the vital force

MILK
the vital force

PROCEEDINGS of the
XXII International Dairy Congress
The Hague, September 29 - October 3, 1986

Edited by

Organizing Committee of the
XXII International Dairy Congress

Springer-Science+Business Media, B.V.

Library of Congress Cataloging in Publication Data

International Dairy Congress (22nd: 1986: The Hague, the Netherlands)

Milk — the vital force.

1. Dar Dairying—Congresses. 2. Milk—Congresses. I. International Dairy Congress (22nd: 1986: The Hague, the Netherlands). Organizing Committee. II. Title.
SF223.I6   1986a   637'.1   86-31508

ISBN 978-94-017-5573-3   ISBN 978-94-017-5571-9 (eBook)
DOI 10.1007/978-94-017-5571-9

All Rights Reserved
© Springer Science+Business Media Dordrecht 1987
Originally published by D. Reidel Publishing Company, Dordrecht, Holland in 1987
Softcover reprint of the hardcover 1st edition 1987

No part of the material protected by this copyright notice may be reproduced or utilized in any form or by any means, electronic or mechanical, including photocopying, recording or by any information storage and retrieval system, without written permission from the copyright owner

PREFACE

From September 29 to October 3, 1986, the XXII International Dairy Congress was held in The Hague, The Netherlands. It was attended by approximately a thousand participants from 56 countries. The scientific and technical programme of this congress was composed in a way that differed from the set-up of some of the preceding international dairy congresses. Thanks to the imaginative contributions of the members of the organizing committee - but in particular that of Drs. H. Schelhaas and Prof.Dr.Ir. P. Walstra, the members of the scientific and technical programme committee - a varied programme could be drawn up. This programme consisted of five plenary presentations, nine seminars on generic subjects, nine round-table discussions on subjects of topical interest, a great many poster presentations and a large number of technical excursions. Seminar programmes of broad interest were drafted with the help of nine seminar committees. This opportunity is taken to acknowledge once again the contribution of the members of these seminar committees.

The scientific and technical programme was completed between an opening session attended by H.M. Queen Beatrix and a closing session. In these sessions words of welcome and appreciation were expressed by the chairman of the Netherlands National Committee of the International Dairy Federation, Drs. H. Schelhaas, the president of the International Dairy Federation, Ir. W.M. Dijkstra, and the chairman of the organizing committee. Also, the Minister of Agriculture and Fisheries, Ir. G.J.M. Braks, addressed the congress during both the opening session and the closing session.

Although these speeches and addresses are of historical significance, it was decided for practical reasons to confine the congress proceedings to the opening address delivered by the Minister of Agriculture and Fisheries, the plenary presentations, and the papers and the summaries of discussions of the nine seminars.

May the reading of the proceedings be as useful to those who attended the XXII International Dairy Congress as those who did not.

W.IJ. Aalbersberg
chairman of the organizing committee

TABLE OF CONTENTS

Preface	v
G.J.M. BRAKS / Opening Address	1
M. WILLIAMS / Agriculture, World Economic Development and Prospects for Food Security	5
H. SCHELHAAS / The Dairy Industry in a Changing Economy	15
R. JARRIGE / Evolution de la Production Laitière Bovine et des Caractéristiques du lait dans les Pays Temperes	27
E.W. SPECKMANN / Present and Future Health Issues and Milk and Dairy Products	47

Seminar I: PROGRESS IN CHEESE TECHNOLOGY

P. WALSTRA / Introductory Remarks	57

Session 1: Rennets and Starters

P.F. FOX / Coagulants and Their Action	61
M. TEUBER / Starter Cultures: Fundamental Aspects	75
A.V. GUDKOV / Starters: As a Means of Controlling Contaminating Organisms	83
C. DALY / Starters: Application in the Dairy	95
DISCUSSION	105

Session 2: From Curd to Cheese

R.C. LAWRENCE and J. GILLES / Cheese Composition and Quality	111
J. KOROLCZUK, J.-L. MAUBOIS and J. FAUQUANT / Mecanisation en Fromagerie de Pates Molles	123
G. VAN DEN BERG / Trends in the Mechanized Manufacture of Semihard Cheese Types	129
N.F. OLSON / Mechanization: Dry Salted Cheeses	137
R. AHLSTRÖM / Automation from Curd to Cheese	145
DISCUSSION	153

Session 3: Properties of Cheese

P. WALSTRA, H. LUYTEN and T. VAN VLIET / Consistency of Cheese	159
J. ADDA / Les Mecanismes de Formation de la Flaveur dans les Fromages	169
E. RENNER / Nutritional Aspects of Cheese	179
H.A. MORRIS and S.R. TATINI / Progress in Cheese Technology - Safety Aspects with Microbiological Emphasis	187
DISCUSSION	195

Seminar II: MILK PRODUCTION AND MILK PRODUCTS IN DEVELOPING COUNTRIES

Session 1

H. BAKKER / Milk production by Indigenous Cattle	201
V.N. TRIPATHI / Milk of Other Animals	209
J.J. MOL / Collection, Transport, Compositional and Quality Assessment	223
J.C.T. VAN DEN BERG / Composition and Quality of Milk as a Basis for Payment of Farmers	233
SUMMARY OF DISCUSSION	239

Session 2

M.R. BACHMANN / Specific Aspects of Milk Processing in Developing Countries	243
A. SJOLLEMA / Recombination of Dairy Ingredients into Milk, Cream, Condensed and Evaporated Milk	251
S.E. BØJGAARD / Recombination of Dairy Ingredients into Fermented Products Incl. Cheese, Butter and Ice Cream	259
T.N. MALETNLEMA / Nutritional Aspects of Milk Products in Developing Countries	269
DISCUSSION	279

SEMINAR II: MILK PRODUCTION AND MILK PRODUCTS IN DEVELOPING COUNTRIES
AND
SEMINAR III: MARKET AND MARKETING OF DAIRY PRODUCTS

Session 3

D.E. DE ROON / Market and Marketing of Dairy Products in Developing Countries	285

F.E. JOLLIET / Marketing Aspects of Dairy Products in Developing Countries	299
L.A. BARRØN DEL CASTILLO / Social and Economic Aspects of Recombination and Indigenous Milk Production	305
F. PRONK / Dairy Food Aid	317
M.J. WALSHE / Criteria for Success or Failure of Dairy Development	329
SUMMARY OF DISCUSSION	341

Seminar III: MARKET AND MARKETING OF DAIRY PRODUCTS

Session 1: Enterprise-Oriented

R. RISTOLA / Planning the Dairy Enterprise to Meet the Challenges of the Market Place	345
R. HILKER / Das Milchwirtschaftliche Unternehmen und die Herausforderungen des Marktes	353
V.N. SERGEEV / Production and Marketing of Milk and Milk Products in East Europe	359
A.R. DARE / Changes in Food Distribution Systems and Their Implications for Dairy Enterprises	363
A.J. KRANENDONK / Major Problems in the Marketing Management of Modern Dairy Enterprises	369
H.R. FELIX / Major Problems in the Marketing Management of Modern Dairy Enterprises	381
SUMMARY OF DISCUSSION	389

Session 2: Product-Oriented

R. HALL / Liquid Milk Market to Year 2000	393
B.A. JOYCE / The Butter Market	415
L. RAUN / The Cheese Market - Demand trends and future challenges	423
K.J. KIRKPATRICK / Demand Trends in the Last Decade and Challenges to be met to the Year 2000 - For Preserved Milk Products	435
E.H. McCONNELL, II / United States Dairy Farmer Promotion	441
E. SCHMEKEL / Market and Marketing of Dairy Products - Do advertising, sales promotion and nutrition education pay?	447
DISCUSSION	453

Session 4: Global Aspects

G. HAYDOCK / The Changing Role of Government in Dairy Policy	457

TABLE OF CONTENTS

T. O'DWYER / The Changing Role of Government in Dairy Policy — 467

W. KROSTITZ / Trends in World Trade in Dairy Products and Future Prospects — 473

F. HÜLSEMEYER / Milk Supply Management Programmes: Their Strengths and Weaknesses — 487

G. SYRRIST / Milk Supply Management Programmes: Their Strengths and Weaknesses — 495

SUMMARY OF DISCUSSION — 503

Seminar IV: MODERN METHODS OF ANALYSIS OF MILK AND MILK PRODUCTS

R.J. BROWN / Modern Methods of Analysis of Milk and Milk Products: Rapid Determination of Main Components — 507

F. O'CONNOR / Determination and Evaluation of Hygienic Quality — 513

DISCUSSION — 523

Seminar V: NEW METHODS OF CONCENTRATING AND DRYING

B.S. HORTON / Reverse Osmosis: Its Technical, Technological, Economical and Legal Achievements and Limitations — 527

H.G. KESSLER / Multistage Evaporation and Water Vapour Recompression with Special Emphasis on High Dry Matter Content, Product Losses, Cleaning and Energy Savings — 545

W.B. SANDERSON and A.G. BAUCKE / New Drying Techniques for Improved Processing and for a Wider Product Versatility — 559

SUMMARY OF DISCUSSION — 569

Seminar VI: CONVERSION OF FEEDSTUFFS IN THE RUMINANT UNDER DIFFERENT CONDITIONS

A.J.H. VAN ES and S. TAMMINGA / Intake and Composition of Tropical Feeds — 573

T.R. PRESTON / Supplementation of Tropical Feeds — 585

K. ROHR and H.J. OSLAGE / Feed Conversion and Nutrient Partitioning — 595

SUMMARY OF DISCUSSION — 605

Seminar VII: FERMENTED MILKS

Session 1

E. LIPIŃSKA / Différents Laits Fermentés, Leur Signification et Leur Caractéristique — 609

TABLE OF CONTENTS

S.E. GILLILAND / Characteristics of Cultures Used for the Manufacture of Fermented Milk Products	623
V.F. SEMENIKHINA / Interrelations Between Strains in Starters for Cultured Dairy Products	631
M.I. GURR / Nutritional Aspects of Fermented Milk Products	641
DISCUSSION	657

Session 2

V.T. MERILÄINEN / Yoghurt and Cultured Buttermilk	661
J.Lj. RAŠIĆ / Other Products	673
C.H. KEHAGIAS / Fermented Milk Products in Developing Countries with Emphasis on Those Produced from Ewe's and Goat's Milk	683
F. WINKELMANN / Yoghurt – Legal Aspects	691
DISCUSSION	703
CONCLUSIONS from Session 1 and Session 2	705

Seminar VIII: MILK AS A SOURCE OF INGREDIENTS FOR THE FOOD INDUSTRY

Session 1

B.K. MORTENSEN / The Use of Milk Powder in Food Products	709
R.A.M. DELANEY, N. DESAI, and V. HOLSINGER / The Use of Milk Powders in Confectionary and Bakery Products	719
J.G. ZADOW / The Use of Lactose in Food Products	737
L.A.W. THELWALL / Derivatives of Lactose and Their Applications in Food Products	749
SUMMARY OF DISCUSSION	759

Session 2

C.V. MORR / The Use of Milk Proteins in Food Formulations	763
S.S. GULAYEV-ZAITSEV / The Use of Caseinates in Foods	769
J.N. DE WIT / Empirical Observations and Theoretical Considerations on Whey Protein Functionality in Food Products	779
J.E. KINSELLA and D.M. WHITEHEAD / Modification of Milk Proteins to Improve Functional Properties and Applications	791
W. BANKS / The Use of Milk Fat and Milk Fat Components in Food Products	805
SUMMARY OF DISCUSSION	811

Seminar IX: INTEGRATED QUALITY CONTROL AND ASSURANCE

Session 1

M.G. VAN DEN BERG / The Quality World and Its Terminology	815
H. WAINESS / Raw Milk Quality, Prerequisite for Excellent Milk and Dairy Products Farm Inspection and Quality Control	829
F. HARDING / Milk Inspection and Quality Payment	837
D.I. JERVIS / The Effect of Raw Milk Hygienic Quality on the Quality of Dairy Products	845
P. VAN DE WETERING / Total Quality Control	855
DISCUSSION	861

Session 2

K. SALMINEN / Quality Management and Management Information	867
J.H.B. CHRISTIAN / Hazard Analysis and Critical Control Points	871
G. ODET / Quality Control and Quality Assurance of Processing a Product at Different Locations	879
A.M. DUKE / Quality Assurance and Image in Relation to the Consumer	887
D.W. ATKINSON / British Food Marks	895
K.J. KIRKPATRICK / Australian & New Zealand Quality Marks	901
J.M. VAN DER BAS / Scandinavian and Dutch Quality Marks	907
DISCUSSION	911
E. MANN / Synopsis of the Congress	913

OPENING ADDRESS

Ir. G.J.M. Braks
Minister of Agriculture and Fisheries
le Van den Boschstraat
             's-Gravenhage
The Netherlands

Your Majesty, Mr. President, Ladies en Gentlemen,

Milk, the vital force.
To my mind this is a well-chosen theme for the XXIInd International Dairy Congress. It shows great similarity to the theme of our national promotion campaign for liquid milk: "Melk, de witte motor". Both phrases posit that milk is a health factor and a source of energy. The international dairy industry can at this moment certainly use vital force. Vital force to deal creatively with a number of major problems.

I have noticed that future developments in agriculture and especially in the dairy industry will be discussed by prominent speakers. Developments which my colleagues and I are discussing at the moment at an informal Council meeting in the Lake District. As Minister of Agriculture and Fisheries of an important dairy exporting country I would like to give you some food for thought on this subject. This is, in fact, the reason why I have briefly deserted my colleagues in the Lake District this morning.

In this gathering I need not mince words. I consider the present situation in the dairy industry on a global scale extremely alarming. There is an ever-widening gap between world milk production and the demand for dairy products. Medium to long-term scales prospects are not overly favourable. For example: at this moment stocks in the EC have risen to the record levels of 1.5 million tonnes of butter and more than 1 million tonnes of skimmed milk powder. Imagine, 1.5 million tonnes of butter means more than 1 packet of butter for every human being on earth. And you must not forget that measures to limit production have been in operation in the ED for over two years. And that, between 1984 and the present day, they have caused production to fall by 4 to 5 percent compared with 1983.

There are also marketing problems. The sharply reduced exchange rate for the dollar, the substantial reduction in income of oil-producing countries and the lack of foreign currency in many countries and the lack of foreign currency in many countries are at the root of those problems. The result is intensified competition between the

dairy exporting countries and trading blocks. That battle is fought
with the help of extensive price support measures.

The gap between supply and demand must be narrowed. No other
conclusion is possible. One way of doing this is to curb production.
And in view of the almost unlimited technical possibilities to
increase production, drastic curtailment of production is inevitable.
Not only in the EC. For I must emphasize that surpluses of dairy
products are not solely an EC problem. They are a world-wide problem.
In spite of measures limiting production in some parts of the world
the world milk output rose 0.5 to 1 percent in 1985. To this rise
dairy producers in the rich part of the world have also contributed.

Only two weeks ago the EC Ministers of Agriculture held a first
discussion about a new package of drastic measures to curb production
even further. We hope to take decisions on this matter in the short
term. It will give the EC the chance to assume its responsibility.
But for the world milk output to be actually reduced and the stocks
to be disposed of, other important dairy producers in the world will
also have to cut their production. Fortunately those countries
recognize the need for such action.

Whether or not the more balanced market situation I have called
for can be achieved in the dairy sector will also be determined to
a great extent by the outcome of the new world trade round within the
framework of GATT that has just started. A world trade round in which
this time the international trade in agricultural products will
participate. I attach great value to these negotiations, even though
I realize that it will be some time before concrete results can be
expected. But I am convinced that GATT is and will remain an important
forum for the solution on a global scale of the problem of dairy
surpluses.

We are talking about the surpluses existing in parts of the world,
while at the same time in other parts of the world an adequate food
supply is still far from being a matter of course. In broaching the
problemens connected with the dairy surpluses I have mentioned one
significant aspect of the problem of the disturbed market equilibrium
for many agricultural products. Another aspect is the adverse effects
for developing countries, since the imbalance on the market has led
to a fall in prices and to destabilizations of international trade.
It is of course obvious that the development of prices on the world
market is of vital importance to the own agricultural production of
developing countries. Developing countries would, accordingly, benefit
greatly by the restoration of market equilibrium. This fact makes the
start of the eighth GATT world trade round even more important.

Dutch policy is aimed above all at stimulating the ability of
developing countries to produce their own food on a permanent basis.
It implies strengthening the local agricultural structure. I realize
all too well that in the short term food aid will remain indispensable.
It should, however, be implemented in such a way that local food
production is not adversely affected. In this context it is of the
utmost importance that your congress will be devoting attention to
dairy production in developing countries. A range of topics will come
up for discussion, from imports of Western dairy product and food aid

to the manufacture of dairy products with locally produced milk. The results of the discussions may possibly prove fruitful for Third World countries. Therefore I greatly appreciate the organization's initiative and have not hesitated to provide the funds to enable a number of experts from Indonesia, Malaysia and Saudi-Arabia to attend this congress.

The programme of your congress further evidences a lot of attention to topics such as quality management and new technologies. And you are obviously not letting the opportunity to discuss product innovation pass unused. Together with cost control these are important elements, which in the future will be a decisive factor in the maintenance of enlargement of market shares. To a great extent the Netherlands owes its prominent position as a dairy produces to these elements, and I know that the dairy industry in this country assigns top priority to them. The manifestation Holland Dairy Land in Utrecht affords ample proof that the Dutch dairy industry has every intention of continuing to do so.

I express the hope that the conference theme "Milk, the vital force", will have an inspirational effect on your discussions. In view of the size and scope of the problems facing the dairy sector in the coming years such inspiration will not come amiss. In that sense I wish you a fruitful congress.

I herewith declare the XXIInd International Dairy Congress open.

AGRICULTURE, WORLD ECONOMIC DEVELOPMENT AND PROSPECTS FOR FOOD SECURITY

Maurice Williams
World Food Council & Overseas Development Council
United Nations S-2955
New York, N.Y. 10017

ABSTRACT. Farmers and agricultural policies have institutionalized productivity gains for staple foods in all regions except Africa, where special efforts are required. The outlook is for future security of supply, adequate to meet the overall food needs of the world's expanding population. However, technology-driven productivity increases are outpacing effective market demand and there are large surpluses in a world with many undernourished people. The solution is to reform the faulty economic policies which constrain the growth of incomes and trade. The farm and trade policies of the major food exporters are not adequately responsive to changing circumstances. What is needed is a political initiative to launch a programme for world farm and trade development which would include: (i) a cease fire in the food subsidy war between the United States and the European Community in the interest of fair market and supply management practices, and (ii) measures to stimulate the growth of agriculture, incomes and trade which would meet the needs of farmers in the developed and developing countries and raise food consumption standards worldwide.

1. THE WORLD FOOD PROBLEM IN PERSPECTIVE

Humankind has had a long history - extending over a hundred centuries - of obtaining dairy and meat products from ruminent livestock. However, scientific dairying with its institutionalized and sustained productivity increases is an achievement of only the last few decades. This achievement is a tribute to the research, high technical standards and managerial skills made possible by the co-operative endeavors of the farmers and their organizations represented at this International Dairy Congress.

Dairy products today are still largely a food of the middle class, mainly in the developed countries, for they consume almost 80 per cent of the world's output. That, however, has been changing as incomes improve and people in the developing regions upgrade the adequacy of their diets. The latent potential is very large for expanding future consumption of meat and dairy products, as well as of cereals which are the staple food of the masses in developing regions.

There is a linkage and similarity of outlook for both meat and cereal products. Livestock consume, in addition to grasses and forage, almost half of the world's cereal production. Significantly, for both cereals and dairy products, technology-induced production increases are outpacing effective market demand -- particularly since 1981 when international markets for food became relatively static. As a result, there are large surpluses of food in a world which has many seriously undernourished people. This is a tragic anomaly. Who is to blame and what should be done about this world food situation?

Some would blame the farmers and agricultural policies for producing so much excess food, large parts of which must be stored or given away and bearing subsidies which are a heavy burden on public treasuries. Few in the OECD countries are pleased with this situation.

But the the problem has to be seen in perspective. Farmers and agriculture are doing their part to contribute to world food security. They are producing on a sustained basis the food that is actually needed to feed adequately all of the world's population. The agricultural sector should not be blamed for the imbalances in the maldistribution of food. That responsibility must be placed where it belongs -- namely on the faulty economic policies and the depressed world economy. These are responsible for constraining the growth of incomes and trade which result in millions of people being undernourished.

It is problems of food distribution which must be more directly addressed. In this regard, the farm and trade policies of the major food producers and exporters in the U.S. and European Community are not adequately responsive to changing food market and development needs. What is needed is a political initiative in support of a world farm and trade development programme. We will propose the essential elements of such a programme.

But first, let us signal the importance of the contribution which farmers all over the world have been making to food security.

1.1 Agriculture's Contribution to A Food Secure World

The decade of the 1970s witnessed achievement of a new stage in the transformation of world agriculture. Sustained productivity gains have been institutionalized on a broad basis. The rate of food production is now almost certain to keep pace with population needs. This process became firmly rooted during the 1970s, not only in the OECD countries, but among most Asian and Latin American countries as well. As a result, the outlook is favorable for adequate production to supply the food needs of the world's expanding population well into the 21st Century. That is a remarkable agricultural achievement.

This food-secure outlook as to overall supply is in marked contrast to earlier concerns that agriculture would not be able to keep pace with growing world demand for food, and that there would be recurrent food crises and a harsh market allocation of scarce supplies.

The dramatic change in the world's food outlook has been the results of strategies which combine policy incentives for farmers with science-driven technologies and the higher priority for agricultural investment recommended by the 1974 World Food Conference. It is the widespread application of the these strategies that have achieved large and sustainable increases of food production in the developing countries of Asia and Latin America, and that placed them on the path toward more adequately meeting the nutritional needs of their expanding populations.

The progress of agriculture in the developing world is constrained in some regions by seasonal climatic instability affecting production and by the ambivalence of some governments in adopting appropriate food strategies.

In particular, productivity-enhancing farm policies and technologies have not yet taken root in most of Africa. This region has been living out earlier fears of faltering food systems and increased vulnerability in a downward spiral of environmental deterioration and a menacing food-population imbalance.

Yet the largely agrarian African countries clearly have the potential for self generating food and agricultural productivity growth, a potential which has been delayed by a complex of conflicting interests and institutional weaknesses. These constraints on African agriculture can be overcome by the measures for economic recovery and development adopted by the United Nations Special Session on Africa in June 1986. The United Nations Programme calls for well directed, sustained and larger efforts by African Governments and international development assistance agencies.

1.2 Assured Expansion of Food Demand and Markets

Agriculture's contribution worldwide to food security on the supply side must now be matched by assured expansion of food demand and markets - both nationally and internationally.

There is a generally mistaken notion that institutionalizing the process of agricultural growth in developing countries leads inevitably to an appropriate balance between food supply and food demand. Some developing country governments espouse this view of food self-sufficency as a political objective. However, the experience of most developing countries is for higher rates of food and agricultural production to stimulate even more rapid income-driven demand for food improvements, and thus for increased imports of meat, feed grain, and dairy products.

A principal reason for the latent propensity of developing countries to increase food imports is their very low level of average food consumption. We know that chronic hunger is life's reality for some 800 million people; however, with better incomes almost all of the people of developing regions would increase their daily caloric consumption by at least 50 percent almost immediately, according to one estimate. Although this would provide them with a reasonably satisfactory level of food consumption, it would still be well below that of developed countries.

Consequently, with development and the growth of incomes, the demand for food rises more rapidly in most developing countries than can be met by even priority attention to their domestic production capacities, with a concomitant need to increase food imports.

This was the experience of most developing countries in the 1970s when accelerated food and agricultural production was accompanied by a surge of farm imports - rising from $45 billion in 1972 to $165 billion by 1980. Developing country markets accounted for half of the world's food imports in 1980-81 (the USSR and socialist countries accounted for about 25 per cent of world imports). Thus the food situation of developing countries - with the exception of Africa - greatly improved during the 1970s, both by their increased production and by accompanying larger food imports.

This situation changed dramatically after 1981. While most developing countries have been able to sustain their farm production programmes, their imports of food fell off as a result of the world recession. It was not the success of developing world farm programmes which dampened their imports of food - as some observers have claimed. Rather it was the collapse of their income growth rates, high world interest rates, and the large international debt burden which put a stop to the farm trade expansion. Unable to earn or borrow additional foreign exchange, and obliged to service past debts at much higher rates of interest, developing countries had no alternative but to reduce food and other essential imports.

The results of these economic dislocations have been to increase the numbers of undernourished people, as the United Nations World Food Council has emphasized.

2. CURRENT FARM AND TRADE POLICIES CALLED INTO QUESTION

The recession and faltering world economic recovery of the early 1980s with the prospects for continuing slow economic growth has called into question the farm and trade policies of the major food producing and exporting countries.

The world economic recovery to date has been uneven geographically, uncertain as to duration, and limited in its capacity to overcome serious imbalances among the major countries. Inflation has been overcome at a cost of high unemployment, severely depressed commodity prices and generally low levels of investment. Massive international debt continues to weaken confidence in the financial system and to limit the capacity for trade revival. International trade is further threatened by fiercely defensive protectionist pressures - which if unrestrained would mean an equilibrium at even lower levels of world production and trade.

While there is broad general agreement that world economic and trade recovery requires a better alignment of national policies by the major industrial countries, there is little consensus on how this can best be achieved.

In these circumstances, the outlook over the next few years is for slow economic growth and relatively static levels of international trade. This period of slow growth coincides with large and sustained increases in food production, and a situation of simultaneous over-supply and under-demand in world markets. Prices are at low levels, surpluses are high, and they are currently being disposed on international markets with minimum concern for longer-term economic and food security ramifications.

The farm policies of the major OECD producing countries are not well adapted to the current situation of changing technology and market development. Their farm economies during the past decade became increasingly dependent on international markets to support their large investments in agriculture, sustained productivity improvements and high farm incomes. Until recently the output of half of cultivated farm acres in North America was exported. And the European Community has become increasingly dependent on farm exports to support its Common Agricultural Policy. Over one-third of the Community's agricultural expenditures have been for export refunds.

The international food market outlook is grim. Over the next three to five years, the prospects are that the projected increases in farm productivity will exceed market demand. Prices are almost certain to remain depressed. And the current fierce competition of subsidized exports among the principal producers will further reduce earnings for all food exporters.

When one considers the prospects for further productivity increases the prospects become even more worrisome. The long term trend of increases in the volume of agricultural production in the European Community is 1.5 to 2 per cent a year. For the U.S. the total factor productivity is about 2 per cent annually. Since the price elasticity of demand for food is low in the high income economies, the growth in food consumption averages between 0.5 and 1 percent annually. Consequently, their farm economics are export driven.

The rate of technologically induced productivity increases appears likely to continue and even accelerate as a result of further advances in the biological sciences. Already there are a number of technologies in various stages of research which could be introduced commercially in the next few years. For example, a synthesized growth hormone has increased milk production by 15 to 40 percent in field trials. Other new developments would stimulate plant yields. And the intensive research now being devoted to biotechnology appears certain to yield a stream of commercial applications in the years ahead.

No one would seriously suggest discouraging advances in science and technology as a means for restraining future production increases. What then are the alternative policy options?

2.1 Options for Global Farm and Trade Recovery

Some believe that the current crisis in developed countries' agriculture provides an opportunity to break away from past farm policies and to restructure world agriculture toward more liberal market policies. They propose treating agriculture in the future GATT negotiations on the same basis and in the context with other trading sectors. This would be a major reversal of past national policies of special protection for agriculture. Many agricultural economists see such policy reforms as necessary in order to reduce the waste of over-investment and to assure more efficient production and marketing as essential to improving world food consumption standards.

Others believe that the application of free trade and a restructuring of world agricultural production according to criteria of international comparative advantage would be disruptive of past social and productive achievements. They point to the difficulties of shifting resources out of agriculture in a period of high unemployment. Also they have expectations that the currently high agricultural capacities in the major producer and exporting countries will be needed in the next few years to improve food consumption in the developing world.

Is it possible to envision a revival of farm trade in the developing countries in the latter years of this decade? Essentially, this would mean a return to the expanding international food trade of the 1970s. We know that the potential need for improved food consumption is large in the developing world, and that their rapidly expanding urban populations will expand future markets. It is a potential which exceeds their capacity to meet from indigenous production. How could this latent market potential be best developed to the mutual advantage of developed country food exporters and low food consumption developing countries? What is needed is a world farm and trade development programme to promote economic growth, income and trade as a basis for raising food consumption standards in the developing world. What would be the essential elements of such a programme?

3. A PROPOSED WORLD FARM AND TRADE DEVELOPMENT PROGRAMME

The first essential element of a world farm and trade development programme should be a negotiated cease fire in the export subsidy trade war between the U.S. and the European Community. Export subsidies in the current circumstances are not an effective means for market development. Nor will they enhance returns from trade or increase market shares. For each side is able to counter subsidy with subsidy with the result that international prices fall even lower, and shares will remain essentially unchanged in the currently static international food market.

All that is being accomplished by the subsidy fight for market shares is to convert even larger portions of the international trade in food to concessionary sales. This is a poor means to support and enhance incomes of farmers in both developed and developing countries.

Developing country food importers may gain some short-term savings, but for the most part export subsidies on the present scale tend to undermine the farm programmes of most developing countries. And the trade of a number of low cost agricultural food producers is being unnecessarily and seriously damaged.

U.S. and European policy makers should cease this trench type warfare of trade subsidy attrition, so reminiscent of the stalemate of trench tactics in the First World War. It is not only excessively costly, in financial and political terms, but is an entirely counter-productive means of resolving current farm and trade problems. What is essential is a high level political understanding among the main participants to manage their respective farm and trade policies at least for a two year period in such a manner as to:

- restrain the use of export subsidies,

- adjust their respective farm policies to moderate production increases, by supporting farm incomes with policies which do not directly enhance production increases, and

- work together to restrain agricultural trade restrictions and expand international food markets.

Following political agreement on these key policy elements, it would then be possible to elaborate specific mutual undertakings concerning export credits and subsidies, and other appropriate measures.

An agreement along these lines would seek to align at least for a temporary period the domestic farm policies of both the U.S. and the European Community in the interest of coordinated supply management and fair market export practices. Such an agreement is almost certainly necessary to set the stage for more intensive formulations and negotiations in the GATT on longer-term trade and marketing standards in agriculture. It would greatly improve the environment and prospects for constructive trade negotiations.

On the basis of such a two year cease fire on export subsidies, the U.S. and the European Community would then be able to launch a broader global effort to promote growth and income-generating farm policies in the developing regions in order to raise food consumption standards. As the experience of the 1970s has demonstrated, there is no inconsistency between the growth of developing world agriculture and developed country farm exports. Agricultural development can lead to broad-based domestic growth of incomes in developing countries and, along with open trade policies, to the expansion of both food production and international food markets.

This programme for world farm and trade development can be envisioned as a farmer-to-farmer mutually supportive one, with farmers in both developed and developing countries allied in their common concern for raising food consumption standards throughout the world.

Farmers generally should oppose actively protectionist trade measures. It is a fact of international life that exports must be balanced by imports over the longer term, and export market development must allow for import repayments in products for which the developing countries are low cost producers. Merchantilist export expansion policies provide only illusionary market gains, as is being demonstrated by the current farm trade policies of the U.S. and European Community.

Also a more growth-oriented pattern of debt rescheduling is essential in order to revive mutually profitable farm trade relations. This should be accompanied by higher levels of economic assistance in support of redirected food and development programmes in the least developed countries, particularly those in Africa.

What is envisioned is a partnership of farmers the world over, from developed and developing regions, for an expansion of income and trade which would match the world's agricultural production potential in order to raise and enrich the diets of the world's people. We would no longer be talking about cutting back farm production in a world of hungry people, but rather of expanding food production and trade to meet these real needs. Then world agriculture and economic development would truly contribute to lasting world food security.

The Hague, Holland
29 September 1986

THE DAIRY INDUSTRY IN A CHANGING ECONOMY

Drs. H. Schelhaas
Dutch Dairy Commodity Board
P.O. Box 5806
2280 HV RIJSWIJK ZH
Holland

ADDRESS BY MR. H. SCHELHAAS, CHAIRMAN OF THE COMMODITY BOARD FOR DAIRY PRODUCTS, IN THE PLENARY SESSION OF THE INTERNATIONAL DAIRY CONGRESS TO BE HELD FROM 29th SEPTEMBER TO 3rd OCTOBER 1986 AT THE HAGUE

SUMMARY

The world dairy industry is currently going through a period of rapid transition. The factors which are instrumental in bringing about changes in the dairy sector are identified as:

1) rapid technological advances;
2) changes in the pattern of demand;
3) declining relative importance of agriculture in general;
4) changing views on the role of government;
5) internationalization of the world economy.

After an extensive analysis of the forces at play the following five conclusions are drawn which might serve as guiding principles for future policy in these turbulent times:

1) the rate and character of agricultural transformation can be influenced but not halted by policy;
2) there is large scope for further development of the dairy sector in the Third World;
3) the West could employ its food surpluses considerably better than has been done so far;
4) worldwide agreements on production and marketing of dairy products are urgently required;
5) there remains considerable scope for improvement of international cooperation.

1. INTRODUCTION: A TIME OF RAPID CHANGE

The world dairy industry currently finds itself in a period of tran-

sition. This is caused by a number of complex factors. In this address
I intend to consider these causes in detail.
But first let me briefly review the present state of the world dairy
industry. Of the world milk production - which amounts to about 500
million tons - 46 per cent takes place in Western Europe, North America
and Oceania, another 27 per cent in Eastern Europe and the Soviet
Union, and most of the remaining 27 per cent in the Third World: admittedly that share is still small, but we should remember that 15 years
ago it was only 20 per cent.

Other major changes have taken place during the past 15 years.
For example, there are still approximately 2.5 million dairy farmers in
the western world today, against 4.5 million in 1970. Altogether the
western dairy industry provides jobs to around 3.5 million persons,
against 6 million in 1970.

That means that in a span of 15 years the number of jobs has fallen by 2.5 million. In the countries of Eastern Europe and the Third
World, there are on a relative basis at least as many persons employed
in the dairy sector. In India alone, the dairy cooperatives in 1984 had
3 million mostly small-scale dairy farmers.

In this paper I will deal with the problems of the dairy industry.
But let us not forget its power and its glory.

The dairy industry possesses a sound base because there is a very
stable demand for dairy products in western and eastern European countries, and in the Third World this demand - although small - is growing
strongly. The industry supplies natural and healthy products of excellent quality, which provide a prime necessity and, what is more, are
highly appreciated by consumers for their nutritional value and flavour. Milk is the nearest approach we possess to a perfect and complete
kind of food and no single food is known that can be used as a substitute.

But, to come back to my main subject, the factors which are currently resulting in substantial changes in the dairy sector are:
- rapid technological advances in dairy farming and in milk processing;
- changes in the pattern of demand for dairy products, partly structural in nature and partly cyclical;
- the declining relative importance of agriculture in general, among
  other things due to the declining number of farmers and changed economic priorities;
- changing views in the task of government, leading to a smaller role
  for agricultural policy and a considerably larger role for the market;
- the internationalization of the world economy.

First of all, I will now discuss each of these factors individually.
Then I will present five conclusions which may be drawn from that set
of interrelated factors.

2. RAPID TECHNOLOGICAL CHANGES

Technological progress is the main cause of changes in society. In the
1960s and 1970s, which have now passed into history, an important technological advance already took place in keeping dairy cattle, leading

to what has been called the first technological revolution. It was featured by among other things the introduction of the milking machine and of improved breeding methods made possible by artificial insemination and an improved understanding of genetics, the universal introduction of the tractor, the introduction of the loose house, cooling tank and milk collection tankers. Technological development may become even more remarkable in the years to come.

Society as a whole is passing through a period of rapid technological progress. In this context we may think of the enormous progress made in the field of micro-electronics (for example, the rise of micro-processors), the automation which this helped to make possible, the tremendous progress in computing, in information technology and in communications, and the rapid progress in biotechnology and microbiology. According to Alvin Toffler *), the changes which society is currently experiencing, both in the technological field and in social respects, are at least as extensive as those which took place during the industrial revolution.

This technological progress wil, as always, find its way to the dairy sector as well. Dairying is a part of this dynamic world and cannot be isolated from changes elsewhere. As a matter of fact the impact on dairy farming will be more spectacular than on the dairy industry. A second technological revolution in milk production is even now presenting itself, while the effects of the first have not yet fully been realised. The important trends this time are far-reaching automation of many activities on the dairy farm, individual cow identification and treatment made possible by computerization, embryo transplantation, considerable and constant improvement of breeding and feeding techniques, introduction of natural growth hormones and the introduction of the milking robot.

As a matter of fact, no more than the first, the second technological revolution will not necessarily lead to any erosion of the family farm as the cornerstone of western dairy farming, bearing in mind the growing scope for making use of the new technology in a small-scale way. Nevertheless, we are going to see considerable changes in the nature of farm management and the demands imposed on the farmer's skills as a manager.

There are two trends which may prove to be particularly spectacular in this second technological revolution, which in the short term may well lead to a considerable increase in milk production, namely application of the recently developed natural hormone preparations and introduction of the milking robot.

The use of natural hormones in dairy farming is capable of leading to a substantial increase in milk yields per cow by around 30% without being accompanied by any significant increase in feed consumption. People in the United States and Canada are convinced that, as a result of this, the next 5 years are going to see more changes in the dairy sector than the past 20 years. In many European countries, however, the application of these hormones is prohibited or may be prohibited shortly, because of by possible consequences for milk quality (health as-

*) Alvin Toffler: The adaptive corporation, London 1985, page 3.

pects) and possible reactions on the part of consumers. I will stress that there are many experts who deny that the use of natural hormons will have any influence on the quality of milk and milk products.

The introduction of the milking robot should be regarded as theprovisional end of the current process of automating of many parts of the dairy farmer's work. In the Netherlands, it is expected that the milking robot - which makes it possible for cows to be milked without any direct human assistance - will be ready for practical application well before the year 2000. According to experts, the robot may make it possible to reduce the number of man-hours per cow per year from 50 down to 30-35 hours. Bearing in mind that his would make it an easy matter to carry out milking 3 or 4 times daily, the milk yield per cow could at the same time increase by a further 15-20% as a result.

In at least two advanced dairy countries, namely the United States and the Netherlands, an average milk yield of 10000 kg per cow is considered feasible by the year 2000. What is more, the Dutch forecast does not take account of the utilization of hormone preparations in dairy farming.

A moderate assumption is that a productivity increase of at least 50% is feasible in almost all milk-producing countries by the year 2000, whilst in many countries a still considerably larger increase is possible.

3. THE CHANGING PATTERN OF DEMAND FOR DAIRY PRODUCTS

3.1. Demand for dairy products in the developed countries

Although milk production has a potential for a strong increase in the next few years, demand for dairy products in the western countries can at best be called stable. It is a fact that in most countries the consumption of butter is showing a downward trend, whereas demand for low-fat milk products is rising at the expense of the higher-fat products. For this reason, a structural reduction of the price of milk fat appears inevitable.

In the longer term, there is the real risk of a further advance of imitation products, both at the expense of the fat component and the protein component of milk. Quite a lot of research is known to be going on aimed at using soya protein to make foods which resemble the traditional dairy products.

Here and there, these research efforts are indeed achieving results (soya milk, a few artificial cheeses, toppings, etc.). There are projections - and I mention them here only as a warning and not as a personal statement - there are projections which show a sharp growth in the market share of these products by the year 2000. From this angle, the use of soya protein might show the same kind of development as soya fat.

In the 1930s (soya was virtually unknown before 1925) soya oil managed to gain a prominent position on the fats market, and it became very important raw material in the manufacture of margarine. In the course of the coming decades, at least according to some experts, soya protein may prove capable of gaining an equally strong position in the world protein market. Of course, for the time being these views are not much more than wishful thinking. Without any risk of contradiction, it may be stated thad substitution of the protein component is far more difficult to achieve whilst maintaining the quality characteristics (flavour, consistency, etc.) at an acceptable price than substitution of the fat component.

Even so, the dairy sector must be alert. Milk is a relatively expensive raw material, compared with soya protein for example, and food technology is advancing with rapid strides - also with regard to the utilization of soya protein. On the other hand, however, milk is a unique raw material, which contains at least about 110 valuable components and thereby offers considerably more scope as base material for a wide variety of products than any other agricultural raw material whatsoever. The practical possibilities in this direction have not yet been more than partly explored. In many countries the range of products made by the dairy industry remains limited and traditionally determined.

What is needed is constant research aimed at finding new products. Modern dairy technology - for example, membrane technology (including ultra-filtration), the progress in microbiology, the possibilities of genetic engineering in dairy micro-organisms and the possibilities of flexible automation - is opening up new prospects for the development of new products.

The dairy industry must in the coming years become more and more a speciality producer or it will gradually lose part of its market share. To summarize it may be stated that in the short term, demand for dairy products will be stable in the developed countries but that in the rather longer term there is the threat that imitation products may be able to conquer a part of the market.

3.2. The demand pattern for dairy products in the Third World.

Unlike that in the developed countries, the demand pattern for dairy products in the Third World is definitely subject to cyclial fluctuations. At present demand for food in general - especially demand for dairy products - in that part of the world is experiencing a setback as a result of
- the fall in oil prices;
- the economic recession in many non oil producing developing countries;
- the effects of the international debt crisis;
- the malaise on the vegetable oils market;
- the fall in prices of many raw materials produced by developing countries.

According to the Director-General of the FAO, during 1986 food consumption will decline in almost half of the poorest countries in the world, and grain imports will fall by 11 per cent. According to my own estimate, dairy imports by Third World countries will fall by around 20 per cent in 1986.

It is particularly sad that all this is taking place in a world where, according to the World Bank, hunger is continuing to increase and where three quarters of a billion people get too little to eat to allow them to lead an active working life, of whom almost half are barely subsisting on a minimum survival diet. There is enough food in the world for everybody except for those who need it most.

For food demand, backed by purchasing power, to recover it is essential that economic growth return in the Third World. The present situation being pretty grim, I am rather pessimistic concerning the short term. On the other hand, the prospects of substantial growth in the longer term are good.

Experience has shown that a period of economic growth in the Third World is accompanied by a more than proportional increase in the level of demand for food and for high-quality food in particular. In the past it has been found that, although in the middle and higher developing countries the local agricultural production is capable of very rapid expansion under such conditions, nevertheless that expansion is far outstripped by the rising demand for food in such a situation. In the longer term, therefore, there are good prospects for a growth in demand for dairy products in the Third World, especially in the traditional regions of dairy consumption such as the Middle East, North Africa, India, Central Asia, the Savanna areas of West Africa, the Highlands of East Africa and many South American countries.

But in other countries, too, the internationalization of consumer habits will generate new marketing opportunities.

4. THE CHANGING ROLE OF GOVERNMENT

In a period of rising surplus problems we see a tendency towards a less large role for governments throughout the world. At the same time a greater responsibility is being given to the citizens and their organizations. In Eastern Europe this has led to a call for greater flexibility in price policy as well, and it has also led to a greater responsibility on the part of the production enterprises. In this way the prices will vary more from region to region, the price bonus on quality can be improved and a greater percentage can be sold direct to local shops and markets.

As regards the dairy industry in the western countries, the less large role of government finds striking expression in the growing resistance to taking over expensive and unmarketable agricultural surpluses. Producing for a market which does not exist has got to stop, is a frequently heard call. This has led to the paradoxical situation where, although strict limits have been set on the financial obligations of governments, nevertheless the instrument used for that purpose

*) Various studies confirm that this applies very evidently in the case of the middle and higher income class of developing countries, among others by a number of papers presented to the 19th International Congress of Agricultural Economists, including those by John W. Mellor: Balancing Overproduction and Malnutrition, Fred. H. Sanderson: World Food Prospects to the Year 2000, and Alex F. MacCalla and Davis and Timothy E. Josling: Agriculture in an Independent and Uncertain World: Implications for Markets and Prices.

- the quota system - is leading to far-reaching intervention in individual farm management. This intervention is incompatible with the strong call for greater flexibility, which is heard in Western countries as well. A quota system has already been introduced in almost all Western European countries and Canada. This is without doubt the biggest change in dairy policy since the Second World War, a change which also causes dairy policy to differ fundamentally from the agricultural policy in other sectors.

Only in the United States are efforts being made to limit supplies by means of a voluntary premium arrangement aimed at encouraging farmers to stop milk production and exerting pressure on prices. In view of the greater flexibility which the American system embodies, I am definitely in favour of that system above the inflexible European quota system.

The power of agricultural policy to improve the dairy farmers' lot is declining, the dairy sector will have to become more self-reliant rather than depend on government.

5. DECLINING IMPORTANCE OF AGRICULTURE

While problems in agriculture are getting more complex agriculture's influence in general has been declining in recent years. Among other things this is due to the sharp fall in the number of farmers, scale enlargement in the processing industry and in marketing (for example, the growing power of the big food chains), urbanization and overproduction in agriculture in combination with a greater priority given to environmental factors, health etc. In this context I should like to emphasise that a policy aimed at avoiding structural surpluses is in the long-term interest of dairy farming, even though in the short term such a policy means less popular measures.

6. GROWING INTERNATIONALIZATION

There is undoubtedly a getting internationalization of the world economy, due among other things to greatly improved communication and transportation systems. The world is getting smaller; the result is a strongly growing inter-dependence.

Many people, especially young people, are strongly convinced that the worldwide progress in the techniques of both communication and destruction have condemned the world to solidarity if it is not to perish *.

Unfortunately, the increasing internationalization has - at least until now - been accompanied by a declining importance of international organizations and international agreements. Nevertheless, the latter situation would appear to be only a temporary development. The increa-

---

* Quoted rather freely from: 'Albert Bressand and Catherine Distler': <u>Le Prochain Monde</u>, Paris 1980.

sing interdependence calls for stronger international organizations and effective international agreements. I am firmly convinced that - for the remainder of this century - we must strive to improve and strengthen international cooperation.

7. CONCLUSIONS DRAWN FROM THESE CHANGES FOR THE DAIRY SECTOR HEADING FOR THE YEAR 2000

7.1. Technological change

A question of crucial importance for the development over the coming 15 years is, what attitude should be taken with regard to the opportunities offered by technological progress.

It is certain - given the situation on the dairy markets - that for the time being there will be no need for the western countries to boost their production further, but rather to bring production levels down. In this case, technological progress may lead to the loss of many jobs. Viewed from the agricultural market, there is therefore no need whatsoever for the present technological progress; the problems in (western) agriculture would be considerably less severe without present-day technological progress. Even so, I believe that opposition to technological progress is not a realistic option. To illustrate this, let me mention the following considerations:
- never before has an effort to stop technical progress been successful, not even where it has cost, or will cost, many more jobs than in agriculture;
- technological progress in dairy farming is capable of leading to a reduced workload for the dairy farmer and to a lower burden to the environment;
- technological progress in dairy farming may lead to a reduction of the cost price. As I have already remarked, the price of milk as a raw material is relatively high, and there is the threat of substitute products to come. If there is a single sector which cannot afford to ignore technological progress, it is the dairy sector;
- as the European Commission has rightly remarked, in agriculture - and among young people in particular - there is a strong desire to participate in the great technological changes of our time.

The views of a number of European agricultural economists on a large number of topical problems of agricultural policy are also relevant in this respect *.

They have been found to agree in large measure about, among other things, the following propositions, which are to be regarded as a concise creed on agricultural policy:
1. "Agricultural policy should not aim at maintaining the number of family farms nor try to maintain agricultural labour force within the sector, not even in a time of unemployment in the economy, when

* Roland Herrman: 'A survey of views of agricultural economists in Europe', in <u>the European Review of Agricultural Economics 12</u> (1985).

the opportunity costs of labour are zero".
2. "The characteristics of the agricultural sector justify government redistribution measures in favour of the farmers, and price fluctuations in agricultural markets should be smoothed by policy measures".
3. "Agricultural price support cannot reduce <u>in the long run</u> the disparity in per capita incomes between agriculture and other sectors".

The <u>conclusion</u> appears inevitable that "the rate and character of agricultural transformation can be influenced but not halted by policy". *

As regards the <u>speed</u> of the process of change, it is above all important that dairy farming should be given time to adapt. That means the need for a certain amount of income protection accompanied by premiums in order to bring the production capacity back within the limits of what can be economically marketed, for example by means of stopping production, early retirement schemes, withdrawing land from productive use, encouragement for other forms of production and activities (e.g. recreation) and forestry.

As regards the nature of the change, we should consider encouraging the development of a modern technology tailored to the smaller farms.

Over the next 15 years, agricultural policy is going to be faced with the difficult tasks of not only controlling the volume of production - and doing so much better than in the past decade - but also the task of providing guidance in carrying out the structural adaptation - amounting to a reduction of the number of farms by at least 50%.

During the period mentioned, the task of agricultural policy is going to be more difficult than in the past 15 years, at least if we want to prevent as much human suffering as possible.

7.2. Dairying as a source of social and economic change

This title: "Dairying as a source of social and economic development" was the theme of the 1974 dairy congress, which was held in India and was so well organized, by the way.

In contrast to the developed world, the coming decades hold out important opportunities for growth in the dairy sector in the Third World. As I have mentioned, when purchasing power picks up again in the Third World, we can expect a vigorous increase in demand for food, especially demand for high-value food.

There are several advantages in the development of dairy farming and dairy industry in the Third World, for example:
- it provides more jobs in rural areas;
- it improves agricultural incomes;
- in improves the quality of the diet.

The role played by the developed countries in this process may be as follows:
- to supply the necessary know-how;

* A somewhat free quotation, taken from: Alex MacCalla, David and Timothy E. Joshing: 'Agriculture in an independent and uncertain world: implications for markets and prices'.

- to supply milk and milk products to countries where, for reasons of
  comparative costs, it would not be logical to stimulate local production;
- to supply raw materials (milk powder and milk fat) to local industry,
  for example in order to make up for seasonal deficits, such within
  the framework of the recipient country's dairy policy;
- to supply speciality products;
- to provide (dairy) food aid, as was done in Operation Flood for
  example.

The conclusion is that there is large scope for further development of the dairy sector in the Third World.

7.3. Food aid must be temporary

With regard to food aid, it may be pointed out that Operation Flood in India has demonstrated that the dairy surpluses of the West can be used in order to build up local dairy farming and dairy industry in a developing country.

For that reason, it is worth considering an enlargement of the dairy food aid programmes in the years to come. Surely it is totally absurd that an organization such as the World Food Programme is unable to obtain sufficient low-fat milk powder and butter oil for its various programmes, while the world is groaning under the burden of dairy surpluses and large quantities of these are being fed at extremely low prices to pigs and poultry.

I should like to point out that the World Bank, too, in its recent annual report has made a plea for a substantial increase in food aid in the coming years, firstly in the form of development projects as e.g. Operation Flood and secondly for the poorest population groups in general and the vulnerable groups in particular, especially children under 5, pregnant and lactating women, and schoolchildren. These groups namely will only in the very long run profit of the hopefully coming economic growth. The majority of ordinary people, - the man in the street - with an excellent spokesman in Bob Geldof, are still unable to understand why it remains impossible to construct a passable road linking the surpluses of the West and the hunger of the Third World. There is food enough that could be transported over such a road and there are more than enough jobless to construct it, and yet nothing happens. To quote Bob Geldof: "From the top of the butter mountain you can see the hunger in Africa".

The third conclusion is that the West could employ its food surpluses considerably better than has been done so far.

7.4. Crisis on the international dairy market

The present situation on the world dairy market - with world butter reserves at about 2 million tons and very low prices for all dairy products - may without exaggeration be called chaotic.

There exists, in fact, a full-scale crisis on this market.

It can scarcely be disputed, and indeed economists generally recognize it to be correct, that liberalization of international trade can lead to a significant increase in prosperity and should offer developing countries in particular an opportunity to increase their exports. Another point is that, given our present-day transport and communication techniques, and unfortunately in view of the nature of any future wars as well, the pursuit of national self-sufficiency is obsolete. However, it will require a great deal of patience and wisdom in order to free the world economy from its present trade barriers.

For the past few years the international climate has not been favourable for the conclusion of international agreements. However, since recently there are signs which may indicate a turn for the better. One such sign has been the unexpected success achieved by the negotiations on the International Cacao Agreement. Another encouraging development is the decision, taken just over a week ago in a GATT Ministerial Meeting, to launch a new round of Multilateral Trade Negotiations. One of the objectives of this new round is to bring all measures affecting trade in agriculture under more effective GATT rules and disciplines.

In this framework it would, I think, be worthwhile to make every effort in order to achieve a broadened and improved form of the present International Dairy Agreement. In doing so, the following particular objectives might be pursued:
- control of milk production in all developed countries;
- reduction of export subsidies, allowing the price level on the world market to improve;
- stimulation of the international specialization principle - this might be a worthwhile attempt to give back to the dairy sector the principle of comparative cost advantages;
- incorporation of a food aid convention in the International Dairy Agreement, in analogy with the International Wheat Agreement.

I should like to see the international dairy problems tackled with a fresh spirit. I would, therefore, like to suggest that a worldwide conference on these problems be organised which could provide new ideas and, possibly, suggestions to the negotiators in GATT.

This brings me to the fourth conclusion, which is that worldwide agreements on production and marketing of dairy problems are urgently required.

7.5. Toward greater inter-dependence

There is undoubtedly - as was said before - a strong tendency to a greater interdependence. An international recession, a recovery of purchasing power in the Third World, a rise or fall in oil prices, and the international debt crisis - all have an impact on the dairy sector.

As regards the world debt crisis - which is therefore also relevant to the dairy sector - I should like to make the following comment:

The present situation, in which the total financial aid given by the West, amounting to 30 billion dollars, represents only one third of the interest due by the developing countries on their debts to the very same West and in which a net export of capital takes place from the poor countries to the rich West, cannot be called anything but paradoxical - certainly if the situation were to go on for any length of time.

It would be far better to introduce a 1986 Marshall plan under which the western governments would take over the Third World debts from the banks and then remit those debts, at any rate at least in part, all this within the framework of a programme of sound economic recovery. In this way, the world would gain greater purchasing power and more employment while at the same time a dangerous time-bomb underneath the whole world economy would be defused. The Third World would benefit from a better economic system - more than from any enlargement of the technical aid wich is also necessary - but in the short term there is only an extremely remote chance of that.

The technicians and the managers do their jobs better than the economists and politicians who ought to be providing the world with a better economic system.

The pursuit of international détente is equally in the interest of the world dairy sector. Of course, a dairy congress such as this can only make a very modest contribution towards that.

But nevertheless I would like to point out that for each tank wich is built it would also be possible to supply several small-scale village cheese factories to the Third World, and for each jet fighter a complete milk factory capable of providing the liquid milk requirements of a medium-sized city. Then - accordingly to Isaiah 2:4: - ... they shall beat their swords into plowshares and their spears into pruning hooks.

The fifth - and final - <u>conclusion</u> is that there remains considerable scope for improvement of international cooperation.

8. CLOSE

The dairy sector heading for the year 2000!

It is clear that the dairy industry in the developed world is passing through a period of turbulence and has difficult years ahead. In the western countries, rapid technological progress and stable demand will result in declining numbers of jobs.

The jobs that remain, however, will be highly qualified ones. In Eastern Europe there seems to be scope for some further growth of milk production, whilst in the Third World a recovery in purchasing power may well lead to great opportunities for the dairy sector.

Major aims to go for are: reinforcement of the position of dairy farmers, which can help to bring about fair incomes for them, a flourishing world economy, a fair international division of work, and a reduction of the international tensions within and across the world. I should like to close by expressing the wish that this congress may contribute towards the attainment of those objectives.

EVOLUTION DE LA PRODUCTION LAITIERE BOVINE ET DES CARACTERISTIQUES
DU LAIT DANS LES PAYS TEMPERES

R. JARRIGE
I.N.R.A. - Centre de Recherches Zootechniques et Vétérinaires
de Theix
63122 Ceyrat
France

RESUME. La production laitière par vache traite va continuer à augmenter, plus particulièrement dans les populations Pie noires, par suite 1°) de la diffusion massive et irréversible des gènes des races nord américaines (Holstein Friesian et autres), 2°) d'une utilisation plus intense des programmes de sélection des taureaux d'insémination artificielle et 3°) des améliorations continues dans l'alimentation, la protection sanitaire et le management des vaches. Les modifications de la composition des laits traités par l'industrie seront d'importance limitée et peuvent être plus ou moins prévues. Il est indispensable de prendre en compte la richesse du lait, en protéines notamment, dans les programmes de sélection. L'application des règles d'hygiène bien connues permet de limiter la contamination microbienne du lait. La fréquence des mammites sub-cliniques peut être ramenée jusqu'à un seuil acceptable. Le paiement du lait suivant ses qualités prendra une importance croissante.

1. EVOLUTION RECENTE

Depuis une trentaine d'années la production laitière est en voie continue d'intensification, de spécialisation et de concentration, en Amérique du Nord, Europe de l'Ouest et Océanie. Le nombre de troupeaux a diminué au rythme moyen de 3 à 5 % par an dans la plupart des pays au cours des 10 dernières années, surtout du fait de la disparition des plus petits. Les autres se sont agrandis par une augmentation de la surface de l'exploitation et (ou) de la charge à l'hectare. La taille moyenne des troupeaux et sa dispersion sont cependant très variables entre pays et entre régions. La productivité du travail a été multipliée grâce aux bâtiments en stabulation libre et à la mécanisation de la traite, de l'alimentation des vaches et de l'évacuation des déjections. Le capital investi s'est alourdi.
   Cette évolution a été permise, facilitée ou stimulée par l'augmentation continue de la production laitière par vache (de 50 à 130 kg par an selon les pays entre 1974 et 1984), conséquence de l'amélioration conjuguée du potentiel laitier des vaches et des conditions de leur

exploitation.

Le potentiel laitier, sous contrôle du patrimoine génétique, est fixé par le nombre et l'activité des cellules sécrétrices. Il peut être exprimé dans un premier temps par la quantité de lait car celle-ci rend compte pour environ 80 % des variations des quantités sécrétées de matières grasses, de protéines et d'autres constituants, aussi bien entre vaches d'une même race qu'entre régimes alimentaires. L'augmentation continue du potentiel laitier a été réalisée par le remplacement des races locales à deux fins par des races laitières spécialisées, surtout du type Pie noir, et par une sélection de plus en plus efficace : extension du contrôle laitier qui concerne maintenant de un tiers à deux tiers des vaches des grands pays laitiers ; testage des taureaux d'insémination artificielle, plus de 70 % des vaches laitières étant inséminées ; accouplements des meilleurs géniteurs pour engendrer les taureaux de la génération suivante.

La Holstein Friesian nord américaine avait acquis un potentiel laitier d'environ 20 % supérieur à celui des populations Pie noires européennes (cf Cunningham, 1983, 1985). Depuis une dizaine d'années, elle a donc envahi progressivement ces dernières. La proportion des gènes Holstein dans les jeunes taureaux Pie noirs mis en testage, qui est une des mesures de cette pénétration, atteint en 1983-4 de 80 à 90 % en Allemagne et en France, 70 % au Danemark et aux Pays-Bas, etc. De son côté, la Brown Swiss est en voie d'absorber les populations Brunes de l'Europe alpine (Suisse, Italie, Autriche, Allemagne) et la Red Holstein est utilisée sur certaines populations Pie rouges. Cette invasion nord américaine touche aussi les troupeaux de l'Europe de l'Est. Les vaches de type Pie noir représentent aujourd'hui près de la moitié des vaches traites de l'ensemble des pays développés ; elles sont environ 11,5 millions en Amérique du Nord (90 % du total), 18 millions en Europe de l'Ouest (55 %), 20 millions en Europe de l'Est (35 %) et quelques millions pour les autres pays : Australie, Japon, Nouvelle Zélande ... Elles ont de plus en plus de sang Holstein.

Les vaches spécialisées exigent une plus grande quantité d'aliments, et de meilleure qualité, pour exprimer leur potentiel laitier accru. Cela a été réalisé par la distribution plus libérale d'aliments concentrés de mieux en mieux équilibrés (sources d'azote, minéraux) et dont le coût relatif diminuait, l'extension vers le Nord de la culture du maïs pour l'ensilage, une meilleure exploitation des pâturages, la conservation des excédents d'herbe de printemps par l'ensilage, l'application des progrès des connaissances sur l'ingestion, la digestion (rumen) et le métabolisme des vaches laitières.

Le rendement biologique de la transformation en lait de l'énergie et des éléments nutritifs des aliments, ou efficacité alimentaire, augmente avec le potentiel laitier (revue de Gravert, 1985) ; une proportion plus élevée de l'énergie ingérée est en effet exportée dans le lait puisque les dépenses d'entretien ne sont pratiquement pas modifiées. Malgré un coût plus élevé des aliments supplémentaires, la marge (recettes laitières - prix de l'alimentation ou des aliments achetés) augmente dans la plupart des situations. Les vaches à haut potentiel laitier sont celles qui répondent le mieux à l'amélioration de l'alimentation, du management et de la protection sanitaire.

La production laitière totale a augmenté beaucoup plus vite que les besoins des consommateurs de l'Amérique du Nord, puis de la plupart des pays de l'Europe de l'Ouest. Pour la maîtriser, les gouvernements ont pris des mesures conduisant à réduire le nombre des vaches laitières. Celui ci a régressé de moitié en Amérique du Nord en une trentaine d'années (de 21 à 11 millions aux U.S.). Il était resté à peu près stable en Europe. Il vient de diminuer brutalement dans la C.E.E. (3.6 % en 1984 et 2.9 % en 1985) à la suite de l'introduction des quotas. C'est seulement en U.R.S.S. (43 millions de vaches traites actuellement) qu'il devrait continuer à augmenter.

2. POURSUITE DE L'ACCROISSEMENT DE LA PRODUCTION LAITIERE PAR VACHE

2.1. Possibilités d'accroissement continu du potentiel laitier

En 1981, les spécialistes (Crowley et Niedermeier, 1981) prévoyaient la poursuite de l'évolution du troupeau laitier des U.S.A. pour les 25 années à venir avec un accroissement minimum de 25 à 30 % de la production par vache, un triplement de la taille moyenne des troupeaux et sans doute une nouvelle réduction du nombre de vaches. Qu'en sera-t-il dans la C.E.E. ? Pour produire leur quota de lait, les éleveurs choisiront des solutions qui se situeront entre deux extrêmes : a) avoir un nombre de vaches minimum de la production la plus élevée, en comptant diminuer le coût total et en libérant des surfaces pour d'autres productions ; b) conserver le même nombre de vaches, mais en réduisant le coût de la ration et en prêtant plus d'attention aux aptitudes pour la production de viande. L'étude sur modèle de Van Arendonk et al. (1985) aux Pays-Bas est en faveur de l'orientation (a) mais en accordant plus de poids à la richesse du lait en matières grasses et protéines qu'avant les quotas ; celle de Niebel et Fewson (1985) est en faveur de l'orientation (b) dans le cas des vaches à deux fins de la Bavière. La première orientation devrait être dominante dans le prolongement de l'évolution récente, tout au moins dans les troupeaux Pie noirs parce qu'ils sont engagés irréversiblement dans la conversion à la Holstein Friesian.

Le potentiel génétique laitier des populations Pie noires va continuer à s'accroître pendant bon nombre d'années aussi rapidement qu'auparavant, sinon plus, du fait de la poursuite de la conversion à la Holstein et d'une sélection plus efficace. Il faut bien voir que la généralisation des programmes de sélection laitière performants est relativement récente, et d'abord en Amérique du Nord.

Le potentiel laitier des animaux Holstein inscrits a stagné dans les années 1960 (mais il s'améliorait chez les vaches non inscrites) ; il ne s'est mis à augmenter qu'au cours des années 1970 (en moyenne de 84 kg de lait par an pour les mâles et de 51 kg pour les femelles) à la suite de l'adoption, en 1968, d'une méthode d'indexation plus précise des taureaux (Lee, Freeman et Johnson, 1985). En France, l'accroissement annuel du potentiel laitier moyen de la population Pie noire est passé de 75 kg par vache entre 1975 et 1980 à 125 kg depuis 1981 à la suite d'une utilisation plus intense des programmes de sélection et de l'infusion croissante des gènes Holstein ; il devrait se maintenir à ce niveau dans les années à venir si les objectifs et les efforts de sélection ne sont pas changés (Bloc, Guerin et Mocquot, 1986). Le progrès génétique semble être du même ordre dans les autres pays européens, leurs programmes de sélection ayant des potentialités équivalentes. L'augmentation de la production par vache est généralement supérieure à celle du potentiel laitier par suite de l'amélioration simultanée des conditions d'exploitation.

A l'utilisation optimale des programmes de sélection, viennent s'ajouter les potentialités de la manipulation des embryons : superovulation, collecte, congélation, transplantation (en moyenne à l'âge de 7 jours), division, sexage (cf Seidel, 1984). Elle permet d'obtenir en moyenne de 5 à 6 embryons de bonne qualité par collecte, dont 50 à 60 % donneront naissance à des veaux. L'éleveur individuel peut l'utiliser pour produire de meilleures femelles pour l'amélioration génétique de son troupeau et/ou pour la vente. C'est sans doute la raison majeure de l'extension de la technique : le nombre de veaux de race laitière issus d'embryons transférés est évalué pour l'année 1984 à 25 000 aux U.S.A., à 3 000 en France, etc. Mais c'est la procréation des futurs reproducteurs mâles par les vaches d'élite (qui n'avaient parfois jamais de veaux mâles) inséminées par les meilleurs taureaux, qui peut apporter une accélération du progrès génétique de la population (de 20 à 30 %), à la condition d'utiliser à plein le raccourcissement de l'intervalle entre générations en transférant des embryons de génisses ou de jeunes vaches. Cela serait réalisé dans des noyaux de sélection dont des modèles ont été proposés (Nicholas et Smith, 1983 ; Colleau, 1985). La division des embryons à l'âge de 7 à 8 jours (Ozil, Heyman et Renard, 1982) permet d'obtenir en moyenne plus de veaux par embryon récolté ; elle est intéressante lorsque les vaches donnent peu d'embryons en réponse au traitement de superovulation ou que l'éleveur ne veut pas appliquer ce traitement pour ne pas risquer de perturber la production ou la reproduction de ses vaches d'élite. De premiers succès ont été obtenus pour déceler le sexe des embryons dans l'espèce bovine, ce qui permettrait de réduire le nombre des embryons transférés. Il est trop tôt pour connaître les applications possibles des travaux actuels sur la culture des embryons associée à leur division pour créer de véritables clones de vaches d'élite ainsi que sur le transfert de gènes.

Comme le souligne Kennedy (1984), on ne voit pas de limite, ou de plateau, à l'amélioration du potentiel laitier dans un futur proche. La production moyenne des vaches Pie noire contrôlées dépasse 8 000 kg aux U.S.A., 6 800 kg au Canada et varie de 5 000 kg à 6 300 kg dans la C.E.E. ; elle augmente de 50 à 150 kg par an selon les pays. La barre des 10 000 kg est dépassée par un nombre important de troupeaux en Amérique du Nord ; il est atteint ou presque par les meilleurs troupeaux européens. Deux vaches ont produit près ou plus de 25 000 kg en une seule lactation.

2.2. Stimulation de la réalisation du potentiel laitier

La réalisation du potentiel laitier peut être limitée par une traite incomplète et par le nombre de traites. La troisième traite journalière, qui avait pratiquement disparu pour économiser de la main d'oeuvre, retient à nouveau l'attention depuis quelques années. Elle a été introduite dans environ 10 % des troupeaux contrôlés de la Californie afin de mieux rentabiliser les installations modernes dans des grandes unités. Elle s'accompagne d'une augmentation moyenne de la production de 12 % (de - 2 à 32 % selon les troupeaux) mais aussi d'une légère diminution du taux butyreux en première lactation et de la fertilité (Gisi, de Peters et Pelissier, 1986). En station expérimentale, les vaches adultes traites trois fois ont produit 17 et 13 % (dans le même sens que la quantité de concentré reçue) de lait de plus que celles traites deux fois, (de Peters, Smith et Acedo-Rico, 1985), cela sans consommer plus d'aliments mais en prenant moins de poids.

L'administration sous-cutanée ou intra-musculaire d'hormone de croissance ou somatropine bovine (bGH) est la voie la plus spectaculaire pour extérioriser le potentiel sécrétoire de la mamelle. Cette action est connue depuis les essais de Brumby et Hancock (1955) en Nouvelle-Zélande à l'aide d'une préparation naturelle relativement impure. La possibilité récente de faire produire l'hormone par des bactéries "implantées" a permis deux essais de longue durée, dont celui de Cornell University (Eppard et Bauman, 1984 ; Bauman et al., 1985) a fait grand bruit. Des vaches Holstein de très bon potentiel laitier ont reçu, à partir du 84ème jour après le vêlage, une injection journalière de bGH bactérienne (méthionyl) à différentes doses. Leur production laitière journalière a immédiatement augmenté jusqu'à un plateau plus élevé que le pic normal atteint au début du 2ème mois. Pour les 188 jours de l'expérience, elle a été accrue en moyenne de 6,5 (23 %), 10,1 (36 %) et 11,5 kg (41 %) selon la dose administrée (13,5 ; 27,0 et 40,5 mg/j respectivement) par rapport à la production du lot témoin qui était pourtant très élevée (27,9 kg). Cette augmentation a été plus forte que celle déterminée par l'injection de bGH naturelle (4,6 kg soit 16 %). Elle a été couverte par une augmentation correspondante de l'ingestion et n'a entraîné aucune modification de la teneur du lait en matières grasses, protéines et lactose et aucun trouble de santé et de reproduction. Ces résultats ont été entièrement confirmés (+ 3,5 kg de lait soit 18 %) dans l'autre essai de longue durée (22 semaines) qui a été réalisé en Australie avec des injections journalières de bGH naturelle (Peel et al., 1985).

Des essais de courte durée apportent quelques précisions complémentaires. La réponse au traitement est beaucoup plus limitée au tout début de la lactation (Peel et al., 1983 ; Richard, Mc Cutcheon et Bauman, 1985). Quand la vache est en déficit énergétique et azoté, la teneur en protéines du lait diminue sensiblement et la teneur en matières grasses augmente nettement (Tyrrell et al., 1982) ; la proportion des acides gras à longue chaîne dans ces graisses (Eppard et al., 1985) et la concentration des acides gras non estérifiés dans le sang augmentent aussi, ce qui traduit la mobilisation des lipides du tissu adipeux au bénéfice de la mamelle.

L'accroissement de la bGH circulante à la suite de l'administration de bGH exogène accroît l'activité sécrétoire de la mamelle, sans doute par les mêmes voies que la multiplication des traites ou des tétées. Elle orchestre un accroissement de l'ingestion et un ensemble de processus métaboliques au service des besoins accrus de la mamelle (revue de Bauman et Mc Cutcheon, 1986).

Les spectaculaires accroissements de production obtenus à Cornell font de la GH un produit rentable pour les fermes laitières de l'Etat de New-York, sous réserve de son coût de production (Kalter, 1984). Ils ont stimulé la production bactérienne de cette hormone par plusieurs firmes et de nombreuses recherches en cours. Avant d'envisager son agrément et sa mise en application, beaucoup d'aspects restent cependant à préciser et à améliorer. Ils concernent : 1°) les modalités d'administration car l'injection journalière prend du temps. Or la stimulation de la production laitière est liée directement au maintien d'un taux sanguin élevé de GH ; avec une injection un jour sur deux, elle est déjà deux fois plus faible qu'avec une injection quotidienne ou une instillation continue (Mc Cutcheon et Bauman, 1986), 2°) l'influence de ce forçage sur la santé de la vache à l'échelle de plusieurs lactations consécutives. Des essais en ferme sont nécessaires. L'utilisation de la GH, implique des rations d'une valeur nutritive et d'une ingestibilité très élevées et un management optimum ; 3°) le passage éventuel dans le lait de l'hormone avec son support et, dans ce cas, son devenir dans le tube digestif (où elle est probablement digérée comme tout polypeptide) et sous l'action des microbes au cours de la conservation et de la transformation du lait.

Des réponses favorables ayant été apportées à ces questions biologiques, la mise en application de la GH dépend de l'autorisation par les services compétents mais aussi d'une politique technique et économique de l'élevage. En face de ses potentialités, il faut mettre toutes les conséquences négatives de l'intensification et de la concentration de la production laitière qu'elle accentuerait (cf plus loin), la désorganisation de la sélection, les problèmes éthiques soulevés par ce traitement et un risque de dévalorisation de l'image de marque du lait dans l'esprit des consommateurs.

2.3. Facteurs biologiques limitant l'accroissement du potentiel laitier

2.3.1. Capacité d'ingestion et alimentation insuffisantes.

En moyenne, la capacité d'ingestion des vaches augmente en même temps que le potentiel laitier pour faire face à l'accroissement des besoins nutritionnels (cf Jarrige, 1986). Par exemple, les Holstein ont consommé plus que les Frisonnes hollandaises dans la comparaison réalisée aux Pays-Bas (Oldenbroek et Van Eldik, 1980). Cependant la capacité d'ingestion est fondamentalement limitée par le volume du rumen qui, lui, dépend de la taille de la vache plus que du potentiel laitier. En outre, au début de la lactation, elle augmente beaucoup plus lentement que les quantités de protéines et d'énergie exportées. De ce décalage résulte un déficit inévitable, dont l'intensité et la durée augmentent en même temps que le potentiel laitier, et qui ouvre la porte à des désordres digestifs et métaboliques (cétoses) et à un retard dans la fécondation. Heureusement, on sait de mieux en mieux constituer des rations qui sont consommées en quantité maximum, digérées sans problèmes et qui apportent les produits terminaux de la digestion dans les proportions les mieux adaptées aux besoins de la mamelle. On sait aussi orienter et optimiser l'activité de la population microbienne du rumen. On peut enrichir la ration en substances (protéines traitées par le formol ...) qui échappent à la dégradation dans le rumen et apportent dans l'intestin grêle les quantités complémentaires d'acides aminés, d'acides gras longs et de glucose nécessaires. Les progrès vont se poursuivre dans ces différentes voies. Ils seront mis en application grâce à l'amélioration des machines (rations complètes), des dispositifs de distribution programmés des aliments concentrés et de la gestion informatisée. Le régime alimentaire pourra accompagner l'accroissement du potentiel laitier aussi longtemps que cela sera rentable. Les éleveurs des troupeaux d'élite savent déjà alimenter des vaches d'un potentiel laitier supérieur à celui des stations expérimentales !

2.3.1. Fertilité et santé. Au plan génétique, la fertilité des génisses semble liée positivement à leur production ultérieure. La liaison devient négative lorsqu'elles entrent en production, mais son amplitude est limitée et décroît quand le nombre de lactations augmente (revues de Philipsson, 1981 et de Freeman, 1984). Le milieu hormonal, le stress de lactation et le déficit nutritionnel des deux premiers mois expliquent cet antagonisme. La fertilité a une héritabilité faible (< 0,10) ou nulle et dépend avant tout des conditions d'alimentation au début de la lactation et du management (détection des chaleurs, etc) ; c'est pourquoi elle peut être meilleure dans les troupeaux les plus productifs que dans les plus mauvais (Laben et al., 1982 ; O'Connor et al., 1985). En plus des difficultés de reproduction, les vaches laitières souffrent de mammites et de troubles de la locomotion, de la digestion et du métabolisme (fièvre vitulaire, cétoses), surtout au début de la lactation. La fréquence et le coût total (pertes, médicaments, travail) de toutes ces maladies augmentent en même temps que le potentiel laitier mais beaucoup moins que le bénéfice apporté par la production supplémentaire lorsque ces troubles sont correctement maîtrisés (Shanks et al., 1978 ; Hansen et al., 1979).

Dans tous les pays, les troupeaux les plus productifs ont à la fois les vaches du potentiel le plus élevé et les meilleures techniques d'alimentation et de conduite. L'analyse des résultats de plus de 6 000 troupeaux du Minnesota (Steuernagel, 1982) montre que les troupeaux d'élite (38 avec une moyenne de 9 500 kg et un pic de lactation de 44 kg pour les adultes) ont le même taux de fécondation que les troupeaux dits moyens (1 110 à 6 600 kg), un intervalle entre vêlage et une durée de tarissement sensiblement plus faibles mais un taux de renouvellement un peu plus élevé (34 % de primipares au lieu de 31 %) ; la valeur de leur production après déduction du coût de l'alimentation est plus élevée de moitié, dans les conditions économiques locales.

2.3.3. Sélection simultanée sur d'autres aptitudes. La prise en compte d'autres caractères retarde obligatoirement l'amélioration du potentiel laitier en raison d'une diminution de l'intensité de sélection laitière et parfois d'un certain antagonisme avec le potentiel laitier. C'est le cas pour la richesse du lait, qui sera examinée plus loin, et pour l'aptitude à la production de viande. Celle-ci reste sans intérêt en Amérique du Nord parce que la viande bovine y est produite par le troupeau de vaches à viande qui est 4 fois plus nombreux que celui des vaches laitières. La situation est différente en Europe où le troupeau laitier fournit de 60 % (France) à la totalité de la viande bovine. Le nombre de vaches laitières régresse et la conversion à la Holstein entraîne une diminution du rendement en carcasse, de la conformation et de la charnure de la carcasse, ainsi que de l'efficacité alimentaire des mâles à l'engrais.

Les conséquences à en tirer pour les programmes de sélection ont fait l'objet d'un récent séminaire de la C.E.E. (Krausslich et Lutterbach, 1985). L'intérêt national conduit, dans la plupart des pays, à donner au moins une information sur les aptitudes bouchères des taureaux à partir de leurs propres performances ou de celles de leurs descendants veaux de boucherie. Cependant, dans les conditions de prix actuelles, l'éleveur spécialisé de vaches Pie noires ne freinera probablement pas l'accroissement du potentiel laitier pour améliorer, ou simplement pour maintenir la valeur commerciale de ses veaux mâles, parce que le premier accroît infiniment plus ses recettes. En revanche, l'éleveur de vaches laitières à deux fins peut accorder plus de poids aux aptitudes bouchères.

2.4. Facteurs économiques limitant l'accroissement du potentiel laitier

Le potentiel laitier optimum d'un troupeau doit être en harmonie, mais en le dépassant d'une certaine marge (disons de 20 à 25 %), avec le niveau de production qui peut être atteint de façon rentable. Ce dernier dépend de 3 catégories de facteurs :
- la compétence de l'éleveur au plan technique (traite, alimentation, surveillance, soins) et aussi au plan économique (gestion, organisation). Si elle est insuffisante par rapport au potentiel laitier du troupeau, la production sera plus faible, les risques et dépenses sanitaires accrus, la reproduction mauvaise et le renouvellement plus rapide, même si les dépenses alimentaires sont aussi élevées, voire plus ;
- le coût des facteurs de production, en premier lieu de l'alimentation puisqu'elle représente couramment de 60 à 70 % du coût total de production du lait. Le coût des fourrages varie selon les milieux. Le rapport du prix du lait à celui de l'aliment concentré a une importance déterminante puisque les aliments concentrés couvrent la majeure partie des besoins nutritionnels supplémentaires consécutifs à l'accroissement du potentiel laitier. Il varie de façon importante selon les conditions du marché, les pays et les régions, par exemple de 0,8 à 1,4 (Pays-Bas) dans les pays de la C.E.E. Les valeurs élevées de ce rapport ont stimulé jusqu'ici l'augmentation de la production laitière par vache et par hectare. Sa réduction, notamment à la suite d'une diminution relative du prix du lait, conduirait à accroître la part de l'herbe pâturée et des fourrages conservés et à limiter l'augmentation du potentiel laitier, tout au moins dans la plupart des pays européens ;
- les caractéristiques et l'orientation de l'exploitation. D'une façon générale, l'accroissement de la taille du troupeau, surtout par

un nombre de vaches à l'hectare plus élevé, diminue le poids des charges de structure dans le coût de production. Le niveau de production laitière optimum varie aussi avec l'importance des autres recettes du troupeau (veaux, etc) et de l'exploitation, ainsi qu'avec la quantité de travail disponible.

L'ensemble de ces facteurs explique que le coût de production du lait soit minimum dans les troupeaux produisant environ 8 000 kg de lait à 36 g/kg de matières grasses dans l'Etat de New-York (Allaire et Thaen, 1985) et de 40 % plus faible en Nouvelle Zélande (3 475 kg au pâturage toute l'année complété par de l'herbe conservée) qu'en Californie (7 140 kg en stabulation permanente avec des aliments achetés (Scott, 1981). Dans les grands pays laitiers ont été, ou seront proposés des modèles qui tiennent compte de tous les facteurs intervenant dans le revenu et de leurs interactions pour aider l'éleveur à adapter le potentiel et le management de son troupeau aux conditions économiques.

L'accroissement du potentiel laitier et de son extériorisation semble correspondre aux besoins des grandes unités hors sol d'Amérique du Nord et de tous ceux des éleveurs spécialisés qui se sont modernisés ont accru leurs effectifs et disposent de ressources alimentaires au prix le plus bas. Il apparaît donc inexorable pour la population des vaches Pie noires et pour les autres races laitières spécialisées européennes. La limitation de la production offre cependant une chance de se maintenir aux plus laitières des races à deux fins, surtout si les prix de la viande par rapport au lait devenaient plus favorables.

La poursuite de cette intensification et de la concentration de la production laitière est-elle utile à la communauté des éleveurs et à l'économie des états ? C'est une autre histoire. Disons simplement que cette évolution, en général nécessaire jusqu'à un certain niveau, a contribué à marginaliser ceux qui n'avaient pas les moyens de la suivre à engager la plupart des autres dans un endettement exagéré, à accentuer le dépeuplement des régions les moins propices, à gonfler les excédents de beurre et de lait écrémé en poudre en même temps que le nombre des chômeurs. Ce n'est certainement pas un modèle à proposer aux nombreux pays en développement qui ont besoin d'accroître leur production laitière pour mieux alimenter leur population, d'autant plus qu'ils n'ont généralement pas de céréales à distribuer aux vaches.

3. COMPOSITION ET QUALITE DES LAITS

3.1. Conséquences des changements génétiques

Les différences entre races à l'intérieur d'un pays sont bien connues. Ainsi la substitution des Pie noires aux races Jersey, Guernesey ou Normande entraîne un appauvrissement du lait. Le lait des Holstein Friesian des U.S.A. contient en moyenne 36 g matières grasses et 31,5 de protéines totales (N x 6,38) par kg. Il est plus pauvre que celui des Pie noires européennes, par exemple de 4,0 et 1,5 g respectivement par rapport à celles des Pays-Bas (Oldenbroek, 1985). L'absorption de la Pie noire européenne par la Pie noire américaine, risque donc d'entraîner un appauvrissement du lait. Elle est probablement en partie responsable de la tendance à la diminution de la teneur en protéines observée entre 1981 et 1984 chez les vaches soumises au Contrôle laitier dans plusieurs pays européens (Danemark, Royaume Uni).

A la suite des travaux italiens sur la transformation en Parmigiano (Mariani et al., 1976) on sait que les variants B de la caséine κ et de la caséine β ont une action favorable sur l'aptitude à la coagulation. Or leur fréquence est beaucoup plus faible dans les populations Pie noires (Losi et Mariani, 1984 Mc Lean, Graham et Ponzoni, 1984, F. Grosclaude, communication personnelle). Si le variant B de la caséine β est beaucoup trop rare (<1 %) chez les Holstein, la fréquence de celui de la caséine κ pourrait être améliorée par la sélection, d'autant mieux qu'il est est en relation positive avec la production laitière en première lactation (Lin et al., 1986).

Les teneurs en matières grasses et protéines du lait ont une héritabilité environ deux fois plus élevée que les quantités de lait, de matières grasses et de protéines par lactation (Maijala et Hanna, 1974, Gaunt, 1980). Elles sont en corrélation génétique positive entre elles, nulle ou légèrement positive avec les quantités de matières grasses et de protéines, et négative avec la quantité de lait. La sélection sur la teneur en matières grasses (et corrélativement sur la teneur en protéines) a été très efficace dans la Pie noire européenne mais au détriment de l'amélioration de la quantité de lait. La sélection exclusive sur la quantité de lait accroît simultanément les quantités de matières grasses et de protéines et elle en diminue les teneurs, sauf si on veille à ne pas les laisser tomber au dessous d'un certain seuil.

L'accroissement des quantités a toujours été jusqu'ici le plus rentable pour les éleveurs quels que soient les modes de calcul du prix du lait, même après l'introduction de la teneur en protéines. Cependant, au cours des dernières années, on a envisagé d'accorder un poids plus important aux teneurs pour mieux tenir compte des souhaits des transformateurs, voire des besoins nationaux. Un enrichissement du lait permet des économies d'énergie dans le refroidissement, le transport et le séchage. La part croissante du lait qui est transformée en fromages conduit à privilégier la teneur en caséines. La C.E.E. encourage un accroissement du rapport protéines : matières grasses ; les quotas ne portant que sur la quantité de lait, ajoutent de l'intérêt à l'enrichissement du lait, etc. Les index des taureaux d'I.A. commencent à tenir compte de ces modifications dans le calcul du prix du lait. Celui utilisé aux Pays-Bas depuis 1980, l'I.N.E.T. combine les valeurs génétiques pour la quantité de lait (kg) et les teneurs en matières grasses et en protéines (%) avec des coefficients de pondération de 0,316, 260 et 500 respectivement (Dommerholt et Wilmink, 1986). L'Allemagne et le Danemark envisagent une orientation semblable. En France, l'indexation des taureaux d'I.A. depuis deux décennies sur la quantité de matière utile = quantité de lait x 0,5 (teneur en matières grasses + 1,21 teneur en protéines) a permis de maintenir la richesse du lait des vaches Pie noires soumises au contrôle laitier malgré l'accroissement des gènes Holstein. Il est envisagé un index synthétique : quantité moyenne de matière utile x 2 taux moyen de matière utile (Bonaiti, 1985, Gastinel et Mocquot, 1986). Il faut à plus forte raison tenir compte de la richesse du lait en protéines dans la sélection des races autres que les Pie noires (Montbéliarde, Normande, etc). L'amélioration génétique de la composition du lait a de l'intérêt pour tous, à la différence de celle du potentiel laitier.

3.2. Variations d'origine alimentaire

3.2.1. **Apport énergétique**. Chez les vaches de potentiel élevé, le déficit énergétique risque d'être plus long en début de lactation et aussi plus fréquent par la suite, car elles essayent de maintenir leur production laitière à partir de leurs réserves corporelles lorsque la ration est insuffisante. Le déficit énergétique modifie peu la teneur en matières grasses et peut même l'augmenter en début de lactation, mais il diminue systématiquement la teneur en protéines. Cette diminution est en moyenne de 0,5 g par unité fourragère manquante (correspondant à l'énergie nette nécessaire à la production de 2,2 à 2,5 kg de lait) d'après la moyenne des 38 essais récapitulés par Rémond (1986) ; elle semble s'accroître au fur et à mesure que s'allonge la période de déficit intense ; elle peut ainsi dépasser 2 g/kg, à la fin du 2ème mois de lactation. Le déficit énergétique vient donc diminuer la teneur en protéines du lait à deux périodes critiques pour elle : d'une part, au 2ème et au 3ème mois de lactation, alors qu'elle est à son minimum physiologique ; d'autre part pendant la sécheresse d'été (juillet) (revue de Journet et Rémond, 1980) où elle est déjà diminuée par la longueur des jours et, à un moindre degré, par les températures élevées. La teneur en matières grasses diminue aussi à cette période.

La quantité d'aliments concentrés qui est nécessaire pour couvrir le plus tôt possible les besoins énergétiques des fortes productrices représente couramment plus de 40 % de la ration hivernale (et beaucoup plus si on lui ajoute le grain de l'ensilage de maïs). La teneur en matières grasses du lait a tendance à diminuer au fur et à mesure que cette proportion augmente - la teneur en protéines étant maintenue ou accrue - en relation avec l'augmentation de la production d'acide propionique dans le rumen à partir de l'amidon des céréales (revues de Sutton, 1981 et de Journet et Chilliard, 1985). On connaît les moyens de prévenir ces accidents. Il s'agit de distribuer à volonté des fourrages d'excellente qualité qui soient consommés en quantité maximum, de choisir la composition des concentrés (sous produits à la place d'une partie des céréales ; substances tampon) et d'étaler leur ingestion.

L'introduction de graisses dans le concentré permet elle aussi de diminuer la proportion d'amidon tout en augmentant, ou en maintenant, la concentration énergétique, et d'apporter des acides gras longs à 18 C dont certaines rations sont assez pauvres. L'influence sur la teneur en matières grasses est toujours négative dans le cas des huiles végétales riches en acides gras insaturés mais variable dans celui des graines oléagineuses (surtout de coton) et des graisses animales (saindoux) (revues de Palmquist et Jenkins, 1980 et de Storry, 1981). Deux essais en ferme confirment ces tendances : diminution de 2 g ‰ dans l'Ohio avec un mélange de graisses animales et végétales, amenant la teneur en graisses de la ration à 6 ‰ (Heinrichs, Hoyes et

Palmquist, 1981) ; maintien ou même augmentation au Danemark quand la quantité de saindoux ajoutée passe de 200 à 600 g/jour, en complément d'une ration pauvre en matières grasses (betteraves, ensilages d'herbe), (Ostergaard et al., 1981). La teneur en protéines n'a pas été modifiée dans le premier cas ; elle a le plus souvent diminué dans le deuxième, ce qui a été le cas dans les trois quarts des essais sur les troupeaux expérimentaux (revues de Emery, 1978 et de Morand-Fehr, Chilliard et Bas, 1986).

Les graisses enrobées dans des protéines selon une méthode australienne ne perturbent pas les fermentations dans le rumen et y sont protégées de l'hydrogénation. Leur incorporation dans la ration entraîne une augmentation importante de la teneur en matières grasses du lait en même temps qu'une diminution systématique de la teneur en protéines (revue de Storry, 1981). Elle permet, plus sûrement que les graisses non protégées, de modifier la composition des graisses du lait, à des fins diététiques (enrichissement en acides gras polyinsaturés) ou technologiques (beurre plus facile à tartiner). Le coût élevé et la fiabilité insuffisante du procédé ont jusqu'ici empêché son utilisation.

3.2.2. Régimes alimentaires. D'après les nombreuses comparaisons réalisées dans les stations expérimentales, (revue de Hoden, Coulon et Dulphy, 1985), le remplacement du foin par l'ensilage d'herbe ne modifie pas la teneur en matières grasses ou l'augmente légèrement ; il diminue sensiblement la teneur en protéines (de l'ordre du g par kg) sauf pour les ensilages d'excellente qualité. Il accroît considérablement les teneurs en carotène et vitamine A. Par rapport au foin ou à l'ensilage d'herbe, l'ensilage de maïs correctement pourvu en grains, accroît la teneur en matières grasses de 3 à 4 g/kg et la teneur en protéines de 1 à 2 g ; cela, par l'intermédiaire d'une augmentation de la production d'acide butyrique dans le rumen et du niveau d'alimentation énergétique respectivement.

Les laits des vaches consommant des ensilages d'herbe, et à un degré moindre, des ensilages de maïs, contiennent des nombres de spores de Clostridium tyrobutyricum beaucoup plus élevés, et plus variables, qu'avec les rations de foins et de betteraves, ce qui a des conséquences désastreuses pour certains fromages. La chaîne de cette contamination depuis le sol de la prairie est bien connue, de même que les règles pour la réduire : d'abord limiter la contamination de l'ensilage (et par là même dans l'atmosphère) en récoltant une herbe non souillée par de la terre et en lui assurant une bonne conservation ; limiter ensuite la contamination du lait par les matières fécales, et aussi par l'air (lactoducs), en appliquant rigoureusement les règles d'hygiène et de propreté, notamment lors de la traite. La majorité des producteurs peut ainsi produire régulièrement des laits contenant moins de 500 spores, seuil au-dessous duquel les altérations des fromages à pâte cuite sont rares et faibles. Mais il suffit de peu de laits fortement contaminés pour dépasser rapidement ce seuil. A l'exception de l'addition d'oxydants tels que les nitrates, qui sont autorisés seulement aux Pays-Bas, les industriels ne disposent que de palliatifs. L'interdiction de l'utilisation des ensilages dans les zones de fabrication traditionnelles de Gruyère et d'Emmental (Suisse, France, Allemagne) reste la seule méthode sûre pour protéger la qualité de ces fromages ; l'autorisation d'emploi existant en Finlande est accompagnée d'une discipline et d'un encadrement très étroits.

3.3. Altérations dues aux mammites

Les mammites sont devenues le principal fléau des élevages laitiers dans tous les pays. Elles sont un des plus difficiles à combattre en raison de la multiplicité des espèces bactériennes en cause comme des causes qui les favorisent. Elles entraînent de nombreuses modifications de la composition du lait (revue de Kitchen, 1981) dont certaines sont utilisées pour déceler ces infections subcliniques : diminution des teneurs en lactose et en K +, accroissement des teneurs en Na+, Cl- et surtout du nombre de cellules somatiques. Les teneurs en matières grasses et en caséines sont sensiblement diminuées, cette dernière diminution étant masquée par une augmentation de la sérum albumine et des immunoglobulines. De plus, les micelles de β caséine sont en partie solubilisées sous l'action de la plasmine venue du sang, et à un moindre degré, des protéinases des cellules somatiques (Barry et Donnelly, 1981). Ces modifications altèrent les propriétés technologiques du lait, plus particulièrement pour la fabrication de fromages. Quand le nombre de cellules somatiques dans le lait augmente, le temps de coagulation pour la présure s'accroît, la fermeté du caillé et le rendement en fromage diminuent. Cela a été observé pour la transformation en Cheddar aussi bien au laboratoire (Ali, Andrews et Cheesman, 1980) que dans les fromageries du Wisconsin (Everson, 1984). Cette détérioration du rendement en fromage apparaît pour des nombres de cellules somatiques inférieurs à 500 000 /ml, seuil qu'approchent ou dépassent encore 30 % ou plus des troupeaux dans de nombreux pays : Angleterre et Pays de Galles, France, U.S.A., Australie ...

Les deux programmes de lutte, scandinave et anglais (Dodd et al. 1969) dirigés principalement contre les microbes responsables habituellement de la plupart des mammites (Str. agalactiae, Str. dysgalactiae et Staph. aureus) ont fait leurs preuves sur le terrain (revue de Bramley et Dodd, 1984). Mais ils sont malheureusement sans action sur les autres bactéries impliquées dans les mammites, St. uberis, E. coli, et les autres coliformes, dont les litières sont le principal réservoir. La fréquence des mammites aigues dues aux coliformes n'est pas modifiée ou même se trouve parfois accrue (Coliform Subcommittee, 1979), ce qui décourage les éleveurs. Les risques de contact (et d'infection) des trayons avec ces bactéries du milieu sont accrus par la forte concentration des vaches dans les stabulations libres, la diminution de la litière, ou son renouvellement insuffisant, et par de mauvaises dispositions pour l'évacuation des déjections. Ces facteurs de risques sont démontrés par les enquêtes épidémiologiques (Brochart, Barnouin et Fayet, 1984, Bakken, 1982).

La lutte contre les mammites comporte l'application rigoureuse, simultanée et continue des règles d'hygiène bien connues concernant les animaux, la traite et les bâtiments, ainsi qu'une utilisation correcte des machines à traire (voir Kingvill, Dodd et Neave, 1977). Elle est guidée par la numération des cellules somatiques ; depuis qu'il a été automatisé, ce test est l'outil de choix pour évaluer le niveau d'infection des troupeaux (environ 20 % de quartiers infectés pour 500 000 cellules/ml), déceler les vaches infectées (de 25 à 35 % le plus souvent), et sensibiliser l'éleveur à une perte de production qui est proportionnelle au nombre de cellules (Jones et al., 1984).

Dans les pays où on applique depuis plusieurs années des programmes de lutte appuyés sur des variations du prix du lait, les valeurs moyennes de la numération cellulaire sont à des niveaux plus bas que dans les autres : Suisse (165 000), Allemagne (235 000), Norvège (245 000), Danemark (280 000) ; cependant, elles diminuent assez peu désormais. L'introduction de ces programmes dans d'autres pays devrait être facilitée par la mise au point attendue d'appareils de détection lors de la traite. A plus long terme, d'autres progrès pourraient venir de l'accroissement de la résistance des vaches aux mammites (revues de Poutrel, 1983 et de Miller, 1984).

3.4. Récolte et conservation du lait

La traite mécanique est de plus en plus automatisée et le sera entièrement dans un proche avenir avec la pose automatique des gobelets trayeurs. La conservation du lait à 4° C à la ferme s'est généralisée. Si elle empêche le développement des bactéries lactiques et autres mésophiles, elle permet celui des bactéries psychrotrophes. Les enzymes protéolytiques (revue de Fairbarn et Law, 1985) et lipolytiques produites par ces dernières ont en général des conséquences défavorables sur les aptitudes technologiques (fromages) des laits et parfois sur la qualité des produits laitiers. On connaît les principales règles pour maintenir le nombre des psychrotrophes à un faible niveau : d'abord limiter la pollution microbienne initiale par la propreté de la machine à traire et de la traite ; ne pas dépasser la température de 4° C et un séjour de 48 heures à la ferme ; collecter le lait dans de bonnes conditions et le traiter rapidement à la laiterie.

L'action de la lipoprotéine lipase sécrétée par la mamelle se développe aussi au cours de la conservation du lait à 4° C, d'autant plus que les globules gras ont été rendus plus fragiles par les chocs mécaniques et thermiques et par l'aération subis par le lait au cours de la traite mécanique et des manipulations (revue de Chilliard et Lamberet, 1984). Cette lipolyse augmente constamment au cours de la 2èma moitié de la lactation, surtout au cours des trois derniers mois sous l'action de l'avancement de la gestation et de la réduction de la quantité de lait (Chazal et Chilliard, 1986). Elle est aussi favorisée par certains facteurs du milieu, tels que les régimes alimentaires à base d'ensilages d'herbe (Chazal, Chilliard et Coulon, 1986).

C'est la diversification des produits laitiers - produits frais et fromages - qui a permis jusqu'ici d'accroître, ou tout au moins de maintenir la quantité de lait consommée par tête dans les pays développés. Pour persévérer avec succès dans cette voie, les industriels doivent

disposer de la matière première qui soit la mieux adaptée dans ses
caractéristiques physico-chimiques aux produits fabriqués et la moins
altérée par les microbes et les produits contaminants. Le moyen le
plus efficace pour y parvenir est de payer le lait suivant sa qualité
(matières grasses, protéines, contamination totale, cellules somatiques,
etc) afin que les producteurs soient correctement rétribués pour leurs
efforts. Le système du paiement du lait doit être obligatoirement pris
en compte dans l'orientation de la sélection. Le plafonnement des productions nationales et, à plus forte raison, leur diminution dans la
C.E.E. renforcent cette orientation.

Références

Ali A.E., Andrews A.T., Cheeseman G.C., 1980. 'Influence of elevated
somatic cell counts on casein distribution and cheese-making. J. Dairy
Res. 47, 393-400.
Allaire F.R., Thaen, C.S., 1985. 'Prospectives for genetic improvement
in the economic efficiency of dairy cattle. J. Dairy Sci., 68, 3110 -
3123.
Bakken G., 1982.'The relationship between environmental conditions and
bovine udder diseases in Norwegian dairy herds'.Acta Agric. Scand.,
32, 23-31.
Barry J.G., Donnelly W.J., 1981. 'Casein compositional studies . II.
The effect of secretory disturbance on casein composition in freshly
drawn and aged bovine milks'. J. Dairy Res., 48, 437-446.
Bauman D.E., Eppard P.J., De Geeter J., Lanza G.M., 1985. 'Responses
of high-producing dairy cows to long-turn treatment with pituitary
somatotropin and recombinant somatotropin'. J. Dairy Sci., 68, 1352-
1362.
Bauman D.E., Mc. Cutcheon S.N., 1986. 'The effect of growth hormone
and prolactin on metabolism'. In : L.P. Milligan, W.L. Grovum et A.
Dobson (Eds), Control of digestion and metabolism in ruminants. 436-
455, Prentice-Hall, Englewood Cliffs, NJ 07632.
Bloc N., Guérin J.L., Mocquot J.C., 1986. 'Les index laitiers : un
outil d'analyse du passé et de maîtrise de l'avenir génétique des races.
In : Annuel pour l'éleveur de bovins. 5-22, ITEB. Paris.
Bonaiti B., 1985. 'Composition du lait et sélection laitière chez les
bovins. Bull. Tech. CRZV Theix, INRA, 59, 51-56.
Bramley A.J., Dodd F.H., 1984.'Reviews of the progress of dairy science:
Mastitis controle progress and prospects'. J. Dairy Res., 51, 481-512.
Brumby P.J., Hancock J., 1955.'The galactopoietic role of growth hormone in dairy cattle'. N.Z.J.Sci. Technol. (A), 36, 417-436.
Brochart M., Barnouin J., Fayet J.C., 1984. 'Les mammites dans
l'enquête éco-pathologique continue en élevages-observatoires'. Bulletin
des G.T.V., 5, 7-24.
Chazal M.P., Chilliard Y., 1986. 'Effect of stage of lactation, stage
of pregnancy, milk yield and herd management on seasonal variation in
spontaneous lipolysis in bovine milk'. J. Dairy Res. (in press).
Chazal M.P., Chilliard Y., Coulon J.B., 1986. 'Effect of nature of forage
on spontaneous lipolysis in milk from cows in late lactation'. J. Dairy
Res.(in press).

Chilliard Y., Lamberet G., 1984. 'La lipolyse dans le lait : les différents types, mécanismes, facteurs de variation, signification pratique'. Lait, 64, 544-578.
Coliform Subcommittee of the research committee of the national mastitis council., 1979. 'Coliform mastitis'. A review. J. Dairy Sci., 62, 1-22.
Colleau J.J., 1985. 'Efficacité génétique du transfert d'embryons dans les noyaux de sélection chez les bovins laitiers'. Genet. Sel. Evol., 17, 499-538.
Crowley J.W., Niedermeier R.P., 1981. 'Dairy production 1955 to 2006. J. Dairy Sci., 64, 971-974.
Cunningham E.P., 1983. Structure of dairy cattle breeding in Western Europe and comparisons with North America'. J. Dairy Sci., 66, 1579-1589.
Cunningham E.P., 1985. 'Recent genetic evolution of E.E.C. cattle populations'. In : H. Kraüslich et A. Lutterbach (Eds), Adapting E.E.C. cattle breeding programmes to market realities. p. 3-11, Office publications C.E.C. Luxembourg.
De Peters E.J., Smith N.E., Acedo-Rico J., 1985. 'Three or two times daily milking of older cows and first lactation cows for entire lactations'. J. Dairy Sci., 68, 123-132.
Dodd F.H., Westgarth D.R., Neave F.K., Kingwill R.G., 1969. 'Mastitis and the strategy of control'. J. Dairy Sci., 52, 689-707.
Dommerholt J., Wilmink J.B.M., 1986. 'Optimum selection responses under varying milk prices and margins for milk production'. Livestock Production Science, 14, 109-121.
Emery R.S., 1978. 'Feeding for increased milk protein'. J. Dairy Sci., 61, 825-828.
Eppard P.J., Bauman D.E., 1984. 'The effect of long-term administration of growth hormone on performance of lactating dairy cows'. Proc. Cornell Nutr. Conf. Feed Manuf., 5-12.
Eppard P.J., Bauman D.E., Mc Cutcheon S.N., 1985. 'Effect of dose of bovine growth hormone on lactation of dairy cows'. J. Dairy Sci., 68, 1109-1115.
Everson T.C., 1984. 'Concerns and problems of processing and manufacturing in super plants'. J. Dairy Sci., 67, 2095-2099.
Fairbarn D.J., Law B.A., 1986. 'Proteinases of psychrotrophic bacteria: their production, properties, effects and control'. J. Dairy Res., 53, 139-177.
Freeman A.E., 1984. 'Secondary traits : sire evaluation and the reproductive complex'. J. Dairy Sci., 67, 449-458.
Gastenel P.L., Mocquot J.C., 1985. 'Choisir les reproducteurs : sur les quantités ou sur les taux ? sur la matière utile ou sur les protéines ?'In : Annuel pour l'éleveur de bovins, p. 9-20, ITEB. Paris.
Gaunt S.N., 1980. 'Genetic variations in the yields and contents of milk constituents'. In : FIL-IDF Bulletin, 125, 73-82.
Gisi D.D., De Peters E.J., Pelissier C.L., 1986. 'Three times daily milking of cows in California dairy herds'. J. Dairy Sci., 69, 863-868.

Gravert H.O., 1985. 'Genetic factors controlling feed efficiency in dairy cows. Livestock Production Science, 13, 87-99.
Hansen L.B., Touchberry R.W., Young C.W., Miller K.P., 1979. 'Health care requirements of dairy cattle. I. Response to milk yield selection'. J. Dairy Sci., 62, 1922-1931.
Heinrichs A.J., Noyes T.E., Palmquist D.L., 1981. 'Added dietary fat for milk and fat production in commercial dairy herds'. J. Dairy Sci., 64, 353-357.
Hoden A., Coulon J.B., Dulphy J.P., 1985. 'Influence de l'alimentation sur la composition du lait. 3. Effets des régimes alimentaires sur le taux butyreux et protéique'. Bull. Tech. CRZV Theix, INRA. 62, 69-79.
Jarrige R., 1986. 'Voluntary intake in dairy cows and its prediction'. International Dairy Federation Bull., N° 196, 4-16.
Jones G.M., Pearson R.E., Clabaugh G.A., Heald C.W., 1984. 'Relationships between somatic cell counts and milk production'. J. Dairy Sci., 67, 1823-1831.
Journet M., Rémond B., 1980. 'Influence de l'alimentation et de la saison sur les fractions azotées du lait de vache'. Le Lait, LX, 140-159.
Journet M., Chilliard Y., 1985. 'Influence de l'alimentation sur la composition du lait. 1. Taux butyreux : facteurs généraux'. Bull.Tech. CRZV Theix, INRA, 60, 13-23.
Kalter R.J., 1984. 'Production cost, commercial potential and the economic implications of administering bovine growth hormone'. Proc. Cornell Nutr. Conf. Feed Manuf., 27-41.
Lee K.L., Freeman A.E., Johnson L.P., 1985. 'Estimation of genetic change in the registered Holstein cattle population'. J. Dairy Sci., 68, 2629-2638.
Lin C.Y., Mc Allister A.J., Ng-Kwai-Hang., Hayes J.F., 1986. 'Effects of milk protein loci on first lactation production in dairy cattle'. J. Dairy Sci., 69, 704-712.
Losi G., Mariani P., 1984. 'Significato technologico del polimorfismo delle proteine del latte nella caseificazione a formaggio grana'. Industria del latte, 20, 23-53.
Maijala K., Hanna M., 1974. 'Reliable phenotypic and genetic parameters in dairy cattle'. In : Ist World Congress on genetics applied to livestock productions. Madrid. Vol. 1. 541-563.
Mc Cutcheon S.N., Bauman D.E., 1986. ' Effect of pattern of lactational performances of bovine growth hormone on lactational performance of dairy cows'. J. Dairy Sci., 69, 38-43.
Mc Lean D.M., Graham E.R.B., Ponzoni R.W., Mc Kenzie H.A., 1984. 'Effects of milk protein genetic variants on milk yield and composition' J. Dairy Res., 51, 531-546.
Mariani P., Losi G., Russov V., Castagnetti G., Grazia L., Morini D., Fossa E., 1976. 'Parmigiano-Reggiano cheesemaking experiments with milk characterized by κ casein variants A and B'. Scienzae Tecnica Lattiero Casearia, 27, 208-227.
Miller R.H., 1984. 'Traits for sire selection related to udder health and management'. J. Dairy Sci., 67, 459-471.

Morand Fehr P., Chilliard Y., Bas P., 1986. 'Répercussions de l'apport de matières grasses dans la ration sur la production et la composition du lait de ruminant'. Bull. Tech. CRZV Theix, INRA, 64, 59-72.
Niebel E., Fewson D., 1985. 'Improving milk and beef production potential within the purebred dual purpose cattle population'. In : H. Krauslich et A. Lutterbach (Eds) Adapting E.E.C. cattle breeding programmes to market realities. p. 201-210. Office publications C.E.C. Luxembourg.
Nicholas F.W., Smith C., 1983. 'Increased rates of genetic change in dairy cattle by embryo transfer and splitting'. Anim. Prod., 36, 341-353.
Oldenbroek J.K., 1984. 'A comparison of Holstein Friesians, Dutch Friesians and Dutch Red and Whites. I. Production characteristics. Livestock Production Science, 11, 69-81.
Oldenbroek J.K., Van Eldik P., 1980. 'Differences in feed intake between Holstein Friesian, Dutch Red and White and Dutch Friesian cattle'. Livestock Production Science. 7, 13-23
O'Connor, Baldwin R.S., Adams R.S., Hutchinson L.J., 1985. 'An integrated approach to improving reproductive performance'. J. Dairy Sci., 68, 2806-2816.
Ostergaard V., Danfaer A., Daugaard J., Hindhede J., Thysen I., 1981. 'The effect of dietary lipids on milk production in dairy cows'. Beretn Statens Husdyrbrugsforsøg, n° 508.
Ozil J.P., Heyman Y., Renard J.P., 1982. 'Production of monozygotic twins by micro-manipulation and cervical transfer in the cow'. Veterinary Record, 110, 126-127.
Palmquist D.L., Jenkins T.C., 1980. 'Fat in lactation rations'. Review. J. Dairy Sci., 63, 1-14.
Peel C.J., Sandless L.D., Quelch K.J., Herington A.C., 1985. 'The effects of long-term administration of bovine growth hormone on the lactational performance of identical-twin dairy cows'. Anim. Prod., 41, 135-142.
Philipsson J., 1981. 'Genetic aspects of female fertility in dairy cattle'. Livestock Production Science, 8, 307-319.
Poutrel B., 1982. 'Susceptibility to mastitis': a review of factors related to the cow. Ann. Rec. Vet., 13, 85-99.
Remond B., 1986. 'Influence de l'alimentation sur la composition du lait de vache. 2. Taux protéique : facteurs généraux'. Bull. Tech. CRZV Theix, INRA, 62, 53-67.
Richard A.L., Mc Cutcheon S.N., Bauman D.E., 1985. 'Responses of dairy cows to exogenous bovine growth hormone administered during early lactation'. J. Dairy Sci., 68, 2385-2389.
Scott J.D.J., 1983. 'Efficiency of dairying under contrasting feeding and management systems in North America, Israel, Europe, and New Zealand'. In : Proc. XIV International Grassland Congress. Lexington U.S.A., 15-24, juin 1981, p. 843-846. Westerview Press. Boulder-Colorado.
Seidel G.E., 1984. 'Application of embryo transfer and related technologies to cattle'. J. Dairy Sci., 67, 2786-2796.

Shanks R.D., Freeman A.E., Berger P.J., Kelley D.M., 1978.'Effect of selection for milk production on reproductive and general health of the dairy cow'. J. Dairy Sci., 61, 1765-1772.
Steuernagel G.R., 1983. 'How top Minnesota dairymen differ from others'. Hoard's Dairyman. June 25.
Storry J.E., 1981. 'The effect of dietary fat on milk composition'. In : W. Haresign (Ed) Recent advances in animal nutrition.1981, 3-34, Butterworths. London.
Sutton J.D., 1981. 'Concentrate feeding and milk composition'. In : W. Haresign (Ed) Recent advances in animal nutrition. 1981, 3-34, Butterworths. London.
Tyrrell H.F., Brown A.C.G., Reynolds P.J., Haaland G.L., 1982. 'Effect of growth hormone on utilization of energy by lactating Holstein cows'. In : A. Ekern et F. Sundstol (Eds) Energy metabolism of farm animals 46-49. Publ. 29 EAAP.
Van Arendonk J.A.M., Wilmink J.B.M., Dijkuizen A.A., 1985. 'Conséquences of a restriction of the herd milk production for the selection on milk, fat and protein'. In : H. Kräuslich et A. Lutterback (Eds) Adapting EEC cattle breeding programmes to market realities. p.211-220. Office publications CEC. Luxembourg.

PRESENT AND FUTURE HEALTH ISSUES AND MILK AND DAIRY PRODUCTS

Elwood W. Speckmann, Ph.D.
National Dairy Council
6300 North River Road
Rosemont, Illinois 60018
United States of America

ABSTRACT. Increased public awareness of food, nutrition and health issues and federal initiatives in health promotion and disease prevention offer a significant opportunity for the dairy industry to reinforce consumer knowledge, attitudes and behavior that dairy foods are "protective foods" and should be consumed regularly as part of a nutritionally-balanced diet to maintain and improve the excellent quality and longevity of life we have achieved to date. Emerging research is documenting the value of milk and milk products, particularly their principal nutrient calcium, in bone health, regulation of blood pressure and decreasing risk of some types of cancer. Additional research is identifying new facts about the nutritional contributions of dairy foods to a healthful diet. The success with which the dairy industry addresses these societal opportunities will have a marked influence not only on the health of the population, but on the health of the dairy industry as well.

In the early part of this century infectious, communicable and nutrient deficiency diseases were prominent in most of the world. Advances in medicine and public health since then have resulted in a remarkable reduction of these maladies in the developed world. As research improved our understanding of the relationship between food, nutrition and health, nutritional guidelines were developed recommending intake of a wide variety of foods to insure adequate intake of the known (and unknown) nutrients. Dairy foods were referred to as "protective foods" (1) because of their excellent nutritional value and because they provided the nutrients supplied in limited amounts, or not at all, by plant foods. The healthy population was encouraged to regularly consume a nutritionally-balanced diet, including dairy foods, to prevent nutrient-deficiency diseases and to promote optimal health. As a result, the health of the developed world has never been better than it is today.
During the the past few decades, there has been a growing awareness and understanding of the relationship between our environment and our quality of life. Diet is a component of our environment which we can modify to achieve and/or maintain health or "wellness." The

consumer is taking responsibility for his/her personal health and is willing to change behavior (including food purchases) based on real and/or perceived health benefits. Today, coronary heart disease, cancer, stroke and accidents represent the four leading causes of death for Americans. Smoking, diet, alcohol use and lack of exercise are controllable risk factors associated with these leading causes of death. Public education to minimize these risk factors are believed to have a measurable reduction on today's spiraling health care costs.

The 1969 White House Conference on Food, Nutrition and Health (2) brought together leaders from academia, government and the private sector to examine the nutritional status and health of Americans. It stimulated a second public health revolution in the United States which is focusing on chronic disease and behavior-related health problems and which is positioning nutrition as preventive medicine. The report "Healthy People" in 1979 (3) and subsequent reports in 1981 (4) and 1984 (5) outlined an ambitious commitment by the U.S. government to improve the health and quality of life in America.

Despite the fact that longevity and quality of life in America has never been better than they are now, the U.S. government is espousing the view that future gains in our health will come, not from increases in medical care, but from a national commitment to health promotion and disease prevention--a commitment that can be achieved only through the combined efforts of all segments of society, from the federal government to the individual citizen (6).

The Surgeon General of the United States set down five major goals designed to improve the health status of all age groups by 1990. These goals have been translated into 15 health priority areas and 226 measurable objectives to reduce health problems in each of these areas (4). Nutrition is one of these priority areas. The government has encouraged those in health care settings, health professions, business and industry, voluntary organizations, and schools to be "partners for health promotion" and participate in a coordinated effort to achieve these objectives by 1990 (5).

I'm confident that many developed countries are formulating and implementing similar policies for health promotion and disease prevention. The NACNE (7) and COMA (8) reports, for example, are having a profound effect on public health policies and practice throughout Europe. In the United States, the Department of Health and Human Services and the Department of Agriculture jointly issued in 1985 "Nutrition and Your Health. Dietary Guidelines for Americans" (9) directed to the general public.

The U.S. Food and Drug Administration is expected to publish proposed guidelines governing public health claims on food labels this fall. And a Surgeon General's report on nutrition and health is expected to be published in 1987. Government initiatives regarding food, nutrition and health are now influencing, and will continue with increasing dimensions to influence, the kinds of dairy foods consumers want, as well as the way we market dairy foods to the consumer.

While nutrition may be one modality in the treatment of individuals affected with, or at high risk of, developing chronic diseases such as cardiovascular disease and cancer, the causal relationship of specific nutrients or foods, or their alleged role in the prevention, mitigation or cure of these maladies is the subject of considerable scientific debate. The result is a confusing array of dietary advice to the general public. There are those who are impatient with our present limited knowledge of science and medicine and who are promoting public health practices to the general healthy population without convincing scientific evidence that such advice will, in fact, improve overall quality of life. As a result of these attitudes, their enhancement by some segments of the food industry and their sensationalism by the media, consumers are being advised constantly that consumption of animal foods are "bad for you." Many are recommending avoiding or at least markedly reducing dairy foods to prevent coronary heart disease, cancer, hypertension and strokes, as well as diabetes, because of their fat, cholesterol, sodium and/or sugar content irrespective of their excellent nutrient profile and nutritional quality. I submit to you that THE PROBLEM IS THE DINER, NOT THE DINNER!

In the latter part of this century, we have allowed dairy foods to fall from a position of strength wherein public health authorities were encouraging consumption of dairy foods to promote health, to a position of weakness wherein public health authorities now are recommending reduction in consumption of dairy foods to prevent disease. There's a popular saying, "IF YOU ARE NOT PART OF THE SOLUTION, YOU ARE PART OF THE PROBLEM." Government initiatives currently are positioning the dairy industry as part of the problem in their efforts to improve the health of the nation. The health promotion/disease prevention initiatives of the federal government offer both a threat and an opportunity for the dairy industry. It threatens our market for dairy foods the consumer perceives as high in fat, cholesterol, sodium and/or sugar. But it also offers an opportunity for the dairy industry to successfully market the excellent nutritional value of dairy foods and their role in health promotion. The dairy industry can be a leader or a follower. For all too long, we have been following in a defensive mode trying to protect our food and our market.

I firmly believe that we must focus our research and marketing strategies to recognize and address the changing needs of society. An industry that recognizes consumer wants and needs and makes its products and marketing strategies part of that trend will, in my opinion, continue to be successful. I submit to you that THE BEST DEFENSE IS A GOOD OFFENSE. The dairy industry must take the offense and embark on a sustained research program to expand and strengthen the data base linking dairy foods with health promotion and thus part of the solution in improving world health. We know very well that the most progressive and successful industries traditionally have a strong and sustained commitment to research. In 1986, the dairy industry in the United States is taking the offense by investing over 4 million dollars in approximately 100 nutrition research projects to generate data which will expand and strengthen the position of dairy foods in health promotion. Nutrition research is providing a competitive edge in the marketplace for dairy foods while concomitantly keeping in perspective consumer health concerns. And this blue chip investment in the future already is yielding significant dividends.

In order to develop nutrition research strategies for the future it is essential to delineate the major nutrition/health issues which influence our current consumption of dairy foods.

Opportunities for increased consumption of dairy foods include:
- overall nutrient density and quality of dairy foods
- role of calcium in reducing risk of osteoporosis, hypertension and colon cancer
- value of cultured and culture-containing dairy foods in promoting optimal gut ecology
- cariostatic properties of dairy foods
- value of dairy foods in protecting against stomach ulcers in adults and acute gastrointestinal illness in children, promoting optimal growth and development in children, value in diabetic diets and in diets of women exercising or losing weight.

Obstacles to increased consumption include:
- concern about fat, cholesterol, sodium and sugar
- concern about lactose intolerance and milk allergy
- increased use of nutrient supplements and fabricated products
- misperceptions about dairy foods
- chemical and microbiological contamination

Milk, cheese, yogurt and ice cream are grouped into one of the four major food categories by nutrition educators because of their excellent nutrient density and quality. In 1984, for example, milk group foods provided the following percentage of nutrients available in the U.S. food supply: calcium - 76, phosphorus - 36, riboflavin 35, protein - 21, vitamin B-12 - 20, zinc - 20, magnesium 19, vitamin A - 12, and vitamin B-6 - 11 (10). Because milk protein is an excellent source of the amino acid, tryptophan, a precursor of the vitamin niacin, and most fluid milk in the U.S. is fortified with vitamin D, milk group foods are a significant source of niacin equivalents and vitamin D. Milk group foods provide all these nutrients while supplying only 10% of the food energy, 12% of the fat and 14% of the cholesterol available in the U.S. food supply (10). On the other hand, high energy-low nutrient foods such as vegetable oils, salad dressings, table spreads, and soft drinks provided little in the way of nutrients but collectively represented 37% of the energy, 46% of the fat, and 39% of the sugar available in the U.S. food supply (10). If individual adjustments in fat, sugar and/or energy are necessary, modifications can be made in consumption of high-energy, low-nutrient foods without affecting the nutritional integrity of the diet.

One of the most encouraging dividends of research funded by the dairy industry in recent years is the emerging evidence that calcium plays an important role in promoting optimal bone health (11-16), in regulating blood pressure (17-20) and in decreasing the risk of colon cancer (21-24).

Calcium intake is particularly important to the dairy industry because calcium links dairy foods with health promotion. Dairy foods are an excellent source of calcium and other nutrients (25-26). In

1984, 76% of the calcium in the U.S. food supply came from dairy foods (10). A national survey of the population by the U.S. government from 1976-1980 (27) revealed that more than 50% of males age 12-18 years and above 34 years of age failed to consume recommended amounts of calcium on any given day. In females more than 50% age 12 years and over failed to consume recommended amounts of calcium on any given day. Between 18 and 34 years of age, the period of peak bone mass development, 67% of females were not consuming recommended amounts of calcium. By age 35 years and over, this percentage increased to 75%.

Metabolic studies of normal perimenopausal women revealed that the calculated mean intake required to attain calcium balance in premenopausal and estrogen treated postmenopausal women was 989 mg/day, whereas for untreated postmenopausal women it was 1504 mg/day (28). This is higher than the current Recommended Dietary Allowance (RDA) in the U.S. of 800 mg for women over 18 years (29).

The National Institutes of Health (NIH) Consensus Development Conference in 1984 (30) concluded that "an increase in calcium intake to 1000 to 1500 mg a day beginning well before the menopause will reduce the incidence of osteoporosis in postmenopausal women." The American Society for Bone and Mineral Research (31) and the National Institute of Arthritis, Diabetes and Digestive and Kidney Diseases (32) have issued statements consistent with the Consensus Development Conference recommendations. These data are significant to the dairy industry in that they support the recommendation that adult women consume 3-4 servings of milk group foods daily to obtain recommended amounts of calcium as well as other nutrients. But will increased calcium intake throughout life improve bone density and reduce risk of fractures? Matkovic and colleagues (33) compared a population typically consuming a low calcium diet with a population consuming about twice this amount. They found that the lower bone mass and higher incidence of hip fractures of the population consuming the low calcium diet was related to a reduced peak bone mass, most likely reflecting a prolonged inadequate calcium intake during the years of bone development, growth and maturation.

The primary dietary approaches to control hypertension are weight control and sodium restriction. There is emerging scientific evidence that dietary calcium intake, renal disposition of calcium, and circulating levels of calcium-regulating hormones, as well as of calcium itself, are related to disorders of blood pressure regulation such as hypertension and thus may have a significant role in the etiology of this malady (18). Hypertensives have a wide spectrum of blood pressure sensitivities to antihypertensive drug therapy. Likewise, some but not all, hypertensive individuals respond to lower sodium or higher calcium intake with a reduction in blood pressure. Preliminary evidence suggests that higher dietary salt intake, lower plasma renin activity and lower serum ionized calcium levels may help identify persons with the greatest hypotensive response to increased calcium intake (18).

To date, at least five studies have increased dietary calcium intake of normal (34), pregnant (35) or hypertensive subjects (36-38) and reported significant decreases in blood pressure. Expectant mothers consuming an extra 2000 mg calcium daily had a 7-10% drop in blood pressure that persisted throughout the last trimester of pregnancy (35). Implications for reducing the risk of preeclampsia are obvious. Researchers at Cornell University Medical College report that 1/3 to 1/2 of their hypertensive patients responded well to calcium therapy (39). McCarron and Morris (37) found that 44% of hypertensive and 19% of normotensive persons achieved a therapeutically significant reduction (10 mm Hg or greater) in standing systolic arterial pressure when consuming an extra 1000 mg calcium daily for 8 weeks. Increasing dietary calcium rapidly is becoming a valuable nonpharmacologic option in the treatment of hypertension. More clinical research is needed to identify those individuals for whom calcium would be most beneficial and to more clearly understand the clinical role of calcium and its hemodynamic effects.

The association between diet, nutrition and cancer continues to be actively investigated and hotly debated (23). While total fat intake has been implicated in breast and colon cancer, the evidence is by no means consistent. Dairy foods continue to receive adverse publicity because of their fat and cholesterol content. Calcium and other factors in dairy foods, however, may play a role in reducing the risk of certain forms of cancer. Garland et al (21) found from a 19-year prospective study that the risk of colorectal cancer was inversely correlated with vitamin D and calcium. This association remained significant after adjustment for age, daily cigarette consumption, body mass index, ethanol consumption, and percentage of calories obtained from fat. Individuals who drank no milk had nearly three times the colorectal cancer risk of those who drank a couple of glasses a day. Lipkin and Newmark (22) also reported that increasing dietary calcium by 1250 mg daily for 2-3 months induced a more quiescent equilibrium in epithelial-cell proliferation in the colonic mucosa of subjects at high risk of colon cancer, similar to that observed in subjects at low risk. Abnormal cellular proliferation is a hallmark of neoplasia.

Goldin and Gorbach (40-41) have made significant contributions to our understanding of the role of lactobacillus organisms on risk of colon cancer. These investigators measured dietary induced changes in several fecal bacterial enzymes (nitroreductase, azoreductase, β glucuronidase, and steroid 7-α- dehydroxylase) which have been associated with the conversion of chemical procarcinogens to proximal carcinogens in the large bowel. These investigators have found that in humans consuming Lactobacillus acidophilus milk or oral supplements of viable L. acidophilus, activities of these fecal bacterial enzymes were significantly reduced. In addition, lactic cultures have been shown to reduce the incidence of tumors in laboratory animals given the chemical carcinogen DMH (1,2 dimethylhydrazine dihydrochloride). Further research is warranted regarding the role of cultured and culture-containing dairy foods on the nutrition and health of human subjects.

Dairy food consumption also has been associated with a reduced risk of stomach cancer. Hirayama (42), in studying the epidemiology of stomach cancer in Japan, reported that daily intake of green-yellow vegetables, soybean paste soup and milk were considered as risk reducing factors for this malady. The Diet, Nutrition and Cancer report by the National Research Council (43) cited earlier studies by

Hirayama and concluded that protective factors for stomach cancer may include consumption of milk, raw green or yellow vegetables, especially lettuce, and other foods containing vitamin C.

Cancer is a complex disease in which diet and/or specific nutrients affect different cancers in different ways. Emerging evidence that polyunsaturated fatty acids are immunosuppressive and thus increase the risk of tumorigenesis warrants further investigation (44). Overall energy intake appears to be most consistently associated with cancer and specific nutrients appear to act as modifiers rather than initiators of tumorigenesis (45). Various components of dairy foods have been identified as reducing several forms of cancer.

The two most prevalent, but preventable, oral health diseases in industrialized nations are dental caries and periodontal disease. Although these chronic disorders are multifactorial, there is increasing interest as to the role of diet in general, and specific foods in particular, regarding how they modulate the development and progression of these maladies (16).

Most foods containing fermentable carbohydrate cause dental plaque acidity to increase to a level conducive to dental caries. Recent findings from both laboratory animal and human dental plaque investigations reveal that dairy foods, including milk and certain cheeses, may protect against caries (46-49). At least seven cheeses--aged Cheddar, Swiss, Blue, Monterey Jack, Mozzarella, Brie and Gouda--produce little or no plaque acid. Moreover, three of the cheeses--aged Cheddar, Monterey Jack and Swiss--effectively prevented sucrose-induced increases in plaque acidity when consumed 30 minutes before sucrose intake (46).

It generally is held that periodontal disease is caused by plaque bacteria resulting, in advanced stages, in loss of tooth supporting alveolar bone. An alternative hypothesis is that loss of alveolar bone resulting from long-term inadequate intake of calcium and/or loss of estrogen at menopause may result in loose teeth and supporting gum tissue facilitating the penetration of Streptococcus mutans thereby initiating and/or exacerbating periodontal disease. The relationship between dietary calcium and periodontal disease at present, however, awaits further research (14,15). The value of consuming recommended amounts of calcium throughout life to achieve optimal alveolar bone density thereby minimizing tooth loss and after tooth loss maintaining proper prosthetic fit and stability is worthy of continued research.

Fat and cholesterol in the diet, and in animal foods in particular, continue to receive public attention. It is my belief that public policy encouraging all healthy Americans over the age of 2 years to markedly decrease consumption of higher fat dairy foods goes beyond scientific documentation of risk/benefit. Data on serum cholesterol levels collected during a major government survey between 1976-1980 (50) reveal that the median serum cholesterol value for males 20-74 years of age was 206 mg/dl. Almost 85% of males had serum cholesterol values below 250 mg/dl, the level above which blood cholesterol is associated with increased risk of heart disease (51).

Closely controlled metabolic studies reveal that individuals differ in their response to changes in fat and cholesterol intake.

Some patients are sensitive to changes in dietary cholesterol, others are sensitive to fat quality, but the majority are relatively insensitive to changes in fat quality or cholesterol quantity (52,53). At present, there is no way to predict which individuals will experience a decline in blood cholesterol levels as a result of diet, or to what degree, or whether such changes will be sustained. Milk fat is about 66% saturated fatty acids, 30% monounsaturated fatty acids and 4% polyunsaturated fatty acids. Approximately 14% of the saturated fatty acids are short chain and do not influence blood cholesterol levels. Mattson and Grundy (54) reported that oleic acid, the principal monounsaturated fatty acid in several fats, including milk fat, is as effective as the polyunsaturated fatty acid, linoleic acid, in lowering low density lipoprotein cholesterol levels in patients with normal blood triglyceride levels, and oleic acid seemingly reduces high density lipoprotein cholesterol levels less frequently than does linoleic acid. These results are encouraging for the dairy industry. Metabolic studies are needed to more precisely study the effect of diet on blood cholesterol level, factors regulating cholesterol homeostasis, and when blood cholesterol is reduced the extent to which it is excreted and/or redistributed within the body.

Major intervention studies conducted during the past 30 years have failed to demonstrate that dietary modifications of fat and cholesterol in healthy people with moderate to low blood cholesterol levels prevent disease and improve overall quality and longevity of life. Nevertheless, the American Heart Association (AHA) recently has published __An Eating Plan for Healthy Americans__ (55) over the age of 2 years which lists "okay foods" as "milk products containing only 0-1% milk fat...cheeses containing not more than 2 grams fat per ounce... vegetable oils...margarines" and "foods to avoid" as "milk products containing more than one percent milk fat...cream, all kinds...non dairy cream substitutes...all cheeses containing more than 2 grams of fat per ounce...butter." Such an extreme position not only adversely affects the organoleptic quality of our diet but also discriminates against nutritious foods on the basis of a single nutrient. An AHA statement to physicians which AHA believes embodies the current medical thinking regarding an optimum diet for "healthy adult Americans" will be published December, 1986 (56). The basic tenets of nutrition education, namely variety, moderation and balance, can provide dietary patterns, including dairy foods, to achieve any lifestyle or health objective without scaring people with the "good food" vs. "bad food" tactic.

Milk fat appears to reduce gastrointestinal illness. Koopman *et al* (57) reported that children over age one taking lowfat milk as their only milk source in the three weeks prior to illness had five times the risk of a doctor's visit for acute gastrointestinal illness as did children over age one taking only whole milk during the same time period.

Also, Dial and Lichtenberger (58,59) reported that milk has a potent antiulcer activity in rats that may be attributable to its phospholipid constituents and/or protein. Milk's buffering capacity is an important adjunct in the treatment of ulcers. Emerging research is revealing how milk may protect against or help treat ulcers.

Pugliese (60) reported earlier this year that parental concerns that their children would become obese and/or develop atherosclerotic disease, junk-food dependency or unhealthy eating habits caused them to severely restrict animal fat and whole milk consumption. This resulted in non-organic failure to thrive. With counseling, all dietary restrictions were removed, whole-fat dairy products were encouraged and the energy intake was significantly increased. This resulted in resumption of weight gain and growth. The medical controversy regarding across-the-board diet reform for coronary artery disease was featured in Medical World News (61). Pediatrician Alvin Mauer, M.D., immediate past chairman of the American Academy of Pediatrics' Committee on Nutrition, expressed concern that blanket recommendations about diet, based primarily on studies of middle-aged men are affecting other groups for which intervention may not be necessary or wise, such as growing and developing children and women of childbearing age. He further commented that the suggestion that healthy children over age 2 should forsake foods they like for the American Heart Association diet might pose a threat to the best nutritional status that has ever been achieved among children (61). The American Academy of Pediatrics reassessed their 1983 position statement "Toward a Prudent Diet for Children" and concluded that "Thus, it would seem prudent not to recommend changes in current dietary patterns in the United States for the first two decades of life without first assessing the effects on growth, development, and such measures of nutritional adequacy as the status of iron" (62).

Recognition in the last 20-25 years that intestinal lactase activity begins to decline after early childhood for a large majority of the world's population has raised questions about the nutritional implications of recommending lactose-containing foods for these individuals. The prevailing medical and scientific opinion today (63-65) is that although most of the individuals with low lactase activity will develop symptoms of intolerance to a large load of 50 mg lactose (equivalent to that in one quart of milk) as used in diagnostic tests, the majority of these individuals can consume an 8-oz. serving of milk (12 gm lactose) without discomfort. Lactose intolerance is not milk intolerance!

For those who do experience difficulty tolerating the amount of lactose in one glass of milk, various strategies are available. Consuming recommended amounts of milk in smaller portions more often throughout the day and taking milk with solid foods enhance the effectiveness of residual lactase activity. Also, whole milk is tolerated better than low fat milk and tolerance for chocolate milk is better than for unflavored milk. Most aged cheeses with their low lactose content and unpasteurized yogurt with its intrinsic lactase activity are well accepted. However tolerance to pasteurized yogurt, cultured buttermilk and sweet acidophilus milk is similar to that for milk for persons with low lactase levels. Lactose-hydrolyzed milk and exogenous lactases added at mealtime are other means of improving tolerance to milk (66).

Research by Nathan et al (67) demonstrated that a modest amount of ice cream may be included in weight-maintaining diets of insulin-dependent diabetics. This is significant because the American Diabetes Association had deleted ice cream from their food exchange list in 1976 thus eliminating this nutritious dairy food from the diet of diabetics. Dr. Nathan's research reaffirms the relatively low glycemic index of ice cream and demonstrates that a medium-size portion of ice cream can be included in the diet of insulin-dependent diabetics because it causes a relatively small rise in blood glucose levels.

Belko and associates (68,69) reported that the requirement for riboflavin is increased by physical activity and during weight reduction. Weight reducing young women required about two times the RDA for riboflavin during both nonexercise and exercise periods. With a large number of women attempting to lose weight, many of whom reduce dairy food intake, and the fact that dairy foods supply 35% of the riboflavin in the U.S. food supply (10), the dairy industry should continue research initiatives in this area.

LIST OF REFERENCES AVAILABLE FROM AUTHOR UPON REQUEST.

Seminar I: Progress in Cheese Technology

INTRODUCTORY REMARKS

Prof.Dr.Ir. P. Walstra, chairman of the seminar committee

Cheesemaking is changing from an art into a science. This may be considered progress, though not everybody would agree. Indeed, something is lost by this transition, but - in my view - more is gained.

The transition from art to science is not the same as the increase in the number of auxiliary science-based activities surrounding cheesemaking, involving mechanics, analytical chemistry and microbiology, process engineering, electronics and systems engineering, to mention some. What I mean is that the technology of cheesemaking itself is getting a more fundamental, scientific basis. This does not only imply that explanations have been found for what happens during cheesemaking which satisfy the academic mind. There is emerging a science-based processing, directed to obtain products of a desired pre-determined quality in an efficient manner. This constitutes important progress in cheese technology. It is a main reason for selecting it as topic number one of this congress. In addition, the progress made now can lead to a more fruitful interchange of ideas between cheese technologists involved in various types of cheese, often markedly different in manufacture and properties, but still being based on the same major principles. I hope that this will also be your conclusion at the end of this seminar.

Seminar I: Progress in Cheese Technology

Session 1: Rennets and Starters

Chairman: Dr. Ir. J. Stadhouders (The Netherlands)
Secretary: Dr. Ir. M. D. Northolt (The Netherlands)

COAGULANTS AND THEIR ACTION

P.F. Fox
Department of Food Chemistry
University College Cork
Ireland.

ABSTRACT. The primary function of rennets is to specifically hydrolyse micelle-stabilizing κ-casein with the minimum of general proteolysis (which reduces cheese yield). Gel strength, and consequently cheese yield, are influenced by rennet action. Rennet retained in the curd plays a key role in the ripening of low/medium-cooked cheese. Secondary rennet proteolysis is primarily responsible for textural changes and resulting large peptides are hydrolysed by bacterial proteinases/peptidases to small peptides and amino acids which contribute to cheese flavour; amino acid catabolism leads to several sapid products. Proteolysis is important for the release of such compounds.

The manufacture of most cheese varieties essentially involves concentrating the casein and fat of milk 6 to 12 fold, depending on variety; in a few varieties, the whey proteins are also concentrated and this is becoming more common through the use of ultrafiltration to concentrate the colloidal phase of milk to the level present in cheese, i.e. to pre-cheese.
    Concentration is achieved by coagulating the casein either by:
1. Limited proteolysis by a crude proteinase (rennet); this method applies to the vast majority of ripened and some fresh cheeses.
2. Isoelectric precipitation, which is used mainly for fresh cheeses, usually by <u>in situ</u> production of lactic acid by a starter culture but direct acidification with pre-formed acid, usually HCl, or acidogen, usually gluconic acid-δ-lactone, is becoming increasingly popular, or
3. Acid plus heat, i.e. acidification to a pH value above the isoelectric point with acid whey, acid milk, citrus juice, vinegar or acetic acid at elevated temperatures, e.g. Ricotta, Quesco Blanco.
    If present, the fat is occluded in the coagulum. In this presentation, I will deal exclusively with the rennet cheeses.

Enzymatic Coagulation of Milk

It has been known in a general way since the pioneering work of

Hammersten in the 1880's that the rennet coagulation of milk involves proteolysis with the formation of para-casein and non-protein-nitrogen. The work of Linderstrøm-Lang (1) and Berridge (2) clearly showed that rennet coagulation is a 2-stage process, the first involving the enzymatic production of para-casein and peptides, the second involving the precipitation of para-casein by Ca^{2+} at temperatures > 20°C. Both stages, especially the primary phase, are now fairly clearly understood. Before describing the current views on rennet coagulation, a brief review of the milk protein system seems appropriate.

1. The Milk Protein System

Bovine milk, and apparently milks of all other species also, contains two markedly different groups of proteins: the caseins, i.e. phosphoproteins insoluble at pH 4.6 and 20°C, and the whey (or non-casein) proteins. In bovine milk, casein represent $\sim$ 80% of the total nitrogen and consists of 4 principal proteins, α_{s1}-, α_{s2}-, β- and κ-, in the approximate ratio of 40:10:35:12, and several minor proteins many of which originate via from post-transcriptional proteolysis of the primary caseins by indigenous alkaline milk proteinase, plasmin. These include γ^1, γ^2 and γ^3-caseins and proteose peptones 5 and 8 fast, all derived from β-casein, λ-casein derived from α_{s1}-casein and at least 20-30 as yet unidentified peptides, most of which are probably derived from the principal caseins.

Because of their high phosphate contents, α_{s1}- (8 or 9 P), α_{s2}- (10-13 P) and β- (5 P) caseins bind Ca^{2+} strongly and precipitate at $[Ca^{2+}]$ > 6 mM. However, κ-casein, which has only 1 phosphate residue, does not bind Ca strongly and is soluble at high $[Ca^{2+}]$. It also reacts hydrophobically with α_{s1}-, α_{s2}- and β-caseins and can stabilize up to 10 times its weight of these Ca-sensitive caseins against precipitation

In milk, > 95% of the casein exists as micelles which consist of $\sim$ 94% protein and 6% small species, mainly Ca and PO4 with some Mg and citrate, which are usually collectively called colloidal calcium phosphate (CCP). The micelles are spherical, 50-300 nm (mean $\sim$ 100 nm) in diameter, with molecular weights of $\sim$ 10^8 Da; they are hydrated to the extent of $\sim$ 2 g H2O/g protein but values ranging from 0.5 to 4 g/g have been reported.

The structure of the micelles is not yet fully established but there is a general consensus that they consist of sub-micelles (spherical particles, M.W. 5 x 10^6 Da) which are held together by CCP bridges, hydrophobic interactions and hydrogen bonds. The micelles dissociate on removing CCP, e.g. by Ca-chelators or acidification, by increasing the pH > $\sim$ 9, or by SDS, urea, guanadine HCl and similar agents. The precise structure of the submicelles is unknown. A widely supported view is that the Ca-sensitive α_s- and β-caseins interact hydrophobically to form the core of the submicelles. Some, and possibly all, sub-micelles have patches of κ-casein on the surface such that the hydrophobic N-terminal segments react hydrophobically with the core leaving the hydrophilic C-terminal regions more or less projecting into the surrounding environment. The submicelles aggregate such that

the κ-casein-rich sub-micelles are concentrated on the surface of the
micelles with the κ-casein-deficient sub-micelles buried within. It is
generally considered that the micelles are stabilized by a zeta
potential of ∿ -20 mV and by steric hindrance caused by the protruding
C-terminal segments of κ-casein which give the micelles a "hairy"
appearance and which prevent close approach.

Recent reviews on the milk protein system include references 3-8.

2. Primary Phase of Rennet Action

With the isolation of κ-casein, the demonstration that it is responsible
for micelle stability and that its micelle-stabilizing properties are
lost on renneting (9), it became possible to define the primary phase
of rennet action precisely. Only κ-casein is hydrolyzed during the
primary phase of rennet action (10), the primary cleavage site being
Phe_{105}-Met_{106} (11). This particular bond is many times more
susceptible to hydrolysis by acid proteinases (all commercial rennets
are acid proteinases) than any other in the milk protein system.

The unique sensitivity of the Phe-Met bond has aroused interest.
The dipeptide, H-Phe.Met-OH, is not hydrolyzed nor are tri or tetra-
peptides containing a Phe-Met bond. However, this bond is hydrolyzed
in the pentapeptide, H.Ser-Leu-Phe-Met-Ala-OMe and reversing the
positions of serine and leucine in this pentapeptide, to give the
correct sequence of κ-casein, increases the susceptibility of the Phe-
Met bond to hydrolysis by chymosin. Both the length of the peptide and
the sequence around the sectile bond are important determinants of
enzyme-substrate interaction. Serine appears to be particularly
important and its replacement by Ala in the above pentapeptide renders
the Phe-Met bond very resistant to hydrolysis by chymosin but not by
pepsin. Extension of the above pentapeptide from the N and/or
C-terminal to reproduce the sequence of κ-casein around the chymosin-
susceptible bond increases the efficiency with which the Phe-Met bond
is hydrolyzed by chymosin. Taking the pentapeptide Ser_{104}-Ile_{108} as a
standard, studies at pH 4.7 show that extending the peptide toward the
C-terminal by 3 residues to give Ser_{104}-Lys_{111} causes a 6-fold increase
in the catalytic ratio, kcat/Km, while addition of Leu_{103} to the
pentapeptide increases the ratio 600 fold. Addition of His_{102} and
Pro_{101} increases kcat/Km a further 5-fold and the peptide representing
the sequence His_{98}-Lys_{111} of κ-casein is hydrolyzed 66,000 times faster
than the original pentapeptide and at a similar rate to intact κ-casein.
κ-Casein and the peptide His_{98}-Lys_{111} are readily hydrolyzed at pH 6.6
(and at pH 4.7) but the above small peptides are not hydrolyzed at pH
6.6. Thus, this sequence His_{98}-Lys_{111} appears to contain the necessary
determinants for rapid cleavage of the Phe-Met bond by chymosin and
presumably by other acid proteinases.

The Phe_{105}-Met_{106} bond of κ-casein is the only Phe-Met bond in the
milk protein system; however, these two residues are not intrinsically
essential. There are numerous Phe residues in all milk proteins and a
substantial number of Met residues. Replacement of Phe by Phe (NO_2)
reduces the catalytic ratio by a factor of about 3 while substitution
of cyclohexylamine for Phe reduces the rate of hydrolysis ∿ 50 fold.

Susbstition of Nle for Met increases kcat/Km ~ 3 fold. Neither porcine or human κ-caseins possess a Phe-Met bond (the former has a Phe-Ile bond at this position), yet both are readily hydrolyzed by calf chymosin although more slowly than bovine κ-casein; in contrast, porcine milk is coagulated more effectively than bovine milk by porcine chymosin, indicating that unidentified subtle structural changes influence chymosin action. Thus, the sequence around the Phe-Met bond, rather than the bond itself, contains the important determinants of hydrolysis. The particularly important residues are Ser_{104}, at least one of the 3 histidines (98, 100 or 102, as indicated by the inhibitory effect of photooxidation), some or all of the 4 prolines (99, 101, 109, 110) and Lys_{111} and/or Lys_{112}.

The importance of the proline residues may reside in their significance in secondary structure. Loucheux-Lefebvre et al. (12) propose that the Phe-Met bond of κ-casein is situated between 2 β-turns which may cause this sequence to protrude from the κ-casein molecule and enable it to fit into the active site cleft of acid proteinases (rennets). Theoretically, the peptide around the sectile bond is capable of forming α-helical or β-sheet structures; the latter would facilitate hydrogen bonding with the active site of the enzyme. For references to the original work in this area, see Refs. 13-15.

While the hydrolysis of κ-casein by acid proteinases other than calf chymosin has not been thoroughly studied, it is generally assumed that all the commercially used rennets hydrolyse the Phe-Met bond. However, as discussed below, the aggregation characteristics of micelles altered by rennet substitutes differ from those of chymosin-treated micelles, suggesting differences in the extent and/or specificity of κ-casein hydrolysis. Furthermore, the common rennets differ in their action on synthetic peptides and this may be used to quantify the individual enzymes in a mixed rennet or as the basis for defining an absolute unit for rennet activity (cf. Fox, 15).

Rennet Substitutes

Most proteinases will coagulate milk under suitable conditions, especially pH, but most are too proteolytic relative to their milk clotting activity; consequently, they hydrolyse the coagulum too quickly causing reduced cheese yields and/or defective cheese which has a propensity to bitterness. Although plant proteinases appear to have been used as rennets since pre-historic times, gastric proteinases from young mammals (calves, kids or lambs) have been used traditionally as rennets for most cheese varieties. In western countries, the only exception is the use of a rennet from species of the genus Cynara in the manufacture of Sierra cheese in Portugal.

Owing to increasing cheese production internationally (~ 4% p.a. over the past 20 years), concomitant with a decreasing supply of calf rennet due to an increasing tendency to slaughter more mature calves, the supply of calf rennet has been inadequate over the past 20 years. This has led to increased prices and to a search for rennet substitutes. In spite of the availability of numerous potentially useful milk coagulants, only six rennet substitutes have been found more or less acceptable for one or more cheese variety: bovine, porcine and chicken pepsins and the acid proteinases from Mucor meihei, M. pusillus and Endiothia parasitica.

Chicken pepsin is the least suitable of these and is used widely only in Israel. Bovine pepsin is probably the most satisfactory and many commercial "calf rennets" contain 50-60% bovine pepsin. The proteolytic specificity of bovine pepsin is similar to that of calf chymosin and it gives generally satisfactory results with respect to cheese yield and quality. The principal disadvantage of porcine pepsin is its sensitivity to moderately high pH (> 6.6); 50:50 mixtures of porcine pepsin and calf rennet gave generally acceptable results but porcine pepsin has been withdrawn from most markets because of more attractive prices for pig stomach for use in pet foods. The proteolytic specificity of the 3 commonly used fungal rennets is considerably different from that of calf rennet but the acceptability of most cheese varieties made with fungal rennets has been fairly good. Microbial rennet substitutes are now widely used in the U.S. but "calf rennet" is still predominant in most European countries and in New Zealand; rennet substitutes are prohibited in many European countries. The extensive literature on rennet substitutes has been reviewed (e.g. 16-21).

Chymosin and all the commercially successful rennet substitutes are acid proteinases, i.e. proteinases with 2 aspartate residues at the active site which is buried within a deep cleft in the molecule. Animal acid proteinases are excreted as inactive precursors which are activated autocatalytically or intermolecularly by specific proteolysis. Microbial acid proteinases, which are mainly of fungal origin, are secreted as active enzymes. The molecular and catalytic aspects of the acid proteinases have been reviewed by Foltmann (22, 23). Acid proteinases have relatively narrow specificaties with a preference for peptide bonds to which bulky hydrophobic residues supply the carboxyl group; this narrow specificity is significant for the success of these enzymes in cheese manufacture. The fact that in cheese, these enzymes operate at a pH far removed from their optima, which in the case of pepsins is $\sim$ 2, is probably also significant. It is surprising that they are active at all at the pH of milk, $\sim$ 6.6. However, not all acid proteinases are suitable as rennets because they are too active even under the prevailing relatively unfavourable conditions.

Although they are relatively cheap enzymes, rennets represent the largest single industrial usage of enzymes with a world market of $\sim 25 \times 10^6$ litres of standard rennet ($\sim £100 \times 10^6$). It is not surprising then that rennets are attractive to industrial enzymologists and biotechnologists. The gene for calf chymosin has been cloned in selected bacteria leading to the secretion of chymosin, although apparently at low yields. One of the principal problems encountered is that while the genetically engineered microorganism is capable of expressing chymosin, most of it remains intracellular, which increases recovery and purification costs [cf. Foltmann (23) for references].

In modern practice, most cheesemaking operations are continuous or nearly so; the actual coagulation step is the only remaining major batch operation although the use of small batches of milk, as in the

Alpma process for Camembert, makes coagulation, in effect, a continuous process. The feasibility of continuous coagulation using cold renneting principles has been demonstrated but not used commercially. The use of immobilized rennets would facilitate continuous operations. Several rennets have been immobilized but their efficacy has been questioned; other problems, e.g. hygiene, also exist. Immobilized rennets have not been commercialized. (cf. Refs. 14 and 24 for reviews).

Secondary Phase of Coagulation

Hydrolysis of κ-casein during the primary phase of rennet action, with the release of highly-charged, hydrophilic macropeptides, decreases the zeta potential of the casein micelles from -10/-20 mV to -5/-7 mV and removes the protruding peptides from their surfaces, thereby reducing the intermicellar repulsive forces (electrostatic and steric) and the colloidal stability of the caseinate system. When $\sim$ 85% of the total κ-casein has been hydrolyzed, the casein micelles begin to form a gel but an individual micelle cannot participate in gelation until $\sim$ 97% of its κ-casein has been hydrolyzed. Reducing the pH or increasing the temperature from the normal values ($\sim$ 6.6 and $\sim$ 31°C, respectively) permits coagulation at a lower degree of κ-casein hydrolysis (cf. 13-15, 34).

Coagulation of rennet-altered micelles is absolutely dependent on a critical concentration of Ca^{2+} which may act by cross-linking micelles, possibly via serine phosphate residues, or simply by charge neutralization. Colloidal calcium phosphate (CCP) also plays an essential role in coagulation: if the level of CCP is reduced by $\sim$ 20%, renneted micelles will not coagulate unless the $[Ca^{2+}]$ is increased. Removal of CCP causes dissociation of the micelles but considerably more than 20% can be removed before micellar dissociation occurs. Presumably, the removal of CCP, which is believed to be attached to the casein via serine phosphate residues, increases micellar charge. However, partial enzymatic dephosphorylation of casein, which decreases micellar charge, increases RCT although interaction of casein micelles with various cationic species predisposes them to coagulation and may even coagulate unrenneted micelles. It would be interesting to investigate the influence of a range of cationic species on the coagulability of micelles with reduced CCP content; increased $[Ca^{2+}]$ offsets the effect of reduced $[CCP]$. Chemical modification of histidine, lysine or arginine residues inhibits coagulation, presumably by reducing micellar positive charge. Thus, reducing the positive or negative charge on the renneted micelles appears to cause opposite effects. It has been suggested that coagulation occurs via electrostatic interactions between a positively charged cluster toward the C-terminal of para-κ-casein, which is exposed on removal of the macropeptide, and an unidentified, negatively-charged cluster on neighbouring micelles.

The apparent importance of micellar charge on the coagulation of rennet-altered micelles would suggest that pH should have a major influence on the secondary phase of rennet coagulation. However, Pyne (25) claimed that pH has essentially no effect on the coagulation process although Kowalchyk & Olson (26) showed that the rate of firming of rennet gels was significantly increased on reducing the pH; the methods used were different and perhaps the definition of "secondary phase" also differed.

The coagulation of rennet micelles is very temperature-dependent ($Q_{10°C} \sim 16$) and normal bovine milk does not coagulate $< \sim 18°C$ unless $[Ca^{2+}]$ is increased. The marked difference between the temperature dependence of the enzymatic and non-enzymatic phases of rennet coagulation has been exploited in the study of the effects of environmental and other factors on the rennet coagulation of milk, in attempts to develop a system for the continuous coagulation of milk for cheese or casein manufacture and in the application of immobilized rennets in cheese manufacture. Interestingly, porcine milk is readily coagulated by calf rennet at 5°C (27) but the factor(s) responsible for the different behaviours of porcine and bovine milks in this regard has not been established. The very high temperature-dependence of rennet coagulation might suggest that hydrophobic interactions play a major role.

Para-κ-casein flocculates in the absence of Ca^{2+} but the rate of flocculation depends on the rennet used, presumably reflecting differences in proteolytic specificity which have not yet been elucidated (28, 29). The rate of firming of rennet gels is also influenced by the type of rennet, presumably for similar reasons (29-31).

The rennet coagulation time (RCT) of milk is not correlated with casein concentration within the range normally encountered. Concentration of milk by ultrafiltration causes a slight increase in RCT at a constant level of rennet although the rate of firming of the gel increases with increasing concentration (32). However, diluting milk with ultrafiltrate or synthetic milk serum causes a marked increase in RCT (32, 33).

Gel Assembly

The assembly of renneted micelles into a gel has been studied using various forms of viscometry, electron microscopy and light scattering. The micelles remain discrete until $\sim 60\%$ of the visual coagulation time, after which the rennet-altered micelles, i.e. those with sufficient of their κ-casein hydrolyzed, begin to aggregate steadily, without sudden changes, into chain-like structures, in which the micelles are bridged by an as yet unknown material, which eventually link up to form a network (cf. 14, 15, 34). Aggregation of the rennet-altered micelles can be described by the Smoluchowski theory for diffusion-controlled aggregation of hydrophobic colloids when allowance is made for the need to produce, enzymatically, a sufficient concentration of particles capable of aggregating, i.e. micelles in which $\sim 97\%$ of the κ-casein has been hydrolyzed. A number of more or less sophisticated formulae have been developed to describe the kinetics of aggregation (cf. 13-15, 35). We do not need to concern ourselves here with this aspect of rennet coagulation except to note that the rate of micelle aggregation and development of gel strength are influenced by the type of rennet used, presumably reflecting subtle difference in the specificity and/or extent of proteolysis in the primary phase.

The hydrophobic amino terminal segment (residues 14-24) of α_{s1}-casein appears to be important in the structure of rennet curd (36); it has been suggested (37) that the structural matrix of young cheese consists of a network of α_{s1}-casein molecules linked together via hydrophobic patches. Softening of cheese texture during ripening is considered to be due to break-up of this network on hydrolysis of α_{s1}- to α_{s1}-I casein by rennet (see below).

Effect of Pre-Renneting Treatment of Milk on Rennet Coagulation

The overall RCT of milk increases markedly with increasing pH > $\sim$ 6.4 to an extent characteristic of the type of rennet; porcine pepsin is the most pH-dependent of the commonly used rennets. Addition of 1.5-2% starter decreases the pH of milk by $\sim$ 0.15 units and cognisance must be taken of this when neutralized or concentrated starters are used. It was common practice to "ripen" milk after starter addition, and it still is for some varieties, e.g. Camembert; ripening allows starter to enter its log growth phase before renneting and the acid developed accelerates rennet action. Improved starter technology has obviated the need for ripening which increases the risk of phage infection. The microbiological quality of milk has improved very considerably, especially with respect to mesophilic lactic acid bacteria; consequently, the pH of cheese milk is normally higher than previously. Since the level of calcium in curd, which is determined primarily by the pH at draining, has a major influence on cheese texture (38), the pH of milk at renneting must have some influence on cheese texture unless counteracting action is taken. The amount of chymosin retained by cheese curd increases as the pH is decreased and this influences the rate of proteolysis during ripening, thus affecting flavour and textural changes; retention of pepsins and microbial rennets in curd appear to be independent of pH (39-40).

It is fairly common practice, standard for many varieties, to add $CaCl_2$ ($\sim$ 0.04%) to cheese milk. This practice causes 3 changes in milk all of which reduce RCT: increased $[Ca^{2+}]$, increased $[CCP]$ and reduced pH. These 3 factors also increase gel strength, which may improve cheese yield. The amount of rennet required for a standard RCT may be reduced if $CaCl_2$ is added. Milk for most cheese varieties is renneted at $\sim$ 31°C which is far below the optimum for rennet coagulation. While RCT or the amount of rennet required may be reduced by increasing the renneting temperature, only a small increase is permissable due to inhibition of mesophilic starters and the adverse effect of higher temperatures on gel development (21).

Both the primary and secondary phases of rennet coagulation are retarded in cheese milk heated to temperatures which cause denaturation of whey proteins; coagulation is completely inhibited following severe heat treatment. This effect is due to disulfide-linked polymerization between heat-denatured β-lactoglobulin (and possibly α-lactalbumin) and κ-casein which apparently makes the rennet-susceptible bond inaccessible and also renders renneted micelles incapable of gelation; consequently, the strength of renneted milk gels is weakened by prior heating (cf. 13-15, 34). The adverse effects of heat treatment on the rennet coagulation of milk, both coagulation time and gel strength, may be counteracted by acidification, with or without reneutralization, apparently due to increased $[Ca^{2+}]$ (41). Cooling and cold storage of milk following severe heat treatment exacerbates the adverse effects of heat treatment, an effect known as rennet hysteresis. The effect is apparently due to reestablishment of the calcium phosphate equilibrium during which some indigenous CCP redissolves (42).

Cold storage of raw milk also adversely affects rennet coagulability due to one or more of the following: 1. Solution of CCP, which is readily reversible by HTST pasteurization or by heating at 40°C x 10 min. 2. Dissociation of some micellar casein, especially β-casein, which is also reversible. 3. Proteolysis by indigenous milk proteinase plasmin, especially of dissociated β-casein. 4. Proteolysis by proteinases secreted by psychrotrophs but this becomes important only when these organisms exceed 10^6-10^7 cfu/ml.

Curd Tension (Gel Strength)

The strength of the rennet gel is as, or perhaps more, important, especially from the viewpoint of cheese yield, as the rennet coagulation time. We will not concern ourselves with this subject, which has been reviewed by Fox (15) and Green & Grandison (34). Suffice it to say that curd tension is influenced by the type of rennet; calf rennet gives a more rapid increase in CT than microbial rennets, standardized to equal RCT, presumably reflecting differences in the extent or specificity of proteolysis (29-31).

Curd (Gel) Syneresis

The first and second stages of rennet coagulation and gel assembly process are common to all rennet cheeses, although the rate and extent of both may vary with numerous compositional and processing variables. However, the subsequent treatment of the rennet gel is variety-specific. The objective of the immediate post-gelation operations is to remove sufficient whey to give cheese with the desired, characteristic composition, which in turn determines the rate, extent and direction of ripening. Dehydration is achieved by the marked tendency of rennet milk gels to contract (syneresis) when cut or broken. Again, we will not review this subject and refer to Fox (15) and Walstra et al. (43); syneresis is slightly influenced by the level and type of rennet used.

Cheese Microstructure

Presumably, cheese microstructure is related to the gel assembly process

and the nature of the gel network. The type of rennet is among the
factors which affect cheese microstructure; Cheddar cheese made using
calf rennet has a more compact, organized structure than Cheddar cheese
made using bovine or porcine pepsins (44).

Cheese Ripening

Most ($\sim$ 90%) of the rennet added to cheese milk is removed in the whey;
the amount retained by the curd is influenced by the type of rennet,
e.g. porcine pepsin is extensively denatured during cheesemaking, pH
[low pH favours retention of chymosin but not pepsins or microbial
rennets (39, 40) and reduces denaturation] and cooking temperature, e.g.
very little, if any, coagulant survives the cooking conditions used for
Swiss cheeses (45). The proteolytic specificity of calf chymosin and
the principal rennet substitutes on isolated α_{s1}- and β-caseins and on
sodium caseinate is now fairly well established and these findings can,
largely, be extended to cheese.

β-Casein in solution is sequentially hydrolyzed at bonds 192-193,
189-190, 163-164 and 139-140 to yield the peptides β'I, β''I, β-II and
β-III, respectively; the bonds 165-166, 167-168 may also be hydrolyzed
to yield peptides indistinguishable electrophoretically from β-II. At
low pH (2-3), bond 127-128 is also hydrolyzed to yield β-IV. The
hydrolysis of β-casein by chymosin is strongly inhibited by 5% NaCl and
completely inhibited by 10% NaCl. The reasons for this inhibition are
not clear but a similar effect is produced by sucrose or glycerol or by
high protein concentrations. Presumably, the effect is related to water
activity, A_W, but as far as we are aware, the influence of varying A_W
directly on the proteolysis of individual caseins has not been studied,
although Al-Mazein (29) investigated the effect of A_W on the proteolysis
of sodium caseinate by rennet.

β-Casein is quite resistant to proteolysis in bacterially-ripened
cheeses throughout ripening and in mould-ripened cheeses until fungal
proteinases become dominant after mould growth. Although the
concentration of β-casein in bacterially-ripened cheeses decreases
during ripening, the β-peptides normally produced by rennet, i.e. β-I,
β-II, do not appear, suggesting that plasmin and/or bacterial
proteinases are responsible. The NaCl in cheese is undoubtably an
inhibitory factor but even in the absence of NaCl the extent of β-casein
hydrolysis is slight.

α_{s1}-Casein in solution has several chymosin-susceptible bonds,
hydrolysis of which is highly dependent on the environment. In contrast
to β-casein, NaCl up to 5% stimulates the hydrolysis α_{s1}-casein and
significant proteolysis occurs in the presence of 20% NaCl. Consequently
α_{s1}-casein is readily hydrolyzed in cheese, initially to α_{s1}-I and later
to α_{s1}-V, α_{s1}-VII and small amounts of α_{s1}-II. In Cheddar and Dutch-
type cheeses, α_{s1}-casein is completely degraded to α_{s1}-I and some further
products by the end of ripening. In mould-ripened cheeses, α_{s1}-casein is
completely degraded to at least α_{s1}-I prior to the mould-ripening phase
and very extensive degradation occurs thereafter. α_{s2}-Casein and para-
κ-casein appear to be quite resistant to rennets and remain largely

intact in bacterially-ripened cheeses. For reviews on proteolysis in cheese see Refs. 46-48.

Significance of Secondary Coagulant Proteolysis

Studies on aseptic curd and cheese with controlled microflora have shown that the coagulant is responsible for the level of proteolysis detected by gel electrophoresis and for most of the nitrogen soluble in water or at pH 4.6; however, little TCA or PTA-soluble N is produced by the coagulant (49). Proteolysis by rennet is believed (36, 37) to be responsible for the softening of cheese texture early during ripening via the hydrolysis of α_{s1}-casein to α_{s1}-I which is sufficient to break the continuous protein matrix. Undoubtably, further proteolysis by both coagulant and bacterial proteinases modifies the texture further. Even in surface mould ripened cheese, and probably in smear cheeses, coagulant is considered to be essential for the development of proper texture, e.g. in Camembert, although the very marked increase in pH (to 7) caused by the catabolism of organic acids and the production of ammonia (by deamination of amino acids) is also essential (50, 51). The proteinases excreted by the mould diffuse into the cheese to only a slight extent and contribute little to proteolysis within the cheese although peptides produced by these enzymes in the surface layer may diffuse into the cheese.

The texture of high cooked cheeses, e.g. Swiss and hard Italian varieties, changes relatively little during ripening. In these cheeses, little if any coagulant survives the cooking process (45) and plasmin plays a significant role in primary proteolysis. The high pH of these cheeses at drainage ensures that much of the plasmin is retained in the curd, in contrast to more acid cheeses, e.g. Cheddar and Cheshire (52).

The secondary proteolytic action of the coagulant influences flavour in 3 separate ways:
1. Some of the rennet-produced peptides are small enough to influence flavour. Unfortunately, some of these peptides are bitter and excessive proteolysis, e.g. due to too much or excessively proteolytic rennet or unsuitable environmental conditions, e.g. too much moisture or too little NaCl, leads to bitterness.
2. The large rennet-produced peptides serve as substrates for microbial proteinases and peptidases which lead to increased levels of small peptides and amino acids. These contribute at least to background flavour, and perhaps, unfortunately, to bitterness if the activity of such enzymes is excessive. Catabolism of amino acids by microbial enzymes and perhaps alterations via chemical mechanisms, leads to a range of sapid compounds - amines, acids, NH_3, thiols - which contribute markedly to characteristic cheese flavours.
3. Alterations in cheese texture appears to influence the release of flavourful and aromatic compounds, arising from proteolysis, lipolysis, glycolysis and secondary metabolic changes, from cheese during mastication (53). This may be the most significant contribution of proteolysis to cheese flavour.

References

1. Linderstrøm-Lang, K., Z. Physiol. Chem. 176, 76 (1928).
2. Berridge, N.J., Nature, 149, 194 (1942).
3. Fox, P.F., "Developments in Dairy Chemistry - 1", Applied Science Publishers, London (1982).
4. Fox, P.F. & Mulvihill, D.M., J. Dairy Res., 49, 679 (1982).
5. Fox, P.F., Proc. 6th Intern. Congr. Food Sci. & Technol., 5, 177 (1983).
6. Fox, P.F. & Mulvihill, D.M., Proc. IDF Symposium 'Physico-chemical aspects of dehydrated protein-rich milk products', p. 188; Helsingor, Denmark (1983).
7. Eigel, W.N., Butler, J.E., Ernstrom, C.A., Farrell, H.M. Jr., Harwalkar, V.R., Jenness, R. & Whitney, R.McL., J. Dairy Sci., 67, 1599 (1984).
8. McMahon, D.J. & Brown, R.J., J. Dairy Sci., 67, 499 (1984).
9. Waugh, D.F. & von Hippel, P.H. J. Am. Chem. Soc., 78, 4567 (1956).
10. Wake, R.G., Aust. J. Biol. Sci., 12, 479 (1959).
11. Delfour, A., Jolles, J., Alais, C. & Jolles, C. Biochem. Biophys. Res. Commun. 19, 452 (1965).
12. Loucheux-Lefebvre, R.-H., Aubert, J.P. & Jolles, P., Biophysical J., 23, 323 (1978).
13. Dalgleish, D.G., 'Enzymatic Coagulation of Milk' in "Developments in Dairy Chemistry - 1 - Proteins", P.F. Fox, ed., p. 157, Applied Science Publishers Ltd., London (1982).
14. Dalgleish, D.G., 'The Enzymatic Coagulation of Milk' in "Cheese: Chemistry, Physics and Microbiology", Volume 1, P.F. Fox, ed., in press, Elsevier Applied Science, London (1986).
15. Fox, P.F., 'Proteolysis and protein-protein interactions in cheese manufacture' in "Developments in Food Proteins - 3", B.F.J. Hudson, ed., p. 69; Elsevier Applied Science, London (1984).
16. Sardinas, J.L., Adv. Appl. Microbiol. 15, 39 (1972).
17. Ernstrom, C.A., 'Milk clotting enzymes' in "Fundamentals of Dairy Chemistry", 2nd edn., B.H. Webb, A.H. Johnson, & J.A. Alford, eds., p. 662; AVI Publishing Co., Inc., Westport, CT (1974).
18. Nelson, J.H., J. Dairy Sci., 58, 1739 (1975).
19. Sternberg, M., Adv. Appl. Microbiol. 20, 135 (1976).
20. Green, M.L., J. Dairy Res., 44, 135 (1977).
21. Phelan, J.A., 'Milk coagulants: an evaluation of alternatives to standard calf rennet', Ph.D. Thesis, National University of Ireland (1986).
22. Foltmann, B., 'Gastric Proteinases' in "Essays in Biochemistry", Vol. 17, R.N. Campbell & R.D. Marshall, eds.; Academic Press, London (1981).
23. Foltmann, B., 'General and molecular aspects of rennets' in "Cheese: Chemistry, Physics and Microbiology", Vol. 1, P.F. Fox, ed., in press, Elsevier Applied Science, London (1986).
24. Fox, P.F., 'Proteinases in Dairy Technology' in "Proteinases and their Inhibitors: Structure, Function and Applied Aspects", V. Turk & Lj. Vitale, eds., p. 245; Pergamon Press, Oxford (1981).
25. Pyne, G.T., Dairy Sci. Abstr., 17, 531 (1955).

26. Kowalchyk, A.W. & Olson, N.F., J. Dairy Sci., 60, 1256 (1977).
27. Hoynes, M.C.T. & Fox, P.F., J. Dairy Res., 42, 43 (1975).
28. Lawrence, R.C. & Creamer, L.K., J. Dairy Res., 36, 11 (1969).
29. Al-Mzaien, K.A., 'Chicken pepsin: isolation, characterization and assessment of its suitability as a rennet substitute', Ph.D. Thesis National University of Ireland (1985).
30. Richardson, G.H., Gandhi, N.R., Divatia, M.A. & Ernstrom, C.A., J. Dairy Res., 54, 182 (1971).
31. Kowalchyk, A.W. & Olson, N.F., J. Dairy Sci., 62, 1233 (1979).
32. Dalgleish, D.G., J. Dairy Res., 47, 231 (1980).
33. Fox, P.F. & Morrissey, P.A., J. Dairy Res., 39, 387 (1972).
34. Green, M.L. & Grandison, A.S., 'Secondary (non-enzymatic) phase of rennet coagulation and post-coagulation phenomena' in "Cheese: Chemistry, Physics and Microbiology", Vol. 1, P.F. Fox, ed., in press; Elsevier Applied Science, London (1986).
35. Payens, T.A.J. & Wiersma, A.K., Biophys. Chem., 11, 137 (1980).
36. de Jong, L., Neth. Milk Dairy J., 30, 242 (1976).
37. Creamer, L.K., Zoerb, H.F., Olson, N.F. & Richardson, T., J. Dairy Sci., 67, 1632 (1984).
38. Lawrence, R.C., Heap, H.A. & Gilles, J., J. Dairy Sci., 67, 1632 (1984).
39. Holmes, D.G., Duersch, J.W. & Ernstrom, C.A., J. Dairy Sci., 60, 862 (1977).
40. Creamer, L.K., Lawrence, R.C. & Gilles, J., N.Z.J. Dairy Sci. & Technol., 20, 185 (1985).
41. Singh, H., Shalabi, S.I., Fox, P.F., Flynn, A. & Barry, A., unpublished data.
42. Morrissey, P.A., J. Dairy Res., 36, 333 (1969).
43. Walstra, P., van Dijk, H.J.M. & Geurts, T.J., 'The syneresis of curd' in "Cheese: Chemistry, Physics and Microbiology", Vol. 1, P.F. Fox, ed., in press; Elsevier Applied Science, London (1986).
44. Eino, M.F., Biggs, D.A., Irvine, D.M. & Stanley, D.W., Can. Inst. Food Sci. Technol. J., 12, 149 (1979).
45. Matheson, A.R., N.Z.J. Dairy Sci. & Technol., 15, 33 (1981).
46. Grappin, R., Rank, T.C. & Olson, N.F., J. Dairy Sci., 68, 531 (1985).
47. Rank, T.C., Grappin, R. & Olson, N.F., J. Dairy Sci., 68, 801 (1985).
48. Law, B.A., 'Proteolysis in Cheese' in "Cheese: Chemistry, Physics and Microbiology", Vol. 1, P.F. Fox, ed., in press; Elsevier Applied Science Publishers, London (1986).
49. O'Keeffe, A.M., Fox, P.F. & Daly, C., J. Dairy Sci., 45, 465 (1978).
50. Noomen, A., Neth. Milk Dairy J., 37, 229 (1983).
51. Lenoir, J., IDF Bulletin 171 (1984).
52. Lawrence, R.C., Gilles, J. & Creamer, L.K., N.Z.J. Dairy Sci. & Technol., 18, 175 (1981).
53. McGugan, W.A., Emmons, D.B. & Larmond, E., J. Dairy Sci., 62 398 (1979).

Starter cultures: Fundamental aspects

Michael Teuber
Institute of Microbiology
Federal Dairy Research Centre
D-2300 Kiel 1
F. R. Germany

ABSTRACT. The function of microbial starter cultures in cheese manufacture is the initiation of biochemical processes (e.g. lactic fermentation, proteolysis, aroma production) which contribute essentially to the overall ripening and final quality of cheese. Some of the necessary enzymes as well as some bacteriophage resistance systems are genetically linked to plasmids. This opens the possibility to stabilize microbial functions in cheese making by genetic manipulation of the involved cultures. Progress and prospects will be discussed.

1. Historical aspects

The invention of starter cultures for the fermentation of milk is about 100 years old since V. Storch in Copenhagen and H. Weigmann in Kiel succeeded independently to isolate the mesophilic lactic acid streptococci responsible for lactic acid and diacetyl production during manufacture of sour milk, sour cream and cheese (1). Since then the list of microorganisms recognized as important for different functions in the dairy industry has been considerably expanded to include thermophilic streptococci, lactobacilli, propionibacteria, brevibacteria, yeasts and moulds like Penicillium caseicolum and P.roqueforti.
   Due to their dominant role in the production of worldwide distributed fermented milk, cream and cheese varieties (e.g. sour milk, sour cream, lactic butter, fresh, soft, semihard and hard cheese), fundamental research has mainly focussed on the mesophilic group N lactic acid streptococci Streptococcus lactis, its subspecies diacetylactis and S.cremoris. The amount of such products is estimated to be between 1 and $2 \cdot 10^{10}$ kg per year in the world (1).
   It is the scope of this paper to summarize the most recent developments in fundamental research areas such as molecular biology, genetics and biochemistry of the lactic streptococci with emphasis on functions vital for a successful cheese technology.

2. Functions and dysfunctions of lactic streptococci in cheese making
   (2)

The main early function is a sufficiently fast and quantitatively defined conversion of lactose into lactic acid. Later functions include formation of diacetyl, gas (CO_2) and other aroma compounds, some possibly generated by a controlled proteolysis during cheese ripening. In the beginning, these processes are surely driven by living bacteria and their enzymes. During ripening and after salting, most of the cells will die, but further biochemical changes could occur due to remaining bacterial enzymes. The number of live cells included in cheese curd during coagulation has been estimated to be at least 10^8 per ml (3), enough to induce substancial biochemical changers in cheese even at low temperatures but during long ripening periods of up to several months. The bacterial enzyme systems are thereby severely influenced by the salt concentration, the pH-value, the water activity and the ripening temperatures. Dysfunctions - if not caused by such inproper biophysical conditions or handling - may have one or more of three microbiologcal reasons:
1. Irreversible loss of technologically important functions from the culture.
2. Inactivation of the culture by bacteriophages.
3. Inactivation of the culture by residual antibiotics and desinfectants.

3. Manipulation of functions and dysfunctions of starters

This attempt is only possible if we understand the functioning of the starter microorganisms, it means the genetic and biochemical rules governing the performance of the bacteria during cheese making.

3.1. Genetics of starters (4)

The last 10 years have witnessed an explosion of activities in the field of streptococcal genetics due to the discovery of plasmids in group N streptococci by L. L. McKay in 1972. From 1976 to 1986, the number of laboratories engaged in genetics of lactobacteria has increased from about 5 to at least 50 all over the world.
   Plasmids are small genetic elements composed of double stranded deoxyribonucleic acid (DNA) replicating independently of the bacterial chromosome which carries most of the genetic information of the cells.
   Plasmids have been detected in all investigated strains of lactic streptococci (several hundred), varying in size from about 1 to 80 megadalton and in number from 1 to more than 10 (5, 6). The plasmid profile of a single strain which can be easily obtained by agarose gelelectrophoresis, is strain specific and quite stable if the culture is kept in milk at ambient temperature during cultivation. The determination of plasmid pattern therefore allows the identification of single, multiple and undefined mixed strain starter cultures and gives

an exact account of their composition. By this method, at least 50 different strains have been recognized in certain mixed strain starters (5).

More important, however, was the discovery that practically all the necessary technologically functions of lactic streptococci were found on plasmids including fermentation of lactose and citrate, ß-casein-specific proteolytic acitivity and formation of slime in ropy strains. In addition, further technologically useful functions like production of nisin and other bacteriocins, resistance to bacteriophages and antibiotics were identified on plasmids (table 1). The observation that growth or treatment of streptococci at higher temperatures (39-42 °C) induces loss of plasmids explains the accumulation of functionally defect strains in whey and artificial media. As a consequence, starter propagation has to be performed under controlled conditions to avoid degeneration caused by plasmid curing. The coding of technologically important functions on plasmid DNA in group N streptococci opened the way to genetic transfer between starter strains.

Several methods have been developed for that purpose. Two naturally occuring pathways are available: conjugation and transduction. For conjugation, donor and receptor bacteria have to be brought into direct physical contact. Lactic streptococcal strains are to a certain percentage (10-35 %) able to use this route of gene exchange at frequencies between 10^{-2} to 10^{-8} on solid media and between 10^{-8} to 10^{-9} in coagulated milk (7). Some plasmids are selftransmissible like lactose-, bacteriocin- and phage-resistance plasmids. Others like protease- or citrate-plasmids can be mobilized by the mentioned selftransmissible plasmids. Sometimes, the complete plasmid complement of a strain or cell is transfered, somtimes only one, most frequently several of them. It may be speculated that conjugation does occur during cheese making and milk fermentations explaining at least in part the complexity of plasmid profiles in mixed strain starter cultures. This view is supported by the fact that plasmids often are changed in size (usually enlarged) during conjugation (8).

The second natural transfer of genes is by transduction: plasmid DNA is encapsulated in the head of bacteriophages instead of phage DNA. If such a phage infects a sensitive bacterium, the plasmid DNA will be injected into that organism and eventually replicate. The size of transducible plasmids is limited by the size of the phage head and DNA (13 to 34 megadalton, see below). Lactose plasmids have been mobilized with this technique (6).

Transformation (=uptake of isolated DNA by bacterial cells) has been recently achieved with lactic streptococci, however, only when the cell walls of the bacteria had been digested with lytic enzymes and the resulting protoplasts had been treated with polyethyleneglycol and phospholipid vesicles. The efficiencies of transformation vary between 10^1 and 10^6 transformants per microgramm of DNA depending mainly on the strain specific rate of regeneration of protoplasts into intact cells. Nevertheless, it is an important prerequisite to successful genetic engineering (6, 9).

Fusion of protoplasts has also been occasionally used for transfer of genes in lactic streptococci (4).

3.2. Biochemistry of starter fuctions

The biochemistry of lactose fermentation has been completely elucidated (10). This disaccharide is transported into the cells via the phosphoenolpyruvate dependent phosphotransferase system. The accumulated lactose-6-phosphate is cleaved by a phospho-ß-galactosidase into glucose and galactose-6-phosphate which is converted into triosephosphates by the tagatose-pathway. Most of the involved enzymes are coded on plasmids.

The conversion of citrate to diacetyl starts with transport by a plasmid coded citrate permease. Further citrate metabolism follows the classical pathway through oxaloacetate, pyruvate and -acetolactate.

The protease system which is necessary to obtain high cell numbers in milk has recently been elucidated in more details (11). The plasmid coded enzyme has a molecular weight of about 145.000 (12). It is bound to the bacterial cell wall, specific for ß-casein and degrades rapidly itself once solubilized from the cells. It is inhibited by sodium chloride concentrations above 2 % which might influence its possible role in cheese ripening. Some of the breakdown products of ß-casein have been characterized. They seem to be generated from the carboxyterminal end including a know bitter peptide. Recently, a cell wall bound peptidase has been described which is able to degrade the peptides produced by the cell wall protease from ß-casein (12). The peptidase products have not been further analysed. It is expected that this will be possible within the next few years. A few streptococcal strains seem to have a protease which can also cleave -and k-caseins (13).

4. Bacteriophages of lactic streptococci (14, 15, 16)

Since cheese making can not be performed under sterile conditions with sterilized milk, bacteriophages are the main cause of dysfunction of starter bacteria. Raw milk and whey seem to be the main source of infection since many phages do survive pasteurization. The many morphologically different virulent phage types described and found all over the world have been classified into 4 genetically distinct phage groups on the basis of DNA-DNA hybridization (17, 18):
1. phages with prolate heads and a genome size of about 13 megadalton.
2. phages with small isometric heads and a genome size of about 20 megadalton.
3. phages with small isometric heads and a genome size of about 25 megadalton.
4. phages with large isometric heads and a genome size of about 34 megadalton.

There seems to be a wide range of genetic diversity and evolution within these 4 genetically distinct groups. However, the existence of further groups can not be excluded (18).

In addition, many starter strains carry prophages in their chromosome which may be released spontaneously or by induction with ultraviolet light or mitomycin C (14). However, only few indicator bacteria are found for these temperate phages. This appears logical since all sensitive strains in a mixed strain culture would be constantly eliminated by spontaneously released temperate phages. The most exciting development in the area of bacteriophage is the discovery of several, plasmid coded phage resistance mechanisms (19, 20, 21):
1. restriction/modification
2. inhibition of phage adsorption
3. inhibition of phage replication.

All three kinds of resistance plasmids can be transfered by natural conjugation to phage sensitive strains rendering them phage insensitive. Some of such strains are already with promising results in testing and practical use demonstrating the great potential of fundamental genetic work for every day cheese technology.

5. Genetic engineering of lactic streptococci (4, 22)

Plasmids offer themselves as vectors for genes to be expressed in lactic acid streptococci. Since these bacteria are generally regarded as safe (GRAS) organisms they could be suitable to produce genetically engineered products for human consumption e.g. food grade enzymes (chymosin, lysozyme, proteases and others). The basic work for the construction of these vectors has just been completed within a joint research project of several laboratories, within the 'Biomolecular Engineering Research Programme' in the European Community (23). The structures of these vectors are shown in figure 1. Most interestingly, these vectors - based on small cryptic plasmids and other plasmids from lactic streptococci - are also able to replicate and be expressed in Bacillus subtilis and Escherichia coli, two genetically much better characterized microorganisms. This will greatly facilitate the elucidation of gene structure and function in starter bacteria. Some of their genes (lactose, protease) have been cloned and are currently investigated regarding detailed structure and function.

These vectors - in a proper form - should enable the incorporation of desired functions (if biochemically well characterized) into lactic acid streptococci.

It should be mentioned at the end that many laboratories are now trying to transfer the experience made with streptococci onto lactobacilli, propionibacteria, yeast and moulds. It can therefore be predicted that similar natural ways of manipulation will become available for these microorganisms in the near future.

Table 1. Plasmid coded functions in lactic streptococci

Function	plasmid size (megadalton)
1. Lactose fermentation	20 to 60
2. protease activity	9.5 to 25
3. citrate use	5.3
4. bacteriocin production	37 to 75
5. slime formation	17 to 30
6. restriction of phages	10
7. inhibition of phage adsorption	34
8. inhibition of phage replication	39 to 41
9. cryptic plasmids (plasmid replication)	1 to 5

copy numbers per cell may vary widely between a few copies and up to 100.

Figure 1. Examples of cloning vectors constructed for S.lactis and S.cremoris: pNZ122 (M. de Vos), pGK12 (G. Venema), pCK21 (M.J. Gasson). These vectors may also be used in conjunction with B.subtilis and E.coli (23).

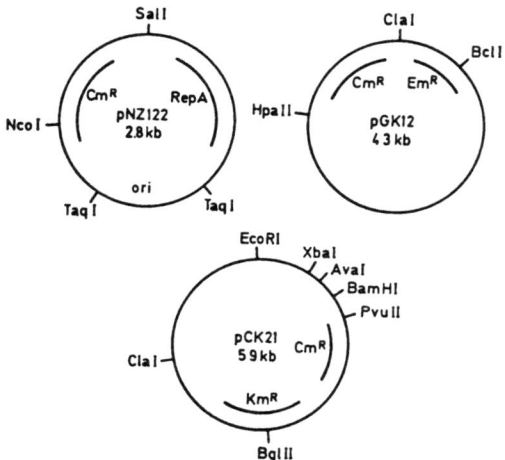

References
1. Teuber, M., Geis, A., 1981. The family Streptococcaceae (Nonmedical aspects). In: M. P. Starr et al. The Prokaryotes, Vol. 2, Springer-Verlag, Berlin-Heidelberg-New York, pp. 1614-1630.
2. Lawrence, R. C., Thomas, T. D., Terzaghi, B. E. 1976. J. Dairy Res. 43: 141.
3. Hugenholtz, J. 1986. Neth. Milk Dairy J. 40:129.
4. Gasson, M. J., Davies, F. L. 1984. The genetics of dairy lactic-acid bacteria. In: Davies, F. L. and Law, B. A. (eds.). Advances in microbiology and biochemistry of cheese and fermented milk. Elsevier, London-New York, pp. 127-152.
5. Andresen, A., Geis, A., Krusch, U., Teuber, M. 1984. Milchwissenschaft 39:140.
6. McKay, L. L. 1983. Antonie van Leeuwenhoek 49:259.
7. Neve, H., Geis, A., Teuber, M. 1984. J. Bacteriol. 157:833.
8. Neve, H., Geis, A. Teuber, M. 1987. System. Appl. Microbiol. in press.
9. Simon, D., Ronault, A., Chopin, M.-C. 1986. Appl. Environm. Microbiol. 52:394.
10. Kandler, O. 1983. Antonie van Leeuwenhoek 49:209.
11. Kok, J., van Dijl, J. M., van der Vossen, J. M. B. M., Venema, G. 1985. Appl. Environm. Microbiol. 50:94.
12. Geis, A., Bockelmann, W., Teuber, M. 1985. Appl. Microbiol. Biotechnol. 23:79.
13. Simons, A. F. M., Exterkate, F. A., Visser, S., de Vos, W. M. 1985 Antonie van Leeuwenhoek 51:565.
14. Teuber, M., Lembke, J. 1983. Antonie van Leeuwenhoek 49:283.
15. Klaenhammer, T. R. 1985. Adv. Appl. Microbiol. 30:1.
16. Davies, F. L., Gasson, M. J. 1984. Bacteriophages of dairy lactic-acid bacteria. In: Davies, F. C. and Law, B. A. (eds.) Advances in the microbiology and biochemistry of cheese and fermented milk. Elsevier, London-New York, pp. 127-152.
17. Jarvis, A. W. 1984. J. Bacteriol. 47:343.
18. Loof, M. 1985. Ph. D. Thesis, University of Kiel.
19. Jarvis, A. W., Klaenhammer, T. R. Appl. Environm. Microbiol. 51:1272.
20. Wetzel, A., Neve, H., Geis, A., Teuber, M. 1986. Chem. Mikrobiol. Technol. Lebensm. 10:86.
21. Daly, C. 1986. in Biomolecular Engineering in the European Community.
Martinus Nijhoff, Dordrecht-Boston-Lancaster, pp. 453-463.
22. de Vos, W. M. Neth. Milk Dairy J. 40:141.
23. de Vos, W. M. 1986. pp. 465-472
Gasson, M. J. 1986. pp. 489-496
Venema, G. 1986, pp. 549-558
in: Biomolecular Engineering in the European Community.
Martinus Nijhoff, Dordrecht-Boston-Lancaster.

STARTERS: AS A MEANS OF CONTROLLING CONTAMINATING ORGANISMS

A.V. Gudkov
All-Union Research Institute of Butter and Cheesemaking
152620 Uglich, Krasnoarmeiskii Boulevard, 19
USSR

ABSTRACT. Starter organisms inhibit the growth of contaminating organisms in cheese by the production of organic acids, competition for lactose, pH and Eh reduction. Additionally some species and strains of lactic acid bacteria produce H_2O_2 and antibiotics. Lactic streptococci and lactobacilli producing such inhibitors were used for semihard cheese manufacture from milk inoculated by up to 100 c.f.u. ml^{-1} of Escherichia coli or Clostridium tyrobutyricum. There were no differences in the quality of these cheeses and the cheeses made from unseeded milk.

Starters play the dominant role in the inhibition of contaminating organisms in cheesemaking. Antimicrobial factors that are due to starter action may be divided into three groups, nonspecific, specific and toxic derivatives of oxigen (Table I).
All starter organisms exhibit the nonspecific inhibitory activity against contaminating organisms. Specific inhibitors are formed by a few lactic acid bacteria. They have limited action spectra. H_2O_2 and other O_2 derivatives are cytotoxic substances but lactic acid bacteria accumulate only small amounts of these substances that inhibit a few contaminating organisms. Therefore toxic derivatives of O_2 were placed into the separate group.
Nonspecific inhibition is caused mainly by the production of lactic and acetic acids and pH reduction. Many pathogenic and spoilage organisms can grow in media at pH values normally found in cheeses (pH 5.0 to 5.5). However, the inhibitory effect of organic acids is higher than that of inoorganic acids at the same pH (1, 10, 62). Lactic acid inhibits the Escherichia coli development at pH 5.1 in the same extent as hydrochloric acid does it at pH 4.5 (1). Acetic acid produced by starter organisms in cheese possesses still more inhibitory activity toward

contaminating organisms (62). Rubin et al. have shown lactic acid entered Salmonella typhimurium cells in the undissociated state, irreversible altered the intracellular conditions that caused the death of cells (62).

Table I. Antimicrobial factors of lactic acid bacteria

Inhibitory factors	Spectrum activity in cheese	References
1. Nonspecific		
1.1. Organic acids and pH	Majority of contaminating organisms	1,28,24,27,57,62,75
1.2. Absorption of O_2, Eh reduction	Aerobes and facultative anaerobes	51,72
1.3. Competition for lactose	Saccharolytic organisms	
2. Specific		
2.1. Bacteriocins	Lactic streptococci, Lactobacilli	12,13
2.2. Nisin	Clostridia, Bacilli, staphilococci etc.	2,38,65
2.3. Other antibiotics (peptides and indefinite substances)	Enterobacteria, staphilococci, clostridia, psychrotrophs etc.	5,6,10-12,15,18,25, 27,28,30,33,35-37, 46,48,49,53,55,56, 58-61,65,67-72
3. H_2O_2 and other derivatives of O_2	Clostridia, staphilococci, micrococci, psychrotrophs etc.	7,9,13,14,16,19,20, 27,44,46,49-51,55, 61,64

Starter organisms rapidly uptake oxygen dissolved in milk and reduce Eh of cheese (51, 72) that suppress the growth of obligate aerobes and strongly inhibits the multiplication of facultative anaerobes within cheese. Stadhouders et al. (72) have shown that anaerobic conditions at pH 5.2 normally found in many ripening cheeses are sufficient to give a complete inhibition of Staph. aureus growth.

Lactose is the limiting nutrient in cheese as it is rapidly used by starter organisms. This is also one of the reasons of ceaseing the growth of staphilococci and enterobacteria in ripening cheese.

The effect of the pH reduction rate in Rossiiskii cheese on the growth of staphilococci is shown in Table II. The number of Staph. aureus generations closly correlated with pH of cheeses at the end of pressing. The pH of many cheeses with low acidity at pressing achieved the normal value after 1 to 3 days of ripening. In spite of this the numbers of Staph. aureus generations in these cheeses were higher than those in cheeses with normal pH at pressing. It means that starter activity during the early stages of cheesemaking plays the most important role in the inhibition of staphilococci growth in cheese.

Table II. Effect of pH on the growth of Staph. aureus in Rossiiskii cheese (24) and E. coli in Dutch-Type cheese (75)

pH at end of pressing[2]	Number of cheeses	Number of generations[1]		
		minimal	mean	maximal
A. Staphilococci in Rossiiskii cheese				
5.15-5.25	51	no growth	2-3	4-5
5.26-5.40	18	no growth	3-4	6
5.41-5.60	13	1	5-6	8
> 5.6	10	2-3	8-9	11-12
B. E. coli in Dutch-type cheese				
5.31-5.42	2		3-4	
5.65-5.82	4	6-7	7	8
5.94-6.48	3	10-11	13	14-15

1) The number of generations was calculated accepting a 10 - fold mechanical concentration of cells incurd (59)
2) Normal values of pH at the end of pressing are 5.15 to 5.25 for Rossiiskii cheese and 5.6 to 5.8 for Dutch-type cheese

In Cheddar and Gouda cheeses made under normal acidification conditions staphilococci enterotoxins were detected when the numbers of enterotoxogenic Staph. aureus strains reached 1.5 to 3.3 x 10^7 g^{-1}; in starter failure cheeses they were detected when the numbers of staphilococci were not less than 3 x 10^6 g^{-1} (3,76). The number of staphilococci in Rossiiskii cheese with normal rates of acidification might reach the dangerous levels if initial counts of staphilococci in cheese milk exceeded 5 x 10^4/ml. So heavy post-pasteurization contamination of cheese milk by staphilococci does not occur in practice. In starter failure Rossiiskii cheese enterotoxins might develop if an initial count of staphilococci in cheese milk exceeded 10^2 c.f.u./ml. In starter failure Cheddar cheese staphilococcal enterotoxins might be formed if cheese milk is contaminated by 5 c.f.u./ml of staphilococci (26). Postpasteurization contamination of milk by 5 to 100 c.f.u./ml of staphilococci is quite possible under poor hygienic conditions.

Similar relation exists between the rate of pH reduction and E. coli growth in Dutch-type cheese (Table II).

Pasteurization of milk, the good hygienic conditions of manufacture and high starter activity protect cheese from the most pathogenic and spoilage organisms. Unfortunally these measures are not always sufficient to prevent the cheese defects caused by lactate-fermenting clostridia, mainly Clostridium tyrobutyricum. Cl. tyrobutyricum grows in media with lactate as an energy source at pH ≥ 5.0, but only at pH > 5.2 it produces butyric acid and hydrogen caused the cheese defects (23). Defects in cheeses with pH < 5.2 (Cheddar, Rossiiskii etc.) due to clostridia do not occur if a starter produces sufficient acidity at a satisfactory rate. The hard cheeses and the most semi-hard cheeses with higher pH values are sensitive to clostridia spoilage. It is possible to lower the pH of cheese but hard and semi-hard cheeses with pH lower than 5.2 do not possess the typical quality (32).

Cheesemakers do not always succeed in keeping of the desirable rate of acidification. This fact and the necessity to protect cheese from clostridia induce a search for additional means of controlling contaminating organisms. The use of lactic acid bacteria strains possessing a specific antimicrobial activity is one of such means (Table III).

Tabl III. Specific antimicrobial activity of lactic acid bacteria

Species	Spectrum of activity			
	Cl. tyrobutyricum	Staph. aureus	Enterobacteria	Psychrotrophs
S. lactic	+ /2,38,65/	+ /27,67/	+ /3,40,54,68/	
S. lactis subsp. diacetylactis		+ /10,15, 34,70/	+ /3,6,12,19, 34,40,48,54, 61,70,77/	+ /6,18, 48,61, 70/
S. cremoris		+ /27,60,67/	+ /3,40,54/	
Leuconostoc		+ /41/	+ /3,6,41,54, 61,77/	
L. plantarum	+ /4,33,35,69/		+ /39,49,54,71/	
L. casei	+ /4,33,35,69/	+ /68/	+ /8,37,39,54,68/	+ /8/
L. fermentum	+/35/		+/68/	
S. thermophilus	+/35/	+ /5/	+ /5,40/	+/56/
S. lactis	+/33/			
L. helveticus	+ /4,33,35/			
L. bulgaricus	+ /4,33/	+ /58/	+ /47,58/	+/58,42/
L. acidophilus	+ /4,33/	+	+	+
Cheese starter		/11,30,66/ /15,29/	/11,25,30,66/ /71/	/30,66/

Numbers of lactic streptococci and lactobacilli produce bacteriocins (12,13) but it is doubtful whether the bacteriocinogenic strains will find use in starters as bacteriocins affect only the organisms closly related to bacteriocinogenic strains. Probably lactic streptococci producing bacteriocins might be used to prevent the cheese defects caused by lactobacilli (12).

Nisin is the sole antibiotic produced by some S. lactis strains that is commercially available and used in food industry. Neither nisin nor nisin-producing strains are used in cheesemaking as nisin inhibits the starter organisms while the acidification capacity of nisin-producing strains is too low to control Gram-negative microorganisms in cheese that are not sensitive to nisin.

Nisin-resistant strains of lactic streptococci with required rate of acid production were selected to foster the acidification of cheese made with nisin-producing starters (38, 65). The combination of nisin-producing and nisin-resistant starters prevents the late blowing of semi-hard cheeses (38, 65). At the sixties these starters were commercially used in the Soviet Union; then they were given up because of difficulties of selection the nisin-resistant cultures with required acid-producing capacity that do not inactivate nisin.

Recently it was found that the lactose-fermenting ability, the resistance to nisin and phages of lactic streptococci are located on the same plasmid (31,45,74). Nisin-production and nisin-resistance can be transfered from S. lactis to S. lactis subsp. diacetylactis (17). The increacing knowledge of the genetics of lactic streptococci could create exciting possibilities for genetically constraction nisin and phage resistant strains with nigh acid-producing capacity.

A number of studies have reported on ability of lactic acid bacteria to produce the other than nisin antibiotic-like substances (Table III). A few lactobacilli strains possess the specific activity against clostridia (4,33,55, 69). Bergere et al. (4) heve found that this inhibitory activity is not sufficient to prevent late blowing in Emmental cheese, but Kundrat (35) successfully used thermophilic lactic acid bacteria strains possessed the specific inhibitory activity toward clostridia in vitro to control butyric acid bacteria in hard cheeses.

Strains with specific inhibitory activity toward enterobacteria are found more often. Screening test of a large number of lactic streptococci cultures for antimicrobial activity against E. coli has shown that medium activity (12 to 30 mm zones of inhibition) was exhibited by 7.5 % of S lactis, 17.7 % of S. cremoris, 23.4 % of S. lactis subsp. diacetilactis with high acid-producing capacity, 26.4 % of S. lactis subsp. diacetylactis with low acid-producing capacity in milk and 81.3 % of Leuconostoc (3). About 0.5 % of S. lactis and 2 % of S cremoris cultures formed the inhibitory zones 32 to 44 mm.

Freshly isolated cultures of lactic streptococci and lactobacilli possessed the higher antimicrobial activity than laboratory ones (40). There are cells with high and low antimicrobial activity in microbial populations (34). Cells with high specific antimicrobial activity usually have a low acid-producing capacity and therefore during long-term cultivation in milk the proportion of cells with high specific antimicrobial activity in a population could decrease.

The most of specific inhibitors produced by lactic acid bacteria are active at acid conditions (6,11,18,30, 41,58). Therefore the lack of inhibition in a medium buffered to the neutral pH does not mean that investigated cultures do not produce specific antimicrobial substances.

Some lactic acid bacteria also inhibit the growth of clostridia, staphilococci, psychrotrophs with the production of hydrogen peroxide and other O_2 derivatives. The maximal amounts of H_2O_2 accumulated in milk by lactic streptococci, L. plantarum, L. acidophilus, L. lactic and L. bulgaricus in depending on strain were resp. 0.03 to 0.19 mM (51), 0.06 to 2.06 mM (22,51), 0.4 to 1.6 mM (7,9), 0.44 mM (64) and 0.32 mM (64). S. lactis C 10 accumilated 0.15 mM of H_2O_2 in medium with glucose and 0.4 mM of H_2O_2 in the same medium after glucose has been substituted with galactose (21).

Bacteriostatic concentrations of H_2O_2 for staphilocicci and Pseudomonas were resp. 6 μg/ml (~0.17 mM) (9) and 22 to 35 μg/ml (0.65 to 1.0 mM) (55). Growth of butyric acid bacteria in depending on strain was reduced by 50 % in the presence of 0.08 to 0.26 mM of H_2O_2, the inhibitory concentrations of H_2O_2 for starter organisms were 2 to 10 fold greater (22). Consequently there are opportunities to select lactic acid bacteria strains that would inhibit some pathogenic and spoilage organisms with H_2O_2 production without self-inhibition.

Lactobacilli produce more H_2O_2 than lactic streptococci. The H_2O_2 content in milk cultures of lactic streptococci usually reached the maximal level in the late lag phase or the early exponential phase followed by a drastic decrease (14,50,51). Lactobacilli in depending on strain and cultural conditions accumulate H_2O_2 in medium during 3 to 15 days (7,9,50,51,55). After the H_2O_2 concentration has reached a maximal level it dropped sharply or slowly (42,51). H_2O_2 is formed by growing and resting cells of lactobacilli at lower temperature-limit for growth, at pH 4.0 to 8.0 and under a low oxygen tension (7,9,16,19,88,42). The resting L. plantarum cells can form H_2O_2 in the medium without carbohydrates (19).

These studies show that conditions in ripening cheeses could not prevent the H_2O_2 accumulation by lactobacilli. Indeed in extracts of 3-months old Manchego cheese Staph. aureus growth was stimulated by catalase.(43).

The antagonistic action of lactic acid bacteria toward the pathogenic and spoilage organisms in associative cultures results from specific and nonspecific factors which affect sinergistically (27,29).

Perfil'ev et al. (53) have devided the lactic streptococci activity against E. coli in associative milk culture into two parts. The first part of this activity was due to nonspecific inhibitory factors; the second one correlated with the sizees of inhibitory zones formed on buffered medium plates ($r = +0.91$). The strains that did not form inhibitory zones supressed E. coli growth in associative milk culture when the initial number of E. coli was not more than 10 c.f.u. per ml of milk (milk was inoculated by 1 % of streptococci cultures). The strains forming the largest inhibitory zones supressed the E. coli growth when 10^5 or more c.f.u. of E. coli were added to ml of milk. Only lactic streptococci strains prodicing high levels of H_2O_2 caused significant inhibition Staph. aureus in sterile milk supplemented with catalase (27). Apparently these strains produce toxic radicals of O_2 in addition to H_2O_2.

One and the same strain of lactic acid bacteria can produce several types of inhibitors. The inhibitory effect of L. plantarum strains isolated from high quality Kostroma cheese toward C. tyrobutyricum was fairly well correlated with H_2O_2 production but the effect against E. coli was not associated with H_2O_2 formation (49). There are differences in mode of lactic streptococci and lactobacilli action on E. coli (54,71). Streptococci decreased the rate of E. coli growth while lactobacilli increase the rate of E. coli dying off in milk and cheese. Combined lactic streptococci and lactobacilli starters were more effective in inhibiting the growth of E. coli and Staph. aureus than streptococcal starter (5,51).

In Kostroma cheese made with starter composed of lactic acid bacteria strains with specific activity against enterobacteria the maximum count of E. coli was 10 to 100 fold less than in cheeses made with starter composed of lactic streptococci without specific antimicrobial activiry (71).

There is evidence of strong inhibition of butyric acid bacteria in cheeses by lactobacilli possessing antimicrobial activities against these organisms in vitro (20,22, 35,36,52). Kostroma cheeses were made from milk inoculated by 10 to 100 spores of C. tyrobutyricum per ml with usual streptococcal starters (control) and with starter containing lactic acid bacteria possessing the antimicrobial activity against clostridia. Experimental cheeses were assessed as of high quality while the control cheeses had the typical for butyric acid fermentation defects and mostly were substandard. Many strains of lactobacilli, although inhibiting butyric acid fermentation, produced off-flavours in cheese.

The starters with antagonistic effect on enterobacteria and butyric acid bacteria are successfully used in the Soviet Union. With these starters good quality cheese can be made from milk contaminated by up to 100 c.f.u./ml of enterobacteria or lactate-fermenting clostridia spores. However, the efficiency of inhibition of contaminating organisms in cheese by starter organisms depends on the pH, NaCl and moisture content, temperature of ripening of cheese (27,73).

REFERENCES

1. Amster,H.; Jost,R. 1980. Experimentia, 36(6),764.
2. Babel,F.J. 1977. J.Dairy Sci., 60(5),815.
3. Belova,G.A.; Shergin,N.A.; Trofimova,T.I.; Bausheva,A.L. 1982. Molochnaia Promyshlennost'. 48(5),19.
4. Bergere,J.L.; Sus,T.; Vassal,L. 1978. Lait, 58(575-576), 215.
5. Bielecka,M.; Melan,K.; Ruszkiewicz,A. 1982. In XXI Intern. Dairy Congr., vol.1,Book 2,Moscow,USSR,281.
6. Branen,A.L.; Go,H.S; Genske,P.R. 1975. J.Food Res. 40(3),416.
7. Collins,E.B.; Koichiro,A. 1980. J.Dairy Sci.,63(3)353.
8. Choi,C.S.; Chung,J.B.; Chung,S.I.; Yanh,Y.T. 1984. Korean J. Veterinary Public Health, 8(1),49.
9. Dahiya,R.S.; Speck,M.L. 1968. J.Dairy Sci.,51(10),1568.
10. Daly,C.; Sandine,W.E.; Eliker,P.R. 1971. J.Dairy Sci., 54(5)755.
11. Gandhi,D.N.; Nambudripad,V.K.N. 1981. Indian J.Dairy Sci.; 34(1),98.
12. Geis,A.; Singh,J.; Teuber,M. 1982. In XXI Intern.Dairy Congr., v.1,Book 2,Moscow,USSR,307-308.
13. Geis,A.; Singh,J.; Teuber,M. 1983. Appl. and Environmental Microbiol. 45(1),205.
14. Gilliland,S.E.; Speck,M.L. 1969. Appl.Microbiol., 17(6)797.
15. Gilliland,S.E.; Speck,M.L. 1974. Appl.Microbiol., 28(6),1090.
16. Glaeser,H.; Sigmaringen,F. 1981. Ernährungs-Umschau, 28,Heft 6, 199.
17. Gonzales,C.F.; Kunka,B.S. 1985. Appl. and Environmental Microbiol.,49(3)627.
18. Goseco,M.A. 1980. Diss.Abstacts Intern.,B,40(10),4724.
19. Götz,F.; Sedewitz,B.; Elstner,E.P. 1980. Arch.Microbiol.,125(3),209.
20. Goudkov (Godkov),A.V.; Sharpe,M.E. 1965. J.Appl.Bacte-

riol.,28(1),63.
21. Grufferty,R.C.; Condon,S. 1983. J.Dairy Res., **50**(4) 481.
22. Gudkov,A.V.; Perfil'ev,G.D. 1979. Trudy VNIIMSP,Uglich,USSR,**30,75.**
23. Gudkov,A.V.; Perfil'ev,G.D. 1980. Molochnaya Promyshlennost',N.12,18.
24. Gudkov,A.V.; Kandrina,S.I., Slipchenko,S.N.; 1981. Molochnaya Promyshlennost', N.1,39.
25. Hosono,A.; Yastuki,K.; Tokita,F. 1977. Milchwissenschft, **32**(12),727.
26. Ibrahim,G.F.; Radford,D.R.; Baldock,A.K.; Ireland,L.B. 1981, J.Food Protection,**44**(3),189.
27. Ibrahim,G.F. 1978. Austr.J.Dairy Technol.,**33**(3)102.
28. Jakuczyk,E. 1978. In **XX** Intern.Dairy Congr., v.E,535.
29. Khalid,A.S.; Harrogan,W.F. 1984. Lebensmittel-Wissenschft und-Technologie, **17**(3),137.
30. Kim,D.S. 1984. Korean J. Veterinary Research,24(2),149.
31. Klaenhammer,T.R.; Sanozky,R.B. 1985. J.General Microbiol.,**131**(6),1531.
32. Kleter,G.; Lammers,W.L.; Vas,E.A. 1984. Neth.Milk Dairy J.,38(1)34.
33. Korhonen,H.; Ali-Yrukö,S.; Ahola-Lutila,H.; Antila,M. 1978. In XX Intern.Dairy Congr., v.E,536.
34. Kopylova,N.V. 1979. In Biologia microorganismov i ich ispol'sovanie v narodnom khozaistve. Irkutsk,USSR, Gosuniversitet,70.
35. Kundrat,W. 1971. Alimenta,**10**(4)133-141;(5),167-180.
36. Laudoniu,A; Dănilă,V.; Stoica,A. 1982. In Intern.Dairy Congr.,v.**1**,Book 2,Mockow,USSR,332.
37. Lee,S.H.; Yoon,G.H;Lee,D.S.; Kim,H.U. 1981. Korean J. Dairy Sci.,3(1)25.
38. Lipinska,E. 1973. Ann.bull.IDF,N.73,37.
39. Makarova,A.P.; Grinevich,A.G. 1975. In Biologia microorganismov i ich ispol'sovanie v narodnom khozyastve. Irkutsk,USSR;Gosuniversitet,60.
40. Makarova,A.P.; Kozlova,M.G. 1981. In Biologia microorganismov i ich ispol'sovanie v narodnom khozyastve. Irkutsk,USSR;Gosunoversitet,21.
41. Marth,E.H., Hussong,R.V. 1963. J.Dairy Sci., **46**(10), 1033.
42. Martin,D.K.; Gilliland,S.E. 1979. J.Food Protection, **43**(9),675.
43. Martinez Moreno,J.L. 1977. Revista Espanola delecheria. N.**103**,3.
44. Matilla,T. 1985. J.Dairy Res.,**52**(1)149.
45. Mc Kay,L.L.; Baldwin,K.A. 1984. Appl. and Environmental Microbiol.,**47**(1)68.
46. Mehta,A.M.; Patel,K.A; Dave.P.J. 1983. Microbios,**38** (152-,73.

47. Mitchell,De.G.; Kenworthy,R. 1976. J.Appl.Bacteriol., **41**(1)163.
48. Mitič,S.; Otenhajmer,I.; Milosavljevic,V. 1978. In XX Intern. Dairy Congr., v.E,540.
49. Perfil'ev,G.D.; Cudkov,A.V.; Matevosyan,L.S.; Sorokina,N.P.; Khandak,R.N. 1979. Trudy VNIIMSP,Uglich, USSR;**30**,22.
50. Perfil'ev,G.D.; Gudkov,A.V. 1979. Trudy VNIIMSP,Uglich,USSR;**30**,35.
51. Perfil'ev,G.D.; Gudkov,A.V. 1982. In XXI Intern. Dairy Congr., v.I,Book 2,Moscow,USSR,353.
52. Perfil'ev,G.D.; Gudkov,A.V. 1982. In XXI Intern. Diry Congr., v.I,Book 2,Moskow,USSR;544.
53. Perfil'ev,G.D.; Sorokina,N.P.; Matevosyan,L.S. 1982. Trudy VNIIMSP, Uglich,USSR;**38**,63.
54. Perfil'ev,G.D.; Sorokina,N.P; Eliseeva,N.N. 1982. Trudy VNIIMSP,Uglich,USSR; **36**,65.
55. Price,R.I.; Lee,S.H. 1970. J.Milk and Fd Technolog., **33**(1)13.
56. Rao,D.R.; Reddy,B.M.; Sunki,G.R; Pulusani,S.R. 1981. J.Food Quality,**4**,247.
57. Rash,K.E.; Kosikowski,F.V. 1982. J.Dairy Sci., **65**(4)537.
58. Reddy,G.V.; Shahani,K.M. 1971. J.Dairy Sci.,**54**(5),748.
59. Reiter,B.; Gillian Fewins,B.; Frayer,T,F.; Sharpe,M.E. 1964. J.Dairy Res.,**31**(3),261.
60. Richardson,G.H.; Divitia,M.A. 1973. J.Dairy Sci., **56**(6),706.
61. Ross,G.D. 1981. Austral.J.Dairy Technol.,**36**(4),147.
62. Rubin,H.E.; Herad,T.; Vaughan,F. 1982. J.Dairy Sci., **65**(2)197.
63. van Schouwenburg-van,A.W.I.; Stadhouders,J.; Witsenburg,W.W. 1979. Neth.Milk Dairy J.,**33**(1)49.
64. Schunz,M.; Coulibaly,O.; Nogai,K.; Wiesner,H.U. 1982. Milchwissenschaft,**37**(4),202.
65. Senesi,E.; Emaldi,G.C.; Caserio,G. 1975. Latte,**3**,84.
66. Shahani,K.M.; Vakil,J.R.; Kilara,A. 1976. Cultured Dairy Products J., **11**(4),14.
67. Shelaih,M.A.; Winkel,S.A.; Oxigbo,O.N.; Richardson, G.H. 1985. J. Dairy Sci.,**68**(3),609.
68. Simonetti,P.; Cantoni,C. 1982. Industrie Alimentary, **21**,784.
69. Simonetti,P.; Milas,N.; Cantoni,C. 1982. Industrie Alimentary,**21**(7),536.
70. Singh,J. 1983. Egypt.J.Dairy Sci.,**11**(2),321.
71. Sorokina,N.P.; Perfil'ev,G.D.; Gudkov,A.V.; Klimova, E.A. 1985. Trudy VNIIMSP,Uglich,USSR;**42**,36.
72. Stadhouders,J.; Cordes,M.M.; van Schouwenburg-van, A.W.I. 1978. Neth.Milk Dairy J.,**32**,193.
73. Stadhouders,J.; Kleter,G.; Lammers,W.L.; Tuintes,J.H.

1983. *Voedingsmiddelentechnologie*,**16**(26)20.
74. Steenson,L.R.; Klaenhammer,T.R. 1985. *Appl. and Environmental Microbiol.*,**50**(4),851.
75. Sukhotskene,I.I.; Stascuitite,I.I.; Gudkov,A.V. 1984. *Molochnaya Promyshlennost'*,N.12,8.
76. Tatini,S.R.; Jezeski,J.J.; Morris,H.A.; Olson,J.C.Ir.; Gasman,E.P. 1971. *J.Dairy Sci.*, **54**(6)815.
77. Zaborkich,E.I. 1975. *In Biologia microorganismov i ich ispol'zovanie v narodnom khozyaistve*, Irkutsk,USSR, Gosunivercitet,50.

STARTERS: APPLICATION IN THE DAIRY

Charles Daly,
Dairy and Food Microbiology Department,
University College,
Cork,
Ireland.

ABSTRACT: The importance of predictable consistent starter activity to the production of high quality cheese is stressed. The selection of stable phage-insensitive strains either as mixtures e.g. Gouda cheese production in the Netherlands or as defined strains e.g. Cheddar cheese production in Ireland is described. The strategy to overcome the phage problem is outlined and involves the use of carefully selected strains, the use of concentrates to avoid variant accumulation in cultures, the production of bulk culture free of disturbing phage and prevention of phage build-up in the factory environment. Recent useful innovations in the preparation of active bulk cultures e.g. the use of external or internal pH control of the medium and the impact of starter culture systems on important aspects such as cheese flavour and yield are mentioned. The conclusion is that stable predictable starters are available to help maximize process control and ensure optimum product quality in dairy fermentations.

INTRODUCTION

The purpose of this contribution is to examine developments in the use of starter cultures in the dairy industry. In attempting to do this one has to be conscious of the great diversity of the dairy fermentation sector in terms of the size range of processing units and the variety of flavourful nutritious end-products. In addition, the starter culture practices of those involved - scientists, culture producers and users - are often deep-rooted and the introduction of changes may require an input from the science of psychology as well as microbiology and technology. Despite this background, however, it is this author's belief that the selection and use of starter cultures today can be much more of a science than an art and thus help to optimize process control of the production of high quality fermented foods. This theme will be developed here. Due to space limitations, only developments in the use of mesophilic cultures containing Streptococcus cremoris, S. lactis, S. lactis subsp. diacetylactis and Leuconostoc species are examined in detail. However, many of the findings also apply to thermophilic cultures, although a more concentrated research effort is needed to maximize their controlled use.

FUNCTIONS OF THE STARTER

The starter culture used has a key role in the production of high quality cheese i.e. with the desired composition and organoleptic properties. The basic reaction is acid production at the correct rate and amount specific to each cheese type - this property is usually used as the measure of activity of the starter and provides the correct environment for cheese ripening in terms of pH, Eh, mineral content, retention of rennet and plasmin, moisture control and curd syneresis. Proper starter activity also ensures maximum inhibition of spoilage microorganisms and pathogens. Citrate utilisation by the starter bacteria is important for eye formation and flavour in some products while the correct proteolysis pattern (including the contribution of starter enzymes) is essential for the desired flavour and texture attributes of ripened cheeses.

FACTORS THAT RETARD STARTER PERFORMANCE

The variety of factors that can retard the performance of starter cultures in dairy fermentations have been well documented and include:
- intrinsic characteristics of the starter bacteria; genetic properties, physiological status, accumulation of variants.
- contamination with bacteriophages from sources such as milk, starters, equipment, air, whey.
- inhibitory substances; antibiotics, sanitizers, natural milk inhibitors, free fatty acids, bacteriocins.
- process conditions; level of starter inoculum, temperatures used, salt content.

The adoption of good manufacturing practices by the dairy industry has helped overcome the problems listed with the possible exception of bacteriophage attack which remains the most serious inhibitor in commercial practice, especially in large production units processing up to 200,000 gallons of milk per day using up to 6 multi-fills of the cheese vats (16). The widespread use of complex rotations of phage-sensitive mixed-strain starters has hindered proper process control and caused variations in manufacturing schedules and product quality. In contrast, stable phage-insensitive starter cultures have been selected and used to overcome these problems and this aspect will be examined in some detail.

STRATEGY USED TO COUNTERACT PHAGE

The strategy that has been developed and used to essentially eliminate the phage problem in large-scale dairy fermentations is based on:
(i)   the selection of phage-insensitive strains - these may be mixed

strain starter cultures (whose precise strain composition is not known) as used in the Netherlands (34, 36) or defined strain blends of known composition, traditionally used in New Zealand (18, 20) and Australia (13, 14) but recently finding increasing use in the USA (29, 30, 39) and Ireland (8, 9);
(ii) supply of cultures to the dairy as frozen or freeze-dried concentrates to eliminate the accumulation of variants in the strains used;
(iii) propagation of bulk-starter free of 'disturbing' phage i.e. phage from outside the culture that have an adverse effect on performance during cheesemaking. It should be noted that the direct addition of super concentrated cultures to the cheese milk is finding increasing use, thus avoiding the need to prepare bulk-starter;
(iv) diminished phage contamination levels in the factory.

Use of phage-insensitive mixed strain cultures

The studies of Stadhouders and co-workers in the Netherlands have clearly demonstrated that the phage sensitivity of mixed strain starter cultures is highly dependent on the conditions of propagation and, as a consequence, the strain balance present (34, 36). This is obvious from the comparative performance of so-called 'P' (Practice) and 'L' (Laboratory) mixed strain cultures used in commercial cheesemaking. The 'P' starters were traditionally propagated in the factory environment i.e. exposed to phage. Thus they have evolved with the optimum balance of phage-insensitive strains and a rapid ability to 'repair' activity following phage attack. They may show fluctuation in performance when used without protection of the bulk culture against phage but they do not fail, in contrast to 'L' cultures - propagated in the laboratory and, hence, predominantly phage-sensitive. The best 'P' cultures, in terms of phage-insensitivity, rate of acid production and ability to produce top quality cheese with the desired composition and flavour have been taken by the NIZO and stored as frozen stocks at neutral pH in the presence of cryoprotective agents at $-196°C$. When supplied as frozen concentrates, to minimize propagation, they can be used continuously, without rotation, for Gouda cheese production in large modern multi-fill factories. Consistent performance is ensured by protection of the bulk culture against phage and by hygienic practices to prevent phage build-up in the factory. The inherent mechanisms of phage insensitivity of these mixtures are quite complex but it is significant that some component strains are not attacked by any phage and the starter must be contaminated with a variety of different phage before all the sensitive strains, with their different phage-host interaction patterns, are attacked (34, 36).

Use of phage-insensitive defined strain cultures

While defined strains have been used in New Zealand since the 1930's new approaches to the selection of phage-insensitive strains have recently been taken in response to the pressures of large-scale fermentations (18, 19, 20). This has led to the introduction of the 'single-pair' system i.e. the use of 2 strains continuously. Defined strain systems for Cheddar cheese manufacture have also been developed in Australia (13, 14), the USA (29, 30, 39) and Ireland (8, 9). Only the key aspects will be highlighted here with emphasis on the experiences gained in Ireland in a collaborative project between the author and Dr. T.M. Cogan of the Agricultural Institute, Moorepark. A key task was the selection of phage-insensitive strains using the Heap and Lawrence test (11) in which each strain (isolated mainly from mixed strain starter cultures) was repeatedly challenged against all available phage (purified isolates and cheese factory wheys) under simulated cheesemaking conditions. Activity in milk at cheesemaking temperatures, gas production. flavour, bacteriocin production, lysogeny, salt tolerance and plasmid DNA profiles (to facilitate strain differentiation) were other attributes examined.

Out of ~300 strains screened, about 30 (mainly S. cremoris but including some S. lactis) proved satisfactory for industrial use. Multiples of 6 strains were used initially but now a blend of 3 strains is favoured. The system is simple to operate; the strains are supplied individually in frozen pellet form and grown together only at the bulk culture stage. The same blend is used in each vat day after day i.e. no rotations are used. This contrasts with the previous use of long awkward rotations by the factories e.g. one large factory propagated 70 and rotated 6 cultures (one for each fill) each day while another factory rotated 12 mixed strain cultures. The new system which incorporates bulk-culture protection and good factory hygiene was quickly accepted by the industry following initial factory trials in 1981 and was used for the production of >46,000 tonnes of Cheddar cheese (~80% of total production) in 1985. The key attribute of the defined strain system is the consistent predictable acid production leading to uniform high quality product and the elimination of starter failures. The economic benefits arising from the introduction of a defined strain starter system in the USA have been documented (38).

MONITORING STARTER PERFORMANCE

In modern fermentations the objective is to achieve consistent starter performance. This is measured as the pH of young cheese and should not fluctuate between fills or from day to day. A key aspect of quality assurance is to monitor the activity and phage status of the bulk starter. This is done using simple milk inhibition tests. In the case of defined strains plaque assays may be used to provide specific phage titres if desired. Similar tests are also used to monitor phage levels in the cheesemaking process. It should be noted that some troublesome phage show temperature specificities and may grow better in pasteurised milk than in sterile milk (14, 15) - these factors must be taken into account in the design of phage tests.

The continued use of a small number of defined strains (e.g. 2 or 3) leads to a major reduction in the diversity and levels of phage present in cheese factories. When factory hygiene at all levels is good, phage are often absent but it is not essential to have a phage-free environment because some phage have a low multiplication factor under cheesemaking conditions. If high levels of phage interfere with the activity of an individual strain, it may be replaced by another well characterized strain or by a so-called Bacteriophage-Insensitive-Mutant (BIM). BIMs can usually be isolated by continued propagation of the strain in the presence of whey (containing the disturbing phage) but must be carefully characterized before use in cheesemaking. BIM isolation is done in the cheese factories in Australia (13, 14) but since some isolates are not satisfactory this task may best be provided by a central laboratory. The concept of BIM isolation is worthy of further research as a possible approach to extending the range of stable phage-insensitive strains available.

SUPPLY OF CULTURES

Major technological developments have taken place in the growth, harvesting, concentration and storage of cultures (2, 27). As a result cultures are now supplied in a variety of forms including frozen or freeze-dried concentrates for bulk culture inoculation and super-concentrates for the direct addition to the cheesemilk. The use of concentrates helps to avoid variability which can occur even in defined strains subjected to excessive propagation.

The convenience and reliability of modern direct-to-vat cultures have led to their increased use in several countries and this trend may be facilitated further by continued research into the optimum conditions for the concentration, preservation and resuscitation of phage-insensitive S. cremoris and thermophilic Lactobacillus strains.

PREPARATION OF BULK CULTURE AND FACTORY HYGIENE

As mentioned above, the preparation of phage-free bulk culture is an important aspect of the strategy to eliminate the phage problem, especially since even a low phage infection at this stage can have serious consequences. The principles involved in physical protection of the bulk culture are well established (41) but often ignored and their effectiveness questioned. However, the effort involved is well worthwhile when used in conjunction with carefully selected starter cultures. Key aspects include heating the culture medium to 95°C to inactivate phage, steaming of the head-space of the tank, use of positive pressure and the introduction of sterile air immediately before the medium is cooled. The air may be sterilized by the use of High Efficiency Particulate Air (HEPA) filters for which an efficiency of phage removal to 1 in 10^8 is claimed (7, 21). The importance of an aseptic culture inoculation of the bulk tank has long been stressed (22, 31) but tanks may have to be modified to facilitate the use of concentrates. A well designed bulk tank system, featuring the attributes mentioned here, has recently been described (34).

Another aspect of bulk culture preparation which is often ignored is the need to control the times and temperatures of incubation and cooling and thus avoid irregular acid production due to the use of over-ripe starters. An important development in this context has been the introduction of pH-control of the medium to provide an increased number of more active cells with enhanced storage characteristics. This may involve external pH-control by the addition of ammonia to a whey-based medium as developed by Richardson and co-workers at Utah State University. The advantages, including cost benefits, of this system have led to its widespread use in the USA (4, 29, 37) and it is being examined with interest in other countries. The use of Internal-pH-Control Media containing special buffering materials to keep the bulk starter pH above 5.2 was pioneered by Sandine and Ayres at Oregon State University (24, 32, 42). Another benefit of pH-controlled media is that they allow the use of relatively slow strains (which often possess marked phage insensitivity) for cheesemaking. The benefit of pH adjustment to enhance culture activity is also recognised in New Zealand where a one-shot neutralisation step is used. This involves neutralisation of the coagulated bulk starter with sodium hydroxide followed by a 2h growth period to allow an approximate doubling of the viable cell count (23).

The reduction of phage levels in the cheese factory is another essential aspect of consistent culture performance - measured by uniform pH reached in all vats of cheese from first to last fill. There is a need for a definitive analysis of the contribution of various factors to this aspect of the phage problem. However, strict attention to quality and hygiene at all levels is required e.g. use of best quality milk to minimize the presence of raw milk streptococci and their phages, sanitation of milk and starter pipelines and, in particular, careful handling and disposal of whey. Cheese vats should be thoroughly cleaned and, preferably, sanitized with chlorine between fills. Whey should be carefully removed, separated with hermetically sealed units and not returned to dairy farmers as this risks recycling phage. The use of whey-cream for standardization purposes represents another source of phage and the pasteurization treatment used should reflect this (34). Controlled air changes and chlorine sprays in the factory may also contribute to reduced phage levels.

IMPACT OF CULTURE SYSTEMS

The developments in the selection of strains and bulk culture preparation (or direct vat addition) of starters described has led to improved predictable culture performance in the dairy industry. This in turn allows an assessment of the impact of culture systems on aspects of major economic importance such as the development of cheese flavour and cheese yield. The importance of manufacturing conditions, cheese composition and ripening parameters to protein degredation and, therefore, flavour development in cheese has been documented (1, 17, 19). While the increasing tonnage of several cheeses being manufactured worldwide with defined strains is testimony to their ability to produce top quality products, at least two aspects of flavour development have

been raised. A concern that a bitter flavour defect may be encountered
with defined strains has often been expressed. Work in New Zealand has
shown that the propensity to cause bitterness in cheese is at least
partly related to the cell population reached in the cheese and may be
controlled by proper strain ratios in the bulk culture and by taking
into account temperature and salt tolerances (18). The level of
starter proteinase has a role in bitterness development (25) while
some fast strains are prone to causing bitterness in Gouda cheese (35).
The bitterness defect has not been encountered with the use of defined
strains in Ireland as described above. Another aspect that is
receiving some attention is the rate of flavour development during
cheese ripening. In this regard the tendencies towards very rapid make
times and marketing of young cheese should be noted. A recent study
(3) concluded that the flavour of Cheddar cheese made with S. cremoris
strains grown under external pH-control in a whey-based medium lagged
slightly behind that of cheese made with conventional bulk starter until
about 9 months, after which flavour differences were minimal. Further
studies are needed with particular attention being given to inoculum
levels, over-selection of Prt$^-$ cells under some growth conditions and
ensuring that proper manufacturing schedules are achieved.

There have been several recent studies on the impact of culture
systems on cheese yield - this aspect may be more important than
differences in direct costs of the cultures and media themselves,
although these may also be significant (12, 37). A major area of
controversy concerns the contribution of milk solids in the culture
medium to cheese yield. Banks et al. (6) claimed that, on average,
40% of starter solids was retained in cheese and that yields were
significantly greater with a traditional skim-milk bulk starter culture
than with a direct-to-vat culture. This conclusion was supported by
studies in a model cheesemaking system (5). In contrast, Hicks and
co-workers claim increased yields from direct-to-vat cultures and from
cultures grown in whey-based and internally pH-controlled media (12, 40).
Increased yields have also been claimed for Prt$^-$ cultures (28) and
cultures propagated in UF retentate (26). When extrapolated to large
scale cheese production, say 10,000 tonnes per annum the cost benefits
are very significant e.g. £500,000 p.a. These aspects need to be
clarified and an agreed approach to yield determinations is needed to
provide accurate reliable information and allow the maximum benefits
of alternative starter culture systems for cheesemaking.

CONCLUSION

In conclusion it is worth highlighting again the key factors in the
success of starter systems in overcoming the phage problem and providing
improved process control of cheesemaking - careful strain selection
based on phage insensitivity; elimination of genetic variability by
the use of frozen storage and concentrates; use of the minimum number
of cultures without rotation; careful bulk starter preparation and
cheese factory hygiene. The philosophies behind the use of mixtures
(34) and defined strains (18) have recently been reviewed and these

reports highlight the principles fundamental to both systems. With either, properly managed, the phage problem is now under control. The demonstrated phage insensitivity of carefully selected defined strains has removed one of the advantages long claimed for the use of mixtures. Perhaps future research will confirm the ability of defined strains to produce the many fine flavoured cheeses desired throughout the world - it is noteworthy that they are presently used in New Zealand for the production of cheeses such as Gouda, Camembert and Blue (18). The more widespread use of defined strains will facilitate the industrial exploitation of advances being made in the genetic manipulation of starter bacteria. Progress in the elucidation of phage-insensitivity mechanisms associated with some defined strains of lactic streptococci has been very rapid (10, 16, 33). Thus it is likely that genetically manipulated strains with superior phage resistance will soon be available for commercial cheesemaking and thus extend the range of suitable strains.

REFERENCES

1. Adda, J., J.G. Grippon and L. Vassal. 1982. Food Chem. 9, 115.
2. Anonymous. 1983. Dairy Ind. Int. 48, 19.
3. Amantea, G.F., B.J. Skura and S. Nakai. 1986. J. Food Sci. 51, 912.
4. Ausavanodum, N., R.S. White, G. Young and G.H. Richardson. 1979. J. Dairy Sci. 60, 1245.
5. Banks, J.M. and D.D. Muir. 1985. Milchwissenschaft 40, 209.
6. Banks, J.M., A.Y. Tamine and D.D. Muir. Dairy Ind. Int. 50, 11.
7. Boelle, A.C., G.J.M. Leenders and J. Stadhouders. 1985. Rapport R121 NIZO.
8. Daly, C. 1983. Irish J. Food Sci. Technol. 7, 39.
9. Daly, C. 1983. Antoine van Leeuwenhoek 49, 297.
10. Daly, C. and G. Fitzgerald. 1986. Proceeding 2nd ASM Conference on Streptococcal Genetics, A.S.M. Washington D.C. USA.
11. Heap, H.A. and R.C. Lawrence. 1976. N.Z.J. Dairy Sci. Technol. 11, 16.
12. Hicks, C.L., F. Marks, J. O'Leary and B.E. Langlois. 1985. Cultured Products J. 20, 9.
13. Hull, R.R. 1983. Aust. J. Dairy Technol. 39, 149.
14. Hull, R.R. 1985. Aust. Soc. Dairy Technol. Technical Publication No. 27, 20.
15. Hull, R.R. and A.R. Brooke. 1982. Aust. J. Dairy Technol. 37, 143.
16. Klaenhammer, T.R. 1984. Advs. in Appl. Micro. 30, 1.
17. Lawrence, R.C., J. Gilles and L.K. Creamer. 1983. N.Z.J. Dairy Sci. Technol. 18, 175.
18. Lawrence, R.C. and H.A. Heap. 1986. I.D.F. Bulletin No. 199, 4.
19. Lawrence, R.C., H.A. Heap and J. Gilles. 1984. J. Dairy Sci. 67, 1632.
20. Lawrence, R.C., H.A. Heap, G.K.Y. Limsowtin and A.W. Jarvis. 1978. J. Dairy Sci. 61, 1181.
21. Leenders, G.J.M., A.C. Boelle and J. Stadhouders. 1983. Rapport R119, NIZO.

22. Lewis, J.E. 1956. J. Soc. Dairy Technol. 9, 123.
23. Limsowtin, G.K.Y., H.A. Heap and R.C. Lawrence. 1980. N.Z.J. Dairy Sci. Technol. 15, 219.
24. Mermelstein, N.H. 1982. Food Tech. 36, 69.
25. Mills, O.E. and T.D. Thomas. 1980. N.Z.J. Dairy Sci. Technol. 15, 131.
26. Mistry, V.V. and F.V. Kosikowski. 1986. J. Dairy Sci. 69, 1484.
27. Porubcan, R.S. and R.L. Sellars. 1979. In Microbial Technology, H.J. Peppler and D. Perlman Eds. 1, 59, Academic Press.
28. Richardson, G.H. 1985. Cultured Dairy Products J. 20, 20.
29. Richardson, G.H. and C.A. Ernstrom. 1985. Aust. Soc. Dairy Technol. Technical Publication No. 27, 12.
30. Richardson, G.H., G.L. Hong and C.A. Ernstrom. 1980. J. Dairy Sci. 63, 1981.
31. Robertson, P.S. 1966. 17th Int. Dairy Congress D2, 439.
32. Sandine, W.E. and J.W. Ayres. 1983. U.S. Patent No. 4,382,965.
33. Sing, W.D. and T.R. Klaenhammer. 1986. Appl. Environ. Microbiol. 51, 1264.
34. Stadhouders, J. 1986. Neth. Milk Dairy J. 40, 155.
35. Stadhouders, J., G. Hup, F.A. Exterkate and S. Visser (1983). Neth. Milk Dairy J. 37, 157.
36. Stadhouders, J. and G.J.M. Leenders. 1984. Neth. Milk Dairy J. 38, 157.
37. Thunell, R.K. 1986. Seventh Biennial Cheese Industry Conference, Utah State Univ. Logan, Utah, USA.
38. Thunell, R.K., F.W. Bodyfelt and W.E. Sandine. 1984. J. Dairy Sci. 67, 1061.
39. Thunell, R.K., W.E. Sandine and F.W. Bodyfelt. 1981. J. Dairy Sci. 64, 2270.
40. Ustinol, Z., C.L. Hicks and J. O'Leary. 1986. J. Dairy Sci. 69, 15.
41. Walker, A.L., W.M.A. Mullan and M.E. Muir. 1981. J. Soc. Dairy Technol. 34, 78.
42. Willrett, D.L., W.E. Sandine and J.W. Ayres. 1982. Cultured Dairy Products J. 17, 5.

DISCUSSION

B. Hahn-Hägerdal (Sweden) : By which methods were the kinetic constants for the various peptides obtained?
P.F. Fox (Eire) : By determining the formation of new amino groups. Most of the work was done by S. Visser of NIZO and has been published.

J. Visser (the Netherlands) : I have reasons to believe that your casein micelle model is too simple (see I. Heertje, J. Visser and P. Smits in Food Microstructure) in the sense that I do not believe in spherical submicelles. My question concerns the casein micelle during the first stage of cheesemaking. Does it retain its original structure and size despite the loss of colloidal calcium phosphate?
P.F. Fox (Eire) : The majority view on casein micelle structure favours the sub-micellar hypothesis. The micelles retain their structure in young pressed cheese.

A.A. Ayerbe (France) : Prof. Teuber has suggested that some plasmids of *Streptococcus* can be involved in the production of bacteriocins which might be active against *Listeria* .
M. Teuber (Fed.Rep. Germany) : Nisin production is coded on a plasmid. Nisin and most of the other bacteriocias have not been tested against *Listeria* . I urgently suggest to do that, because some of these compounds are active against Gram-positive bactaria.

P. Walstra (the Netherlands) : How easily do starter organisms lose plasmids, especially if the plasmid codes for a property that is not used by the bacteria under the conditions of propagation of the starter?
M. Teuber (Fed.Rep. Germany) : Plasmids are lost at high frequencies by 1. growth at high temperature; 2. growth in artificial media without lactose or casein; 3. freeze drying.
C. Daly (Eire) : The variability of natural strains can be prevented by using a minimum of transfers, i.e. by the use of deep-frozen stocks. Strains of *S. cremoris* being phage-resistant by uptake of a plasmid have proved stable under industrial conditions for one year (T.R. Klaenhammer, personal communication).

J. Hart (Switzerland) : 1. Does plasmid born phage resistance allow resistance to naked phage DNA introduced by transfection?
2. Does screening using phage lysates for phage resistance allow detection of lysogens as a form of phage resistance?
M. Teuber (Fed.Rep. Germany) : 1. Experiment not yet done.
2. There is no evidence yet, because no strong correlation between resistance and lysogeny could be found.

J. Stadhouders (the Netherlands) : Dr. Gudkov mentioned that the antagonistic effect of lactobacilli against butyric acid bacteria is due to hydrogen peroxide. I think that in cheese a low redox potential is reached soon. As a consequence no oxygen is present to be converted to hydrogen peroxide.

A.V. Gudkov (U.S.S.R.) : Lactic acid bacteria accumulate hydrogen peroxide at low oxygen tension. We have found that there is enough oxygen to accumulate hydrogen peroxide.

J. Stadhouders (the Netherlands) : The strains of the starter of Prof. Daly are vere rarely replaced. But if it does, what is the effect on the quality of the cheese?

C. Daly (Eire) : If a build-up of phages in the system occurs, it is usually due to poor hygienic practices. This can be corrected in response to the information provided by the phage monitoring system. When well characterized replacement strains are used, there is no effect on the quality of the cheese.

N. Tofte Jespersen (Denmark) : Is it correct that in the Irish system the strains are not propagated in the factory?

C. Daly (Eire) : The indivudual strains are not propagated in the factory. Each strain is supplied as a frozen concentrate to inoculate the bulk culture medium. This guarantees a consistent inoculum and limits the accumulation of variants in the culture.

P. Walstra (the Netherlands) : Is not it essential in the Irish and other methods of starter handling to rigorously prevent contamination of cheesemilk with whey or curd from previous batches?

C. Daly (Eire) : Yes, factory hygiene at all stages is very important and I have mentioned this in more detail in the text. I have stressed the importance of cleaning and chlorination of vats between fillings.

F.M. Driessen (the Netherlands) : For the manufacture of Gouda type of cheese it is considered important that a part of the starter bacteria is of the "fast type". Are such types in your set of strains used for Cheddar cheese?

C. Daly (Eire) : The majority of phage-insensitive strains have a medium rate of acid production compared to "fast types". However, blends of 2 or 3 strains give a consistent performance. There is an increasing evidence from various sources that a considerable proportion of so-called Prt$^-$ variants can be present in a culture without adverse effects on acidification or ripening.

R.J.M. Crawford (UK) : Do the strains in the starter of Prof. Daly belong to *Streptococcus lactis* or *Streptococcus cremoris* and do the starter combinations produce CO_2 and diacetyl?

C. Daly (Eire) : The strains used are mainly *Streptococcus cremoris* with a few *Streptococcus lactis*. *Streptococcus cremoris* strains appeared to be more insensitive to phage attack. Up till now the system has been used only for Cheddar cheese. There is no significant production of CO_2 and diacetyl.

J.L. Maubois (France) : What are your comments regarding the recent proposal of Prof. Kosikowski to use UF retentate as a propagating medium for starters, thus using its buffering capacity and the higher cheese yield of this liquid?

C. Daly (Eire) : I have no further information than the details in the publication (cited in the text). The effect on yield that is claimed would be very significant. However, there is a need for an agreed approach to measure the cheese yield so that conclusions can be assessed properly.

Seminar I: Progress in Cheese Technology

Session 2: From Curd to Cheese

Chairman: Prof. Dr. Ir. P. Walstra (The Netherlands)
Secretary: Dr. D. Muir (U.K.)

CHEESE COMPOSITION AND QUALITY

R C Lawrence & J Gilles
New Zealand Dairy Research Institute
Palmerston North
New Zealand

ABSTRACT. The manufacture of any cheese variety involves four main factors which influence quality during ripening: (a) the pH of the curd at whey separation, since this determines the mineral content and the proportions of residual calf rennet and plasmin in the cheese, (b) the ratio of moisture to casein (which is related to the moisture to non-fat substance, MNFS), (c) the final pH of the cheese after salting and (d) the salt to moisture ratio (S/M) which, together with the MNFS and ripening temperature, controls the rate of casein breakdown in the cheese. These four factors can be varied to give specific cheese types with particular chemical compositions and enzyme activities. Each cheese variety develops a characteristic texture and flavour during ripening, which is largely dependent upon the extent of casein breakdown and the change in pH. The calcium content of a cheese has less effect on quality than pH, S/M or MNFS but nevertheless serves as an index of acid production in the curd at whey separation. Cheese varieties can, therefore, be differentiated by their characteristic ranges of pH and of calcium to solids-not-fat ratios.

1. INTRODUCTION

Cheesemaking is a relatively simple matter, the removal of moisture from a rennet coagulum. The single most important variable involved is the rate and extent of acid production but the cooking temperature used, the size of the curd particles and the proportion of fat in the curd are significant (Figure 1) in particular cheese varieties. In dry-salted cheeses such as Cheddar and Cheshire, much of the acid production occurs before the curd is separated from the whey, whereas in brine-salted cheeses most of the required acid production takes place after whey separation. To compensate for the relatively low acid production at the vat stage in Gouda and Swiss manufacture, moisture removal is enhanced by cutting the rennet coagulum into smaller curd particles and pressing the curd in the vat.

In general the need for a high cooking temperature is inversely related to the extent of acid production in the cheese. Thus

Cheshire, Camembert and Feta curds are not normally cooked at all whereas in Swiss cheese manufacture a high cooking temperature is used. Lowering the fat content of the cheese milk is also important for low-moisture cheeses such as Parmesan and Romano. The less fat present in the milk used for cheesemaking, and therefore in the rennet coagulum, the easier it is to remove moisture for the same manufacturing conditions, since the presence of fat interferes mechanically with syneresis. As one would expect, dry-salting tends to result in greater moisture expulsion from the curd than brine-salting.

It should be noted that some of the above factors are inter-related. Thus moisture expulsion is mainly due to the acid-producing activity of the starter bacteria, but the cooking temperatures used may adversely affect this activity. The size of the pressed cheese also affects its rate of cooling and thus the activity of the starters. When thermophilic starters are used, for instance, care must be taken not to allow the temperature of the cheese to fall below about 35°C. Similarly, starter activity in dry-salted cheeses is markedly affected by salt to moisture percentages greater than about 4.5.

The successful development of continuous mechanized cheesemaking systems has been dependent upon the availability of reliable starter cultures since this has allowed the cheesemaker to control the rate of acidity increase, and thus the required expulsion of moisture, in a given time. In view of the importance of acid production in cheese manufacture it is not surprising that the pH of the curd at whey separation and the final pH of the cheese should both be important quality control parameters.

2. EFFECT OF PH ON CHEESE QUALITY

As the pH of curd decreases there is a concomitant loss of colloidal calcium phosphate from the casein sub-micelles and, below about pH 5.5, a progressive dissociation of the sub-micelles into smaller casein aggregates (1). At pH's below about 4.8, the aggregates appear to exist only as short strands. These are up to 15 nm in length and 3 to 4 nm in width, roughly the diameter of a spherical casein molecule (2). The texture of a cheese depends primarily upon its pH and can range from springy through plastic to non-cohesive. This is irrespective of whether the cheese has been brine-salted or dry-salted. Historically cheeses which possessed a certain texture were given specific names, such as Cheddar or Gouda, after the localities in which they were first made. Figure 2 has been constructed from an interpretation of electron micrographs of different cheese varieties. While this diagrammatic representation is undoubtedly simplistic it does help one to visualize what is occuring. Thus Gouda tends to become increasingly more Cheddar-like in its texture during ripening, presumably because the sub-micelles are degraded to give a range of casein aggregates similar to those normally associated with Cheddar.

Overlaps occur between texture types (Figure 2) since the effect of pH can be modified by other compositional factors, particularly the moisture, salt and calcium contents. Between pH 5.5 and 5.1 much of the colloidal phosphate and a considerable part of the casein is dissociated from the sub-micelles. These changes in the size and characteristics of the sub-micelles significantly increase their ability to absorb water (1, 3). It is not surprising that various types of texture can be obtained between pH 5.3 and 5.1 since a wide range of casein aggregates is present and differences in the sodium and calcium ion concentrations, as well as the proportion of water to casein, markedly affect the extent of swelling of the sub-micelles.

2.1. Factors Affecting the pH of Cheese

2.1.1. <u>Dry salted cheese</u>. The pH of dry-salted cheese is determined primarily by the curd acidity at salting, which in turn is to a large extent controlled by the acidity developed at whey separation (4). Any further decrease in pH after salting is dependent upon the residual lactose in the curd and the starter activity (Figure 3). In a cheese with a S/M of 6%, the activity of all <u>Streptococcus cremoris</u> strains is inhibited and a high proportion of residual lactose remains unmetabolized. In a cheese with a S/M of 4.5%, however, the starter is not inhibited and the lactose will be rapidly metabolized. This explains why the pH values of one-day old Cheddar cheese may range from 5.3 (which is about the pH of the curd at salting) down to pH 4.9. In general, the higher the pH the greater the proportion of lactose initially left unmetabolized. This residual lactose is not, however, detrimental to the quality of the cheese provided that the cheese is cooled immediately after pressing.

The pH is also dependent upon the buffering capacity of the curd, which is largely determined by the concentrations of protein and phosphate present, and to a much lesser extent by calcium. The proportion of phosphate and calcium retained in the cheese is mainly influenced by the extent of acidification prior to the separation of the whey from the curd. Mineral losses after the draining stage are small under normal circumstances, despite the further decrease in pH.

2.1.2. <u>Washed curd cheese</u>. The pH of cheeses such as Gouda can be readily controlled by adjusting the concentration of residual lactose in the curd. Since the lactose content of the cheesemilk varies during lactation, it is normal commercial practice to drain off the same proportion of whey throughout the season but to vary the proportion of water used to wash the curd. This ensures that the lactose content of the cheese curd before brining is no greater than that required to ensure that the pH after brining is between 5.15 and 5.25. It is possible therefore to manufacture Gouda-type cheeses to relatively precise pH specifications. If too much lactose is present after washing the curd, the pH will be lower than normal and the quality of the cheese will be markedly affected.

2.1.3. <u>Swiss-type cheeses</u>. In the manufacture of Swiss-type cheeses, the pH at draining is determined by the proportion of <u>Streptococcus</u>

thermophilus added. The final pH is controlled independently by the Lactobacillus strain in the starter culture. The Strep. thermophilus hydrolyses the lactose to glucose and galactose but further metabolizes only the glucose moiety to lactic acid. The use of Strep. thermophilus is therefore a somewhat elegant method of controlling the metabolism of lactose since the proportion of easily fermentable sugar available in the curd is effectively halved. The galactose moiety is subsequently metabolized relatively slowly by the Lactobacillus strain. It is the fermentation of residual sugar in the curd to lactic acid which largely determines its final pH. The pH of the cheese at the time of transfer to the hot room is critical for normal eye development since it affects both the texture of the cheese and the growth of propionibacteria.

3. MAIN FACTORS INFLUENCING CHEESE QUALITY

The manufacture of all cheese varieties involves four main factors which influence the quality of the cheese during ripening: (a) the pH of the curd when it is separated from the whey, since this determines its mineral content and the proportions of calf rennet and plasmin in the cheese, (b) the salt to moisture ratio which influences the activity of the residual rennet and plasmin in the cheese, (c) the final pH of the cheese after salting and (d) the ratio of moisture to casein (Figure 4). Each of these four factors can be varied so as to give specific cheese types of particular chemical composition and enzyme activity. Each cheese variety develops a characteristic texture and flavour during ripening, which is largely dependent upon the pH of the cheese and the extent of casein breakdown (5).

The importance of pH in texture development has been emphasized but a second factor, the ratio of moisture to casein, is also involved. Clearly the lower the ratio of moisture to casein, the firmer will be the casein matrix of the cheese. In addition, the ratio of residual rennet to casein will be lower and the rate of change in texture will therefore be less marked. Thus low-fat cheeses, which tend to contain a relatively low ratio of moisture to casein, are harder than those with a higher fat content, after the same period of ripening. It is difficult, however, to measure the casein content of cheese accurately and at commercial plants it is usual to analyse only for for fat and moisture. A practical compromise, therefore, is to calculate MNFS rather than measure the moisture to casein ratio. The non-fat substance (NFS) is not the same as the casein in the cheese but is equal to the moisture plus solids-not-fat. The major part (about 85%) of the solids-not-fat in cheese consists of casein, and a significant part of the minerals, which constitute about another 10%, is also associated with the casein. The relationship between the casein in cheese and the NFS value is not strong but changes in MNFS for any particular cheese variety correlate well with changes in the ratio of moisture to casein. The MNFS value gives a much better indication of potential cheese quality than the moisture content in the same way

that the S/M ratio is a more reliable guide to quality than is the
absolute salt content of the cheese.

Recent developments in marketing have resulted in a demand for cheese
of greater uniformity of composition than in the past, and the
relationship between cheese composition and quality is now well
established. Ranges in pH, MNFS and S/M have, for instance, been
suggested (4) for Cheddar cheese (Figure 5), assuming that the cheese
is to be stored for at least 6 months. The fat in the dry matter (FDM)
content must also be included for regulatory purposes but plays no
direct role in the quality of the cheese. Variations in FDM between
48 and 58% have no significant effect upon cheese quality. As long as
acid production during manufacture has been normal, Cheddar cheese
with a chemical composition that falls within the inner square is
almost certain to result in a premium cheese that is acceptable to the
vast majority of customers. Cheese with a composition falling in the
outer square is less certain to mature well, but will still be first
grade cheese. It is important to recognize that MNFS, S/M and pH are
inter-related and that these three parameters must be controlled as a
group to ensure premium grade cheese.

Ranges of composition are necessary in any quality control system
since it is not yet possible to produce a completely uniform line of
cheese within a day's manufacture, no matter what system of
cheesemaking is used. This does not necessarily mean, however, that
some of the cheese produced is of poorer quality than the rest. The
rate of ripening will differ but all of it is likely to be acceptable
as long as the composition of the cheese is within the required
compositional range. In mechanized cheese plants, a significant
relationship has been found to exist between the FDM and MNFS values
in cheese, probably as a result of the relative inflexibility of the
procedures available for the control of moisture. This is of
commercial interest since changing the FDM is an effective way of
adjusting the MNFS in the cheese as the composition of the milk
changes.

The actual MNFS percentage for which a cheesemaker should aim depends
upon when the cheese is required to reach optimum quality. Experience
has shown that if Cheddar cheese is to be stored at 10°C, and the
cheese is to be consumed after six to seven months, then the MNFS of
the cheese should be about 53%. The higher the MNFS percentage the
faster the rate of breakdown. Thus if one anticipates that the cheese
will be consumed after three to four months, the MNFS percentage can
be increased to about 56%. However, the higher the MNFS the more
rapidly Cheddar cheese will deteriorate in quality after reaching its
optimum. It might seem surprising that an increase of only 3% in MNFS
should affect the rate of breakdown so markedly but much of the
moisture in cheese is bound to the caseins and their degradation
products, and also to the calcium lactate and sodium chloride present.
A small increase in MNFS apparently results in a relatively large
increase in available moisture. Even small increases in water activity

have been shown to affect markedly the rate of proteolytic activity in cheese (6). It follows that the higher the MNFS, the more rapidly Cheddar cheese will deteriorate in quality after reaching its optimum.

The major difficulty in achieving cheese of uniform quality in mechanized Cheddar cheese plants usually results from the relatively wide variation in S/M levels that occurs in the cheese. Variations in the acidity of the curd before salting, in the quantities of salt delivered by salting equipment and in the dimensions and physical structure of the milled curd all contribute to differences in salt uptake. The main source of variation, however, is the moisture content of the curd that is presented to the salter. This is partly unavoidable where large vats are used, since up to 30 minutes may elapse while curd is discharged on to the draining conveyers. Consequently, S/M values within a single 20-kg block of Cheddar cheese may vary by more than 1%. As long as the titratable acidity at salting is normal, however, a salt to moisture ratio between 4.5 and 6.0% will tend to result in acceptable cheese.

4. FACTORS INFLUENCING THE PROPORTIONS OF RENNET AND PLASMIN IN CHEESE

It must be stressed that it is not possible to predict with complete accuracy the quality of a cheese during ripening from its chemical composition only, since a further factor, the extent of acid production in the curd prior to whey separation, is also important. When calf rennet is used as the coagulant, most is lost in the whey at draining but some remains in the curd, depending upon the pH and the proportion of whey retained in the curd. The distribution of calf rennet between curd and whey is pH dependent whereas that of milk clotting enzymes from the Mucor spp. is not (7). The pH at draining also determines the proportion of plasmin in cheese. Plasmin, a native milk proteinase, is associated with the casein micelle in fresh milk but dissociates as the pH decreases. The proportion of plasmin is therefore greatest in cheeses such as Emmenthaler, in which the whey is drained from the curd at a relatively high pH, and least in cheeses such as Cheshire. In Mozzarella and Swiss-type cheese, the high scalding temperatures used inactivate most, if not all, of the chymosin, the active proteolytic component of calf rennet. Plasmin, however, is relatively heat resistant and casein breakdown in these varieties appears to be largely due to plasmin.

Since much of the acid production occurs at the vat stage for dry-salted cheeses such as Cheshire and Cheddar, the activity of the starter up to draining is more important than for brine-salted cheeses. The pH at draining is relatively low and chymosin retention therefore is high. Chymosin activity is much more important than plasmin activity during the ripening of Cheddar. In the manufacture of brine-salted cheeses, most of the acid is developed after the whey has been drained from the curd and the pH at draining is relatively high. One would therefore expect slightly more plasmin and less

chymosin to be retained in Gouda than in Cheddar. Nevertheless, the
contribution of plasmin to casein degradation in Gouda is reported to
be small in relation to that of rennet and bacteria (8).

The pH of Cheshire and Cheddar cheese normally decreases slightly in
the first 14 days of ripening as the residual lactose is metabolized
and thereafter increases only slightly, usually less than 0.2 pH
units, after six months. The texture of these cheeses therefore
normally changes relatively slowly during ripening. In contrast, the
pH of Gouda and Swiss-type cheeses rises relatively rapidly during
ripening. Rindless Gouda cheese therefore becomes less firm in
texture during ripening as the pH rises. Rinded cheeses, however,
lose moisture continually through the rind during ripening and this
will tend to off-set the softening effect due to the rise in pH. It
may be coincidental that the increase in pH is greatest towards the
end of the lactational season when the plasmin content of the milk is
at a maximum (5).

5. CLASSIFICATION OF CHEESE VARIETIES

It should be noted that variations in calcium content within any
specific cheese variety have a much smaller effect on quality than pH,
S/M or MNFS. Nevertheless, the calcium level serves as an index of the
extent of acid production before the curd is separated from the whey and
can therefore be used to distinguish one cheese type from another.
Indeed cheese varieties are best classified by their characteristic
ranges of pH and ratio of calcium to solids-not-fat (4). It would be
preferable to know the proportions of calcium and phosphate directly
associated with the casein but these are difficult to measure. In
general there is a good correlation between the ratios of calcium and
phosphate retained in different cheese types. One therefore usually
needs to determine only the calcium level to obtain an assessment of the
overall mineral content of a cheese.

Swiss, Gouda and Cheshire traditionally have fairly narrow ranges of
calcium and pH but the range for Cheddar is relatively wide. There is,
however, considerable overlap between the various cheese types. Thus a
Cheddar cheese with a relatively high calcium content will tend to be
similar in texture to a Gouda cheese.

Most brine-salted cheeses such as the Gouda- and Swiss-types, possess
'eyes' but this is a result of the relatively high pH and moisture of
these cheeses and not of brine-salting per se. The curd of a
brine-salted cheese is pressed under the whey to remove pockets of air
before brining and its texture is therefore closer than that of
traditionally made dry-salted cheese. The prerequisite to eye formation
is a close texture, which can be achieved in dry-salted cheese by
removing the air from between the particles of salted curd by vacuum.
The development of vacuum pressing, therefore, now makes possible the
manufacture of dry-salted cheese with 'eyes', provided that the chemical

composition is similar to that of traditional brine-salted cheeses and
the starter culture contains gas-producing species. The activity of the
residual rennet and plasmin in the curd will be more rapidly affected by
dry-salting than by brining and the rate of ripening of dry-salted
cheese will be slightly slower than in a brine-salted cheese of the same
composition. There is no evidence to suggest, however, that the
mechanisms by which the casein is degraded are affected by the changes
in salt concentration as the salt diffuses into the curd.

6. THE MANUFACTURE OF CHEESE FROM ULTRA-FILTERED MILK

The pH of cheese made from ultrafiltered (UF) milk can be controlled
relatively easily by removing appropriate proportions of lactose from
the milk. It is still nevertheless difficult to duplicate exactly the
texture and flavour of traditional hard and semi-hard cheese varieties
when UF milk is used. UF cheese tends to have a smoother consistency
than that made from normal milk, presumably because an increased
proportion of calcium is present and/or the presence of whey proteins.
The texture of UF Cheddar, for instance, tends to be similar to that of
traditional Gouda. This may be a consequence of the difficulty involved
in achieving the ratio of minerals to protein required for a texture
that can be characterized as cheddary. In traditional Cheddar
manufacture the mineral/casein ratio is dependent only upon the extent
of acid production in the vats. When using UF milk, however, there are
four distinct stages in the process (pre-acidification of milk;
ultrafiltration; diafiltration and acid production in the coagulum) at
which mineral loss can occur and control is by no means easy.

UF cheese also ripens more slowly than traditionally made cheese,
apparently because β-lactoglobulin, the major whey protein, inhibits
plasmin activity. In addition, undenatured whey proteins are resistant
to the action of chymosin and the other proteinases normally found in
cheese. Most UF cheeses are marketed under traditional names even
although the normal pattern of casein breakdown is changed and the
textures and flavours therefore tend to be different to those of
traditional cheeses. These new characteristics may well prove to be
acceptable commercially and it may be preferable to develop a new range
of cheese varieties rather than to attempt to duplicate exactly the
properties of traditional cheeses.

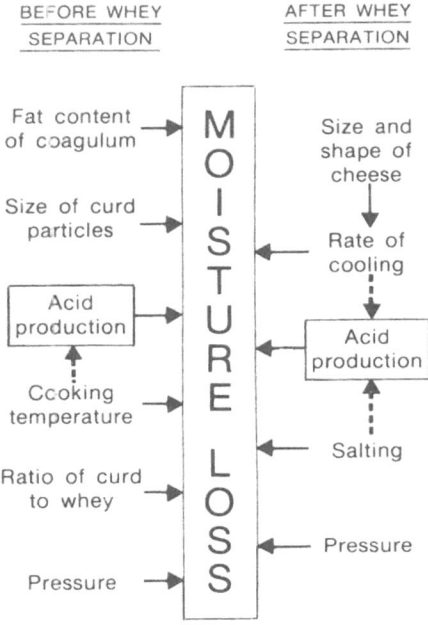

Figure 1. Main factors that determine the expulsion of moisture from a rennet coagulum.

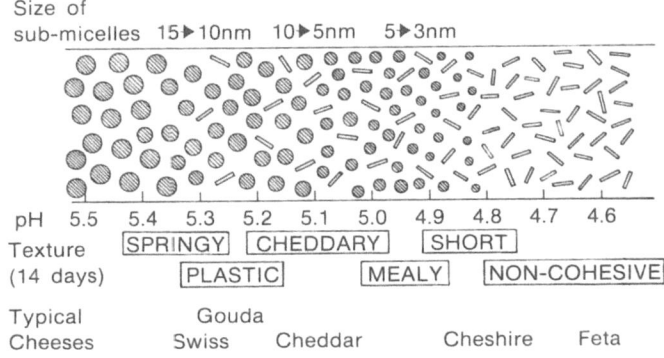

Figure 2. Diagrammatic representation of the effect of pH on cheese microstructure and texture.

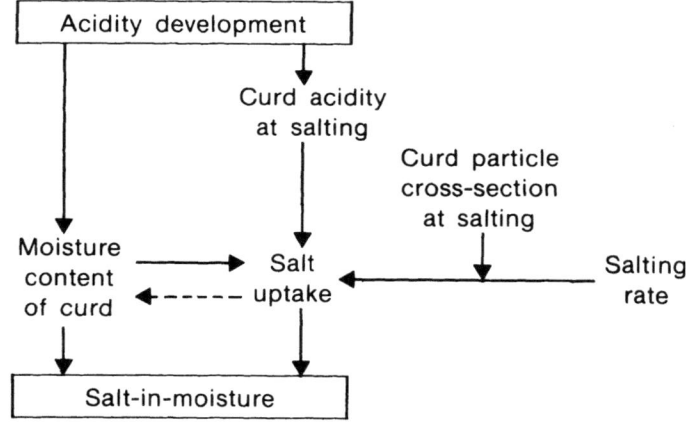

Figure 3. Main factors that determine the pH of dry-salted cheeses such as Cheddar.

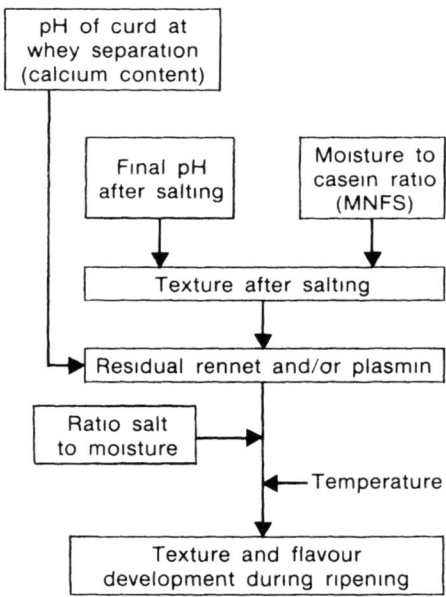

Figure 4. The relationship between the compositional parameters (pH, MNFS, S/M, Ca) of a cheese variety and the development of texture and flavour during ripening.

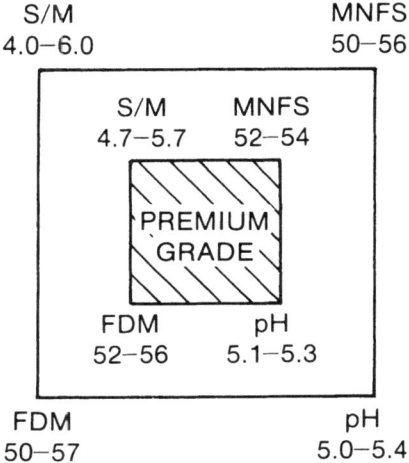

Figure 5. Ranges of pH, MNFS, S/M and FDM suggested for Premium (shaded) and First grades of Cheddar cheese.

REFERENCES

1. Roefs, S.P.F.M., P. Walstra, D.G. Dalgleish, and D.S. Horne. 1985. Neth. Milk Dairy J. **39**:119.

2. Hall, D.M., and L.K. Creamer. 1972. N.Z. J. Dairy Sci. Technol. **7**:95.

3. Creamer, L.K. 1985. Milchwissenschaft. **40**:589.

4. Lawrence, R.C., H.A. Heap, and J. Gilles. 1984. J. Dairy Sci. **67**:1632.

5. Lawrence, R.C., L.K. Creamer, and J. Gilles. 1986. J. Dairy Sci. In press.

6. Creamer, L.K. 1971. N.Z. J. Dairy Sci. Technol. **6**:91.

7. Holmes, D.G., J.W. Duersch, and C.A. Ernstrom. 1977. J. Dairy Sci. **60**:862.

8. Visser, F.M.W., and A.E.A de Groot-Mostert. 1977. Neth. Milk Dairy J. **31**:247.

MECANISATION EN FROMAGERIE DE PATES MOLLES

J. Korolczuk, J.-L. Maubois et J. Fauquant
Dairy Research Laboratory - I.N.R.A.
65, rue de Saint-Brieuc
35042 Rennes Cedex
France

RESUME. La mécanisation en fromagerie de pâte molle s'est orientée ces dernières années autour de deux démarches principales : - la mise en continuité des opérations classiques des technologies traditionnelles par le biais d'équipements mettant en oeuvre soit des macrovolumes (coagulateur Alpma), soit des minivolumes (chaîne Cartier) avec dans ce dernier cas, une simulation parfaite de la technologie artisanale grâce à la robotique - l'utilisation des avantages résultant de la mise en oeuvre de technologies nouvelles telle que l'ultrafiltration sur membrane pour la conception d'équipements totalement automatisés produisant de façon continue des fromages de forme et de texture variées. Des exemples choisis pour illustrer ces deux tendances, il peut être envisagé pour l'avenir le franchissement de nouveaux paliers pour l'automatisation dans les grandes fromageries de pâtes molles. La robotique devrait permettre de sauvegarder, voire de développer, la fabrication de fromages à caractéristiques organoleptiques tranchées. L'ultrafiltration sur membrane devrait amener, quant à elle, outre une standardisation généralisée de la teneur en protéines des laits de fabrication à la création de nouveaux fromages de caractéristiques organoleptiques très variées.

INTRODUCTION

Au sein de la grande famille des fromages, ceux dits à pâte molle se caractérisent par un faible poids (100 g à 1000 g), une teneur en eau voisine de 50 %, une très grande diversité de forme, de texture, de surface (croûtes fleurie, lavée, salée, sèche, etc) et un affinage de quelques semaines. Leur fabrication artisanale requéraient une main d'oeuvre abondante pour les opérations de formation et de travail du caillé (moulage à la louche - soins en hâloir).

L'industrialisation des unités de production qui s'est réalisée ces dernières années repose sur la mise en place d'actions de mécanisation. Celles-ci se sont orientées autour de deux démarches principales :
- une mécanisation pouvant aller jusqu'à la robotisation des opérations des techniques traditionnelles de fabrication ;

- l'utilisation des avantages résultant de la mise en oeuvre de technologies nouvelles telle que l'ultrafiltration sur membrane pour la conception d'équipements totalement automatisés produisant de façon continue des fromages de forme et de texture variées.

A l'origine de ces deux démarches, se trouvent les nécessités économiques de réduire le coût de la main d'oeuvre et d'accroître l'homogénéité des productions, particulièrement importante pour nombre de ces fromages vendus à la pièce. Une motivation supplémentaire distingue la seconde démarche, celle d'accroître le rendement par rétention dans le fromage des constituants protéiques du lactosérum.

Pour illustrer ces deux tendances de la fromagerie actuelle de pâtes molles, nous avons choisi de vous présenter quelques exemples concrets de réalisations industrielles récentes.

1. MECANISATION DES TECHNOLOGIES TRADITIONNELLES

Soucieux de réduire les écarts de poids constatés au niveau des fromages égouttés (σ = 27 g selon Maubois et Mahaut, 1974, pour un poids moyen de 270 g), les équipementiers se sont d'abord orientés vers le concept de répartition moule par moule de la quantité de caillé correspondant à un fromage. Des deux voies possibles : la microbassine telle que proposée par Rematome ou l'insertion d'un "diviseur" dans une minibassine telle que proposée par Cartier, seule la dernière a connu, à notre connaissance, plusieurs réalisations industrielles. Un de ces unités fonctionne actuellement à Vire, dans l'usine de l'ULN, pour la fabrication de 150 000 Camemberts/jour (Anonyme, 1980).

Remplies de lait emprésuré, les bassines sont amenées par convoyeur lent jusqu'à un poste de décollement par électrolyse et tranchage (horizontal et vertical). Dans la bassine est alors inséré un diviseur délimitant 24 portions. Gerbées par 4 pendant le temps de la synérèse (30 à 40 minutes), les bassines toujours en mouvement sur le convoyeur sont ensuite désempilées, recouvertes d'un répartisseur, d'une rehausse, d'un bloc moule et d'un plateau. L'ensemble est retourné dans le poste de moulage ou bloc-moule et plateau d'un côté, bassine et diviseur de l'autre sont séparés ; ces derniers retournant pour un recyclage constant vers le poste de lavage. Empilés par 10, les ensembles plateaux-blocs-moules-rehausses sont encore convoyés pendant 90 minutes, temps auquel la rehausse sera enlevée. De nouveau gerbés par 12, les plateaux de 24 fromages sont ensuite dirigés vers les salles d'égouttage.

Les possibilités de standardisation de la teneur en protéines des laits de fabrication par emploi de la technique d'ultrafiltration sur membrane ou par addition de caséinate (jusqu'à un maximum de 5 g/l) ont amené ensuite les équipementiers à envisager d'autres solutions pour réaliser une mécanisation à la mesure de l'accroissement de capacité des entreprises productrices de pâte molle (500 000 l à 600 000 l de lait traité par jour). Cette standardisation en protéines (37 à 45 g de protéines au litre) rend, en effet, beaucoup moins cruciaux les écarts de poids en raison de la diminution du volume de caillé à répartir par moule et de l'augmentation de fermeté de ce même caillé, ce qui facilite son transfert et diminue les pertes en fines dans le lactosérum. C'est dans cette démarche que s'insère la technologie proposée par la société Alpma dont les coagulateurs équipent maintenant les grandes unités françaises de pâte molle. L'installation la plus récente est celle mise en place dans l'entreprise Bridel sise à Retiers (France).

Le lait de fabrication dont les teneurs en protéines et en matière grasse, le pH ont été amenés aux valeurs souhaitées, la première par ultrafiltration et/ou addition de caséinate, la seconde par écrémage, le troisième par développement fermentaire et addition de lait décationisé (Riallant et Barbier, 1980) est déversé à l'extrémité du coagulateur de 54 m de long. La capacité hemicylindrique délimitée par la bande plastique transporteuse est séparée à l'aide de cloisons appliquées de façon étanche à cette bande en volumes de 800 l dans lesquels est injectée et mélangée la présure. En fin de coagulation, le caillé est décollé de ces cloisons par application d'un courant électrique continu. La cloison est enlevée et nettoyée en N.E.P. Quant à la masse de caillé, elle est découpée en cubes de 1,5 cm d'arête et transportée jusqu'au poste de moulage se trouvant à l'autre extrémité du coagulateur ou après une évacuation sur filtre du lactosérum de synérèse, la répartition se fait dans des ensembles plateaux-blocs-moules et rehausses. La mécanisation de l'ensemble est poursuivie puisque aussi bien les différentes phases de l'égouttage en tunnel à cinétique de température contrôlée par microprocesseur que les retournements sont réalisés de façon dynamique. Il en est de même pour l'opération de salage qui est réalisée en piscine de saumure (durée : 45 minutes).

Les progrès récents réalisés en matière de robotisation ont tout récemment amené à envisager la mise en place d'équipements simulant les technologies artisanales traditionnelles. La première réalisation industrielle illustrant cette nouvelle démarche de la mécanisation en fromagerie est celle récemment réalisée à la Coopérative d'Isigny (France) (Morel, 1986). Le lait cru maturé, additionné de levains lactiques et de présure est versé en quantités prédéterminées en fonction de sa teneur en protéines, dans des bassines ayant une capacité maximale de 230 l à fond alvéolé (Sté Maisonneuve). Lorsque le coagulum atteint la fermeté désirée, les bassines sont amenées par transporteur sous un robot de moulage comportant 20 louches commandées par automate (Sté Staimsima). En forme de sphère mais avec les deux quarts inférieurs articulés et comportant à l'intérieur des couteaux de découpe, ces vingt louches viennent prendre le caillé en surface des bassines et le déposer délicatement dans un bloc de 20 moules. Le remplissage de ces derniers est effectué en six passages. Cette robotisation simule donc tout à fait les gestes dit de "sabrage" et de moulage à la louche réalisés en fromagerie traditionnelle de Camembert et les fromages obtenus à Isigny ont droit au label d'appellation d'origine : Camembert de Normandie. La productivité du robot est de 1200 fromages/heure ce qui autorise une production maximale en deux équipes de 24 000-25 000 fromages/jour. Elle peut être rapprochée de celle d'un mouleur expérimenté qui, toutes choses étant égales, moule en moyenne 200 fromages/heure. Le robot de moulage conduit, toutefois, à des produits plus homogènes en poids en raison de la plus grande précision des volumes de caillé prélevé. De l'économie de main d'oeuvre ainsi réalisée, il est espéré un maintien à un seuil acceptable pour le consommateur de l'écart de prix entre fromage traditionnel et fromage dit industriel (du simple au double actuellement)

2. MECANISATION DES TECHNOLOGIES NOUVELLES

Si la transformation du lait en fromage Feta a représenté et représente toujours la principale application du procédé dit MMV (Maubois et al., 1969), procédé basé sur l'emploi de la technique d'ultrafiltration sur membrane, de nombreux autres fromages sont maintenant fabriqués selon cette technologie, que ce soit dans le domaine des pâtes fraîches ou dans celui des pâtes demi dures ou encore dans le domaine des pâtes molles. Pour illustrer les possibilités de mécanisation dans ce dernier type de fromage, trois réalisations récentes ont été également choisies.

Le système Camatic développé par la société Alfa Laval a été mis en place en France et en RFA pour la production continue de Camemberts. Le lait pasteurisé est ultrafiltré jusqu'à une teneur en protéines de 14-15 g p. 100 soit un facteur de concentration en volume de 5,1-5,2 à la température de 50°C. Le rétentat refroidi à 30°C est additionné de 2 % de levains lactiques mésophiles et de 0,75 g p. 100 de NaCl. Lorsque son pH atteint la valeur de 5,2, le rétentat est injecté dans les têtes de moulage de l'ensemble Camatic où il est emprésuré et versé à la dose de 360 g/moule dans les moules ayant une soupape en fond. La coagulation et le durcissement du caillé sont effectués en dynamique avec gerbage et dégerbage en chaîne (durée 70 minutes). Pour le démoulage, une quantité de courant électrique continu égale à 1 Coulomb est appliquée par fromage, ce qui provoque le décollement du caillé de la paroi du moule acier inoxydable. Les blocs moulés sont retournés et une pression d'air comprimé est appliquée à l'endroit des soupapes. Les blocs moulés retournent automatiquement en tête de l'équipement Camatic en subissant au passage un nettoyage dans une machine à laver placée sur le convoyeur. Les fromages démoulés sont ensuite maturés et salés en dynamique comme dans la fromagerie classique. La capacité de l'installation mise en place à la fromagerie Gallais est de 5000 fromages/heure. Elle ne requiert pour son fonctionnement que la présence d'une personne.

Un autre système de mécanisation, certes plus simplifié mais qui démontre bien les possibilités de nouveaux formats offertes par la technologie MMV, est celui installé à la fromagerie Guilloteau (France). Le lait pasteurisé additionné de 0,50 % de NaCl est ultrafiltré à 50°C jusqu'à une teneur en protéines de 13,0 g p. 100. Le rétentat additionné de levains lactiques thermophiles et de présure est ensuite versé dans des bacs parallélépipédiques ayant la hauteur du fromage (5 cm). Ces bacs sont ensuite placés, une fois empilés dans des étuves à 43°C où s'effectue la coagulation et maturation du caillé. Le bloc de caillé qui une fois développée la moisissure de surface donnera le fromage Pavé d'Affinois résulte de la simple action sur chaque plaque de caillé d'un couteau diviseur (96 pavés par plaque).

Les caractéristiques de résistance mécanique des membranes minérales d'ultrafiltration, membranes dites de troisième génération, leur "nettoyabilité" même après mise en contact avec des fluides très

visqueux nous ont amené récemment à proposer l'ultrafiltration de lait
de chèvre acidifié et coagulé pour la fabrication de fromages à pâte
molle, de caractère lactique, de type Ste Maure (Mahaut et al., 1986).
Le lait entier de chèvre traité thermiquement à 85°C pendant 15 s est
additionné de levains lactiques mésophiles (2 %) et de présure (7 ml/
100 kg) en cuve de coagulation à la température de 20-25°C pendant 15
heures. Le coagulum obtenu est alors brisé mécaniquement, réchauffé
doucement à 40°C et ultrafiltré. L'équipement d'ultrafiltration utilisé
(SFEC, France) comporte trois étages en série : le premier équipé d'une
pompe de recirculation centrifuge permet d'atteindre une teneur en protéines dans le rétentat de 7,5 g p. 100, le second équipé d'une pompe
de recirculation volumétrique conduit à l'obtention d'un rétentat ayant
une teneur en protéines de 10,5 g p. 100, le troisième équipé d'une
pompe de même type mais alimentant des cartouches dites courtes (0,6 m)
permet d'atteindre une teneur en protéines de 16 à 17 g p. 100 dans le
rétentat soit une teneur en matière sèche de 40 g p. 100. Après refroidissement statique à 4°C pendant 15-16 h pour raffermissement de la
pâte, le caillé est moulé en forme Sainte Maure dans un système de
boudineuse. Salage et affinage sont ensuite conduits dans des conditions classiques.

3. CONCLUSION

Les deux tendances illustrées à l'aide des exemples évoqués ci-dessus
devraient vraisemblablement se développer au cours des prochaines
années. Il est probable, en effet, que grâce aux moyens maintenant disponibles ou qui le seront pour standardiser la matière première laitière destinée à la transformation fromagère (matière grasse - protéines
pH - vitesse d'acidification - degré de protéolyse, etc) de nouveaux
paliers seront franchis pour une automatisation complète en fromagerie
de pâtes molles. De nouvelles percées de la robotique sont à attendre,
grâce aux progrès de la microélectronique, elles devraient permettre
de sauvegarder voire de développer des fromages à caractéristiques
organoleptiques tranchées qui autrement étaient condamnés à disparaître
étant donné le coût de la main d'oeuvre nécessaire à leur obtention.
Les technologies nouvelles outre leur généralisation de leur emploi
pour la standardisation de la matière première devraient quant à elles,
par les extraordinaires possibilités technologiques qu'elles offrent,
amener à la création de toute une palette de nouveaux fromages aux
goûts, formes, textures, saveurs et arômes très diversifiés.

4. REFERENCES

. Anonyme, 1980. 'La nouvelle unité de fabrication de pâtes molles de
l'Union Laitière Normande à Vire'. Technique Laitière, **948**, 27-34.
. Mahaut M., Korolczuk J., Pannetier R. et Maubois J.-L., 1986.
'Eléments de fabrication de fromage de type pâte molle de lait de
chèvre à caractère lactique par ultrafiltration de lait acidifié et
coagulé'. Technique Laitière et Marketing, **1011**, 24-28.

. Maubois J.-L., Mocquot G. et Vassal L., 1969. 'Procédé de traitement du lait et de sous-produits laitiers'. <u>Brevet français</u> n° 2 052 121.
. Maubois J.-L. et Mahaut M., 1974. 'Application de l'ultrafiltration sur membrane dans l'industrie fromagère'. <u>Revue Laitière Française</u>, **322**, 479-484.
. Morel F., 1986. 'La robotisation du moulage à la louche : fruit de trois années de recherches à la Coopérative d'Isigny'. <u>Technique Laitière et Marketing</u>, **1006**, 28-31.
. Rialland J.-P. et Barbier, 1980. 'Procédé de traitement du lait par 1 résine échangeuse de cations en vue de la fabrication de la caséine et du lactosérum'. <u>Brevet français</u> n° 2 480 568.

TRENDS IN THE MECHANIZED MANUFACTURE OF SEMIHARD CHEESE TYPES

G. van den Berg
Netherlands Institute for Dairy Research
P.O. Box 20
6710 BA   EDE
The Netherlands

ABSTRACT. This review deals with the modern developments in the mechanisation of the manufacture of mainly semihard cheeses. Trends in designing of the equipment and their relation to the underlying technology are treated in the following sections.
Introduction: mechanisation as part of modern management systems
Curd preparation: the design of curd preparation tanks
Draining and moulding: developments in the vertical automatic draining systems
Cheese moulds: various new trends
Cheese storage: conditioning systems and cheese carrying materials

INTRODUCTION

Cheesemaking is, in spite of all research, a subtle and time-consuming process and is concerned with a complicated raw material. Its aim is still to convert the largest possible amount of casein and fat from milk into a good-quality product. The almost autonomous reactions that underlie the cheesemaking process are mainly controlled by feed-forward procedures and conditioning measures.

A great many technical facilities are available to a modern cheese factory. But even when the equipment is well chosen, a lot of process analysis work is necessary for accurate adjustment of all operation steps, in order to derive maximum profit from the investments. Another point is that attention should be given beforehand to the cleanability and ease of maintenance of the equipment. The possibility of making accurate measurements in the proper places is also an important desideration in modern process control.

As regards this last point, automation should have a sound basis in mechanisation, and both should be parts of one concept. An important aspect is the system of data processing by means of which the manager, the quality officer, the operator and the technical staff can be supplied with selected information. Process control and quality assurance should also be integrated.

In this paper, the most important developments in mechanisation of the various stages of the cheesemaking process will be dealt with, and for the most part they will be concerned with Dutch cheese varieties. A number of general items, such as saving in energy consumption and noise abatement, though they certainly have their effects on the development of machinery, will not be considered in detail.

CURD PREPARATION

Curd preparation is still a batch process, since the most promising approach to a fully continuous system failed some ten years ago at the stage of scaling-up (1). Nowadays the curd is usually prepared in closed tanks which have been designed for automatic cleaning and provide better protection against contamination than the formerly used open cheese vats. In these tanks the milk is coagulated and cut, after which part of the whey is sucked off. Then the curd is washed, if necessary, and stirred to promote further syneresis. The capacity of a present-day curd making tank may be larger than 20 000 l; for the manufacture of softer and smaller cheese types these tanks should have smaller capacities, so that batch times can be limited and the cheese composition can be better controlled. For Gouda cheese the batch time should not exceed 30 - 40 minutes.

Various dairy equipment manufacturers offer curd making tanks in the shape of a "Double-O", which are equipped with a power unit mounted on the superstructure and with two wide cutting/stirring assemblies. The individual cutting blades are now a larger distance apart than in previous types. Freely swinging blades are used to increase the stirring effect. Sometimes these assemblies are supported by stays on the bottom of the tank to make lighter constructions possible. Another interesting type is the horizontal cylindrical tank of the OST IV concept. It has a horizontal shaft with the cutting/stirring assemblies, mounted in the heart of the cylinder and supported at both ends of the cylinder.

Several curdmaking tanks have been tested, under practice conditions, for their suitability for the manufacture of Gouda cheese in the Experimental Dairy of the Netherlands Institute for Dairy Research (2-6). Losses of fat and curd fines in the whey and the formation of curd lumps are important issues which are influenced by the design of these tanks. But the results obtained in practice are also affected by the manufacturing procedures used. In this respect, it has been found that the effective capacity of the whey pump should be so large that after one minute of curd sedimentation the proper amount of first whey (40 - 50 % of the cheese milk) is sucked off in 5 - 6 minutes via a horizontal tubular strainer. The risk of curd lumps being formed has been reduced by setting the bottom of some makes of tank at a smaller angle and by designing appropriate and efficient stirring blades. To keep losses of fat and curd fines at a low level, the design and sharpness of the cutting blades should be optimal. The whey strainer should be sufficiently dimensioned and the size of its inflow area should permit a smooth drain-off (7).

Some makes of tank are provided with a stop-cock at one end and are mounted or can be brought in a tilted position to facilitate emptying. This is effective in particular for the OST IV tank. When a tank is horizontal and has one or even two drain cocks in the bottom, the bottom (and the cutting assembly) is so designed that a smooth flow to the cock(s) is ensured.

Recently, equipment has been introduced by means of which the strength of the renneted gel can be measured. It enables the cheesemaker to start cutting at the correct point of time, and signals whether there is actually any coagulum to be cut or not. The many aspects of this matter concerning process control systems have been laid down in a monograph which comes up for discussion in the Expert Group B12 of the IDF (8).

The cleanability and protection against contamination of a curdmaking tank, and of the auxiliary equipment, is, of course, a matter of the utmost importance. For example, the cock in the whey discharge pipe should be mounted as closely as possible to the tank, to avoid that any whey flows back into the tank. It is also recommended to equip every tank with a separate starter-dosing system, which, in combination with an effective cleaning system, can give adequate protection against problems with disturbing bacteriophages (9). It also renders separate aseptic rooms for the curdmaking tanks unnecessary (10).

To sum up, the most important recent developments concerning curd making tanks are:

Shape	- "Double-O" shaped tanks
	- short, horizontal, cylindrical tanks;
Cutting assemblies	- greater distance between cutting blades;
	- freely swinging stirring blades;
Whey strainer	- general use of a large, horizontal, perforated tube;
Bottom	- one valve or two valves at lowest point of bottom (when tank cannot be tilted);
Gel strength	- indicator on the tank.

DRAINING AND MOULDING

This is a crucial unit operation in the manufacturing process, because it determines the structure and the weight of the cheese and, to some extent, its moisture content. For cheeses having a closed texture or round holes, draining should proceed in such a way that any mixing with air is avoided after the curd preparation has finished, because every small air bubble that remains in the curd causes a hole in the cheese. In addition, care should be taken that the whey is drained evenly from all parts of the curd block. Randomly distributed quantities of whey between the curd particles may not only cause iregular (nesty) holes in the cheese, but also acid, white spots. Too rapid drainage, particularly when fairly soft curd is concerned, presses the curd into the perforated plates; as a result, these become clogged and the drainage becomes insufficient. Also, a certain lapse of time and a slight pressure on the curd are necessary between the first drainage and the cutting stage, in order that the fusion of the curd particles is sufficient and a firm curd mass is obtained that can easily be cut and will be able to withstand the subsequent, mechanical treatment.

The draining and pre-pressing vat (sometimes called "strainer vat"), which once was the start of mechanization in the modern cheese factory, has been improved by equipping it with automatic curd distribution (11) and mechanized delivery of the curd blocks. However, this equipment still has disadvantages, such as the risk of irregular draining, and insufficient control of weight and composition of the cheese (12). The most important development in this field has been that of the automatic draining and moulding machines. The Casomatic, developed from the curd-portioning section of the continuous Nicoma cheesemaking system (13), is well known. It consists of vertical pipes divided into three draining zones (14). The whey-and-curd mixture enters the machine at the top - under the liquid level - and is drawn from a jacketed buffer tank in which the whole batch of curd suspension is kept homogeneous. Control devices are mounted on the draining machine to keep the whey and curd independently at their proper levels. The whey level is maintained during drainage by supplying fresh whey. A certain whey flow through the curd is necessary to compress the curd. Overflow levels in the whey drainage pipes are used to maintain a constant pressure drop in order to control the flow rate of the whey through the perforations. Flow-restrictions are also mounted in the whey discharges. During cutting and discharging a block of curd this whey flow is interrupted. These regulations are particularly necessary when cheese types are made with a diameter of more than 22 to 24 cm. Ultimately, the machine places the blocks of curd in cheese moulds. The Stork curd portioning machine can be used for the same types of cheese (15). To make cheeses with an open texture, the curd can be drained with rotary strainers before entering the pipes.

An earlier approach was the Holvrieka curd-portioning machine, now mentioned the "Conomatic R" consisting of one pipe with the section for draining and cutting the curd blocks at the top. The construction is such that a preconcentrated curd and whey mixture is led in at the bottom of the pipe (16). In the manufacture of smaller cheese types the curd blocks can be drained with pipes without complicated systems for controlling whey drainage being necessary. Well-known is the Stork machine that is capable of producing Edam, baby-Gouda and other cheese types (17).

The contents of the whey-and-curd buffer tank, which I mentioned before, are cooled slightly during draining of the batch, in order to decrease syneresis of the remaining curd. This measure, in conjunction with the proper starting time of pressing and the discharge frequency of the curd blocks, is an important means of controlling composition and weight of the final cheese.

The dimensions of the draining pipe should be only slightly smaller as those of the cheese mould, because before being pressed the curd has to flow and fill the mouls smoothly and without any force being used. This is more obvious when the curd is firm. On the other hand, the flow of curd may not be extended over a large distance, because it would cause a higher moisture content in that part of the cheese. This is the reason why this modern draining equipment is not very flexible with respect to various cheese types. Therefore attempts have been made to design machines with one or more exchangeable perforated inner pipes of various sizes, round or rectangular in shape (18). The knife and discharge parts are also exchangeable. For smaller cheese, when using several inner pipes, the moulds are mounted in one frame (the so called multimoulds) (19).

Another approach is the division of the curd block into smaller units after it has been discharged from the draining equipment (20).

To summarize the new developments in this stage of the process:
- pre-pressing vats with automatic curd distribution;
- various types of automatic draining pipes;
- exchangeable draining pipes.

CHEESE MOULDS

The aim of pressing the curd in a mould is to give the cheese its typical shape and a closed rind. The latter is particularly necessary for naturally ripening cheese to prevent microorganisms from entering the cheese during brining and ripening (21). Therefore the curd block should be well drained and the temperature during pressing well controlled. Nevertheless, particularly when cheese types that require prolonged natural ripening are concerned, the type of cheese mould should be carefully selected.

It is often assumed that more effective pressing results in more whey being squeezed out. But it has been found that more whey is liberated when there is no pressing at all. The rind formation, which inhibits expulsion of whey from the inside of the cheese, is based on compression and more intensive fusion of the curd particles in the outermost layer. The drainage of whey from this layer is facilitated by the deformation of the curd around the fine perforations and woven threads of the liners or the inner surface of the moulds themselves (22). It may be expected that there is hardly any difference between the various modern cheese moulds in the amount of whey expelled during pressing, with the preconception that the whey can easily be drained from the mould.

The holding of the cheese between pressing and brining is necessary for sufficient acidification. Originally the shape of the cheese had also to be improved at that stage by placing the cheese, after trimming, upside down in the empty mould again. But nowadays, the shape of the mould and particularly of the lid are so well designed that, after pressing, the cheeses leave the mould, in a nearly perfect shape. The advantage is that trimming after pressing is no longer necessary, but on the other hand the lid is sharp-edged and has to be lifted perpendicularly to avoid damage to the cheese rind. Another consequence of this finely designed mould is the need to remove the cheese from the mould by blowing or applying a vacuum. When using moulds with loose liners, automatic emptying is possible after loosening the lid. After separate cleaning of mould, liner and lid, the liner is automatically replaced in the mould and, after filling, also the lid.

To simplify handling, moulds have been developed with fixed lining material from plastics or woven stainless steel threads and even finely perforated moulds with a special surface structure to ensure the formation of a normal rind. However, the latter were not successful, in those cases where they could not meet the high demands imposed on them in connection with cheese rind formation.

To be noted are the attempts to introduce self pressing cheese moulds to avoid the use of presses (23).

The new developments in moulds can be summarized as follows:
- improved shape of moulds for round cheeses;
- auxiliary equipment to empty the moulds;
- multimoulds;
- moulds without loose lining material;
- self-pressing cheese moulds.

BRINING

Mixing of salt with curd, and even injection of a salt solution into the curd just after draining (24, 25), causes damage to the cheese texture. So the cheese types concerned have still to brine so they can take up the desired amount of salt.

Apart from an integrated system of brining, ripening and transport of Edam cheese, no developments concerning this stage of the process have been reported recently. A particular point worth mentioning is that, not long ago, a plea was made for improving the hygienic design of the equipment (26).

TREATMENT OF THE CHEESE DURING RIPENING

For naturally maturing cheese air conditioning is very important. The actual weight loss as a result of drying is crucial and should not be too small, to prevent visible growth of microorganisms. On the other hand, too high weight losses cause cracks in the plastics layer and mould growth in the rind in spite of the use of natamycin. In addition, the cheeses should always be placed on clean and dry wooden shelves. After removing the cheeses they should be cleaned at one side and used again at the other side.

The causes of any variance in weight loss during storage were studied (27), because weight loss contributes to the final composition of the cheese. This investigation resulted in a further improvement on the distribution of pre-conditioned air over the cheeses. Each shelf with cheeses is separately supplied with air, usually by means of injection pipes (28, 29). The flow of the air can be reversed and the cheeses relocated after every treatment (28). Further treatments of

the cheeses, such as turning, plasticizing and cleaning of the
shelves, are mostly carried out outside the store. These manipulations
are usually combined with automatic transport and loading systems.

There is an increasing interest in the use of metal plates
instead of wooden shelves. Steel plates can be cleaned more
effectively and do not impair the drying process. Their disadvantage
of not absorbing moisture from the cheese can be met by a well chosen
type of perforation. The original idea came from Switzerland (30), but
it has now been adapted to the Dutch cheese varieties (31).

For small cheese types, such as the classic Edam cheese, systems
have been developed to hang them in open plastic nets during ripening
(32, 33).

To summarize, the latest developments in cheese storage are:
- improved automatic systems for loading, unloading, turning of
  cheeses and cleaning of shelves;
- improved air supply systems;
- perforated metal plates instead of wooden shelves;
- ripening nets for smaller cheese types.

REFERENCES

1. P. Zwaginga, 61 st Ann. Meeting IDF Comm. B, Stockholm (1977).
2. E. de Vries, G. van den Berg & W. van Ginkel, Test-Report NIZO (1976)R105.
3. E. de Vries, Test-Report NIZO (1979)R110.
4. E. de Vries & W. van Ginkel, Test-Report NIZO (1980)R113.
5. E. de Vries & W. van Ginkel, Test-Report NIZO (1983)R118.
6. E. de Vries & W. van Ginkel, Test-Report NIZO (1984)R120.
7. G. van den Berg, E. de Vries, A.G.J. Arentzen, Off. Orgaan FNZ 65(1973)825; Voedingsm. tech. 4(1973)(36)127.
8. A.C.M. van Hooydonk & G. van den Berg, Monograph IDF Expert Group B12 (1986).
9. J. Stadhouders, Neth. Milk Dairy J. 40(1986)155.
10. R. Hansen, North Europ. Dairy J. 52(1986)12.
11. K. Gasbjerg, Danish Dairy Ind. Worldwide 4(1984)31.
12. 'Vochtgehaltespreiding in kaas', Kon. Ned. Zuivelbond FNZ, Report 23 (1975).
13. A.G.J. Arentzen, Dtsche Molkerei Ztg. 91(1970)2251.
14. R. Hansen, North Europ. Dairy J. 45(1979)66.
15. E. de Vries & W. van Ginkel, Test-Report NIZO (1985)R123.
16. R. Hansen, North Europ. Dairy J. 49(1983)218.
17. R. Hansen, North Europ. Dairy J. 43(1977)88, 93.
18. R. Hansen, North Europ. Dairy J. 51(1985)275.
19. R. Hansen, North Europ. Dairy J. 47(1981)208.
20. R. Hansen, North Europ. Dairy J. 49(1983)33.
21. G. Hup, J. Stadhouders, E. de Vries & G. van den Berg, Zuivelzicht 74(1982)270.
22. P. Walstra & T. van Vliet, Neth. Milk & Dairy J. 40(1986)241.
23. R. Hansen, North Europ. Dairy J. 50(1984)207.
24. R. Hansen, North Europ. Dairy J. 51(1985)248.

25. V.D. Surkov et al. USSR Pat. (1981) SU 884634.
26. J. Stadhouders et al., Zuivelzicht 77(1985)892.
27. S. Bouman, Zuivelzicht 69(1977)1130.
28. G.J. van Elten, North Europ. Dairy J. 45(1979)98.
29. NN, Zuivelzicht 69(1977)1130.
30. V. Oehen, Schweiz. Milchztg. 95(1969)99.
31. M.S., Zuivelzicht 74(1982)872.
32. Dutch Patent Appl. 7313365.
33. Dutch Patent Appl. 7700794.

MECHANIZATION: DRY SALTED CHEESES

Norman F. Olson
Department of Food Science
University of Wisconsin-Madison
1605 Linden Drive
Madison, Wisconsin   53706   USA

ABSTRACT: Mechanical handling of dry-salted cheese, i.e. Cheddar has progressed at all stages of manufacturing to enhance efficiency, sanitation, and cheese quality and yield. Future mechanization will emphasize flexibility and will interface with the use of ultrafiltration of milk. Computer-mediated process monitoring and control are integral parts of mechanized systems. These controls include milk standardization, temperature, time, pH and coagulum firmness. Methods of pressing salted curd and of cheese packaging are rapidly evolving technologies.

1. EVOLUTION OF CHEESE MECHANIZATION

Mechanization of dry salted cheeses involved substantial innovations in equipment from 1950 through the 1970's. These developments coincided with consolidation of cheese plants and larger processing capacities within single plants. Mechanical cutting of curd allowed the use of larger traditional open horizontal vats and closed vats. Whey drainage and subsequent handling of the curd was done in large covered, conveyer systems or in open or closed vats equipped with stirring devices to maintain the curd in the granular form. The conveyers are typically used for Cheddar cheese that is cheddared and milled; the finishing vats with stirrers are used for granular (stirred-curd) Cheddar or varieties like Colby, Monterey, Muenster and brick cheeses.
 Fusing the curd after salting has evolved in two divergent routes. In the United States, 218 Kg cylinders of cheese are formed in barrels, or 290 Kg blocks of cheese are pressed in rectangular draining hoops. Curd fusing towers are used to a greater extent in other countries for form 18 Kg blocks of cheese. The two modes of pressing cheese evolved because of the methods of utilizing the cheese. Barrel and large block cheeses are often converted into process cheese. Efficiencies in handling and packaging large pieces are obvious for this utilization method. The large blocks may be cut, after pressing, into 18 Kg blocks that are packaging and matured.

They can also be matured intact and cut into pieces for retail use. Continuous curd fusion and formation of 18 Kg blocks in the tower block-former is a commonly used process throughout the world. Multiple towers with mechanized vacuum packaging of blocks and mechanized cartoning of the packaged blocks provide a continuous labor-saving process and produce blocks of cheese that are convenient for shipping.

2. FUTURE TRENDS IN EQUIPMENT AND PLANT DESIGN

Radical changes in design of cheese vats and curd handling equipment are not likely to occur in the near future. The use of ultrafiltration in the cheese making process will channel capital investments and development efforts to that technology and to curd-handling systems that will retrofit to coagulation chambers for ultrafiltered milk.

In addition, flexibility of equipment and plant design is sought by the cheese industry (Elliot, 1984; Hansen, 1983). Consumer demand for a wider variety of cheeses appears to be the driving force in the United States. Flexibility can be either in the stirred-curd system (Elliot, 1984) or by modification of operating procedures of the continuous curd-handling belt systems (Hansen, 1983).

Previous innovations in mechanization have replaced manual tasks with machines without altering the basic cheesemaking procedures. This even applies to the use of ultrafiltration in cheesemaking since the same control of pH, calcium and lactose levels and whey syneresis are essential (Lawrence, et al., 1984). Future developments probably will focus initially on use of computers to control, monitor and evaluate various stages of cheesemaking. Further mechanization, acceleration or modification of processes will require better understanding of the fundamental changes occurring in milk and cheese curd during manufacturing. The remainder of this paper will discuss the various stages of cheesemaking from a process standpoint and relate that to the potential for mechanization.

3. MILK COMPOSITION STANDARDIZATION

The demand for more variety in cheese and for lower fat levels will necessitate greater emphasis on standardization of milk composition. Techniques for removing fat or adding solids not fat to adjust milk composition are well known. The ability to carry this out with greater precision and more careful evaluation of economic value has been made possible by rapid, automated compositional analyses and interactive computer programs (Kerrigan, 1985). Computer programs such as the Milk Resource Allocation Decision Support System allow cheese manufacturers to use milk resources to maximize net returns, maximize cheese yields or minimize cost. The operator enters compositional and cost data for all available milk resources such as milk from various storage tanks, condensed skimmilk and nonfat dry milk, into the program. After one of the three objectives are chosen, the program will determine the best system for milk standardization, which of the various milk sources can be used most effectively and the economic value of using the selected milk resources and selected method of standardization. The program will also perform a sensitivity analysis to determine what the value of various resources, such as nonfat dry milk or cream, must be to allow their use as a component in milk standardization. This particular program is one example of the powerful interactive computer programs that can be utilized by the cheese industry.

Increased production from existing equipment is attained in numerous Cheddar cheese plants by concentrating milk. This is typically done by vacuum evaporators, especially with energy-efficient turbo-fan or mechanical vapor recompression evaporation (Zimmer, 1986). Concentration of milk before cheesemaking can also be done by reverse osmosis or ultrafiltration. Typically, the milk volume is reduced about 15% by these processes before cheese is manufactured in traditional vats.

Reverse osmosis has been evaluated in a commercial plant to concentrate milk 5, 10, 15 and 20% before manufacturing Cheddar cheese in traditional vats (Barbano, 1986). Comparisons were made with regular milk from the same storage tank and made into cheese in similar vats as used for the concentrated milks. The amount of lactic starter and cooking temperatures had to be reduced and curd size increased to obtain pH values and moisture levels close to the controls. The enhanced starter activity in concentrated milks has been observed in several studies. Increased concentrations of non-protein nitrogen in these milks and in the serum phase of curd may cause this enhancement.

Yield of cheese per kilogram of milk increased in proportion to the concentration factor. This increased throughput would improve manufacturing efficiencies. Recovery of casein in cheese was slightly higher with concentration of milk in one series of trials but not in the second. Recoveries were not significantly different from those in control cheeses in either series.

Recovery of fat in cheese made from concentrated milk was significantly higher than from control milk. The effect was traced to homogenization of milk by a pressure relief valve during the reverse osmosis process. Modifying the valve avoided homogenization and resulted in similar recoveries of fat in the control and experimental cheeses. Homogenization can also occur in thermal evaporators if excess shear occurs in valves and other constrictions.

Lipolysis was greater in cheese from concentrated milk that was partially homogenized. This may cause rancid flavors in cheese if heat treatment of milk does not inactive milk lipases or if lipases are produced in cheese by psychrotrophic bacteria. Small fat globules produced by shear during concentration are difficult to recover from whey by centrifugal separation creating loss of whey cream and adverse effects on whey powder.

The effects of milk concentration on cheese yield, when adjusted for concentration effect, have been variable between studies. Theoretically, increased yields should only result from the higher concentrations of lactose, or its metabolites, whey proteins and non-protein nitrogen in the cheese serum plus the additional water to compensate for these added solids. Other factors could influence yield. Higher protein (casein) concentrations in concentrated milk should improve clot formation especially in milk that is low in protein before concentration. Using thermal evaporation to standardize seasonal variations in protein content may be beneficial for uniform clotting and attendant cheese yield increases.

Thermalization of milk before ultrafiltration on farms is claimed to increase cheese yields (Zall, 1986). A 7.5% increase in yield over control cheese was observed when skimmilk was thermalized. Similar comparisons have not been made with Cheddar cheese.

Lower than expected yields can occur from milk concentrates if the concentration factor is too high (Bush et al., 1983, Jensen et al., 1984). The high ratio of curd to whey and slower whey expulsion after cutting do not provide a lubricant layer around the curd grains. Stirring causes abrasion and loss of yield.

The degree of milk concentration is limited also by residual lactose levels in cheese. Quality of cheese may be affected by excess lactose when milk is concentrated over 15%. Defects such as browning during heating, undesirable fermentations and calcium lactate crystals on cheese surface have been observed in cheese made with concentrated milk (Barbano, 1986; Bley et al., 1985; Johnson, 1986).

4. PROCESS CONTROLS

4.1 Curd Firmness Measurement.

Automated, mechanized equipment protects milk and cheese curd from contamination but makes these products less accessible for observation and testing. Consequently, remote, automated sensing, monitoring and controlling of process equipment becomes essential.

Firmness of the milk gel at cutting traditionally was assessed subjectively by the cheesemaker. Alternatively, cutting was done on a time schedule with some adjustment of milk-clotting enzyme levels. A variety of instruments have been developed to objectively measure curd firmness and more accurately indicate the proper point to cut the milk gel. Most of the instruments create an oscillatory motion that is transmitted through the gel to a receiver or the force required to create the motion is measured (Kawalchyk and Olson, 1978; Richardson et al., 1985). The output of the oscillations is directly proportional to curd firmness.

The curd firmness sensing devices can be used to activate curd cutting or to signal the operator to evaluate the curd firmness versus the time that is prescribed for cutting. It should be feasible to use a computer program that would arrive at the best solution for the compromise between optimum curd firmness and time.

Information on curd firmness values can be a useful management tool. Data on firmness, rate of curd firming would give a quick review of process controls, precision of personnel in carrying out manufacturing functions, and enable management to correlate this variable with quality, composition and yield of cheese.

4.2 Monitoring pH

Enclosed, mechanized equipment also poses some problems in sampling and relating the acidity of the sample from the mechanized system with measurements from a traditional system. Computerized process control in which pH and temperature during Cheddar cheese manufacturing are being monitored is under development in Australia (Linklater and Hall, 1986). Samples of whey are siphoned from the curd-whey slurry in a vat, the pH is measured and the data fed into a computer for storage, analysis and display.

At any stage up to whey drainage, the cheesemaker can have an instantaneous display of the stages of manufacturing that have been reached in a series of vats, current pH, and temperatures and the time intervals elapsed from renneting. considerably more information can be displayed on individual vats such as target time intervals, target pH values and temperatures and pumping times.

Computer programs to evaluate and display the large amount of data should be extremely valuable for production management. Displays of the data over the time course of manufacturing individual vats should easily allow managers to pinpoint those vats that followed prescribed procedures and those that deviated.

A computer program was developed that analyzed the data over a designated time period and listed those lots of cheese in which manufacture procedures or other data fell outside of specified limits. This allows management to quickly determine whether these deviations affected cheese quality and allowed management the option to use these lots of cheese for purposes other than extended maturation. Analysis of data also provided frequency distributions of pH values, process time intervals and cooking temperatures. Improved sensing devices and imaginative computer programs should make remote sensing a more powerful production management tool in the future.

5. Control of Syneresis of Curd

Automation and control of the cheesemaking process has advanced substantially with processes to measure curd firmness, pH and temperature. These techniques will aid in many stages of cheese manufacture including whey expulsion or syneresis of cheese curd. Controlling the process of removing serum from curd is critical in developing any mechanized system and in accelerating cheesemaking procedures.

Temperature and pH effects on syneresis well known (Walstra et al., 1985). Over the ranges of temperatures and pH values attained during manufacturing of Cheddar, pH gradients had a slightly greater effect on syneresis. However, pressure applied to curd had an even greater effect than pH on whey expulsion. The magnitude of pressure is critical since slight increases in pressure caused considerable whey expulsion but maximum expulsion is reached at low pressures with further increases having little effect. This may explain the substantial effects of stirring curd in whey on syneresis since stirring imposes slight pressures on the curd particles. Similar to pressure effects, moisture expulsion is more rapid during initial stages of stirring before reaching an asymptotic level of expulsion.

The above information and recent studies on permeability of curd (Van Dyk and Walstra, 1986) indicate that numerous factors must be considered in developing mechanized systems. This would be especially pertinent if deviations are to be made from the traditional temperature, acid production and curd-handling regimes. A good understanding of the basic physical properties of cheese curd will make process modifications more systematic and consistent.

6. PRESSING CURD IN FINAL FORM

Cheddar cheese curd is usually pressed into 18-Kg blocks, 220-Kg cylinders (barrels) or 290-Kg blocks (Everson, 1984; Honer, 1985). The block-forming towers are widely used throughout the world to replace substantial manual handling of the 18-Kg cheeses. The 290-Kg cylinder and 290-Kg block are almost exclusively used in the U.S. Substantial mechanization has been applied to filling, pressing, handling and packaging of the 290-Kg blocks.

Moisture variation within the 220-Kg cylinder and 290 Kg block is a major problem to be resolved. The moisture content may be 1.5 to 4.5% lower in the center of these cheeses as compared to the outside (Everson, 1984; Blattner, et al., 1985). The moisture gradient is established during or within 24h after pressing (Reinbold, and Ernstrom, 1986). Subsequent equilibration is very slow because of the rate of moisture diffusion through cheese curd.

Moisture distribution changes radically during pressing and the first 24h of maturation because of the effects of pressing and temperature differentials within the block (Reinbold, and Ernstrom, 1986). After pressing, the moisture content of the center is about 3% higher than the outside. The applied pressure is thought to seal the outside region of the cheese block and retain water within the cheese (Walstra et al., 1985). Within 24h; the distribution had reversed with the moisture content of the center being about 5% lower than the outside. It is proposed that the higher temperature in the center of the block increased the vapor pressure of the water in that region. The moisture vapors migrated through channels between the incompletely fused curd and condensed on the cooler curd grains at the edge of the block.

Although the problem of moisture distribution has not been solved, an economic solution was developed for cylinders of cheese that are typically used to make process cheese. The moisture distribution in a large number of cylinders was analyzed with a computer to develop moisture profiles. Computer simulation experiments were made to obtain a cheese trier sample that would most closely match the true moisture content of the barrel (Blattner et al., 1985). This has permitted better estimation of the solids content of the cheese cylinders which is the basis of payment.

7. PREDICTIONS FOR FUTURE MECHANIZATION

Further development of equipment to handle curd for traditional manufacturing procedures will be relatively modest. Exceptions will focus on flexible equipment to handle specialty cheeses. Most efforts will be directed towards equipment that can process curd from ultrafiltered milk. Considerable research and development will be done on procedures to obtain uniform moisture contents in larger blocks of cheese. Techniques for temperature control, filling the forms, pressing and vacuum treatment will be evaluated. Developments in packaging and maturing the large blocks will also be actively pursued. All of the developments will have to rely on a combination of biological, chemical and physical research procedures and innovative engineering.

8. REFERENCES

Barbano, D.M. 1986. "Reverse osmosis prior to cheese making". Proc. IDF Seminar on New Dairy Products via New Technology. Internat. Dairy Fed., Brussels, Belgium. p.31.

Blattner, T.M., N.F. Olson and D.W. Wichern. "Sampling barrel cheese for moisture analysis: Comparison of methods". J. Assoc. Off. Anal. Chem. 68:718.

Bley, M.E., M.E. Johnson and N.F. Olson. 1985. "Factors affecting nonenzymatic browning of process cheese". J. Dairy Sci. 68:555.

Elliot, R. 1984. "Cheese plant gains flexibility". Dairy Field. (8):58.

Everson, T.C. 1984. "Concerns on problems of processing and manufacturing in super plants". J. Dairy Sci. 67:2095.

Hansen, R. 1983. "Flexible plant for the production of Cheddar and English Territorials in the Express Creamery in Ruyton XI Towns". North European Dairy J. 49:233.

Honer, C. 1985. "The Europeanization of U.S. Cheesemaking:. Dairy Record 86(3):84.

Jenson, L.A., C.S. Bush, N.F. Olson and C.H. Amundson. 1984. "Use of membrane processing of milk in cheese manufacturing". Rept. Res. Activities, W.V. Price Cheese Res. Inst., University of Wisconsin, Madison, Wisconsin, U.S.A.

Johnson, M.E. 1986. Personal communication.

Kerrigan, G.L. 1985. "A linear programming approach to the milk resource allocation problem faced by cheese manufacturers". M.S. Thesis, University of Wisconsin, Madison, Wisconsin, U.S.A.

Kavalchyk, A.W. and Olson, N.F. 1978. "Firmness of enzymatically-formed milk gels measured by resistance to oscillatory deformation". J. Dairy Sci. 61:1375.

Lawrence, R.C., H.A. Heap and J. Gilles. 1984. "A controlled approach to cheese technology". J. Dairy Sci. 67:1632.

Linklater, P.M. and R.J. Hall. 1986. "Computerized process control for Cheddar cheese manufacture". Presented at 7th bienn. Cheese Ind. Conf., Utah State Univ., Logan, Utah, U.S.A.

Reinbold, R.S. and C.A. Ernstrom. 1986. "Temperature, pH and moisture profiles in 290 Kg stirred curd Cheddar cheese blocks". J. Dairy Sci. 69 (Suppl. 1):76.

Richardson, G.H., L.M. Okigbo and J.D. Thorpe. 1985. "Instrument for measuring milk coagulation in vats". J. Dairy Sci. 68:32.

van Dyk, H.J.M. and P. Walstra. 1986. Syneresis of curd. "One-dimensional syneresis of rennet curd in constant conditions". Neth. Milk Dairy J. 40:3.

Zall, R.R. 1986. "On-farm ultrafiltration". Proc. IDF Seminar on New Dairy Products via New Technology. Internat. Dairy Fed., Brussels, Belguim. p.9

Zimmer, A.G. 1986. "Mechanical vapor recompression evaporators". Proc. IDF Seminar on New Dairy Products via new Technology. Internat. Dairy Fed., Brussels, Belguim. p. 163.

AUTOMATION FROM CURD TO CHEESE

       R. Ahlström
       Alfa - Laval
       Lund
       Sweden

Principles in automation of cheese production

When choosing to automate a large scale cheese factory, a great attention is paid at selecting the most reliable system. Certain functions in cheese making are still to be handled by the cheesemaker and the automation system is only a "tool" for the cheesemaker. The automation package consist of an operation desk, a control and transmitter system, and all have to be at an equal level of reliability and easy to operate. If we start in the plant most transmitters used are flow-level-pressure-vacuum- temperature and which continuously send information to the operator via the control system. Further a number of remote controlled valves, distributing the product in a closed pipe system, are equipped with microswitches and feed their signals to the control system. Any production status can now be controlled so that valves as well as pumps operate in accordance to the logic in the control system. The most common systems to handle the logic and the entire cheese making instructions are computers or advanced PLC-systems. The computer system will inform any deviation or malfunction to the operator via informative systems as a typewriter or other types of displays.

Further the computer logic acts immediately to the reported fault and takes necessary action steps in order to prevent loss of valuable product. This, if I may say, emergency type logic, is often a large part of the computer system and it is also the one which differ widely when comparing systems from various suppliers. The safety aspect is essential in a control system and utilizing the transmitter and control system in a sensible way could save a lot of money.

The quality of the final product is influenced very much of the automation system. A poorly programmed control system will give a poor quality cheese, and the same effect if the transmitter system gives false information. A certain number of important checkpoints give the operator possibilities to make the final judgement to proceed in the manufacturing or to prolong or shorten certain steps. The size of the computer memory should be of a size handling various types of cheese to be produced and further the seasonal variations of the various parameters.

I have chosen to illustrate a cheddar cheese plant, which is fairly easy to follow and to understand what the automation package can do. We will soon follow the different steps from curd to cheese but first we have to enter into the control room and to see how the main operator works.

The control desk consists of a process display and a keyboard. The microprocessored controlled keyboard comprises colour coded keys in four groups, each with a special connection to the process. New commands are activated by entering keys from left to right. The depression of a key is immediately confirmed on the alphanumerical display or on the process display. The keyboard is used to activate new functions, for messages to the operator and for special commands.

The process display shows functions in operation. It is designed as a symbolic flowchart or, as here, as a coordinated system. Product routes for example are indicated by rows showing sources and columns showing destinations. A steady light in an intersection indicates a function in operation, a flashing light can be an alarm.

At a display unit the following information showing the actual cheese making process, could be given to the main operator.
- Vat number
- Total volume
- Temperature from filling to emptying
- Filling volume/Time Cheese milk
- Filling volume/Time starter
- Temperature step 1. Time when achieved
- Rennet adding. Volume/Time
- Agitation. Different steps/speed
- Temperature step 2. Time when achieved
- Cutting when and how long
- Temperature step 3. Time when achieved
- Whey drainage. Volume, number of steps
- Curd emptying time start finish
- CIP. Eventual flow - temperature - time.

The operator has to give the computer the following information:
- Rennet, if manually added when it has been done and at what volume.

For the continuous cheese making program, type Alf-O-Matic information about the speed of the various belts, depths of the curd, and which agitators are activated, are of importance.

In the plant a number of operators are working.

- Reception and storage area
- Pasteurization
- Cheese vat area
- Continuous cheese making system
- Blockforming area
- Cold room and storage

Strategically placed control panels ensure good communication between the various suboperators and the main operator.

The growing power of microprocessors now makes it a practical proposition to utilize them for improving productivity, not only on the shop floor but also at management level. They can be made to study and analyze the data they generate, and present it in a form on which rational management decisions can be based.

Data logging, i.e. retrieval of data from the process, is a preliminary step in all management tasks.

Production planning. Logged data can be processed to provide records of material consumption, stock inventories, ets. as a guide to prediction of future requirements. Information can also be obtained on how much of the plant's available capacity is being utilized.

Maintenance planning can be made much more efficient if the maintenance manager has access to records showing haw many hours each machine has run and how many times each valve has operated since last serviced.

Optimization of operations. Computer-generated records of power consumption, for example - especially if broken down by consumption sources - can be an invaluable aid to identifying soft spots in operational economy.

Quality assurance. A bad run can easily be traced to its source with the help of information from the ALERT system.

Total plant supervision. All the information mentioned above can be logged, collated and printed out as periodical summary reports for the guidance of top management.

Sensing devices of transmitters

Flowmeters

Flowswitches

Fixed level transmitters

Level measuring system

Pressure switches

Pressure gauges

Temperature transmitters

Before we enter into the part "from curd to cheese", i.e. from the cheese tanks to the storage room, I should stress the importance of having a gentle treatment of the product from the reception and through the whole system. Secondly to standardize the fat, preferable automatically in order to achieve a high accuracy.

Nowadays many customers start talking about standardizing the protein as well, which I believe will give a good return. Also standardization with lactose will be possible in the future. Now we have a good product as a base in order to produce a cheese quality of high standard.

Back to the cheese tank. This picture shows the necessary transmitters such as high level which prevents overfilling, low level which prevents filling if last rinse water still remains after CIP, and further informs the computer when it is time to change from one tank to the other after finished emptying.

Starter addition is done automatically by means of injecting the starter directly into the milk line filling the vat. The amount could be measured by flowmeaters or by time-controlled positive pumps. Poor accuracy has forced suppliers to look into other systems in order to get as accurate as possible volume of starter addition. The possibility to use a load cell system could give somewhat better accuracy.

The heating of the cheese milk is taking place, by hot water or steam in the jacket. This is controlled by the computer which operates the heating regulating valve. Temperature switches inform the operator when final cooling temperature is achieved.

The dosing of rennet is preferably done through the distribution nozzles, and the closed rennet dosing system is controlled by accurate positive pumps.

A signal from a recently designed transmitter could inform the operator when the coagel is ready for cutting. PH and moisture control are very interesting areas for measuring acidity and moisture level during the chees making process in the vat. Today the existing manual methods are not acceptable due to the delay in getting the result.

When the cheese making process in the enclosed tank is ready, the computer system starts emptying the vat into the continuous cheese making system, and a positive pump gently brings the curd from the tank to the whey screen and further to the first belt. A level sensor situated at this first belt controls the speed of the pump.

The speed of the belt is preset in accordance with the cheese making programme. Normally a manual test of the water content of the curd is taken at the whey screen, but I understand that it may be possible to also automate this function and to let the computer system decide how many of the agitators at the belts should be activated.

The speed of the second belt, matting and fusing belt as well as the cheddaring belt are also controlled by the computer programme.

A manual test of the water content of the product at the chip mill could be automated and this signal could be treated by the computer and control the speed of the mellowing and salting unit. Sensors at the salt distribution system ensure that salt is fed to the distribution system, and further, a fork system measures the curd depth on the mellowing conveyor and controls the amount of salt delivered to the curd.

The discharge to the blockforming system is done by vacuum. The vacuum level is controlled by a vacuum control system. The curd is now forming a coloumn in the blockformer, continuously drained throughout the emptying phase. Level transmitters regulate the input of curd from the continuous cheddaring machine.

In the blockformer base unit the function of the portioning coloumn, controlling the weight and injecting the block into the plastic bag is taking place. A number of transmitters and safety devices are used to secure each movement at the right time.

Immediately after the discharge from the blockformer, a semiautomated bag system ensures that the ready block cheese is well protected.

Micro-switches at the conveyors automatically stop the conveyor when a block is forced from the blockformer to the conveyor, by a piston.

An automated weighing scale prints the exact weight and could as well print numbers to indicate which cheese vat and which blockformer the ready cheese came from. The operator at the nearby vacuum sealing machine could also manually adjust the blockformer weight control if any mayor deviation of the weight, 40 Lbs should occur.

After the vacuum treatment, the block of cheese enters the
wrap-around-cartoning system or the shoe-box-system, further to the
rapid cold store and then to the storage room. In both areas
temperatures and humidity will be regulated. The velocity of the air
is also important.

Now overlooking the entire process there are often thousands of signals
from the process, which are treated by the computer system. The
transmitters used today are safe and we have to avoid transmitters of
low quality or of a <u>too</u> sensitive character. PH measurements
especially for cheese making have always been of great interest, but it
is difficult to rely on their signal.

<u>Principle Computer System</u>

The operator is always in over-all charge of the process at his
terminal. His commands are received by the control system, which
executes them in accordance with the logic conditions stored in the
program memory.

Operators or maintenance staff do not have to undergo computer training
to be able to run and supervise the process. One reason is that the
program language is basic version of their everyday language.
Furthermore the operator terminal is customised to the application. In
the program mermory has been loaded an application program for the
process concerned. Pumps and valves of the plant are actuated at he
precised moment in time demanded by the process cycle. Simultaneously
and continuously the operator and if required, the management, receives
information on process and equipment status and progress via the main
operator terminal and any other terminals provided. The application
program also incorporates a number of safety interlocks to prevent
erroneous operation, to supervise the process parameters, and safeguard
the quality of the product

Even the best computer system will fail one day and it is therefore
important to have a system which is self-diagnostic, which means that
the downtime for the system will be as short as possible. Skilled
servicemen are a must to handle the system in the right way. But also
a munual back-up system placed close to the individual major components
is important in order that the operator can take control in case of
emergency.

<u>PROBLEMS ARISE MAN/MACHINE</u>

In cheese-making one parameter could basicly affect may other areas and
it is very important to look into the consequences of a program change.
I believe most of the problems occur due to wrong or faulty information
through the transmitter system or bad commumnication between the
operator and the computerized information system.

Training of operators is essential for trouble free operation. It is very important that todays system is designed in a way so it motivates the operators to keep fully operational control of the plant, and that the system gives complete and clear information to the people involved in running the plant from day to day business.

The first step to get the operator committed to the system is well planned theoretical and practical training before and after the system is installed.

FUTURE IMPROVEMENTS OF TRANSMITTER/AUTOMATION SYSTEM

Transmitters

If possible, the transmitters have to be even more reliable, sometimes doubling.

- The recently developed transmitter measuring the firmness in the coagel has to be further tested for different types of cheese.

Desirable Types of Future Transmitters

- Continuous measuring of water in finished block of cheese, maybe measured just before the block enters the plastic packing.

- Continuous measuring of lactic acid bacteria and the moisture level trough the system.

- Continuous measuring of salt in the product in order to ensure an even distribution as possible.

ADVANTAGES OF AUTOMATING A SYSTEM

Automation done in a sensible way is, of course, profitable. Labour savings in the plant just described compared to a manual system show a saving of about 7-10 men. For a plant as described with an output of 7500 pounds cheese an hour, following people is needed.

one person at the reception and storage.
 "   "   " the pastuerizing section.
 "   "   " the cheese making tanks.
 "   "   " the continuous cheese making system.
 "   "   " the blockforming system.
 "   "   " the control panel.

Total minimum 6 operative persons. Plus service people.

A further advantage of great importance is, of course, that the enclosed system has a high yield return and low product losses, and with a big output of high quality cheese day after day.

A very comprehensive management feed-back system is obtainable from todays computers, limiting human mistakes, and gives a good return in production reports, energy, service intervals etc.

A disadvantage with an automated system in earlier days was in flexibility, but that has changed as the described plant can now undertake to produce products such as gouda type for example Egmont, Mozzarella etc., and it is now possible to get the cheese blocks in different sizes. A collator after the blockformer could handle 640 Lbs blocks by stacking 16x40 Lbs blocks.

The restriction on whether to automate a system or not today is mainly the size of operation and the economical aspects.

DISCUSSION

J.F. Boudier (France) observed that, with brine-salted cheese, the pH of some green cheese rose during the first 48 hours after immersion whilst others did not. As this seems at odds with the speakers comments for hard cheese he asked for comment.
R.C. Lawrence (New Zealand) replied that brine salted cheese usually had little lactose for fermentation by the time the brining took place, hence a decrease in pH is not to be expected. Why sometimes an increase occurs, is not well understood. Possibly, buffer capacity may change due to release of phosphate.

G.W. Jameson (Australia) remarked that he could not agree with the speaker's negative attitude to the use of ultrafiltration for improving the manufacture of Cheddar cheese. He and his co-workers at CSIRO had carried out many trials at a substantial scale and had found no effect on cheese grading (consistency and flavours). He suggested that the process they had developed did release whey and that the residual whey protein in the cheese had no deleterious effect.
R.C. Lawrence (New Zealand) replied that commercial confidentially agreements made it hard for an outsider to evaluate many new processes. In experiments in New Zealand, 3-4 % whey protein could be incorporated in a cheese within the range of Cheddar cheese normally produced but with a softer, smoother body. He felt that whey protein was a foreign protein in cheese - like soya protein - and that much remained to be learned about the technology of such cheeses.

H. van Dijk (the Netherlands) asked Maubois to comment on the differences in quality of Camembert cheese produced by ultrafiltration.
Maubois (France) stated that if the mineral content of a cheese produced by ultrafiltration is controlled, the cheese is organoleptically identical to that made by a conventional process. He was sure that whey protein incorporation in soft cheese did not affect ripening. He also pointed out that the most expedient way of making soft cheese was to carry out the ultrafiltration process after coagulation.

H. van Dijk (the Netherlands) asked if, when ultrafiltration is used for cheesemaking, there is a problem with damage of milk fat globules leading to oiling-off.
Maubois (France) : There is no problem if the equipment is correctly designed. The equipment must have pumps, bends and valves specially designed to reduce turbulence. In addition the flow rate can be reduced by a factor of two (with previously coagulated milk) without a significant reduction in flux.

R. Peters (Canada) commented that, in France, Camembert cheese could be purchased at a price which could vary by a factor of two. Was cheese prepared by ultrafiltration realising a premium price?

J.L. Maubois (France) replied that "real" Camembert could not be made by ultrafiltration. As he had stated in his lecture, traditional cheese was less uniform and because of the inclusion of imperfections (holes) less dense. However, the consumer in France will only pay a limited premium for traditional cheese. All large manufacturers of Camembert cheese use ultrafiltration to some degree. Improvements in performance result from such practices.

R. Peters (Canada) stated that there was a reluctance to standardize cheese because milk-fat realised a higher price in cheese then in butter: would prof. Olson comment?

N.F. Olson (USA) replied that standardization is not widely used in the manufacture of Cheddar cheese. Solids-not-fat or protein is added to achieve uniformity in fat percentage, moisture content and milk-non-fat-solids. The Italian cheese industry practise standardization in USA for economic reasons.

P. Walstra (the Netherlands) suggested that the development of new sensors was a critical fact of progress towards further automation. He invited the speakers to comment.

R. Ahlström (Sweden) replied that cost consideration had slowed the development of new systems. Key areas for new equipment were for measurement of salt level, moisture level and the distribution of these components in cheese.

N.F. Olson (USA) asked if the moisture content of curd could be measured during the cheddring process?

R. Allström (Sweden) replied that such measurements had been successfully carried out in trials and that their extension into commercial practice would come with time.

N.F. Olson (USA) commented that the heterogeneity of cheese curd could cause problems for sensors.

R. Ahlström (Sweden) agreed.

R. Peters (Canada) noted that the pH of cheese curd before salting is a critical factor in the control of cheese quality: how was this measured automatically?

R. Ahlström (Sweden) replied that this measurement could only be done at great expense with automatic equipment.

R. Peters (Canada) then inquired how pH of cheese at salting was evaluated.
R. Ahlström (Sweden) stated that the pH of the whey could be monitored or a conventional measurement made on curd. Automatic monitoring was not yet very reliable and was currently very expensive.

P. Walstra (the Netherlands) asked Daly (Eire) if starter could be injected into milk flowing in a milk-line without causing phage problems.
C. Daly (Eire) replied in the affirmative.
G. van den Berg (the Netherlands) commented that there were no problems with bacteriophage contamination if starter is injected close to the cheese vat.
P.S. Robertson (New Zealand) commented that problems might arise if the milk is ripened after starter addition rather than set immediately.

Seminar I: Progress in Cheese Technology

Session 3: Properties of Cheese

Chairman: Dr. J. E. Auclair (France)
Secretary: G. Steiger (Switzerland)

CONSISTENCY OF CHEESE

Pieter Walstra, Hannemieke Luyten and Ton van Vliet

Department of Food Science
Agricultural University
Wageningen
the Netherlands

ABSTRACT. Cheese is a visco-elastic material with a very low yield stress and a very high apparent viscosity. Although rheological parameters vary considerably in magnitude, qualitatively the behaviour is the same, at least for small deformations and long times. Greater differences are experienced if deformation is rapid and large, as is the case during eating, cutting, grating or spreading of cheese. The application of fracture mechanics opens new and better ways for understanding these phenomena. The effect of several compositional and external variables, including proteolysis, is discussed.

1. INTRODUCTION

The consistency of cheese - defined as its resistance to lasting deformation - affects some important properties:
- eating quality, i.e. the consistency as perceived in the mouth;
- usage properties, e.g. ease of cutting, spreading or grating and melting characteristics;
- handling properties, including shape retention, also and of particular importance for curd;
- hole formation, i.e. whether eyes or slits are formed.

The consistency as perceived by the consumer is an important quality mark and it markedly changes with maturation in most types of cheese. Consequently, endeavours to accelerate cheese ripening have led to a renewed interest in the development of cheese consistency.

Curd and cheese are visco-elastic materials (1); they have a very low yield stress, i.e. almost any stress applied leads to a lasting deformation, how ever slow. Cheeses vary in having different elastic moduli and different (apparent) viscosities, and the application of rheological theory seems straightforward. There are, however, some hard complications:

a. Already for quite weak stresses, the deformation is not any more proportional to stress (this is designated non-linear behaviour), which makes interpretation of results difficult. High stresses may cause widely different phenomena: flow, yielding or fracture. This greatly depends on the type of cheese.

b. Results depend on the geometry and on the time scale of deformation. Consequently, one should do rheological measurements under similar conditions as occur during the use of the cheese. These are widely different if one compares, e.g. chewing and stand-up properties. For instance, Shama and Sherman (2) found that firmness of two cheese types as perceived by panelists during eating, agreed more or less with the results of compression tests only if compression rate was of the order of 1 cm/s and relative deformation about 0.5; otherwise the softer cheese was rated as the firmer one.

c. It is very difficult to perform tests that can be interpreted in unequivocal rheological terms. The intricacy of the underlying theory may be a main reason for the lack of progress made in cheese rheology.
d. Cheese contains numerous inhomogeneities (1) and this hinders interpretation of rheological results in terms of differences in cheese structure.

Despite these problems, some progress has been made, and some results will be briefly reviewed, without going into theoretical detail, and without trying to cover the whole field.

2. DYNAMIC MEASUREMENTS

These measurements permit separate determination of an elastic or storage modulus and a viscous or loss modulus, at least for small deformations (linear behaviour) and for a wide range of deformation rates. Here, a modulus means the ratio of stress to relative deformation, as long as this ratio is constant (independent of stress). Several studies have been done on curd (e.g. 3 - 6) and some of the results have been recently reviewed (7). Apart from the relevance of this work for curd making processes, two interesting conclusions can be drawn.

a. The moduli are widely variable, and the differences mainly stem from differences in the unevenness of the network; i.e. not the nature but the number of bonds varies with conditions. In other words, it is the geometry rather than the chemistry that varies. One manifestation of this is that the moduli are proportional to protein concentration to the power 2.6, rather than a power of about 1.2, to be expected if the network were even.

b. The ratio between viscous and elastic modulus, called the loss tangent, is, however, independent of the evenness of the network. It depends on deformation rate, but it is otherwise remarkebly constant, except when varying pH. Figure 1 gives an example. The striking change from a more liquid

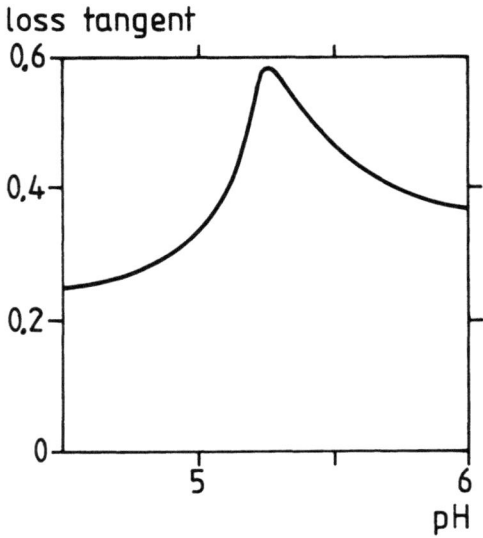

Figure 1. Loss tangent (= loss modulus/storage modulus) of renneted milk at various pH; 20 °C, frequency 0.01 rad s^{-1}, aging time 10^4 s. After Roefs (6).

material (high loss tangent) to a more solid one when going from pH 5.2 to 4.6, may well be related to the clear change in cheese consistency over the same pH range, but we have not yet been able to find similar results in cheese. In semihard cheese, the loss tangent is about 0.4 and it hardly alters with maturation. But it may be that there are two changes at work, the one increasing and the other lowering the loss tangent. We probably must study these phenomena at a slower deformation rate.

3. FRACTURE MECHANICS

The theory of fracture mechanics of biological materials has shown interesting developments in recent years (8 - 10) and may be fruitfully applied to food materials, including several types of cheese. We are extensively studying Dutch-type cheeses and similar work on Cheddar cheese is being done by Green et al. (11). Interesting considerations have also been given by Peleg and coauthors (e.g. 12). It would lead too far to give here an outline of the theory. Suffice it to say that one usually studies the relation between stress (σ) and strain or relative deformation (ϵ) when applying an increasing stress or an increasing strain; the modulus $E = \sigma/\epsilon$, as long as ϵ is small. It is more interesting to find σ_f and ϵ_f, i.e. stress and strain at which fracture occurs. An ideally elastic (and brittle) material shows that σ/ϵ is constant until the material breaks; hence, $\sigma_f = E \epsilon_f$.

As is shown in Figure 2, cheese shows a different behaviour. The so called compression curve (measured force against applied compression) has a strange shape (Fig. 2A), but only for very small compression is σ proportional to force and ϵ to compression. Recalculating to true σ and ϵ results in a curve that can easier be understood. It is seen that σ/ϵ decreases with increasing ϵ; consequently $\sigma_f/E\epsilon_f < 1$ (assuming that fracture occurs at the maximum in the curve). We found this ratio to vary from 0.2 - 0.7 for various kinds of cheese. Furthermore, it is seen (Fig. 2B) that the relation between stress and strain and the fracture behaviour may greatly depend on type of deformation: *tension* (i.e. uniaxial pulling at a piece of cheese that is usually notched to predetermine the breaking site), *bending* (by pressing on a cylindrical test piece supported on two beams) or *compression* (uniaxial, between parallel plates). The true stresses given in Figure 2B are average stresses in the test piece. Now, the cheese will fracture where the stress is highest, and if one calculates the maximum stress (assuming the sample to be homogeneous), it turns out that the actual breaking stress i.e. the stress at the site where fracture starts) does not significantly depend on the test mode, at least for reasonably firm samples. This is a gratifying result. Note, however, that ϵ_f markedly depends on deformation mode.

There are more complications. Fracture may, and often does, occur before the maximum in the curve. It can even be observed that a test piece contains major cracks when cutting it in half before it shows any outward signs of breaking. Moreover, behaviour depends on the rate of deformation $\dot{\epsilon}$ (= $d\epsilon/dt$). We generally found E to be proportional to $\dot{\epsilon}^{0.15}$; the explanation is presumably that part of the stress applied to the sample relaxes during compression, and that this part is larger for longer times, hence slower deformation. In accordance with the dependence of E on $\dot{\epsilon}$, the fracture stress likewise increases with deformation rate. It has also been found that somewhat deforming a piece of cheese a few times before doing a full compression test, lowers the fracture stress (13).

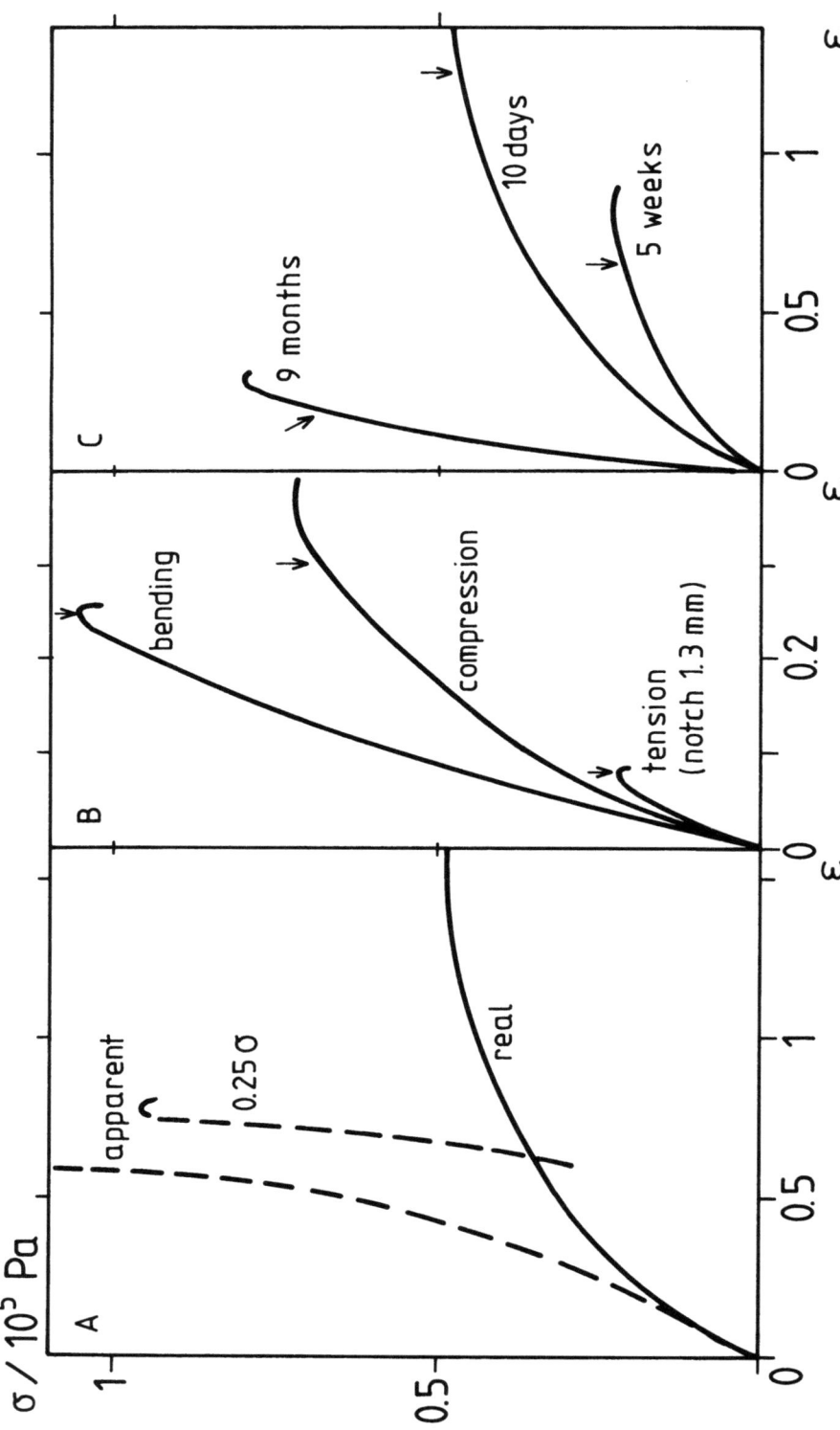

Figure 2. Fracture curves (true stress σ against true strain ε) of Gouda cheese at 20 °C. Arrows indicate beginning fracture. A. Compression, rate 5 cm. min^{-1}; broken line gives the uncorrected stress and strain; cheese 10 days old. B. Various mpdes of deformation, cheese 10 months old, $\dot{\varepsilon} = 0.0013$ s^{-1}. C. Compression; cheese of various age; $\dot{\varepsilon} = 0.028$ s^{-1}.

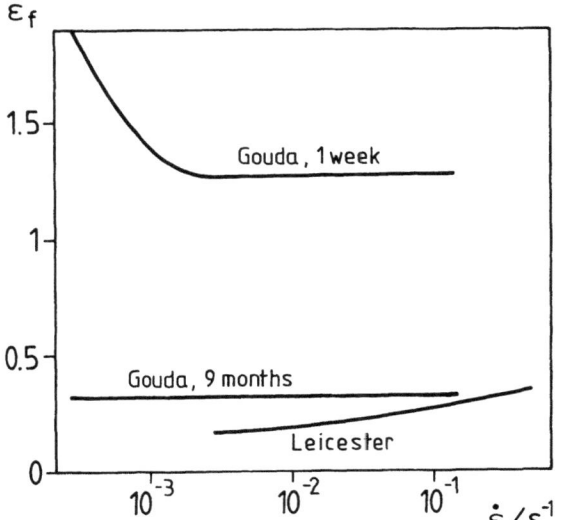

Figure 3. Deformation at fracture (ε_f) in compression as a function of strain rate ($\dot{\varepsilon}$) for three types of cheese. Results on Leicester cheese after (13).

Figure 3 shows that ε_f may also depend on $\dot{\varepsilon}$, especially if the cheese is soft and immature. In the latter case slow deformation does not cause fracture at all and an analysis of elongational flow while compressing a sample, as done by Casiraghi et al. (14), may yield interesting results.

Except in some cheeses with high ε_f, we never observed a significant dependence of ε_f on $\dot{\varepsilon}$. But in a crumbly cheese like Cheshire or Leicester, this is different: see Figure 3. We have as yet no explanation for this; it may be that the actual fracture occurs before the ε at maximum stress if $\dot{\varepsilon}$ is high, but less so if $\dot{\varepsilon}$ is low.

Often, firmness or hardness of cheese is defined as the fracture stress or some closely related parameter. Clearly, σ_f depends on (1) number and type (strength, relaxation time) of the bonds in the network and (2) on the presence of irregularities or even minor cracks in the sample. At the tip of a crack or the periphery of a weak spot a considerable stress concentration occurs. The simplest theory predicts (8, 10) that fracture occurs once a crack has attained a critical length, and this critical crack length is proportional to σ^{-2}. A similar relation will hold for the size of weak spots. But the occurrence of stress relaxation (the extent of which is not evenly distributed throughout the sample) during deformation makes explanation of fracture mechanisms difficult. For the time being it appears useful to determine ε_f besides σ_f and also note the modulus E at small deformation. It appears that a small ε_f corresponds with what is commonly denoted as a "short" texture. When comparing widely variable cheeses, it also is useful to determine the effect of $\dot{\varepsilon}$ on these parameters. Other useful parameters have - in our opinion - not yet been identified.

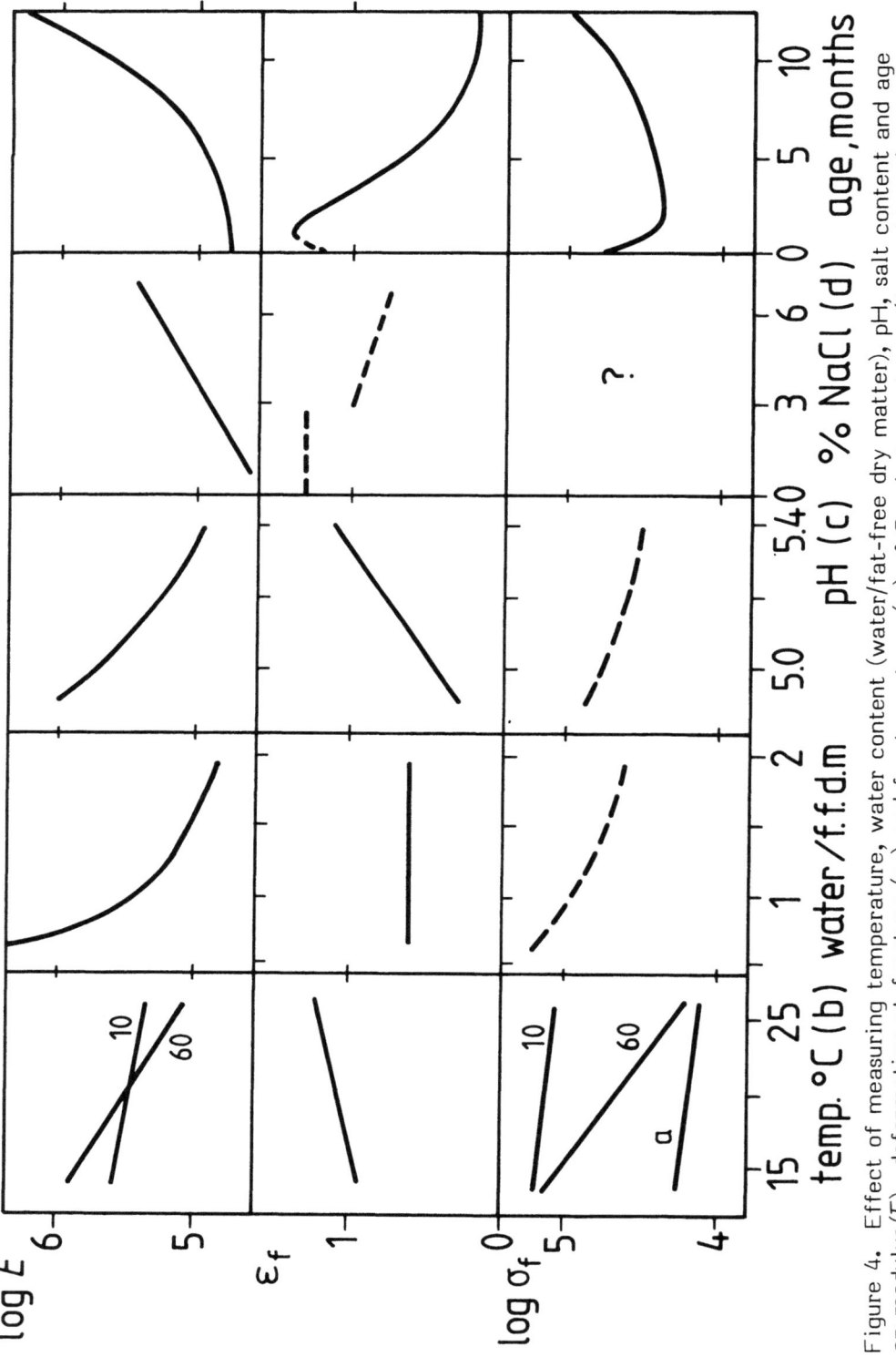

Figure 4. Effect of measuring temperature, water content (water/fat-free dry matter), pH, salt content and age on modulus (E), deformation at fracture (ε_f) and fracture stress (σ_f) of Gouda-type cheese (free of holes). Fat content in dry matter 48 %, unless denoted otherwise (10 and 60 %). Notes a: cheese from filled milk (soy oil in stead of butterfat); b: water/ffdm $\approx$ 1.3; c: 4 weeks old, water content $\approx$ 41 %; d: pH $\approx$ 5.2, water content 50 %

4. EFFECT OF CHEESE PROPERTIES

It is common knowledge among cheese technologists that composition and manufacturing procedure of the cheese greatly affect its consistency. But reliable data are hard to come by. The main reason is that varying only one parameter mostly is impossible or nearly so. One may try to determine compositional and rheological parameters of a large number of samples of widely varying properties and calculate a correlation matrix (15, 16). This poses two problems. The first is that the rheological parameters cannot be determined with great accuracy (a variation coefficient of 20 % between "identical" samples is quite common) and since there are a great number of variables, statistical significance often cannot be attained. Second, even if a significant correlation is found, it is uncertain whether it is causative. We therefore tried to deliberately vary one parameter, leaving the others constant. The problem is now to achieve this and we cannot claim to always have been completely successful. Experimental details will be published later and the results given below are to some extent tentative and only semi-quantitative. Figure 4 illustrates some trends found by us; some of these have been observed also by others.

Temperature of measurement (3, 16, 17). The general tendency is that E and σ_f decrease with increasing temperature, but the effect primarily depends on melting of the fat. In a low-fat cheese, and also if the milk fat has been replaced by an oil, the decrease is only some 40 % over 10 °C.

Fat content (15, 16, 18). Generally, a higher fat content gives a lower E and σ_f, but a higher fat content in the dry matter usually goes along with a higher water content in the fat-free cheese and this will be the main causative factor. To obtain the results of Figure 4, the latter ratio was kept constant. As is seen, the effect greatly depends on temperature. It may also depend on water content.

Water content (16, 19, 20, 21), or rather water in the fat-free dry matter, has a very strong effect on E : see Figure 4, remembering that the scale is logarithmic. The effect of, say, a 1 % change in water content is greater as water content is lower. Water content did nod clearly effect ε_f, but may affect the way of fracturing; consequently σ_f appears to vary less than does E.

Acidity (16, 20, 22, 23). It is difficult to determine the sole effect of pH, since other factors usually vary along with pH, notably the calcium phosphate content and the extent of proteolysis. It turns out that ε_f increases markedly with pH; in other words, a more acid cheese is shorter. Moreover a higher pH tends to favour fracture of the cheese in tension rather than in shear (see Section 5). Also firmness, i.e. the modulus and to a lesser extent σ_f, vary with pH. The effect of pH on consistency is quite different in curd (6); in other words, proteolysis affects the relationship. Table 1 summarizes some of the interesting results of Noomen (24) on soft cheese.

Calcium phosphate content. As yet, a clear effect has not been found. At low pH, very little influence is indeed to be expected, since colloidal calcium phosphate is fully dissolved; see also the work on curd (5, 6). At higher pH this may be somewhat different (25). It has been shown that in soft cheeses at not too low pH, more calcium phosphate gives a firmer consistency (23, 26).

Salt content. Several workers have observed that longer brining goes along with a firmer cheese, but, then, this causes a lower water content along with a higher salt content. Figure 4 shows that also at constant water content, more salt gives a distinctly higher E ; this will partly be due to the (volumetric) protein content in the fat-free cheese still being higher for a higher salt content. It appears as if ε_f suddenly decreases around 2.5 % salt (= 5 % salt in water). This may correlate with the chalky appearance of the body of a high-salt cheese as compared to the more yellowish, smooth and glossy looking low-salt one.

Proteolysis. The effect of proteolysis, keeping other factors constant, has hardly been studied. De Jong (19) has shown for soft cheese, and Creamer and Olson for Cheddar (20), that especially the splitting of α_{s1} casein causes a softer (more liquid-like) or shorter texture, respectively. See also Table 1.

Table 1. Consistency of soft cheese (water/fat-free dry matter 2.3) as a function of pH and proteolysis (depending on presence or absence of rennet; action of microbial enzymes negligible). After (23, 24).

pH	no rennet	rennet present
4.8	firm, fairly short	firm, short
5.4	rubber-like	liquid

5. MATURATION

Several authors have studied changes in cheese consistency with aging (e.g. 16, 19, 20, 26, 27). Some of our results are in Figures 2c and 4. Interpretation of the results is not easy, since several properties change with maturation. A greatly simplified scheme of relations is:

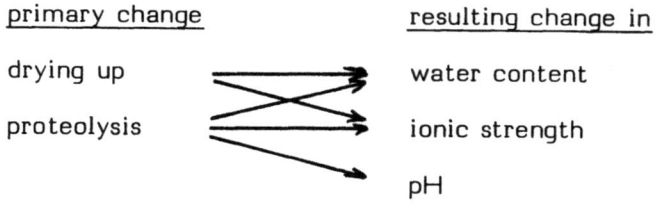

All these factors affect the consistency. Moreover, the rate of either primary change depends on conditions: loaf size and shape, curing conditions (notably temperature), amount of rennet included, number and type of bacteria (or bacterial enzymes) present, ionic strength, pH, etc. (see e.g. 28). Several of these parameters are correlated.

Consequently, we need not be surprised by the change in consistency with maturation being greatly dependent of the kind of cheese. See for instance Figure 3. We also found that in compression tests of some Dutch-type cheese, the way of fracture altered with aging. An immature cheese fractures in tension (the test piece bulges out and develop vertical cracks at the outside), while an older cheese fractures in shear (diagonal cracks develop inside the test piece). The most striking difference is between soft and (semi) hard cheeses. The former mostly show fairly rapid proteolysis, often a considerable increase in pH and the lowering of water content is in a range where it moderately affects consistency. Harder cheeses show less change in pH and a decrease in water content has a greater effect.

We have formed the impression (but this is admittedly only a tentative conclusion that we may have to modify after accumulation of more experimental data) that for (semi)hard cheese the considerable increase in E is almost exclusively due to a decrease in water content (and the concomitant increase in ionic strength). Proteolysis as such beyond its early stages seems to have little effect, except in as far as it causes a decrease in water content (higher E) and an increase in pH (lower E). Wodecki et al. (21) found slight differences in the effect of water content on E according to these differences being due to variation in the green cheese or to change during aging; but these results do not materially disagree with our tentative conclusion. We feel that neither Eberhard's results (16) are at variance with ours, although he gives a slightly different interpretation. Sensory evaluation showed (29) that panelists rated a piece of cheese from the outside of a loaf to be more mature than a sample from the inside of the same loaf; the former piece did certainly not show more proteolysis but had a lower water content. ε_f clearly decreases with maturation: see Figure 4. We also found for a wide range of samples of Gouda and Edam cheese, a fair negative correlation ($\tau = -0.86$, $N = 26$) between ε_f and age between 1 and 6 months, but especially for very young cheese, also pH considerably affects ε_f. The increase of pH with age may, to some extent diminish the effect of proteolysis per se on ε_f. The effect of aging on σ_f is less clear, and different authors have somewhat different results; the problem may be that the fracture mode alters with age and that this works out differently on σ_f, according to the test procedure.

For many soft cheeses, the most important aspect of consistency is the change from firm and short to soft and smooth, or - somewhat exaggerating - the liquifying of the cheese. This only occurs if the following conditions are all met (23):

- pH at least about 5.3;
- ratio of water to fat-free dry matter over about 2;
- at least about 60 % of the α_{s1} casein split;
- content of calcium phosphate not too high.

Naturally, the critical value of each variable depends somewhat on the magnitude of the others.

Accelerated ripening of cheese is topic of great practical interest. Characterization of the consistency of the cheese (is it as after normal ripening?) then becomes important. The above considerations lead to the conclusion that it is still difficult to do so reliably by means of instrumental tests, although we are decidedly making progress. It is hoped that the newly installed IDF working party on cheese rheology will provide at least some answers. But sensory evaluation should still play a central part.

REFERENCES

1. P. Walstra & T. van Vliet, *IDF Bulletin*, Doc. **153**(1982)22
2. F. Shama & P. Sherman, *J. Texture studies* **4**(1973)344
3. H.J.M. van Dijk, 'Syneresis of curd', Ph.D. thesis, Wageningen, 1982
4. L. Bohlin, P.O. Hegg, E.H. Ljusberg-Wahren, *J. Dairy Sci.* **67**(1984)729
5. S.P.F.M. Roefs & T. van Vliet. *Proc. 9th Int. Congr. Rheology* **4**(1984)249
6. S.P.F.M. Roefs, 'Structure of acid casein gels', Ph.D. thesis, Wageningen, 1986
7. P. Walstra & T. van Vliet, *Neth. Milk Dairy J.* **40**(1986)241
8. J.E. Gordon, 'Structures, or why things don't fall down' Penguin, 1978
9. J.F.V. Vincent, 'Structural biomaterials', MacMillan, 1982
10. A.G. Atkins & Y-W. Mai 'Elastic and plastic fracture', Horwood, 1985
11. M.L. Green, K.R. Langley, R.J. Marshall, B.E. Brooker, A. Willis & J.F.V. Vincent, *Food Microstructure* **5**(1986)169
12. N. Pollak & M. Peleg, *J. Food Sci.* **45**(1980)825
13. E. Dickinson & I.C. Goulding, *J. Texture Studies* **11**(1980)51
14. E.M. Casiraghi, E.B. Bagley & D.D. Christianson, *J. Text. Studies* **16**(1985)281
15. A.H. Chen, J.W. Larkin, C.J. Clark & W.E. Irwin, *J. Dairy Sci.* **62**(1979)901
16. P.P. Eberhard, 'Rheologische Eigenschaften ausgewählter Käsesorten', Ph.D. thesis, Zürich, 1985
17. J. Culioli & P. Sherman, *J. Texture Studies* **7**(1976)353
18. D.B. Emmons, M. Kalab, E. Larmond & R.J. Lowrie, *J. Texture Studies* **11**(1980)15
19. L. de Jong, *Neth. Milk Dairy J.* **30**(1976)242
20. L.K. Creamer & N.F. Olson, *J. Food Sci.* **47**(1982)631
21. E. Wodecki, J. Budny, K.W. Hoppe, A. Gryzowska & J. Turowski, *J. Food Eng.* **3**(19847)295
22. C.W. Raadsveld & H. Mulder, *Neth. Milk Dairy J.* **3**(1949)222
23. A. Noomen & J.C.M. Jacobs, unpublished (1984)
24. A. Noomen, *Neth. Milk Dairy J.* **37**(1983)229
25. P. Zoon & T. van Vliet, unpublished (1986)
26. H. Mulder, *Versl. Landbouwk. Onderz.* **51**C(1946)467
27. C.W. Raadsveld & H. Mulder, *Neth. Milk Dairy J.* **3**(1949)117
28. B.A. Law, in P.F. Fox, Ed. *Cheese: Chemistry, Physics and Microbiology*, Vol. 1, Ch. 10, Appl. Sci., in press.
29. H. Oortwijn, Unpublished report 84.40 of the Rijkskwaliteitsinstituut voor Land- en Tuinbouwprodukten (RIKILT), 1984

LES MECANISMES DE FORMATION DE LA FLAVEUR DANS LES FROMAGES

J. Adda
Institut National de la Recherche Agronomique
78350 Jouy-en-Josas
France

RESUME. La flaveur des fromages apparaît au cours de l'affinage à la suite de la dégradation des divers constituants du caillé suivant des mécanismes dont seules les grandes lignes sont connues. La fermentation lactique joue un rôle essentiel de par la chute de pH qu'elle provoque et qui a notamment pour effet de limiter l'action des divers systèmes enzymatiques susceptibles d'agir sur les lipides et les protéines. L'action des lipases sur les glycérides provoque la libération d'acides gras libres qui seront transformés en composés odorants (esters, lactones méthylcétones, alcools primaires...). Les protéines seront source de peptides de poids moléculaire varié et d'acides aminés. Ces derniers seront source de molécules odorantes. Les connaissances que nous avons de tous ces phénomènes apparaissent encore insuffisantes lorsqu'il s'agit de réguler l'apparition de la flaveur notamment en vue d'accélérer l'affinage.

1. INTRODUCTION

La flaveur des fromages apparaît au cours de l'affinage à la suite d'une série de réactions, pour la plupart enzymatiques, qui transforment les constituants du caillé, donnant ainsi naissance à de nombreuses molécules odorantes ou sapides, dont les proportions et parfois même la nature varieront suivant la technologie utilisée.[1,2,3]

2. LA FERMENTATION LACTIQUE

L'étape d'acidification par fermentation du lactose apparaît elle-même déterminante pour la qualité sensorielle des fromages en fin d'affinage[4]. La production d'acide lactique provoque une chute de pH, qui exerce une influence importante, de par les conditions physico-chimiques qu'elle crée au sein du substrat, régulant l'activité des microorganismes et des enzymes, qui interviendront ultérieurement dans la modification des constituants du caillé. L'acide lactique produit servira de substrat à certaines flores secondaires, comme les moisissures ou les bactéries propioniques, étant dans ce dernier cas métabolisé en acide propionique. La croissance des bactéries lactiques s'accompagne en outre d'un abaissement du potentiel d'oxydo-réduction, qui crée les conditions favorables à certaines réactions ultérieures[3].

La fermentation du citrate[5] conduit à la production de diacétyle qui pourra être ultérieurement réduit enzymatiquement en acétoïne puis en 2,3-butylène glycol que certains lactobacilles (*Lactobacillus plantarum* et *L. brevis*) pourront transformer en 2-butanone et 2-butanol[3].

3. LE ROLE DES LIPIDES

Le rôle de la matière grasse dans la génèse de la flaveur des fromages semble être double. La matière grasse joue tout d'abord un rôle essentiel pour la solubilisation et la rétention d'un certain nombre de molécules hydrophobes et il semble même que l'interface eau/matière grasse ait une influence importante sur la perception des arômes.

L'importance de la matière grasse découle ensuite de la nature des molécules formées par sa dégradation. Ces molécules importantes pour l'arôme mais sans doute aussi pour la saveur apparaissent par deux mécanismes distincts : la lipolyse et l'oxydation.

La lipolyse est le mécanisme enzymatique qui conduit à la libération d'acides gras libres. Plusieurs systèmes enzymatiques d'origine différente sont susceptibles d'intervenir au cours de l'affinage. En premier lieu la lipase naturelle du lait qui est activée par les traitements mécaniques et qui résiste à la pasteurisation attaque les triglycérides avec une spécificité intramoléculaire (relative à la position de l'acide gras) et une spécificité intermoléculaire (relative à la nature de l'acide gras). En second lieu interviennent les différentes lipases produites par les flores de contaminations au premier rang desquelles viennent les psychrotrophes et les différentes flores d'affinage.

Les lipases des bactéries psychrotrophes sont, comme la lipase naturelle du lait, très actives sur les triglycérides et conduisent à la formation de mono et diglycérides.

Les bactéries lactiques ont pour leur part une activité lipasique faible et assez spécifique étant surtout actives sur les mono et les diglycérides. Des activités lipasiques plus élevées se rencontrent dans les flores secondaires : *Penicillium caseicolum* produit une lipase exocellulaire alcaline, alors que *P. roqueforti* produit pour sa part une lipase alcaline et une lipase acide. *Geotrichum candidum* possède une lipase qui hydrolyse préférentiellement l'acide oléïque.

Sauf dans les cas assez rares où l'action des lipases, et notamment celle des psychrotrophes, conduit à une rancidité perceptible, la lipolyse n'est un phénomène important dans la génèse de la flaveur que du fait de la transformation des acides gras ainsi libérés en composés ayant un impact élevé (lactones, esters, méthylcétones, alcools secondaires).

La présence de *delta* lactones peut résulter de la cyclisation spontanée des *delta* hydroxy acides normalement présents en faible quantité dans la matière grasse du lait et libérés par lipolyse mais il semble que d'autres mécanismes puissent également intervenir.

La présence d'esters est à attribuer à des mécanismes enzymatiques comme cela a été démontré chez Pseudomonas fragi, reconnu comme responsable de l'apparition d'une note fruitée, parfois constatée dans le cheddar, ou chez les levures par exemple.

Les acides gras libres servent également de substrat aux moisissures pour la formation de méthylcétones par un mécanisme enzymatique qui a été parfaitement décrit chez P. roqueforti[2]. Les acides gras sont, tout d'abord, oxydés en position beta puis l'hydroxy acide ainsi formé subit une deshydrogénation conduisant au β-oxoacide qui sous l'action d'une décarboxylase est transformé en méthylcétone ; celles-ci seront pour une part réduites en alcools secondaires.

Le mycélium comme la spore sont également capables de réaliser ces conversions et les différents facteurs technologiques influant sur le taux de formation des cétones ont été bien étudiés.

La présence de méthylcétones insaturées peut s'expliquer en faisant appel à un mécanisme similaire qui ne porterait pas directement sur des acides gras libérés par la lipolyse, mais sur des acides insaturés, qui prendraient naissance par oxydation enzymatique des acides gras polyinsaturés[6].

Les mécanismes oxydatifs peuvent également être invoqués pour expliquer la présence d'un certain nombre d'autres composés, parmi lesquels les alcools primaires et notamment l'octen-1-ol-3, caractéristique des fromages à croute moisie. Un mécanisme de nature enzymatique, permettrait ainsi à P. caseicolum de produire ce composé à partir de l'acide linoléique, tandis que des composés diinsaturés comme les 1,5-octadienol-3 et 1,5-octadienone-3 également mis en évidence dans les fromages à croutes moisies proviendraient de l'acide linolénique[6].

4. LE ROLE DES PROTEINES

Les protéines apparaissent de plus en plus comme jouant un rôle essentiel dans la formation des composés sapides et des composés d'arôme déterminant les qualités et les défauts des fromages.

Le phénomène le plus apparent et le mieux connu, du moins dans ses étapes initiales, est celui de la protéolyse et plus spécialement la dégradation enzymatique des caséines[7]. Peuvent y participer les protéinases natives du lait mais surtout les enzymes coagulantes et les enzymes provenant des différentes flores microbiennes qui se succèdent au cours de l'affinage. Ce phénomène de protéolyse est sous la dépendance d'un certain nombre de paramètres (pH, température, humidité, teneur en Na, Ca, Mg, ...) qui influent sur l'activité des sytèmes enzymatiques ou sur l'état du substrat.

Plus difficile à décrire, mais sans doute tout aussi importants, sont les phénomènes, encore mal connus, qui semblent se produire au cours de la maturation enzymatique des laits et par lesquels l'état d'oxydation des protéines serait modifié[8], ce qui permettrait à un stade ultérieur l'apparition de composés d'arôme par un mécanisme qui pourrait être alors purement chimique[2].

Parmi les protéinases du lait, seule une métalloprotéine, la plasmine, semble jouer un rôle, qui dans la plupart des fromages reste néanmoins mineur par rapport à celui des enzymes coagulantes. Son rôle est néanmoins à prendre en considération lorsque les conditions qui prévalent (température de chauffage du caillé, pH) limitent l'activité protéolytique des enzymes coagulantes présentes dans le caillé, ce qui est notamment le cas dans les fromages à pâte pressée[4,7]. Possédant une bonne stabilité thermique, la plasmine a un maximum d'activité vers pH 7,5 et paraît dégrader surtout la caséine β qui semble jouer un rôle particulier dans l'apparition de l'arôme.

Le mécanisme d'action des enzymes coagulantes d'origine gastrique et notamment celui de la chymosine est aujourd'hui bien connu, alors que celui des enzymes d'origine fongique est moins bien établi bien que l'on sache qu'elles ont un mode d'action différent. Outre son action initiale sur la caséine K, la chymosine scinde la caséine αs_1 dès les premiers moments de l'affinage et son action combinée à celle de la pepsine se poursuit tout au long de cette période. Comparativement son action sur la caséine β reste modérée. La coupure des chaines ne se fait pas au hasard mais de façon préférentielle au niveau des liaisons contenant le groupement carboxyle d'une phénylalanine ou d'une leucine. Ces coupures conduisent à l'apparition de peptides à haut et bas poids moléculaire mais ne provoquent pas l'apparition d'acides aminés car les très courts peptides ne sont pas scindés.

Parmi les peptides libérés, certains à caractère hydrophobe présentent une amertume certaine et sont susceptibles de conduire à des défauts de goût dans la mesure où ils sont produits en trop grande quantité ou ne sont pas ultérieurement scindés en peptides à plus courte chaîne et en acides aminés par les systèmes peptidasiques des flores successives au premier rang desquelles viennent les bactéries lactiques. Ces dernières sont douées d'activité relativement faibles. Les protéases associées à la paroi sont néanmoins capables de dégrader les caséines et les gros peptides, provoquant ainsi l'apparition de peptides amers. L'intensité du phénomène apparaît lié au nombre de cellules présentes mais aussi au caractère propre des souches, l'existence de souches amères paraissant maintenant bien établie. Ce caractère amer pourrait être lié, à la quantité de protéases liées aux parois (comme le suggèrent notamment des essais réalisés avec des variants protéase négatif de *Streptococcus cremoris*) mais aussi à la présence ou à l'absence de peptidases membranaires et intracellulaires capables de scinder les peptides amers en peptides de taille plus réduite et en acides aminés. Des peptidases ont ainsi été mises en évidence chez *Str. diacetylactis*, *Str. thermophilus* et *Lactobacillus casei*. Elles apparaissent comme des métalloenzymes dont le pH optimum est généralement proche de la neutralité.

Des systèmes protéolytiques très complets ont également été mis en évidence chez d'autres microorganismes d'affinage comme Penicillium : ces systèmes comprennent des protéases exocellulaires (métalloprotéase et aspartylprotéase) dont l'action sur les caséines αs_1 et β se traduit par l'apparition de peptides de haut et bas poids moléculaire, ainsi que plusieurs exopeptidases.

L'apparition d'amertume dans les fromages à croute fleurie semble ainsi pouvoir s'expliquer par un excès d'aspartylprotéase découlant notamment du développement trop abondant de la moisissure de surface. Cette amertume peut toutefois être évitée si la quantité d'enzyme produite est limitée par une augmentation du pH obtenue en utilisant une atmosphère ammoniacale ou en favorisant la croissance d'un microorganisme capable comme G. *candidum* de provoquer une remontée du pH.

D'autres systèmes protéolytiques intra ou exocellulaires et notamment ceux des microorganismes de surface tels que microcoques, corenyformes et levures interviennent également dans l'affinage.

Au total l'action des divers systèmes protéolytiques conduit non seulement à la formation d'acides aminés libres importants en eux-mêmes pour la saveur et qui vont comme nous le verrons être à l'origine de produits volatils odorants mais aussi à celle d'un ensemble de peptides, de poids moléculaire très différent, parmi lesquels les peptides amers déjà mentionnés, mais aussi tout un ensemble de petits peptides, dont la structure reste encore mal connue et qui apparaissent pourtant comme jouant un rôle important au regard de l'intensité totale de la flaveur des fromages ou dans l'apparition de certaines notes de saveur très caractéristiques : la saveur sucrée de certaines pâtes pressées cuites résulterait de l'interaction de Ca et Mg avec des petits peptides, alors que la note bouillon de viande serait due à la présence de petits peptides et d'acides aminés. La protéolyse pourrait en outre avoir des conséquences indirectes importantes sur la perception de l'arôme puisqu'il apparaît que, de façon générale, les interactions entre molécules odorantes et les protéines diminuent au fur et à mesure que le niveau de protéolyse augmente et les acides aminés eux-mêmes pourraient former des complexes instables avec des molécules odorantes.

Le profil en acides aminés libres d'un type de fromage donné diffère de celui d'un hydrolysat de caséine, ce qui laisse penser qu'il y ait soit des mécanismes de libération préférentielle soit des mécanismes de dégradation préférentielle. Cette seconde hypothèse, qui n'infirme pas la première, peut être étayée par quelques exemples de mécanismes qui ont pu être reliés à l'activité de microorganismes caractéristiques d'un type de fromage ou à l'apparition d'un type de molécule d'arôme particulier, encore qu'il soit difficile dans un système en évolution, où anabolisme et catabolisme se superposent, de toujours bien établir le bilan de transformation d'un acide aminé, à plus forte raison lorsque les produits d'arôme formés sont eux-mêmes remétabolisés.

Un exemple de dégradation préférentielle, liée aux phénomènes de décarboxylation et de désamination, peut être celui de la dégradation de l'histidine et de la tyrosine dans les fromages du type Comté. On constate en effet qu'en dépit d'une teneur en histidine libre supérieure à celle en tyrosine libre, il y a toujours dans les fromages une teneur en tyramine supérieure à la teneur en histamine. On peut soit alors supposer que la décarboxylation de la tyrosine se produit plus facilement que celle de l'histidine ou que l'histamine est plus facilement désaminée que la tyramine. Une telle spécificité pour un substrat a été par ailleurs observée chez *Bacterium linens* où de fortes quantités d'ammoniac sont formées par une désamination portant surtout sur la sérine. Par ailleurs l'existence d'une aminotransférase active sur les acides aminés aromatiques a été démontrée également chez *B. linens*. C'est par cette voie qu'est notamment dégradée la phénylalanine.

Au-delà de mécanismes particuliers, l'étape clef de la transformation des acides aminés reste la réaction de transamination, qui a notamment été décrite chez les bactéries lactiques du groupe N et certains streptocoques du groupe D. Sur ce mécanisme se greffent ensuite des réactions diverses comme celles de décarboxylation oxydative susceptibles de conduire à des acides gras volatils.

Les acides aminés servent aussi de substrat pour la formation d'aldéhydes, par une réaction qui débute aussi par une transamination enzymatique suivie d'une décarboxylation. C'est par un mécanisme de ce type que seraient notamment formés les 2 et 3 méthylbutanal et 2-méthylpropanal par les streptocoques lactiques var. maltigenes, responsables du défaut décrit comme malté qui se rencontre parfois dans le cheddar. Il semble que dans les conditions normales d'affinage les aldéhydes ne représentent qu'un stade transitoire et que ces composés soient réduits en alcools correspondants. Ceci explique la présence de très larges quantités de 3-méthylbutanol dans la plupart des fromages où les aldéhydes ne sont normalement pas présentes.

Une série de réactions similaires, conduisant au phényléthanol a été démontrée chez les levures. Ce composé semble être ultérieurement catabolisé avec ouverture du cycle benzénique. Ces réactions permettant le passage de l'acide aminé vers l'aldéhyde correspondant sont à rapprocher du mécanisme purement chimique, dit de Strecker qui diffère cependant du précédent par le fait que l'étape de transamination est remplacée par la réaction d'une acide aminé avec un composé dicarbonyle, ce qui conduit à un aldéhyde. Cette réaction de Strecker a notamment été invoquée pour expliquer la formation de méthanethiol à partir de la méthionine mais il apparaît qu'elle n'est qu'une des voies non enzymatique pouvant conduire à ce composé considéré comme important pour l'arôme d'un certain nombre de fromages.

L'autre mécanisme le plus souvent invoqué est celui qui permet la production de méthanethiol à partir de methionine libre ou d'un résidu methionine, lorsque les conditions d'oxydo réduction sont favorables : il y a alors libération de H_2S à partir des groupements HS des protéines ce composé réagissant sur la méthionine pour libérer du methanethiol. Ce mécanisme, invoqué pour expliquer la présence de methanethiol dans le cheddar, est à rapprocher des observations faites par ailleurs sur ce même type de fromage et qui tendraient à prouver que l'état d'oxydoréduction des protéines dans le lait de fabrication soit l'élément déterminant l'apparition d'un arôme convenable au cours de l'affinage.

Il est par ailleurs bien connu qu'un certain nombre de microorganismes soient capables de produire du méthanethiol. Les voies enzymatiques qui comportent une étape de transamination suivie d'une coupure du composé intermédiaire ainsi formé par une demethiolase.(L-methionine-alpha - deamino *gamma* mercaptomethanelyase) ont été décrites et les caractéristiques de cet enzyme sont connues pour un certain nombre de microorganismes dont *P. putida*[9] et *B. linens*[10].

L'enzyme isolé de *B. linens* apparaît comme un multimère. Son activité, induite par la L-methionine atteint un maximum à pH 8,5 et augmente en présence d'ions Na. Elle est inhibée par les ions Cu. L'enzyme utilise le L-methionine comme substrat mais aussi, et avec un rendement supérieur, les dipeptides Me-Ala et Ala-Me.

Du fait de sa très grande réactivité le méthanethiol est lui-même à l'origine de plusieurs composés soufrés importants : par simple condensation, lorsque les conditions d'oxydo-réduction sont favorables, il donne du disulfure de méthyle et par un mécanisme encore inconnu conduit également au trisulfure. Il est susceptible d'être estérifié par des acides gras volatils, aussi bien par un mécanisme enzymatique que par simple réaction chimique. Il apparaît enfin vraisemblable qu'il puisse par cétalisation avec le formaldéhyde donner le bis (methyl-thiomethane) mais l'hypothèse reste à vérifier.

Un certain nombre d'autres composés reconnus comme importants pour l'arôme comme le phénol, l'acétophénone, l'indole ou susceptible de provoquer dans certains cas des défauts comme le crésol, dérivent de la dégradation d'acides aminés aromatiques. On peut postuler une voie enzymatique car leur présence est liée à celle de certaines flores, sans qu'il soit possible d'en mieux préciser le mécanisme.

La même remarque peut être faite pour les pyrazines qui semblent également avoir une origine microbienne. Celle-ci n'a été mise en évidence avec certitude dans les fromages que pour la 2-methoxyisopropyl pyrazine, responsable d'un défaut rappelant la pomme de terre dans les fromages à croute lavée et ou sa présence est liée à la croissance de *P. taetrolens* avec la valine comme précurseur. Dans le Brie, la présence de cette pyrazine a été reliée au métabolisme de *P. caseicolum* et il a été démontré que des alkylpyrazines peuvent être produites par *Lactobacillus helveticus*. Il a été proposé que le schéma métabolique supposé exister dans les végétaux puisse également être retenu pour expliquer la formation des méthoxypyrazines par les microorganismes mais aucune preuve de l'existence d'un tel mécanisme n'a jamais été apportée. Le même constat d'ignorance pourrait être formulé quant à l'origine de nombreux autres composés.

5. L'AFFINAGE ACCELERE

On peut donc voir qu'au total nous n'avons encore que des connaissances très fragmentaires sur les mécanismes qui conduisent à l'apparition de la flaveur dans les fromages et que seules les étapes initiales de dégradation des composants du caillé sont relativement bien connues. Ceci explique que les tentatives qui ont pu être faites pour accélérer l'apparition de la flaveur au cours de l'affinage n'aient en général été fondées que sur l'accélération de ces étapes initiales sans que l'on en soit pour autant capable d'en maîtriser parfaitement les effets[11].

Sachant que l'on se trouve en face d'une série de mécanismes qui ne peuvent être dissociés que lorsqu'il s'agit d'en obtenir une meilleure compréhension, il semblerait à priori logique de tenter d'augmenter simultanément la vitesse de toutes les réactions ; ceci semble pouvoir être obtenu en augmentant la température d'affinage. Quelques résultats ont pu ainsi être obtenus, limités dans leurs effets avec parfois même l'apparition de défauts qui s'atténuent si l'on prend la précaution de freiner l'étape initiale de fermentation lactique en commençant par abaisser la température d'affinage avant de la relever ultérieurement[12].

D'autres tentatives reposent sur l'ajoût de systèmes enzymatiques susceptibles de libérer une plus forte quantité de précurseurs de composés d'arômes. A cette fin diverses préparations destinées à augmenter le taux de protéolyse ou de lipolyse ont été utilisées.

L'ajoût de lipases a parfois été signalé comme bénéfique surtout pour les fromages où les acides gras libres apportent traditionnellement une contribution importante.

L'utilisation de protéases microbiennes s'est révélée être une arme à double tranchant car si elle permet une dégradation plus rapide des protéines elle se traduit fréquemment par une apparition de peptides amers avec souvent la formation d'un arôme déséquilibré ou bien ne se traduit que par une modification de la texture sans qu'un arôme ne se développe pour autant. Ces distorsions s'expliquent aisément car si on arrive bien à augmenter la teneur du caillé en précurseurs on n'a pas pour autant accéléré les mécanismes qui auraient permis d'utiliser ces précurseurs devenus disponibles en plus grandes quantités pour la production de molécules responsables de l'arôme. Un moyen qui a été utilisé pour obvier à cet inconvénient est l'ajoût de cellules ayant subi un choc thermique suffisant pour inactiver les systèmes acidifiant mais respectant l'activité des protéinases et des peptidases ce qui permettrait d'augmenter la quantité d'acides aminés libres et peut être d'atteindre ainsi le stade de transformation de ces acides aminés en produits d'arôme. Dans le même esprit l'utilisation de levain, lactose déficient, à haute concentration a été envisagée, mais ces deux techniques même si elles permettent de mieux dégrader les peptides n'apportent pas les solutions désirées.

Les mécomptes recontrés dans l'accélération des phases initiales de la protéolyse expliquent que certains se soient plutôt attachés à essayer d'augmenter le rendement de réactions situées plus en aval et contrôlant plus directement l'apparition de composés d'arômes[13]. Un exemple de ce type d'approche est l'incorporation d'un enzyme et de son substrat avec éventuellement les cofacteurs nécessaires, sous forme encapsulée à l'intérieur du caillé. C'est ainsi que des essais indépendants ont démontré la possibilité de produire des quantités plus importantes, de diacétyle, 3-méthyl-1-butanol, méthional ou de methanethiol dans le cheddar. On peut toutefois douter que la production d'un seul métabolite, en quantité plus importante, suffise à obtenir l'effet désiré.

6. CONCLUSION

C'est au moment où il apparaît nécessaire de mieux contrôler l'apparition de la flaveur ou lorsqu'il semble intéressant de racourcir la période d'affinage, tout en conservant au produit fini les mêmes propriétés sensorielles, que l'on ressent les limites de nos connaissances. La plupart des unités de recherche qui travaillaient sur la flaveur, et dont les résultats ont souvent été considérés par le monde laitier avec plus de curiosité que d'intérêt réel, ont été reconverties à d'autres fins. On voit donc mal comment sortir de l'empirisme qui préside le plus souvent aux tentatives faites pour contrôler la flaveur, tant que l'on n'aura pas mieux explicité à quels produits est réellement due la flaveur des fromages et éclairci les mécanismes de leur formation et de leur transformation, c'est-à-dire en reprenant le problème à la base.

REFERENCES

1. LAW, B.A., 1981, 'The formation of aroma and flavour compounds in fermented dairy products', Dairy Sci. Abstr., 43, 143-154
2. LAW, B.A., 1984, 'Microorganisms and their enzymes in the maturation of cheeses' in Progress in Industrial Microbiology, 19, 246-283.
3. ADDA, J., GRIPON, J.C. and VASSAL, L., 1982, 'The Chemistry of Flavour and Texture Generation in Cheese', Food Chem., 9, 115-129.
4. LAWRENCE, R.C., HEAP, H.A. and GILLES, J., 1984, 'A Controlled Approach to Cheese Technology', J. Dairy Sci., 67, 1632-1645.
5. COLLINS, E.B., 1972, 'Biosynthesis of Flavor Compounds by Microorganisms', J. Dairy Sci., 55, 1022-1028.
6. KARAHADIAN, C., JOSEPHSON, D.B. and LINDSAY, R.C., 1985, 'Contribution of *Penicillium* sp. to the Flavors of Brie and Camembert Cheese', J. Dairy Sci., 68, 1865-1877.
7. GRAPPIN, R., RANK, T.C. and OLSON, N.F., 1985, 'Primary Proteolysis of Cheese Proteins During Ripening A Review', J. Dairy Sci., 68, 531-540.
8. KRISTOFFERSEN, T., 1985, 'Development of flavor in cheese', Milchwissenschaft, 40, 197-198.
9. LINDSAY, R.C. and RIPPE, J.K., 1986, 'Enzymic Generation of Methanethiol To Assist in The Flavor Development of Cheddar Cheese and Other Foods' in "Biogeneration of Aromas", Am. Chem. Soc. Symp. Series. T. Parliment and R. Croteau Eds.
10. FERCHICHI, M., HEMME, D. and NARDI Michèle., 1986, 'Induction of Methanethiol Production by *Brevibacterium linens* CNRZ 918', J. Gen. Microbiol., (sous presse).
11. LAW, B.A., 1984, 'Accelerated Ripening of Cheese' in Advances in the Microbiology and Biochemistry of Cheese and Fermented Milk. Davies and Law Eds. Elsevier Applied Sci. Pub.
12. FEDRICK, A., ASTON, J.W., DURWARD, I.F. and DULLEY, J.R., 1983, 'The Effect of Elevated Ripening Temperatures on Proteolysis and Flavour Development in Cheddar Cheese', New Zealand J. Dairy Sci. Technol., 18, 253-260.
13. BRAUN, S.D., 1984, 'Microencapsulated Multi-Enzyme Systems to Produce Flavors and Recycle Cofactors', Thesis The University of Wisconsin-Madison.

NUTRITIONAL ASPECTS OF CHEESE

E. Renner
Dairy Science Section
Justus Liebig University
Bismarckstr. 16
D-6300 Giessen
Federal Republic of Germany

ABSTRACT. Cheese is a product rich in essential nutrients like protein, minerals and vitamins, as it can be derived from the nutrient density of protein, calcium and vitamin B_2 in different cheese varieties. Because of protein hydrolysis during cheese ripening the digestibility is increased. Cheese is suitable for persons suffering from lactose malabsorption and for diabetics, as it has a very low lactose concentration. Calcium, phosporus and magnesium in cheese are as well utilized as those in milk. During ripening, B vitamins are both used and synthesised by the cheese microflora. By heating or ultrafiltering the cheese milk, whey proteins pass into fresh cheese, thereby the biological protein value is increased. The utilization of the proteins of processed cheese is thought to be better than that of the proteins of natural cheese.

1. NUTRIENTS OF CHEESE

1.1. Milk fat

Although very often it is recommended that, because of the too high intake of fat and energy, low-fat cheese should be prefered, consumers generally favour high-fat cheeses because a high fat content contributes significantly to the flavour quality. Such a typical aroma develops only when the fat-in-dry matter content is at least 40 - 50 %, because the aroma is due mainly to the breakdown products of fat formed during cheese ripening.

Cheese contributes to the total fat intake only to a small extent: although cheese consumption in Germany is rather high (more than 15 kg per caput per annum) cheese contributes less than 5 % of the fat intake.

As to the cholesterol content of cheese which is another objective against the cheese lipids it has to be emphazised that the cholesterol content of cheese is rather low (0 - 100 mg/100 g, depending on the fat content) and that, therefore, cheese contributes only 3 - 4 % of total cholesterol intake (RENNER 1983). Furthermore, the cholesterol in the diet has only a limited effect on the level of blood cholesterol, because the body has a control mechanism which ensures that the synthesis of cholesterol by the body is reduced when the amount of cholesterol consumed increases (FLAIM et al. 1981).

1.2. Protein

The nutritional importance of cheese mainly arises from its high content of biologically valuable proteins. Table I shows that the protein content of different varieties of cheese varies between 20 and 35 %, for fresh cheese, cottage cheese and Feta between 10 and 18 %. In Table I, also the nutrient density is given which is a good measure for the content of nutrients related to the energy content. Here, the relative nutrient density is presented, where a value of more than 1.0 means that this specific food contributes to the supply with this specific nutrient in a higher extent than to the energy supply (at a nutrient density of 3.1 in a 3.1-fold extent). The nutrient density (related to the recommended dietary allowances for adults in the age of 36 - 50 years) for the protein content of different types of cheese is in the range between 2 and 8. A 100 g portion of soft cheese will provide 35 - 45 % of the daily protein requirements of an adult and 100 g of a hard cheese will supply 50 - 60 %.

Table I: Content and nutrient density (ND) of protein, calcium and vitamin B_2 in some cheese varieties (RENNER and RENZ-SCHAUEN 1986) (FDM = fat-in-dry matter)

Cheese variety	FDM %	Protein g/100g	ND	Calcium g/100g	ND	Vit. B_2 mg/100g	ND
Parmesan	35	36.0	4.1	1.3	9.2	0.60	2.1
Emmental	45	27.4	3.1	1.2	8.5	0.34	1.2
Cheddar	50	25.4	2.8	0.8	5.5	0.45	1.6
Edam/Gouda	48	21.8	2.8	0.7	5.6	0.35	1.4
Tilsit	45	23.2	3.1	0.75	6.3	0.35	1.5
Blue cheese	50	22.0	2.7	0.7	5.4	0.40	1.5
Brie	60	16.8	2.0	0.3	2.2	0.58	2.2
Camembert	45	20.1	3.2	0.4	3.9	0.58	2.8
Limburg	40	22.4	3.6	0.5	5.0	0.35	1.8
Feta	40	17.6	3.5	0.65	8.2	0.50	3.1
Cottage cheese	20	14.0	5.6	0.1	2.5	0.28	3.5
Fresh cheese	40	10.1	3.0	0.1	1.9	0.25	2.3
Fresh cheese	skimmed	13.2	8.4	0.1	4.0	0.30	6.0

The nutrient density is related to the recommended dietary allowances for adults (36 - 50 years)

In cheese manufacture the casein of milk is incorporated into the cheese while most of the biologically valuable whey proteins pass into the whey. Since the whey proteins are nutritionally superior to casein, which is somewhat deficient in sulfur containing amino acids, the biological value of the proteins in cheese in somewhat lower than that of the total milk protein, but is still higher than that of casein alone. The biological value of the proteins is not impaired by the action of rennet or of other enzymes active during cheese ripening, nor is it affected by acid formation. The Maillard reaction does not occur during cheese manufacture so that the availability of lysine in cheese is almost the same as in milk. Ripening periods of 16 - 20 weeks produce no significant changes in the NPU and PER values of the proteins of Tilsiter and Gouda cheeses, in fact, in some cases the NPU and PER values of cheese proteins are higher even than those of milk proteins (STAUB 1978).

Cheese can contribute significantly to the supply of essential amino acids. In Table II, where the amino acid composition of milk and cheese proteins is compared to the reference protein, which indicates the ideal concentration of essential amino acids in a dietary protein, it can be seen that cheese protein meets the requirements to the same extent as milk protein, except those for methionine plus cystine.

Table II: Concentration of essential amino acids in milk and cheese protein, compared to the reference protein

Essential amino acid	Content (g per 100 g protein)		
	Reference protein	Milk protein	Cheese protein
Tryptophan	1.0	1.4	1.4
Phenylalanine+Tyrosine	6.0	10.5	10.9
Leucine	7.0	10.4	10.4
Isoleucine	4.0	6.4	5.8
Threonine	4.0	5.1	4.8
Methionine+Cystine	3.5	3.6	3.2
Lysine	5.5	8.3	8.3
Valine	5.0	6.8	6.8
Total	36.0	52.2	51.6

During cheese ripening part of the water-insoluble casein is converted into water-soluble nitrogenous compounds which include the intermediate products of protein hydrolysis as well as free amino acids. Cheese ripening can be looked upon as a sort of predigestion whereby the digestibility of the proteins is increased. The true digestibility of a number of cheese varieties is almost 100 %. Small peptides can pass through the walls of the intestine and it is possible that they penetrate even cell membranes so that they become directly available to the cell. An experiment with rats demonstrated that the rate of utilization of cheese protein was higher than the rate for casein. The mean degree of utilization of the essential amino acids of cheese protein is 89.1 %, i.e. greater than the corresponding value for milk protein (which is 85.7 %) and almost equal to the value for egg protein, which is 89.6 %. The free amino acids of cheese, particularly aspartic and glutamic acid, are said to promote the secretion of gastric juices. It should be noted that a food allergy to cheese protein has never been described (DILLON 1984).

The decarboxylation of free amino acids during cheese ripening produces amines. The principal amines found in cheese are histamine, tyramine, tryptamine, putrescine, cadaverine, and phenylethylamine. The concentrations of individual amines in cheese show great variations and depend on the ripening period, on the intensity of flavour development and on the microbial flora. Average values of the contents of tyramine and histamine in different types of cheese have been determined and are shown in Table III. It is evident that Cheddar cheese contains an astonishingly high concentration of tyramine, and that blue cheeses have high concentration of amines, tyramine and especially histamine.

Table III: Average tyramine and histamine contents of some cheese varieties (RENNER 1983)

Cheese variety	Content of tyramine µg/g	histamine µg/g
Cheddar	910	110
Emmentaler, Gruyère	190	100
Blue cheese	440	400
Edam, Gouda	210	35
Camembert, Brie	140	30
Cottage cheese	5	5

Physiologically active amines can affect the blood pressure, with tyramine and phenylethylamine having a hypertensive and histamine a hypotensive effect. However, mono and diamine oxidases convert the biogenous amines that are consumed in foods relatively quickly into aldehydes and finally into carboxylic acids by oxidative deamination. Although opinions on the toxicity threshold values of amines vary widely it is concluded that healthy persons are able to metabolize the biogenic amines ingested even when large amounts of cheese are consumed, without adverse physiological reactions (BINDER and BRANDL 1983). It is, however, possible that some sensitive persons who suffer from a genetically-determined lack of monoamine oxidase may be subject to attacks of migraine as a result of eating cheese.

1.3. Lactose and lactic acid

There is no lactose in many cheeses or only a very low concentration (1 - 3 g/100 g) because most of the lactose of the milk passes into the whey and that retained in the cheese curd is partly converted to lactic acid during cheese ripening. Therefore, like other cultured milk products, cheese is suitable for the diets of persons suffering from lactose malabsorption and of diabetics (BLANC 1982).

Cheese usually contains both lactic acid isomers, L(+)- and D(-), the relative proportion of the D-isomer depending on the type of starter culture used and on some other ripening factors. The content of D(-) lactic acid in different types of cheese can be very different (fresh cheese 4 - 14 %; ripened cheeses 10 - 50 %). The human organism has only a limited capacity to metabolize D(-) lactic acid but from the data available in the literature, a toxic effect of D(-) lactic acid cannot be derived for the adolescent or the adult. As a logical conclusion in a revised statement, the WHO has not limited the admissible intake for adults while for infants (up to 1 year of age), a D(-) lactic acid-free diet is recommended (BARTH and DE VRESE 1984).

1.4. Minerals

The calcium and phosphorus contents of cheese are as important as that of milk, since 100 g of soft cheese will supply about 50 % of the daily Ca and P requirement and 100 g of a hard cheese will meet the daily Ca and P requirement completely.

The average concentration of Ca in a number of cheese varieties is shown in Table I. The nutrient density for Ca in different types of cheese varies between 2 and 9. Cheeses produced by rennet coagulation usually have higher calcium contents that those made from acid-coagulated milk.

The calcium, phosphorus and magnesium in cheese are as well utilized by the body as those in milk (KANSAL and CHAUDHARY 1982). The ratio of calcium to phosphorus in cheese is also thought to be desirable nutritionally.

There is a wide range in the Na content which is due to the different amount of NaCl added to cheeses; the following are average values for the salt content (%) of different cheeses:

Fresh cheese, Cottage cheese	0.8
Emmental	0.8
Tilsit, Camembert, Cheddar, Gouda, Edam, Brie	1.7 - 2.0
Parmesan, Roquefort, Feta	2.5 - 3.0

Probably, the Na requirement is less than 500 mg per caput per day, 5 g of NaCl are considered to be sufficient for the adult. Since a high sodium intake can induce hypertension, a restricted sodium intake is recommended to accomodate the diet of consumers under medical management for hypertension. Although even in countries with a high consumption, cheese contributes only for about 5 - 8 % to the total sodium intake, the manufacture of low-sodium cheese was investigated by using a brine containing mainly KCl. Taste panel results showed that cheese prepared to contain up to 75 % less sodium than traditional cheese was acceptable to consumers (KARAHADIAN et al. 1985). It should be considered also that hypertension may be due to a deficiency of dietary calcium rather than to an excessive intake of sodium, since it has been observed that patients suffering from hypertension consume about 25 % less Ca than normotensive persons, because of a low consumption of milk and dairy products (McCARRON et al. 1982).

1.5. Vitamins

The concentration of fat-soluble vitamins in cheese depends on its fat content. Most (80 - 85 %) of the vitamin A contained in milk passes into the cheese. The figure is naturally lower for the water-soluble vitamins. The values for thiamine, nicotinic acid, folic acid and ascorbic acid are 10 - 20 %, for riboflavin and biotin, 20 - 30 %, for pyridoxine and pantothenic acid, 25 - 45 % and for cobalamin, 30 - 60 %; the rest remains in the whey (REIF et al. 1976). However, milk contains such high concentrations of some B vitamins that cheese still contributes significantly to the supply of these vitamins. This is especially true of vitamin B_{12}.

Table I lists the average concentrations of Vitamin B_2 in a number of cheese types. In spite of the low transfer rate, there is a positive nutrient density for vitamin B_2 in all cheeses with values between 1.5 and 3.

The concentration of B vitamins changes during ripening since these vitamins are both used and synthesised by the cheese microflora. The concentration of several of the B vitamins depends on the type of starter culture used and increases with time of storage. After a long ripening period, the concentration of these vitamins in cheese therefore may be increased. By isolating individual microorganisms from cheese it could be shown that they are able to synthesise nicotinic acid, folic acid, biotin and pantothenic acid. The synthesis of vitamin B_{12} by propionic acid bacteria in hard cheese, especially in Emmentaler, has aroused great interest. Propionic acid bacteria have therefore been added experimentally to cheese milk in the manufacture of Edam, Tilsiter and a number of other types of cheese with the result that in some cases the cobalamin content was doubled.

2. FRESH CHEESE (QUARG)

Table I also includes values for the concentrations of protein, minerals and vitamins in quarg (fresh cheese). Milk destined for quarg production is nowadays often strongly heat treated (at 95°C for 10 minutes). This leads to complex formation between casein and whey proteins so that a large part of the whey proteins is precipitated with the casein on acidification and passes into the quarg. The percentage of the total nitrogen precipitated increases from 77 - 79 % to 88 - 89 %. This product has, therefore, a higher content of essential amino acids and a higher biological protein value (RENNER et al. 1983). The whey proteins are also incorporated in a high extent into the fresh cheese by ultrafiltering the cheese milk.

From the point of view of nutrition, quarg, which is usually produced by means of a lactic acid culture, is similar to other cultured milk products. Because low-fat quarg is rich in biologically valuable proteins, calcium and phosphorus, and because its calorie content is relatively low, it is recommended for all sections of the population, but particularly for older people and as part of slimming diets. Quarg is also easily didestible and this makes it valuable in therapeutic diets, expecially in cases of liver disease.

3. PROCESSED CHEESE

Processed cheese contains roughly the same proportions of nutrients as the cheese from which it was made. The casein is hydrated and peptized by the action of the emulsifying salts and the proportion of water-soluble protein therefore increases considerably. Except for the Na and K content, which are higher, the mineral concentration is also similar to that in the original cheese. Polyphosphates have the widest range of application as emulsifying salts, but citrates and lactates are used also. The addition of polyphosphate does not increase the phosphate content significantly; the natural variation in the phosphate content of cheese is 0.4 - 2.7 % and of processed cheese 0.8 - 2.7 %. Some losses of vitamin B_1, B_2, nicotinic acid, panthothenic acid and vitamin B_{12} occur during the manufacture of processed cheese. The free amino acid content of the cheese and the in vitro digestibility of the proteins are increased by processing and the utilization of the proteins of processed cheese is thought to be better than that of the proteins of natural cheese. No change in the availability of lysine could be detected (LEE and ALAIS 1981).

Polyphosphates ingested with food are unable to exert a physiological effect because they are quickly broken down by enzymes to monophosphates which are then absorbed. They are therefore no danger to health. Experiments with rats have shown that polyphosphates are well tolerated, even when administered over long periods of time. The phosphate ingested as part of processed cheese has to be considered in the context of the total phosphorus intake as it might even contribute to meeting the P requirement.

References

BARTH,C.A., M.DE VRESE: D-Laktat im Stoffwechsel des Menschen - Fremdstoff oder physiologischer Metabolit? Kieler Milchw.Forsch.Ber.**36**, 155-161(1984)

BINDER,E., E.BRANDL: Über das Vorkommen und die Bedeutung biogener Amine in Lebensmitteln. Österr.Milchw.**38**,257-259(1983)

BLANC,B.: Die Biosynthese des Käses als Grundlage seines Nährwertes. Alimenta **21**,125-134(1982)

DILLON,J.C.: Le fromage dans l'alimentation. in: Eck,A.(edit.): Le fromage.p.497-510. Paris:Lavoisier 1984

FLAIM,E., L.F.FERRERI, F.W.THYE, J.E.HILL, S.J.RITCHEY: Plasma lipid and lipoprotein cholesterol concentrations in adult males consuming normal and high cholesterol diets under controlled conditions. Am.J.Clin.Nutr.**34**,1103-1108(1981)

KANSAL,V.K., S. CHAUDHARY: Biological availability of calcium, phosphorus and magnesium from dairy products. Milchwiss.**37**,261-263(1982)

KARAHADIAN,C., R.C.LINDSAY, L.L.DILLMAN, R.H.DEIBEL: Evaluation of the potential for botulinal toxigenesis in reduced-sodium processed American cheese foods and spreads. J.Food Protect.**48**,63-69 (1985)

LEE,B.O., C.ALAIS: Etude biochimique de la fonte des fromages. III. Evolution des acides aminés libres et de la digestibilité in vitro des protéines. Lait **61**,140-148(1981)

McCARRON,D.A., C.D.MORRIS, C.COLE: Dietary calcium in human hypertension. Science **217**,267-269(1982)

REIF,G.D., K.M.SHAHANI, J.R.VAKIL, L.K.CROWE: Factors affecting B-complex vitamin content of cottage cheese. J.Dairy.Sci.**59**,410-415(1976)

RENNER,E.: Milk and dairy products in human nutrition. Munich: Volksw. Verlag 1983

RENNER,E., U.KARASCH, A.RENZ-SCHAUEN, A.HAUBER: Untersuchungen über Qualitätskriterien von Speisequark. Deut.Milchwirtsch.**34**,1410-1417(1983)

RENNER,E., A.RENZ-SCHAUEN: Nährwerttabellen für Milch und Milchprodukte - Energie- und Nährstoffgehalt von 500 Produkten. Giessen: Verlag B. Renner 1986

STAUB,H.W.: Problems in evaluating the protein nutritive value of complex foods. Food Technol.**32**(12),57-61(1978)

PROGRESS IN CHEESE TECHNOLOGY - SAFETY ASPECTS WITH MICROBIOLOGICAL EMPHASIS

H. A. Morris and S. R. Tatini
Department of Food Science and Nutrition
University of Minnesota
1334 Eckles Avenue, St. Paul, Minnesota
55108, USA

ABSTRACT. Milk has been preserved as cheese for centuries. Improvement in sanitation and milk treatment has minimized contamination by pathogens. Inhibition by starter bacteria, high acid, high salt, low moisture, metabolites and curing conditions inhibit growth and survival of pathogens. Progress is being made in monitoring and eliminating possible health problems associated with cheese. Public health safety issues concerning cheese are discussed briefly in relation to Staphylococcus aureus toxin, enteropathogenic Escherichia coli, Salmonellae, Listeria monocytogenes, mycotoxins, and amines.

INTRODUCTION

Judging from the literature and commercial practices, the cheese industry is highly concerned about product quality and safety. Particularly, the concerns are to insure the microbiological and toxicological safety of cheese. Advances have been made in prevention of contamination and growth of pathogenic and toxigenic microorganisms in and on cheeses. Some of these advances include: 1) elimination of many pathogens from the cow population, better milk quality, and an increased use of pasteurized milk in cheese making; 2) improved equipment designs such as enclosed vats and curd handling equipment; 3) more effective cleaning and sanitation procedures; 4) more reliable starter culture systems with better control of acid development and cheese composition; and 5) more effective control of the growth of molds on cheese by improvement in packaging materials and procedures along with the use of antimycotic agents such as natamycin and sorbic acid. Other advances have been made in analytical procedures for monitoring the presence or absence of pathogenes or toxins that might possibly be associated with cheeses. Along with the improved procedures, hazard analysis critical control point (HACCP) concepts are being utilized to assure the safety of cheese.

Natural cheeses, in general, represent hostile environments for the growth and survival of pathogenic bacteria. Initial pH values of near 5.0, cold curing temperatures, relatively high salt contents of about 4 to 5% NaCl in the free water of the cheese, relatively low moisture content, the presence of metabolites such as lactic and acetic acids and competitive inhibition from the growth of starter bacteria, all contribute to a inhibition of growth and death of pathogenic bacteria. However, there is evidence to the contrary from recent outbreaks and experimental studies indicating that salmonella may survive during aging and that E. coli and Listeria monocytogenes may grow in soft cheeses.

The most common pathogenic bacteria of concern in natural cheeses are Staphylococcus aureus and Salmonellae in hard cheeses, and enteropathogenic Escherichia coli and, more recently, Listeria monocytogenes, in semisoft and soft cheeses. In addition to toxins produced by S. aureus, the presence of amines in cheese and the potential for mycotoxins to be present in moldy cheeses are of concern. We will present views about the above mentioned concerns following a HACCP analysis format.

PATHOGENIC BACTERIA

Of the microorganisms mentioned above, S. aureus represents a hazard in hard cheeses (Cheddar and Swiss) due to accumulation of heat and proteolysis stable enterotoxin mostly during manufacturing of cheese. Viable S. aureus which really do not represent a food poisoning hazard decrease during ripening of cheese at 5-10°C. The most common contributing factor to enterotoxins is contamination of pasteurized milk with enterotoxigenic S. aureus followed by some faulty fermentation (slow acid development) due to antibiotics, inhibitors or bacteriophage allowing growth to high levels (> 1-10 million per g) with accumulation of enterotoxins (1,2,3). In cheeses with high levels of starter bacteria and active fermentation, S. aureus do not produce high levels of toxin because of competition (4). In terms of HACCP, cheese milk is expected to be contaminated with low levels of enterotoxigenic S. aureus depending on the level of sanitation; but critical control points are proper fermentation and pH of cheese, the failure of which can easily be monitored by milling acidity and/or final pH. In the case of starter failure or slow acid development, the cheese should be tested for enterotoxin. Water activity, pH and ripening temperatures have no effect on staphylococcal enterotoxins which persist under these conditions for several years.

Significant progress has now been made in developing a dip stick procedure (a 4-hour test) for enterotoxins A-E (Igen, Inc., Maryland, USA). Control includes sanitation, monitoring fermentation and, when failure of fermentation is noted, test for toxins. Sampling for S. aureus, thermonuclease and enterotoxins should be done on multiple samples taken from outside and inside of several blocks representing each lot (refer to ICMSF, 1980).

Salmonellae represent a more widespread and moderately severe food poisoning. Salmonellae are widespread in the environment and there are more than 2,000 serotypes, each of which is capable of causing food poisoning. Therefore, they are likely to be found in raw milk in low numbers. They do not grow at refrigeration temperatures in milk, and are killed by pasteurization. They are expected to die in hard cheeses during aging at 5 to 10°C due to low pH's, A_W of about 0.95, and free fatty acids released during ripening. However, data from recent outbreaks traced to Cheddar cheese suggest that pH 5.0 and salt levels of 4-5% in the water of cheese with low A_W may not be adequate to control Salmonella (5,6); therefore, pasteurization of cheese milk and prevention of post pasteurization contamination followed by good fermentation should be utilized. Also, testing of cheese following the ICMSF scheme should be utilized. Significant progress is being made in developing rapid and specific methods for salmonella analysis; namely DNA-probes by integrated genetics - Gene Tech, and enzyme linked immunoassays (Enzabead, Litton Bionetics). While this may add cost to the product, it does assure safety. From a HACCP standpoint, salmonella will be present in raw milk on a regular basis (4%) and this may result in contaminating cheese plant environment. Thus, prevention of post pasteurization contamination along with proper fermentation are necessary control measures. It is claimed that some strains may persist in cheese during ripening for a long period (8 months) (6). Therefore, testing before release may be the final checkpoint for consumer safety.

Coliforms, _Escherichia coli_ and enteropathogenic strains of _E. coli_ may be infrequently found in raw milk and may find entry into cheese during manufacture; however, these organisms do not grow well during cheese making and during ripening of hard cheeses. Because of low pH and relatively high salt content (low A_W) they decrease in numbers and eventually die (7). In soft ripened cheeses, however, due to higher pH's during ripening, they may grow to high levels depending on temperatures of ripening (15°C or >) (8,9). Control consists of pasteurization of milk, good starter fermentation and in soft cheeses testing for enteropathogenic _E. coli_ (EEC).

An organism of recent concern, especially in soft ripened cheeses, is _Listeria monocytogenes_, which may cause fatal illness in susceptible population, namely prenatal and neonatals and elderly undergoing chemotherapy or immuno-compromised individuals. Surveys in the USA indicate the prevalence of this organism to be about 4% in raw milk. According to Bradshaw et al. (10) current pasteurization process guidelines are adequate to destroy _Listeria_ in whole milk. It is a psychrotroph and can grow at 3-5°C. Recent reports presented at the International Association of Milk, Food and Environmental Sanitarians at the 1986 annual meeting in Minneapolis, Minnesota, indicate that _Listeria_ can survive in cheese during manufacture, can persist in some hard cheeses, and can grow in some soft cheeses (11). A considerable amount of research is in progress concerning _Listeria_ including developing immunoassays for rapid and specific detection of the organism.

To aid in controlling _Listeria_ in cheese, pasteurized, low somatic cell count milk should be used along with good sanitation practices.

MYCOTOXINS

There are three possible sources of mycotoxins in cheese: 1) injestion of contaminated feed by dairy animals and subsequent passage of the injested mycotoxin(s) or metabolites into the cheese milk; 2) growth of adventitious toxigenic fungi on cheese that might result in formation of toxins; 3) mold starter cultures used in the manufacture of mold ripened cheeses.

The main mycotoxin of concern that might occur in milk is M1. It is a metabolite of the highly carcinogenic aflatoxin B1 (12). Because of the structural resemblance of M1 to B1 toxicologists consider that M1 also has carcinogenic properties (13). However, Hsieh (14) showed that M1 isn't carcinogenic. Nevertheless, tolerance levels have been set in milk ranging from 0.01 to 0.5 µg/kg (13). Presumably cheese levels would be similar.

The amount of M1 when found in cheese from three surveys ranged from 0.02 to 1.3 µg/kg (12). About one-half of the cheeses contained no M1. What does this mean in reference to safe levels in cheese. As Hamilton (15) writes, "The time honored concepts of cause and effect are not as clear and simple with mycotoxins as we would wish and frequently assume. We have nondescript diseases with multiple nondescript causes. It is a rare field outbreak in which everything equals what is seen in the laboratory." He presented the following postulates in regards to mycotoxins in feed that might be applicable to cheese. "Koch's postulates in mycotoxicology. 1) Find the mycotoxin in suspect substrate from the toxicosis outbreak; 2) Find in the substrate a fungus that produces the toxin; 3) Induce the toxicosis in experimental animals by ingesting or contacting the toxin." Postulate 1 hasn't been met. As Bullerman (16) wrote, "The significance of mycotoxins as causes of human diseases is difficult to determine because there is no direct evidence of such involvement in terms of controlled experiments with man."

However, it is to the consumers and the industries best interest to minimize the presence of mycotoxins in cheese. Control measures to reduce the amount in milk include: eliminating or reducing mold growth on dairy animal feed (13), ammonia treatment of feed (17), and the use of hydrogen peroxide plus riboflavin plus heat treatment of milk (18).

A few molds, isolated from various cheeses and grown on special media, have produced material toxic to chicken embryos or that have been identified as mycotoxins (19,20,21). The predominant molds isolated by Bullerman's group were of the Penicillium species with a few Aspergillus and Fusarium species. They found that 25/349 isolates obtained from Cheddar cheese produced toxins (20) and 10/183 from Swiss cheese (19). Four moldy Swiss cheeses out of 33 contained penicillic acid in trace amounts (19). Northolt et al. (21) also found a predominance of Penicillium species in Dutch hard cheese but in addition found Aspergillus versicolor. Analysis of 39 cheeses molded with A. versicolor found sterigmatocystin in the outer 1 cm thick layer of 9 of them. Sterigmatocystin is similar in structure to aflatoxin B1 and is reported to be a carcinogen (13). As intimated by Zerfiridis (22), the presence of a particular mold on or in a cheese doesn't mean that it has produced toxins or will grow and produce toxins. Also, toxin production in a special medium doesn't mean that toxins will likewise be produced in cheese, even though mold growth occurs on the cheese. For example, Penicillium species known to produce patulin and penicillic acid were grown extensively on Swiss and Mozzarella cheeses; however no patulin or penicillic acid were detected in the cheeses as a result of the mold growth (23). Likewise, Olivigni and Bullerman (24) using an isolate of P. roqueforti that produced patulin and penicillic acid on laboratory media, found that no toxins were produced when the mold was grown on Cheddar or Swiss cheeses. Indications are that the lack of production of patulin and penicillic acid on cheese is because cheese is low in carbohydrate and high in protein (23,24,25) or that if it is produced it is rapidly degraded (26).

Since in all of the above cases the cheeses examined had extensive mold growth on them, it is highly unlikely that consumers would eat them. Issues raised earlier about mycotoxins and health risk are also pertinent in the case of toxins produced by adventitous fungi.

Controls include prevention or minimizing mold contamination of cheeses and inhibition or prevention of growth of potentially toxin producing molds on cheeses. Improvement in recent years in cleaning and sanitation methods, air filtration of cheese plants and packaging rooms, packaging materials and methods that exclude air from cheese, have been dramatic in reducing contamination. The use of antimycotic agents, particularly of natamycin, to treat cheese surfaces by direct application or through incorporation in cheese coatings, has resulted in more effective mold control (27,28,29). Also, as reported by Ray and Bullerman (27), sorbate and natamycin inhibit mycotoxin production to a much greater extent than they inhibit mold growth.

Another control measure that should be further encouraged is the education of consumers concerning care of cheeses in the home and potential hazards of consuming moldy cheeses. Bullerman (30) has summarized guidelines for consumers in this regard.

Mold starter cultures used in the manufacture of mold ripened cheeses are capable of producing toxins (31). Those of greatest concern are penicillic acid and PR toxin but they are unstable in cheese (32). Cyclopiazonic acid may be produced by some strains of Penicillium camemberti and it has been found in small amounts in the crust of Camembert cheese but not in the inner part (33). Schock (34) found that 3/13 cheeses ripened with P. camemberti contained cyclopiazonic acid. Schock (34) further found that in Ames tests that there was no evidence that extracts of P. camemberti or P. roqueforti were mutagenic. There were no adverse effects on mice given large quantities of Blue or Camembert cheese (equivalent to 100 kg/person daily). Additionally, Schock et al. (35) injected crude extracts of four strains of P. camemberti intraperitoneally into mice at high doses (5-10 mg). No toxic effects were found.

A useful conclusion concerning possible risks from consuming mold ripened cheeses is Scott's (31): "Of fungal metabolites or mycotoxins that originate from penicillia used to process cheese and that have been detected in the cheeses themselves at µg/g concentrations are limited to roquefortine, isofumigaclavine A, mycophenolic acid, ferrichrome and cyclopiazonic acid. Subject to any new findings that might indicate carcinogenic activity of any of these compounds to animals, their presence in cheese would appear to pose no hazard to the consumer. Contamination could always be limited, if necessary, by proper selection of a non- or low-toxigenic strain." Similar conclusions have been made by Egmond (13) and Schoch et al. (35).

AMINES

The public health significance of tyramine, histidine and tryptamine found in some cheeses has been well reviewed by Edwards and Sandine (36). Apparently from the various surveys of these amines in cheese (36,37,38) there is negligible health risk in eating cheese for all but the rare individuals lacking monoamine oxidases. However, persons on monoamine oxidase inhibitor therapy should not consume cheese.

The amines in question are produced by bacterial amino acid decarboxylases acting on the free parent amino acids during cheese maturation. Some of the bacteria implicated are strains of Streptococcus faecium, Streptococcus mitis, Lactobacillis bulgaricus, Lactobacillus plantarum, and streptococci of the viridans group (36).

One control measure is to minimize contamination of cheeses with such bacteria and also restrict their growth in cheese. Also, according to Edwards and Sandine (36): "Whether the amines accumulate and persist in cheese depends on a number of factors. These are the presence of capable bacteria or enzymes; availability of substrate; presence of the suitable cofactor; existence of a proper environment in the cheese as influenced by pH, temperature, salt, and water availability; existence of potentiating compounds such as diamines; and amine catabolism."

Since nitrates are used to aid in controlling butyric acid fermentation primarily in Edam and Gouda cheeses in Europe (39) and elsewhere, there is concern over the possibility that carcinogenic nitrosamines might be formed in these cheeses. As discussed earlier, there are no cases of toxicity arising from cheese. The occurrence of nitrosamines in cheeses made from milk to which nitrates have been added is minimal.

All of these studies showed no relationship between nitrate level added or the residual nitrate and nitrosamine content. Because of this apparent lack of correlation and insignificant amounts of nitrosamines in cheese no control measures are indicated.

There is also concern about the amount of nitrate in cheese. The quantity of nitrate in cheese when added at the rate of 15 g $NaNO_3$/100 liters of milk, decreased in the cheese from an average of 56 mg/kg initially to about 30 mg/kg after 6 weeks. Nitrite levels were 1 mg/kg at about two weeks and then dropped to 0.5 mg/kg or lower. Munksgaard and Werner (42) found from 6 trials with ^{15}N-labelled nitrate after 16 weeks storage that the cheeses contained 13-55% of the added ^{15}N, mainly present in the NPN-fraction. Only 0.3-10% was still present as nitrate.

Again, the significance of added nitrate in some cheeses to human health risks appears to be minimal.

REFERENCES

1. S.R. Tatini, J.J. Jezeski, H.A. Morris, J.C. Olson, Jr., and E.P. Casman, J. Dairy Sci. 54(1971)815.
2. J. Stadhouders, M.M. Cordes, and W. Ivan Schouwenburg-van Foeken, Neth. Milk Dairy J. 32(1978)193.
3. V.L. Zehren, and V.F. Zehren, J. Dairy Sci. 5(1968)645.
4. S.R. Tatini, W.D. Wesala, J.J. Jezeski, and H.A. Morris, J. Dairy Sci 56(1973)429.
5. D.S. Wood, D.L. Collins-Thompson, D.M. Irvine, and A.N. Myhr, J. Food Prot 47(1984)20.
6. J.-Y D'Aoust, D.W. Warburton, and A.M. Sewell, J. Food Prot.48(1985) 1062.
7. M.W. Yale, and J.C. Marquardt, New York Agr. Exp. Sta. Tech. Bul. No.270(1943).
8. H.S. Park, E.H. Marth, and N.F. Olson, J. Milk Food Technol.36 (1973)543.
9. J.L. Kornacki, and E.H. Marth, and N.F. Olson, J. Food Prot.45 (1982)310.
10. J.G. Bradshaw, J.Z. Peeler, J.J. Corwin, J.M. Hunt, J.T. Tierney, E.P. Larkin, and R.M. Lwedt, J. Food Prot.48(1985)743.
11. E.T. Ryser, and E.H. Marth, J. Food Prot. 49(1986) in press.
12. P.M. Scott, J. Food Prot.41(1978)385.
13. H.P. van Egmond, Food Chemistry 11(1983)289.
14. D.P.H. Hsieh, The mutageniety and carcinogenicity of mycotoxins. Proceedings V. International symposium on mycotoxins and phycotoxins, Vienna, Austria, Sept. 1982, 228-31.
15. P.B. Hamilton, J. Food Prot.41(1978)404.
16. L.B. Bullerman, J. Food Prot.42(1979)65.
17. R.L. Price, O.G. Lough, and W.H. Brown, J. Food Prot.45(1982)341.
18. R.S. Applebaum, and E.H. Marth, J. Food Prot.45(1982)557.
19. L.B. Bullerman, J. Food Sci.41(1976)26.
20. L.B. Bullerman, and F.J. Olivigni, J. Food Sci.39(1974)1166.
21. M.D. Northolt, H.P. van Egmond, P.S.S. Soentoro and W.E. Deyll, J. Assoc. Off. Anal. Chem.63(1980)115.
22. G.K. Zerfiridis, J. Dairy Sci.68(1985)2184.
23. F.Y. Lieu, and L.B. Bullerman, J. Food Sci.42(1977)1222.
24. F.J. Olivigni, and L.B. Bullerman, J. Food Sci.42(1977)1654.
25. W.T. Stott, and L.B. Bullerman, Appl. Microbiol.30(1975)850.
26. W.T. Stott, and L.B. Bullerman, J. Food Sci.41(1976)201.
27. L.L. Ray, and L.B. Bullerman, J. Food Prot.45(1982)953.
28. W.G. de Ruig, and G. van den Berg, Neth. Milk Dairy J.39(1985)165.
29. A. Reps, S. Poznanski, L. Jedrychowski, A. Babuchowski, and A. Fetlinski, XXI Int. Dairy Congress Brief Communication 1(1)(1982) 519.
30. L.B. Bullerman, J. Dairy Sci.64(1982)2439.
31. P.M. Scott, J. Food Prot.44(1981)702.
32. P.M. Scott, and S.R. Kanhere, J. Assoc. Off. Anal. Chem.62(1979)141.
33. J. LeBars, Appl. Environ. Microbiol.38(1979)1052.

34. U.W. Schoch, Ph.D. Thesis abstracted in Milchwissenschaft 38(11)(1983) 683 and Dy. Sci. Abs.47(1985)446.
35. U. Schoch, J. Luthy, and C. Schlatter, Milchwissenschaft 39(10) (1984)583.
36. S.T. Edwards, and W.E. Sandine, J. Dairy Sci.64(1981)2431.
37. M.N. Voigt, R.R. Eitenmiller, P.E. Koehler, and M.K. Hamdy, J. Food Technol.37(1974)377.
38. M.N. Voigt, and R.R. Eitenmiller, J. Food Prot.41(1978)182.
39. T.E. Galesloot, Neth. Milk Dairy J.18(1964)127.
40. H. Werner, XXI Int. Dairy Congress Brief Communication 1(1)(1982)540.
41. K. Goodhead, T.A. Gough, K.S. Webb, J. Stadhouders, and R.H.C. Elgersma, Neth. Milk Dairy J.30(1976)207.
42. L. Munksgaard, and H. Werner, Fate of nitrate in cheese. Statens Mejeriforsg, Hillerod (The Danish Government Research Institute for Dairy Industry) beretning 263(1985).

DISCUSSION

H. van Dijk (the Netherlands) asked whether the often assumed relation between calcium content and cheese consistency could in fact be due to pH, because of the fair correlation between cheese pH and calcium content via the pH at the end of curd making.
P. Walstra (the Netherlands) considered this to be likely.
R.C. Lawrence (New Zealand) confirmed that pH and water content were more important than Ca content and that the latter mostly varied little within one type of cheese; between widely varying types of cheese, the Ca content may, however, be partly responsible for the difference in consistency.

H.T. Badings (the Netherlands) asked whether comparisons had been made between the results of experimental tests and sensory evaluation.
P. Walstra (the Netherlands) said that he had no facilities for such comparisons, but that they would be most useful.

Name and country unknown : Is anything known about the relation between cheese structure (as e.g. evaluated by texture measurements) and flavour, for instance the rate of release of flavour components?
P. Walstra (the Netherlands) : It is very likely that a relationship exists, but I am not aware of any experimental work in this field.

B. Bianchi-Salvadori (Italy) : Is it possible to predict from consistency measurements on an immature cheese (say, 2 weeks old) the development of texture during ripening?
P. Walstra (the Netherlands) : Not yet, but it may be that a combination of consistency and chemical analysis (pH, water content) may eventually prove to be useful.

P. Walstra (the Netherlands) : Is it known to what extent the rates of the various reactions producing flavour components depend on conditions such as temperature and pH? This would be most useful information in relation to accelerated cheese ripening.
J. Adda (France) answered that such knowledge is not yet available.

H.T. Badings (the Netherlands) : Would it be possible to greatly accelerate cheese ripening without significantly altering the flavour profile of the cheese.
J. Adda (France) considered this to be unlikely for most types of cheese. This would not imply that accelerated ripening could not lead to good-tasting cheese varieties.

J. Visser (the Netherlands) : A French study has shown that polyphosphates in a rat feeding trial induce hypocalcaemia. Other have shown that for that reason processed cheese should not be given to infants and elderly people.
E. Renner (Fed.Rep. Germany) : I do not think that in humans an effect has to be feared, since polyphosphates are quickly broken down by the body to monophosphates, such a physiological mechanism can not exist.

J. Visser (the Netherlands) : An Australian patent claim an anticaries effect of $a\hat{s}_1$-casein. Since cheese is rich in $a\hat{s}_1$-casein, are there indications that cheese also has an anticaries effect?
E. Renner (Fed.Rep. Germany) : I remember a publication 1 or 2 years ago, where such an anticaries effect of cheese was described. At least, it is generally accepted that cheese does not have any cariogenic effect.

J.L. Maubois (France) : You suggested that the consumption of quarg must be recommended to people suffering from liver diseases. Can you comment on that? Is this general for fresh cheeses or special for fresh cheeses enriched with whey proteins. In the latter case is that related to the need for amino acids such as valine of isoleucine?
E. Renner (Fed.Rep. Germany) : It is already a long tradition to give quarg to patients recovering from liver diseases, since quarg fulfils the demands of those patients for a palatable, easily digestible protein with a high biological value, which is necessary for the regeneration of the affected liver cells. The biological value of thermo-quarg or UF-quarg is still higher due to the presence of whey proteins. Therefore the significance of these products for this purpose is still greater.

G.G. Shiler (USSR) : What is the significance of the steadily increasing cheese consumption for the nutritional status of the consumer?
E. Renner (Fed.Rep. Germany) : The increasing cheese consumption, which we can observe in all parts of the world, can be considered as very positive from a nutritional point of view, because of the high nutritional quality described.

G.G. Shiler (USSR) : Why have whey proteins a higher nutritional quality than casein?
E. Renner (Fed.Rep. Germany) : The whey proteins have a higher concentration of some of the essential animo acids, in particular of lysine, tryptophan, thronine, isoleucine and cysteine, which results in a higher biological value as compared to casein.

N.J. Tofte Jespersen (Denmark) : The proportion of lactic acid in the D(-)form in cheese was mentioned as 4 to 14 % for fresh cheese and 10 to 50 % for ripened cheese. Are these figures correct? We use lactic streptococci producing L(+) for most of the semi hard cheeses and for Cheddar, and we use *S. thermophilus* and *Lactobacillus helerticus* for Emmental, and none of them produce D(-) lactic acid.
E. Renner (Fed.Rep. Germany) : Some types of lactic streptococci also produce a small amount of D(-) lactic acid. Therefore, we often find a small proposition of this isomer in cheese. The figures which I gave on the content of D(-) lactic acid in different cheese varieties have been obtained in investigations performed in Switzerland. I also was surprised on the high concentration of D(-) lactic acid in some cases.

G.G. Shiler (USSR) : Do you consider the excess of certain essential amino acids in cheese protein as compared with the reference protein as positive, neutral of negative?
E. Renner (Fed.Rep. Germany) : Clearly as positive, because in a mixed diet, which includes proteins - like cereal protein - with a low concentration of essential amino acids, this excess in cheese protein makes it possible to achieve a good balance in the concentration of essential amino acids of the whole diet.

M.D. Northolt (the Netherlands) was surprised at the very high level of tyramine in Cheddar cheese reported.
E. Renner (Fed.Rep. Germany) answered that this was just an example from the literature. It was further mentioned in the discussion that results obtained by H. Joosten (the Netherlands, not present) on formation of biogenic amines in Gouda cheese had shown that (a) in a cheese made from pasteurized milk under strictly hygienic conditions almost no biogenic amines are found, and (b) that if contaminating organism are present that can decarboxylate amino acids to amine, the production of free amino acids rather than the decarboxylation determines the rate of amine production.

M.D. Northolt (the Netherlands) further said that he had done experiments in Gouda cheese made of milk contaminated with *Listeria monocytogenes* ; unlike the behaviour of *Salmonellae, Listeria* did not grow in the cheese, but neither died off during the first six weeks.
H.A. Morris (USA) considered it likely that future legislation in the US would compel pasteurization of all milk used for making milk products, including cheese. However, post-pasteurization infection may in some instances be of greater concern.

P. Walstra (the Netherlands) remarked that sterigmatocystine, - although similar in structure to aflatoxin B - is nevertheless about a thousand times less toxic. He also recalled the lecture of Professor J. Koeman during the Annual Sessions, where he showed that consumption of cheese (even if containing nitrate) most probably diminishes the formation of nitrosamines in the stomach.

Seminar II: Milk Production and Milk Products in Developing Countries

Session 1

Chairman: Prof. Dr. H. A. Jasiorowski (Italy)
Secretary: Dato Osman Bin Din (Malaysia)

Indigenous production of cow's milk and milk of other animals

MILKPRODUCTION BY INDIGENOUS CATTLE

H. Bakker
Hendrix International b.v.
P.O. Box 1
5831 MA Boxmeer
Holland

INTRODUCTION

In many tropical countries dairy production is of considerable significance to the human population:

- Milk is a highly valuable nutrient, contributing to a better balanced diet for people in both rural and urban areas.
- Dairy production can be an important specialized enterprise and can also contribute to the economics of mixed farming by the small holder, providing gainful self-employment and/or sale of products.
- Ruminants create the possibility of converting low quality roughage and by-products into products which are digestible by humans. Areas which are unsuitable as arable lands may thus be used indirectly for food production.
- Owning a herd of livestock means status and security against risks.

Circumstances for animal production in tropical countries are often vastly deviating from conditions in temperate zones. Climatic conditions are severe by high temperatures sometimes in combination with a high relative humidity.
    These climatic conditions have a direct impact on the animals and an indirect effect via feed availability and feed quality. Especially in the arid and semi-arid zones feed is scarce and of a very low digestibility during the greater part of the year. In many areas infectious diseases e.g. anaplasmosis and parasites, e.g. ticks add to the already difficult circumstances of animal production.
    Dairy production systems, under these circumstances, are generally extensive: feed and capital inputs are very limited; production levels of milk, meat and reproduction (age at first calving, calving interval) are low. Indigenous cattle and buffaloes are well adapted: they have a certain amount of tolerance to climatic conditions resistance to diseases, and their often low production potential is not a limiting factor.

From FAO (1983) figures we may obtain an impression of the major ruminant livestock species in industrialized, developing countries and those with a centrally planned economy.

	developed	developing	centrally planned
cattle	276	735	214
buffaloes	-	102	22
sheep	351	482	303
goats	18	369	89

These figures show that a large proportion of the world population of these species is kept in the developing countries. They play a very important role by converting low quality roughages, which otherwise would be almost useless, into valuable products or functions to the human populations.

Feed

The largest part of these roughages is obtained by grazing systems on marginal lands. Other important sources are crop-residues (like straw and stalks) and agro-industrial by-products from milling, sugar production, vegetable oil extraction etc.

Winrock Research Centre (1978) estimated the quantities of metabolisable energy (ME) from different resources for ruminants in developing countries:

- Grazing areas and permanent pastures   43 %
- Areas with non-agricultural functions
  (roadsides, forests etc.)   11 %
- Green fodder produced on crop lands   21 %
- Crop residues   24 %
- Cereals   0,5 %
- Oil seeds   0,5 %
- Agro-industrial by-products   1 %

It may be concluded from these figures that direct competition in land use between ruminants and mankind is very limited.
Obviously, the milkproduction of cattle or buffaloes on low quality rations will be low.

Very often milkproduction will be no more than 1.000 liters per lactationperiod of which a substantial part will be consumed by the calf. An age at first calving of 3 to 4 years, a calving interval of 1½ to 2 years and a low growth rate are some of the characteristics reflecting the low input levels in cattle and buffalo husbandry systems in developing countries.

Functions

Dairy products play a very important role in human nutrition. However, dairy husbandry is seldom a single-purpose operation. In many tropical countries cattle and buffaloes are kept for three purposes: milkproduction, meatproduction and labourproduction.

Producing bullocks as draught animals for agricultural operations or rural transportation is often a major reason for keeping cattle. Manure for fertilizing the crop land and as a fuel is an important by-product of animal production.

The illustration of this point can be given by figures of Ehrlich et al (1977) who reported that in India, of the 800 million tons of dung annually produced, more than 300 million tons is used after drying as a fuel for cooking and heating.

Farming systems

Larger ruminants in developing countries are kept in different farming systems. Payne (1976) presented figures about the distribution of cattle and buffaloes in different farming systems in the tropics:

- Nomadism   5 %
- Transhumance   10 %
- Ranching   29 %
- Mixed farming with crops   56 %
- Specialised and intensive farms   1 %

Of course dairy animals will not be distributed evenly over these different systems. The large proportion of cattle and buffaloes kept on mixed farms is striking. This farming system is prevalent in Asia.

Ranching is of great significance in South America and partly in Africa. Nomadism and transhumance herding systems are mainly found in Africa. It is also clear that specialized intensive production systems in the tropics are in numbers almost negligible.

DAIRY DEVELOPMENT

Many countries aim to further develop dairy production. As was mentioned in the Introduction section the main arguments are rural development, better balanced diet to the human population, saving foreign currency and utilizing available roughages. The individual farmer considers increase of milkproduction as a tool to generate a regular flow of income.

To increase the production levels per cow in a dairy development program a large number of points should be taken into consideration as is illustrated in Figure I.

- Increase of the feed quantity and quality.
- Improvement of management, housing, husbandry methods and veterinary care.
- Genetic improvement production and reproduction capacity of the cattle.

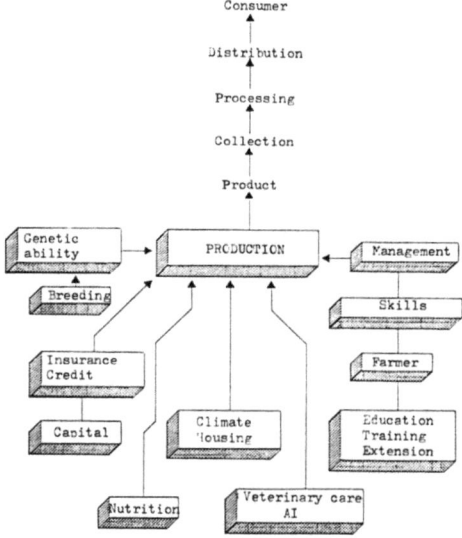

Figure I.: Some factors of importance in dairy development

However, it is also necessary to create an infrastructure for dairy cattle husbandry. Important components are: reliable and attractive marketing outlets for the product, possibilities to purchase raw materials such as feed, effective extension, education, practical training, veterinary and A.I.-services.

In addition to these points attractive loan schemes and adequate cattle insurance are necessary if investments in dairy animals have to be made. A cattle development policy has to cover all these points to avoid failures of the program.

Some examples of dairy development program

One of the largest examples of dairy development in the tropics is the Operation Flood program in India. Milk co-operatives in the villages are collection points for locally produced milk. The milk is transported to dairy plants at district level, where it is further processed and then supplied mainly to the urban population.

A number of services are supplied to the dairy farmer via this organization such as veterinary care, A.I., concentrates etc. Payment is done to the farmer very frequently. This scheme is adopted in India as the national dairy development strategy and applied in more than 150 districts. It is also used as a model for dairy development in other Asian countries such as Indonesia and Sri Lanka.

More detailed information will probably be presented in the contribution of Dr. Kurien to this Congress.

In the state of Kerala in the South of India a similar program is executed. In an Indo-Swiss dairy development program cross breeding of local cattle with Brown Swiss bulls or Jersey bulls was applied to upgrade the local cattle population. In subsequent stages of the program extension activities are combined with A.I., pedigree registration and milkrecording under field conditions, to select superior cows to be used as bull dams and to select a group of bulls on the basis of their progeny's performance under field conditions. Young male offspring will then be used as the next generation of breeding bulls.

These selection cycles may result in a genetic progress of 1-2 % per year in the cattle population, which has 50-75 % exotic blood of Brown Swiss or Jersey origin due to the previous cross breeding.

The Kerala dairy development scheme is an example of the application under field conditions of a scientifically designed breeding program. A description was presented by Chacko et al (1985).

An example of a dairy development program in Kenya is given by Voskuil (1984). It is a co-operation of the Kenyan and the Netherlands' governments. Its target group are smallholder farmers in seven districts scattered over Kenya. As the project has its target group in the high potential areas it adopted mainly the most intensive system of zero grazing. The average farm size of a project farmer is $\pm$ 3 hectares, of which almost 1.25 hectares are utilized to keep 2.5 dairy cows and followers.

In the Kenyan zero-grazing system the cows are almost entirely fed on Napier grass (Pennisetum purpureum). Legumes are added of which a mixture of Napier and Desmodium gives variable, but often good results. A carrying capacity of 1 cow plus followers per 0.5 hectares is recommended.

Under zero grazing conditions animals are kept indoors, making the availability of the manure a favourable side effect. Economic analyses have shown that milkproduction contributed 62 % of the total farm income by using 42 % of the farm size. So dairying contributed 1.5 times as much as the other farm activities per hectare.

Some technical results achieved on the farms are given. They show that still much room for improvement of the management is left:

calving-interval	455	days
interval calving-1st insemination	128	days
interval calving-conception	173	days
cow-mortality	4,3	%
calf-mortality	12	%
milkproduction (305 days)	2.032	kg

Another effect of dairy development is the creation of employment in the rural areas, not only within the farms, where at smaller farms more excessive labour is available without alternative opportunities, but also external labour is employed, as a survey within the Dairy Development Project shows.

The family contributes 70 % of the required labour on the dairy activities, of which the women execute about 25 %. The women expressed their views in several cases that less of their time is involved in dairy activities than before. When the cows were taken for grazing in the traditional system, women were looking after them most of the day, leaving little time for their other agricultural and household duties.
Casual labourers carry out 15 % of the work and also 15 % is carried out by permanently employed workers. So labour opportunities are created and income generated to the landless, poor people.

LITERATURE

- C.T. Chacko, F. Bachmann, F. Schneider, W. Kropf   1985
  Results of dairy cattle improvement programmes under
  field conditions in India: Kerala as a case study.
  Paper presented at E.A.A.P. Conference Thessaloniki,
  Greece 1985.

- P.R. Ehrlich, A.H. Ehrlich, J.P. Holdren   1977
  Ecoscience; population, resources, environment.
  W.H. Freeman and Co., San Francisco.

- W.J.A. Payne   1976
  Systems of beef production in developing countries.
  In: Beef Cattle production in developing countries.
  Ed. A.J. Smith. Centre for Tropical Veterinary Medicine,
  Edinburgh.

- Winrock International   1978
  The role of ruminants in support of man.
  Winrock Intern. Livest. Research and Training Centre,
  Morrilton, U.S.A.

- Food and Agriculture Organization   1983
  Production yearbook Vol. 37.
  FAO, Rome.

- Ed. A.J. Smith   1985
  Milk production in developing countries.
  University of Edinburgh, CTVM, Edinburgh.

- G.C.J. Voskuil   1984
  Small scale milkproduction and its role in rural development.
  Paper presented at FAO seminar on dairying for rural
  development, Nairobi.

MILK OF OTHER ANIMALS

V.N. Tripathi
Division of Dairy Cattle Genetics,
National Dairy Research Institute,
Karnal 132 001 (Haryana) India

ABSTRACT. Besides cattle, the species of livestock contributing to the supply of milk are buffaloes, goats, sheep and camel. The total population of cattle, buffaloes, goats and sheep in the world during the year 1985 was 1267.9, 128.7, 461.0 and 1122.6 million, respectively. The amount of milk produced by cows, buffaloes, sheep and goats during the year 1985 in the world was 455.3, 32.6, 8.6 and 7.6 million metric tons, respectively. The amount of milk produced by cows, buffaloes, sheep and goats in the developing countries during the year 1985 was 76.4, 32.5, 4.8 and 5.5 million metric tons, respectively. In this article, an attempt has been made to review the aspects of breeds, feeding, management, milk production and composition and production efficiency of buffaloes, sheep, goats and camel in the developing countries.

1. INTRODUCTION

The population of cattle, buffaloes, sheep and goat in developed countries, developing countries and the world is presented in Table 1. It is evident from the Table 1 that most of the buffaloes and goats are found in the developing countries of the world. There was a decline in the population of both cattle and buffaloes in the developed countries of the world and its magnitude was 4.85 and 14.30% in the two species, respectively. The increase in the population of cattle, buffaloes, goats and sheep during the last decade in the developing countries was 13.00, 11.93, 10.77 and 8.75%, respectively.

The amount of milk produced by cows, buffaloes, sheep and goats, during the year 1985 in the world was 455.339, 32.638, 8.604 and 7.605 million MT, respectively (Table 2). The amount of milk produced by cows, buffaloes, sheep and goats in the developing countries during the year 1985 was 76.432, 32.543, 4.841 and 5.595 million MT, respectively. The percent increase in the amount of milk fom cows, buffaloes, sheep and goats during the last decade in the world was 16.49, 39.51, 26.59 and 21.52, respectively. The percent increase in the amount of milk from cows, buffaloes, sheep and goats during the last decade in the developing countries was 30.79, 39.66, 37.65 and 27.199, respectively. The amount of the total milk produced in the world during the year 1985 was 504.206 million MT and the share of developed and developing countries was 384.497 and 119.411 million MT, respectively (Table 3). The contribution of cows, buffaloes, sheep and goats in the total supply of milk in the world based on the estimates for the year 1985 was 90.31, 6.47, 1.77 and 1.51%, respectively. The corresponding values for the four species was 98.48, 0.02, 0.98 and 0.52% in the developed countries and 64.01, 27.25, 4.05 and 4.69%, respectively in the developing countries of the world.

It is evident from the Table 4 that buffaloes are the second most important source of milk in the world. The proportion of buffaloes milk vis-a-vis cow milk in the world was 7.17% while it was 0.03% in the developed countries and 42.58% in the developing countries. The proportion of the sheep milk in comparison to cow milk was 1.89, 1.00 and 6.33% in the world, developed and developing countries, respectively. The contribution of the goats in comparison to cow milk was 1.67, 0.53 and 7.32% in the world, developed and developing countries, respectively.

It is evident from the preceding discussion that buffaloes, goats and sheep are important source of milk for the people of the developing countries of the world and there was continuous increase in the population of all the four species of farm animals.

An attempt has been made to review on the aspects of breed, feeding, management, milk production and composition and production efficiency of buffaloes, sheep, goats and camels in the developing countries.

2. BUFFALOES

2.1 Breeds of Buffaloes

There are two types of water buffaloes namely Swamp buffaloes and River buffaloes (Cockril, 1977).

The characteristics of Swamp buffaloes are: (i) heavy bodied, (ii) stockily built, (iii) the body is short, (iv) the belly large, (v) the forehead is flat, (vi) the eyes prominent, (vii) the face short, (viii) the muzzle wide, (ix) the withers and croup are prominent, (x) the neck is comparatively long, (xi) the dorsal ridge extends backward and ends abruptly just before the end of the chest.

The characteristics of river buffaloes are: (i) comparatively longer face, (ii) smaller girth and (iii) bigger limbs,

(iv) the dorsal ridge extends further back and tapers off more gradually.

The differences between males and females are more marked in river breeds than in Swamp buffaloes.

Breeds of buffaloes of Indian subcontinent are: Murrah, Nili-Ravi, Kundi, Surti, Mehsana, Jafarabadi, Bhadawari, Tarai, Nagpuri, Pandharpuri, Manda, Jerangi, Kalahandi, Sambalpur, Toda and Malabari.

2.2 Feeding

The main objectives of feeding lactating buffaloes are provide sufficient nutrients for the maintenance, growth (in case of primipara) and milk production. In a normal feeding practice at a farm the ration of a dairy buffalo consists of two parts viz.,(i) maintenance, and (ii) production part. The maintenance part of the ration depends upon the body weight and production part is dependent upon quantity and composition (Specially the fat percentage) of the milk. Till 1972, both starch equivalent (SE) and total digestible nutrients (TDN) systems were followed in expressing the energy value of the feeds but presently the ME system is being followed widely.

2.3 Milk Production

The lactation yield was highest (2272 kg) in Murrah and lowest (956 kg) in Surti breed of buffaloes (El-Arian, 1986).

2.4 Milk composition

Sharma et al. (1980) studied the milk composition of different breeds of buffaloes at Bombay. Fat percentage was 7.40 ± 0.594, 7.40 ± 0.570, 7.40 ± 0.689 and 7.30 ± 0.477 in Jafarabadi, Mehsana, Murrah and non-descript breeds respectively. The fat percentage was 7.6 ± 1.063, 7.6 ± 0.860, 7.4 ± 0.707 and 7.4 ± 0.58 in monsoon, autumn, winter and summer seasons, respectively in Murrah buffaloes.

A total of 5069 milk samples were analysed since June 1980 to December 1984 in Murrah herd of NDRI, Karnal (India). The overall average fat, protein and SNF contents were 7.54, 4.43 and 11.13%, respectively (Chawla and Tomar, 1986).

The effects of year and month on milk constituents were found to be significant. The fat content was highest in April (8.20%) and lowest in February (7.32%). The protein content was the highest in February (4.73%) and the lowest in March and November (4.30%). The average percentages of milk constituents in different breeds of buffaloes are presented in Table 5.

2.5 Efficiency of Feed Conversion

An investigation on the efficiency of feed conversion of high yielding crossbred cow, the high yielding Indian dairy cow, high and low yielding buffaloes and goats at the National Dairy Research Institute has made interesting revelation (Sundaresan, 1980). A high yielding cow, averaging 3000-3500 litres of milk per lactation, has an efficiency of converting nutrients of feed taken into milk nutrients, energy 25 to 30 per cent and protein 25 percent. A buffalo averaging 1500-2000 litres of milk per lactation has an efficiency of about 20 percent in case of energy and and about 15 percent in case of protein. A buffalo with lesser production of about 1000 litres per lactation but maintained on coarse forage and grain byproducts gave 35% energy conversion and 26% protein conversion. A local cow yielding about 400 litres of milk in a lactation has an efficiency of conversion of energy and protein of only 4-6%. The high yielding cow with greater efficiency will need lesser amount of land when put to high quality animal feed. Whereas a low yielding animal with less efficiency in feed conversion will need greater amount of land but of coarser feed material including byproducts. The medium yielding buffalo on coarse feed gave the best efficiency. To the extent that byproducts of human feed like cereal straw and crop waste can support a population of low yielding buffaloes, they would be ideal to situation where land cannot be allotted to high quality forages for high yielding cows, where animal husbandry is essentially a subsidiary of the agricultural enterprise.

3.0 GOATS

3.1 Breeds of Goats in the Tropics and Sub-tropics

The following are the important breeds of goats which are considered medium milk yielders:

Breed	Country of origin
Barbari	India, tropical, dry
Beetal	India, tropical, dry
Black Bedouin	Egypt, Istreal, tropical, very dry
Damani	Pakistan, tropical, dry
Damascus	Syria, Lebananon, sub-tropical dry
Dera Din Panah	Pakistan, tropical, dry
Jamnapari	India, Tropical/sub-tropical, dry
Kamori	Pakistan, sub-tropical, dry
Kilis	Turkey, sub-tropical, dry
Malabari	India, tropical, humid
Marwari	India, tropical, dry

Sudanese Nubian	Egypt and Sudan, tropical, dry
Zeribi	Egypt, tropical, dry

3.2 Feeding

Goats have feeding habits which differ from those of other ruminants and lead to the destructiveness of the former when uncontrolled. By means of their mobile upper lip and very prehensile tongue, goats are able to graze on very short grass and to browse on foliage not normally eaten by other domestic livestock.

Goats are inquisitive feeders with a feed range from herbage to tree bark. Goats relish eating aromatic herbs in area of sparse feed supply and hence can penetrate deep into the desert. While goats will accept a wide variety of feeds, they have contrary to popular opinion fastidious feeding habits. Food that is acceptable to one goat is sometimes not acceptable to another, and goats usually refuse anything which has been spoiled by other animals. Goats can distinguish between bitter, sweet, salty and sour tastes and show a high tolerance for bitter taste than cattle.

3.3 Milk Yield and Lactation Length

Average milk yield and lactation length in different breeds of goats are presented in Table 5. Damascus and Jamnapari are the two highest milk yielding breeds of goats in the developing countries.

3.4 Composition of Goat Milk

The composition of goat milk differs from breed to breed and between individuals within breeds as well as due to stage of lactation and environmental factors such as nutrition. It can be seen that there is a tendency for the milk of tropical breeds to be higher in total solids, mainly due to higher fat and protein contents, within a breed, and also within lactation, total solids tend to be inversely related to milk yield. However, in tropical breeds, the lactose level is generally similar to that of temperate breeds and ash percentage does not vary greatly.

4.0 SHEEP

4.1 Breeds

There are a large number of breeds of sheep in developing countries which can be classified as to whether they are hairy coated or wooled, thin tailed or fat-tailed/fat rumped, and

whether they are horned or polled (Williamson and Payne, 1978).
Some of the important milch breeds of sheep are Sudanese Desert, Lohi, Kathiawari and Awassi.

4.2 Milk Yield and Milk Composition

In most of the contries milk production is for domestic consumption. For instance, the population in many Indian cmmunities and most nomadic sheep owner throughout western Asia derive an important part of their diet from sheep milk. In India and in western Asia, ghee is sometimes prepared from sheep milk. Different types of cheeses and some form of Yoghurt are commonly made from sheep milk. Ewes are usually milked once a day by hand. Ewes respond to good management and feeding in the same way as milking cows.

In western Asia, the Awassi is the most productive milch breed. Finci (1957) reported that in Israel the average production in the 11 best flocks was increased by 266 kg in 18 years, the average animal yield in 1955-56 being 359 kg. Yield was maximum at the fourth lactation. The fat content was 6 to 8%.

Ewes of the Lohi breed yield upto 3.6 kg of milk a day (Kaura, 1941) while the Kuka breed is said to produce 1.8 to 3.6 kg of milk per day (Kaura, 1942). The Sonadi breed produce 0.9 to 1.4 kg of milk daily (Anon, 1953).

In Africa, Dwarf west African ewes managed under poor conditions, average 30 to 40 kg of milk in 120 to 135 days of lactation and when well fed can produce 75 to 85 kg. Ewes belonging to Sudan desert breed produce upto 2.3 to 2.7 kg daly. Priangan ewes from Indonesia weighing 30 to 40 kg can produce 21 to 53 kg of milk with a 5% fat content per lactation (Atmadilaga, 1958).

Marawari breed of sheep on an average yielded 36.79 ± 9.05 kg of milk during average lactation length of 17 weeks and milk had fat and total solid content of 5.07 ± 0.04 and 15.95%, respectively (Patel et al., 1984). Patanwadi breed from Saurashtra (India) produced 37.63 ± 8.52 kg of milk during average lactation length of 17 weeks and the fat and total solid contents of the milk was 6.08 ± 0.08 and 15.91%, respectively (Patel et al., 1984).

Parity, frequency of milking, age of ewes, levels of nutrition were responsible for causing significant variation in milk production of sheep (Labussiere et al., 1984; Akcpinar et al., 1984; Abou-Naya, 1981; Vitkov, 1985).

4.3 Feeding

Practically, all sheep in developing countries are maintained on unimproved grazing. In Africa and Asia, they are grazed extensively on grazing lands as well as on crop stubbles after its harvesting. They are selective grazer and prefer short grasses, legumes and wide variety of low growing herbs. When they are transferred to a new locality, they have little knowledge of what forage is suitable and often they do not thrive well for a long period. Average daily intake of dry matter was reported to be 1.14, 0.93 and 1.27 kg per 100 kg body weight for sheep of the Merino, Black head, Prussian and Dwarfer breeds, respectively (Williamson and Payne, 1978). The feed consumption of the tropical breeds was somewhat lower than that of temperate type of sheep. As sheep tend to thrive best on dryer climate where feed supply fluctuates both in quantity and quality from the wet to dry season, supplementary feeding is often of importance during the dry season. Sheep can be maintained through drought period on hay and if this can be supplemented with 0.11 kg of protein concentrate per head per day, normal growth and development can be maintained.

Where sheep encounter a long dry season, its productivity can be increased by supplementing their rations while they are on grazing and by managing those sheep destined for slaughter in a feed lot.

Good feeding is very important before the onset of the breeding season. Supplementary feeding prior to the onset of breeding season helps to improve prolificacy. Good feeding is also required during the later half of the pregnancy. Well fed wool sheep will usually produce a heavier but coarse fleece. Sheep should be offered common salt at a rate of 7 g/day. It may be necessary to provide trace minerals such as copper and cobalt in a suitable form, if a deficiency is known or suspected. Sheep should be given access to water all times. On an average, sheep will drink 4 to 5 litres of water under semi-arid tropical conditions daily. When sheep are fed feed indoors they should be provided water free-choice. When sheep are fed indoors, the feed may consist entirely of succulent fodder or of a mixture of forage and concentrates. If suitable forage or browse is not available, mature sheep may be fed upto 450 gms of concentrates daily. A suitable concentrate mixture would consist of 38 parts cereal grains; 20 parts wheat bran; 20 parts rice bran; 10 parts oil cake (groundnut, cotton seed, sesame meal); and 2 parts mineral mixture.

Feeding of good legume hay or fresh green forage or browse wth this type of concentrate would improve the ration and meet the requirement of all the nutrients.

5.0 CAMELS

5.1 General

Camels are used for the transport of people and goods in arid and semi-arid regions; milk, meat, fibre and hides are the by-products.

There are more than 14.5 million camels in the world and

nearly 70% of these animals are to be found in Africa (FAO, 1974). More than 70% of the total world population of camels are raised in the tropics. Nearly 91% of camels in Africa are found in the tropical areas, whereas in Asia two-thirds of camel population is found outside tropical regions. Tropical countries with large camel populations are Sudan, Somalia, Ethiopia, Mauritania, Saudi Arabia and India with 23.4, 20.6, 6.8, 4.8, 4.0 and 3.9% of the total world camel population, respectively.

5.2 Breeds

Camels belong to order <u>Artiodactyla</u>, family <u>Camelidae</u>, and genus <u>Camelus</u>. They are of two kinds, viz. the Arabian or one-humped camel (<u>Camelus dromedarius</u>) and the Bactrian or two humped camel (<u>C. bactrianus</u>) of Turkestan. Neither the Arabian nor the Bactrian form exists any longer in wild state, though there are some semi-wild herds, which have escaped from captivity. Bactrian camels are distinguished by a thick wooly coat that is usually reddish brown in colour, by a deep fringe of hair under the neck and have shorter limb bones than dromedaries. The dromedary is slightly larger than the Bactrian camels. Its coat is also woolly but shorter than that of the Bactrian and the coat colour is usually fawn. It has very broad feet, long eyelashes, trapdoor nostrils that can be closed and extremely thick lips.

Only dromedary is used in tropics and there exists a large number of different types of dromedary, many of which are spoken of as ´breeds´. These breeds may be classified into two general types, the riding camel and the baggager. Breeds within these types are not marked by many pronounced functional or conformational traits. The conformation and performance of the plains camel vary in all degrees between the light, fine-boned, thin-skinned, alert, desert, riding type and the massive but rather mean looking phlegmatic baggager type from the riverine areas, accustomed to good living and regulated activities.

5.3 Productivity and milk composition

A good riding camel can cover about 130-160 km at a speed of about 56 km per day. It is capable of carrying a load of 224-261 kg (CSIR, 1970). The baggage camel travels at a walking pace at just over 4 kg an hour and can carry a full load for 2.4 km a day for an indefinite period.

Bactrian camels may produce upto 5,000 kg of milk per lactation which may vary in length from 6 to 18 months. However, the average production is only 800 to 1200 kg. The fat content in the milk of Bactrian camels varies from 5.8 to 6.6% (Kulaeva, 1964). Properly cared dromedaries will yield 2,722 to 3,629 kg of milk in a 16 to 18 months lactation while under desert conditions it will yield 1,134 to 1,588 kg of milk in 9

Table 1. Population of cattle, buffaloes, sheep and goats (million)

	Cattle			Buffalo			Goats			Sheep		
	1974-76	1985	Change %	1974-76	1985	Change %	1974-76	1985	Change %	1974-76	1985	Change %
World	1192.449	1267.997	+ 6.34	115.247	128.769	+11.73	416.415	461.056	+10.72	1053.554	1122.616	+6.56
Developed countries	444.919	423.325	- 4.85	0.860	0.737	-14.30	24.473	26.895	+ 9.90	518.313	540.559	+4.29
Developing countries	747.529	844.672	+13.00	114.386	128.032	+11.93	391.942	434.160	+10.77	535.242	582.057	+8.75

Source: FAO (1986)

Table 2. Milk production from cow, buffaloes, sheep and goats (million, MT)

	Cow milk			Buffalo milk			Sheep milk			Goat milk		
	1974-76	1985	Change %	1974-76	1985	Change %	1974-76	1985	Change %	1974-76	1985	Change %
World	390.908	455.359	+16.49	23.394	32.638	+39.51	6.797	8.604	+26.59	6.258	7.605	+21.52
Developed countries	332.471	378.928	+13.97	0.093	0.095	+ 2.15	3.280	3.764	+14.76	1.859	2.010	+ 8.12
Developing countries	58.437	76.432	+30.79	23.301	32.543	+39.66	3.517	4.841	+37.65	4.399	5.595	+27.19

Source: FAO (1986)

Table 3. Contribution of cow, buffalo, sheep and goat in the supply of total milk in the year 1985.

	Total milk (million MT)	Cow %	Buffalo %	Sheep %	Goats %
World	504.206	90.31	6.47	1.71	1.51
Developed countires	384.497	98.46	0.02	0.98	0.52
Developing countries	119.411	64.01	27.25	4.05	4.69

Scource : FAO (1986)

Table 4. The percentage of buffalo, sheep and goat milk to cow milk in the year 1985

	% of Buffalo milk to Cow milk	% of sheep milk to Cow milk	% of Goat milk to Cow milk
World	7.17	1.89	1.67
Developed countires	0.03	1.00	0.53
Developing countries	42.58	6.33	7.32

Source : FAO (1086)

Table 5. Lactation yield and length in different breeds of goats.

Breed	Country	Lactation yield Kg.	Daily yield Kg.	Lact. length days
Barbari	India Pakistan	150-228	1.6	180-252
Beetal	" "	140-228	1.2	208
Black Bedouin	Israel	-	1.3-2.0	-
Damani	Pakistan	104	1.0	105
Damascus	Cyprus	500-560	2.0	190-290
Dera Din Panah	Pakistan	200	1.5	130
Jamnapari	India	200-562	1.5-3.5	170-200
Kamori	Pakistan	228	1.8	120
Kilis	Turkey	280	1.0	260
Malabari	India	100-200	1.0	181-210
Marwari	India	90	0.9	106
Sudanese Nubian	Egypt, Sudan	70	1.0-2.0	-

Source : Devendra and Burns (1983)

months (Yasin and Wahid, 1957; Iwema, 1960). A good animal can attain peak yield of 9 kg. The content of fat, protein and lactose in dromedary milk are 3.8, 3.5 and 3.9%, respectively (Nawita et al., 1967).

Average dromedaries weigh from 454 to 900 kg while average Bactrian camels may weigh somewhat heavier. Average dressing percentage of Bactrian camels varies from 56 to 70 (Kulaeva, 1964).

Camels in India clipped in March or April yield on an average 0.90 to 1.35 kg of hair (CSIR, 1970). The average yield of hair of the Bactrian camel is 4.5 kg (Williamson and Payne, 1978).

5.4 Feeds and feeding

Camels like to browse rather than graze. They should be allowed to forage for at least 6 hours. In many countries, camels depend entirely on browsing while in other countries for example Somalia, grass forms a main part of the ration. Camels can also be reared by stall feeding on grain and fodder. An adult camel requires at least 142 g of salt in its ration daily. Riverine camels must be watered daily. Indian desert camels can be maintained in good condition if they are watered on alternate days. Somali camels will maintain their condition even when watering is infrequent.

5.5 Management

Females must have access to good grazing for at least 2 months prior to calving and 3 weeks after calving. Many new born die before they are 3 weeks old. New born must be fed colostrum. Only one quarter of the available milk is fed to young one upto 3 weeks of age and it is enhanced gradually thereafter. Desert fed dam dries off after 9 months of lactation while well fed animal dries off after 18 months of lactation (Williamson and Payne, 1978).

Acknowledgements: I am extremely grateful to Dr. R. Nagarcenkar, Director, N.D.R.I., Karnal for his valuable help in preparing and presenting the article on my behalf. Thanks are also due to Mr.M.N. El-Arian and Dr. P.U. Gajbhiye for their assistance in the preparation of this article.

REFERENCES

1. Abou-Naga, A.M., El-Shobokshy, A.S., Marie, I.F. and Moustafa, M.A.(1981) Milk yield of Rahmani, Ossimi and Barki local ewes. Alexandria J. Agri. Res., 29: 480-493 (Anim. Breed. Abstr., 53: 5776).

2. Akcapinar, N., Aydin, I. and Kadak, R. (1984) Lamb production and milk yield of Red Karaman sheep under private farm conditions in Erzurum region. Ankard University Veteriner Fakultesi Dergisi, 31: 114-127 (Anim. Breed. Abstr., 53: 4952).

3. Anon (1953). Indian Farming, No.5, 3: 10 and 26.

4. Atmadilaga, D. (1958) Study on the milk yield of Indonesian sheep with special reference to the Priangan breed. Hemera Zoa, 65: 3-14.

5. Chawla, D.S. and Tomar, O.S. (1986) Effect of year and month on milk yield and its components in Murrah buffaloes. Terminal Progress Report of Buffalo Project, NDRI, Karnal, India, PP. 12.

6. Cockrill, W.R. (1977) The water buffalo. FAO, Rome

7. CSIR (1970) The Wealth of India (Livestock supplement).

8. Demiruren, A.S., Beheshti, R.D., Salimi, H., Salesh, B.H. and Djaferi, A. (1971) Comparison of the reproductive and productive capacities of sheep of the Kellakul, Kizili, Bakhtiari and Baluchi breeds in Iran. Tech. Rep. No.1. Anim. Husb. Res. Inst., Teheran.

9. Devendra, C. and Burns, M. (1983) Goat production in the Tropics. Commonwealth Agricultural Bureaux.

10. El-Arian, M.N. (1986) Genetic analysis of Murrah buffalo herds. Ph.D. Thesis, Kurukshetra University, Kurukshetra.

11. F.A.O. (1974) Production Yearbook, Vol. 27, FAO, Rome.

12. F.A.O. (1986) Monthly bulletin of statistics. 9 : 19-28.

13. Finci, M. (1957) The improvement of the Awassi breed of sheep in Israel. Bull. Research Council, Israel, 6: 1-106.

14. Iwema, S. (1960) The ship of the desert in (Dutch). Veeteelt-en Zuivelberichten, 3: 390-394 (Anim. Breed. Abstr., 30: 98).

15. Jenness, R. (1980) Composition and characteristics of goat milk: Review 1968-1979. J. Dairy Sci., 63: 1605-1630. International Symposium : Dairy Goats (ADSA Diamond Jubilee meeting), Baton Rouge, June 28-July 1, 1980.

16. Kaura, R.L. (1941) Some common breeds of Indian sheep. Indian Farming, 2: 175-179.

17. Kaura, R.L. (1942) Some common breeds of Indian sheep. II. Indian Farming, 3: 122-125.

18. Khosla, S.K. and Mathur, A.K. (1984) National herd book, Bull. Ministry of Agri., Dept. of Agri. and Co-op., New Delhi.

19. Kulaeva, V. (1964) The production of bactrian camel (in Russian). Konevod. Konnyi Sport, 34: 9-10 (Anim. Breed. Abstr., 32: 535).

20. Labussiere, J., Bennemederbel, B., Compaud, J.F. and Chevalerie, F. Dela (1984) The principal milk production traits under morphology and milk ejection in Lacaune ewes milked once or twice daily with or without stripping. In III Symposium International de ordeno Mecanico de pequenos Puminates Valladolid (Espana) May 1983. Editorial Sever Cuesta, 625-652.

21. Nawito, M.F., Shalash, M.R., Hoppe, R. and Rakha, A.m. (1967) Reproduction in the female camel. Bull. Anim. Sci. Res. Inst. (Cairo) (2).

22. Patel, K.S. and Dave, A.D. (1984) Milk production ability of Marwari and Patanwadi ewes. In milk composition. Indian Vet. J., 61: 323-326.

23. Sharma, U.P., Rao, S.K. and Zariwala, I.T. (1980) Composition of milk of different breeds of buffaloes. Indian J. Dairy Sci., 33: 7-12.

24. Sundaresan, D. (1980) Buffalo and dairy development. Summer Institute on buffalo management systems. 2-30 June 1980 pp. 254-262.

25. Tothill, J.D. (1948) Agriculture in Sudan. Oxford Univ. Press. London.

26. Vitcov, V.T. (1985) Effect of different planes of feeding on the productivity of pleven Black-head dairy sheep and their East Friesian and Awassi crosses. Feed intake, milk yield and body weight, Zhivot. Nauk. (1985) 22: 3-9. (Anim. Breed. Abstr., 54: 265).

27. Williamson, G. and Payne, W.J.A. (1978) An Introduction to Animal Husbandry in the Tropics. Third Edition, the English Language Book Society and Longmans, pp. 436-462.

28. Yasin, S.A. and Wahid, A. (1957) Pakistan Camles - a preliminary survey. Agric. Pakistan, 8: 289-297.

COLLECTION,TRANSPORT,COMPOSITIONAL AND QUALITY ASSESSMENT

Ir. J. J. Mol
Private consultant dairy development projects
Boslaan 135
6741 KG Lunteren
the Netherlands

ABSTRACT. An efficiently operating milk-collection is of great value for the success of dairy development.Retention of milk-quality is to a great extent a matter of limiting bacteriological spoilage.Milking utensils are the most important source of contamination by microorganisms,which determine the keeping quality of milk.The milk-collection in most developing countries is in a disadvantageous situation due to the fact that cooling of milk at farmlevel is not possible.To limit not only the development but also the activation of the bacterial flora in milk a minimal time interval between milking and collection should be pursued.Milk with undesirable bacteriological changes has to be rejected at the receiving platform.To quarantee an undisturbed intake of milk,quick tests,such as those based on appearance,odour and alcohol stability,must be applied.Various methods for preserving collected milk are used,such as cooling and adding hydrogen peroxide.The thermization process of milk to make the system of milktransport from the collecting centre to the processing plant more flexible deserves attention.The same holds true for the application of the lactoperoxidase system as a preservation method.A successfully operating system of milk-collection depends on the availability of competent milkhygienists.Practical training in the field should be emphasized.

1. Introduction
This contribution will deal with supplying milk-processing plants with fresh farm milk in a way that is economically justified and that quarantees the retention of the milk quality.A poorly operating system of milkcollection results in manufactured products with an inferiour quality or even in loss of this expensive raw material owing to its limiting keeping capacity.Table 1 shows that farm milk is an expensive raw material in developed as well as in developing countries(1).

Table 1. Farmgate prices in US-cents/kg

Jamaica	18(1983)	Spain	24(1982)
Holland	23(1983)	Panama	24(1982)
France	22(1983)	Colombia	28(1983)
Switzerland	43(1983)	Brazil	20(1982)
Australia	16(1983)	Trinidad	59(1983)
Ecuador	28(1982)		

2. Milkcollection systems

Although the specific goals of milkcollection in developed and developing countries are identical, the circumstances under which it takes place can differ considerably. In many developed regions, like countries in Western Europe, North America etc, the system has been developed where the milk is collected immediately after milking, cooled in farm-cooltanks and transported by roadtankers to the processing plant. With this system, at least in the Netherlands, 4-6 milkings are collected before the milk is transported to the plant. Before the development of this system the milk was transported directly to the dairy plant in 30-40 litre cans, at least in the summer months, twice daily. Now this can-collection system has nearly completely been abandoned.

However the manner of collecting milk in most of the developing countries is totally different. Farms do not have cooltanks, apart from some exceptional dairy farms with a large number of milking animals. The great majority of the individual farmers in developing countries deliver only a small quantity of milk, which makes the purchase of a cooltank unjustified - apart from the fact that the necessary electrical power is not generally available on those premises. Consequently it is inevitable that the milk under those conditions has to be delivered as soon as possible after milking. How the milk has to be collected depends on the local conditions, such as the distance to the dairy plant, the availability of all-weather roads etc. Due to the fact that in many regions the milk-density is low and consequently the milkprocessing units are too far away, the need of intermediate collection-centres with or without chilling-facilities is obvious. The time-interval between milking and cooling is the most essential difference between the aforementioned systems of milkcollection. Whatever system of milkcollection might be applied the basic condition in all these circumstances is identical: the raw milk must arrive at the receiving platform of the dairyplant as fresh as possible, at least without flavour defects and sufficiently stable to withstand the necessary heat-treatment in the processing.

By far the most important prerequisite for good milkquality at delivery is keeping the contamination and the multiplication of the microflora in the product during the collection at a low level(2).

3. Contamination of the milk

Both the type of the microflora and the number of viable microorganisms as a result of the contamination at farmlevel are of importance for the resistance of the milk to unfavourable conditions during transport from the farm to the dairyplant.
Microorganisms enter the milk from various sources during milking and handling at the farm,such as the udder,the skin of the cow,the air,the milking utensils etc.There are,however,distinct differences regarding the numerical contribution of these sources to the total number of the microorganisms.During the training of farm-inspectors a worthwhile exercise is calculating to what extent the contaminant at the farm contribute to the total load of microorganisms.Based on the available approximate figures of microorganisms concerning the previously mentioned sources of contamination,the result of these calculations always indicates insufficient cleanliness of the milking utensils and the milkcans as the main cause of the high bacterial populations in the milk at farmlevel.This statement is valid provided that the high bacterial counts are not caused by the incidence of mastitis in the herd.
According to an IDF-Morograph(3) the total viable count of ex-farm milk,produced under good hygienic conditions,should not exceed 10,000 bacteria per ml.Although we found these low counts in farm-milk last year under the tropical conditions of a country like Thailand,the platecounts were usually much higher,up to one million and more(4).The poor cleaning and disinfection of buckets and milkcans at the farm is a well-known fact.Without doing an injustice to the good farmers,we can state that the majority of farmers in developing countries,especially those whose milk production is a sideline,do not pay attention to this vital factor in milk-hygiene.In any case instruction for these farmers about the necessity of brushing the utensils with hot water and an alcaline detergent cannot be emphasized too much.

4. Microflora of raw milk

Regarding the keeping quality of raw milk during collection,the type of the microflora which enters the product by contamination is of even more importance than the total number of microorganisms(5). Table 2 gives a survey of the groups of microorganisms that occur in raw milk(6).These groups of microorganisms are most likely to be identical in various types of milk,produced in tropical regions as well as in countries with a more moderate climate.The dominant flora in these milks however can be presumed to be different.Under very good hygienic conditions of milking,with a low level of total plate counts in the fresh milk,visible dominant flora do not generally occur.However this situation will change if the milk is contaminated due to lack of cleanliness of the milking utensils.In such cases a dominant flora of Gram-negative bacteria will appear,along with higher numbers of non-thermoresistant lacticacid bacteria.

Table 2. Composition microflora of raw milk

```
pathogenic bacteria
non-thermorisistant micrococci
non-thermoresistant coryneforms
yeasts and moulds
Gram-negative rod-shaped bacteria
non-thermoresistant lactic acid bacteria
thermorisistant bacteria,
          e.g. some species of Streptococci
               some species of Micrococci
               Microbacterium lacticum
               sporeformers
```

5. Milkcollection

The collection of farmers' milk is the most critical stage maintaining the raw milk in such a condition that it can be processed into good dairy products. This is especially the case if the milk is heavily contaminated and cooling at the farm before transport is not possible. In these circumstances the development of microorganisms will start; at first the multiplication of Gram-negative rods, including the coliforms, will usually be dominant. In a continuing process of deterioration, not so much the coliforms as the strickt aerobic Gram-negative rods will decrease in activity due to lack of oxygen, while the lactic acid bacteria will gradually take over this dominant position. The other representatives of the raw milk microflora have no significance with respect to the limited keeping quality of raw milk. In the light of such knowledge about the sensitivity of raw milk to quality defects due to the growth of bacteria, it is obvious that the locations of the centres for milkcollection have to be chosen very carefully. Although the level of milkhygiene at the farms is an important factor, it is to be recommended under the mostly severe climatic conditions in most developing countries that the time-gap between milking and collection should not exceed 2 hours. Other local factors, such as the means of transport and the roads, determine the acceptable distance from the farms to the point of collection. The "Dairy Farming Promotion Organization"(DFPO), responsible for the intake of raw milk in Thailand, has reduced the opening -time of their collection centres to one hour in the morning and one hour in the evening. In their system of payments late delivery of the milk results in a heavy penalty. Such stimulation to deliver the milk as soon as possible after milking is undoubtedly a matter of great importance for the quality of the milk.

Certainly such a well-organized collection system will limit the development of the bacterial flora in the milk before collection. However we have to realize that the interval between milking and cooling at the centre, even in these circumstances, will always be too long to avoid the so-called pre-incubation period of the bacterial flora.It is a well known fact that this delayed cooling always result in a greater increase in the bacterial flora during storage,even at temperatures of approximately 4 degrees Celcius(7), than the increase in milk that is cooled immediately after production.This unfavourable situation in most of the collection systems in developing dairy industries can only partly be compensated by extreme care in the hygiene of milking.

6. Milkcontrol on reception

Upon the arrival of the milk at the receiving platform the quality of the milk has to be checked immediately.The essential objective of the tests is to determine quickly the suitability of the raw milk for heattreatment and processing into products.Milks with undesirable bacterial changes have to be rejected.The degradation of milkfat and proteins due mainly to large numbers of Gram-negative bacteria and the production of acids from lactose by the lactic acid bacteria are the main changes that make the raw milk unfit for processing.The first test should be that of the appearance and the odour of the milk immediately after the lid of the can is removed. Milk with incipient souring or undesirable off-flavour has to be rejected.Cans with milk which pass this first test are often checked for their suitability for heattreatment.For this second check the alcohol stability test with 68% ethanol has proved useful,although there is much confusion about the interpretation of the results.The basic aim of this test is to check the reduced stability of the casein micelles as a result of lactic acid development in the milk.There are however considerable differences in stability among lots of milk,and there are influences of season,perhaps of th stage of lactation and individual cows(8).Even spontaneous curdling after milking may occur,which is probably due to an excessive concentration of colloidal calcium phosphate.This phenomenon of casein in-stability is known as the "Utrecht milk abnormality".It is obvious that the instability of milk detected by means of the alcohol stability test is not always caused by the development of lactic-acid.Questionable results require retesting by means of a titration to dtermine the acidity of the milk.

Batches of farmers'- milk that passed the quality check at the firs point of delivery are collected in a recipient for weighing at the platform.At that moment samples have to be taken for determination of composition and quality.

7. Preservation of milk

Whatever system of milkcollection is used immediate preservation of the product is needed.Several procedures have been designed to preserve the food value of milk but local factors determine which system is to be recommended.

For centuries surplus milk on the farms has been converted into products with a longer shelf life than that of raw milk.

This practice of preserving milkconstituents is based either on a reduction of the wateractivity(some types of cheese,cooked butter, etc) or on a decrease of the pH by lactic acid bacteria(fermented milkproducts).These traditional methods of preserving milk,still practised,are not recommended,at least not fresh products,for hygienic reasons(9).

There is no question that a heat-treatment of milk is an efficient process for preserving milk,but this technique is mostly applied as part of the milkprocessing in dairy plants.We have to realize however that the level of the temperature needed to preserve the milk can be much lower than that applied during the usual pasteurization process.To destroy the group of Gram-negative psychrotrophic bacteria a heat-treatment for 10-20 s. at 60-65 degrees Celcius is effective.Such a heat-treatment is known as the thermization process(10). Applying this heat-treatment in a centre of collection before cooling could make the system of transport of the milk to the dairy plant much more flexible.The feasibility of such treatment of milk depends on local factors,but an electrical heating system as thermization unit in combination with a plate heat exchanger as regenerative and a cooling tank could be an interesting subject for a study.

Probably in most regions the dairy industry has been advanced by the application of cold in the preservation of milk.When there is a lack of low-temperature groundwater in tropical countries the cooling of milk depends completely on the availability of artificial cold.

Engineering developments in the field of refrigeration have resulted in several systems for cooling of milk.The stationary tank cooling system and the instant cooling system by plate heat exchangers have received the widest application.Designing the layout of a collection system,including the equipment for the cooling of milk,is always a matter of making compromises.Various local factors,including financial ones,determine,with sufficient consideration for the milk-quality aspects,the most advisable layout of a system.It is recognized that the organization of a milkcollection system very substantially affects the price of the raw milk at the processing plant.With this knowledge care must be taken to avoid over-investment with high fixed costs for the system.

At collecting points without the above-mentioned refrigeration facilities not seldom hydrogen peroxide is used as an alternative method of preservation.This treatment was approved by the FAO in 1957 to be used "when technical and/or economic reasons do not allow the adoption of cooling facilities for maintaining the quality of raw milk".Milk of poor quality can be preserved satisfactorily with the addition of 400-800 ppm hydrogen peroxide(11).Although this chemical additive might be the least objectionable of all possible preservatives it oxidizes in too high concentrations worthwhile milkconstituents(12).

One of these milk constituents is the enzyme lactoperoxidase,which belongs to the most active natural antibacterial systems of milk. Due to its relatively high concentration this enzyme is also active after the milk has been discharged from the udder.Particularly in

milks with a low level of contamination the multiplication of the
bacterial flora is postponed considerably.During the last decade
research has focused on strengthening this enzyme-system so that
it can be utilized practically for the preservation of raw milk
(13,14).Lactoperoxidase catalyses the oxidation of thiocyanate in
the presence of hydrogen peroxide into an antimicrobial agent.Although thiocyanate and hydrogen peroxide are natural constituents
of milk,their low concentration is a limiting factor in the activity of this enzyme system.It has been found that this system can
be activated considerably by the addition of small amounts of sodium thiocyanate(10-12 ppm) and hydrogen peroxide(8-10 ppm) to the
milk.Field experiments under tropical conditions(15) have proved
that this system can be used successfully to extent the keeping
quality of raw milk.These experiments have provided the approximate relation between milk temperature and length of the lag-
phase before bacterial multiplication starts:

Temperature in degrees Celcius	Lag-phase in hours
30	7-8
25	11-12
20	15-16
15	24-26

We can only follow the development of this milk-preservation-system with great interest and concern because it offers an attractive alternative to the hydrogen peroxide treatment of milk.

8. Transportation of the milk to the dairy plant
The handling of the milk at the farm and at the milkcollection-
centre has consequences for the system of milktransport to the
dairy plant.In developing countries this system of bulktransport
is in general not flexible,which means that preferably the milk of
only two milkings has to be collected daily.This greater frequency
of collection in situations of developing dairy industries compared
with those in advanced dairy countries has to be attributed to the
quality of the milk at the centres.As we have discussed before the
delayed cooling of the milk after production gives rise to a higher
development rate of the psychrotrophic bacteria in the coolingtank
compared with situations in which the milk has been cooled immediately.A thermization treatment of the milk or the application of the
lactoperoxidase-system in combination with cooling of milk could
make the bulk-collection of the milk much more flexible.

9. Training of milkhygienists
Whatever system of milkcollection may be applied,a successful operation depends not least on the availability of competent milkhygienists who have the confidence of the farm community.The FAO-Regional
Dairy Development and Training Teams have,for many years,made a
worthwhile contribution to the training of milk-hygienists and support of such activities must be stressed.The success of dairy development projects depends to a large extent on the availability of

training in the field, which is one of the most worthwhile investments that can be made. Sufficient priority to this training-component has to be given in the budgets of development projects.

References

1. Homewood J.T, Technology for the Developing Countries, in: Modern Dairy Technology, volume 2, R.K. Robinson

2. FIL-IDF, Document 120(1980), Factors influencing the bacteriological quality of raw milk

3. FIL-IDF, Document 83(1974), Bacteriological quality of cooled bulk milk

4. Chuaprasert S, Quality aspects of raw milk in Thailand, Asian Institute of Technology, AE-84-13(1984)

5. Cousins Ch.M, Milking Techniques and the microbial flora of milk, International Dairy Congress, Paris 1978

6. Stadhouders J, Microbes in milk and dairy products, An ecological approach, Neth. Milk and Dairy Journal 29(1975) 104-126

7. Stadhouders J, Cooling of raw milk immediately after production as a main factor for controlling the bacterial growth during storage at 4 degrees Celcius, Neth. Milk and Dairy Journal 22(1968) 173-178

8. Walstra P. and Jenness, Dairy Chemistry and Physics

9. Joint FAO/WHO Expert Committee on Milk Hygiene, FAO Agricultural Studies no 83

10. Berg M.G. van den, The thermization of milk, FIL-IDF Document 182(1984)

11. Siegenthaler E.J, Possibilities and limits for the use of hydrogene peroxide to facilitate milkcollection under tropical conditions, Milchwissenschaft 22(1967),

12. Lück H, The use of hydrogene peroxide in milk and dairy products, Milk Hgiene, WHO Monograph Series 48(1968)

13. Reiter B, Antimicrobial systems in milk, J. Dairy Res. 45, 131-147

14. Björck L, the Lactoperoxidase System, Seminar Natural Antimicrobial Systems, University of Bath(England), September 1985, pg. 18-30

15. Schmekel J. and Harnulv G, Activation of the Lactoperoxidase System as a means of saving milk in tropical countries from early spoilage, *Seminar Natural Antimicrobial Systems, University of Bath(England)*, September 1985, pg. 75-79

COMPOSITION AND QUALITY OF MILK AS A BASIS FOR PAYMENT OF FARMERS.

J.C.T. van den Berg

Department of Food Science
Agricultural University
Wageningen
the Netherlands

1. THE SYSTEM OF PAYMENT

A proper system of price setting is the only way to improve the compositional and the hygienic quality of raw milk, and to prevent adulteration. Systems for payment may be very simple, or very complicated, because many factors have an influence on the choice to be made, such as:
- the stage of development of the dairy industry. If it is still in its beginning, a simple system should be preferred.
- the educational level of the farmers. It is very important that the farmers understand the system of payment, otherwise they will feel themselves too much left to the mercy of the buyer of the milk.
- the dairy legislation, if any. The system of payment should be in line with the legislation. Unfortunately legal requirements are sometimes too severe for the majority of the farmers, whilst it also occurs that no facilities exist to enforce the legal standards. In these cases, the adjustment of the system of payment to the official legislation will be extremely difficult and seems unrealistic.
- farmers are not prepared to submit themselves to payment on quality, or they are not satisfied with the system as such, or with the classification of their milk. In such cases they may prefer to sell to less critical or less quality conscious buyers.
- last but not least, it also depends on the collection system and the requirements of the dairy plant.
This all means that the choice of the system for payment and the way it should be introduced and implemented, largely depends on local conditions. It has technological, but particularly organizational and psychological implications.
The expert panel who presented in 1972 the FAO publication "Payment for milk on quality" stated already: "Payment on quality is primarily a human problem with technological implications". And little has changed in these - almost - 15 years.

The system of payment should satisfy two conditions:
- the price to be paid must correspond with the value of the milk, which depends on quantity and composition, and on hygienic quality.

- the farmer must not only receive a fair price, but he should also be convinced that the price is fair.

2. PAYMENT ON COMPOSITION AND QUANTITY

The following possibilities exist:
- payment on the quantity of milk exclusively. This is easily understood by the farmer, requires no laboratory testing, but does not comply with the condition that payment should depend on composition. Moreover, the system usually leads to adulteration by fat extraction (creaming) and water addition.
- payment for fat only. The farmer is paid for his efforts to increase the fat content of the milk, whilst - seemingly - adulteration of milk does not pay. Seemingly, because it pays to skim the milk partly and to sell the part of the milk with the high fat content to the dairy plant, and the remainder with the low fat content directly to the consumer or elsewhere.
- payment for quantity and fat. This system is usually well understood by the farmer, rewards him for compositional quality (high fat content), whilst creaming of milk does not pay. A proper setting of the price for fat on the one side and for skimmed milk (or "quantity") on the other side may meet some difficulties, whilst adulteration with water is still profitable.
- payment for quantity, fat and protein. Many countries introduced payment on protein, because protein is not only an important nutritive component in milk, but it largely contributes to the yield in the manufacture of so called concentrated products, like milk powder, condensed milk and cheese. If a system for payment on quantity ànd fat content ànd protein content is introduced, the price that goes to the fat must precisely correspond with the value of the fat, for instance the value of the fat in butter or ghee. If not, adulteration by creaming will pay. The remainder of the milk price must - in some way or another - be divided over quantity and protein. Apart from being complicated, this system still makes addition of water financially attractive.
- payment for fat and protein. If farmers are exclusively paid for fat and protein content of the milk, the quantity they supply is only indirectly of importance, because more milk means more fat and protein and there is no payment for the quantity of milk as such. With this system, the farmer receives the full benefit for his efforts to increase milk production and to improve the compositional quality, whilst adulteration of milk by fat extraction or water addition does not pay. However, under conditions, partly creaming of milk and selling the high fat part to the plant and the remainder to others, may pay. A proper price setting of the fat and the protein may be found difficult.
- payment on fat and protein with a quantity deduction. Dairy plants producing concentrated products like milk powder, condensed milk and cheese are more interested in large quantities of milk fat and protein than in large quantities of milk, or actually water. The latter must be handled and processed, without directly contributing to the yield in dairy product. In this case a system may be followed by which the milk price consists of two elements, namely a gross price for fat and protein, and a deduction per 100 kgs of milk or part thereof to meet several expenditures of the dairy plant, such as costs of milk collection and processing, and miscelleneous levies. Sometimes allowances are given for quantity, lean-season supplies, etc., but the total of the allowances is always lower than the total of the deductions. This system is followed in the Netherlands. An imaginary example is given in the table.

Payment of milk on composition
(imaginary example)

Gross price per kg of fat	Dfl 10.00
Gross price per kg of protein	Dfl 11.50
Quantity deduction per 100 kg of milk	Dfl 12.50
Fat percentage	Dfl 4.40
Protein percentage	Dfl 3.35

Calculation of price per 100 kg of milk

Gross price fat	4.40 × 10.00 =	Dfl 44.00
Gross price protein	3.35 × 11.50 =	Dfl 38.53
Total gross price		Dfl 82.53
Quantity deduction	1.00 × 12.50 =	Dfl 12.50
Farmer's payment		Dfl 70.03

Farmers supplying the same quantities of fat and protein, but different quantities of milk will receive a lower price as the quantity of milk is higher. With this system a large number of advantages is achieved:
1. the farmer receives indirectly the full benefit of his efforts to increase the quantity of milk, and directly to increase the fat and protein content.
2. extraction of fat by creaming does not pay, because it lowers the amount received for fat.
3. since it does not change the quantities of fat and protein, addition of water will not be profitable. On the contrary, it will increase the kilogrammes of "milk", which results in a higher quantity deduction and ultimately a lower price to the farmer.

The gross fat value or price can be based on the whole sale price of butter, taking the legal fat content (in most cases 84 per cent) into consideration, and the gross protein value or price by deducting the gross fat value and the whey value from the whole sale price of cheese. Time is too short to go more into detail.

Analyses for protein content may be found too complicated. Therefore, a system similar to the payment for fat and protein content can be followed, in which system milk is exclusively paid for fat and solids-non-fat, with or without quantity deduction. Estimation of the fat content is sufficiently accepted and not too difficult, whilst the solids-non-fat content can be calculated if the density is known. The density test is a simple test as well.

Proof of the adulteration of milk fat extraction is - apart from exceptional cases - not possible, and consequently it must be avoided by proper price setting. Proof of water addition is difficult. The density test is unreliable in this respect; it may leave additions of 5 to 10 per cent of water undetected. Most accurate is the freezing-point test, but this test is less suitable for use in collection centres and small industries. Moreover, raising of the freezing point by the addition of water can be avoided by the addition of some sugar to the water.

Therefore, the system with the quantity deduction certainly has advantages. It also punishes the farmer who adulterates the milk by negligence, such as incomplete drainage of rinsing water from the milking equipment after sanitation.

3. PAYMENT ON HYGIENIC QUALITY

The processing of high quality milk is one of the conditions for the manufacture of high quality products.

Milk which passes the platform test is not necessarily milk of high hygienic quality. To test milk for hygienic quality, more elaborate and time consúming tests must be performed. Since daily testing is not economically justified, random samples can be taken at irregular intervals, e.g. once every two weeks. The results of these tests must have financial consequences for the farmers, i.e., farmers who pay much attention to the hygienic quality of the milk should be rewarded. In this way the payment will stimulate hygienic milk production.

The introduction of an adequate system, and even the introduction of new tests, often causes difficulties, because the farmers may not understand the purpose of the tests, let it be the interpretation of the results. Therefore, the system of payment for hygienic quality and the tests to be performed, with their financial consequences, should be cautiously and gradually introduced. The performance of a quality test should start well in advance (e.g. half a year) of the attachment of financial consequences. During this period, the farmer will get acquainted with the system and the tests, and he can be taught how better results can be obtained.

The interpretation of the tests on which the milk is graded and the financial consequences should be modest in the beginning, but can be tightened gradually as required.

For psychological reasons it may be found advisable to avoid designations like "bad" and "poor" in the system of classification, whilst deductions for poor quality milk should be avoided as well. In most cases it is better to distinguish a few classes only, for instance "standard" (never "normal") for the lowest quality and "good" and "excellent" for the better qualities. Standard quality receives the basic price, whilst good quality receives a small and excellent quality a substancial premium. It is not a good practice to advise the farmer of the classification of his milk exclusively. He should also know the grading of all tests separately, which is necessary to trace the cause of low classification.

Major tests that can be introduced for the evaluation of the quality of the milk refer to purity, bacteriological quality and smell; in addition a large number of other tests can be included in the system. Some of these tests are of general importance, some are important for the manufacture of particular products. Consideration of all tests would be beyond the scope of this paper.

Nevertheless, I like to draw your attention to the fact that for the ensurance of a correct system of payment tests may have to be adjusted to new developments in milk handling and milk processing. Adjustments which, in turn, depend on the possibilities created by the advancement of dairy science. However, the local situation plays a very important role as well. A typical example is the test for the bacterial quality of milk. Originally, the bacterial quality was exclusively assessed by smelling the milk; a sour flavour being an indication that the milk was not fresh. With the acidity test a more accurate method was introduced, and for routine purposes the alcohol test became available. Usually alcohol of 68 per cent, and sometimes more, is used to sort out milk in which too much production of lactic acid took place. I am referring to cow's milk. If it is merely the objective to sort out milk which cannot stand normal pasteurization the clot-on-boiling test can be performed. However, this test is laborious and requires much glassware. The same objective can be obtained with the alcohol test, using weaker alcohol, for instance alcohol of 65 per cent. Although the value of these tests for estimating the bacterial quality of milk cannot be denied, they rather are platform tests for sorting out poor quality milk, than tests that fit in a system for payment on quality. For this purpose the dye reduction test was and is widely used. The flora which generally develops in raw milk that has not been cooled to low temperatures, has the ability of reducing and thus discolouring the methylene blue or resazurin added to the milk in the performance of the test. This flora mainly consists of gram-positive bacteria, like lactic acid bacteria. Such milk will generally become sour after prolonged storage.

As a result of the introduction of milk handling and milk collection systems, whereby the raw milk is stored at very low temperatures over long periods, the dye reduction test has lost its importance, because a different type of flora, principally consisting of gram-negative bacteria with no or little reducing capacity develops, whilst the gram-positive bacteria are present in a dormant state as a result of the low temperature, which also results in a low reduction capacity. This means that it is advisable to replace the reduction test by the germination test (plate court test) for the bacteriological grading of deep-cooled milk.

Bacteria which grow in deep-cooled milk, so called psychrotrophic bacteria, may produce lipolytic and proteolytic enzymes, which can cause defects like rancidity, and bitterness and occasionally gel formation in milk. Since many of these enzymes are not destroyed in the UHT-sterilization process, they may cause defects in so called long-life milk after long storage periods. However, psychrotrophic bacteria grow slowly and start producing significant amounts of enzymes if their number amounts to about 10^6 per ml. As a result, the danger of the presence of these enzymes in UHT-milk should not be overestimated. Production of the enzymes will only be possible in case of excessive contamination, insufficiently deep cooling, or storage of the milk over very long periods. If contamination is heavy and temperature of cooling is just low enough to retard or suppress growth of lactic acid bacteria a rapid development of gram-negative bacteria may take place and storage of milk for periods over 24 hours will involve a risk.

The occurrence of bacterial enzymes can be expected in raw milk as a result of contamination with a fully grown culture. In such cultures, which may be found in insufficiently sanitized milking utensils, like milk cans, storage tanks, etc., an accumulation of the bacteria and their enzymes may be found, especially if the temperature of the utensils is quite high between two milkings or storage periods. A contamination by such an accumulation may make milk unsuitble for the manufacture of UHT-products, even if the total bacterial count of the milk is still comperatively low. For these reasons, the introduction of a special test for the determination of the enzymatic quality would be worth while for milk to be used in the manufacture of UHT-products.

To the psychrotrophics in milk belong species of Pseudomonas, Achromobacter and others; that is, species of a group often indicated "water-bacteria", because they are found in large numbers in surface water. Although not all water-bacteria are psychrotrophic, it would be interesting to consider the possibility that fresh milk is inoculated with psychrotrophics if milk is adulterated by adding water.

There are other examples that the value of the dye reduction test and the plate count test is limited. They merely give an idea about the extent of contamination and growth of certain micro-organisms. They do not give an insight into the kind of bacteria which may be important in the manufacture of certain products, like thermoresistant bacteria for pasteurized milk and butyric-acid bacteria for certain cheese varieties. For this purpose additional tests must be implemented.

Although the system of payment should be adjusted to new developments, it will be a good point for discussion to what extent tests necessary for the determination of the usability of fresh milk for the manufacture of specific products should find reflection in this system.

The implementation of a system of payment with the objective to promote the production of milk of high compositional and hygienic quality can only be successful if there is a direct contact between the dairy plant and the farmers. If farmers sell their milk to middle men who,.on their turn, sell the milk in bulk to the plant, such a link does not exist, and the results of the system of payment will not or hardly reach the individual farmer.

To conclude, I should like to emphasize the necessity of a well organized extension service if the payment for quality is introduced. It is not fair to punish farmers for the supply of poor quality milk, if they do not know how they can improve the quality. Particularly the small farmer may suffer in such a situation. Moreover, without an extension service it will be more difficult to reach the objective of the system of payment, that is a better quality milk for better quality products.

REFERENCES

Driessen, F.M. (1983), Lipases and proteinases in milk, thesis, Agricultural University, Wageningen.

SUMMARY OF DISCUSSION

On the subject of increasing productivity in developing countries through cross-breeding and/or upgrading to European type breeds, mention was made of the rapid increase in the first cross, followed by a slower rate of increased production in subsequent generations. It was agreed that the dramatic increase in the first cross was by hybrid vigour, coupled with genes from improved breeds.

The question of genotype-environment interaction in terms of ecological, climatic, nutritional, management and economic factors was discussed. It was agreed that the situation in different developing countries has to be examined on the basis of their own circumstances, before decisions were made as to the types of exotic breeds, to be introduced and the level to which the indigenous breed should be graded to. Research and experience in India indicate that the temperate blood level should not exceed 50 - 60 %.

The role of other ruminants was discussed. It was emphasised that the dairy breeds of water buffaloes contribute significantly to milk production in many countries. The seminar was informed about the prospect of application of embryo transfer methods in propagation of high producing buffaloes.

Although the total contribution from goats is small, the goat has a high milk yield per unit of liveweight compared to buffaloes and cows, and this is of importance to small farmers.

The importance of milk quality of small farmers, through collection, handling, storage, transportation, processing and distribution of milk was emphasized. The peculiar circumstances of different countries as a result of different socio-economic conditions were discussed. The need of simple and low-cost tests to determine milk quality in terms of hygiene, fat and solids-not-fat at farm gate level was highlighted, as it was felt as these were key factors to initiate the proper and controlled development of the small holder sector in such countries.

The importance of extension and training in areas of feeding, management and health control were emphasized, as without such concurrent action the benefits of genetic upgrading will not be realized.

In his concluding remarks, the session chairman stressed the great need for dairy development in developing countries. While the countries are pursuing their programmes, a great need exists for assistance from highly industrialized countries, through bilateral and multilateral programmes. In this respect the FAO, through its various programmes such as the semen donation scheme, has and is, assisting such countries in the efforts to develop the dairy industry. More assistance is needed to help these countries in the context of the overall economic and social developments in such countries. Such support is essential to ensure the otherwise widening gap between the developed and developing countries.

Seminar II: Milk Production and Milk Products in Developing Countries

Session 2

Chairman: L. L. Muller (Australia)
Secretary: L. Sandberg (Finland)

Recombination of dairy ingredients

Specific Aspects of Milk Processing in Developing Countries

M.R. Bachmann
Laboratory of Dairy Science
Swiss Federal Institute of Technology
ETH-Zentrum
CH-8092 Zürich
Switzerland

Abstract: Characteristics of milk processing in developing countries are the poorly developed technological and economic environment, in many cases the warm climate and the long distances as well as the poor quality of roads between milk production areas and consumption centres. Milk processing has to be well adapted to these specific conditions in order to survive. Every technological improvement must, therefore, be adapted to the local technological and economic environment. The appropriateness of traditional local dairy products and of products of industrial countries is discussed. A few guidelines for better adapted milk processing in developing countries are given. Emphasis is put on the consideration and use of sound development principles. Examples from several countries are cited.

1. INTRODUCTION

The fact that, 40 years after the foundation of FAO and nearly 40 years after the start of dairy development cooperation with Third World Countries [1], we are still discussing special problems of dairy development, shows that there are difficult problems to be solved. Problems, which are different from those we encounter in industrial countries. If we had to deal with the same problems they would have been solved long ago and we wouldn't talk about them now. It is striking that numerous difficulties and many failures can be found precisely in the field of milk collection, milk processing and marketing of dairy products in developing countries. Are dairy people in developing and developed countries less inventive and clever than other specialists? We do not think so. The reasons for the mentioned situation have to be found elsewhere. On one hand they result from a misconception of technology transfer, on the other hand they have to do with the extremely delicate and perishable nature of the raw material "milk". Unlike grain, sugar beets, cotton or other raw materials milk does not tolerate any delay or interruption concerning its processing. Either a breakdown of a processing line or an interruption of energy supply immediately leads to losses. That perishableness of milk together with the problems of technology transfer makes milk processing in developing countries an art. Fortunately, there also exist well functioning dairy projects in the Third World. We, therefore, can learn from both, failures and success stories. While studying both, we may find out how to deal with the collection, transformation and marketing of milk. The problems of technology transfer and milk processing are closely interwoven. We will, therefore, first look at some basic problems of technology transfer and later at their effect on milk processing.

2. THE BASIC PROBLEMS OF TECHNOLOGY TRANSFER

Recently Prof. Hoelscher of the Centre for Applied Technology of the University of Houston, Texas, very aptly said: "It is known that a new technology, or a system of technologies (e.g. a factory or a manufacturing plant), in a given location must be adapted to the local technological 'culture' if it is to prosper following its absorption to that culture; otherwise the exercize would ultimately end in failure." [2] That means, that a "local technological culture" has to be able to assimilate the new technology or the technological improvement. If such an assimilation does not take place, the new technique will remain a foreign body and will quickly or slowly die. The necessity to make technological improvements absorbent for the local technological environment is often the stumbling block for development.

The necessity to adapt technology to the "local" technological culture means furthermore that techniques which have proven successful in a certain environment will not automatically be successful in a different technological setting. To make this clear, let me cite a case from the recent history of dairy development. The big success of "Operation Flood" and the techniques used for developing the milk scheme of the four biggest Indian cities is by no means a guarantee that "Operation Flood" will be the same success when applied to other areas of the Indian subcontinent or Africa or Latin America. The "technological environment" of these other areas may not be capable to assimilate the techniques which have been used for "Operation Flood". On the other hand dairy projects and their techniques which have been successful in Nepal or in the Highlands of Latin America would not have made use of the technological potential which exists in certain parts of India and, therefore, would also have been, a failure.

However, in many countries, the existing "technological culture" would not be able to absorb the high technology which has been applied for "Operation Flood". In these cases, dairy development must start on a much lower technological level, choosing simpler methods for milk collection and processing. In these situations it is usually important to adapt the technology also to the "economic environment". A poor adaptation to the economical environment is probably just as often the cause for the death of a dairy-plant as a poor adaptation to the local "technological culture". Dairy development is rich in examples which demonstrate how the chosen techniques make dairy products expensive, put them out of reach of the average consumer and thereby push the company out of business.

Some of the most important special problems of milk processing in developing countries are therefore:

- adaptation of the technical improvement to the local "technological environment"

- selection of techniques and products that fit into the "economic environment"

- adaptation of techniques and products to the geographical and climatic conditions

and finally,

- choice of technologies according to sound development principles.

Let me now discuss these specific aspects of milk processing in developing countries with the help of a few examples.

2.1. Adaptation of the technical improvement to the local "technological environment"

What does the local "technological environment" consist of? A highly developed technological environment consists of a proper network of communications including good roads and reliable post- and telephone connections. It further consists of suppliers for every type and size of industrial hardware and software, of services for maintenance, repairs and waste disposal. It further consists of a reliable energy and water supply system, of skilled and well trained labourers, technicians, engineers, scientists and managers. Schools, training centres, universities and research centres also make part of the well developed "technological environment".

In a poorly developed "technological environment" most or all of the above mentioned elements are lacking. In such an environment division of labor does hardly exist, that means the entrepreneur builds his own workshop, makes his own tools and produces the goods. Sometimes he even is his own merchant. He uses locally available energy sources such as human labor, animal power, hydraulic power, solar energy and wood as fuel. The raw materials to be processed such as food crops, milk, vegetable and animal fibres, wood, clay and so on are ready available locally. The necessary techniques and skills are handed over from father to son or mother to daughter and pass on from generation to generation. The production processes are relatively simple and clear. However, that does not exclude that great skill and craftsmanship are used.

Let me give an example for such an environment from my own country, Switzerland. On the Swiss alps there exist in a poorly developed technological environment small cheese factories. They produce dairy specialities such as different types of hard cheese, semi-hard cheese, butter, double cream and "ziger". These cheese factories used to be built - and some are still today - by the cheese maker and the milk producers themselves. The cheese maker knows how to make his tools, including pails, stirrer, harp and cheese press from locally available material. The only item which is made by a specialist is the copper cheese vat. For making cheese vats, an early division of labor has set in.

During the last century the above mentioned specialization in making cheese vats has led to the creation of specialized workshops and industries which produce equipment and machinery for milk processing. Beside of the cheese vat it was above all the invention of the cream separator at the end of the last century which helped to develop the equipment industry in the field of dairying. There exist today huge firms such as Alfa-Laval, APV, Westphalia, Silkeborg, Stork etc., which developed originally from modest beginnings and in a poorly developed technological environment.

We need in fact not much imagination to guess what happens if a technology or a whole system of technologies is transferred from a surrounding with a high technological culture into a surrounding with a low technological culture. Most probably the exercise will end in failure and lead to another "development-wreck". Judging from the numerous development-wrecks in the field of dairying scattered all over the world, dairy experts seem to have very little imagination.

It is true that not only the so-called dairy experts are responsible for this sad state of affairs, but also the agencies that finance dairy development and the counterparts in the developing countries themselves. The tragedy of dairy development is the great amount of money available compared with the small amount of imagination or - in other words - the small amount of common sense involved. This tragic situation makes it possible that technologies and whole systems of technologies are exported into developing countries which are not in the least adapted to the technological environment of these countries. In order to illustrate this let me give an example:

There is a country in Africa that printed on one of its current banknotes a picture showing school children drinking pasteurized milk from one-way packages, symbolizing a recent achievement of the country, that is the school milk programme. After some time it has been found that the milk packaging used in that programme did cost more than the milk inside. Furthermore it proofed to be difficult to distribute pasteurized milk in a tropical country where no reliable cooling facilities exist. To set up and organize a closed chain of cooling facilities and to keep it working is costly and makes the product which is being marketed through it very expensive indeed. The end of the exercise was not a very happy one. The school milk programme still exists .... but only on the banknote.

2.2. Selection of techniques and products that fit into the economic environment

The above mentioned example shows that a system and a product which is badly adapted to the economic environment has not a big chance to survive even in a fairly well to do country. Milk products which have been developed in and for an economic environment with high

purchase power are not suitable for an economic environment where
consumers have an extremely low purchase power. For such an environment
low cost products have to be produced with the help of low cost technologies. What are low cost products? The characteristics of low cost
products are:

- They can be manufactured with simple means.

- They keep well under natural climatic conditions until they are
  consumed.

- They do not need expensive packages. One does not have to throw
  away part of what has been paid for.

- They provide essential nutritional elements.

- They are complementary to the traditional local diet.

The economic environment includes also the foreign exchange situation of a country. If there exist difficulties to obtain foreign
exchange, the production process should not depend too much on imported
equipment and materials. Delays in obtaining foreign exchange for
spare-parts or auxiliary material risk to hamper or even to stop production for weeks or months. Such interruptions are more harmful for
dairies than for other industries because of the necessity to process
the milk daily and immediately after arrival. Unprocessed wheat or
sugar beets or wool can easily be kept until a processing line is
repaired or a piece of equipment installed. Milk can not even wait two
or three hours.

It is a rule that the traditional dairy products of developing
countries - and there exist such products in all countries - are well
adapted to the technological and economic environment and, therefore,
are relatively cheap whereas poorly adapted products produced with
poorly adapted technologies always risk to be expensive.

However, there is an exception from this general rule. Sometimes
there exist in Third World countries small, but interesting markets for
luxury products. These markets are usually connected with the foreign
community or with tourism. They absorb luxury products such as fresh
butter, ice cream, soft and fresh cheese etc. One should always try to
use such opportunities in order to lower the prices of those products
which are sold to the poorer population.

2.3. Adaptation of techniques and products to geographical and clima-
     tic conditions

This specific aspect of milk processing in developing countries is
closely linked with technological and economic considerations. Most
developing countries are tropical or subtropical countries. Furthermore, many of them are subdivided into costal regions and inland
regions and lowlands and highlands. In many developing countries all
these geographical and climatic differences exist. However, the most

important common denominator is the relatively high temperature. In certain countries such as Nepal, Bhutan, parts of Afghanistan, Bolivia, Peru etc. the milk producing areas are situated in the highlands and, therefore, not in a hot climate. In these cases it is not the temperature, but the poorly developed technological and economic environment which is the common characteristic.

Why should these geographical and climatic conditions be taken into consideration? Because it is expensive to ignore them. Trade with products which are not or poorly adapted to these conditions can be detrimental. Let me explain this with the help of an example. We pointed out before that pasteurized milk must be kept cool from the moment of production till consumption. Without cooling its shelf life is extremely limited. However, the installation of chains of cooling facilities linking the dairy factories with the retail shops or even with the individual households and the maintenance of such cooling chains in good working order is extremely costly and difficult. This is true for all countries, but particularly for developing countries. Those dairy specialists who are familiar with the maintenance of chilling centres in developing countries will certainly agree. There exist at least two alternatives to the production and marketing of pasteurized milk. The first alternative is the production and marketing of cultured milk. Cultured milk or sour-milk has been made and consumed in tropical countries for thousands of years. Sour-milk has a naturally improved shelf-life. The other alternative is sterilized milk. Sterilized milk can be produced with simple means and in small set-ups as we could proof by experimental work carried out in East Africa during the last four years [3].

The geographical situation of many developing country is such that milk production takes place far from the main consumption centres. These centres are often near or at the coast in a hot and humid climate. Milk production usually takes place in less humid areas, on irrigated planes or in the highlands. In all these cases the milk or milk products have to be brought to the consumers over long distances and usually over bad roads. For this reason, it would be advisable not to transport perishable raw milk which contains 87 % of water - or with other words - dead weight, but a milk concentrate which keeps well under the existing climatic conditions. Such concentrates which can be transported without expensive cooling are e.g.: Ghee, different types of hard cheese, sun-dried butter milk or sour-milk, Khoa and "Dolce de leche" which is well-known in Latin America.

All traditional milk concentrates are "natural" preserves. Hard cheese e.g. has excellent keeping qualities because of its high content on lactic acid and due to its reduced water activity. "Dolce de leche" is a natural preserve because of its low water activity and its high osmotic pressure. Sun-dried buttermilk which is known in Nepal under the name of "Zurpi" and in Afghanistan under the name of "Kurud" keeps well for years because of its high content on lactic acid and low moisture content. It may be interesting to have more information on this traditional but little known dairy product. We analyzed "Kurud"-samples from Afghanistan and found moisture contents between 5 and 6 %. The average composition of "Kurud" is shown in table 1.

Table 1: Average composition of "Kurud"

Protein	62,3 %
Ash (including impurities)	13,8 %
Fat	8,7 %
Lactose	6,7 %
Lactic acid	3,7 %
calorific value	20,1 MJ/kg

The pepsin-hydrochloric-acid-solubility of the proteins is 93 % or more, which indicates an amazingly high digestibility of the proteins.

The mentioned examples of traditional products make it clear that it is absolutely possible to adapt dairy products on one hand to the geographic and climatic conditions and on the other hand to a poorly developed technological and economic environment. Without any doubt there exist in developing countries alternatives to huge milk powder works, to condensed milk factories or to automated pasteurizing plants.

2.4. Choice of technologies according to sound development principles

Dairy development is an important factor of agricultural development. Ruminants transform very efficiently energy- and protein-sources which are not directly accessible to man. Milk can be produced on small-holdings or even by landless peasants. In many cases, there is no need for additional land for fodder production. Milk production, collection and transformation create additional working opportunities and income. Peter Timmer of the Harvard Business School said: "A healthier food production sector is also needed because that is where much of the world's poverty is. It is where most of the world's productive jobs can be created in the next ten, twenty, or thirty years" [4]. However, in order to make full use of the development potential of dairying, it must be well located and well adapted to the abilities of the users.

Dairy activities should take place as far as possible in rural areas. This could create additional working and earning opportunities where they are lacking, adding to the attractiveness of rural life and counteracting rural exodus. It goes without saying and has been explained before that milk collection and processing has to be adapted to the technological culture of this rural environment. The jobs must be tailor-made for unskilled or marginally trained labor. The rural location also asks for products which are well adapted to their geographic and climatic environment. Products should not contain much dead weight and must keep well during transport without artificial cooling. The factories must be relatively small and easily to manage. Service for maintenance and replacement of machinery and tools must be available within the country and should be paid with local currency.

The energy sources used must be available the whole year round. Preference should be given to those energy sources which are produced within the country and which can easily be kept in stock (coal, wood).

If electric current is used, frequent interruptions and severe fluctuations in tension must be taken into consideration. For all processes which use electric current, an alternative power source (manpower, combustion-engine, etc.) has to be foreseen as a stand-by. The highly perishable raw material "milk" does not allow the slightest interruption during processing. For this reason, the well-functioning of a dairy plant must be much more reliable compared with a plant which transforms less perishable raw material. Recommendations for building and running such rural dairy plants in warm countries exist [5].

The use of local construction techniques, of tools and equipment which have been built in the country itself, helps to reach a further development goal, that is, the creation of jobs and income in other sectors of the home economy. We have mentioned earlier on, renowned international companies which emerged from small workshops making tools for milk processing. True development is a result of economic exchange and interactions between different sectors within the country and is not based on imports of technologies from outside. Development cannot be bought, but must grow from the country's own resources. Well chosen imported equipment and systems can, at the most, serve as examples or prototypes. However, they should never replace indigenous development activities. Development means improvement of the existing technique which is perfectly adapted to the local technological and economic environment. If the money which has been spent for imports of poorly adapted and often unsuitable equipment would have been used for improving existing, indigenous technologies, many a country would nowadays be much better off technologically and economically.

References

[1] Winkelmann, F. (1985) 'Beitrag der FAO zur Entwicklung der Milchwirtschaft in der Dritten Welt', Welt der Milch, **39** (46) 1323-1232

[2] Hoelscher, H.E., (1985) 'Development - The Issues Aehead', Int. J. DevelopmentTechnology, **3** (4), 237 - 239

[3] Kurwijila, R.L. (1986) Flame Sterilization and Marketing of Milk in 20 Liter Containers, Diss. ETH Zürich

[4] Timmer, C.P., (1985) 'Realistic apporaches to world hunger: How can they be sustained?' Food and Nutrition Bulletin, **7** (1), 2

[5] Bachmann, M.R., (1981) 'Technology Appropriate to Food Preservation in Developing Countries' in: Development in Food Preservation-1, S. Thorne, ed., Applied Science Publishers, London and New Jersey

RECOMBINATION OF DAIRY INGREDIENTS INTO MILK, CREAM, CONDENSED AND EVAPORATED MILK

A. Sjollema
ccFriesland, Cooperative Company
P.O. Box 226
8901 MA LEEUWARDEN
The Netherlands

ABSTRACT. A survey is given of the most important general quality specifications of dairy ingredients for the production of recombined milk, cream, evaporated and condensed milk. Supplementary to the general specifications, specific requirements are needed for the skimmilkpowders used as ingredients for recombined products. For recombined milks and creams, low or low-medium heat powders are recommended. Specially for recombined UHT milk and cream, the powder has to be free of heat resistant enzymes. For recombined evaporated milk a high heat powder is needed which has a good heat stability. For sweetened condensed milk a medium-heat powder is recommended, having specified viscosity properties.
Tests to check the powders on these points are mentioned.

1. INTRODUCTION

In 1979 the IDF published a monograph (1), covering the technological and engineering aspects of recombination. In 1980 an IDF-seminar on recombination was held in Singapore. Proceedings of this seminar were published in 1982 (2).
In these publications, the technology of recombining was reviewed extensivily. For that reason the present paper will not cover the technology of recombining in general, but a special, however very important, part of it, namely the quality specifications for the dairy ingredients and the influence of ingredients on the quality of the final products.

2. GENERAL QUALITY REQUIREMENTS OF DAIRY INGREDIENTS

2.1. Skimmilkpowder

Without any doubt, skimmilkpowder (SMP) is the raw material with the greatest influence on final product quality. For all products, SMP of good general quality is a necessaty. Internationally accepted are the quality specifications of the American Dry Milk Institute.

For recombining, extra grade quality according to A.D.M.I. is necessary.
Based on these extra grade A.D.M.I. specifications, the following
requirements are recommended (table 1).

TABLE 1: General quality requirements skimmilkpowder for recombination.

Moisture	max. 4.00 %
Milkfat	max. 1.25 %
Titratable acidity (as lactic acid)	max. 0.15 %
Solubility index	max. 1.25 ml
Scorched particles	max. 15 mg (Disk B)
Bacterial estimate	max. 50.000/g
Coliforms	absent in 0.1 g
Yeasts and moulds	max. 50/g
Extraneous matter	absent
Lumps	absent
Flavour	normal

Apart from these general requirements, specific requirements depending on the recombined milkproduct for which the SMP is intended, are necessary. Very important in this context is the heat treatment of the SMP, or more correct of the skimmilk from which the SMP is made.

Classification of SMP for heat treatment is usually based on the amount of undenatured whey protein, present in the powder. It is expressed as Whey Protein Nitrogen Index (WPNI) in mg N/g powder. The higher the heat treatment, the lower the WPNI-value. SMP's are classified as follows:
Low heat powders   : WPNI $\geq$ 6.0 mg
Medium heat powders : WPNI $\overline{1.51}$ - 5.99 mg N/g
High heat powders   : WPNI $\leq$ 1.5 mg
This classification is a rather rough one. WPNI-values depend not only on heat treatment, but also on the original amount of whey protein in the powder, which is not constant.

Other, probably more reliable heat treatment test have been proposed, such as the "casein-number", that is the amount of Nitrogen from casein + denatured whey protein, expressed as a percentage of total nitrogen. In international trade however, the WPNI is still universally used as an indication of heat treatment.

2.2. Buttermilkpowder

Part of the SMP (10 - 15 %) in recombined products can be replaced by sweet-cream buttermilkpowder (BMP). Due to its relatively high phospholid content, BMP aids in the emulsification of the added fat. In some products like pasteurised milk and sweetened condensed milk it can give a fuller, creamier taste, but addition of more that 15 % BMP can have a detrimental effect on flavour. In heat sensitive products, such as evaporated milk, BMP gives an improvement of the heat stability.

Quality requirements for BMP are given in table 2. Most items are identical with SMP, except for the higher fat content and requirements for lactate and free fatty acids, which are meant as an assurance against the presence of (neutralized) fermented buttermilk.

TABLE 2: General quality requirements buttermilkpowder for recombination.

Moisture	max. 4.0 %
Milkfat	min. 4.5 %
Titratable acidity	max. 0.15 %
Solubility index	max. 1.25 ml
Scorched particles	max. 15 mg (disk B)
Bacterial estimate	max. 50.000/g
Coliforms	absent in 0.1 g
Yeasts and Molds	max. 50/g
Extraneous matter	absent
Lumps	absent
Flavour	normal
Fatty acids (BDI-method)	max. 1 maeq/100 g fat
Lactic acid	max. 40 mg/100g solids non fa

It is not usual to differentiate BMP's on heat treatment because in the final product it comprises only a small part of total non fat milk solids. Therefore, the influence of the heat treatment of BMP on the quality of the final recombined products can be considered as small.

Care should be taken with storage of BMP. As it contains up to 10 % milkfat it is much more susceptible to oxydation than SMP. Therefore storage should be at low temperature if possible and it should be used within one year from production.

2.3. Milkfat

Milkfat for recombined dairy products should fulfill the requirements for IDF-standard 68 A for anhydrous milkfat (table 3).

TABLE 3: Quality requirements for anhydrous milkfat for recombination.

Milkfat	min. 99.8 %
Moisture	max. 0.1 %
Free Fatty Acids	max. 0.3 % (as oleic acid)
Cu	max. 0.05 ppm
Fe	max. 0.2 ppm
Peroxide Value	max. 0.2 maeq/kg
Coliforms	absent in 1 g
Taste and Odour	clean, bland
Neutralizing Substances	absent

Physical structure of the fat is not important, because for recombining purposes, the fat is always melted.

The fat should be packed in air-tight and light-proof containers and be kept at normal ambient temperature.

An alternative milkfat source is unsalted, sweet-cream butter. It is sometimes recommended for recombined products where a fresh milk flavour is very important, such as pasteurized milk. The difference with a high quality anhydrous milkfat is however small and a big disadvantage of butter is, that it has to be transported and stored under refrigeration.

3. RECOMBINED MILK AND MILK-PRODUCTS

3.1. Recombined milk

Three types of milk have to be considered: pasteurised, sterilised in containers and UHT-sterilised, followed by aseptic packaging.

Especially for pasteurised milk a good milk flavour, resembling that of fresh pasteurised milk as close as possible, is important. Therefore ingredients should be of excellent quality. To avoid a cooked flavour, skimmilkpowder should be of low heat or low-medium heat specification, with a WPNI of at least 4.0 mg/g.

For milk, sterilized in containers, the heat treatment of the milkpowder is less critical, because a cooked flavour of the powder is masked by the much stronger sterilisation flavour of the final product.

For UHT-milk, a low level of cooked flavour is of importance again and most specifications for SMP for this purpose ask for a low or low-medium heat powder with WPNI at least 4.0 mg/g.

An important quality deviation in UHT-milk is enzymatic proteolysis, resulting in the development of a bitter flavour during storage or even gelation of the milk. It is caused by traces of heat-resistant enzymes, present in the SMP as a result of growth of psychrophylic bacteria in the milk before processing into skimmilkpowder.

It is therefore important that the SMP is free of heat-resistant enzymes. A direct check on the presence of very low levels of proteolytic enzymes in powders is not easy. However by determining the amount of pyruvate in the powder, an impression can be obtained of the bacteriological quality of the milk from which the powder is made. A pyruvate level of less than 9 mg/100 g powder gives a reasonable guarantee of absence of heatstable enzymes.

For in-container sterilized milk and for pasteurised milk, the absence of traces of enzymes is less important. For in-container sterilized milk because the more severe heat treatment inactivates the enzymes, for pasteurised milk because within the short shelflife of this product, off-flavours or gelation do not develop.

Sometimes heat stability requirements are included in specifications for SMP for recombined milks. This does not seem to be a necessaty because heat stability in non-concentrated products is generally no problem, provided the powders are of good general quality.

3.2. Recombined cream

In general, all points mentioned for recombined milk are valid for recombined creams too. Because of its high fat content, even more emphasis should be laid on the flavour quality of the milkfat used. Quality requirements for skimmilkpowder do not differ from these for SMP for recombined milks.

3.3. Recombined evaporated milk

Contrary to recombined milk, the heat stability is for recombined evaporated milk of utmost importance. Heat stability is determined to a large extend by the skimmilkpowder used. Therefore, specially prepared heat-stable skimmilkpowders have to be used for recombined evaporated milk.

Heat stability is achieved by a high preheating of the skimmilk before concentrating and drying. Therefore, powders for recombined evaporated milk always have to be of the high heat type (WPNI < 1,5). This alone is however not a guarantee for sufficient heat stability. Natural heat stability is dependent on many factors, such as breed of cows, season, area and climate. Moreover the effect of preheating on heat stability differs from plant to plant, depending among other things on the type of heating. Generally, a high temperature short time type of heating (for example 1 minute 120°C) gives a better heat stability than a low temperature, long time type (for example 30 minutes 85°C), but this too is dependent on season.

It is therefore necessary to test the heat stability of a powder in a more direct way. Several tests have been published, all of which are based on the heating of a 20 % solution of the powder in a sealed tube in an oil bath of 120°C and observing the moment of coagulation. Well-known versions of this tests are developed in Australia (3) and Ireland (4).

Coagulation times according to these tests of 20 minutes for American Standard evaporated milk and 30 minutes for English Standard give a reasonable guarantee that the powder is suitable for the production of evaporated milk.

Heat stability of recombined evaporated milk can be increased by addition of 10 - 15 % sweet-cream buttermilkpowder. A high heat stability of the BMP itself is probably not essential for an increase of the heat stability of the evaporated milk in which it is incorporated. For reasons of security it is however usual to specify BMP's for recombined evaporated milk to be of the high heat type.

Other quality aspects of recombined evaporated milk, like viscosity and fat separation are little dependent on the ingredients. There control does not differ from that in normal evaporated milk from fresh milk.

Colour tends to be slightly darker than that of conventional evaporated milk, due to the fact that a highly preheated milkpowder is the main ingredient. It is, at least partly, compensated by the fact that the production site is near the consumer's market. So, the product reaches the consumer shorter after production than imported products do.

Another aspect which is influenced by the use of milkpowder as ingredient is an increased sensitivity to age thickening. This is most probably caused by the high concentration of the milk before drying. Although there is no known method to counteract this higher susceptibility to age thickening, it rarely gives problems in practice, probably due to the relatively short time between production and consumption.

3.4. Recombined sweetened condensed milk

The most important quality aspect of this product is the viscosity. The main factor determining the viscosity is the milkpowder used. The powder has to be of the medium heat type, WPNI roughly between 3.0 and 6.0. The WPNI however, gives not more than a first indication that the powder could be acceptable for recombined sweetened condensed milk.

Viscosity properties of a powder depend on so many other things (breed of cows, season, feed conditions, weather conditions, manufacturing process etc.) that a specific test for viscosity properties is a necessity. Some tests have been developed by powder manufacturers with the purpose to classify their powders in viscosity classes. These tests are based on mixing on laboratory scale of the appropriate amounts of water, SMP and sugar, applying a programmed heat treatment to the obtained mix and measuring the viscosity. One of these tests, developed by CSIRO-Australia is published as an Australian Standard (5).

None of the existing tests are fully satisfactory. One of the problems is, that in the manufacture of recombined sweetened condensed milk a pasteurisation step of a few minutes at 80 - 90°C is necessary. As this heat treatment has a big influence on the final viscosity and this influence is dependent on the powder used, a comparable heating step has to be incorporated in the viscosity test. For reliable test results an exactly reproducable heat treatment is necessary. This is difficult to realise by external heating a small batch of a highly viscous fluid. In the Australian test this has been solved by using a high shear mixer and heating the water-powder-sugar mixture internally by the friction heat of the mixer. This way of heating is difficult to control and only very experienced technicians can obtain reproducable results. Nevertheless, as this is the only published method, it is frequently used to specify milkpowders for recombined sweetened condensed milk.

Together with collegues of the New Zealand Dairy Research Institute we are working on a new viscosity test in which we have tried to avoid some of the problems of the existing tests. Heating is done internally by direct electrical heating, which makes is possible to follow exactly a pre-programmed temperature profile. Other improvements are quick cooling by vacuum-flashing, control of lactose-crystallisation by seeding and an improved method of measuring the viscosity. In the coming year the test will be evaluated in practice.

Although the powder used is the main factor determining the viscosity, some correction is possible by process factors. Increasing or decreasing the pasteurisation temperature increases or decreases the viscosity. The range is however small, because on the one side an effective pasteurisation must be effected, on the other side a too high temperature can give flavour deviations.

Viscosity can also be influenced by homogenisation. Normal in recombined sweetened condensed milk processing is a light homogenisation at 2 - 4 MPa to ensure a good fat dispersion. By increasing the homogenisation pressure, the viscosity is raised. Summarizing: The viscosity range is determined by the choice of SMP. Within that range, a further adjustment is possible by variations in pasteurisation temperature and homogenisation pressure.

Equally important as the initial viscosity is the change in viscosity during storage, the "age-thickening". A certain age thickening is normal, but a too pronounced increase in viscosity is inacceptable. While the initial viscosity can be influenced to a certain extent by processing, the age thickening is entirely determined by the SMP. Therefore, in the viscosity tests mentioned, a test for age thickening has to be included. This is done by storing the test mix for some time, for example 1 week at 40°C and measuring the viscosity again. Dividing the viscosity after storage by the initial viscosity gives the age thickening ratio (ATR). A limit for this ATR has to be included in the powder specifications, together with a range for the test values for initial viscosity.

It is not unusual to incorporate a small amount of buttermilkpowder in recombined sweetened condensed milk. Main function is an improvement of flavour but a small increase of viscosity can also be achieved. This effect is however too small to be of practical importance for the contro of viscosity.

Other quality aspects of recombined sweetened condensed milk are not directly dependent of specific SMP-properties. Their control does not differ principally from that in conventional condensed milk from fresh milk.

4.   CONCLUSION

In the production of recombined milk and milk products, the selection of dairy ingredients and particularly of skimmed milkpowder is of utmost importance for the quality of the final products. Depending on the product, the heat treatment and in most cases one or more functional properties have to be specified. As these powder properties are determined by raw milk properties and powder production conditions, much of the success of recombining depends on the professional knowledge of the milkpowder manufacturer and on the reliability of functional tests.

REFERENCES

(1) International Dairy Federation: 'Monograph on the recombination of milk and milkproducts' - Doc 116 (1979)
(2) International Dairy Federation: 'Proceedings of IDF Seminar on recombination of milk and milkproducts, Singapore 7 - 10 Oct. 1980' Doc 142 (1982)
(3) Standards Association of Australia: 'Determination of the heat stability of skimmilkpowder' Australian Standard 1629.3.4 (1978)
(4) P.M. Kelly. Irish Journal of Food Science and Technology 1, 129-135 (1977)
(5) Standards Association of Australia: 'Determination of the viscosity index of dried skimmilk' Australian Standard 1629.3.5 (1978)

RECOMBINATION OF DAIRY INGREDIENTS INTO FERMENTED PRODUCTS INCL.
CHEESE, BUTTER, AND ICE CREAM

Svend Erik Bøjgaard
Pasilac-Danish Turnkey Dairies Ltd.
Europaplads 2
DK-8000 Århus C
Denmark

1. RECOMBINED FERMENTED MILK PRODUCTS

The manufacture of recombined fermented milk products is very widespread in the Middle East. This area produces a vast number of various fermented products, and the most well-known of these are: set yoghurt, Laban drink, Ayran, yoghurt drink, stirred yoghurt (flavoured), Lassi, buttermilk, Labaneh, cottage cheese, strained yoghurt, and quarg.

Products can be divided into groups in many different ways - for instance, according to whether the product is intended for drinking or eating, whether the fermenting culture used is a yoghurt culture or a mesophile culture, and whether the fermented product has been concentrated or not after the fermentation.

If we are to take a closer look at the manufacture of some of the above products in different groups, it would be natural to look at the most well-known, such as set yoghurt, Laban drink, and Labaneh.

In connection with set yoghurt there are three questions that I find relevant, namely:

a) How much do we demand in terms of quality?

b) Should investments be low?

c) Do we aim at the most economical solution in the slighty longer term?

In addition to these questions, there are naturally many others, but my reason for focusing slightly on these three questions is two of them are almost always overlooked, and the third almost always dominates totally.

Question a):

The first comment always is that the product should be good, but later on it often turns out that there is always room for some degree of compromise with regard to quality if the manufacturer can make more money.

Question b):

The price is often the all-important aspect and is often totally decisive of the composition of the plant, which is naturally dissatisfactory for all parties concerned.

Question c):

It is rare that investors decide in favour of a plant which is capable of producing the very best quality, and which, at the same time, provides the most economical solution over a period of 5 years, for instance.

I should like to present a suggestion of how such a solution might look in the case of medium-sized and large manufacturers of set yoghurt.

Figure 1 shows a recombined set yoghurt plant. Pre-heated water is mixed with skimmilk powder and emulsifier in the mixing unit.

The mixture is then further preheated, the butter oil is added, and the mix is homogenised and pasteurised. The milk is then stored in buffer tanks for a minimum of 20 hours and is subsequently pasteurised again.

The milk is inoculated with the starter culture and the mix is heated to 44°C before it is filled into containers. Next, the containers are placed in the incubator until the desired pH has been reached, and the containers are then moved to the cold store where the yoghurt is cooled down quickly.

This processing method will give the best quality product with the lowest possible TS. The investment is high, the running costs are slightly high too, but on the whole the final quality product is cheap.

There may be some of you who might say that this could have been done in a different way, and of course it could. It could be done in many ways with many small modifications here and there, depending on what other purposes the processing lines are to be used for.

Basically, I think it is fair to say that the best solution is to tailor the processing line to the specific job in question and to give the three above-mentioned questions thorough consideration.

2. CHEESES MADE FROM RECOMBINED MILK

Today a wide range of soft and semi-soft cheeses are produced from recombined milk with good results.

Especially the production of various white cheeses, such as Domiati and Feta, from recombined milk has proved very successful. If we look at the raw materials, we find that especially the quality of the milk powder is highly important.

In general it is true to say that the higher the TS content we want in the final product, the higher the demands we make on the milk powder. In addition, these demands are even stricter if the cheeses are to possess stretching and melting properties.

For the manufacture of Domiati cheese there is a choice between a traditional cheesemaking method and the ultrafiltration method.

If the traditional method is used it is essential to make sure that the loss of TS - especially proteins - is minimised. One of the important things to remember is that the coagulum must be firm before cutting, and on the whole it is important to have equipment designed for manufacture of cheese on the basis of recombined milk.

The ultrafiltration method for manufacture of cheeses on the basis of recombined milk is not particularly widespread as yet, but there are so many advantages involved in this process that this method would naturally be chosen if a new plant for Domiati cheese manufacture were to be built.

Figure 2 shows a flow sheet for the manufacture of ultrafiltered Domiati on the basis of recombined milk. As appears from the figure, the recombining and pasteurising sections are identical to the ones indicated in Figure 1. The recombined, standardised, and pasteurised milk is led to a buffer tank. From the buffer tank the milk is pumped to the ultrafiltration plant via the pre-treatment unit. in the ultrafiltration plant, the milk is concentrated approx. 1 : 3.5.

The concentrate from the ultrafiltration plant is pasteurised and patent blue phosphoric acid, salt, GDL, and rennet are added. this is followed by thorough mixing and filling into bags, cups, or cartons.

The containers are left for incubation for a number of hours before being moved to cold storage.

A comparison between one UF Domiati made from fresh milk and another made from recombined milk will show very little difference between the two.

The next product we are going to look at is the Laban drink. The traditional method is to produce a yoghurt to which we add the same quantity of slightly salted cold water at the desired pH. We then mix it properly and get a fresh sour drink. Today this method is still used, but an increasing number of variations is being introduced, and we find major variations in total solids, viscosity, salt content, and taste. On top of that, UHT treatment is now used for a variety of the products. The reason for this is to give the products a long shelf life. When fermented products are UHT treated it is necessary to add stabiliser in order to protect the protein during the heat treatment. This stabilising together with the heat treatment produces other variations.

The last of the fermented recombined products for consideration is Labaneh. Labaneh is a concentrated fermented product, which is often referred to as a cheese product on account of the whey drainage that takes place during manufacture.

Labaneh consists of 24-30% TS, of which 4.5-10% is fat. pH in Labaneh is most often in the range from 3.9 to 4.3. There are many different ways of producing Labaneh, and I should like to describe these methods by capacity by dividing them into two groups: small scale production and large scale production.

a. Small scale production

1. direct mixing to final TS before fermentation

2. draining in a trolley table or on a drainage belt

3. draining in a bag

4. separation in a nozzle type separator

b. Large scale production

5. thermo methods (including self-cleaning nozzle type separator')

6. ultrafiltration of milk

7. ultrafiltration of yoghurt

In terms of product quality, most of the above methods give extremely good products. However, the quality of number 1 does not quite come up to that of the others, and number 6 produces a product, which is slightly inferior to the rest.

Looking at the recombined fermented products as a whole, we can safely say that we are today capable of manufacturing these products in a quality which is fully comparable to products made from fresh milk.

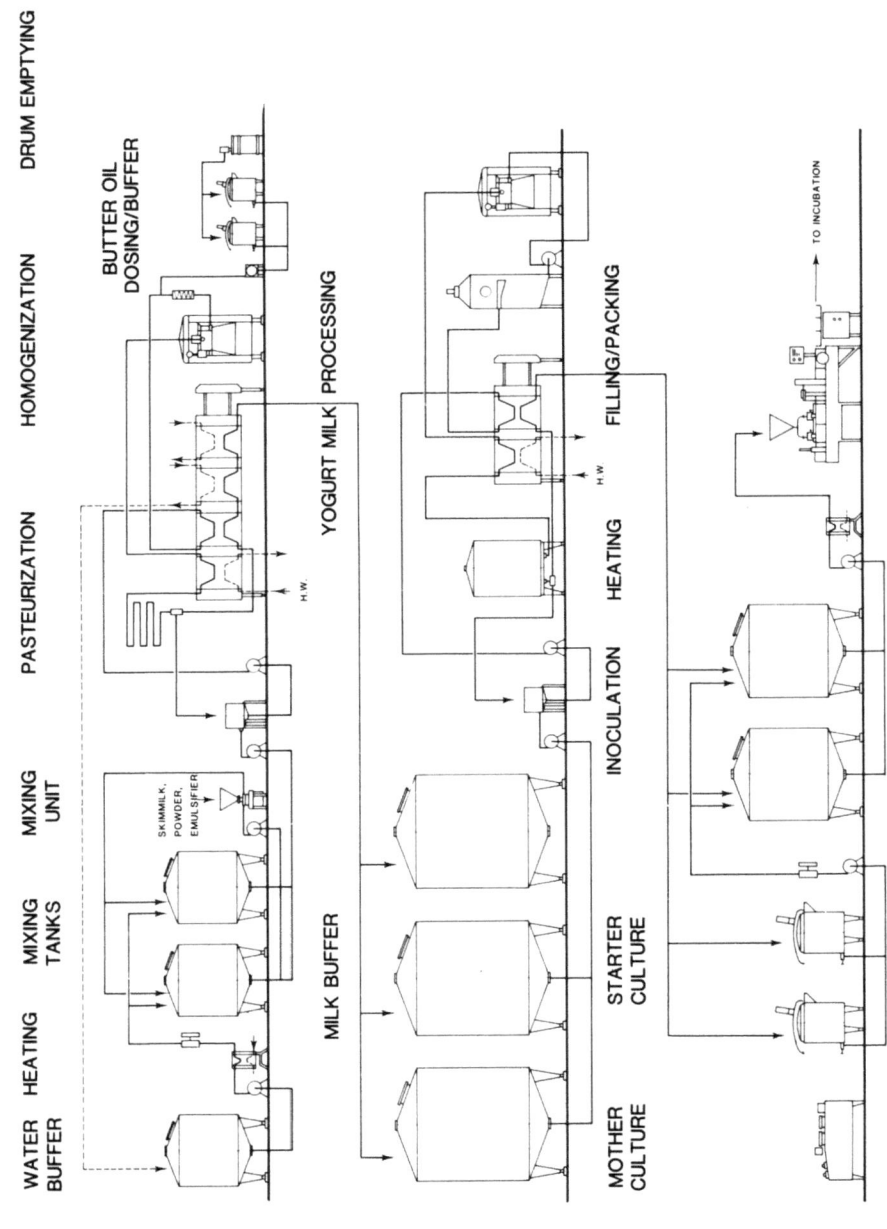

Figure 1: Recombined set yoghurt plant

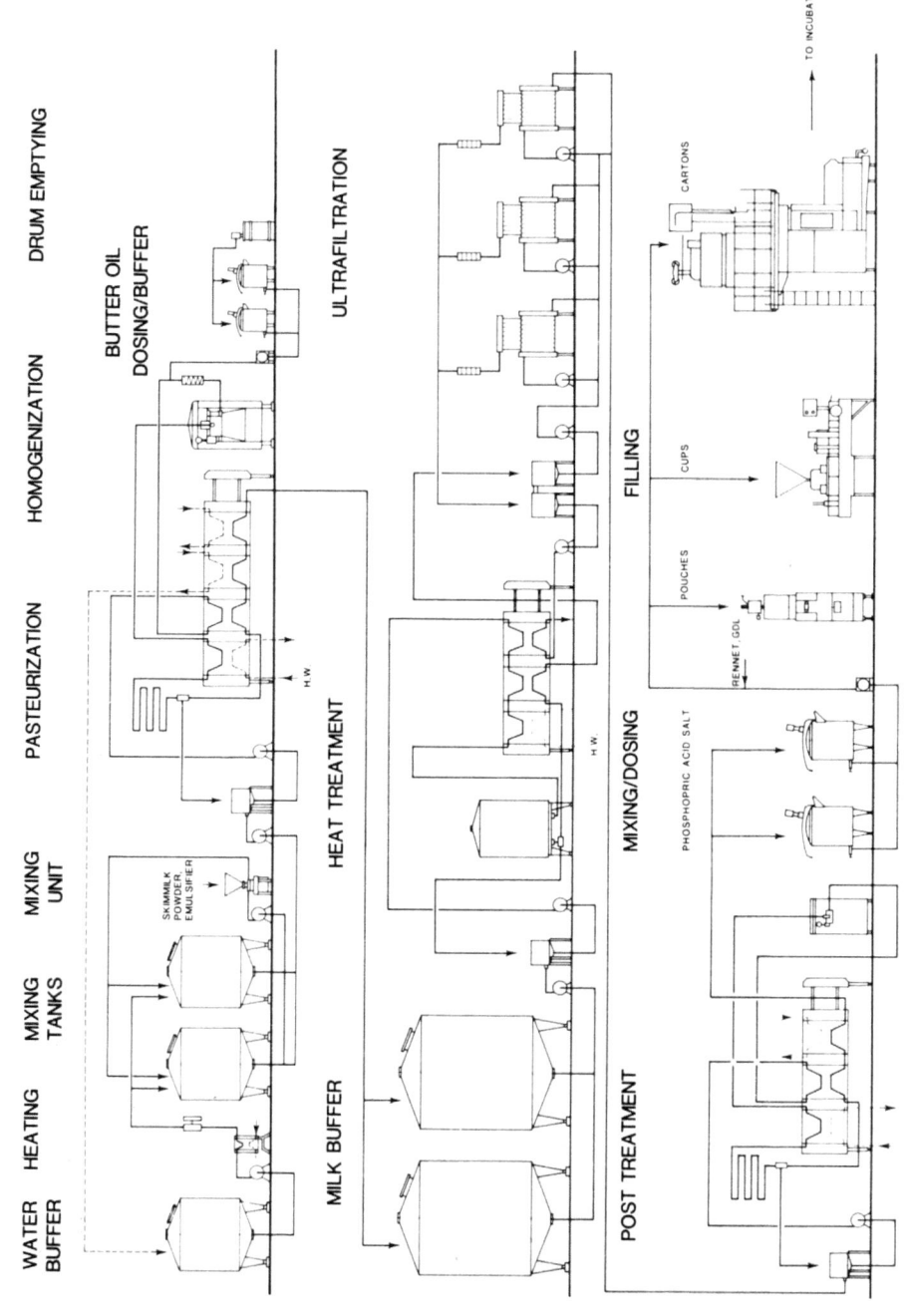

Figure 2: Flow sheet of the manufacture of ultrafiltered Domiati on the basis of recombined milk

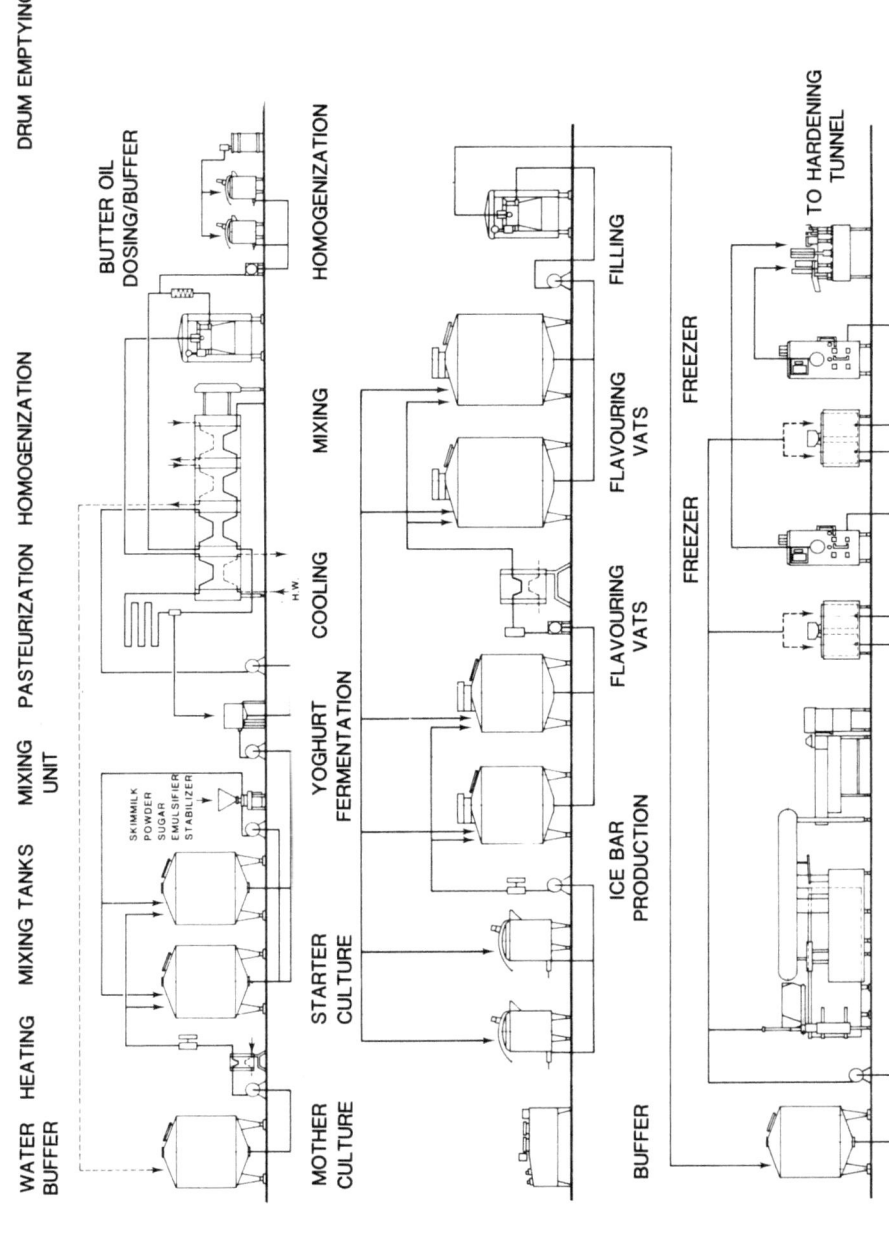

Figure 3: Flow sheet of yoghurt ice cream manufacture

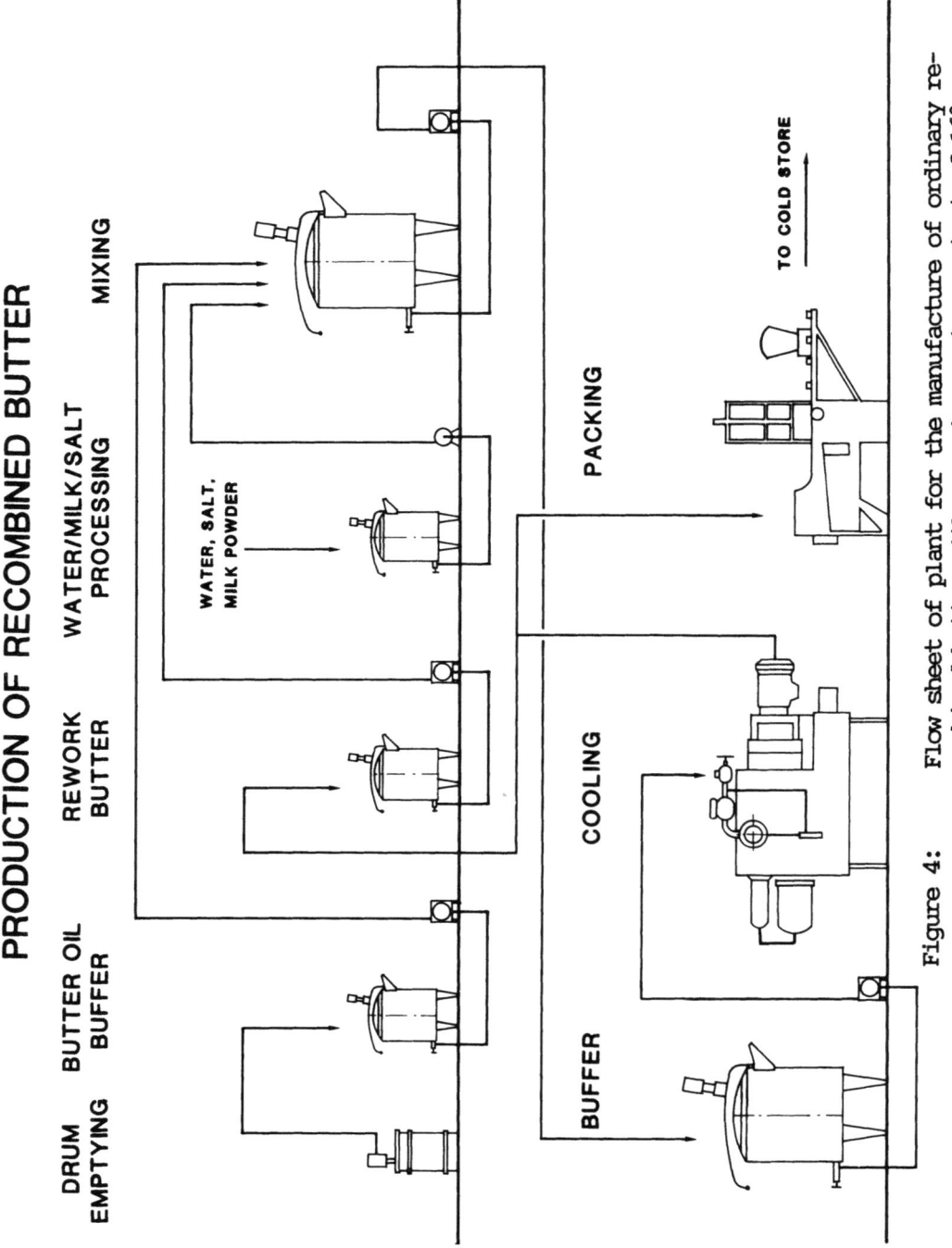

Figure 4: Flow sheet of plant for the manufacture of ordinary recombined butter with a maximum water content of 16%

3. ICE CREAM

Ice cream manufacture from recombined milk mixes is an established method, but manufacture based on fermented milk is a fairly new thing. Ice cream made from yoghurt, for instance, is a product, which is gaining increasing ground in various parts of the world.

If we compare the yoghurt ice cream with a traditional ice cream, the yoghurt ice cream has a fresher flavour and a much higher nutritional value.

Figure 3 shows a flow sheet of yoghurt ice cream manufacture. In this process two different mixes are produced. One of these is fermented with yoghurt culture. When the yoghurt fermentation has been completed, the yoghurt is cooled and mixed with the other mix. This mixture is homogenised and is then ready for manufacture of yoghurt ice cream, here in the form of ice bars and cups.

4. RECOMBINED BUTTER

Recombined butter, like all other recombined products, will only be of good quality if it is made from good raw materials. The manufacture of recombined butter provides us with a number of possibilities, which traditional butter manufacture did not give us, namely provision for varying structural properties, flavour, and water content.

The development and commercial manufacture of butter oil fractions with different melting points now makes for the manufacture of recombined butter with different melting points, i.e. it is possible to make recombined butter which at refrigerator temperature 5°C is soft and spreadable or which at 25-30°C is still firm but spreadable. As far as flavour is concerned, we can also make adjustments, partly by changing pH, partly by adding various flavours, such as cream flavour, and partly by changing the salt admixture. Last but not least it is possible to change the water-fat ratio and make low-fat butter with a fat content as low as 40%.

It is probably appropriate to mention that it is of course possible to replace the butter oil either wholly or partly with various vegetable fats, but that is a different subject altogether.

The manufacture of the different kinds of recombined butter also calls for various types of equipment.

Figure 4 shows a flow sheet of a plant for the manufacture of ordinary recombined butter with a maximum water content of 16%.

The basis for this manufacture is three products, the oil phase (butter oil), water phase (water, salt, and milk powder), and re-work products.

These phases are pumped to a mixing tank in a specific ratio. In the mix tank, the three phases are thoroughly mixed. The mix is pumped via a buffer tank to the surface scraped heat exchanger, where the product is cooled and crystallised. The butter product is passed to a butter packing machine for packing.

The process equipment shown on the flow sheet does not look very complex, but there is a lot of know-how involved in the manufacture of recombined butter of a quality that can measure up to that of traditionally manufactured butter.

NUTRITIONAL ASPECTS OF MILK PRODUCTS IN DEVELOPING COUNTRIES

Dr. T. N. Maletnlema
Tanzania Food and Nutrition Centre
P.O. Box 977
Dar es Salaam
Tanzania

ABSTRACT.
   Many mammals yield milk suitable for human consumption, but of all these animals the cow is the most popular. Milk is a very well balanced food in terms of nutrient content and except for slight deficiency in iron and some vitamins, milk can provide all nutrients and energy required by man. Some of its contents, especially proteins and minerals can be concentrated to a very high level by processing the milk. In the developing countries where nutrient deficiency is a major problem malnutrition, in the form of protein energy deficiency, anaemia, iodine deficiency diseases and hypovitaminosis A, is rampant; affecting mainly children and women. In treating and preventing malnutrition, milk and in particular, dry skimmed milk remains exceptionally useful since it is digestible even by those severely malnourished and it can easily be mixed with local starchy staples. Unfortunately milk production and handling remains grossly defective in the developing countries thus maintaining a state of milk scarcity, high prices and a high degree of dependence on milk surplus western countries. How should these developing countries deal with their milk problem?

Mr. Chairman, Ladies and Gentlemen.

Thank you for giving me an opportunity of sharing our experiences with you and learning from your enormous wealth of knowledge on dairy and dairy products. I cannot speak for all developing countries because they differ a lot, but I speak for a few countries in Africa where milk is a highly treasured commodity.

Whenever one talks of milk for human consumption we think of the cow's milk, but there are several other animals that provide nutritious milk for human consumption. The most common being the buffalo, reindeer, donkey, moose, camel, goat, sheep, lemur and the yak. [1] In terms of protein content the reindeer and the lemur produce the most concentrated milk with a content of 11.5 gm% and 7.3 gm% protein respectively as compared to 3.2% in cow's milk and only 1.1% in human milk. In terms of fat content the reindeer again has the highest concentration of 16.9 gm%, followed by the buffalo and sheep with 7.4 gm% each as compared to 3.7 and 3.8 gm% in cow and human milk respectively.

TABLE I   BIOLOGICAL DATA ON COW'S MILK

Food Item	Biological value	Digestibility	Determined NPU	PER	Hhemical Score
Untreated Cow's milk	84.5	96.9	81.6	3.09	60
Skim milk powder				3.11	64
Casein	79.7	96.3	72.1	2.85	58

Data source reference 3 FAO (1970)

This concentration of proteins and fat can be multiplied 8-10 times by drying the milk to a powder form ending up with full cream powder milk, which is a common constituent of human diets. Further concentration of proteins, minerals and vitamins can be achieved by removing the fat content of the milk before drying it to powder and obtaining dry skimmed milk. Cow's milk is particularly suitable for human consumption because of its high biological value averaging about 84.5 for whole cow's milk. [2] (see Table I) and 79.7 for milk protein casein. It is also highly digestible with a digestibility of 97% in addition to the fact that it can be taken easily since it is usually available in the liquid form and can be mixed in all sorts of other foods. Whole cow's milk contains good amounts of Vitamin A, Carotene, Vitamin D, B group, and Ascorbic acid in just enough quantities for human infant. Milk is also very rich in calcium, but although cow's milk contains three times more calcium than human milk, infants cannot utilise the calcium from cow's milk efficiently (see Table 2).

Contrary to expectations, milks of all types have deficiencies of iron, but in man, infants born with adequate iron stores can maintain normal haemoglobin levels up to six months of age using only the human milk iron. Goat milk is very deficient in Vitamin B_{12} and therefore not suitable alone for feeding infants in whom it may cause anaemia.

TABLE II COMPOSITION OF MATURE WHOLE HUMAN AND COW'S MILK COMPARED

Nutrient	Human Milk	Cow's Milk
Energy kcal/100 ml	75	66
Protein (g/100 ml)	1.1	3.5
Fat (g/100 ml)	4.5	3.7
Sugar (Lactose) (g/100 ml)	6.8	4.9
Essential Amino Acids (mg/100 ml)		
Leucine	100	350
Methionine	25	88
Tryptophan	18	49
Minerals per litre		
Calcium (mg)	340	1170
Magnesium (mg)	40	120
Iodine (ug)	30	47
Zinc (mg)	3-5	3-5
Selenium (ug)	13-50	5-50
Iron (mg)	0.5	0.5
Vitamins per litre		
Vitamin A (I.U.)	1898	1025
Thiamin (ug)	160	440
Riboflavin (ug)	360	1750
Niacin (ug)	1470	940
Folic acid (ug)	52	55
Cynocobalamine (ug)	0.3	4
Ascorbic acid (mg)	43	11
Calciferol (I.U.)	22	14

Modified from D. B. Jelliffe and E.F.P. Jelliffe [1]

The Nutritional Status of Communities in the Developing World
The general state of nutrition in the developing world is that of a large number of people, especially children and women, suffering from severe malnutrition syndromes. Kwashiorkor and Marasmus in children, Anaemia, Goitre, Hypovitaminosis A and fewer cases of hypovitaminosis B like Beriberi and Pellagra [3] in all age groups. This kind of severe malnutrition affects 5-10% of the population, but following the 1980s economic depression the situaton has grown worse and upto 20% of cases may now be found affected in some developing countries. However the severe cases are only a small part of the problem as there are many more, I would estimate up to 60% of the population affected by undernutrition due to inadequate intake of foods, especially the energy foods. This particularly affects children who in underfives clinics appear under weight for age and may not be as active in physical work or classroom work as normal children would be. In women clear indication of this undernutrition presents in their tendency to lose or gain little weight during pregnancy and deliver small infants weighing less than 2.5 kg at term

(see Table III). After delivery their milk output is low, ranging from 300 to 500 ml per day as compared to an average of 700 ml for well fed healthy women. 1/

In men and older children apathy and low physical output for work is the greatest manifestation of undernutrition. This then results in low production evident particularly in the agriculture sector.

TABLE III   FOOD AVAILABILITY AND REPORTED MEAN BIRTH WEIGHTS (1970)

Country	Food available/cap/day		Mean birth weight
	Kcals	Proteins, g.	Kg.
Ghana	2084	43.5	2.89
Uganda	2157	55.9	2.92
Tanzania	2144	60.2	2.94
South Africa	2116	58.7	2.95
Nigeria	2732	77.0	2.98
Senegal	2299	64.0	3.08
Mozambique	2127	40.4	3.15
U.S.A. (Negro)	3156	93.7	3.10
Russia	3182	92.2	3.32

Data modified from Meredith & FAO

Table III demonstrates the relationship between food availability and birth weight of infants. Several field studies show that the more satisfactory the dietary intake by the mother, the higher the birth weight, but in most of the developing countries, 10-15% of infants are born underweight today due to undernutrition and other factors affecting the mothers during pregnancy.4/ Supplimentary feeding projects 5/ have demonstrated beyond doubt that adequate protein and energy intake could improve the situation significantly. In many of these studies dry skimmed milk has been added to local starchy foods like cassava, maize, yam and rice to provide the missing proteins.

Suitability of available foods for feeding children.

In all the developing countries the problem of food intake by the children is compounded by the fact that not only is the food not available in sufficient quantities, but even the food available is often not suitable for consumption by children. The main problem is that of bulk, which is a common property of starchy foods, especially unprocessed starchy foods which have a tendency of absorbing a lot of water and thus diluting the little nutrients that are in the cereal or root flour. Since the children have a small stomach capacity the little food taken only includes few nutrients and parents have no supply of a concentrate that could be added to the cereal porridge to bring up the concentration of nutrents in the children food. The second problem encountered is that of nutrient inbalance in the tropical foods available. For example, the common maize has a deficiency of the amino acid, tryptophan and therefore niacin and the very common root crops like cassava, potatoes and yams are very rich in carbohydrates, but contain very little protein and therefore many diets based on these require a considerable addition of proteins and other nutrients to make a balanced diet. Milk and in particular powder milk provides a ready food to suppliment these starchy diets. 6/

Suitability of milk as a diet suppliment.

From time immemorial communities all over the world have used milk either as a main food among the herdsmen or as a supplimentary food among the non-herders. In many communities milk is an accepted food for feeding the children and infants. However many communities find it difficult or impossible to obtain milk for their children, but still even in these communities few parents manage to obtain milk for their children and families at high costs because of the recognition that by adding milk to the diet you improve the nutritional quality. 6/ Thus many traditional weaning diets based on cereals must by customary practice have milk added in small or large quantities. In some communities of Africa one cannot imagine feeding a lactating women or an orphan without plenty of milk around. Custom for some also demand that lactating women be fed on a diet rich in meat and milk with the conviction that by taking milk the lactating woman will also secrete more milk for the infant.

In food processing too, dried milk powder is used considerably, especially in the dry skimmed form in the formulation of weaning foods and formulae for infants. Processing of milk to infant formulae has enabled parents and doctors to nurse to childhood full-term neonates and prematures who would otherwise have died of starvation or inappropriate feeding. However, much more research is still required to bring the infant formulae to the human milk composition level. In the meantime legislation and nutrition education must be carried out to prevent misuse of formulae all over the world. Where necessary orphans and others who cannot get foster mothers or "wet" nurses must continue to rely on commercial formulae, but poor or remotely placed communities who cannot procure such formulae have to rely on artificially made formulae. Cameron and Hofvander 7/ give considerable details on how one should go about making formulae from various types of milks available. One rule they stress is absolute cleanliness.

Milk as fresh liquid or as a powder added to other foods like cereals, roots or vegetables is much more easily acceptable than other protein concentrates like fish powder or soya flour. In commercial undertakings several developing countries have manufactured relatively cheap weaning foods based on local staples enriched with among other things, milk. Examples include the Algeria made Super-amine based on wheat and precooked peas to which 10% dry skimmed milk powder is added; Faffa made in Ethiopia is also based on wheat, peas and soya flour to which 5% dry skimmed milk is added. In Tanzania attempts have been made to manufacture Lisha based on maize and soya with 5% dry skimmed milk added. Vitamins and minerals are also added to these foods.

The milk industry in Africa.

The milk industry in most of the developing countries has not evolved from local conditions and demands. In a few African countries studied the milk industry is just a copy of the western industry and depends heavily on the west for technology, machinery and of late even basic raw materials like milk powder and fat. The industry produces milk products for consumption in urban areas mainly as pasteurized milk, powder, butter, cheese, condensed or evaporated milk (one country out of 9) and sour or fermented milk (one out of 9 countries). Fermented milk is very common in the rural areas where fresh milk cannot keep for long. Unfortunately the milk fermenting technology is not accepted by the urban community and therefore it has not been developed, despite the fact that it is much more suited to the hot tropical climate. 8/

TABLE IV   MILK PRODUCTION AND PROCESSING STUDY IN A FEW AFRICAN COUNTRIES, 1980s

Country	Production '000 litres per day	Types of Processing Carried Out
Ethiopia	40	Pasteurization Butter extr. and Drying
Burundi*	1.8	Pasteurization Butter extr., Cheese and Drying
Botswana*	300	Pasteurization
Kenya	2470	Pasteurization Condensing and Evaporating Butter extr. and Cheese Drying and Ghee
Malawi*	9.5 (small holder only)	Pasteurizing Butter extr. and Ghee Cheese
Swaziland*	10	Pasteurizing Fermenting
Tanzania*	1369	Pasteurizing Butter extr. and Cheese
Zambia*	35	Pasteurizing
Zimbabwe	411	Pasteurizing Butter extr. and reprocess Cheese, etc Drying and Formulae

* These countries import considerable quantities of milk, etc.
Ref. as for Table V.

TABLE V   MILK CONSUMPTION AND PRICE IN A FEW AFRICAN COUNTRIES COMPARED TO THE NETHERLANDS

Country	Consumption kg/h/year (ml/h/day)		Price US$ per litre
Ethiopia	19.6	(54 ml)	0.30
Burundi	13.3 rural	(36 ml)	
	73.0 urban	(200 ml)	0.61
Botswana	109	(299 ml)	?
Kenya	50	(137 ml)	?
Malawi	5.5	(15 ml)	
Swaziland	63.6	(174 ml)	0.31
Tanzania	39.9	(109 ml)	0.50
Zambia	8.1	(22 ml)	?
Zimbabwe	24.3	?	0.23
Netherlands			0.35

Compiled from: Kitegile, J.A. (1984) The Potential for Small Scale Milk Production in Eastern and Southern Africa. IDRC-IR 983 Manuscript Reports.

For many years now dry skimmed milk enriched with Vitamin A has been used in large quantities to save lives of severely malnourished children in the various parts of the developing world and also for refugee's children and famine striken communities. In this way milk has helped save a lot of lives which would otherwise have been lost.

Milk Costs.

Milk and milk products are very expensive in most of the developing world compared to the developed world. A study of a few countries in Africa shows that milk for local consumption is very expensive in Burundi followed by Tanzania and is cheapest perhaps in Zimbabwe, Swaziland and Ethiopia. But even in these areas where it appears cheap, if compared to a developed country like the Netherlands, it is still difficult to get and subsidized by the government and therefore expensive. Tanzania prepared a commercial weaning food based on maize and soya and in order to improve the quality added 5% dry skimmed milk and this shot the price up by 10% see Table VI.

TABLE VI   CHANGE IN PRICE OF A WEANER FOOD AS A RESULT OF ADDING 5% DRY SKIMMED MILK

Product, Lisha, Tanzania

Composition:
1. Maize 70%, Soyabeans 28.3% and Minvit. mix. 1.7%
2. Maize 68%, Soyabeans 25.3% DSM 5% and Minvit.mix.1.7%

Costing in November 1980 (TShs.)*

	1. Plain CSB	2. CSB + DSM 5%
Ex-factory	6.05	6.55
Whole sale	6.20	6.80
Consumer	6.50	7.25

* 1 US$ was then = 12.5 TShs.

Milk Production and Consumption.

A number of interwoven factors influencing milk production and consumption include climate, type of cattle, animal health and care, transport, equipment, market, price, purchasing power of the consumer, food habits and customs. In most of the developing countries hot and dry climatic conditions often accompanied by poor soil fertility, inferior varieties of grass and wide spread animal diseases combine to prevent development of dairy cattle over vast areas. High inputs are often required to render even small areas of land suitable for dairy cattle farming. In the few areas where the climate is suitable animal diseases and shortage of drugs reduce already low milk yield of for example the Zebu and the Sanga types in Botswana, Zambia, Malawi, Tanzania, Kenya, etc. to almost zero yield. Attempts to improve breed type by crossbreeding has so far not been very successful although considerable improvement has been achieved in a few countries. The main problem is that of management.

Thus although milk is a very nutritious and acceptable food in the developing countries, its production and handling remains a major problem minimizing milk consumption (Table V). The major questions to ask ourselves is whether tropical countries should invest heavily (by further debts) into dairy farming or try to find alternatives to cow's milk. Animal experts argue that milk is the cheapest and most suitable protein food, but nutritionists still see it as one of the most expensive.

References.
(i) D.B. Jelliffe & E.F.P. Jelliffe (1978) Human Milk in the modern world. Publ. Oxford Medical Publications.

(ii) FAO (1970) Amino acid content of foods and Biological Data on Proteins.

(iii) Latham, M.C. (1984) Strategies for the control of malnutrition and the influence of the nutritional sciences. F.A.O. Food and Nutrition 10:1, 5-31

(iv) Aebi Hugo, Whitehead Roger (1980) Maternal Nutrition during Pregnancy and Lactation.
Nestle Foundation Publication No.1
Publ. Hans Huber, Bern, Switzerland

(v) Gwatkin, D.R.; Wilcox, J.R.; Wray J.D. (1980)
Can health and nutrition interventions make a difference?
Overseas Development Council, Washington, D.C.

(vi) Aalbersberg W.I.J. (1979) Nutritional Aspects of the export of Dairy Products. Personal communication

(vii) M. Cameron and Yngve Hofvander (1980) Manual on Feeding Infants and Young Children. FAO, Rome. pp. 95-104

(viii) Bachman M.R. Fermented Milk in the Dairy Development of the Third World Countries. IDF Bulletin 179 pp. 143-147

DISCUSSION

J.C.T. van den Berg (the Netherlands) : How to solve the problems of the supply of liquid milk to large urban centres with a population of several million people, e.g. Mexico city? A recombining industry could be important if the local supply of milk is inadequate.
M.R. Bachmann (Switzerland) : A recombining industry established with the purpose to dispose of surpluses from developed countries may hamper the development of local milk production and dairy industry. Large scale milk plants based on recombining have been closed. For instance in Pakistan the Karachi plant was closed in 1980, and liquid milk will, as a new approach be produced in small plants located close to the milk production.

R. Rosenfeld (Israel) : Milk solids for recombination are marketed to developing countries at subsidized prices and competition arises with local dairy products. There is a conflict between Prof. Bachmann's idea of development and Ir. Sjollema's message to use imported milk constituents. It may be difficult for a developing country to put high quality requirements on subsidized imported materials.
A. Sjollema (the Netherlands) : Recombining plants in developing countries are guided by experts from advanced countries and assistance is given to run the plants and to maintain quality control on raw materials and finished products.
L.L. Muller (Australia) : An example of gained benefits from putting up a local recombining industry could be Singapore where annual savings of 15 million USD were achieved some 20 years ago in import expenditure.

J.J. Mol (the Netherlands) : 1. What is the difference between stirred yoghurt and laban?
2. What is the experience of using whole milk-powder (WMP) in the manufacture of fermented milk products?
S. Bøjgaard (Denmark) : 1. Only the name is different, but the fat content of Laban may differ considerably from country to country, e.g. from 4-5 % to up to 10 %.
2. The use of WMP is a good idea for small capacity plants as investment in equipment is lower. Badly stored WMP may, however, cause a rancid taste. If a proper timing of the delivery of WMP is not under control, the use of skim milkpowder is to be recommended from the point of view of an even and high quality of end-products.

P. Rosenfeld (Israel) : 1. What are the aspects at adding salt to Domiati cheeses?
2. Is ultrafiltration "a must" in manufacture of cheese by recombination?

S. Bøjgaard (Denmark) : 1. All Domiati cheeses include added salt. Originally salt was added to the milk at an early stage before treatment in vats due to questionable milk quality and usually high ambient temperature (35°-40°). In controlled conditions of industrial scale manufacture salt is added to the cheese prior to packaging.
2. Ultrafiltration is not "a must". Small quantities of cheese can well be made without UF but in handling, say, 10000 liters or more per day, the application of UF is justified.

Name and country unknown : Why is buffering in tanks for 24 hours important in the manufacture of recombined yoghurt?
S. Bøjgaard (Denmark) : It is important for the manufacture of all liquid products; as proper rehydration at a temperature of 5-10° C will require at least 16 hours. A period of 20 hours is considered as a "safe" keeping time.

F. Pronk (Italy) : Although milk is expensive in developing countries the World Food Programme receives more and more requests from governments for support to dairy development projects. The justification is to diversify and improve the income of smallholder farmers as land is available for livestock. Why is there a growing trend in requests? Do governments assume that purchasing power is available or is the reason mainly to stimulate the development of local agriculture?
T.N. Maletnlema (Tanzania) : The nutritionists are behind the requests as there are at present no comparable alternatives to the valuable nutrients of milk. Soya products are partial alternatives but not so common in Africa. What WFP receives in the form of requests from African governments is supported by nutritionists and urban governments.
E.A. Olaluku (Nigeria) : A major problem of Africa's developing countries is the lack of a firm and consistent policy for dairy development. The governments stress the importance of improved nutrition and the role of milk, but dairy development projects are in difficulties due to lack of support and infrastructure. People are prepared to pay even comparatively high prices for milk and milk products, but due to severe shortages or unavailability of milk, research institutes are developing and promoting production of soya-beans as an alternative source of protein. There is a real need for support from international agencies to make governments to realize the need for coordinated programmes for dairy development.

D.N. Maletnlema (Tanzania) : Let us each put pressure on our respective governments.

P. Shalo (Kenya) : There are problems when applying advance technology and using the high capacity equipment available from industrialized dairy countries. The plants are too big for their environment in developing countries. The economy of scale requires long distance haulage of milk on rough roads in a hostile climate. The milk deteriorates to unacceptable standard and the producers feel disincentives through rejection or deducted payment for their milk. The developing countries appeal for properly designed, easily maintained and low capacity equipment at acceptable prices from manufacturers.

M.R. Bachmann (Switzerland) : The appeal is highly appreciated. Small capacity equipment, in some cases even hand-operated versions, should be available. Manufacturers have not responded adequately to previous requests. However, mini dairies of 1000, 3000 and 6000 litres daily capacity are now available. More should be done to establish manufacture of equipment and milk handling utensils in the developing countries as well.

S. Bøjgaard (Denmark) : It is surprising that this question is raised by a Kenyan representation. The country appears to have a good network for urban supply of milk and an industry with appropriate technology as judged from an economic point of view. A great variety of miniplants is available as well as single pieces of equipment from many manufacturers. It should however be kept in mind that low output unavoidably means higher general costs per unit of milk or product manufactured.

M. Larsson-Raznikiewics (Sweden) : The camel has been neglected as a milch-animal. Its milk has a high nutritional value, a high content of vitamin C and it keeps well in high temperatures: one day without souring and 2 - 3 weeks in the refrigerator. The camel survives long drought periods and demands little for feeding.

M.R. Bachmann (Switzerland) : This neglection is a fact. Camels may produce up to 1500-2000 litres per lactation. The milk varies, however, in content and the processing technology is different. Camels' milk does not, for instance, coagulate by addition of rennet. Fermented milk can be made from it.

J.C.T. van den Berg (the Netherlands) : Breeding of camels for milk production is carried out e.g. in the USSR. There will be problems due to the nomadic nature of the animal. Processing and marketing is difficult. The camel is not competing with cattle and sheep for feed.

S.R. Sukarto (Indonesia) : The main objectives for dairy development in Indonesia are to create job opportunities and more income for farmers. Improved nutrition has lower priotity. Small farmers with 1-2 hectares of land are encouraged and assisted to keep 3-4 cows for supply to the dairy industry and urban citizens.

P. Rosenfeldd (Israel) : About 25 % of the population in many countries are elderly people. Should they necessarily need milk proteins or are there alternatives?

T.N. Maletnlema (Tanzania) : Milk protein are of highest quality and more complete than other available proteins. For infants, butter is not complete as it is deficient e.g. in linoleic acid. Some addition of vegetable fats might be necessary.

L.L. Muller, chairman (Australia) : The seminar has been interesting and rewarding due to excellent lectures and to the contributions from the participants. Nutritional aspects could have been dealt with first as they form the basis of understanding of the role and the importance of the recombining industry. The need to develop small and pragmatic equipment is obvious as well as the need to consider the local traditionally known products when designing local dairy plants.

Dairy development needs a clear policy: Recombination initially can form a basis. At the same time local milk production should be promoted, so that the recombination industry can gradually convert to locally produced milk.

Seminar II: Milk Production and Milk Products in Developing Countries, and
Seminar III: Market and Marketing of Dairy Products

Session 3

Chairman: R. P. Aneja (India)
Secretary: Prof. E. A. Olaloku (Nigeria)

Market and marketing of dairy products in developing countries

MARKET AND MARKETING OF DAIRY PRODUCTS IN DEVELOPING COUNTRIES

Drs. D.E. de Roon
Dutch Commodity Board for Dairy Produce
P.O. Box 5806
2280 HV RIJSWIJK ZH
Holland

QUANTITATIVE ASPECTS

I have been asked to give you an outline of the trend developments which have taken place in our sector in the production and trade spheres, with specific reference to the quantitative aspect a subject that, because of its statistical approach, may perhaps not appeal directly to the imagination! However, in order to achieve a more listener-friendly presentation, I shall imitate the World Bank and classify the member states of the United Nations in 9 categories. This not only allows the facts to be presented in a more easily digestible form but - as we will see later - it also provides a deeper insight.
    The economies are classified by the World Bank by Gross National Product per capita. This classification is useful in distinguishing economies at different stages of development. Many of the economies included are also classified by dominant characteristics - to distinguish oil importers from oil exporters, for instance. Countries with populations of less than a million, however, are not included.
    The classification devised by the World Bank is as follows:

0 low income economies   average G.N.P. per capita $ 260
  - Asia
  - Africa

0 lower-middle income economies   average G.N.P. per capita $ 740
  - oil exporters
  - oil importers

0 upper-middle income economies   average G.N.P. per capita $ 1950
  - oil exporters
  - oil importers

0 high-income oil exporters   average G.N.P. per capita $ 11250

0 industrial market economies   average G.N.P. per capita $ 11430

0 East European non-market economies

When assessing the figures which will be presented later on, it should
be born in mind that there are extremely large differences between the
country groups mentioned above both in terms of population size and
surface area.

	number of countries	population (millions) mid 1984	area (,000 km 2)
0 low income economies			
- Asia	12	2,101.8	16,124
- Africa	22	251.2	15,451
0 lower-middle income economies			
- oil exporters	8	347.3	7,166
- oil importers	29	333.3	11,280
0 upper-middle income economies			
- oil exporters	8	200.3	7,870
- oil importers	14	309.3	14,209
0 high income oil exporters	5	18.6	4,311
0 industrial market economies	19	733.4	30,935
0 East European non market economies	8	389.3	23,421
		4,684.5	
Countries not included	37	18.2	

Statistics concerning the dairy sector have been collected for all
these country groups.

The data concerning the production of the various types of milk
and its processing into cheese, milk powder, condensed and evaporated
milk, butter and butter oil, have been taken from the FAO production
yearbooks, while the foreign trade figures have been taken from the
publications 'World Trade in Dairy Products" by the Dutch Commodity
Board for Dairy Produce. Whilst the figures taken from the Commodity
Board are based on accurate observations at the external frontiers of
the exporting countries, the data collected by the FAO are based in
part on estimates; estimates which, moreover, undergo fairly frequent
correction over the years. We must allow for considerable margins of
error. Accordingly, no absolute value whatever may be assigned to the
figures which will be presented later and the conclusions which are
drawn from those figures. Nevertheless, having familiarized myself

with the many statistics I feel that there is a good case for urging that the collection and processing of statistics on milk production and milk manufacturing should be taken seriously.

My analysis will be performed over the years 1978 - 1984.

MILK PRODUCTION

Total milk production - cow, buffalo, sheep and goat milk - in 1984 was in the region of 495 million tons. During the 7 years under review, this production increased by around 38 million tons, the average annual increase being just over 1.3%.

All the country groups distinguished here made a contribution towards this growth.

Three-quarters of the total milk output is produced in the industrialized world of the east and west. Milk other than cow milk makes a modest contribution to this milk lake, amounting to just under 10%.

Percentage distribution of the produced milk, by type.

	1978	1984
- cow milk	91.4	90.5
- buffalo milk	5.6	6.4
- sheep milk	1.6	1.6
- goat milk	1.5	1.5

I should like to say a few words abouts each different type of milk in the table.

The production of cow milk is concentrated in the industrial market economies and the Eastern European non-market economies. Over 80% of all cow milk is produced in these economies. Under the influence of the restriction on the production of cow milk which has now been introduced in several countries, starting in 1984 we see a decline in production begin to emerge in the industrial market economies. A further decline, especially in the EC member states, is to be expected.

Cow milk production in the Eastern European non-market economies proceeds erratically: a downward trend until 1982, followed by an upward trend.

A substantial growth in cow milk production can be observed in the group of the high-income economies, oil exporters (average annual rate of growth 7.4%). In absolute terms, however, the volumes concerned are negligible, although compared relatively against the population of these 5 countries it represents a very considerable increase in milk production on a per capita basis. From these figures it appears that the establishment in those countries of large-scale dairy holdings is beginning to bear fruit.

The production of buffalo milk is in fact confined to the low-income economies of Asia (in particular India). With an average annual growth rate of over 3.5%, buffalo milk is one of the fastest-expanding products. The production of buffalo milk in the Eastern European non-market economies is on the decline.

For sheep milk, no specific regions of concentration can be indicated. The share of the EC member states is around 21%. World production is growing at an average of 2.2% per annum. The question may be posed whether the production restriction which has been imposed on cow milk in many industrialized countries will form an incentive for sheep farming.

Production of goat milk takes place largely in the low-income economies of Asia and Africa. The biggest growth, on the other hand, can be seen in the high-income economies, oil exporters. However, in this respect we must immediately ask ourselves whether there is a question of any real growth or of changes in the methods of observing production.

Now how is milk production related to population size, in other words how much milk is produced per capita of population? The results of this calculation are by no means surprising. At most, there may be some surprise when learning that the per capita milk production in the non-market economies is higher than that in the industrial market economies, which include such renowned dairy producers like Denmark, France, New Zealand and the Netherlands.

Milkproduction per capita 1984

	kgs
O low-income economies	
– Africa	24
– Asia	26
O lower-middle income countries	
– oil exporters	14
– oil importers	43
O upper-middle income economies	
– oil exporters	70
– oil importers	95
O high-income economies	
– oil exporters	40
O industrial-market economies	315
O East European non-market economies	361

MILK UTILIZATION

An important segment of milk utilization, namely the consumption of treated or untreated liquid milk, lies outside the range of observation and might at best be calculated as a balancing item. There are no worldwide figures available, either on indigenous consumption or on direct ex-farm sales.

This section will therefore have to be limited to the manufacture of milk into cheese, butter and ghee, evaporated/condensed milk, whole and skimmed milk powder and whey powder. A steady increase in production can be observed for all the products just mentioned.

Index figure of the world production of certain dairy products.

(1978=100)

	1979	1980	1981	1982	1983	1984
cheese (all kinds)	104.3	106.7	109.9	111.1	115.2	115.7
butter and ghee	100.7	102.2	101.0	104.1	112.0	110.8
evaporated and/condensed milk	102.3	102.9	104.6	104.9	102.0	103.8
dry whole cow milk	99.3	110.8	116.1	113.7	114.0	121.0
skim milk	99.1	99.0	100.3	108.1	119.0	105.9
dry whey	103.3	126.7	150.9	155.3	161.2	154.8

Developing countries have only a modest share in the production of these dairy products. Indeed, it would not have been justified to expect anything else.

Share of the developing countries in the production of:

	%
- cheese (all kinds)	15.8
- butter and ghee	22.5
- evaporated/condensed milk	22.4
- dry whole cow milk	21.2
- skim milk	2.1
- dry whey	8.0

At this point a few comments are called for. Both for butter and ghee (14.1%) and for evaporated/condensed milk (8.4%), the share of the low-income economies of Aisa are higher than would be expected on the basis of the milk production in this country group. As regards butter and ghee, it may be pointed out that in those countries this dairy product forms a traditional item of the regular diet.

Finally, the production of condensed milk is found to be concentrated in one of the countries (India) which forms part of this group. In terms of production volume of evaporated and condensed milk, this means that India would rank fifth in the world. Indeed, that country has a strongly expanding industry (with an average annual growth rate of 5%). 1)

1) However, the question must be asked whether, as far as India is concerned, the usual definition for evaporated and condensed milk has been followed in the FAO statement of production figures.

WORLD TRADE IN DAIRY PRODUCTS

On the basis of the trade statistics of the main dairy-exporting countries, the Dutch Commodity Board for Dairy Produce compiles an annual review with the not completely accurate name of the world trade in dairy products. Not completely accurate because the exports of, among other countries, the Eastern European non-market economies have to be left out of these reviews due to insufficient information. 1) Furthermore, the incompleteness of the statistical data makes it impossible to identify the trade flows of relatively minor products which are nevertheless extremely interesting in terms of added value, such as powdered foods for infants.

Bearing in mind the dominant importance of the industrial market economies as offerors of dairy products on the world market, the figures which will now be presented give a fairly true picture of reality. It has been assumed that exports to a given country group are equal to the imports by that country group. First of all, an overall picture of world trade in milk and dairy products as derived from the export statistics of the major exporting countries.

1000 m.t.

product / year	1978	1979	1980	1981	1982	1983	1984
cheese	1086.5	1228.8	1284.4	1390.9	1445.5	1480.7	1622.9
butter and ghee	916.9	1188.2	1258.5	1229.7	1222.0	1069.9	1135.3
evaporated and/ condensed milk	688.2	718.6	875.1	913.2	929.6	841.9	865.4
whole milkpowder	517.2	624.7	800.9	792.6	755.0	651.7	754.3
skim milkpowder	1613.3	1685.2	1450.3	1268.4	1391.4	1659.1	1842.0
milkequivalent 2)	33,222	36,411	36,175	35,068	36,575	38,590	42,724

This foreign trade consist of flows amongst the various industrialized countries on the one hand and exports by industrialized countries to developing economies on the other hand.

1) All the author possesses is a single total figure, with no specification by country of destination.
2) It has been assumed that 100 kg of milk yields 12.0 kg of cheese, 12.1 kg of unskimmed powder, 9.2 kg of skimmed powder, and 47.6 kg of condensed milk. For butter, the production weight was taken.

Exports to developing countries show the following developments:

1000 m.t.

product / year	1978	1979	1980	1981	1982	1983	1984
cheese	141.9	177.4	245.9	272.0	289.9	308.4	380.8
butter and ghee	261.4	370.4	429.4	424.5	365.0	347.2	404.7
evaporated and/ condensed milk	551.5	595.1	735.5	739.0	779.2	672.1	724.6
whole milkpowder	384.7	455.4	582.1	575.2	539.1	503.2	586.4
skim milkpowder	637.8	730.6	767.5	749.1	669.0	651.1	817.5
milkequivalent exports to developing countries as % of	12,711	14,801	17,174	17,135	16,141	15,561	18,829
total exports	38%	41%	47%	49%	44%	40%	44%

It should be mentioned that both in 1981 en 1984 food-aid deliveries were much more voluminous than in the other years. 1).

For the purpose of this presentation, we shall confine ourselves to considering the trade flows which go to developing countries. First of all, a few comments of a more general nature.

In the 1950s and 1960s, the world economy went through a period of uninterrupted growth, which was however disrupted in the 1970s. In particular the second increase in oil prices, which took place in 1979, was absorbed less well than it had been in 1973.

In the first half of the 1980's, real GDP growth slowed throughout most of the developing world, and per capita incomes declined in many countries. Developing countries suffered in 1979-84 from a combination of more expensive oil, high real interest rates, prolonged recession in industrial economies, and more trade barriers. But averages conceal wide differences in individual performances. One of the most worrisome aspects of the early 1980's has been the continued decline in low-income African countries.

In general, it may be stated that the low-income South-East Asia countries have withstood the recession better than the low-income sub-Saharan African countries.

In addition, in the middle of the 1980s the oil-exporting countries themselves were also hit by the lower oil prices and the resultant fall in export revenues. Although this contains a measure of relief for oil-importing countries, it is overshadowed by other (negative trends).

1000 m.t.

1) Food-aid deliveries	1978	1979	1980	1981	1982	1983	1984
skim milkpowder	236	256	228	333	268	252	356
butteroil	53	47	29	59	41	23	57

The following table gives an impression of the trends which have taken place in the developing countries.

Real growth of Gross Domestic Product

	average 1965/'73	1973/'80	1981	1982	1983	1984	1985
low-income countries							
- Africa	3.9	2.7	1.6	0.8	0.3	0.7	2.1
- Asia	5.9	5.0	5.4	5.7	8.6	10.2	8.3
middle-income countries							
- oil exporters	7.1	5.8	4.4	1.0	-1.9	3.1	2.5
- oil importers	7.0	5.5	2.1	0.8	0.8	4.1	3.0
high-income countries							
- oil exporters	9.2	7.7	1.6	-1.7	-7.1	1.3	-5.0

(source: World Development Report 1986)

The trends developing in world dairy trade should be seen in the light of the situation which has just been sketched in brief. The following tables show the development of exports 1) to the developing countries for the years 1978 - 1984. It should be born in mind that consignments of food aid are not identifiable as such in the export statistics.

Exports of dairy products to low-income economies Africa

1000 m.t.

	1978	1979	1980	1981	1982	1983	1984
cheese	2.4	1.6	1.9	2.2	2.4	1.8	1.9
butter and ghee	8.7	10.0	9.3	12.8	9.1	7.4	13.9
evaporated and/condensed milk	23.7	17.4	21.6	19.3	21.6	15.8	26.6
whole milkpowder	15.8	12.7	18.8	19.2	19.6	15.2	16.9
skim milkpowder	47.4	56.2	61.1	75.0	67.6	59.5	84.8
milkequivalent	723	776	889	1046	971	828	1147

In relation to population size, the imports are of little significance.
An increase of any importance can only be observed for butter and ghee and skimmed milk powder, products which usually form part of food aid programmes. The virtual unchanging volume of imports of other dairy products could indicate that the imported volumes do not exceed

1) The terms "exports to" and "imports by" will be used interchangeably.

the minimum volume considered necessary. How do imports relate to
indigenous production? For cheese the share is only 2.5%, for butter
and ghee it is 27.7%. The other products are not manufactured locally.

Exports of dairy products to low-income economies Asia

1000 m.t.

	1978	1979	1980	1981	1982	1983	1984
cheese	0.3	0.4	0.3	0.4	0.4	0.5	0.4
butter and ghee	28.3	28.0	22.6	27.1	26.5	8.1	28.1
evaporated and/ condensed milk	9.9	9.4	11.5	11.7	14.2	12.9	17.1
whole milkpowder	33.8	34.7	31.6	33.0	47.4	35.0	30.3
skim milkpowder	86.6	90.7	62.6	104.2	93.5	35.8	91.5
milkequivalent	1723	1324	991	1460	1468	718	1312

This country group includes China and India, both giants in terms of
population size. India accounts for the lion's share of imports of
butter and ghee and skim milk powder. Only for whole and skim milk
powder do imports represent a substantial proportion of domestic supplies,
in the sense that the local production is of little significance.

Exports of dairy products to lower-middle income economies-
oil exporters

1000 m.t.

	1978	1979	1980	1981	1982	1983	1984
cheese	12.6	18.1	18.4	20.9	25.9	42.8	47.0
butter and ghee	43.2	60.2	64.3	63.8	52.0	51.1	66.6
evaporated and/ condensed milk	152.2	142.8	188.2	196.3	212.9	117.4	103.1
whole milkpowder	61.6	61.1	84.5	77.5	56.6	43.7	48.4
skim milkpowder	84.9	89.4	101.0	124.8	130.3	124.5	117.7
milkequivalent	1900	1988	2409	2647	2600	2368	2354

Whereas imports of cheese continue to grow steadily, since 1982 imports
of evaporated and condensed milk have exhibited an unprecedented
relapse. One of the causes of this must be the fact that imports of
cheese (and also those of other products, as a matter of fact) are
spread over all the countries in this group, while a single country
- Nigeria - accounts for the lion's share of condensed milk imports.
This clearly illustrates the vulnerable position of a product, which
for its sales practically depends on one market, which in turn depends

in great measure on the exports of one single commodity (in this case oil) for its foreign exchange revenues.

Exports of dairy products to lower-middle income economies-oil importers

1000 m.t.

	1978	1979	1980	1981	1982	1983	1984
cheese	26.9	19.7	18.3	21.6	20.0	20.4	27.9
butter and ghee	50.7	77.1	99.0	84.9	92.4	64.4	61.5
evaporated and/ condensed milk	73.4	76.9	85.4	84.5	71.1	75.8	80.4
whole milkpowder	57.6	100.4	110.8	96.9	102.3	108.1	92.8
skim milkpowder	172.0	220.0	187.0	160.1	157.3	168.1	195.3
milkequivalent	2775	3623	3379	2983	2964	3113	3352

With the exception of cheese and skim milk powder (partly as food aid), the conclusion must be that the volume of imports reached a (provisional) peak in 1980.

Exports of dairy products to upper-middle income economies-oil exporters

1000 m.t.

	1978	1979	1980	1981	1982	1983	1984
cheese	56.8	73.2	118.6	132.0	145.6	129.8	168.4
butter and ghee	77.5	117.9	160.3	166.2	123.7	146.6	161.3
evaporated and/ condensed milk	111.7	123.2	186.4	192.1	201.1	167.4	185.6
whole milkpowder	123.0	133.2	196.1	239.3	197.0	187.7	253.3
skim milkpowder	175.9	177.6	248.3	221.1	155.4	190.2	230.4
milkequivalent	3714	4018	5859	6049	5075	5198	6551

Here again we see a divided picture. A striking feature is the constant upward trend of cheese imports, in which respect it should be noted that the two warring states, Iran en Iraq, together account for approximately three quarters of imports.

Exports of dairy products to upper-middle income economies-oil importers

1000 m.t.

	1978	1979	1980	1981	1982	1983	1984
cheese	16.3	26.3	43.3	46.6	47.8	49.4	60.2
butter and ghee	28.3	45.9	43.8	38.3	30.0	23.1	26.8
evaporated and/condensed milk	86.1	105.7	119.4	122.7	135.4	142.1	158.4
whole milkpowder	38.8	55.9	72.1	46.6	47.4	45.4	54.9
skim milkpowder	64.5	84.3	88.5	47.2	43.5	52.0	73.6
milkequivalent	1366	1865	2213	1581	1577	1673	2115

Here we see a highly diversified country group, which includes, among others, the EC member states of Greece and Portugal, but also the city states Singapore and Hong Kong. Here again, sharp increases are to be seen in cheese and condensed milk.

It should be noted that within this group Greece accounted for almost 80% of evaporated/condensed milk imports and 65% of cheese imports in 1984.

Exports of dairy products to high income economies-oil exporters

1000 m.t.

	1978	1979	1980	1981	1982	1983	1984
cheese	26.4	38.0	45.1	48.3	47.7	63.8	75.0
butter and ghee	24.8	31.2	30.2	31.4	31.4	46.4	46.5
evaporated and/condensed milk	94.6	119.8	122.9	112.4	123.0	140.6	153.5
whole milkpowder	54.1	57.3	68.3	62.6	68.8	68.1	89.8
skim milkpowder	6.5	12.4	19.0	16.8	21.2	20.9	24.2
milkequivalent	961	1207	1435	1370	1486	1663	1998

Seen in relation to the population numbers of these 5 countries the only conclusion we can draw is that they represent a large import market, although it should immediately be pointed out that the products imported are scarcely manufactured locally. Until 1984 there was practically no question of any decline in import levels.

So far, this presentation has emphasized the developments with time according to country groups. The following table shows an overall summary, giving average annual growth rates per product and per country group. This table makes it possible to make a relative comparison amongst the country groups.

Average annual growth rate (1978/1984)

Country group Product	low-income economies		lower-middle income econ.		upper-middle income econ.		high income
	Africa	Asia	oil exp.	oil imp.	oil exp.	oil imp.	oil exp.
cheese	-4.1	2.4	24.5	0.6	19.9	24.3	19.0
butter and ghee	8.2	-0.1	7.5	3.3	13.0	-0.9	11.1
evaporated and/ condensed milk	1.9	9.5	-6.3	1.5	8.8	10.7	8.4
whole milkpowder	1.1	-1.8	-4.0	8.3	12.8	6.0	8.8
skim milkpowder	10.2	0.9	5.6	2.1	4.6	2.2	24.5

The picture which emerges from the table is highly diversified, which is one of the reasons justifying the differentiation which has been made according to country groups. An important finding is that, by and large, it reveals not inconsiderable growth rates. A further noteworthy finding is that the average annual percentages for cheese exhibits the biggest relative differences. In addition, it is striking that the low-income economies of Africa show high percentages for butter and ghee and skim milkpowder, evidently food aid, and the lowest percentages for other products. Finally, from the figure it can be deduced that the sharp rise in incomes of the oil-exporting countries has led to a substantial growth in imports of dairy products.

It has now become clear that for 1985 we must expect a substantial fall in exports, at least by the European Community to third countries. What is more, the figures for the first 6 months of 1986 are not encouraging either. Unfortunately, this congress is taking place just a few weeks too early for the definitive figures for 1985 to be incorporated in the analysis.

If we finally combine milk output as advised by FAO and milk imports, as calculated by us, we reach the following results:

Milk: outputs + imports

1000 m.t.

	1978	1979	1980	1081	1982	1983	1984
Low income economies Africa	6509	6485	6207	6413	6841	6905	7300
Low income economies Asia	49130	51089	52332	53735	52419	53545	55695
Lower mid.income econ. oil. exp.	6287	6285	6730	7212	7212	7029	7075
Lower mid.income econ. oil. imp.	15666	17051	16727	16520	16226	16386	17598
Upper mid.income econ. oil. exp.	15992	16802	19241	20293	18816	19117	20473
Upper mid.income econ. oil. imp.	29371	30331	30722	30733	32650	32339	31348
High income econ. oil. exp.	1466	1745	2039	1990	2132	2362	2738
Total	124419	129788	133997	136895	136296	137683	142228
Index (1978=100)	100.0	104.3	107.7	110.0	109.5	110.7	114.3

If we relate these results with the number of inhabitants in 1984, 'disappearance' of milk in that particular year presents the following picture.

Disappearance of milk per capita 1984

	kgs
0 low-income economies	
- Africa	23
- Asia	27
0 lower-middle income economies	
- oil exporters	20
- oil importers	53
0 upper-middle income economies	
- oil exporters	102
- oil importers	101
0 high-income economies	
- oil exporters	147

I leave it to the reader to draw conclusions by comparing this table with the one covering per capita milk production at the beginning of this analysis.

References:

World Bank, Development Report, several volumes;
Produktschap voor Zuivel, Wereldhandel in Zuivelprodukten, several volumes.

MARKETING ASPECTS OF DAIRY PRODUCTS IN DEVELOPING COUNTRIES

F.E. Jolliet
NESTEC LTD
Senior Vice-President Milk Products Department
Avenue Nestlé 55
1800 Vevey / Switzerland

ABSTRACT. Although basic marketing rules apply to all markets, either developed or less developed, they have to be adapted to the realities of individual markets; or of a given region, if habits/consumer attitudes towards foods/political/agricultural/economic and infrastructural constraints are homogeneous. Consequently, it is not recommendable to develop an international centralized marketing strategy except in the case of branding policy. Important overall marketing constraints will be discussed.

You are well aware of the fact that, as a group, developing countries are characterized by a total lack of homogeneity; and, obviously, this is also very true of milk consumption; therefore, please be indulgent with me; in effect, the handling of the theme given to me carries a serious risk of misleading errors arising from gross over-simplifications. Milk products consumption levels, the range of products consumed and consumer habits and attitudes in relation to milk products vary considerably from one continent to another : and even within a same country vast differences may be found in the uses of milk or in the image one has of milk. In certain markets, the vestiges of a millenary milk culture are still present, but milk consumption is nevertheless insignificant. Conversely, in other markets with no ancestral milk tradition, a healthy development of milk products consumption has occurred during the 20th century; by the political conviction of its government, the Republic of China is preparing the onset of a similar evolution.

At this point, I will cut short the inventory of differences and contradictions and focus on one of the rules of conduct which the Company I work for has followed for a long time in the commercialization of dairy products. It is a philosophy which is fully in line with my opening remarks; this rule is : flexibility in the choice and application of marketing strategies, and compatibility with local realities. However tempting it might be in economic terms, we do not believe in the theory of "centralized worldwide marketing". We believe in "tailor-made marketing".

This being said, I would like to put the accent on a few essential features which are fairly common to this extremely disparate developing world and which should not be overlooked by the dairy marketer.

THE MARKETING ENVIRONMENT

First, a few words about the <u>marketing environment</u>. There is a good chance that if asked to rank in order the problems of the marketing environment, the Chief Executive of a Company operating in the industrialized world under a liberal economy would first mention : competition and trade concentration. To the same question, his counterpart working in developing countries would answer : competition and government.

Although variable from one country to another, <u>governmental intervention is omnipresent</u>. Let me just mention a few specific points :

- severe price control

- state-owned distribution channels and governmental brands

- lack of political stability and over-frequent policy changes

- direct or indirect pressure on recombining dairy plants for the use of local raw materials of vegetable origin.

SOME HEADLINES ABOUT CONSUMERS

Now let's tune in on the <u>consumers</u> ! A <u>dramatic fact</u> dictates the marketing strategy to be implemented : the <u>enormous disparity of incomes</u> between a small segment of very affluent people, a larger but still relatively small middle-class segment with rather low income, and a huge mass of people representing a minor part of total disposable income, say less than 20 %. What is worse is that much, much time will have to elapse before this situation improves.

This state of affairs has three consequences, as follows :

- <u>Milk and milk products are not accessible to everyone.</u>

    I think one should have the courage to say it. By its very nature, milk is - and will remain - an over-expensive food for a large part of the population in developing countries. Many governments have tried their hand at "social marketing", aimed either at their own citizens or at the needy in third-party countries, via programmes of free - or largely subsidized - distribution of milk products. Few have been successful. They are costly programmes and therefore very often sporadic. And then, to create distribution channels that ensure that the poor populations, the really needy, are the sole beneficiaries of these social programmes, and to obviate illicit misappropriations, represents a challenge which is almost unsurmountable.

- **Milk products consumption will see no major progression in the next decade.**

  Per capita consumption will not change much from the present average level of 30 to 40 kilos of fresh milk equivalent. This is all the more true since the Western world milk surplus conjuncture, and the situation of depressed world milk prices that goes with it, cannot last for ever. Whilst costly for the tax-payers in "surplus" countries, this conjuncture makes a useful contribution to the feeding of certain populations in developing countries - in effect, about 1/4 of the milk demand in these countries is met by international trade.

- **The keystone of any milk products marketing strategy in developing markets is income/price segmentation.**

  This will be the case for a long time to come and a little later on I will refer to the product strategy which this situation demands. Of course, it is not out of the question that, here and there, more sophisticated segmentation principles might be applied, based on psychosocial or other criteria : since a narrow target is involved, product sales generated via this approach would obviously be modest.

I would like to mention yet another note-worthy factor also linked to the disparity of incomes. It has to do with what might be called the duality of attitudes and motivations with regard to milk. For consumers in the high- and middle-income brackets, milk nearly always has the same significance as for the populations of the developed world : it is a basic food product. Such consumers are fairly well conditioned to the use of milk in different forms and for different purposes. On the other hand, the low-income brackets have a completely different image of milk. For these consumers, material difficulties are a fact of daily life and they see milk rather as a food supplementary to the traditional diet of their children; the uses to which milk products are put by this category of consumers are limited; it is a rather conservative category with little or no motivation to widen the range of eating habits.

Just before closing this "consumer chapter", let me quickly mention the lactose intolerance issue. It is distressing that, now and again, certain articles make the rounds of the world press denouncing the "evils" of milk for Black and Asiatic populations suffering from lactose intolerance. These articles reflect scientific conclusions which, more often than not, are valid only for special cases who have been administered doses well above the real daily absorption levels of these populations. Our opinion is that this "problem" does not significantly affect the development of milk products consumption in developing countries.

THE PRODUCTS

- On the subject of price segmentation, I would like to recall an important factor. Price segmentation is not a question of offering the consumer a choice of 1st, 2nd or 3rd quality products. It is a question of offering best possible quality at each price level. In other terms, is is essential that the purchaser of a lower-range product feels fully satisfied with the quality/price ratio, the value-for-money, offered.

    To the question of how to meet this expectation with a range of milk products at clearly differentiated prices, there is one first answer : the panoply of liquid milks and milk conserves sold in developing countries is generally pretty broad and a price hierarchy establishes itself quite naturally. A second answer is to resort, within a same group of products, to differently costed packaging materials and methods : there is a wide choice to be made between cartons, flexible pouches, tinplate and plastics. And finally, in all countries where demand is not satisfied by local milk production, a third answer is provided by the innovative technique of combining dairy ingredients with ingredients of vegetable origin. You will forgive me if, at a dairy Congress, I refer to "imitation milk". I think, though, that we cannot shut our eyes to a reality of the world. Nowadays in the Philippines, more than half (55 % to be exact) of milk consumed is in the form of imitation milk or, more precisely, filled milk. For Thailand and Malaysia, the respective figures are 42 % and 34 %. The onset of this movement in the Philippines in 1955 has spread like an ink-spot and development has now reached Africa and Latin America.

- If milk products consumption in developing countries is fairly widespread in terms of liquid milks or milk conserves, either concentrated or in powder, the same cannot be said of fermented products and cheese. The Middle East and to a lesser extent Latin America are exceptions to the rule ! On the Indian sub-continent, in North Africa (Sahel included) and the high plateaux of East and Southern Africa, part of the milk sold is used for the in-home preparation of curd products, post-drained or otherwise; but the consumption level is rather small. In South-East Asia, a region with no milk atavism, much time and marketing investment will be needed before there is any worthwhile evolution.

- Concentrated milks, sweetened or not, have been on the market for over 120 years and are still finding takers. I am using this example simply to illustrate the problems set by the duality of consumer attitudes in developing countries. Dairy marketers know that, to satisfy the spirit of modernity of the better-off layers of the population and also to ensure product longevity, it is necessary to continually rejuvenate/renovate these products, sometimes by repositioning. This should not be done too quickly, however, for fear of alienating vast numbers of conservative consumers.

- Just a few words about <u>the dairy legislations</u> in force in developing countries. Very largely, these follow the models established in developed countries, especially those of liberal background. And sometimes, I think, the model does not deserve to be followed. To substantiate this thought, I would mention two recent cases, without entering into detail : The "minimum durability" concept (Best before...), adopted by the EEC a few years ago to express the optimal life of a product, is not in my opinion adapted to conditions in the developing world. By the same token, the EEC prohibits the use of stabilizers of vegetable origin, such as carragheenan, in evaporated milk; this is an acceptable measure for temperate climate countries, but it is much less acceptable for countries in the tropical and equatorial zones.

COMMUNICATIONS WITH THE CONSUMERS

I will limit my comments to two facets : brand policy and media advertising.

- <u>Brand policy</u>

    In my introduction, I said that milk products marketing could not follow the direction of "centralized worldwide marketing". I have just stressed that, in product strategy terms, it is desirable to implement a specific policy for developing countries. We will see a little later on, that the responsibility for media advertising is, in principle, a local management responsability.
    Conversely, if there is an area in which a company with international ramifications applies as strict a policy of centralization as possible, it is certainly the brand policy area. Let us make it clear that the brand is not just a word; in the consumer's mind, the brand, associated to its unique visual properties, represents a <u>symbol and guarantee of quality.</u> Quality accidents and counterfeit products are certainly not limited to developing countries alone; nevertheless, may I say that in certain of these, they are rather more frequent than elsewhere. Under such circumstances, a brand enjoying a food franchise is <u>a guarantee against fraud;</u> and in the optic of the enormous segment of conservative consumers, <u>loyalty to the brand is created, a level of loyalty</u> one finds <u>very rarely in the industrialized world</u>.

- <u>Media advertising</u>

    . The mentalities, the cultures and the emotional values of consumers differ so much from one market to another that <u>very seldom does an international campaign suit without local adaptations</u>, and this even when an identical product, positioned in the same way, is involved.

- Channels of communication with the consumer are sometimes extremely limited. Up until recently, commercial television for food products was banned in Saudi Arabia. Elsewhere, TV is inexistant or has very low household penetration. High illiteracy rates in developing countries seriously limit the efficiency of messages carried in the press. The multitude of vernacular languages in a same country - over 100 in Nigeria - is yet another stumbling block for the marketer.

- Radio and cinema are highly important advertising media in developing countries. It is said that every third African is likely to own a transistor and seldom turns it off.

A FEW WORDS ABOUT KEY MARKETING PEOPLE

In many markets of the developing world, the quality of services offered by Market Research and Advertising agencies is not always up to client expectations, and this despite high costs - talent is rare and staff costs high. The rotation of personnel is considerable and this is a serious problem, particularly when local agencies have many expatriates on their pay-rolls; and if the successors are also expatriates, many years will elapse before they get a real feeling for local values. It is therefore up to us, the dairy marketers, to install marketing people, of high talent of course but above all well-adapted to local conditions. Whether local or expatriate, these people should have a deep personal understanding of the mentality and attitudes of the potential consumers of the products they are called upon to develop, promote and sell. In order to tap latent opportunities, we also ask of these marketing people a sense of creative imagination to generate innovations specific to developing countries. It is only at this price - and this will be my conclusion - that these countries will see a growth of milk products consumption in the long term.

SOCIAL AND ECONOMIC ASPECTS OF RECOMBINATION AND INDIGENOUS MILK PRODUCTION

L.A. Barrón del Castillo
Chief, Meat and Dairy Service
Animal Production and Health Division
Food and Agriculture Organization of the United Nations
Via delle Terme di Caracalla, 00100 Rome, Italy

This paper forms a part of a combined seminar grouping referring to milk production and milk products and the marketing of dairy products in developing countries.

We are therefore collectively talking about some aspects of dairy development and I am privileged to share the speaker's platform with a number of authorities on this subject.

1. Why Dairy Development was Promoted in a Number of Developing Countries

Primary emphasis was placed on combatting malnutrition and here UNICEF was in the foreground with its international scheme for the construction of recombining dairy plants for feeding schemes for children in the developing world.

FAO, with its early and close involvement with the UNICEF programme, which was the precursor to the International Scheme for the Coordination of Dairy Development (ISCDD), assisted in the development of the total dairy industry along with the FAO/DANIDA Dairy Development and Training Programme, which was aimed at supplying trained dairy personnel for the dairy infrastructure of developing countries.

As awareness grew in the developing world that they could possess a modern dairy industry, it is natural that the pattern of the industry or the model would be that which prevailed in the developed dairy world.

There were a number of reasons why the dairy industry pattern of the developed world was 'transplanted' to the developing countries. Among these reasons was the fact that the developed world had evolved an efficient, capable and quality oriented dairy industry. The existence of such an impressive industry became known to more and more developing countries and their policy makers.

It was recognized that a dairy industry was needed, was desirable and, through various forms of multilateral and bilateral aid, was attainable. This, together with the availability of significant stocks of skim milk powder and butter oil in the developed world, led again quite naturally, to a linkage between utilization of these dairy products in a recombined form in the dairy processing plants built, often essentially for this purpose, in developing countries. The agressive sales techniques of dairy equipment manufacturers contributed also to the widespread 'transplanting' of modern sophisticated dairy technology in developing countries.

Even though there have been several decades of dairy development in the world there still remains a striking imbalance between the need for milk and its availability. Some statistics illustrate this very significant disparity.

Table 1 – <u>Milk Production, Population, Livestock Resources, Milk Yields and Milk Consumption</u>

Category and Region	% World's Population	% World's cattle & Buffaloes	% Milk Production	Yield of milk per dairy cow metric tons	% growth in consumption of milk 1970-1980
Total Developed countries	26	32	79	3.13	10
Total Developing countries	74	68	21	.66	45
Africa				.36	
Far East				.51	
Latin America				.95	
Near East				.64	
Asian Centrally Planned Economies				.67	

Figures used from FAO Production Yearbooks – 1979-81 period – except as noted.

The virtually inverse relationships shown above are a striking example of the lower productivity of the milking animals in the developing countries as exemplified by the milk cow. The demand (need) for milk is strong in the developing countries, constrained only by income levels and availability of milk.

2. Key Components for Successful Dairy Development

There have been a number of studies made and evaluation reports prepared of dairy development projects over the past two decades. It is not the purpose of this paper to try and survey the successes and the failures on the myriad of projects that range between these two extremes. Nor is it productive to try and attach blame for those projects which were not as successful as they might have been.

It is possible, however, to synthesize some common conclusions which have come from dairy development project evaluations. These conclusions, many of which were frequently repeated, clearly indicate that certain key components are necessary which, if present, will markedly improve the chances of successful dairy development.

Among these components the following bear reiteration:

Dairy development, particularly in the early stages, can seldom be based on complex, sophisticated imported technology, high yielding animals and concentrate feeds. It should begin with locally available resources and follow a resource conscious policy aiming at gradually improved utilization.

The government of the developing country must have a dairy policy as part of the national economic policy and its agriculture component. To ensure that the dairy industry has an adequate supply of trained personnel there must be a national dairy training programme developed for all levels of the industry.

The dairy sector policy must be long-term in nature, in recognition of the time it takes to develop the dairy industry, and should clearly indicate objectives as to the role that indigenous milk production and the small-scale milk producer will fulfill in national dairy development.

Prices paid to producers for indigenous milk production must be established and maintained at adequately remunerative levels.

Prices of imported dairy products including non-commercial imports under bilateral or multilateral aid programmes must be established at levels equivalent to prices paid for locally produced milk so as not to act as a disincentive to such production.

To the extent that governments, in furtherance of nutritional policies, wish to make milk more readily available to certain needy or vulnerable groups in the society, such schemes should be considered and financed as welfare/subsidy schemes and the price structure and margins throughout the dairy sector should be maintained so that all segments are economically viable.

While the role of milk in the diet of consumers is important, nutritionally it should be recognized that for the urban poor milk will likely always remain at price levels which they cannot afford to purchase in significant quantities, if at all. Rural populations will gain nutritionally, not primarily from increased milk consumption but from increased incomes from organized milk production/collection which will enable them to increase expenditures on other foods thus improving their nutritional intake. In many ways dairy development can be the initiator of rural development with wider implications for the well being of rural families.

The important role of the small-scale milk producer must be recognized and policies and programmes should be designed to assist these producers to develop appropriate organizations that will assist them to produce more milk more profitably while strengthening their bargaining power, decision making and self-reliance. The greatest impact is achieved when milk producers create their own organizations for dairy development.

When national dairy policies favour the maximum viable development of local milk production recombination assumes a natural position – probably transitory in nature – as part of a nation's total dairy industry.

We have all become increasingly more aware, through the press and television of the problems associated with human organ transplants and the tragic fact that rejection of the transplanted non-compatible "foreign body" is frequently the result. Similarly, transplanted segments of the dairy industry – in an attempt to copy the developed dairy world – often results in a similar rejection, in that the segment of the transplanted industry fails to flourish in an enviornment where its assimilation is difficult.

3. Some Additional Observations for Viable Dairy Development Projects

As noted earlier in this paper there have been, over the years, a number of studies and reports on dairy development projects in developing countries. In addition to the key components which are needed for success and which were enumerated earlier, these studies lead to three major additional observations.

Firstly, there has not been sufficient attention paid in the past when dairy development projects were being implemented to obtaining solid, hard, reliable information which could be used to determine, in quantifiable terms, how well the objectives of the developmental projects were being met. Not only was valuable management information not available, at the critical time, during project implementation and development, so that programme modification and correction might have been made, but original base-line data was not collected early enough, making an objective assessment of the project's results difficult, if not impossible.

Secondly, it is clear that the often _ad hoc_ implementation of some of the sub-projects in an entire dairy development project and the non-implementation of others, jeopardized the entire project. Key elements which were needed to ensure economic viability were omitted with the result that the dairy development lacked cohesion and unity. Dairy plants were constructed with capacities designed for large milk intakes but the production aspects were under-developed leading to non-viable processing costs. Milk production was promoted but incentive prices were not forthcoming. The problems of a regular supply of foreign exchange to supply packaging material, replacement parts or milk production enhancement inputs was not adequately foreseen or was minimized. In a single word the approach taken - for a variety of reasons - was not _integrated_.

One of the very real problems with dairy development in the Third World is that virtually everything that is done to increase milk production, improve the infrastructure, reduce dependence on imported dairy products or improve the quantity and quality of the milk supply to the consumers of the country appears to be a step forward. When the base upon which an industry is built starts from a relatively low position a number of incremental improvements, however uncoordinated or costly, give the appearance of significant progress. It is only after some years of applying _ad hoc_ solutions to the overall plan that enthusiasm wanes among donor and recipient alike. Expectations of progress which once were high will, because of the uncoordinated non-integrated approach, fall and enthusiasm will wane.

Thirdly, in spite of the inadequacy of the information base upon which to base an assessment of success from the social and economic viewpoint there are a number of examples of solid growth through integrated dairy development in developing countries.

The mission survey study type of approach which was utilized by FAO in the past identified constraints to dairy development and proposed a number of projects to eliminate these constraints and bring about sound dairy development.

The total 'package' of projects represented an integrated approach to eliminating constraints and achieving progress. If a number of sub-projects were not implemented the integrated approach broke down and the continuing weaknesses of parts of the dairy infrastructure often prevented viable dairy development. A measure of potential success, therefore, was the acceptance or implementation rate of project proposals.

For the FAO Expert Consultation on Dairy Development in 1984 the situation was reviewed in four countries where, on average, 93% of the project proposals were implemented. Comparing the rate of progress in these four countries with that of developing countries in general, it was found that milk production had increased by an average of 4.1% annually in these four countries, versus 2.6% in all developing countries while milk powder imports decreased by 23% versus 13% in all developing countries.

Since these two general criteria, milk production increases and decreased milk powder imports are, to a degree, complementary they are frequently expressed as goals in a dairy development plan. To this extent this analysis showed that the integrated or holistic approach was more likely to achieve these goals than was one that was non-integrated and piecemeal.

4. Recombination - A Tool for Dairy Development

The speaker who will follow me on today's programme has undoubtedly the greatest experience of anyone in skillfully utilizing recombination and donated dairy products to develop his country's dairy industry.

The essence of the dairy development strategy in India was to utilize the food aid, i.e. donated dairy products, in such a way so that their sale into the market would not depress indigenous milk products and, at the same time, would generate funds which were utilized in building up the basic dairy infrastructure to the point where the donated commodities could be replaced with indigenous supplies. With the cooperation of the World Food Programme, the EEC and others, this use of food aid has been the key to ending the need for aid.

Indeed the pioneering work done in this regard in India, with its great potential for generating funds for dairy development, has led to major changes in the way that food aid can be used as a tool for rural and dairy development. At the end of 1984, out of an annual total aid programme of some 250 000 metric tons of skim milk powder, approximately 25% were going in support of dairy development projects.

Stated simply, the objective of promoting dairy development from the standpoint of a food aid organization can best be done through the provision of milk powder and butter oil which, in the recombining process, contribute to the throughput and thus to the efficiency and profitability of the dairy plants. The sales proceeds which the recipient government (or their designated dairy industry development authority) receives for the donated dairy commodities through local sales generate a fund that is used to support small-scale milk producers in their efforts to increase milk production and other related aspects such as improved milk collection, provision of technical inputs for milk production enhancement, better marketing, etc.

The goal of this type of dairy development is to gradually replace donated dairy commodities by locally produced milk so that the dairy plants become independent from external supplies.

It may readily be seen therefore that, given the requisite controls over imports of dairy products and their price and the assurance of incentive prices for indigenous milk production, recombination need not present a barrier to the development of a nation's dairy industry. Indeed, if a food aid dairy development scheme can be developed, significant funds can be generated to help make such a scheme a reality.

5. Socio-Economic Aspects of Dairy Development

The cumulative effect of the improvements that can derive from dairy development and the symbiotic influence it can have on so many sectors of the rural enviornment make it a most valuable tool - if properly used - for rural development.

It must be recognized that to a very large degree traditional cattle keeping in many parts of the developing world is, at best, an attempt to establish and maintain a precarious subsistence existence for the cattle keeper and his family. The milk and meat which is derived from this type of agricultural enterprise is utilized by the producer's family. Small periodic surpluses are either preserved - and sometimes in an ineffective and primitive fashion - or sold or bartered for other foodstuffs or basic necessities.

The cattle, often the producer's sole source of wealth or capital, are vulnerable to the exigencies of disease, drought, feed shortages, etc. The market for his 'surplus', which often occurs when supply is at its peak, is uncertain. The producer's bargaining power which, at best, is poor, weakens even further in over-supply situations and exploitation is the norm.

Further, this situation is compounded by two factors which constantly haunt the rural poor. These are lack of alternative employment opportunities either in the rural areas or the cities and the overall depressing outlook for the producer and his family in that there does not seem to be any 'ray of hope' on the horizon that this pattern of spirit shattering subsistence/semi-starvation and grinding poverty will ever change for the better.

The first significant signs of progress as dairy development starts to become a reality in a developing country are the small but vitally important first steps that a milk producer can make away from the precarious subsistence level towards a more market-oriented approach where markets are more secure, prices are remunerative and dependable and decisions can be made about the future which are founded on reasonable confidence.

The producer can begin to be a part of an organized structure and his wishes and his decisions can start to be considered.

Gradually the sale of his production will become regularized. Opportunities for milk production enhancement expand and better feeding, breeding, animal care practices increase the quantity and quality of the milk he sells. Instead of simply struggling to stand still at the level of bare subsistence he is slowly, gradually but surely becoming a viable entrepreneur.

The impact of the socio-economic effects of dairy development on an individual milk producer is great and of tremendous significance for he and his family. The cumulative effect on a village, a rural area and, indeed, on a nation can also be of fundamental importance.

Some of the benefits of dairy development as they affect the rural socio-economic environment are year-around incomes for a significant number of people; increased family viability; reduced exodus of unemployed to urban areas; accelerated modernization process in rural area (water, electricity, roads, telephone, medical[1] and veterinary services, schools, etc.); expanded and more remunerative roles for women in agriculture; increased rural employment and growth of producer-owned/directed cooperatives and associations.

The synergistic effect of these, and other associated benefits which flow from dairy development can bring increased hope and confidence to the rural areas of developing countries. This, associated with increased availability of milk and dairy products for the nation's consumers and decreased dependence on scarce foreign exchange for dairy product imports offer convincing arguments for support of rational, well planned, integrated dairy development which will be an instrument for economic development and social change.

6. <u>Integrated Dairy Development Programmes - The New Direction</u>

FAO has had a long and close involvement with dairy development. During the past three years there has been a major examination of FAO's dairy development activities. This critical review, while revealing much of continuing merit and value, has led to a consensus

---

[1] A concrete example of the services that can be provided is the Indian Integrated Rural Development Project, promoted by NDDB for organizing health care, nutrition and invironmental improvement. The implementing agency in the Kaira District is the Tribhuvandas Foundation. This far-reaching scheme, financed by the milk producers' cooperatives has tremendous potential for improving the quality of rural life and warrants close study and replication elsewhere.

for further action and an associated reorientation of programming required to effect the necessary changes. This new direction is illustrated by reference to the Director-General's FAO Programme of Work and Budget for 1986-87 which states, in connection with Dairy Development, that:

> "Activities will comprise planning, implementation and coordination of integrated programmes for dairy development, with active involvement of potential donor countries and the full commitment of recipient countries. This will involve greater integration of dairy development activities, particularly at field level, by merging the International Scheme for the Coordination of Dairy Development (ISCDD) and the FAO/DANIDA Dairy Development and Training Programme's Regional Dairy Development and Training Teams (RDDTT's) into an International Dairy Development Programme (IDDP) to be supported by FAO's Government Cooperative Programme, in close collaboration with the Regional offices. First priority will be given in these integrated programmes to domestic milk production, particularly small producers."

In practice FAO's efforts will be directed toward dairy development in which the integrated approach is the norm and to developmental activities which are designed to enhance the ability of the small-scale milk producer to profitably produce more milk.

The activities of the FAO RDDTTs will increasingly be those where small dairy development training units (DDTUs) will provide a focal point for basic small-scale milk producers' training and orgnization. These units can also serve as central milk collection points, as sources of supply for milk production service inputs to the producer and as contact points for provision of first-aid animal health, AI services and veterinary care.

On a larger scale, but utilizing the same modular integrated approach, the FAO ISCDD programme will be directed toward Model Projects on Integrated Dairy Development. Model Projects integrate milk production, processing and marketing in order to ensure the most productive use of available resources, higher returns to milk producers and improved supplies of milk at fair prices for urban and rural consumers. The inter-relation between the components of the Model Project, their sub-components and the implementation modules are summarized in Annex A of this paper.

Integrated dairy development allows for the introduction of a wide variety of project activities which can be replicated and ultimately integrated into the total national dairy development plan as circumstances, needs and priorities permit. This type of approach utilizes those parts of the dairy industry infrastructure which exist and which may only need adjustment or correction and focuses on strengthening the weak or missing links in the total system to form a cohesive and viable whole.

The emphasis of the Model Project concept stems from the fact that a Model Project is, by its very nature, a self-contained integrated dairy development project capable of functioning in a viable way as part of the total national dairy industry. As such, it links opportunities for increased milk production with markets and thus provides a 'model' of how integration can work. Model Project Formulation Missions have already reported in Ecuador and Tanzania, with Syria and Indonesia to follow later.

FAO is confident that with this reorientation, which is based on the active invovlement and participation of the small-scale milk producer - the chief beneficiary of the programme - a new vitality and purpose will strengthen dairy development.

A number of you will have shared with me the rewarding experience of observing, first hand, the beneficial impact of properly planned, integrated and implemented dairy development on individuals and communities in the developing world. I hope that with the passage of time even more of you will become involved in this important work. The dairy world knows what to do, it is now up to us to do it. Let us ensure that we work together in the true spirit of cooperation. The potential of our joint actions for the developing world is too great for us to take any other path than that of unity.

Table 2 - INTEGRATED DAIRY DEVELOPMENT - MODEL PROJECT INTER-RELATIONSHIPS

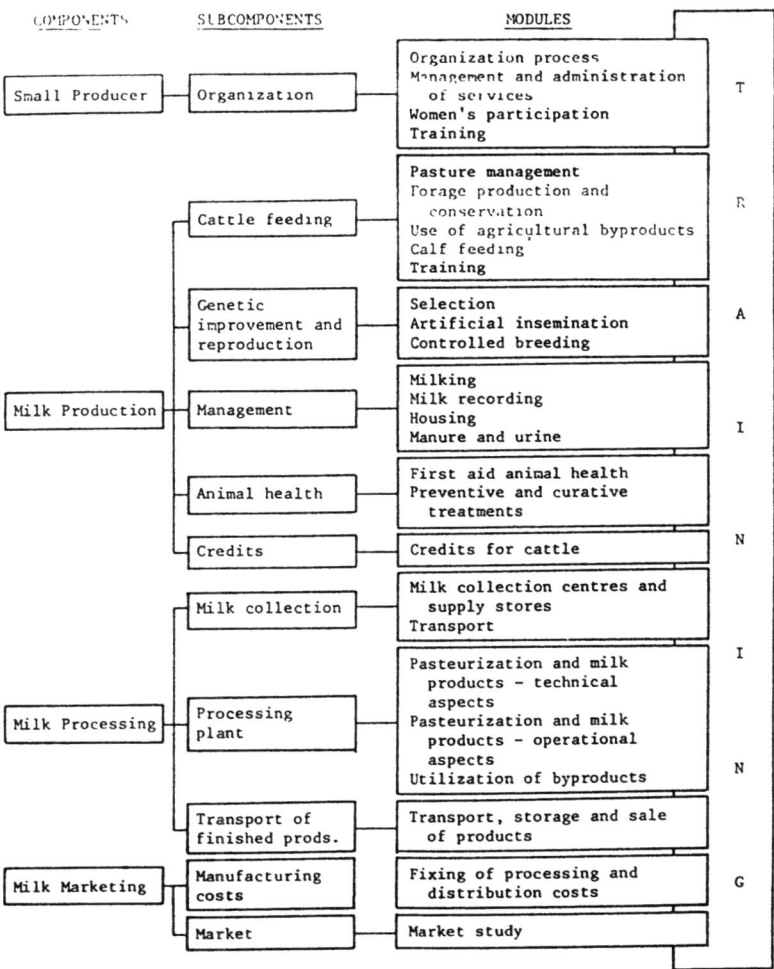

DAIRY FOOD AID

Frits Pronk (Consultant)

World Food Programme
Via delle Terme di Caracalla
00100 Rome
Italy

ABSTRACT. Dairy food aid amounted to an annual average of almost 400,000 tons over the 1981-85 period. More than two thirds of the quantity was committed to projects for human resource development through supplementary feeding of vulnerable groups, and for dairy development. Distribution among recipient countries was uneven: in 1985 over 40 percent of all dairy food aid in the form of skim milk powder went to 11 countries, which also received 60 percent of butter oil food aid. There seems to be scope for more dairy food aid to be used for development purposes in a number of countries, at the same time making a useful contribution to the solution of the major producing countries' surplus problems. The cost should not be higher than that of the most heavily subsidized methods of disposal (e.g., skim milk powder for pig feed). Marketing of recombined or reconstituted skim milk powder and butter oil generates funds in dairy development projects that can strengthen rural economies and improve farmers' incomes in poor areas. Prerequisites for such projects are a sound dairy price policy and, of course, a good development plan ensuring that the funds generated are effectively used to promote animal husbandry and local milk production.

I.   INTRODUCTION

1.   This paper will deal with both aspects of "dairy food aid", i.e., food aid in the form of dairy products and food aid to promote dairy development. Both are provided bilaterally as well as multilaterally. The largest bilateral donor is the EEC, followed by the United States; the largest multilateral donor is the World Food Programme, annually handling, over the last five years (1981-85), about 95,000 to 100,000 of the average total of approximately 395,000 tons of dairy food aid (subject to strong fluctuations). By far the most important commodity is skim milk powder, followed by butter oil. Bilateral donors support various projects for supplementary feeding of vulnerable groups and school feeding and other development projects. They also provide dairy food aid to recipient countries on grant or on long-term credit

arrangements for sales on the open market to generate counterpart funds for a variety of development purposes, or give it free as emergency aid. The World Food Programme provides aid either through development projects or to victims of emergency situations; in all cases, this aid is free, and assistance is also provided to recipient governments to administer and monitor the operations.

2. World dairy food aid flows are in large part directed to the countries listed in Table I. As can be seen from this table, only a few countries received a major share of all aid in skim milk powder and butter oil. In 1985, Brazil and Mexico joined the list of major recipients, with 38,857 tons of skim milk powder and 10,200 of butter oil.

3. WFP dairy food aid goes to 60 to 70 countries every year. During the last two years (1984 and 1985), more than two thirds of this aid was shipped to three types of projects for the feeding of expectant and nursing mothers and pre-school children, and for primary school feeding, as well as for dairy development projects. The last category alone now absorbs almost 40 percent. Approximately 20 percent of WFP's dairy commodities are used for relief feeding in emergency situations.

4. Although the total quantities of dairy products provided as food aid to developing countries (either on grant or credit terms) are by themselves significant and have cost as much as about half a billion United States dollars a year to date, they are small in comparison with the existing surpluses in the EEC and the USA of over 2.5 million tons. Clearly, food aid, provided as a development resource or as humanitarian relief, can put such surplus commodities to good use in needy countries. This question will be further considered in Section VII.

II. ROLE AND FUNCTION OF FOOD AID

5. The role and function of food aid vary from project to project and according to circumstances. The simplest concept is that of food aid in the form of direct feeding to vulnerable groups of the population, especially small children either nutritionally at risk or in various stages of malnutrition. In such cases, the role of food aid is purely nutritional and represents additional consumption. In other feeding schemes, a government budget for the supply of food may be wholly or partly replaced by external food aid; in these cases, food aid releases funds from the government budget for other purposes within the same sector, e.g., improvement of classrooms, canteens or kitchens or for the training of more teachers, etc. In addition to its nutritional role of supplementing generally deficient diets, food aid also represents a useful budgetary support. In food-for-work projects, family rations normally have a higher market value than the part of the wage they

Table 1

Concessionary Exports of Skim Milk Powder and Butter Oil to Major Recipients[1]
(Tons)

Skim milk powder

	1981	1982	1983	1984	1985
Bangladesh	600	500	31	101	9
Bolivia	2 500	3 007	6 847	7 088	5 745
Brazil	1 263	789	694	2 222	12 625
China	2 700	-	-	5 220	7 886
Egypt	24 492	26 018	28 034	31 410	21 724
Ethiopia	4 869	5 032	7 464	8 436	26 897
India	72 710	36 360	13 125	71 109	10 612
Mexico	365	500	500	708	26 232
Morocco	4 006	3 965	11 045	10 029	4 817
Pakistan	10 580	21 036	12 039	11 577	8 832
Somalia	15 297	14 711	4 959	10 074	7 356
Total:	139 382	111 918	84 738	157 974	132 735
% of all food aid in skim milk powder	41.9	41.7	33.7	43.4	41.4

Butter oil

	1981	1982	1983	1984	1985
Bangladesh	3 000	-	2 000	1 500	-
Bolivia	-	6 500	-	481	400
Brazil	-	13	-	-	55
China	-	-	-	2 000	2 677
Egypt	3 160	2 947	1 040	4 500	2 000
Ethiopia	1 200	1 160	964	1 393	8 851
India	23 588	8 200	1 426	27 473	4 188
Mexico	-	-	-	-	10 145
Morocco	200	440	240	980	-
Pakistan	1 873	277	5 522	533	1 757
Somalia	1 815	1 500	-	650	500
Total:	34 836	21 037	11 192	39 510	30 573
% of all food aid in butter oil	58.8	51.9	48.7	70	61.9

[1] Food Aid in Figures, 1986, FAO, Rome.
(Preliminary estimates for 1985)

replace (otherwise they would be less attractive to the workers), and in addition to providing budgetary support to the government, such aid therefore constitutes a transfer of income to the workers. In the various feeding schemes and in food-for-work projects, food aid is always distributed to well defined groups of people, although its main impact need not always be nutritional.

6. It is clear that only through feeding schemes can food aid have a direct nutritional effect, though in many cases this is diluted because the rations intended for the selected beneficiaries (mothers and infants) are shared with other members of the household. Although in food-for-work projects workers receive rations for the whole family, the rations are known to partly replace usual food purchases, often making nutritional additionality limited. The analysis of the effects of projects both for supplementary feeding of vulnerable groups and for food for work is therefore shifting to a determination of the income transfer effects of the food aid, i.e., the effective value of the food aid to the recipient household in terms of equivalent income added or food purchases foregone.

7. In dairy development projects, the commodities, after having been delivered free to the recipient government, are sold to the dairy plant(s), which process them and sell the pasteurized milk or other dairy products on the open market. The primary role and function of food aid is to create a fund from which a dairy development plan or schemes comprising it can be carried out. Secondary roles are to utilize the dairy plant's capacities to a fuller extent, to meet market demand gaps and to reduce foreign exchange requirements for initial imports of dairy products.

III. WFP's APPROACH TO DAIRY DEVELOPMENT: POLICY CONSEQUENCES FOR RECIPIENT GOVERNMENTS

8. The primary objective of dairy development projects is to provide support to the dairy animal husbandry sector in rural areas, thus strengthening the dairy farming sector and the incomes of farmers - especially smallholders. For this to be achieved, a well designed plan is needed, as well as the means to carry out such a plan, i.e. in the first place, funds, both local and foreign currency and also, in some cases, technical assistance. WFP can provide the local currency component through funds generated by the sale of its food aid commodities. WFP skim milk powder and butter oil are supplied free to the recipient government which, by selling them to the dairy plant(s) for reconstitution or recombination, create a fund for dairy development.

9. The price at which the government or project authority should sell the skim milk powder and butter oil to the dairy plants is of crucial importance. Sales at a low price would undoubtedly help the dairy plants to overcome financial difficulties and provide the market with cheaper

milk products. However, it would constitute unfair competition with the
local dairy farmers and would create a disincentive for the very aim of
the project, namely, the promotion of local milk production. Prices
should therefore be set at a level equivalent to a fair incentive price
for the local farmers' milk. This latter price should therefore be
decided upon first. Thus, it may well appear that the market cannot bear
a consumers' price level which -- after allowances for collection,
processing, distribution and a reasonable profit margin for the plants
-- would initially be based upon a remunerative producers' price. If
this is so, and government cannot or does not want to subsidize prices,
the inevitable conclusion is that processed dairy products are still too
expensive for the stage of economic development the society has reached,
and the project should not be started. However, experience in
WFP-assisted projects suggests that this is generally not the case, but
that in most countries promotion of local production can be justified on
the grounds that a large enough segment of the public can afford
pasteurized milk and processed dairy products. Moreover, in certain
cases, a transitional period can be considered during which prices can
be gradually adjusted. On the other hand, it is important that prices
not only be properly established at the outset, but also regularly
reviewed as warranted by inflationary and other developments. This is
important in almost all WFP-assisted projects, together with the
establishment of fair prices for producers at the beginning of the
project. In many countries, increases in food prices can lead to
political problems. The full socio-economic background therefore needs
to be carefully analyzed before a decision is taken, in the case of this
type of project, as well as in other types of market-oriented
policy-linked projects. There should, however, be consensus among donors
in regard to pricing policy. One donor cannot effectively maintain a
certain price level if other donors are willing to provide dried skim
milk and butter oil on softer terms, which take no account of the need
to protect local dairy farmers.

10. The volume of assistance in the form of food aid to dairy
development projects is to some extent determined by both the local
demand at a given price and the need for capital in the form of local
currency for carrying out the dairy development plan. It is, of course,
also determined by the processing capacity of the plant(s) concerned and
by the rate at which local milk production increases. The latter is an
important element for calculating the duration of the external supplies.
These should taper off as local production increases but other factors
may interfere, such as unutilized processing capacity or the need for
more funds. Their upper limit is, however, determined by the capacity to
invest local funds quickly. The next section will deal with this
problem.

IV. UTILIZATION OF FUNDS GENERATED FROM THE SALE OF WFP COMMODITIES

11. The primary function of food aid in dairy development projects is
to create a fund to finance a dairy development plan, or schemes within
such a plan. Yet, not all governments which submit requests for food aid

to dairy projects, have such development plans. Their main concern is
often to utilize their existing processing capacity more fully and to
provide increased quantities of cheap milk to the population. This
concern is in itself perfectly justifiable; it should not, however, be
the only objective of projects which can be a base for effective dairy
development. While the need for a dairy development plan is now widely
recognized, the need for the related policy package is not. WFP's
experience shows that it is difficult to prepare a good dairy
development plan, within a certain time frame, with a budget split
between external and local financing. Without market surveys, it is even
more difficult to select reasonable price levels and to forecast the
reaction of farmers, middlemen and consumers to them. Also,
circumstances can be expected to vary considerably over the years; thus,
a plan for five or more years is difficult to draw up. Consequently,
flexible plans open to annual reviews and adjustments should be
advocated. On the basis of such plans, the need for local and external
inputs can be established and ways and means explored on how to obtain
these inputs. Timing is crucially important. Both bilateral and
multilateral experience show that project implementation, even when
based on a solid plan, is often frustrated because certain essential
items do not become available to the project in good time (equipment,
vehicles, infrastructure, etc.), as their availability depends on action
by external donors if it involves costs in (locally unavailable) foreign
currency. Of paramount importance, as always, is the availability of
qualified staff to manage and execute the plan. Usually such staff are
not available in adequate numbers, and both technical assistance and
training are required. Needs in this respect should be anticipated long
before the project is expected to start. Of course, this principle
applies to any kind of development project.

12. A dairy development plan should include the promotion of local
milk production, expansion of collection of farmers' milk, improvement
and perhaps diversification of processing and improvement of
distribution and marketing. As regards production, in all cases -
according to WFP's experience - due attention should be paid to the
promotion of fodder production, including the use of agricultural
by-products and concentrated feed-mixes, and the establishment or
strengthening of extension services, a credit system and cooperatives.
The last three issues often pose problems to the project authorities,
and long-term technical assistance is sometimes needed. Unfortunately,
technical assistance in these difficult areas is not always readily
available.

13. Increased milk production - and increased income for the farmer -
does not always, and certainly not fully, result in increased deliveries
to the dairy plants. The middlemen may well take more of the farmers'
increased production by paying better prices than the dairy plants.
Without going into the pros and cons of the phenomenon of middlemen, it
may be worth noting here that in some countries their ability to compete
with the dairy plants by offering better prices to the producers and
obtaining higher prices from the consumers (for usually poorer quality

milk) shows that the public may be willing to pay higher prices than governments assume. However, middlemen's activities can also make it more difficult to measure the increased volume of local fresh milk resulting from a dairy development project.

14. It is obvious that the administration of the funds generated for development is critically important. Experience in many projects is not always favourable in this respect. First, dairy plants are normally slow to pay the sales proceeds for the food aid commodities -- delays often amount to several months. Second, project authorities are usually slow to disburse the funds, often because items to be purchased or activities to be carried out have not been identified as part of the project design. Even when they have been identified, actual expenditure authorization or tender procedures or administrative procedures may take too long. Accounting and auditing may pose problems if the necessary skills are lacking. Foreign exchange may not be readily and sufficiently available, holding up execution of parts of the plan, according to which the local funds cannot be utilized. It is not unusual for a year or more to elapse between the date when funds are due to be paid and their actual disbursement. Meanwhile, in most countries, the purchase value of the funds has been seriously eroded as a result of inflation, although the funds may have been deposited in an interest-bearing bank account. The crux here again is not money but the quality of management of the funds, and also the programming for their utilization and the existence of a flexible procedure to revise the original investment plan on a timely basis. Management of funds is the most difficult element of any project, both for the governments and for the donors concerned.

V. PHASING-OUT OF WFP ASSISTANCE

15. Any successful project should, after a period of external assistance, continue on its own. How long should this period of external assistance last in the case of dairy projects? The answer is: when indigenous capacity has been built up to a level that maintains development momentum in the sector. One index of this level, though not the only one, may be the local dairy plants being able to operate economically on the strength of local supplies, hence without food aid. In such cases, external supplies should bring the throughput of the plants to the level of breaking even and be phased down as the supply of local milk begins to push the throughput volume above this level.

16. This break-even point is, however, only *one* of several criteria for phasing out WFP assistance. It might well be that food aid starts in a country where dairy plants already operate economically. The end of food aid has to be determined by its objective. If the latter is to increase per capita consumption of milk, new dairy plants may have to be established, as the existing ones, though economically sound, are no longer sufficient. The economic operation of the plants consists not only in generating adequate supplies, but also in proper management and the availability of equipment.

VI. GENERAL EXPERIENCE WITH WFP-ASSISTED PROJECTS FOR DAIRY DEVELOPMENT

17. WFP is supporting 17 projects in 16 countries at a total cost to WFP, i.e., including transport and some administrative costs, of 265 million dollars, of which 163 million dollars are for skim milk powder and about 102 million dollars for butter oil. In volume, this amounts to about 155,000 tons of milk powder and 45,000 tons of butter oil. About 40 percent of these resources are taken up by two large projects in China and Cuba. These projects are for five years in the case of China and three in Cuba and will probably not need additional time for successful completion. All the other projects are for four or five years, but will very likely need at least one to two more years, after which an expansion may be considered. Thus, annual shipments by WFP to all dairy development projects have amounted to approximately 30,000 tons of skim milk powder and 7,000 tons of butter oil over the last few years.

18. For most projects the greatest difficulties stem from the dairy development plan having been not concrete enough or too optimistically conceived, perhaps underestimating the need for foreign exchange as an indispensable part of its financing, or overestimating the availability of such foreign exchange. Another cause of deficient implementation is the shortage of qualified staff. The underutilization of local funds, including the part generated by the sale of WFP commodities, is a common feature of most projects. This underutilization obviously results in slow development.

19. In half the number of projects, the price issue poses problems. Although a government may have set producers' prices at the beginning of a project at satisfactory levels and agreed to adjustments as necessary, it often is perceived to be politically very difficult to increase these prices later because of subsequently necessary consumer price increases (if the country cannot afford subsidies). Farmers react by either producing less or selling more to middlemen. Factory throughput falls, and WFP - or other donors - are called upon to fill the gap. More local funds may thus become available, but their effect through the plan would be extremely limited as farmers need, for expanding their production, the economic incentive as much as - if not more than - any kind of developmental support.

20. In many projects problems arise in plant operations, due mainly to poor or outdated equipment and lack of trained staff. Dairy equipment is generally not available in the country and has to be imported and paid for in hard currency, which, for various reasons, is not always easily and readily available. There are, of course, other reasons why the progress of dairy development projects may be slower than expected, even though the recipient country may provide all the required inputs. For instance, dairy projects are particularly susceptible to drought, and climatic hazards may well destroy prior achievements. Thus, lack of success may also be the consequence of circumstances outside the government's control. The long-term nature of dairy development projects

must also be taken into consideration. Spectacular results cannot be expected after a few years, and it is often a slow process to provide all the inputs required to assist dairy farmers, particularly the smaller farmers, to obtain more and better cattle, and provide the improved feeding and management necessary to increase milk production. The industry as a whole may require the establishment or improvement of milk collection, transport and processing facilities. The above discussion includes most of the substantive issues arising from the experience of WFP.

VII. DAIRY PRODUCTS IN FOOD AID IN GENERAL

21. Roughly 72 percent of all dairy food aid, or about an annual average of about 285,000 tons over the last five years - but with strong fluctuations - is provided bilaterally. The EEC supplies almost half this quantity, the United States slightly less. Canada is a relatively small donor of dairy products, but of its average donation of almost 14,000 tons a year, only 40 percent is bilateral, while most of it goes through WFP. Of its bilateral aid, the EEC channels about 18 percent through non-governmental organizations (NGOs), i.e., almost 14 percent of its total dairy food aid; for the United States these percentages are 94 and 79, respectively and for Canada 62 and 24, respectively.

22. Most NGOs (as well as WFP) utilize the commodities entrusted to them largely, if not solely, for supplementary feeding programmes and other development projects, while most of the so-called dairy programme aid is provided by the European Community.

23. It would appear that, very roughly, of all dairy food aid something of the order of magnitude of 250,000 tons a year is used for supplementary feeding and other development projects, of which 95,000 to 100,000 tons is provided by the WFP and the rest mainly by voluntary agencies. Of all the skim milk powder and butter oil provided as food aid, almost half goes to only eleven countries (Table 1). Of this, in 1984, almost half of the skim milk powder and 72 percent of the butter oil went to just one country: India. The second largest recipient in 1984 of these commodities was Egypt. India and Egypt alone absorbed one third of all skim milk powder and butter oil in 1984. In 1985, one third of all skim milk powder and half of all butter oil went to only six countries. Thus, there is a very uneven distribution of dairy food aid over the world.

24. Since 1984, WFP has been supporting a dairy development project in China by providing about 45,000 tons of skim milk powder and 13,330 tons of butter oil for a period of five years. The project covers six cities. A recent evaluation mission found the project highly successful as it had exceeded all its quantitative targets in milk production and the increase in the number of dairy animals. Substantial investments in new plants and renovations have been made to meet the rapid increase in the supply of fresh milk to the plants, which grew by 70 percent during the 1981-85 period in the six cities assisted by WFP. The dramatic increase

in the number of private dairy farms participating in the project has had an important effect on incomes and employment, particularly of women. The Chinese authorities have requested similar assistance, but on a larger scale, for another 14 cities, as well as the continuation of the ongoing operations in the six cities.

25. Most of the dairy food aid is utilized in support of feeding programmes for expectant and nursing mothers, young infants and primary school children and for dairy development projects. A substantial percentage is also allocated to assistance to refugees. Dairy products, particularly skim milk powder, are very suitable for these purposes, provided that there is an adequate supply of clean water for reconstitution. There may be ample scope for utilizing much more of the existing dairy surpluses for development in the Third World. Of course, such development costs money, but it may not cost much more than storing butter for three years at 1,300 dollars a ton and then selling it for 450 dollars a ton to the Soviet Union, or than subsidizing skim milk powder at 1,550 dollars a ton for use as animal feed.

VIII. COST

26. The method of calculating the cost of dairy food aid that takes best account of realities, of market and price conditions[1], is clearly based on opportunity cost considerations, i.e., takes into account what the cost of alternative use would be at the moment of designating the commodities for development assistance. Obviously, no recipient - or development budget for that recipient - could be charged more than the price for which the commodity could be commercially imported. But even that price can be obtained only by the exporter because he is prepared to pay other costs, such as those for storage and heavily subsidized disposal methods, to achieve or maintain that level. Also, if the quantities now designated for development aid were added to the export market, prices would decline further.

IX. CONCLUSIONS

27. a) Most dairy food aid is in the form of skim milk powder, an average of 307,000 tons a year and butter oil, an average of 45,000 tons a year. Other dairy products amount to an average of 43,000 tons over the last five years. 72 percent are provided bilaterally, 28 percent multilaterally.

   b) These quantities are small compared with existing surpluses, now of the order of 2.5 million tons, but could be considerably increased if a more even distribution of assistance between Third World countries were achieved and in particular if China were included in bilateral donors' lists of recipients.

---

[1] Production costs are in this context not realistic: for quantities in surplus they are costs of a product that cannot be sold and in that sense are unrealistically high.

c) Unlike cereal food aid, dairy food aid is largely in the form of project aid, the EEC being the largest donor of programme aid (i.e. aid provided in bulk for monetization). WFP provides only project aid, in which mechanisms for use of generated sales proceeds are stipulated in detail and are carefully monitored.

d) For dairy development projects, first of all a proper price policy must be established by the government concerned, and, secondly, dairy food aid commodities - skim milk powder and butter oil - donated free to the recipient government, must be sold by the latter to the dairy plants at prices equivalent to those paid to the local milk producers to avoid undercutting the latter. The funds thus generated are to be used for promotion of local milk production. The prices paid to local producers should provide an adequate incentive, and be regularly reviewed

e) Dairy development projects belong to the category of so-called "policy-linked" development assistance, because their implementation and success, or lack of success, often depend directly on recipient governments' economic policies. Consideration of such projects must therefore include discussion of such policies between donor and recipient, in particular dairy price policies. In this connection, there should be closer cooperation among donors to ensure that well designed projects supported by one donor are not jeopardized by assistance offered by other donors on less rigorous terms. Dairy development projects may be rather difficult to design and implement but can achieve significant impact.

f) More dairy food aid could be usefully given: there is a surplus of absorptive capacity in a number of developing countries, and the cost should not be higher than the cost of the most heavily subsidized methods of disposal, for which development food aid would be a good alternative.

Criteria for Success or Failure of Dairy Development [1]

Michael J. Walshe
World Bank
Washington, D.C., U.S.A.

Abstract. Demand for dairy products is increasing rapidly in the developing world. Although many countries have the potential to substantially increase production progress will be governed their ability to handle biological and technical constraints and economic, social and policy issues. Constraints imposed by climate, animals, feed supplies and feeding are discussed. Institutions and support services to backstop and implement development are considered to be crucial and success or failure is governed by their performance. Modern enterprises involving pasteurized or processed milk have difficulty competing with raw milk in many countries and consequently the evolutionary nature of dairy development is stressed. The potential for smallholder dairying is emphasized and smallholder development is usually more efficient and less risky than efforts which focus on large capital intensive dairy farms.

Introduction

1. Over the last 20 years the rate of growth of milk output (1.3%) exceeded the population growth rate (1.0%) in developed economies but in developing ones the milk and population growth rates were equal at 2.7% [2]. Population growth rate exceeded milk production in Africa (3.0% versus 2.2%) and the Near East (2.6 versus 2.2%) but the milk production growth rate exceeded the population growth in Latin America (3.1 versus 2.6%) and the Far East (3.1 versus 2.4%). The International Food Policy Research Institute (IFPRI) has projected the consumption and production of milk by region to 1990 and 2000, using 1977 as the base year (Sarma and Yeung 1985). The analysis covered 104 developing countries but although China was excluded because of inadequate data, official Chinese policy calls for a dramatic increase in production by year 2,000 (from about 1.1 million cows in 1986 to about 10 million by 2,000). Incremental production to 1990 and 2000 was estimated at 39.8 and 86.3 million ton respectively. These figures are broadly in line with those of FAO (Agriculture: towards 2000) on their medium growth scenario. Despite the large output increases envisaged, the IFPRI analysis depicts substantial deficits in milk for all regions - 26.7, 20.6, 13.6 and 3.5 million ton for

---

[1] The views in this paper are those of the author and should not be interpreted as reflecting those of the World Bank

[2] Source: Calculated from data in FAO production yearbooks 1976, 84.

Asia, N. Africa/Middle East, Sub-Saharan Africa and Latin America respectively. Even under the alternative assumption that income growth might be 25% lower than during the 1966-77 period, the gap between production and consumption could still be 43 million tons or 20% of consumption. Other analysts have arrived at lower estimates of demand (Sanderson 1984) but all agree that substantial increments in production will be required by the year 2000. From this it is clear that demand for milk and milk products, which is a prerequisite for successful dairy development, will not be a constraint in developing countries.

2.   Most countries have the potential to increase domestic production to meet all or part of this demand. Success or failure will depend on each country's ability to exploit its potential by effectively dealing with: (i) biological and technical factors; and (ii) economic, social and policy issues.

3.   Success is important because livestock is a major source of income and employment for rural people. The sub-sector often accounts for as much 20-50% of agriculture value added in developing countries and dairying makes a major contribution in many countries. Since basic food production is often for subsistance only, the sale of livestock products can account for as much as 80% of the cash income for smallholders. Some of this income is usually invested in food crop production via purchases of improved seeds, fertilizers, etc. A source of cash income is essential to unlock the cycle of low inputs - low productivity - low income and dairying may be able to provide it.

4.   The World Bank has provided substantial support per dairy development. From 1963-82, for example, 75 projects were supported in 44 countries which were exclusively for dairying or had a dairy component (Frankel 1982). These projects covered a broad range of climatic and ecologic conditions ranging from the cold arid regions to the humid tropics. Similarly, the production systems, type and size of farms and the economic, marketing and social conditions encountered covered a broad spectrum. An outline of the main deficiencies encountered and the conditions that govern the success or failure of dairy enterprises is presented in this paper.

5.   <u>Management</u> is the single most important factor governing the success or failure of dairy projects. This applies at the farm, institutional (milk collection, processing and marketing) and at the national (industry administration and economic management) level. Management of support services such as credit, input supplies, extension and artificial insemination (AI) is also crucial. Successful dairy industries are characterized by efficient management at all levels. In fact, few agricultural activities require the range or the same degree of sophistication in institutions and support services as dairying. It is not surprising, therefore, that experience with dairy project implementation indicates that if problems arise at the level of the farm, the institution, or the support service they are invariably caused by a deficiency in management.

Biological and Technical

6.   Climate. Tropical climates have well known deleterious effects on the performance of improved dairy breeds both directly and indirectly. High temperatures, exceeding 27°C, combined with high humidity (over 80%) depresses feed intake and milk production. The degree of temperature effects depends on numerous factors: e.g. type of feed, quantity of feed offered, stage of lactation and milk yield. High tropical temperatures determine the type of forages available and their quality. Tropical forages have substantially lower feeding value than temperature climate ones. In general terms, forage digestibility drops by about 1-2 units per 400 km moving from temperate latitudes towards the equator. For example, lactation yields of about 4,000 - 5,000 kg can be supported, without supplementation, on high quality forages grown in temperate regions, but yields of 2,000 - 3,000 kg are the maximum that can be achieved on improved tropical pastures even when dry season irrigation is available. Temperate pastures can sustain peak yields of about 25-30 kg/day compared with about 10 kg on tropical pastures. In fact, a yield of 1,500 - 2,000 kg is a reasonable expectation if cross-bred cows are fed on tropical pastures (mixed grasses and legumes) without supplementation.

7.   **Choice of Genotype (Breed).** The choice of cattle most suitable for milk production in the tropics has received considerable attention because of the importance of genotype-environmental interactions. The superior milk production capabilities of European breeds (Bos taurus) and the heat tolerance and disease resistance of the indigenous Zebu types (Bos indicus) are well recognized. Zebu type cattle prevail in the tropics, not so much far heat tolerance per se, but because of their feeding behaviour. The Zebu has nearly 25% less digestive capacity per unit of size and consequently is forced to be a slower and more selective feeder than, for example, Holsteins (McDowell, 1972). On tropical grass pastures, the Zebu will select a higher quality diet but will utilize less of the total forage dry matter. In the coarse grass areas of the humid tropics it is unwise to replace the Zebu with European breeds; however, Zebus utilize crop residues less efficiently than European types and buffaloes because they have a lower rumen fermentation rate and a faster rate of passage through the digestive tract.

8.   A yield of 4500 kg of milk per lactation is considered to be biologically the most efficient level possible with high quality tropical forages and supplemental concentrate feeds (McDowell 1985). In addition calving internal will exceed 420 days and calf mortality will be 10% or greater. Tropical grasslands and/or crop residues will only support substantially less production and yields may be as low as 1800 kg from a 450 day lactation. Based on experience in Columbia and Puerto Rico, McDowell suggests that, while pure Holsteins are satisfactory at high resource levels, Holstein/Zebu crosses are preferable when resources are more limited. He proposes the following scheme:

Milk Yield Supported by Resources	Type of Animal
> 4,000 kg	Pure Holstein or 75% Holstein cross
3,000 kg	50 - 65% dairy cross
2,000 kg	25 - 50% dairy cross
< 1,500 kg	Dairy cross 25% or native breed

    Based on Bank experience, pure dairy breeds should be used only at higher elevations and cooler climates and cross-breds are prefereble at lower elevations in the humid tropics. The usual feed resources and management levels found in the tropics rarely justify the use of pure-bred dairy cows despite their biological potential for superior milk production. Even in relatively good environments a policy of upgrading native cattle by cross breeding is much less risky than importing pure breds because it affords farmers the opportunity to gradually up-grade feeding and management, and in addition disease problems are more manageable.

9.   Feeding is generally the dominant constraint on dairying at farm level in developing countries. The problem arises from the low quality forage and grasses in the tropics and sub-tropics (para 6) and the cost of supplementing forages or grazing with concentrate feeds. Supplementing forage (silage, hay or grazed grass) or cereal straws with expensive concentrate feeds is complicated by the cows ability to substitute one kind of feed for another. Although nutritionists have shown that about 0.4 kg of concentrate feed contains adequate nutrients for one kg of milk the incremental response from concentrate feeding over silage, for example, is much less because silage dry matter intake is reduced by about 0.5 kg for each kg of concentrate DM consumed. A similar type relationship exists for other forages and the level of substitution is related to the type and quality of forage on offer. The substitution problem has important implications for dairying under all climatic conditions. As a rule of thumb, it is prudent, when appraising dairy projects in developing countries to assume that at least one kg of concentrates will be needed per kg of milk in excess of the estimated yield from grass and forage only. The common mistake of over estimating forage quality and the level of production it can sustain should be avoided. Generally speaking a kg of milk must be worth at least as much as a kg of concentrates before even a low level of supplementary feeding can be justified.

10.   Feed is a major cost for virtually all dairy farmers and the feed/milk price ratio is crucial to profitabililty. Grazed pasture is normally the cheapest feed source, followed by green forage or conserved forage (silage and hay). Concentrate feed is usually much more expensive. In Western Europe, for example, grass silage is about four times as expensive as grazed grass per unit of TDN and dairy concentrate feed is about 8.5 times as expensive. The role that grazed pasture can play in efficient dairy production in the developing world is not sufficiently realized in many countries and consequently there is a tendency to adopt the so called "modern

production systems" which are highly capitalized and rely heavily on concentrates and cultivated forages (e.g. maize silage).

11.   Efficient low cost systems based on grazing are common in temperate, and sub-tropical and tropical climates where grazing land is available. In Latin America, for example, specialized dairy breeds are grazed at higher altitudes and a dual purpose beef-milk system (based on grazing) using beef breeds or beef-dairy breed crosses is widespread and expanding rapidly in the tropical and sub-tropical areas. Farmers must compete with low priced imported milk products and reconstituted milk and little if any concentrate supplementation can be afforded. The Bank's experience with these low cost systems is generally favorable. On the basis of profitability and efficiency they merit strong support.

12.   In countries where rice and wheat are staples large quantities of by-products (rice bran and wheat pollard) are usually available at low cost. In such countries efficient dairying can develop; the feeding regimen can be based on limited forage and/or cereal straws combined with heavy concentrate feeding. Heavy concentrate feeding is usually justified when the milk feed price ratio is greater than about 1.5 to 1.0. Relatively cheap dairy concentrate rations can be formulated from cheap feeds such as rice bran, wheat pollard, sugar beet and citrus pulp, oil cakes and other products. The recent expansion of dairying in a number of developing countries is primarily due to a favorable milk concentrate feed price ratio of about 2 or 3 to 1. Countries like India, China, Turkey and Indonesia come readily to mind. In the countries mentioned pigs are not important except in China and feeds are therefore relatively cheap. It appears that these countries can produce milk virtually as efficiently and as cheaply as countries with much better grazing and forage resources. In contrast, countries where pigs are important like Thailand and the Philippines have much less potential for dairying because there is much more competition for cereal by-products and therefore they must rely more heavily on grazing, forage and feed grains. Unlike Asia, the African and Latin American human food chain is mainly composed of maize, millets and cassava which yield little if any by-products apart from straw for animals.

13.   From an animal nutrition standpoint countries with a climate suitable for high quality forage production have a distinct advantage for dairying. Maize can be grown for silage or green feed over a broad latitude (0 to 50°) and climate range and can contribute a high proportion of a dairy cows diet. The legume alfalfa is a favoured diry feed and can usually be grown between about 25 and 50° latitude. In Egypt and the Northern part of the Indian sub-continent and partss of the Mediterranean climatic region, the winter legume Berseem (<u>Trifolien Alexandrium</u>) is an important forage. It is the corner stone for efficient milk production systems which make maximum use of cereal straws and other by-products in India, Pakistan and Egypt. It is now apparent that forage production on fallow land can make a major

contribution to dairy development in North African and Middle Eastern countries with a Mediterranean type climate.

14.   The economic importance of forage quality is inadequately appreciated in the developing world. The forage quality problem is exacerbated by lack of good applied research in forage production, conservation and supplementation in most situations. Countries do not appreciate the cost of inferior forage in terms of decreased production or increased requirements for expensive concentrates. American research, for example, has shown that if harvesting is delayed until digestability drops from 67% to 65% the ability of forage to contribute to the total TDN ration is reduced from 75 to 58% for milking cows and from 100 to 57% for dairy heifers. It is worth noting that 58% digestability represents a high level for forage in most developing countries.

Economics, Policy, Institutions and Support Services

15.   Labor Costs. The cost of labor has a profound effect on dairying. When labor is cheap, or family labor, with a low opportunitycost, is available, forage production, harvesting, feed storage, feeding and milking can be carried out manually. In predominantly manual systems one person can handle about 5 to 7 cows (i.e. India, China). In fully mechanized systems one person can handle up to 100 to 200 cows under New Zealand type conditions where cows are not housed in winter and supplementary feeding is minimal, or up to 80 to 100 cows in Western Europe and the U.S. where cows are housed in winter.

16.   The point at which a country's dairy industry switches from manual to mechanical milking depends on the cost of labor and the amount and cost of capital to sustain the mechanical system. However, mechanical milking demands a new and more sophisticated technology which developing countries lack and they invariably experience enormous difficulty in making the change from manual to machine milking. Specialized support services for machine testing and maintenance, clean milk production and mastitis control are usually absent. Furthermore investment in more mechanized systems must be justified by an increase in labor productivity sufficient to justify investment costs. In some centrally planned countries the increase in labour productivity may not be sufficient to justify investments in some instances, because of inherent pressures to maintain a large labor force, even though mechanical milking can be justified at the prevailing labor rate. In others e.g. India, Pakistan and China, the low prevailing labour rate does not justify machine milking even on large commercial farms.

17.   Smallholder dairying has the additional important advantage that investment in fixed infrastructure, or machinery and equipment per cow or per kg of milk produced is invariably only a small fraction

of that found on large mechanized farms. Housing and ancillary
facilities (e.g. feed storage and manure storage and disposal) have
simpler designs, incorporate cheap construction materials and can be
built with low cost labor (e.g. family labor).

18.   The Bank's experience with wide scale support for
smallholder dairying has been generally favorable. This statement is
true for all regions as indicated by successful dairy development in
Kenya, Latin American countries, Easter and Western European
countries, North Africa, India and Korea. Smallholder dairying is
much more efficient and less risky, for example, than large capital
intensive and highly mechanized with low production coefficients and
poor labour utilization rates. Dairy development should be an
evolutionary process beginning with smallholders and increasing in
size and level of sophistication as overall development progresses.

19.   <u>Institutions and Support Services</u>. A country's success or
failure in dairy development is closely related to its performance in
building efficient dairy industry institutions and support services.
Institutions or agencies must be established to handle activities at
three levels. At the national level policies relating to dairying as
well as economic political and marketing issues must be efficiently
and effectively handled if success is to be ensured. At the second
level institutions must be developed to handle milk collection,
processing distribution and marketing as well as supplying inputs
(feeds, fertilizer veterinary supplies, detergents, sterilants etc.).
In addition institutions at this level are usually needed to provide
artificial insemination (AI) and specialized services e.g. milking
machine maintenance, mastitis control and clean milk production. At
the third level (farmer level) services such as extension, veterinary
and training must be available to farmers as well as access to
agricultural credit.

20.   In the developed world efficient institutions have evolved
to handle dairy activities at all levels but usually this is not the
case in the developing countries. We are all familiar with the range
of institutions involved - national milk marketing or dairy boards,
large sophisticated cooperatives or commercial companies which provide
a range of support services including processing and marketing. In
contrast institutions and services are invariably weak or non existent
in developing countries and, therefore, when we talk about dairy
development we are, in large measure, talking about institution
building and developing support services. Although it is dangerous to
over generalize one can pinpoint problems and weaknesses which occur
frequently. Before doing so it is useful to consider the evolution of
dairying in the developed world and the relevance of the experience to
developing countries.

21.   <u>Evolution of Dairying.</u> In Europe, Africa, the Indian
sub-continent and North America farmers traditionally milked their
cows for home consumption and limited barter or sale to their

neighbors. For example demand for an all year round supply of liquid milk expanded with the growth of towns and cities in Europe and specialized dairy farms were established to meet the demand. Farmers sold raw milk to consumers or to vendors who distributed it. In time dairy farmers were licensed by veterinary authorities. Until the sale of pasteurized milk became widespread during the thirties and forties loose raw milk was distributed by 'the milkman' in the US and Europe. Pasteurization and the distribution of pasteurized milk in bottles and home refrigerators revolutionized the liquid milk industry after the war. City milksheds could be expanded and licensed producers were usually allocated a winter and summer quota at a fixed price.

22. In the developed world pasteurized milk was quickly accepted because consumers could afford the price and the public health benefits were well publicized. Although pasteurization is relatively expensive in the developed world where dairy industries are well organized and managed (about US$0.1/liter including 'containers' in the US which is equivalent to about 50% of the farmgate price in most developing countries) it is even more expensive in developing countries. In many such countries consumers traditionaly boil milk. They do not discriminate against raw milk and are prepared to pay as much for 'loose' raw milk as for pasteurized milk in containers. Milk collection, cooling, processing and handling charges are expensive in developing countries and farmers who sell milk to processing plants receive, as a rule, about half the consumer milk price. Thus they can almost double their income by delivering direct to consumers or usually realize a higher price by selling to a vendor who delivers loose milk from door to door.

23. The evolution of dairying in the developed world is often forgotten and it is not surprizing that attempts to transpose modern dairying, based on pasteurized milk, encountered serious problems because in most instances they cannot compete with the traditional sector. This is particularly true for countries with a low per capita income, a tradition of milk production and most importantly a tradition of boiling milk before consumption. These include Turkey, India, Pakistan, North Africa, Middle Eastern and some Latin American countries.

24. The lesson is that successful dairy development must cater for the traditional as well as the modern sector and the skill with which this is handled is crucial to success (para 25). Governments at the instigation of dairy technologists often pass legislation prohibiting sale of raw milk which in most instances is fortunately ignored because its implementation would totally disrupt the milk supply for some major cities. Nevertheless this tyoe of legislation leads to some hidden avoidance costs which are very hard to measure.

25. Governments frequently establish public sector processing plants to supply pasteurized milk (para 22). These plants, are often unprofitable, not only because they are usually badly managed and

inefficient, but they have the impossible task of competing against farmers or vendors selling raw milk from door to door. Their continued operation is usually dependent on government subsidies. Given this situation a dairy project which aims to process and market pasteurized milk through a public sector plant may be doomed to failure. The main lesson is that dairy development should not preclude the sale of raw milk.

26. The successful development which has taken place in India under Operation Flood based on the Anand model is noteworthy and instructive. Competition with the private sector within a radius of about 100 km of large cities like New Delhi is studiously avoided. It is fully appreciated by the architects of Anand and Operation Flood that cooperatives cannot compete with the private sector close to the city because individuals can sell raw milk and home made profitable milk products. The cooperatives further away are successful because they provide an efficient milk outlet for small village producers by collecting, processing, transporting and selling in the city. It is not fully appreciated that Operation Flood milk in India is incremental to an existing local supply. As the market expands, Operation Flood will provide a higher proportion of the cities supply because the opportunity to expand supplies, in close proximity to the city, is limited. It is important to realize that a somewhat similar situation prevails in most developing countries.

27. Inadequate and inefficient support services are a major constraint on dairy development. Extension is needed to help farmers improve forage production and conservation as well as providing guidance on the adoption of efficient cost effective feeding systems. More often than not extension is weak and research in applied feeding is weak or non existent. Animal health frequently poses a problem when the veterinary service is unable to provide the necessary vaccinations, drugs and medicines and private veterinary practitioners are not available (in part because the public veterinary service is supposed to undertake the task). AI performance is, more often than not, grossly inadequate and specialized services such as clean milk production, machine milking and mastitis control are in many cases unheard of.

28. Much greater attention should be given to the provision of support services. In doing this greater emphasis should be placed on institution building and on transfering responsibilities for services from the public to the private sector. This has been a feature of the Indian Anand model which should be studied and copied. In India efficient cost effective veterinary, extension and AI services are provided to small village producers and are sustainable because costs are fully recovered by direct payment for the service or a small charge on each liter of milk sold to the cooperative. It is becoming more and more obvious that services should be located in the private sector in most cases. Furthermore, the principle of full costrecovery from beneficiaries not only ensures sustainability but also

performance because farmers will express dissatisfaction if performance is not up to standard.

29. Public sector AI services have a dismal performance record. They are invariably located in the veterinary service and the resources and managerial competence to manage and operate an AI service is usually lacking. Consequently, much more attention should be given to the provision of cost effective AI services in developing countries and more emphasis placed on the most effective combination of AI and natural mating. For example a small concentrated AI service can be used to breed sires which can be widely disseminated for natural mating.

30. <u>Profitability</u>, which is influenced by a host of factors, is of paramount importance to the success or failure of dairy projects . From the farmers standpoint the farmgate milk price and the cost of milk production are crucial. The milk price is governed by the supply and demand for milk and the efficiency with which milk is collected processed and marketed and is therefore closely correlated with the efficiency of dairy institutions. Price can also be determined by the cost and availability of imported products. Reconstituted milk can, for example, compete with domestic milk and, therefore, the policies of exporting countries as well as those of the importing one exert major pressures on the price paid to producers.

31. If demand and price are adequate, successful and profitable dairying will depend on the efficiency of the farmers' milk production system which in turn depends on the farmers management capabilities and the availability and efficiency of support services. If farms are relatively small, or if manual milking and feeding is practised there is usually less risk of poor profitability at the farm level. Futhermore, profitability is usually not an issue if the farmer can distribute raw milk or sell to private sector vendors who distribute raw milk (e.g. Periurban producers in India, Pakistan and Turkey). In the Bank's experience, domestic dairying based on the sale of raw milk or home made milk products (e.g. soft cheese) can compete, without difficulty, with reconstituted milk from products imported at world prices.

32. Dairying based on the sale of pasteurized milk can normally compete with reconstituted milk if the domestic industry is reasonably efficient. On the basis of current world prices for milk powder and butter oil, a farm gate price of 20-22 cents/liter for 3.5% fat corrected milk can be justified using the World Bank's methodology for analyzing projects. A higher milk price can be justified if consumers discriminate against reconstituted milk and this is reflected in a lower price. It is unlikely that a dairy project aimed at milk production for processed products (e.g. cheese, milk powder or butter) can be justified at present because of the prospect that these products will be available at depressed world market prices in the forseable future. Nevertheless, provision for investment in limited

processing to handle seasonal surpluses is justified as part of an
efficient dairy industry. These surpluses or imported products can be
usedin the "valley" season. For example, a policy which enables a
blend of reconstituted and domestic fluid milk to be sold in the "off
season" is common (e.g. India and South America); if properly managed
it can be efficient and avoid disruption.

33.   Governments need to give careful consideration to policies
governing the importation of milk products, especially skim and whole
milk power and butter oil which can be reconstituted. They must
strike a balance between allowing consumers access to "cheap" imports,
which may not be available in the long term and which are a drain on
foreign exchange, and providing legitimate protection to domestic
dairy farmers. Furthermore, governments must develop policies to deal
with milk products which are donated free by bilateral donors under
programs such as the World Food Program (WFP).

34.   Institutions must be developed to implement these policies
and, in this regard India is a good example of effective development.
The Indian Dairy Corporation (IDC), a Government of India Enterprise,
is responsible for administering WFP and other donations of milk
powder and butter oil as well as being responsible for appraising and
financing investment proposals under Operation Flood. In fact a large
proportion of Operation Flood Funds are generated by the sale of
donated milk powder and butter oil. In addition, the National Dairy
Development Board (NDDB), a large organization employing about 850
professional staff (including 250 engineers), was set up to act as a
consultant to IDC to prepare dairy development plans and to organize
and implement services under Operation Flood. These two pivotal
organizations together with village cooperatives, District Cooperative
Milk Producers Unions and State Federations of Unions provide India
with a comprehensive, well planned institutional framework for dairy
development. The striking success of Operation Flood is attributable,
in no small measure, to these institutional arrangements which have
been carefully crafted to meet India's needs. Each country will need
to develop institutions adapted to its own needs – the complete
spectrum from apex institutions to those servicing farmers must be
provided. A country's success or failure will largely depend on its
achievements in establishing effective institutions.

35.   Conclusion. Dairy development embraces a large spectrum of
activities involving production, processing and marketing.
Institutions and support services to backstop and implement
development are crucial and success or failure is, in large measure,
governed by their performance. Development can only take place if a
market exists and if Government policies provide a favorable economic
environment. This should not be interpreted as an advocacy for
protection and control, although some regulation may be called for to
avoid undue disruption caused by surpluses being dumped by developed
countries or by free donations. The paper has not attempted to
provide a blueprint for success or failure but instead has tried to

indicate from the World Bank's experience the main factors and issues which must be taken into account. At the Bank we view dairying as one of the main development sectors for developing countries and look forward to providing continued support for dairying in the forseable future.

BIBLIOGRAPHY

Food and Agriculture Organizations of the United Nations, 1981. Agriculture: Toward 2000. FAO, Rome

Frankel, Jack 1982. A Review of Bank Financed Dairy Development Projects. AGR Technical Note No. 6, World Bank for Reconstruction and Development, Washington. D.C.

McDowell, R.E. 1985. Meeting constraints to Intensive Dairying in Tropical Areas. Cornell International Agriculture Mimeograph No. 108, Ithaca, N.Y.

McDowell, R.E. 1982. Improvement of Livestock Production in Warm Climates. W.H. Freeman and Co., San Francisco.

National Academy of Sciences 1981. Effect of Environment on Nutrient Requirements of Domestic Animals. National Academy Press, Washington, D.C.

Sarma, J.S., Yeung Patrick 1985. Livestock Products in the Third World: Past Trends and Projections to 1990 and 2000 Research Report 49. International Food Policy Research Institute Washington, D.C.

Sanderson, Fred. H. 1984. World Food Prospects to the Year 2000 Food Policy 9(4) pp363-373.

SUMMARY OF DISCUSSION

The discussion highlighted the following main issues:

(a) Dairy development involves a well coordinated programme of production, processing and marketing. In order to provide the required incentives to stimulate production and to ensure the success of the industry in developing countries, concerted efforts must be made in the areas of government policy measures and assured market arrangements.

(b) Dairying is essentially an evolutionary process which invariably starts with the labour-intensive small holder production through to the capital-intensive large scale production of the developed dairy industries. A systematic approach is therefore required for the development of the industry in developing countries. Attempts to transfer the developed industries models to the developing country situations have usually ended up in considerable frustrations.

(c) The market demand for milk in most developing countries can be met either throught increased domestic production or importations or a combination of both. However, in order to stimulate domestic production, the demonstration of profitability of milk production will remain a key factor to the success of dairy development in the developing countries.

(d) Dairy Food Aid, particulary in the form of Dairy Development Projects, has been a useful tool in promoting dairy development with important socio-economic benefits at the small holder level. There is however an urgent need for closer cooperation between donors and recipients for greater emphasis on well-designed projects that will promote an orderly development of local production capacities.

The seminar concluded that:
1. "Milk" has indeed been a "Vital force" in alleviating rural poverty and providing better nutrition in most developing countries where efforts have been directed towards dairy development.
2. Dairying can play a major role in assisting small farmers to optimize their resources, particulary where such resources, in terms of animal numbers, farm size and feed resources, are rather marginal, as is the case in most developing countries. Therefore a need exists to promote the total exploitation of all available resources for milk production, such as for example the feeding of crop residues, to enable the rural small holder to increase his income.

3. A well coördinated dairy development programme can be a "Vital force" in the transfer/redistribution of resources form the urban rich (who is prepared to pay for milk as a highly protective food) to the rural poor (who can produce the milk cheaply) given the good production and marketing environment of an organized dairy industry.

Seminar III: Market and Marketing of Dairy Products

Session 1: Enterprise-Oriented

Chairman: Prof. B. Imbs (Poland)
Secretary: Ing. O. Obermaier (Czechoslovakia)

Planning the dairy enterprise to meet the challenges of the market place

Major problems in the marketing management of modern dairy enterprises

PLANNING THE DAIRY ENTERPRISE TO MEET THE CHALLENGES OF THE MARKET PLACE

Roine Ristola
Valio Finnish Co-operative Dairies' Association
P.O. Box 360
SF-00101 Helsinki
Finland

The title itself is challenging enough and the subject could probably be approached from different angles. However I have tried to reduce it to manageable levels for the simple reason that my somewhat narrow view could not possibly cover the whole wide field.

It is easy and safe to start with a worn cliche' and if I say that we exist in a business environment which is constantly changing I think that we can all agree on that. It has significance for today's subject though because it's the changes that present us with challenges and can even pose us threats unless we are prepared for them. Thus planning is required to anticipate changes and help us take advantage of the opportunities they may open for us.

Before we start planning for tomorrow it is essential that we know where we are today in relation to our environment. In order to establish that let us look first of all at the changes which have taken place over the past years, and what significance we can attach to them. Then we ought to try and spot the trends of development taking place on the market scene and what possible effect they are likely to have. We must assess our own strenghts and weaknesses, also what is expected from us and, what we need to reach our objective.

A.   CHANGES WHICH HAVE TAKEN PLACE IN THE MARKET PLACE

When looking round the market scene today we will notice that the number of retail outlets has gone down from what it was a few years ago and that larger units now account for an increasing share of the market. Multiple chains have grown in size and power at the expence of the independent grocer. The number of articles competing for the same space on the shelves or in the cool cabinets of the retail shops has increased and includes at least in some countries a steadily growing sector of what is known as "own labels". Generic articles have also appeared in our line of business and so have of course substitutes and imitation products.

The consumers tend to change their preferences which reflects itself in their eating habits. The protection of consumers to the extent it is applied today is also a fairly new invention and so is the tendency of governments in some countries to get involved in directing what people should eat.

All the above and probably a number of other factors not listed have led to:

B. CHALLENGES AND DEMANDS

1. Increased competition among our customers for the favour of the consumers.
2. Increased competition among suppliers in our line of business.
3. Increased competition among suppliers of substitututes and imitation products.
4. The predominance of price as the main attraction.
5. Increased bargaining power of the national chains commesurate with their size.
6. Higher price for the place on the shelves or in the cool cabinet of shops.
7. Demands for improved service with little chance to charge for it.
8. Constant requests for more money to be spent in price (and throat) cutting competition.

Those are but some of the challenges we are facing today practically on our doorstep. Further afield in the world market we come up pretty soon against uncontrollable surpluses of basic products like butter and milk powder.

C. WHERE ARE WE AND WHY?

Having established the state of the environment we shall have to try to place ourselves in it. In other words ask ourselves: "Where are we?"

In short we shall have to stop for a while (which is very difficult) and devote some time for self assessment and analyzing the situation.

There are certain areas within our company which will have to be in order to guarantee our success today and our survival in the future:

- profitability
- productivity
- market standing
- resources (financial, know how, manpower)
- management ability.

Are we all right on that score?

The above requires an honest approach including the admission of our limitations due either to factors relating to internal strengths

and weaknesses or then imposed by forces beyond our control.
We may have inherited or accumulated some weaknesses over the years which could have an adverse effect on our efficiency and which we had better do away with. A closer look might reveal:

a) Organisational ones
    - Orientated towards production (unproportionately)
    - Slow to react (top heavy)
    - Development of different functions uneven (structure not balanced but leaning over to one side).
    - Poor co-ordination between different functions (or dept.)

or

b) Operational ones
    - Slow flow of information leading to malfunction within the organisation and to complaints from customers.
    - Indifference towards constructive critisism from the market (customers).
    - Progress uneven. Technically advanced (system orientated) otherwise lagging behind.
    - Low degree of motivation among the staff.

Those are but few of the shortcomings we may come across when honestly assessing the state of our company.

Evidently the fact remains that in the daily rush to rationalize and expand in order to catch the fast disappearing profit we just may have overlooked something obvious.

Having confessed to being guilty to at least some weaknesses it's time to look for the strenghts we possess and must exploit to the limit of course:

- The long standing of our industry with the professionalism it brings with it.
- The high degree of mechanization and automation available at different stages of process if needed.
- The traditional acceptability of milk since ancient times.
- The suitability of milk for a variety of different products.
- The eye- and taste appeal of dairy products.

Assuming now that both the study we undertook first of our environment and then our selfassessment went deep enough and were searching enough, we ought to have located our present position with sufficient accuracy, in other words we know now where we are (ref. to the questions above).

D. COURSE TOWARDS THE OBJECTIVE

...Knowing our present position product/market we will be able to proceed by asking 'where do we want to go; what is our objective? This is a very basic question as it may reveal a difference between our objective and the harsh reality illustrated by where we are going

as we are. This is called the planning gap which we obviously must
fill to get where we want to go.

We may be able to do it at least partly by just eliminating our
internal weaknesses (a job that may well prove a challenge in itself).

As to the remaining part of the gap if we manage to close it we
have successfully completed what the title of the subject suggested
that we should do - we have faced the challenges and got where we wanted
to -.

The KEY to our success no doubt is the customer (being a market
orientated enterprise we accept that) so we must ask ourselves "what
does our customer expect from us?" This very question has been asked
by many and in different ways and a certain pattern of factors affecting the interrelationship between manufacturer and retailer has emerged.
Whatever the order of priority you will normally find most of the following on the list:

- Marketing and the sales organization
- The range of products
- The demand for them in general and/or the novelty value of some particular product
- Pricing and discounting policy
- The suppliers' image
- Good standing with the buyer based on past performance
- Reputation of reliability regarding distribution and delivery
- The strength of the supplier's consumer-directed advertising.

We had better examine each point separately and relate it to our
business environment and to our plans to find out if there is room
for improvement and thereby a possibility to narrow the gap.

1 Marketing and the sales organization

The marketing idea is that you can make a profit more easily by
attending to your customer's requirements than by not doing so. Have
we done everything in our power to try and find out those needs or
is there possibly room for improvement? Is our sales force organized
to achieve our current and future goals and not the past ones? Do they
know what those goals are and are they motivated enough to fill their
part in achieving them? Have we defined our market clearly? Are we
satisfied with the market we supply or should we aim beyond that? Those
are some of the questions we should ask ourselvs and make the people
responsible for planning and running the marketing function in our
organisation contemplate constantly.

2 The range of products

Is our range of products satisfactory? Is it comprehensive from
the customer's point of view? Do we systematically plan what is known
as product portfolio? If we do not, we should. In a balanced portfolio
we need products with capacity to generate a steady flow of cash to

be used to support those which (rapid growth area and poor market share) need financial support. However we must try and avoid areas of excessive vulnerability (problems can be caused by economic fluctuations, uncertainty in supply of raw material, government measures etc...).

3 The demand for our products in general and the novelty value of some of them

The highly professional retailer keeps a constant lookout for the more profitable lines. How do we perform? Use of computers in controlling the stock means that only the best will survive. Do our products turn over quickly enough? Active product development is a must (and requires investment). If we want to guarantee a continuous demand for our products also in the future let us pay attention to young people, be with it, and educate the educators.

4 Pricing and discounting policy

Let us determine our objectives and their relative priorities and adjust our policy accordingly. - High profitability when you are looking for an increase in market share does not necessarily materialize -.

Different customers have different needs. It could be worth while to study in detail the effect segmenting the market would have. Finding the profitable sectors and eliminating the unprofitable ones could mean a lot to our company's profitability in general.

If our strategy is aimed at attaining market leadership, once we have reached our target, our prime task is to try and protect our position. One way of doing it is by sensible price policy. (Let us not deliberately attract competition).

We should be careful to avoid the position where we find ourselves financing the war between rivalling groups.

5 The supplier's image

We can not gain customer loyalty today by product promotion alone, our image counts too. Are we known as the supplier of high quality products and a supporter of good courses?

For us as the manufacturer of branded goods a positive image is most important.

More information than before is expected to flow costantly form the business to the public. Has our information and/or PR-dept. succeeded in building up and maintaining good relationships with public media?

We have to keep testing our image from time to time to find out if we are slipping up and where.

6 Good standing with the buyer

Mutual trust is essential. Have we passed the test?

The realization that we are in this together and should find a basis for permanent co-operation would be ideal. We should try to get this idea across.

The buyer could give us valuable hints as to the packaging, merchandising and even pricing the goods (this works both ways). Let's consult him!

7 Reputation of reliability

This rests on both good quality products and punctual delivery when promised. The distribution must be adopted to varying customer needs and instant attention to complaints is essential.

8 Consumer directed advertising

What is our attitude towards the requests of our advertising manager? Volume and quality of our advertising efforts show that we belive in and are behind our products and serves as a means of convincing our customers as well as keeping the consumers informed.

Retailers can count on our adv. support and time their own promotional efforts accordingly.

E. THE IMPORTANCE OF RIGHT INFORMATION

If on each of the above paragraph we can honestly say that we are all right we should have no problem. - Faced with a bold statement like that however I would suggest a checking of the information system -, because it is essential that we get right information. Moreover when we talk of marketing information speed is essential even at the expence of accuracy. A more or less constant flow of information is needed on:

- products
- sales
- promotions
- customers

and last but not least on competitors of whom we normally know far too little although we should be monitoring their activities all the time.

Once in possession of knowledge indicating the need for action we should be quick to react.

Although we seem set with appropriate corrections to reach our ultimate target let's not forget that a great deal may happen between now and for instance 1988. In order not to be taken by surprise and also to be able to make plans of which we could truly say even in 1990's that we did the right thing in 1986 we must try and work out the trends of today to see where they are leading us.

F. FUTURE BEYOND TOMORROW

We may be fairly certain that:
- Mergers will continue between different groupings in retail/wholesale trade - In GB in 1984/85 twelve multiples had a 60 % share of

the market and it has been estimated that by 1990 the same percentage will be accounted for by just three companies -.
 - The suppliers are also likely to merge into larger units.
 - Cash and carries will in some countries still increase in numbers and in strength. The same applies to some extent to superstores although individual stores will hardly grow anymore (it is estimated that the sales area in Western Europe has grown by about 20 % in the '70's.)
 - Multiple chains and other purchasing groups will develop their warehousing systems and expect increasing quantities of goods go through them.
 - This will mean additional costs (most probably) to the industry which will have to pay the bill for services rendered.
 - The need to reduce costs will force the retailer (and supplier) to look closer at the stock and cut the inventory (fewer articles).
 - The demand for more frequent deliveries will grow.
 - Branded goods will face increasing competition (in some countries) from the part of own labels.
 - We shall probably have to prepare ourselves for more government involvment in business in general than today.
 - The health aspect of food will continue to figure high on the list of priorities.
 - The consumer wants to know more and more of what he/she is purchasing.
 - The size of the family is getting smaller, with a household of 1-2 accounting for an increasing share of the total.

If we try to imagine what kinds of demands these prognoses may bring us in the future and how they will differ from the challenges we are facing today we will probably come to the conclusion that the demands will be practically the same only more so. In addition to what has been discussed before the uncertainties facing us today will still be there tomorrow: cost of energy, cost of raw material, cost of labour, the price a manufacturer will have to pay for the social benefits etc...
Market is there - we do not normally create it out of nothing -. The difficulty is sometimes knowing exactly what our market is and whether we have really covered all that is within our reach.
The task of marketing here is to communicate back what customer research says will sell. Then it is the task for the R&D and manufacturing to make it.
Worldwide, irrespective of the fact that world population is increasing at an alarmingly high rate and is expected to exceed 6 billion already before the year 2000, there seems to be little consolation for countries producing surpluses in the near future.
In fact, if anything the number of importing countries is likely to come down with an increase in their domestic production.
GATT round will gradually open doors for freer trade also in dairy products and that means that we shall have to prepare ourselves for more thorough segmenting of the market than before.
Greater flexibility is also called for by the fact that the retailer is and will keep changing his apperance and image at a fast rate to attract customers. It is increasingly difficult both for the retailer and for us as suppliers to find a faithful customer. It is the quality

- price -relationship together with everything connected with good service on which the decision to buy is based.

Irrespective of what in recent years has been said about small being beautiful I still believe in benefits of scale, provided that with an increase in size you dont lose your flexibility.

Expansion & growth however usually require money and, with no inflation to help us pay our debts, we have to ask ourselves: "Can we afford it?" probably more often than what we are used to, because investment usually pays back in the long term but requires money immediately.

Should you choose the road to expansion through acquisition there are areas where caution is advisable:

- Check and recheck the estimates re: revenue to be expected.
- Apply a generous margin of tolerance to forecasts re: the trends of development (both own and those of competitors').
- Beware of subjective opinions and estimates re: capital needed to make the new investment work.
- Check your resources both financial and personnel needed.
- Constantly keep an eye on competition (they are not likely to remain idle while you are building up your empire).

Whatever we do, the fact remains that we shall always have to fight for our share of the market. - No vacuum is known or allowed to exist in this trade today - neither are we likely to be allowed to explore any potential market alone. However, suppose we do not let our imagination stop there but ask ourselves is there anything beyond? In astronomy there are forces we do not see or detect by ordinary means though we know they exist. Could it be that there are 'black holes' beyond our existing market concept towards which our products would gravitate if only we could get close enough.

FINALLY

I would like to end by summarising in three points a recipe that should see us a long way towards our target.

1. Let us put customer statisfaction above everything else on our list of priorities.

2. Let us be quality conscious in everything we do.

3. Let us be adventurous and bold enough to look beyond the obvious.

Above all let's keep in mind what someone has wisely said: "Best companies attain excellence not by imitation but by developing themselves!"

DAS MILCHWIRTSCHAFTLICHE UNTERNEHMEN UND DIE HERAUSFORDERUNGEN DES MARKTES

Dr. Rudolf Hilker
Meiereizentrale Nordmark e.G.
Waidmannstraße 1o
2ooo Hamburg 5o
Bundesrepublik Deutschland

Aufgrund der Forschungsergebnisse der Natur- und Ingenieurwissenschaften sowie deren Umsetzung in neue Technologien erhöht sich der Investitionsbedarf in der Molkereiindustrie stetig. Um das Risiko möglichst klein zu halten, bedürfen Investitionsentscheidungen stärker denn je einer Absicherung zum Markt. Im Referat wird auf Veränderungen, ausgelöst durch Bevölkerungsentwicklung, Ernährungs- und Einkaufsgewohnheiten, eingegangen.

1. VORBEMERKUNG

Die Welt der Wirtschaft ist in der letzten Dekade aus vielfältigen Gründen in ihren internationalen Zusammenhängen immer komplexer geworden. Verfolgt man die Literatur, die sich mit Management-Theorien beschäftigt, so kann man heute zwei Richtungen erkennen :

Viele Autoren versuchen, möglichst viele Einzelaspekte wirtschaftlicher Datenveränderungen in ihren Auswirkungen auf die Unternehmen zu untersuchen - Einzelaspekte verschiedenster Art, wie zum Beispiel die Bevölkerungsentwicklung, die Kaufkraftentwicklung oder Änderungen in den Technologien; auf einige mir besonders wichtig erscheinende werde ich im Laufe meines Referats zurückkommen. Diesen Untersuchungen liegt überwiegend das Bemühen zugrunde, Daten und Datenauswirkungen zu quantifizieren.

Eine andere Richtung wirtschaftswissenschaftlicher Autoren betont dagegen die Notwendigkeit, trotz oder gerade wegen der Vielzahl der Änderungseinwirkungen eine Gesamtschau, eine Einbeziehung gerade auch der nicht quantifizierbaren Einflußfaktoren in die Untersuchungen vorzunehmen. Es liegt auf der Hand, daß die Ergebnisse solcher Arbeiten sehr allgemein gehalten sind und keine konkreten Empfehlungen beinhalten.

Eine Untersuchung der Zukunftsaspekte des Milchmarktes hat nun eine Vielzahl exogener und endogener Daten zu berücksichtigen. Einige dieser Daten sind rechenbar, quantifizierbar, andere wesentliche Einflußfaktoren dagegen eindeutig unrechenbar, nicht quantifizierbar.

2. DIE ENTWICKLUNG DER MÄRKTE

Die Milchindustrie steht wie kaum eine andere Branche in einer doppelseitigen Marktbeziehung. Dabei ist der Beschaffungsmarkt, der Markt für den Rohstoff Milch, ebenso in unsere Untersuchung einzubeziehen wie der Absatzmarkt, der Markt für die hergestellten Produkte.

2.1. Der Beschaffungsmarkt

Der wichtigste exogene Einflußfaktor auf den Beschaffungsmarkt Milch ist die Agrarpolitik. Es gibt schlechthin keinen Staat und keine Staatengemeinschaft, die nicht - aus welchen Motiven auch immer - in den Milchmarkt eingegriffen haben. Es soll gar nicht erst der Versuch unternommen werden, die Motivation selbst zu untersuchen, lediglich die Folgen sind für unsere Überlegungen wichtig.

Generalisierend ist zu sagen: Die Summe aller Eingriffe führte fort von dem einer marktwirtschaftlichen Wirtschaftsordnung innewohnenden Automatismus, Angebot und Nachfrage zur Deckung zu bringen. Das Angebot an Milch ist weltweit größer als die Nachfrage. Das Ungleichgewicht läßt sich an den Bestandszahlen ablesen.

Ein anderer Indikator ist der Geldbedarf der Staaten, der hierdurch erforderlich geworden ist. Die Marktordnungskosten in der EG haben inzwischen je kg Milch eine Höhe erreicht, die - bezogen auf die überschüssige Milch - in etwa dem Preis gleicht, der den Landwirten bezahlt wird.

Es gibt keinen Zweifel - der Versuch, der Landwirtschaft über eine forcierte Preispolitik ein adäquates Einkommen zu sichern, ist gescheitert, eine Änderung der Politik ist unausweichlich.

Die neue Politik wird mit welchen Mitteln auch immer - seien es marktwirtschaftliche und/oder administrative - die Angebotsmenge erheblich zurückführen müssen. Die Auswirkungen für die Milchindustrie morgen sind ableitbar.

Es sei herausgestellt : nicht die Milchindustrie hat in der Vergangenheit den Anreiz gegeben, die Produktion zu steigern. Sie hatte aber als Folge staatlicher Politik technische Kapazitäten zu errichten, um die produzierten Mengen aufnehmen und verarbeiten zu können. Die Milchindustrie - und diese Aussage gilt insbesondere für den EG-Raum - wird in der kommenden Dekade die Verarbeitungskapazität erheblich abbauen müssen. Hierbei ist von ausschlaggebender Bedeutung, welche Maßnahmen die Administration ergreifen wird.

Es wird folgende These aufgestellt : Je marktwirtschaftlicher die Maßnahme sein wird, desto mehr wird die Milcherzeugung an für die Produktion ungünstigen Standorten aufgegeben und zu günstigeren Standorten verlagert.

Und umgekehrt, je administrativer die Maßnahme sein wird, um so mehr wird die Einschränkung der Milchproduktion losgelöst vom Standort prozentual - das heißt flächendeckend in annähernd gleicher Höhe - erfolgen.

Der Einflußfaktor Agrarpolitik ist in seiner Auswirkung überhaupt nicht zu unterschätzen. Andererseits entzieht sich gerade dieser Faktor jeglicher wirtschaftswissenschaftlichen Voraussage und weitestgehend dem Einflußbereich des Management der Milchindustrie.

Die Milchindustrie steht also vor einem schwierigen Kapazitätsanpassungsprozeß. Dieser Vorgang hat nicht nur ökonomische und technische Dimensionen, sondern natürlich auch eine erhebliche soziale Komponente.

Es erscheint unerläßlich, daß in weiten Teilen der Milchindustrie das Informationsmanagement sowohl zu den Mitarbeitern als insbesondere auch zu den Milcherzeugern verbessert und ausgebaut wird; bei steigender Unternehmensgröße in der Milchindustrie ist dem Informationsmanagement steigende Beachtung zu widmen.

2.2. Der Absatzmarkt

Nachfrage-Boom und -Degression infolge Kaufkraftsteigerungen und Kaufkraftverfall im internationalen Milchmarkt wurden uns in den letzten Jahren in besonders deutlicher Weise demonstriert. Das künftige Volumen des Milchmarktes vorauszuschätzen, ist außerordentlich schwierig. Noch schwieriger ist eine Aussage darüber, aus welchen Regionen der Welt ein sicherlich infolge des Anstiegs der Weltbevölkerung wachsender Bedarf gedeckt werden wird.

Ich messe dem Ausgang der GATT-Verhandlungen in Punta del Este ein außerordentliches Gewicht bei : Der Agrarhandel als Teil des gesamten Weltmarkthandels hat aus vielfachen Gründen eine erhebliche politische Bedeutung erhalten. Es scheint zumindest fraglich, ob in den westlichen Industriestaaten langfristig ein so hohes Produktionsniveau für Milch - das deutlich den Selbstversorgungsgrad überschreitet - gehalten werden kann, daß hieraus erhebliche Mengen für den Weltmarkt bereitgestellt werden.

Leichter greifbar sind Daten für den europäischen und den deutschen Raum. So wissen wir zum Beispiel für Deutschland, daß die Bevölkerung von heute bis zum Jahre 2000 um rund 2 Millionen auf 54 Millionen zurückgehen wird. Dem steht andererseits ein Bevölkerungswachstum in Europa gegenüber.

Genauso wichtig wie diese Kennziffer ist aber die Kenntnis von der strukturellen Zusammensetzung der Bevölkerung : Der Anteil der Menschen über 60 Jahre wird größer sein und fast die gleiche Zahl erreichen wie die der Menschen unter 30.

Die Menschen werden weniger körperliche Arbeit ausüben, sie werden ein höheres verfügbares Einkommen haben, und sie werden in wesentlich kleineren Haushalten leben : die 1- und 2-Personen-Haushalte werden, zumindest in Deutschland, von 47 % in 1960 auf 66 % im Jahre 2000 steigen; die 3- und Mehr-Personen-Haushalte werden im gleichen Zeitraum von 53 % auf 34 % sinken.

Die Lebensgewohnheiten werden sich ändern. So wird zum Beispiel der Außerhausverzehr an Bedeutung gewinnen, der Faktor Qualität - was auch immer darunter subjektiv zu verstehen ist - wird eine noch größere Rolle spielen.

Aus all diesen Gründen heraus wird sich die Nachfrage sowohl nach der Art der Produkte als auch nach deren Quantität und Qualität nachhaltig ändern. Hierbei zeigen alle Untersuchungen, daß die Nachfrage nach Milchprodukten in einem positiven Trend liegen wird. Dennoch ist

festzuhalten, daß die zu erwartenden Absatzsteigerungen in globaler
Betrachtung gering sein werden - so gering, daß sie in keiner Weise
ausreichen, um die Angebotsmenge aufzunehmen.

Hierbei ist zu beachten, daß der ausgewiesene Selbstversorgungsgrad der EG auch die Mengen beinhaltet, die zu Subventionspreisen -
sei es im Bereich der menschlichen oder der tierischen Ernährung -
abgesetzt werden. Hieraus ergibt sich, daß der tatsächliche Selbstversorgungsgrad der EG erheblich über dem veröffentlichten Niveau liegt.

Der Weg zum Konsumenten führt über den Handel, die Zahl der Lebens-
mittelgeschäfte wird weiter rückläufig sein. Gab es im Jahre 1980 in der
Bundesrepublik Deutschland noch 94.000 Lebensmittelhandlungen, so werden
es 1990 nur noch 66.000 sein.

Auch aus diesem Sachverhalt heraus werden sich die Anforderungen
an die Produkte hinsichtlich ihrer Qualität, ihrer Aufmachung, ihrer
Verpackung ändern.

Eine andere Auswirkung dieses Konzentrationsprozesses ist und wird
es verstärkt sein, daß der Zugang zum Konsumenten über den konzentrier-
ten Handel stets schwieriger und damit risikoreicher sein wird. In der
Bundesrepublik entfallen inzwischen auf die 10 nachfragestärksten Ein-
kaufsorganisationen rund drei Viertel des gesamten Warenumschlags.

Die Dominanz der organisierten Nachfrage wird auch dadurch deutlich,
daß der Anteil der Produkte, die ohne markenartikelmäßige Profilierung
auf dem Markt angeboten werden, nachhaltig gestiegen ist. In einzelnen
Produktbereichen entfallen in Deutschland 30 % der Nachfrage auf soge-
nannte anonyme oder auch weiße Produkte. Für den Produzenten ist hierdurch die Gefahr einer problemlosen Austauschbarkeit als Lieferant gegeben. Allein hieraus ergibt sich, daß Konsumentenwerbung als Sicherung
von Marktanteilen in Zukunft stark steigende Werbeetats erfordern wird.

3. INTERNE EINFLUSSFAKTOREN

Der nachhaltigste interne Einflußfaktor auf die Struktur der Milchindustrie ist die Weiterentwicklung der Technologie der Milchverarbei-
tung. Gerade in den letzten Jahren sind in unserer Branche entscheidende
Durchbrüche in neuen Technologie-Bereichen gelungen. Erwähnt werden
können als Beispiele : die Ultrafiltration der Milch; die Umkehrosmose
der Milch; neue Trocknungstechniken; neue Ausfälltechniken des Kaseins;
die Trennung des Butterfetts in verschiedene Fettfraktionen; die Beherr-
schung steriler und aseptischer Arbeitsweisen zur Verlängerung der Halt-
barkeit von Milchprodukten.

Diese Entwicklungen sind durch eine intensive Zusammenarbeit
zwischen der Forschung, der Molkereimaschinen-Industrie und der Milchindustrie eingeleitet worden; die Anwendungsprozesse jedoch müssen aus-
schließlich von der Molkereiwirtschaft erarbeitet werden.

Ziel der Arbeiten ist einmal eine Rationalisierung der Ablauf-
prozesse, eine Verbesserung der Qualitäten und der Haltbarkeiten der
bestehenden Produkte und vor allem - durch Aufgliederung der Milch in
ihre Einzelbestandteile und anschließendes Zusammensetzen in gewünschte
Kombinationen - die Entwicklung neuer Produkte.

Es liegt auf der Hand, daß für den Forschungsaufwand erhebliche Geldmittel benötigt werden und ein großer Kapitalbedarf für die Anschaffung der Maschinen besteht. Rechnete man in den 7oer Jahren mit Investitionswerten von 5oo.ooo DM je Tonne Jahres-Rohstoffverarbeitung, so zeigt der Wert heute bereits 1 Million DM mit weiter steigender Tendenz.

Parallel hierzu ist weiter der Durchbruch neuer Kommunikationstechniken in Produktion und Verwaltung zu berücksichtigen.

4. ERGEBNIS

Welche Schlußfolgerungen sind hieraus für die Entwicklung der Milchindustrie zu ziehen, welches Maßnahmenbündel ist zu treffen, um in den Herausforderungen des Marktes bestehen zu können?

Voraussagen, insbesondere wenn sie in die Zukunft gerichtet sind, sind unsicher - das sagte der Satiriker George Bernard Shaw. Mit meinen Aussagen wollte ich nachweisen, daß insbesondere die nicht exakt vorhersehbaren, nicht quantifizierbaren Auswirkungen veränderter internationaler und nationaler politischer Rahmenbedingungen für die Zukunft von erheblicher Bedeutung sein werden. Aus diesem Grunde ist für die nächste Dekade innerhalb der Milchindustrie eine Unternehmenspolitik angebracht, die ich mit dem Titel "Risiko-Miniminierung" kennzeichnen möchte. Was ist das?

Zur Minimierung der Risiken, die sich aus einer rückläufigen Milchproduktion ergeben werden, ist es zwingend erforderlich, daß Zukunftsunternehmen sich eine größere Rohstoffbasis sichern, eine Basis, deren Größenordnung deutlich über den heutigen Bedarf hinausgeht. Wobei zu beachten ist, daß für diese Zusatzmengen möglichst keine Investitionen vorgenommen werden.

Die Bearbeitung und die Verarbeitung der Milch müssen in einer langfristigen Unternehmensstruktur festgelegt werden; Betriebe, die bei der vorauszusehenden Rohstoffverknappung zu schließen sind, sollten durch systematische Desinvestitionen aufgezehrt werden.

Zur Minimierung der Risiken auf dem Absatzmarkt ist eine Weiterentwicklung und ein Ausbau des gesamten absatzpolitischen Instrumentariums erforderlich. Dieses Instrumentarium besteht aus den Teilstücken Marktforschung, Marketing, Werbung.

Zur Risikominimierung ist es weiter erforderlich, daß das Zukunftunternehmen möglichst in allen bedeutenden Marktsegmenten mit ausreichenden Marktanteilen vertreten ist. Wir haben in der EG in den letzten Jahren wiederholt erleben müssen, wie durch Brüsseler Entscheidungen sich Rentabilitäten verschoben haben, die Werte einzelner Milchinhaltsstoffe verändert wurden und damit Marktverschiebungen eintraten. Genauso haben wir gerade in jüngster Zeit erfahren müssen, wie durch politische Entscheidungen auf dem Weltmarkt bestimmte Abnehmerländer von heute auf morgen mit bestimmten Produkten ausfielen.

Eine optimale Kostenstruktur ist lebensnotwendig, denn nur dann können die Mittel erbracht werden, um das absatzpolitische Instrument mit vernünftigen Etats zu versehen, die technische Innovation zu sichern, den Landwirten einen wettbewerbsgerechten Preis für die Milch zu zahlen und gleichzeitig die notwendige Kapitalbildung vorzunehmen.

In der Vergangenheit wurden in unserer Branche viele Untersuchungen zur Frage der optimalen Betriebsgröße vorgelegt. Diese Untersuchungen beruhten überwiegend auf technischen Daten, stellten Kostendegressionsverläufe in den Mittelpunkt ihrer Betrachtung. Diese Aussagen müssen ergänzt werden um die Komponente optimale Unternehmensgröße. Die Kombination von optimalen technischen Einheiten für verschiedene Produktbereiche mit einer entsprechenden Marktstellung führt erst zu einer optimalen Unternehmenseinheit.

Wir wissen, daß Käsereien über gute Kostenstrukturen verfügen, wenn sie etwa 3oo Millionen kg Kesselmilch pro Jahr verarbeiten; daß Buttereien zwischen 2o.ooo und 25.ooo Tonnen Butter herstellen, Trockungsbetriebe im Durchlauf über 7oo Millionen kg liegen sollten und daß kapitalintensive Konsummilchbetriebe ebenfalls im Zwei- und Dreischichtbetrieb genutzt werden sollten.

Daraus errechnet sich sehr schnell, daß ein Unternehmen, welches sich den Herausforderungen des Marktes von morgen erfolgreich stellen will, über mehrere spezialisierte Betriebsstätten verfügen muß und dementsprechend einen Rohstoffeingang in einer Größenordnung von 1 1/2 Milliarden kg benötigt.

Ungeachtet dieser Aussagen werden auch morgen und übermorgen spezialisierte kleinere Unternehmenseinheiten ihre Chance haben, aber sie werden stets mit größeren Risiken leben müssen.

PRODUCTION AND MARKETING OF MILK AND MILK PRODUCTS IN EAST EUROPE

V.N. Sergeev
State Agroindustrial Committee
27, Kalinin Avenue
G-19 Moscow 121019
Soviet Union

Planned and consistent policy of the Governments of the Council of Mutual Economic Assistance countries directed towards increasing living standards and health improvements of the population comprises measures to enlarge output of milk and milk products in these countries.

After the devastating Second World War when in East Europe farms and dairies were almost completely destroyed it was necessary to invest considerable resources for reconstruction, for dairy cattle breeding, creation of forage basis as well as for building new plants.

Especially high milk production rates are observed in recent 10-15 years. For example, in Hungary and Czechoslovakia milk production grew by more than 40% in the period since 1970 till 1985 and in the USSR - 1,5 times. Now the CMEA countries share in world milk production is 35%, butter - 40% and cheese - 21%. The Soviet Union occupies leading position in the world as to the gross output of milk, fermented milks and products and butter.

In order to more completely meet consumers demands for various dairy products modern technical-industrial basis of dairy industry has been created. New dairies have been built in Moscow, Leningrad, capitals of Federal Republics and large industrial centres. It allowed to provide city population with large variety of high quality products.

At the modern dairies production is carried out according to the schemes giving complete usage of raw and by-products. Simultaneously labour and life conditions as well as medical service of the staff have been improved.

Creation of new plants allowed to concentrate production. So in Czechoslovakia in 1970 there were 231 milk processing plants and in 1980 their number decreased to 140. As it is estimated by 2000 year the country is going to have about 50 dairies with average per day capacity of 380 tons of milk. The majority of these plants has been mechanized and some - automatized. In 1985 in Gambec and Cheska Line highly mechanized dairies were put into operation.

Characteristic feature of dairy industry of our countries is large variety og milk products developed according to nutritionists recommendations including nutritional, age and professional aspects. In the USSR assortment numbers about 500 items of dairy products of which 200 are fresh milk products. About 100 cheese varieties such as Emmental, Edam, Gouda, Chedder, Camamber, Roquefort etc. are being produced in Czechoslovakia.

In order to more completely satisfy population demands for food staff the Food Program had been developed in the USSR. According to this Program milk production in the Soviet Union will reach 107-110 million tons by 2000 year and in Poland it is planned to get about 20 million tons of milk. Similar tasks were also put forward in Bulgaria, Czechoslovakia, Hungary and Rumania.

Governments of the East European countries adopted legislations and approved special programs on dairy cattle breeding. Such programs exist in Hungary under the title "About cattle breeding including milk production" which have been directed towards reaching by 2000 year scientifically recommended milk consumption. This document also envisages creation of large specialized complexes provided with meat and dairy breed cattle. The problem is solved by two ways: importing of high yielding dairy cows and by herd formation with the help of native dairy cows crossbreeding. The actions udertaken allowed to increase milk production mainly due to higher yields. For 15 years average yield per cow grew almost 1,5 times. Fulfilment of the programs mentioned above made it possible to produce more than 440 kg of milk per head per year in German Democratic Republic and Czechoslovakia. Average per head milk production in CMEA countries exceeds 340 kg a year i.e. it is at the level of the European Economic Community. Butter production in these countries amounts to 2 300 thousand tons which is much more higher than in EEC.

Recently much work was done to change dairy products assortment according to recommendations of dietologists based on scientific researches. For example, in Hungary and the USSR greatly increased manufacturing of liquid milks and fermented milks with low fat content (1-1,5%), dietetic sour cream and low fat protein enriched quarg products. Butter creams with 35-40% of fat are being produced in Hungary. Hungarian dairy industry started to put out concentrated and powder milks with low lactose content. Milk proteins received by ultrafiltration process are used as additives to meat, confectionary and bakery products.

Milk products manufactured in Bulgaria are of great demand. They include sour milk, white brine cheese, cachcaval "Balcan". Patents and licences for Bulgarian lactic acid starters are used in 20 countries of the world. Bulgarian white brine cheese and cachcaval are exported to many European and Near East countries, to the USA and Australia.

Dairy industry has become an important branch of the economy in Poland. Government of the country conducts consistent policy in the sphere of milk production and marketing and in fixing prices beneficial for suppliers and retail prices reasonable for consumers. Seim of Poland approved state program of the agriculture development and food supply till 1990.

Milk and milk products manufacturing and consumption in Poland gained high level which is due to the Government care of people health. The share of milk protein in proteins of animal origin consumed approximates 50% and share of milk fat is about 40% of fats consumed. According to the above mentioned program milk production will reach 20 mln tons by 2000.

Special attention in CMEA countries is paid to the increasing of baby foods on milk basis manufacturing. Corresponding programs envisage production of formulas replacing human milk and large variety of foods for feeding children in day nurseries, kindergartens and schools. In the USSR there has been created a new branch of the industry - babt foods production. Five modern specialized plants for manufacture of dry formulas and five sections for fluid and paste baby foods at the existing dairies were built. They are equipped with latest machinery, technological processes are highly automatized and computerized According to the Food program in the USSR dry baby food mixtures production will reach 83,5 thousand tons by 1990 (two times more versus 1985). It will completely satisfy consumers demands. Output of liquid and paste milk baby foods will increase to 100 thousand tons which is 2,5 times more in comparison with 1985. For this purpose in each regional and indusytial center where population exceeds 300 thousand special sections will be organized at dairies.

Dairy industry constantly introduces new baby foods recommended by pediatricians including various kinds of powder and liquid milk products enriched with whey protein, dry fermented mixtures, products on the delactosed basis and also with protective additives.

Large variety of baby formulas is manufactured by dairies in Czechoslovakia. They are known in Europe under the trade names "Sunar", "Sunarka", "Eviko","Feminar", "Lacten", "Relacton". At school children receive milk breakfasts and acquire a habit to drink milk. Schoolchidren are proposed to have vitaminized milk, fermented milks, cream, puddings, cheese, quarg products and butter. In industrial regions where surroundings are worse pupils get such breakfasts free of charge. Here meals include fruits, vegetables and meat.

Similar programs for children and pupils feeding exist in Poland, Hungary, Czechoslovakia, Bulgaria, German Democratic Republic and Rumania.

In all countries-members of the Council of Mutual Economic Assistance workers at the mines, metallurgical works, chemical plants and other harmful enterprises are supplied with milk free of charge.

Further rapid development of dairy industry in these countries is reached by considerable investments increase, radical renewal of the equipment, introduction of electronics, more effective scientific researches in dairy field as well as by skill improvements and labour productivity.

In 1986-1990 the Soviet government will allot 1,7 times more resources for dairy industry than in previous five years. It is planned to supply the industry with new equipment for 1,5 milliard roubles. It will be traditional equipment of higher capacities and robots for working in stores with finished products. Usage of membrane and other energy saving processes will lead to technological schemes excluding any wastes.

In Poland special decision was adopted by the government fulfilment of which will promote more rapid development od the dairy industry in 1985-1995. Considerable resources are alloted for dairy cattle breeding program and modernizing milk processing plants.

Retail system is being constantly improved, actions are taken to provide shops for milk and milk products sale with refrigerators. For example in Poland has increased number of model shops which belong to cooperatives. In 1985-86 there will start to work 150 such shops, in 1986-1990 - 240 and in 1991-1995 - 250. By 2000 year it is planned to have 1000 such shops.

Network of specialized retail milk shops provided with refrigerating equipment has largely increased in Hungary and Czechoslovakia.

In conclusion it should be said that examples given above on measures undertaken in the CMEA countries to grow milk and milk products manufacturing, to improve their quality and to widen variety help to imagine what tremendous programs their governments adopt to improve people well-being.

CHANGES IN FOOD DISTRIBUTION SYSTEMS AND THEIR IMPLICATIONS

FOR DAIRY ENTERPRISES

A R DARE
Managing Director
St Ivel Ltd
Dorcan House
Eldene Drive
Swindon, Wilts, SN3 3TU

ABSTRACT. The UK market has witnessed some major changes:
- Van selling has almost disappeared, being replaced by pre-order distribution systems.
- There has been a major concentration in buying power which has had effects on distribution systems as well as product catalogues.
- There has been significant investments in chiller cabinets by Retailers and improved chill chain facilities by manufacturers.

These major trends have fundamentally changed the nature and the control of the distribution systems in the UK.

MAIN TEXT.

The 1980's are proving to be a decade of change in food marketing and distribution. More crucially it is also a decade of enormous change in consumer habits, attitudes and behaviour: the consumer of 1986 is far more discerning and knowledgeable than the consumer of 1976.

These changes have been particularly marked in a number of key areas which have greatly influenced the way in which the retailer has responded and developed both inside and outside the store. With the 2.4 children, working father, housewife at home 'typical' family now a figment of the past, the decade has seen the growth of individualism within the family unit, leading to the demise of the formal family meal at the expense of individual and snack occasions. The housewife's requirements for food have altered.

Secondly the fact that now more than half the housewives of Britain are in employment has led to a demand for convenience, not just in terms of the foods which are purchased, but also in terms of shopping convenience. The decade has seen the development of a revolution in terms of shopping behaviour:

1. The introduction of one-stop shopping, with all the housewife's needs met under one roof, with convenient parking alongside, and with corresponding savings in terms of time and, hopefully, cost.

2. The move to weekly or even monthly major shopping trips, with the role of local stores restricted to top-up shopping, providing a niche for convenience stores with longer opening hours.

The bigger retailers have reacted to these changes by opening up massive out-of-town superstores, already numbering some 400 in the UK, and expected to reach 750 by the early 1990s.

Thirdly the enormous growth in the ownership of consumer durables has altered cooking and storage habits in the home. In 1976, 75% of homes had a fridge, but only 18% a freezer. By 1985, 97% of homes had a fridge, and 66% a freezer. Microwaves were unknown in 1976, yet by 1985 they were present in 14% of homes and are forecast to reach 25% penetration by 1990. Hence the growth in whole new product sectors, particularly in the chiller and frozen cabinets, initially with products like yogurt, but latterly with ready meals, from the basic to the luxurious. The chiller has seen a plethora of new products and innovations, some of which are here to stay, while others have enjoyed a short but profitable life cycle and have now given way to the next generation of product ideas.

And lastly, but by no means least, the British housewife has become increasingly concerned about the food she eats. We should not imagine that every household is devoted to eating healthier foods, but certainly in general there is a greater concern about the quality of food. People are looking to eat good food, rather than convenient fillers. And the chiller of course, with its fresh natural imagery, is perfectly suited to counteracting concern about processing and the inclusion of additives.

This consumer revolution has opened up enormous opportunities not just for the manufacturer but also for the innovative and forward-looking retailer.

In 1976 the grocery trade was competing almost solely on price. We can all remember Tesco's checkout, and Sainsbury's 1978 price war. Such policies inevitably proved to be self-defeating as they were not only hamstringing retailer development, but were also found to be increasingly out-of-touch with the consumer's needs for better and different foods, individually packaged and well presented. The leading retailers were forced to re-examine their strategies.

Hence the development of the quality fresh imageries of our more influential retailers of today. Fresh food, both from the fish, meat and vegetable counters, and also from the chilled cabinet, proved to not only provide better margins to help fund further development and growth, but also to greatly influence consumers' perceptions of retailers. Sainsburys, of course, had always enjoyed a reputation in the fresh food area, and they have been successful in maintaining this advantage. But the likes of Tesco, Asda and others are not far behind.

So the consumers began to find the products they needed in greater profusion with the more forward-looking retailers. They shopped with them in greater numbers, at the expense of the cooperative and independent sector, so that by 1985, 76% of grocery purchases were made through the multiple trade, compared with only 62% in 1980. And as the big becomes bigger, the small began to fall by the wayside, to be gobbled up by their bigger brethren. By 1985 the concentration of retailer power had already reached such a stage that the top 8 buying points now account for 80% of grocery sector sales.

And as bigger and better stores are built, the retailers are devoting more and more space to fresh foods, which offer them better margins and more profit to fuel further growth. Ten years ago it was said that the chiller cabinet provided some 17% of store profits from a 5% allocation of space within an average store. Today in new stores, the temperature control area occupies up to 50% of total food space, with two thirds of the area devoted to chilled foods. Already an average store is thought to have an allocation of some 20% devoted to chill, accounting for around one-third of total profits.

Moreover this greater space allocation has also pinpointed the need for greater efficiency within the cabinet operation itself. Maintenance of the right temperature reduces write-offs and helps maintain product life. Keeping the cabinet spotless improves consumer's perceptions and store hygiene standards. Line by line profitability signals out the key products and determines the allocation of space within the cabinet.

All of which has led to opportunities for growth. Growth for the retailer from better management and control of the chilled cabinet area. And growth for the manufacturer from greater scope for the national distribution of short life products because of the increased retailer efficiencies.

We have dealt so far with how retailers have responded within the store but it has also been necessary for them to crucially examine any other areas of their operations in which there is any opportunity to improve efficiency and thus sharpen their competitive edge.

Distribution channels and methods was a natural candidate for their attention. Traditionally manufacturers had maintained large expensive networks of vehicles and depots which were substantially under-utilised and this contributed excessively to the overall unit costs of moving products from the factories through to the consumer. Van selling was still widespread with its inefficient use of transport resources, low service levels, inadequate technical standards and the image that it was operated by the manufacturer in a way that did not necessarily have the best interest of the retailer in mind. Added to this the escalating costs of running van sales operations meant it was impossible for some of the smallest manufacturers to operate their own distribution facilities, denying the retailers the chance to purchase the innovative good value products which they badly wanted to extend the product range offered to the ever more selective consumer in the ever more competitive market place.

One obvious alternative was for the retailers to own and operate their own chilled distribution facilities in a similar way to their treatment of longer life products but this was rejected by many who preferred to devote their scarce capital and management time to retailing and building new outlets.

What has happened with many retailers therefore is that they have found that they can best exert pressure on the efficiency and quality of chilled distribution services to their stores by encouraging one or more of their major chilled goods suppliers to become a nominated distributor, collecting orders from the retailers' outlets, laying them off to the various manufacturers, and consolidating the incoming goods from a number of sources of supply through a number of warehouses into a single delivery to each branch using their existing facilities on a marginal basis.

This mode of operation has improved the retailers control over his fresh product business in an immediate and dramatic way. He is now able to use all the desired sources of supply with the lack of transport facilities amongst the smaller manufacturers ceasing to prevent him from having access to their products. His store staff have achieved more effective control over the in-store stock presence and manufacturers have improved their service levels through the measure of stock forecasting and accurate advance ordering that the retailer with his better disciplines can now offer.

Technical standards for refrigeration, product presentation and life can be set and monitored and the retailer is beginning to be able to address and decide upon many service options to improve unit costs, issues which could never have been handled by the manufacturer acting on his own, such things as twilight and night deliveries and demand smoothing for example which potentially have a big impact on final unit costs. In addition the improved 'chill chain' has enabled many new sensitive products to be sold 'fresh'.

In a number of cases these manufacturers who have turned into distributors are large dairy companies for whom the prospect of offsetting the loss of their direct deliveries from Regional warehouses into Retailers branches by revenue for distributing product from other manufacturers is very attractive. In the case of some of the smaller fresh product manufacturers both in the dairy field and in others the reverse has occurred and they have closed down their distribution resources, concentrating solely on manufacture and marketing.

In summary therefore distribution systems in the most dynamic and profitable section of the food industry in the United Kingdom, the fresh chilled food sector, have been in a constant state of change and development over the last ten years. The major features of this change have been:

1. The developing consumer requirements of choice and convenience in fresh food.

2. The desire of a smaller number of more influential retailers to strengthen their competitive position.

3. The requirement for more efficient methods of distribution to maximise product availability and quality at minimum cost.

4. The necessity for manufacturers of fresh products to utilise their distribution resources by acting as nominated distributors to retailers or to disinvest from distribution resources.

In each part of this important change, dairy product enterprises have led the way in responding and adapting to the challenges created by the all-important aspirations of the most important person in the chain - the consumer.

MAJOR PROBLEMS IN THE MARKETING MANAGEMENT OF MODERN DAIRY ENTERPRISES

A.J. Kranendonk
ccFriesland
Postbus 226
8901 MA Leeuwarden
Holland

Ladies and gentlemen,
When I was asked to hold this talk, the situation was rather different than the one we find ourselves in today.
Oil was about $ 18 per barrel, the Dutch florin stood at about f 2,80 per dollar and the butter mountain of the E.E.C. was at its old enormous level.
In the meantime, things have changed rather drastically, even compared to the rather negative conditions I was talking about earlier, which were already bad in relation to the history of milkproducts.

Slide 1.*The Environment is changing.
Obviously the environment is changing very quickly and this will drastically alter the traditional marketing approach to the dairy industry. Let me go into this in more detail:

Slide 2.*Current Account Balances Developing Countries.
If one looks at the developing countries and their balances of payments, one sees that both in oil-exporting and not-oil exporting countr the current accounts tend to be negative.   (alike
The situation, as we are all aware, is getting worse every day.
The oil-exporting countries, for a while, were still plucking the benifits of the oil crisis but in 1986 they are already suffering heavily from the lowering of the oil price.

Slide 3.*Milkimports = Oilexports.
It is interesting to note that there is a strong relationship between the milk imports of these countries and the oil exports.
I hasten to mention that the scale of the two lines is rather different; the oil export value being a 100 times that of milk, unfortunately.

Slide 4.*Development Oil Revenues and US$/Ecu Exchange Rate.
Let me now stop a little bit and discuss with you what these oil exporting countries have suffered in the last few years.
Take the total Opec oil revenues.

In 1980, if one takes the amount produced and the price of the oil
which was then $ 35 per barrel, the index of their revenues can be
put at a hundred, which corresponds to about one billion dollars per
day. Only six years later, in May this year, the index was down to
sixteen, one-six, and revenues were in the order of $165 mln. per
day. It is obvious that the countries are having foreign exchange
difficulties under these conditions.
Now that is for the whole world.
If one takes the European world in particular - who are the main exporters of milkproducts - one sees that in the short period of one
year between May 1985 and May 1986, the exchange rate of the ecu to
the dollar dropped.40 which corresponds to a drop of 28%.
It must be clear that under these conditions it is going to be increasingly difficult to sell milkproducts from the Common Market to
the Opec countries.

Slide 5.* Growth World Population & Development Urbanisation.
Let me now side-step Opec for a moment and maybe infuse this rather
negative start of my presentation with a, potentially at least, positive
note. Let us look at the world population.
Some say, that an outburst of population as depicted here, where this
year in May the worldpopulation crossed the 5 billion mark and is expected.
to go to over 6 billion in the year 2000, is not something to look
forward to.
On the other hand, for a marketeer in milkproducts, it is an enormous
potential of mouths which have to be fed and milk offers a unique
combination of healthy ingredients to fill these hungry mouths.
Another interesting statistic is the projection of the urban population
in the year 2000, which shows that more than half of the world
population is expected to live in urban dwellings by that year.
Living in an urban situation, of course, necessitates the consumption
of industrial products or agricultural products, processed in an
industrial way. The demand is thus there. What is lacking, is the money
to pay for it.
Let us now go back to the milkmarket and look at that in some detail.

Slide 6.* Imports Third Countries.
The milk market consists of a part export and a part local production.
Let us look at exports first. Exports for us, are imports for third
countries. Now the main imports are evaporated milk, sweetened condensed
milk and milkpowder in consumer packs.
Purposely we have left out the bulk milkpowder. If one looks at the
evaporated milk, 3/4 of the total imports are taken up by 6 countries,
practically all of which are oil-producing.
The only exeption here being Greece, which incidentally is also an
E.E.C.-country. These "Group 1 countries", as we have called them, each
have an imported tonnage of over 20.000 tons. There is another group
which takes about 16% of the total and consists of about 12 countries,
each taking more than 3.500 tons and finally there is a rest of 9% of
the total of smaller countries, each taking less than 3.500 tons.

In the "Group 1 countries", one still finds Algeria and Nigeria, two
countries which due to their falling oil exports, have since then,
reduced drastically the imports of evaporated milk.
In the SCM-market, the sweetened condensed, one sees a more or less
similar situation.
Group 1, consisting of the mentioned 7 countries, are accounting for
60% of the total imports of sweetened condensed. The Group 2 countries
which consume over 1400 tons a year each, take up 20% of the total and
the other countries stand at a little less than 20% of the total. This
market is a bit more stable at the moment than the evaporated market.
Also in this market there are less oil-exporting countries. In the
milkpowdermarket one sees once again the same picture with the mentioned
7 countries taking up 52% of the total, each importing over 10.000 tons.
The Group 2 here, does 27% of the total and there are quite a number
of other countries which in total do 21%.
It must be mentioned here that this picture I am showing, is not one
which occurred at one time: it is a development which has been relatively stable over the last few years.

Slide 7.* Production Cow Milk in Developing Countries
If then the imports are lagging, how are the consumers managing to
fill their needs for milk products? Production, one could say. Local
production in their own countries. I show you here the developments
of the local production of cows-milk in developing countries for the
Near East, the Far East (including India) and Africa and apart from the
Far East where the production is indeed growing although at a relatively
slow rate from 16½ million tons in 1978 to less than 20 million tons
in 1984, local production is not seen to take over from imports.

Slide 8.* Self-Sufficiency Rates of Some Milkproducts in the Industrial
Countries.
Unfortunately overproduction in the industrial world does not seem to
follow the market. Here I show you the self-sufficiency rate of some
of the milk-products in our industrialised world. Above you see the milkpowder picture in which production of skimmed milk in 1983 was
almost 500% more than what is consumed; the whole-milkpowder stands
at well over 250% and both these markets have been at these high levels
for the last 6 years, on the lower half of this graph the butter overproduction which leads to the mountain, the condensed milk situation
which is also at high levels and fortunately cheese, which maintains
itself at more or less selfsufficiency level. All in all an extremely
bleak scenario; and what I will now try to do is provide you with
a little optimism in showing you what I feel are possible creative
answers to the current situation.

Slide 9.* Creative Answers.
We at ccFriesland, have been involved in the last few years, in increasing the percentage of local production and local marketing of
milk products in developing countries. To that effect we have built
or acquired quite a number of manufacturing companies in different
parts of the world; we have started trading operations where manufacturing

was not the right solution and we have built sales organisations there
where that was most appropriate.

Slide 10. Manufacturing and Trading Companies ccFriesland.
This is the world of ccFriesland at the moment.
We have divided this world into three regions, each with a separate
directorate and profit responsibility. Region I is the Far East, where
we have factories in Malaysia, Thailand, Indonesia, Guam, Taiwan and
Okinawa where we have trading companies in Hongkong and Singapore.
Region II is our Africa- and America's region where we have factories
in Nigeria, Saudi-Arabia, North Yemen and the one which is not operating
at the moment in Beyruth and where we have a trading company in Athens
and lastly Region III, our European region, which consists of our
factories and head-office in Leeuwarden/Holland, a factory in Groningen
and Sloten and a factory in Germany, trading offices in Holland and
in Brussels.

Slide 11. *Local Production and Marketing of Milk Products.
Thus, as the situation calls for a different set-up, we provide one,
either as local manufacturer, as operator of a trading company or
with our own sales/distribution organisation.

Slide 12.*To Assist or Participate in Local Milk/Farm Developing Projects.
We have also felt that it is often not enough to simply, market locally
and produce locally, the milk products; one should also go further down
nearer to production and assist and, if necessary, participate in local
milk projects and in farm development projects. At the moment we are
heavily involved in three major projects of this kind and we expect
more projects to come. If Mohammed doesn't come to the mountain, the
mountain will go to Mohammed.

Slide 13.*Marketing Approach.
And we are placing more emphasis on marketing as a way to strengthen
our position in our various markets.
One part of that is segmenting the markets in terms of demographics. We
will try to appeal to people of different ages and different incomes,
different life-styles, approach light and heavy users in a different way
and try to supply our consumers with products which are specifically
adapted to their application of those products. This means that we will
go to the consumer and try and find products which will suit the needs
of that consumer best.
It does mean that we will take a step away from the kind of bulk production and bulk marketing we were used to in the past and try and fine-
tune our products to suit the necessarily smaller niches in the market.
Extra emphasis and attention will be placed on the positioning of ourselves (the corporate identity of ccFriesland is a major project within
our Company) the positioning of our products (as with heavy emphasis
on the quality of our products) and more and more attention will be
placed on the quality of the distribution of our products.

Slide 14.*To develop Different Products and Packages.
All this leads to a major emphasis on the production of different products and different packages, more adapted to the market than before.
To this effect, a relatively large research unit is now working out various new product- and package-developments, based on developments in trade where fresh products, which are often distributed in chilled chains, are becoming more important, improved transportation possibilities where paperpacks now become a thin of the present and where, unfortunately our major investments in tin-lines will in the long run, loose out in part to the paperpacks, and developments in new techniques in recombined products and UHT aseptically packed products.

In conclusion, I think we can say that the marketing of milk products is at a major crossroads.
What we have done in the past, may not be good enough for the future.
The situation in the markets where we traditionally have been able to sell our products are such, that a quick change from the current gloomy situation cannot be forseen.
Even so, the potential in the developing markets is such that there must be ways to go about it.
We are organising ourselves in finding ways to use all the creativity availabe in our company and using it productively. The results have been impressive from the start, and I'm sure they will yield even more benefits in the near future.
This brings me to the end of my presentation.
I have tried to indicate to you a number of ways which can be used in finding new paths in a rapidly changing environment, ways which we, at ccFriesland, are actively exploring.

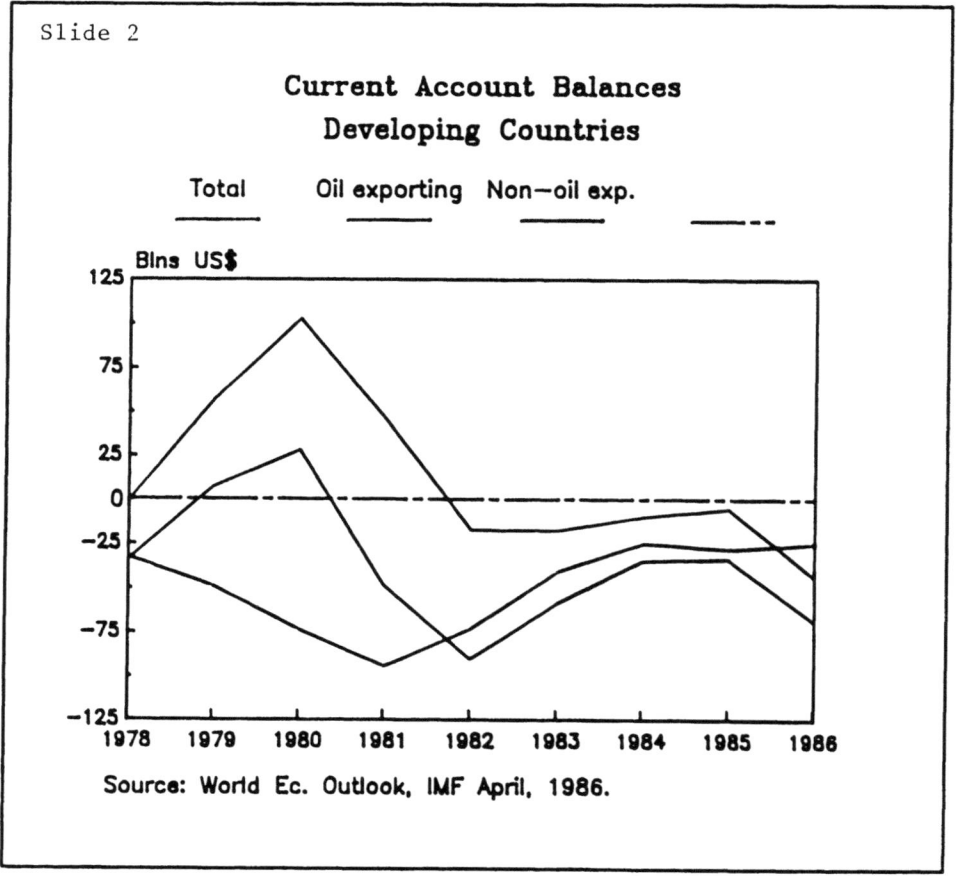

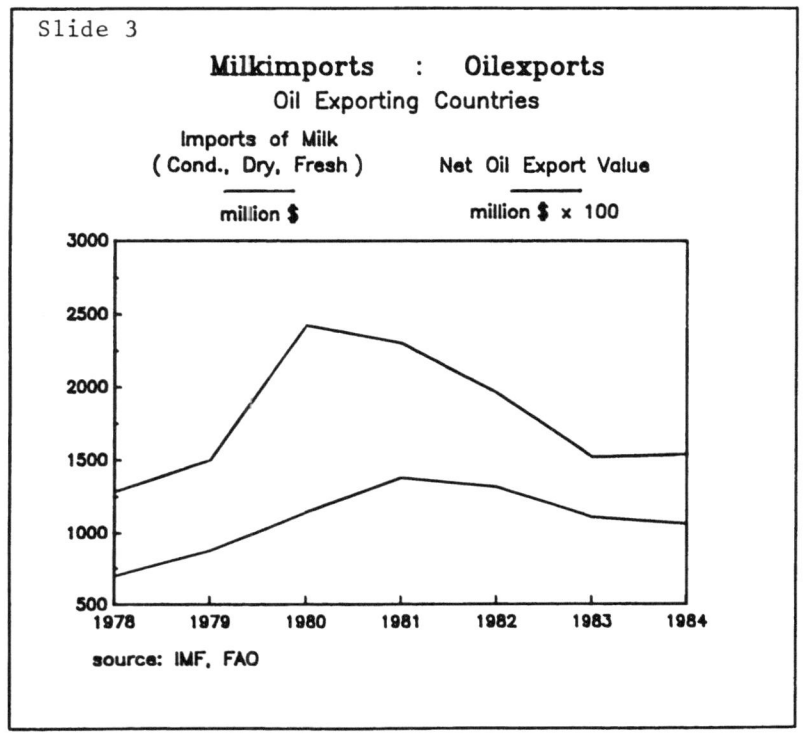

Slide 3
Milkimports : Oilexports
Oil Exporting Countries

Imports of Milk (Cond., Dry, Fresh) — million $

Net Oil Export Value — million $ × 100

source: IMF, FAO

Slide 4

Development of Oil Revenues. (Total Opec)

Year	Mln B/D	$/Barrel	Mln$/Day	%
1980	30	35	1.050	100
1985	15	25	375	36
1986	15	13	195	19
May		11	165	16

Development of US$/ECU Exchange Rate

Date	X-rate
May 7, 1985	1.42
Dec 11, 1985	1.15
May 7, 1986	1.03

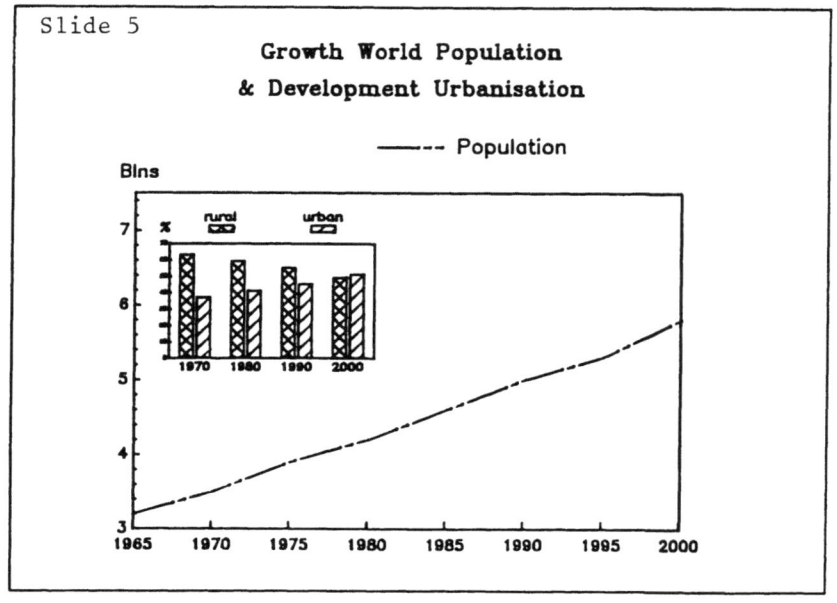

Slide 5

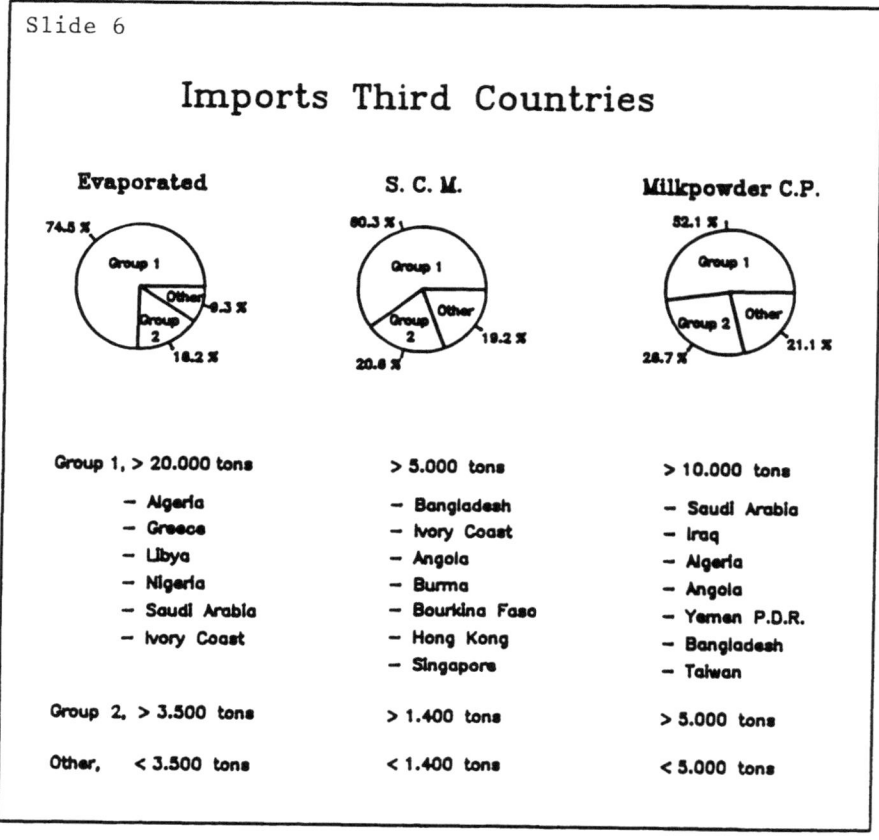

Slide 6

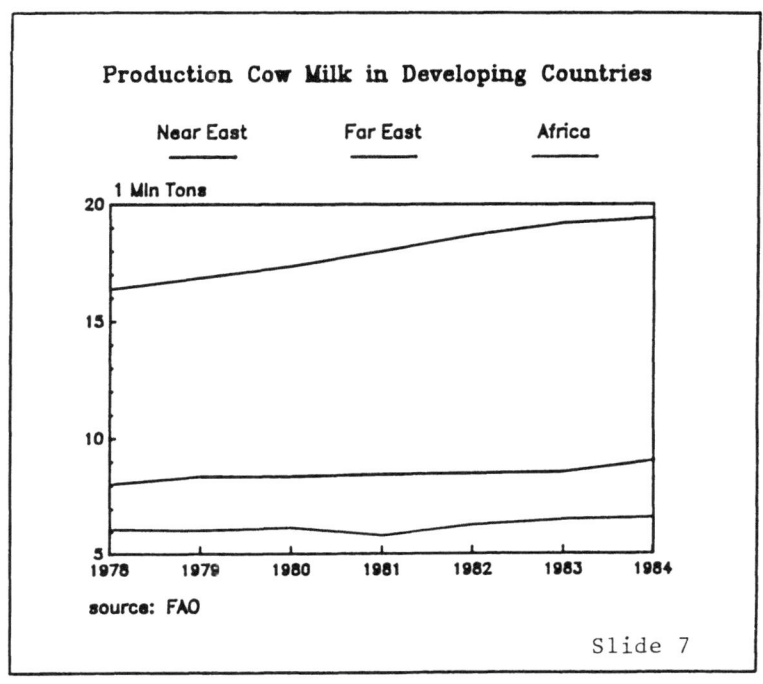

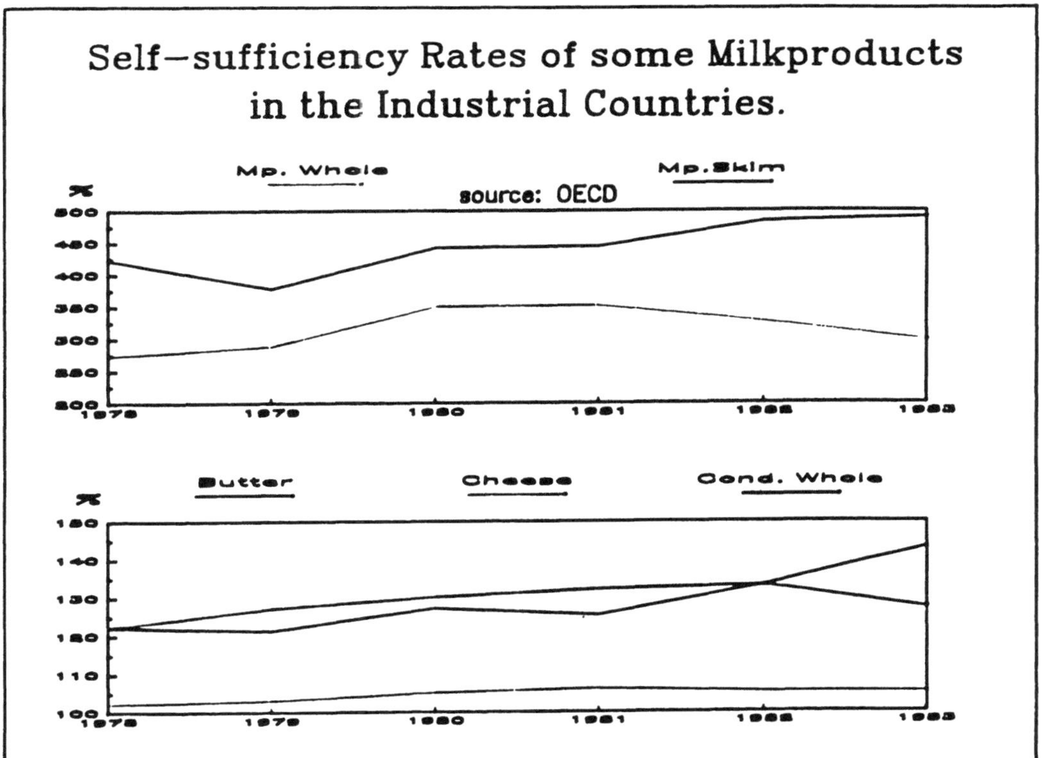

Slide 9

CREATIVE ANSWERS

Slide 10

MANUFACTURING AND TRADING COMPANIES:

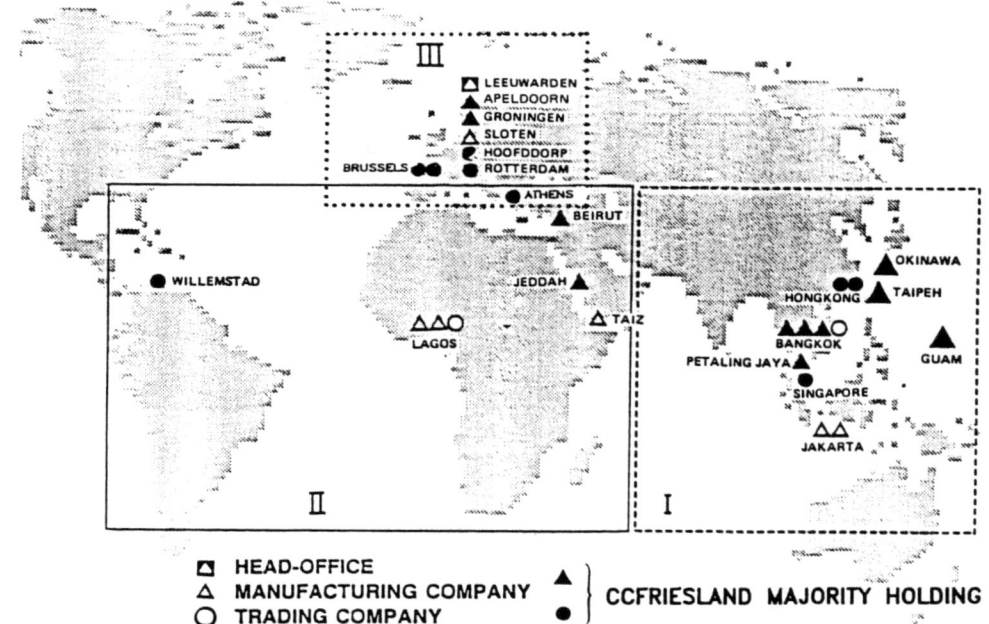

Slide 11

1. Local Production

    &

    Marketing of Milk Products

    — Manufacturing Companies
    — Trading Companies
    — Own Sales Organisation

Slide 12

2. to Assist or Participate in Local Milk / Farm Developing Projects

Slide 13

3. Marketing Approach

* **Segmentation**
    - Demographic Aspects (Age, Income)
    - Lifestyle
    - Light – versus Heavy Users
    - Application

* **Positioning**
    - Corporate Identity
    - Quality (A – Brands)
    - Distribution

Slide 14

4. to Develop Different Products and Packages

Based on :

- Trade Developments   [Chilled Products]

- Improved Transport   [Paper Packs]

- Techniques   [Recombined, UHT]

MAJOR PROBLEMS IN THE MARKETING MANAGEMENT OF MODERN DAIRY ENTERPRISES

Hans Rudolf Felix
TONI-MILCHVERBAND Winterthur
Marketing-Director
Archstrasse 2
8401 Winterthur
Switzerland

Ladies and Gentlemen,

I am very happy indeed to have the possibility of making a contribution to the topic of 'Major Problems in the Marketing Management of Modern Dairy Enterprises'.
   Until not very long ago, marketing was a rather hazy notion within the dairy industry. At the time when the finding 'We have to produce what the market demands' and not 'what is easiest with the aid of the existing equipment' was at last accepted by the Swiss dairy market, this principle had already been practised for a long time by other branches of the consumer goods industry - such as soup, chocolate, oil, soap and detergents.

Now which are the MAJOR PROBLEMS as far·as the Swiss market is concerned, that the dairy industry is confronted with and that have a decisive influence on our actions?

In my opinion these problems are
  - the <u>changes in consumer habits</u> which have effected the total dairy market as well as
  - the constantly <u>growing market position of the trade</u>, in particular of the mighty COOPs and chain stores.

With your permission, I would like to discuss briefly both problems and their aspects. Afterwards, I would like to explain the marketing strategy of TONI Yogurt by means of a case history.

Major problems of the Swiss dairy industry arise from changes in consumer habits, saturated markets, narrow margins, and government controlled retail prices for commodity items such as milk, butter, and cheese.

Therefore, dairies concentrate on specialities in expanding market segments which do not come under government control (yogurt, desserts).

Marketing strategy of TONI yogurt:

- attractive product range
- high quality products and excellent customer service
- responsibility for environmental protection. Unlike all other competitors, TONI sells its yogurt in the brown glass jar, that is in a circulation glass, i. e. no charge of deposit, return on a free basis. With this feature, TONI obtained an exclusive position among all competitors.

The system of returnable packaging applied by TONI proved to be not only economically successful, but also contributes, together with the efforts of the trade and the consumers, to a greater ecological awareness.

1. CHANGES IN CONSUMER HABITS

Today it is quite indispensable for any enterprise within the food industry that its basic strategy takes consumer habits into account.
- Convenience products are gaining importance. Deep frozen products register a marked increase in sales. There is also a wide range of ready-made products, such as yogurt salad sauce, ready-to-serve desserts, etc.
- Increased awareness of nutritional aspects. As a result, preference is given to natural food. This means a boost for health food and low-calorie products.
- Eating out has become more frequent.
- Demand for longer and easier shelf-life with a corresponding boom of UHT-products such as milk, cream, etc.

A look on the market situation as a whole (milk products 1975 - 1985) shows a long-term stagnation in the consumption of milk and dairy products.

Until 1981 a regular increase of the total consumption of milk and dairy products has been noted. In 1982 consumption started to decline, and in 1985 we are once more at the level of 1975. Higher standard of living and the afore-mentioned changes in consumer habits have also resulted in a shift from less expensive to more expensive and processed milk products such as cream, cheese, yogurt, and ice-cream.

Because of this, the general trend differs from one group of products to the other. A comparison of the per capita consumption in the course of the past 10 years shows the following development:
- milk   downward trend, increasing percentage UHT-milk
- <u>butter</u>   stagnation
- <u>cheese</u>   marked growth thanks to large product range
- <u>cream</u>   also pronounced growth owing to coffee cream
- <u>yogurt</u>   considerable growth, very high per capita consumption, growth rate 48 %
- <u>ice-cream</u>   also marked growth, growth rate 31 %

At this point, it should be mentioned that the prices for commodity items such as milk, butter, cheese, are fixed by the Federal Price Control Board while cream, specialities, and ice-cream are not subject to price regulations. For this reason, dairy enterprises are obviously concentrating their marketing activities on the speciality ranges, i. e. on markets with a growth potential that offer a possibility for the profile-making and somewhat better margins.

2. POWERFUL DEMAND POSITION OF THE TRADE

In the past years the structure of the milk product market according to the different trade channels has undergone a basic change. Business is dominated by the 'big ones', i. e. COOPs and the chain stores. While in 1975 the independent dairy outlets still claimed 38 % of the market, its share had declined to a mere 24 % in 1985. The dairies have to make their products available in COOPs and chain stores - if they don't - it means to renounce a priori on a potential of 76 % of the total market. This situation leads necessarily to a stiff competition and price-cutting among suppliers and resulting negative effects on the already small margins, in particular as far as the commodity items such as milk, butter, and cheese are concerned.

Which possibilities do producers have at their disposal in view of the given situation? In my opinion there are basically two possible approaches:
- As supplier close collaboration with the chains on a private label. As we all know, however, this implies that we have to renounce largely on developing our own brands and become therefore more dependent on the trade.
- Profile-making based on specialities which are distributed as branded articles, i. e. outstanding product performance and substantial marketing investments as well as active product innovation.

The TONI Dairy Association has - already years ago - opted for a 'dual-approach strategy' which resulted in a good partnership with chains and COOPs. In some markets, for instance cheese, we produce under private labels - in others - for instance yogurt, we push our own brand. Our marketing strategy for TONI yogurt is rather unusual.

Therefore, I would like to tell you more about it.

TONI YOGURT IN BROWN GLASS JARS

Of the various dairy product markets the yogurt market is the one with the thoughest competition. In 1985 it represents a value of 153 mio US $ and has been showing an upward trend for many years. In 1985 the tonnage was 106'700 tons. For years the TONI Milk Federation has been involved in two brands, namely 'TONI', their own federation brand, and 'CRISTALLINA', the brand of the Central Federation of Swiss Milk Producers. The market is dominated by two large COOPs with their own brands. The rest is divided between milk federation and private dairies. The TONI Federation became deeply involved in that market and - as you can see - could achieve remarkable success. The market share rose from 3.9 % in 1981 to 11.9 % in 1985. TONI's turnover increased from 19 mio units in 1981 to 58 mio units in 1985.

How did this development come about? Please allow me to give you the respective detailed information.

Unlike all other competitors, TONI sells its yogurt in the brown
glass jar, that is in a circulation glass, i. e. no charge of deposit,
return on a free basis. With this feature, TONI obtained an exclusive
position among all competitors. You surely remember the oil shock of
1973. It demonstrated the dependence and the limits of prosperity and
wealth, the euphoria of the sixties and the apparently unshakable
believe in unlimited growth became cracked. Everywhere people became
interested in saving and protection of the environment. We also
started to think about what TONI could contribute to energy saving
and environment protection. One possibility was to replace one-way
packaging by circulation glass and we decided to put our plans into
practice with TONI yogurt.

May I present the product with a slide ?
- Full-cream milk yogurt, 3.5 % fat content
- Natural and 26 flavoured varieties
- Product contains live yogurt cultures
- 100 g = 430 Joules / 103 kcal

Packaging
- light-protected, brown glass jars, 180 g
- cover until 1981 : aluminium foil
  since 1982 : plastic cover for multiple use

Shelf-life
- 26 days

Consumer price
- US $ -.50

We will not be able to get away without looking at least at some
background data.
    The date was January 29th, 1974:
TONI-dairy in Zurich invited to a press conference. A large scale test
was announced in the agglomeration of Zurich to introduce the
deposit-free recycling glass for yogurt, whipping cream, and steril
coffee cream. Duration of the experiment: 6 months. Should 30 % or
more of the glasses be returned, the experiment would be considered
successful. The TONI-dairy promised in that case to purchase a glass
washing machine for the price of about 0.3 mio US $ and thereby
created the necessary conditions for a circulation glass, which could
be reused 30 to 50 times. The participating Journalists were not
particularly optimistic. They reckoned with a returning quota of
5 to 10 %. But they were wrong.

In August 1974, on the occasion of another press conference, we were able to present the proud result: returning quota 40 %. This may be considered sensational. The glass washing machine was purchased, the returning quotas became stabilized at about 35 %.

Problems appeared in the first phase because of insufficient sales. On the one hand, the recycling glass campaign was a big success, on the other hand, yogurt sales in the glass were stagnating; part of the problem was the inadequate marketing support.

The real break-through of the circulation glass concept and with it the commercial success for TONI came only about 8 years after the start of the campaign, in February 1982, thanks to the introduction of the new TONI yogurt line with a new packaging design and the new advertising approach. The three essential objectives were:
  a) Keeping the high quality standard
     (selfexplanatory for branded items)
  b) Consideration of consumer's requests regarding
     packaging design
  c) Continuation and promotion of the concept of
     circulation jars by increased consumer information
     and application of suitable commercial methods

We are pleased to see that our efforts to promote the circulation jars are supported by our major customers. The specially designed rack for returned glasses, distributed to the trade free of charge, has already its fixed place in front of many food stores. The taking back of the empty glasses is handled by the delivery service.

Now a word regarding the communication concept.
It features glass packaging and the new cover. The quality of the TONI product is taken for granted. This product is not advertised directly, but through the vehicle of packaging by using the medias like magazins, the daily press, TV and Radio commercials, billboards, and point-of-sale material.

I will be glad to show you a few examples of the magazine campaign. The rather provocative head-lines are as follows:
  1) Each product has the packaging which it deserves
  2) Mirror, mirror on the wall
     who is in the whole country best of all?
  3) Unmistakably a premier grand cru classé
  4) Why are so many people buying the very yogurt that makes
     so much work?
  5) There are international brands.
     And there are environmental brands.

In addition, every advertisement contains the remark: 'For better quality no packaging is good enough'. At the same time, a 30 seconds TV commercial was aired.

We are particularly pleased about the fact, that partly due to our measures the total yogurt market showed an above average total growth of 5 % in 1982 and continued its upward trend also in 1985.

Today we can say, that it was worthwhile to take that courageous step twelve years ago. The philosophy of the circulation jar has been accepted by a large part of the population. Not only the public but also TONI have been profiting from this fact. We believe that in the long run only oecological solutions can be expected from a company and of course only if there is no disadvantage compared to other solutions.

SUMMARY OF DISCUSSION

IDF should coördinate all efforts and activities regarding the problems discussed. In the future distribution will only take place through the supermarket. The future of the export of dairy products will be of dominant importance. Governmental control by intervention measures, quota fixation etc., is necessary.

Seminar III: Market and Marketing of Dairy Products

Session 2: Product-Oriented

Chairman: L. W. J. Hurd (Canada)
Secretary: Dr. D. E. de Roon (The Netherlands)

Analysis of demand trends in the last decade and challenges to be met to the year 2000 for milk and milk products

Do advertising, sales promotion and nutrition education pay?

LIQUID MILK MARKET TO YEAR 2000

Richard Hall
Liquid Milk Director
Dairy Trade Federation
London NW1 4QP
United Kingdom

One would have thought that the market for a staple commodity like milk, which has helped feed millions of people all over the world for thousands of years, is likely to be fairly stable in a period of relative world peace. In fact, the opposite seems to be the case.

Governments are pulling out of price control. The free market is giving ever greater power to the supermarket chains. Growing health concern is creating huge shifts towards reduced fat milks. Some countries' markets are in decline, while others are expanding rapidly. And there are various conflicting trends in packaging.

To the customer, milk may still generally be thought of as a standard white liquid utility product. But to the industry, almost every aspect of the market is changing and becoming more diverse.

With so much going on, I felt the only way to do justice to this seminar was by conducting a survey of various leading dairying nations to enable an objective assessment of the main developments (Table 1). In consultation with the IDF Liquid Milk Working Group, I sent a questionnaire to 11 countries in the European Community and ten others in the rest of the world at the beginning of June. I have already had over two thirds back and am hoping to publish the full results as soon as they are available. This presentation therefore represents some preliminary findings, but even at this stage there are some important conclusions to be drawn.

The questions were divided into two sections - the first was statistical and the second dealt with issues. If my analysis seems underdeveloped, then that is because many of the answers have thrown up a further set of questions and these have yet to be pursued.

My market definition was intentionally restrictive to assist comparison. It includes all whole, semi-skimmed and skimmed white milk plus vitaminised and flavoured milk, but excludes buttermilk, fermented milks, yogurts and cream.

Total market size

Inevitably, changes in population size have a major impact on overall consumption. In very broad terms, this means that a relatively static European population is seeing only moderate changes in consumption levels, while faster growing populations in other countries surveyed are helping to generate significantly higher consumption in many cases.

The situation in India dwarfs any other worldwide considerations in the survey (Table 2). Their population is forecast to grow by over 50 per cent from 608 million in 1975 to 944 million in the year 2000, with consumption more than trebling from 11 billion tonnes to 36 billion tonnes. That increase alone is greater than the entire European Community market, but even then only takes consumption per person to just over one third of the European Community average.

On the other side of the coin, per capita consumption is falling quite seriously in several countries with a reputation at the top end of the scale (Table 3). Finland, a world leader, is estimating a one third decline from 246 kg per person in 1975 to 164 kg per person in the year 2000. Norway would be down by more than 10 per cent from 170 kg to 152 kg and the United Kingdom is facing a 20 per cent drop from 146 kg to 118 kg. New Zealand has already lost over 20 per cent since 1975 from 144 kg to 112 kg, but is hoping for much greater stability in the next 15 years. Another significant fall is anticipated in the United States, by nearly 20 per cent from 116 kg to 95 kg, though this would be more than offset in total market terms by a population increase in excess of 25 per cent.

Heat treatment

Something I did not expect to find in this survey is that sterilised and UHT milk are almost exclusively a central European phenomenon.

In 1975, virtually 100 per cent of heat treated milk in the countries surveyed outside the European Community was pasteurised (Table 4). By the year 2000, the UHT share is not predicted to be more than 20 per cent in any of these markets - the highest being 20 per cent in India, 15 per cent in Australia and 5 per cent in New Zealand.

Yet, ten years ago in the European Community, UHT milk already accounted for around one third of heat treated sales in Germany and Italy (Table 5). And by the year 2000, the overall EEC average could well be above that. Sterilised milk looks as if it is confined to Belgium and Spain, France, Italy and the United Kingdom on any real scale.

Fat content

Here is where some massive changes are taking place and this appears to be true in the vast majority of countries. Whole milk dominated most markets in 1975, with a 100 per cent share in many of them. By the year 2000, it is expected to hold 50 per cent or less in over one third of those surveyed.

The greatest movement is being experienced by Finland and the United Kingdom, where semi-skimmed and skimmed milk sales have taken off from nothing at all in 1975 and are forecast to reach 70 per cent and 50 per cent respectively by the year 2000 (Table 6). Market share increases of 30 per cent or more are also foreseen in Australia, Canada, the Netherlands, Norway and the United States. The extra residual butterfat will have major implications for dairy policy and capital investment throughout the world.

As between semi-skimmed and skimmed milk, skimmed milk seems to grow early on to capture an eventually steady 5-10 per cent share of the market, mainly for slimmers. Semi-skimmed milk, on the other hand, is more of a straight substitute for whole milk and grows on a long term continuum. One of the most interesting possibilities from this survey is that there is no clear point at which semi-skimmed milk necessarily stops growing.

Packaging

Packaging patterns could hardly be more diverse. Cartons undoubtedly have the edge over other forms of packaging, but their fortunes are rather mixed (Table 7). In Scandinavia and Italy, they accounted for over 90 per cent of sales in 1975 and this is likely to remain the case in the year 2000. Their share is rising rapidly in a wide range of countries, most of all in Ireland, the Netherlands and Spain. But they are confronted by a substantial decline in North America - especially in the United States, where their share is predicted to drop from over two thirds in 1975 to only 25 per cent by the year 2000.

Glass has been traditionally strong in several countries like New Zealand, the United Kingdom, Ireland and Australia, but in each it is declining very fast (Table 8). Plastic is picking up a lot of the slack and is very much an overall beneficiary.

Government price controls

In 1975, most of the countries surveyed had some form of Government price control - the exceptions being Germany, India, Sweden and the United States. By 1985, they had been relaxed in Denmark and the United Kingdom and by the year 2000 they are expected to have been removed as well in Canada, New Zealand and Spain (Table 9). So, there is a distinct move away from Government price control.

Subsidies are on the way out, too. Of the six countries known to have had subsidised milk prices in 1975, only two believe they will still be there in the year 2000 - Norway and Sweden. They have already been removed in Denmark, New Zealand and the United Kingdom and are not reckoned to last in Canada or Ireland.

Milk prices

There is no clear pattern in the relationship between milk prices and inflation from the survey. Inflation rates are also markedly different. In the USSR, milk cost 36 kopecks a litre in 1975 and is forecast to cost exactly the same amount in the year 2000 (Table 10). Milk prices are predicted to rise very little and well below inflation in Germany and the United States. In contrast, the New Zealand price is thought likely to have risen by a factor of 20 times between 1975 and the year 2000.

Strangely, there does not appear to be a close correlation between relative price and consumption levels, though that will need looking at in more detail.

Distribution

Perhaps because of our national preoccupation with doorstep deliveries in the United Kingdom, I sought to find out more about distribution in other countries. Unfortunately, information on this seems particularly sketchy.

Two messages do come through, though. One is that the share of small shops is losing out to the supermarkets, but then that is hardly a revelation.

The second is that home deliveries are unquestionably in decline (Table 11). In Australia, they are expected to have collapsed from 70 per cent in 1975 to a mere 5 per cent by the year 2000. In the United Kingdom, New Zealand and Ireland, falls of 25 per cent or more are anticipated during the same time span. In the Netherlands, they are just about holding on after a severe fall during the 1970s and in the United States they seem likely to disappear altogether. So, for those of us who care about the doorstep service, we have a fight on our hands. Personally, I wonder if armchair shopping by television won't give us an opportunity to improve our home delivery trade again in the longer term future.

....

So that is what is happening, at least what seems to be happening, in the marketplace. If I have misinterpreted any figures, I hope you will let me know. But what of the most important issues facing us in the next 15 years? I identified four in my questionnaire and asked how important people felt they would be - on a scale of very, quite, not much or not at all.

Diet and health debate

The first, maybe predictably, was diet and health. After all, semi-skimmed and skimmed milks are growing almost everywhere. Fat is on the run. But calcium and all the other nutrients in milk are beginning to help the industry recover lost ground. What we do not know and cannot easily tell is how the debate is affecting total sales.

A large number of countries felt that the debate would still be very important over the next 15 years (Chart 1). The rest thought it would be quite important and only one saw it as not very important.

Growth of substitutes

My second issue was the growth of dairy substitutes. In Europe, these are becoming increasingly widespread, particularly in catering uses, but perhaps more seriously for creams and spreads than for liquid milk, where their volume is extremely small. In France and Germany, there are still severe restrictions on substitutes, but these are being challenged in the European Court and the European Commission is looking earnestly at means of regulating their labelling, presentation and advertising.

As to the survey, substitutes drew a non-committal answer, with the greatest concentration viewing them as of not much importance (Chart 2). But almost as many said they would be quite or very important.

Opportunities for premium value

Opportunities for premium value were my third area of enquiry. I wanted to discover to what extent the various countries saw a chance for milk to overcome its virtual imprisonment as a staple white liquid commodity.

Evidently, most people thought it was worth trying, but with limited expectations (Chart 3). Milk is such a basic food and there is extensive competition, the replies explained. There was some hope of more branding, snack packs and so on, but not for very much of the market, it seems.

Maintenance of wide retail distribution

Finally, I asked about retail distribution and the importance of maintaining our home deliveries and small shop sales, which are so clearly under pressure.

Here, there was a clear majority who felt this was very important, more even than in response to the diet and health question (Chart 4). But opinion was sharply divided and some suggested it was not really very important at all.

Conclusion

So, it looks as if all four of these issues will be major challenges for most of us over the next 15 years and we shall have to see how far we can improve on these forecasts of sales. My parting thought is that if anyone had to devise an ideal food product for the 1990s, it would almost certainly be:

* fresh and natural

* high in vitamins, minerals, calcium and protein

* low in fat

* free of additives, preservatives, flavouring and colouring

* versatile, convenient and good value for money

It would probably also be liquid and white. And milk has it all. If we cannot sell that, then we have only ourselves to blame.

RCSH 25.9.86

TABLE 1

COUNTRIES SURVEYED

EUROPEAN COMMUNITY	REST OF WORLD
BELGIUM	*AUSTRALIA
DENMARK	*CANADA
FRANCE	*FINLAND
*GERMANY	*INDIA
GREECE	JAPAN
*IRELAND	*NEW ZEALAND
*ITALY	*NORWAY
*NETHERLANDS	*SWEDEN
PORTUGAL	*USA
*SPAIN	*USSR
*UNITED KINGDOM	

* REPLIES RECEIVED

TABLE 2

TOTAL MARKET SIZE

1. HUGE GROWTH IN INDIA

	1975	2000	% CHANGE
POPULATION IN MILLIONS	608	944	+ 55
CONSUMPTION IN KG PER PERSON	18	36	+ 100
CONSUMPTION IN BILLION TONNES	11	36	+ 230

NOTE: EEC DAIRY SALES OF LIQUID MILK IN 1984 WERE 20 BILLION TONNES WITH AVERAGE OF 95 KG PER PERSON.

TABLE 3

TOTAL MARKET SIZE

2. DECLINES IN HIGH CONSUMING COUNTRIES

KG PER PERSON	1975	2000	% CHANGE
FINLAND	246	164	- 33
NEW ZEALAND	144	110	- 24
NORWAY	170	152	- 11
UNITED KINGDOM	146	118	- 19
USA	116	95	- 18

TABLE 4

HEAT TREATMENT

1. PASTEURISATION DOMINANT OUTSIDE EUROPEAN COMMUNITY

	1975 % PASTEURISED	2000 % UHT
AUSTRALIA	100	15
CANADA	100	2
FINLAND	100	3
INDIA	100	20
NEW ZEALAND	100	5
NORWAY	100	0
SWEDEN	100	0
USA	100	0
USSR	99*	0*

* 1985

TABLE 5

HEAT TREATMENT

2. UHT AND STERILISED STRONGER IN EUROPEAN COMMUNITY

HIGH UHT COUNTRIES

%	1975
GERMANY	31
ITALY	38

HIGH STERILISED COUNTRIES

%	1985
BELGIUM	63*
FRANCE	12*
ITALY	3
SPAIN	31
UNITED KINGDOM	5

* OTHER SOURCES

TABLE 6

FAT CONTENT

MASSIVE RISES IN SEMI-SKIMMED AND SKIMMED MILKS

% SEMI-SKIMMED AND SKIMMED	1975	2000
AUSTRALIA	0	30
CANADA	52	87
FINLAND	0	70
NETHERLANDS	33	64
NORWAY	12	49
UNITED KINGDOM	0	50
USA	30	60

TABLE 7

PACKAGING

1. VARIED FORTUNES FOR CARTONS

%	1975	2000
STAYING HIGH		
FINLAND	93	90
ITALY	90	97*
NORWAY	98	97
SWEDEN	100	100
GROWING		
AUSTRALIA	30	50
IRELAND	15	75
NETHERLANDS	48	85
NEW ZEALAND	0	30
SPAIN	2	45
UNITED KINGDOM	6	30
FALLING		
CANADA	54	38
USA	67	25

* 1985

TABLE 8

PACKAGING

2. GENERAL DECLINE OF GLASS

%	1975	2000
AUSTRALIA	65	0
IRELAND	80	10
NETHERLANDS	44	10
NEW ZEALAND	99	50
SPAIN	37	0
UNITED KINGDOM	93	55

3. GENERAL GROWTH OF PLASTIC

%	1975	2000
AUSTRALIA	0	30
CANADA	43	60
NEW ZEALAND	0	20
SPAIN	7	35
USA	31	75

TABLE 9

PRICE CONTROLS

1. GRADUAL REMOVAL

	1975	1985	2000
CANADA	✓	✓	X
DENMARK*	✓	X	X
NEW ZEALAND	✓	✓	X
SPAIN	✓	✓	X
UNITED KINGDOM	✓	X	X

2. SUBSIDY REMOVAL

	1975	1985	2000
CANADA	X	✓	X
DENMARK*	✓	X	X
IRELAND	✓	✓	X
NEW ZEALAND	✓	X	X
UNITED KINGDOM	✓	X	X

* OTHER SOURCE

TABLE 10

MILK PRICES

1975 = 100	MILK 2000	INFLATION 2000
STATIC		
USSR	100	100
LOW RISES		
GERMANY	120	200
USA	175	250
HIGH RISES		
NEW ZEALAND	2000	1300

TABLE 11

DISTRIBUTION

HOME DELIVERIES IN DECLINE

% OF HOUSEHOLD MARKET	1975	2000
AUSTRALIA	70	5
IRELAND	75	50
NETHERLANDS	45	18
NEW ZEALAND	76	50
UNITED KINGDOM	88	55
USA	7	0

CHART 1

ISSUE 1 - DIET AND HEALTH DEBATE

CHART 2

ISSUE 2 - GROWTH OF SUBSTITUTES

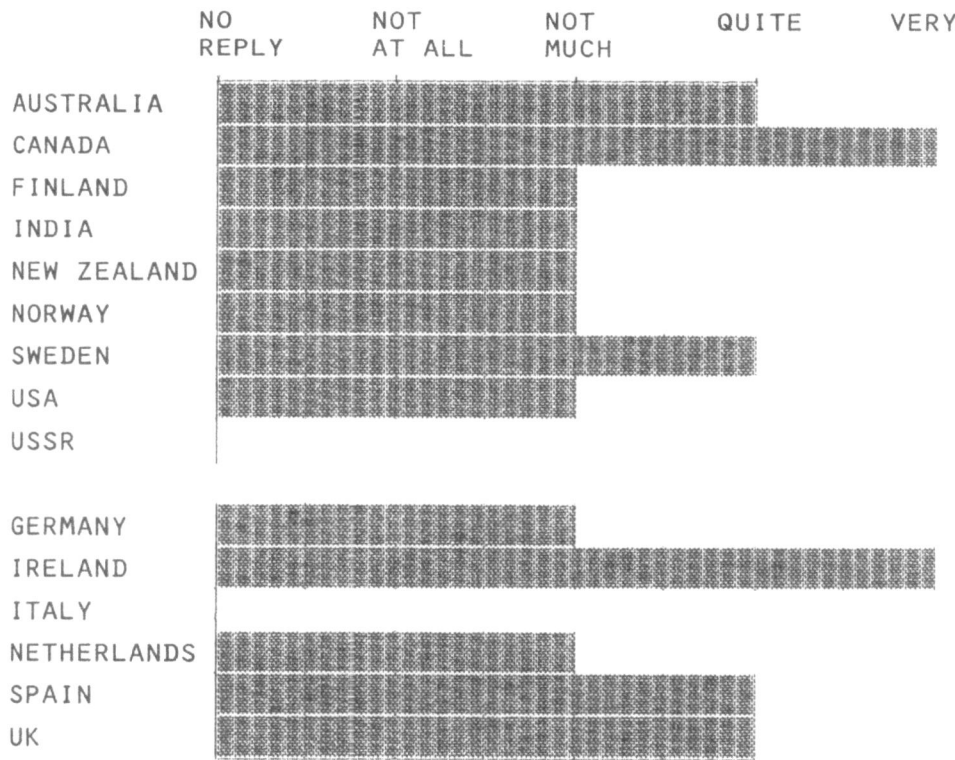

CHART 3

ISSUE 3 - OPPORTUNITIES FOR PREMIUM VALUE

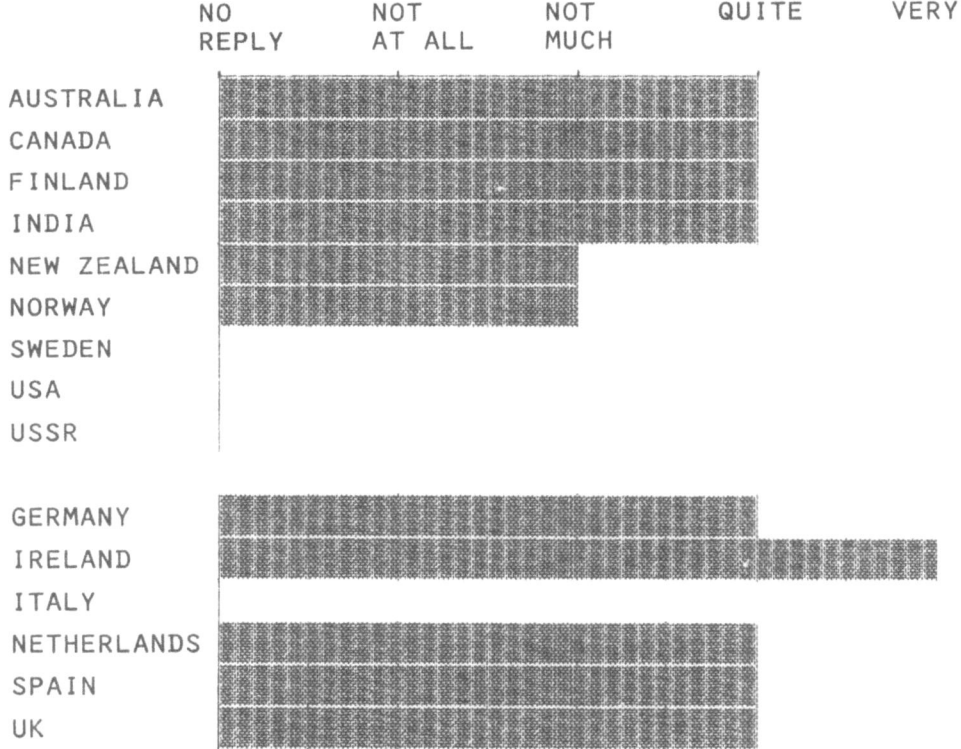

CHART 4

ISSUE 4 - MAINTENANCE OF WIDE RETAIL DISTRIBUTION

THE BUTTER MARKET

Brian A. Joyce
Managing Director
An Bord Bainne/The Irish Dairy Board
Grattan House
Mount St. Lr.,
Dublin 2, Ireland

INTRODUCTION

In many respects I feel I have been handed the "poisoned chalice" in being asked to speak about butter at these sessions of the IDF. I believe that it is entirely reasonable for me to state the accumulation of all planning and management errors in terms of balancing supply and demand in dairy support systems around the world manifest themselves in the international butter market. As a product of last resort, it is utilised to preserve the surpluses of milk over and above the prevailing level of demand.

If there is anyone at the conference who is unaware of the considerable difficulties that continue to dominate the thinking of dairy producing nations in terms of market prospects, they must have been having a long vacation on another planet.

Even a one-handed economist (most economists have usually two - (on the one hand or on the other!) - would tell you that it is possible to balance supply and demand by allowing the pricing mechanism to play its part. The support system in the European Community and indeed support systems elsewhere, was introduced in the first instance to help introduce an element of stability into producer incomes. The interaction of supply and demand was deemed to be so devastating to individual producers as to take them well below an acceptable standard of living. It remains so today. We therefore must find ways of balancing supply and demand and at the same time, preserve producer incomes at acceptable levels. We have an urgent need to find these mechanisms within the European Community. It must be pointed out that we are by no means alone in the predicament of dairy imbalance and it would be grossly unfair to expect the EC to carry the whole burden of trying to restore international balance. In making a contribution to the debate as to how we can improve the management of the dairy support systems in the Community, I do so on the assumption that other producing and exporting regions of the world will play their part in restoring order where none exists.

Before outlining a possible course of action for the European Community to restore equilibrium in the dairy sector, let me put on record that I do not believe that feeding whole milk that has been processed first into butter and powder, to animals, represents a long term solution to the dairy surplus situation. Finding itself in the dilemma that it is, the European Community has little option but to try to do precisely this as a means of ridding itself of past accumulations of stock. There would be little point in continuing to produce these surpluses for recycling in this manner at 100% cost or more of the support prices. Such an extraordinary measure which has now been embarked upon by the EC needs to be coupled with measures that will effectively ensure that a repetition will not be necessary. Continued losses of this magnitude can only ensure lower income for producers in the short to medium term.

Selling surpluses in the international market at 10% or 15% of their original cost does not represent a solution to the more fundamental problems that the EC faces in the dairy sector. This expedient can only be looked upon as a short term measure as it will undermine producers' incomes and stability just as surely as recycling product to animal feed.

Short term expedient measures to make a sale to intervention less attractive will NOT make a contribution to the solution to the problem. The one handed economist will readily agree that if the support threshold is lowered so also will the market and the relative attractiveness of both options (selling to intervention or selling in the market) will remain unchanged.

Hoping that the problem will go away if it is ignored does not represent a solution. The accumulation of surpluses has done much to undermine the CAP and a continuation of a system that continues to receive such unwelcome bad publicity can only be detrimental to the interests of producers everywhere.

SEARCH FOR SOLUTIONS

We must therefore look to achieving market balance by either increasing consumption or reducing production or a combination of these factors. The problem we face today will not go away of its own accord; it will not be solved by tinkering and tampering with the support mechanisms and it is very much in the interests of dairy producers everywhere that the Europeans find solutions of a type that will provide stability in terms of market outlets and stability in terms of producer incomes as a consequence.

If we are to take a medium to long term strategic view of dairy policy, we should be seeking to incorporate certain basic elements into any changes that are needed. It is vital for the long term survival of a Common Agricultural Policy that is favourable to producers that any solution to present problems in the dairy sector should:
(a) provide a plan for the disposal of all existing surpluses in outlets other than the normal commercial or food aid outlets;
(b) it must increase consumption or reduce production to a level where there will be no significant recurrence of the stockpile problems that we are currently experiencing;

(c) it must set the scene for the future in the dairy sector in such
    a way that the cost of dairying within the CAP and within the
    total European Community will be seen to be reasonable in the
    member states and that the dairy industry will cease to be the
    political football of national and Euro-politicians.

These criteria could be difficult enough to satisfy and careful consideration will be required to ensure that we do not take further measures that are yet again perceived to be too little and too late.

Let us therefore examine the possible parameters of any future measures that might satisfy the criteria outlined. In summary we must examine:

(1) Consumption and the near term prospects for improving consumer offtake.
(2) The impact of the Voluntary Cessation Scheme already implemented by the Community.
(3) The necessity or otherwise of further measures that will bring about supply/demand equilibrium.

CONSUMPTION

Many people would agree that the dairy industry has done far too little to promote their product and to counter absurd claims that have been made against some of their products. Such a point of view has more than a little substance to it. Increased promotional expenditure will make a contribution towards increasing demand and/or arresting a decline. There are however certain long-term international trends in food and drink sectors which we cannot afford to ignore. In the food sector, there is a growing recognition of the need for a healthier lifestyle. Indiscriminate eating and drinking combined with sedentary and low energy output occupations make for unhealthy people. Within this general trend, there is a tendency towards a reduction in the total intake of fat, and dairy fat is fighting an uphill battle to maintain consumption of butter, without giving much thought for increasing consumption. If we are realistic, it is probable that butterfat consumption within the European Community will fall in the five to ten year timescale. In certain member states, including my own, new products in the spread category have succeeded in blurring the distinct difference that existed between butter and margarine for many decades and have started to win market share on an alarming basis from butter. Realism would dictate that there is little elasticity of demand in this sector in terms of the price/volume ratio. There is a need for continuing vigilance on the part of the butter industry to reinforce the consumer demand for butter by effective advertising and a strong programme of information to reinforce the perception of butter as a pure and natural product against imitations that contain colouring, artificial preservatives and stabilisers. Consumption of butter as part of a balanced diet and normal lifestyle has been a long tradition in Ireland and other European countries with no ill effects on health. Can the same be said of competitive imitation products?

The cheese market in the European Community has been increasing steadily for many years. The trend is expected to continue and offers the prospect of utilising additional quantities of milk in the medium to long term outlook. Cheese has a much more positive image in the context of the more healthy lifestyle that is perceived to be necessary in the 1980s. There are much more learned people in my audience on the subject of cheese and I will confine my remarks to saying that the growth in cheese consumption will make a SMALL contribution to the problems of surplus, but we would be exceptionally foolish if we engaged in the delusion that the cheese industry will provide us with a solution in an acceptable timeframe to our present problems.

Liquid consumption of milk accounts for 17% of total European Community milk production. There are strongly identifiable trends within this sector. Growth has been evident in the skimmed or semi-skimmed categories at the expense of full cream milk though this trend has been halted in the countries where it has gone furthest. Within the total beverage industry there are identifiable trends away from full strength alcoholic drinks towards lighter drinks. The first segment of this trend is a move towards consumption of more wines at the expense of whiskies and other hard liquor. This trend is now at its next stage in the United States where wine consumption is now suffering and lighter drinks known as "coolers" or west coast coolers, have become extremely fashionable. There are also identifiable trends in the soft drink sector. There is a perceptible movement away from dark coloured drinks to lighter coloured drinks and in general, there is growth in consumption being experienced in low sugar or no sugar drinks. The growth in consumption of pure orange juice across the European Community represents a good example of this overall trend.

The purity of milk and its nutritional values are under-exploited in the context of these trends and considerably more effort needs to be made both in terms of promotional support and in terms of product innovation. I would be optimistic that if the right approach is taken that profitable increases in the European Community consumption can be achieved in the next five to ten years. It must however be said, as in the case of cheese, that such increases in consumption will not constitute a solution to the chronic surpluses that we are currently experiencing in the European Community.

The dairy importing nations of the world have traditionally provided an important outlet for European Community produce and it is important to comment in very general terms on the consumption prospects in these countries. The oil-producing countries of the world have been an extremely important element in the growth up to 1980 or 1981 of total exports from the European Community. This trend has sharply reversed and many of these countries are now facing severe economic and financial difficulties with a consequential decline in overall demand for dairy products. This purchasing power decline in oil-producing countries has coincided with a decline in prices of products competing with milkfat such as palm oil and coconut oil, to all time lows. The short term prospect for palm oil is a continuation of the expansion of production which will leave prices at ludicrously low levels in the short to medium term. You would need to be a super optimist to expect the problems of dairy surplus to be solved in the international arena at any kind of acceptable price level.

In summary, there are things we can do to assist the level of consumption in all product areas within the Community. The continuation of them will not constitute a solution to our problems and we must look elsewhere for a solution.

The Council of Ministers of the EEC agreed on May 6, 1986 to put a Voluntary Definitive Discontinuation Scheme into action in the member states. The scheme provides for:
- 2% reduction in the milk year 1987/88
- 1% reduction in the 1988/89 milk marketing year.

The reduction will become compulsory in the event of the voluntary scheme not providing the necessary response. Assuming that the 3% comes off the current level of production and assuming that the entire reduction were to come from butter and skim milk powder production, it would reduce production by:
- 140,000 tonnes of butter
- 280,000 tonnes of skim milk powder

The quantifiable surplus of the Community at present (over and above domestic consumption, export demand and food aid programmes) is of the order of:
- 400,000 tonnes of butter
- 800,000 tonnes of skim milk powder

Assuming that 1½% on average over next year and the following year comes out of the Community production through the voluntary scheme, we could look forward to a stockpile of:
- 1.3 million tonnes of butter
- 1.1 million tonnes of skim milk powder

by March 1989 which would leave us approximately where we are today. The other assumptions used in arriving at this level of stock are:
(i) annual firesale to the Soviet Union of 150,000 tonnes;
(ii) annual usage in animal feed programmes of 150,000 tonnes of butter and 400,000 tonnes of skim milk powder;
(iii) international demand remains at 1986 levels.

The cessation programme will make a significant contribution to the surplus problem but it does not, of itself, constitute a solution within the criteria set out earlier in this paper.

ADDITIONAL MEASURES REQUIRED

The Community faces the grim prospect of taking additional measures to restore balance between supply and demand in the milk sector. The introduction of the super levy in 1984 was hailed as the solution to the problems of surplus. It is time to say that the introduction of the super levy staved off an even bigger disaster for Community dairy farmers. Regrettably, it was characterised by the same traits of so many instruments of Community policy. It was too little and too late. It was a compromise hammered out that looked at the future through rose-tinted glasses. Since the introduction of the super levy, stocks have grown as follows:-

	Butter '000 tonnes	Skim Milk Powder '000 tonnes
April 1, 1984	907	881
July 31, 1986	1,358	988
September 11, 1986	1,376	1,068

In pointing to the failure of the super levy introduction to arrest the growth of surplus, I am not denigrating the very considerable political achievements of obtaining Community-wide agreement of putting a ceiling on production. The introduction this year of a definitive discontinuance scheme is a recognition by the Commission and the Council that the super levy quota was struck at too high a level.

It is necessary to quantify the "structural surplus" relative to market requirements. Our own studies would suggest that this structural surplus relative to 1986/87 before the definitive discontinuance scheme, is of the order of 8% of Community production. Translated into product, this amounts to 400,000 tonnes of butter and 800,000 tonnes of skim milk powder. The definitive discontinuance scheme, if the 3% ultimately comes off the European Community quota, would leave a residual surplus of about 5% of total production. There will be little relief in the market place - whether within or outside the Community market - until a plan of action to eliminate this structural surplus is in place. This is a formidable task for the Commission and a very unpalatable prospect for the Council of Ministers who will have to carry home such decisions to their constituents. The irony is that the longer such decisions are delayed, the more instability we can expect in the market place. This instability will continue to undermine producers' incomes.

Having only introduced a definitive discontinuance scheme in 1986, there will be understandable reluctance to engage in further measures for the time being. The fact remains however that the Commission must find a palatable way of restoring equilibrium. It is not for me to say how this will ultimately be achieved. The range of options include:
- straight quota cuts
- temporary or medium term buyout programme
- permanent buyout programme
- tightening of existing super levy rules

or a combination of these measures. These matters are a matter for the Commission to propose and the Council to decide. There is no escaping that measures to reduce production by 8% relative to 1985/86 will have to be taken. It would be infinitely preferable that real measures to restore supply and demand equilibrium were taken sooner rather than later. Producer incomes are being eroded at an alarming rate by measures being taken by the Commission which are making no contribution to a longterm solution to the surplus problem.

The dairy support programme was designed to provide an acceptable level of income for dairy farmers in the Community. It was never intended to farm millions of acres on the US and other continents through intensive feeding of imported cereal substitutes. Any solution should recognise this. A buyout programme, for example, should have a strong and identifiable bias to encourage farmers who have an abnormally high density of cows per hectare to leave dairy farming.

The accumulated stockpile acts as a deadweight on the state of the Community and international markets. It needs to be removed from contention. Recycling or the prospect of recycling these surpluses will continue to act as a drag on the market and this fact is widely recognised. Usurping commercial outlets simply to rotate intervention stocks has proved a disastrous policy over the last two years. I would therefore suggest that all accumulated stocks are reclassified and no longer available for export or for the subsidised schemes within the Community for human consumption. We must recognise that the accumulation of these stocks was a major error which should not be perpetuated by allowing them to destroy any chance of market recovery. It gives me no pleasure to say that all these stocks require to be recycled into the animal food sector. In view of the size of the stockpile, all the stock will be far too aged for human consumption on any stock rotation programme. It is better to recognise this now than to continue to make piecemeal decisions that fail to solve the problem.

I would therefore suggest that the financial dilemma facing the Commission in relation to these stocks is solved in the following manner:-

(1) The member states should give their consent to the raising at commercial interest rates about 6 billion ECUs on the Eurobond market.
(2) The Commission should then pay the member states for these stocks and formally declare that they will not be put into recirculation except into animal food programmes.
(3) The bonds which would be interest-bearing at commercial rates could have redemption dates of say 1989/91, at a time when the dairy budget would be in a position to fund such redemptions.
(4) The savings arising from reduced production as suggested would be sufficient in themselves to redeem the bonds.
(5) In taking the present stockpile out of circulation for human consumption, the Commission would be providing sufficient room in the commercial market for fresh production to be absorbed. It would be a reasonable expectation that new stockpiles would not accumulate and that dairy expenditure on a year-to-year basis would become much more modest.
(6) Market firmness would be a consequence of the stock declaration and the reduced production. This would apply to both Community and non-Community markets. It would be reasonable for the Commission to expect that substantial savings would arise in the funding of export refunds and related measures, as a consequence of the new confidence in the market place.
(7) Producers could have a reasonable expectation of maintaining and indeed enhancing their income in contrast to the present position where their incomes are being eroded by piecemeal measures that are related to short term budget priorities.

It all seems like a good idea! But will it happen? If one were to study recent history of the decision-making process in the European Community, it seems more probable that we will have another year or two of debilitating measures before resolute action is taken. This would be a tragedy. There are many people in the audience who are influential in the making of European Community policy and I would urge that they would lend their support to a composite and comprehensive solution to a problem that has been with us for far too long. We will have failed our producers if we do not succeed in this objective.

THE CHEESE MARKET – demand trends and future challenges

Laurits Raun
Danish Dairy Board
22, Frederiks Allé
DK-8000 Aarhus C
Denmark

ABSTRACT. The international trade in cheese is probably the least depressing when discussing trade in dairy products. While a development in the trade in most dairy products has shown a steady decrease over the last years and the outlook at present seems rather grim, the cheese market proven an overall stability even though there are great variations between product types in the world trade for cheese. There has been a shift towards the more value-added and speciality products. EEC is the main cheese exporter in the world, and the cheese market is the only one where the EEC up till now has succesfully defended the position and mainly due to the great variety of the product range. Denmark could, taken as an example, very well illustrate this development. There a number of elements lead to a very high specialisation in Balkan and Middle East types of cheese, and concurrently with that a parallel specialisation in different fancy-type cheeses as Camembert, blue and white mould cheeses and cream cheese. Even if there will always be a basic demand for traditional bulk cheese, the future growth of the cheese sector must stem from the further development of specialities or maybe from some new "cheeselike" products for entirely new markets not yet discovered by the industry.

The cheese market and the product "cheese" differ in a number of ways from all other dairy markets or dairy products, and when assessing that market one could easily be misled into believing that it has a life entirely of its own quite separated from developments for other dairy markets and products.
This is of course not the truth - at least not the full truth.
It would be trivial for a forum like this to lecture on the interrelationship between different dairy products,

fat, protein etc., but on the other hand hardly anybody
would deny that "cheese" is synonymous with a much wider
range of products and product concept than any of the
other main dairy products: liquid milk, butter or
preserved milk. Even though the latter with all the specially adapted varieties for further use in modern industry may prove somewhat of an exception.
Liquid milk and butter have been adequately treated by the
previous speakers; Mr. Graham following me will deal with
the preserved milk products and I will abstain from giving
more than an ultrashort short run-down on developments in
the overall milk production and dairy market situation.
We all know the main scenario by hart: milk production and
dairy product stocks have - except for a short interlude
1981-82 - been steadily increasing over the last decade,
demand has generally speaking been stagnating and has fallen behind in meeting supply.
The cheese market - or the world trade in cheese - has
shown a much better evolution than that of other main products, at least if you do not take into account some important dispose-off sales.
It is well known that the agricultural sector in most
parts of the world benefits from a number of different
subsidies, but in the cheese market these are of much less
importance. The subsidy is generally speaking of a
percentage-wise much less significance than for skimmilk
powder, wholemilk powder, butter or butter oil, and for
many kinds of cheese they are non-existant.
According to FAO, the world cheese production has
developed from some 10 mio. tonnes in 1976 to approx. 13
mio. tonnes in 1985 or an increase of 27% (fig. 1). Most
cheese is produced as well as consumed in the same area,
and international trade in cheese is of limited significance for the sector, but nevertheless the world cheese
market has over the last decade shown a development as indicated on the graph shown in fig. 2, which is based on
the 11 most important cheese exporting countries constituting approx. 95% of the total world trade in cheese.
Trade has grown from 495,000 tonnes in 1970 to 773,000
tonnes after an all-time peak in 1984 of 817,000 tonnes or
60% over this period.
The EEC is treated as one country so consequently intra-EEC trade does not count.
It is clearly demonstrated that trade in cheese as such is
of increasing importance. Not only has trade increased by
the mentioned 60%, but also the world trade/production
ratio has developed from 4.9% in 1976 to 6.1% in 1985.
According to the GATT-secretariat statistics issued in
connection with the International Dairy Arrangement the
price development for cheese shows an evolution which
falls well within the general picture also for skimmilk

powder, wholemilk powder, butter and butter oil (fig. 3-6) even though one can easily notice that the price span for cheese in the low-price periods tends to be somewhat broader.

But one should again not be misled. The GATT statistics are only for socalled "certain cheeses" which are in fact the bulk product Cheddar which is a raw material of an almost industrial nature and to a large extent destined for further production into processed cheeses, pizzas, chips and other products. We shall later see that this is not necessarily representative of the whole trade.

Commercially interesting markets for cheese will mainly be found in the groups of high or medium income countries, i.e.: in the industrialised world and/or oil producing countries which, apart from Europe and North America, mainly includes some countries in Latin and South America, the Middle East, Japan and Australia.

Furthermore some demand originates from the higher social groups in all developing countries where the broader population generally speaking has no tradition of consuming cheese and - what unfortunately is most often the case - does not possess the means to demand cheese.

This variety of demand sources indicates the multiplicity of the cheese market, and that the cheese producer and -trader must follow a policy of diversification and specific adaptation to a multitude of different niches if he wants to flourish.

This is exactly what has been the case over the last decade as at no time before. Maybe forcefully as a result of the depressed general economic environment and increasing surplus milk production and thereby an intensified competition, but the evolution has been distinct.

Not two markets are quite alike. This is a simple lesson which must be learned if the export efforts shall have any chance of success.

I will thus risk arguing that the cheese market of today is not a homogenous market, but a great number of smaller and highly individual markets both as to geography, trade politics and products.

To make progress in the analysis we must realise this, and I believe that we can roughly split the world cheese market into four main categories which each has shown their individual pattern of development, namely:
1. traditional Western (bulk) cheese such as Cheddar or Edam/Gouda.
2. other types of traditional Western cheese such as Emmental, Tilsit, Danbo or Italian types.
3. Western fancy-type cheese as blue, Camembert, blue/white mould, cream cheese etc.

4. Balkan and Middle East types of cheese, the socalled white cheeses such as Feta, Halloumi, Akawi etc.

It is always dangerous to generalise, but as it is not possible on a world wide scale to separate even these four main categories of cheese from the available statistics, and Denmark is one of the major cheese trading nations, I will risk a basis of our own statistics and experience to reflect on the lessons to be learned from the past development about the future which we should prepare ourselves to meet.

In 1976 the Danish dairy industry exported 115,000 tonnes of cheese of which 53,000 tonnes or 46% went to markets outside the European Economic Community. In 1985 - which was a particularly poor year for our cheese sector - Danish cheese exports had grown to 199,000 tonnes of which 130,000 tonnes or 65% went to non-EEC countries (fig. 7 and 8).

The significant increase in Danish exports to third country markets is due not only to the working up of new markets geographically, but also to the development of new production techniques and the introduction of new products.

It is a well-known fact - which is also clearly demonstrated here - that the development must be attributed mainly to the Danish exports to the Middle East area of "white cheese" of which Feta will be the most well-known and also the most important type.

This development has to a great extent been made possible as Denmark, being a member of the EEC, could profit from the Common Agricultural Policy which has the instruments to secure a stable trade policy.

But to make the Danish Feta exports not only a trade, but an adventure, has demanded much more than EEC support.

In the beginning of the 70's the oil crisis hit the Western countries, but to the oil exporting countries this was the beginning of an era of new economic strength which made possible the realisation of a heavy demand for consumer goods, and consequently also for food products.

In the Middle East Feta cheese has traditionally performed an important part of the ordinary diet, and a profound belief by Danish dairymen that it was possible from cows' milk to produce a feta cheese which could meet the latent demands from the market and be compared with the original produced from sheep's milk led to an extensive product development and the finding of new techniques now when that particular market possessed the necessary purchasing power.

This again led to extended and intensive market research, a better understanding of market demands in other countries and a willingness to adapt the products to fulfil the most individual market requirements and preferences.

As a result of all theese elements the Danish Feta exports do today not concern just one product, but Feta in a number of variations and several other white cheeses each adapted to its specific destination.
If for a moment we turn our attention to the fancy-type cheese we can note a positive development much less spectacular, but very steady and very important (fig. 8)
Over the last decade Denmark has just about doubled the production of these cheeses which include a very wide range of cheeses such as Danish Blue cheese, Mycella, Camembert, Brie, blue and white mould cream cheese etc. (fig. 9 and 10)
All of them very high value added products demanding craftmanship and care in the production.
In the same period Danish exports of this group of cheese also almost doubled. If we take 1976 as 100 it was 193 in 1985.
The demand for the white cheese originates almost entirely from one area: the Middle East.
But for the fancy cheeses the demand stems from all over the world, mainly from the "Western" industrialised countries, but increasingly from other markets as well.
Many things separate these two groups of cheese. The one forms an important part of the daily diet of the man in the street. The other is a luxury product for gourmets.
The production techniques and trade customs are all different. But between them they account for almost all the increases and growth of the Danish cheese industry over the past decade.
In my statistics I have taken the two groups of traditional Western types together as the bulk cheese in my country has never played a very important role, but in the same period the traditional Western-type cheese as a group has only developed very little. Even if this may not be quite the same situation for some other countries I venture to say that the demand for at least traditional bulk cheese is largely satisfied, and the other traditional Western cheeses show only a marginal overall growth. Also here demand is satisfied, and only specialities developed within this market show growth and then mainly at the expense of other cheeses within the same market segment.
At this particular moment all dairy markets have recession as a common denominator. The cheese market is in general a somewhat less traumatic experience due to the multinature of the product and the high degree of added value.
But even with an increased economic activity in the traditional cheese consuming market areas it is most unlikely that we will witness another boom as for the Feta adventure from 1974 to 1985 unless some countries with enormous populations and market potential demands, but at present

only with low purchasing power, get part in an economic recovery as for example Egypt, Sudan, Nigeria and Mexico. One point I have so far not touched upon here. But it should also be mentioned that a protective trade policy in many parts of the world creates one of the most important barriers for further growth of the cheese trade. Not only by putting up the barrier as such, but also because experience has taught us that a wider range of cheese offered to the consumer leads to an increase in consumption and demand.

The world has been thoroughly scanned for market possibilities for cheese, and as under the present conditions it is not possible to indicate any new outlets for an increased production this leads to an intensified competition on the existing markets.

The future challenge for the cheese industry in our individual countries will be to establish a niche or platform where it is supplier per excellence.

We have to some extent seen this development within the EEC where each member country so to speak has its own speciality.

Denmark is the master of Feta, the Netherlands of Edam and Gouda, France of Camembert and Brie, Germany of Edam and Emmental, the UK of Cheddar.

But still each country also has a need for a wide range of diversified products to support this position.

Summing up, I will conclude by arguing that the challenge not only to the cheese sector, but to the whole dairy industry is not as much to develop further the existing, traditional and well-known products whatever they may be, liquid milk, butter, condensed milk or cheese, however important this might be both as to quality and product.

The **real challenge** is to develop new products and alternative outlets for milk which after all is the most allround high nutritious raw material nature gave us. I shall keep myself from indicating any such new outlets here, but I shall be astonished if we shall not listen to reports of such in this forum in a few years from now.

Fig 1 World production of cheese.

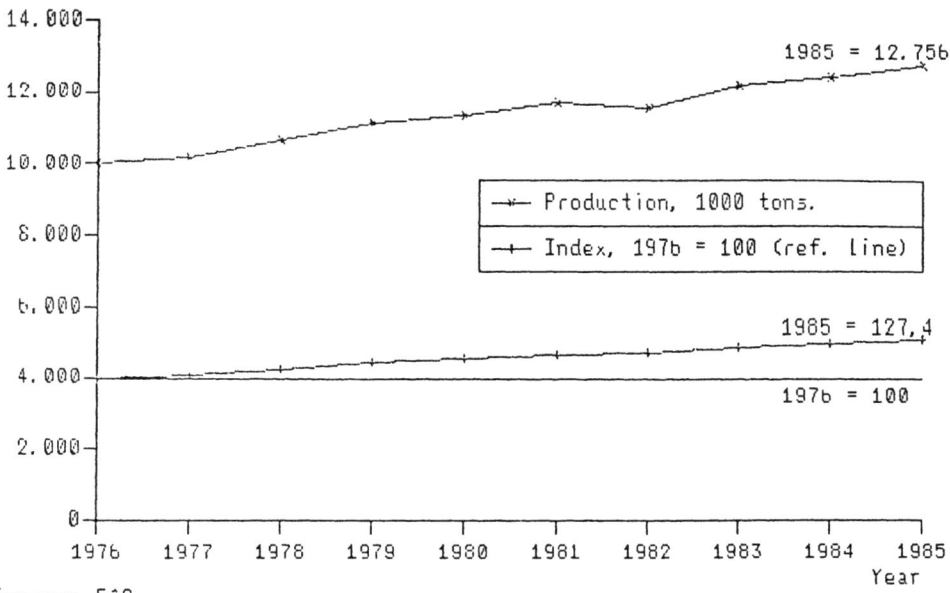

Source: FAO.

Fig. 2 Exportation of cheese from the main exporting countries*.

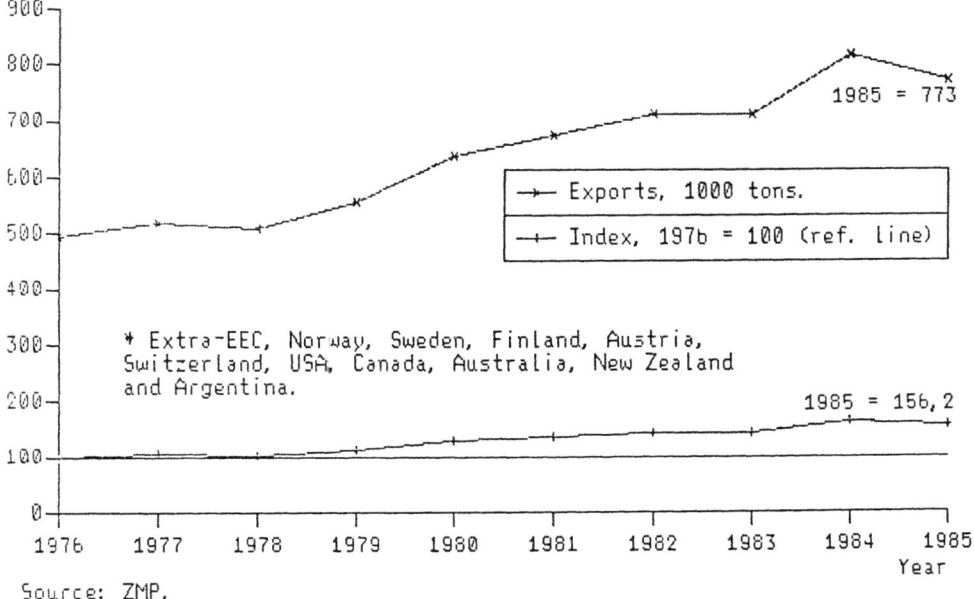

* Extra-EEC, Norway, Sweden, Finland, Austria, Switzerland, USA, Canada, Australia, New Zealand and Argentina.

Source: ZMP.

Fig. 3

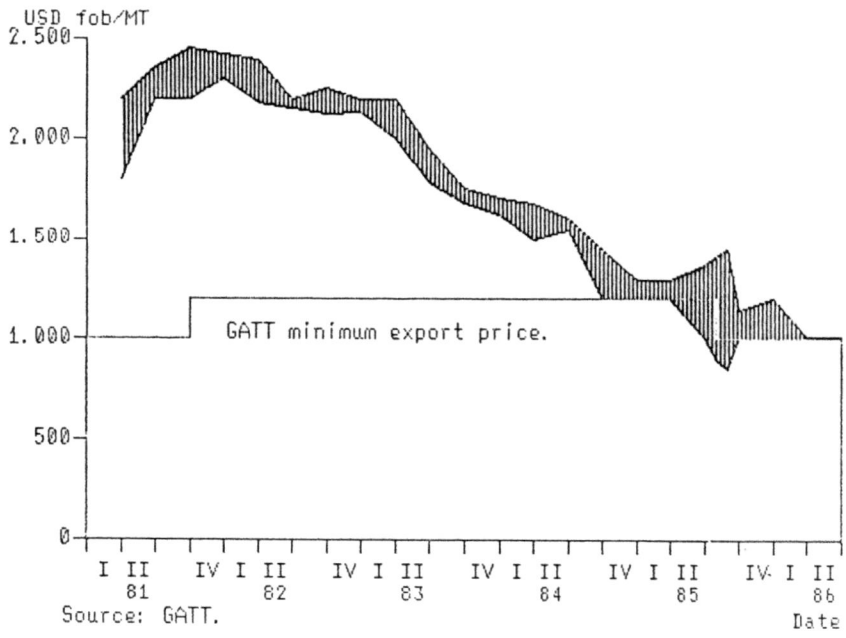

Fig. 4

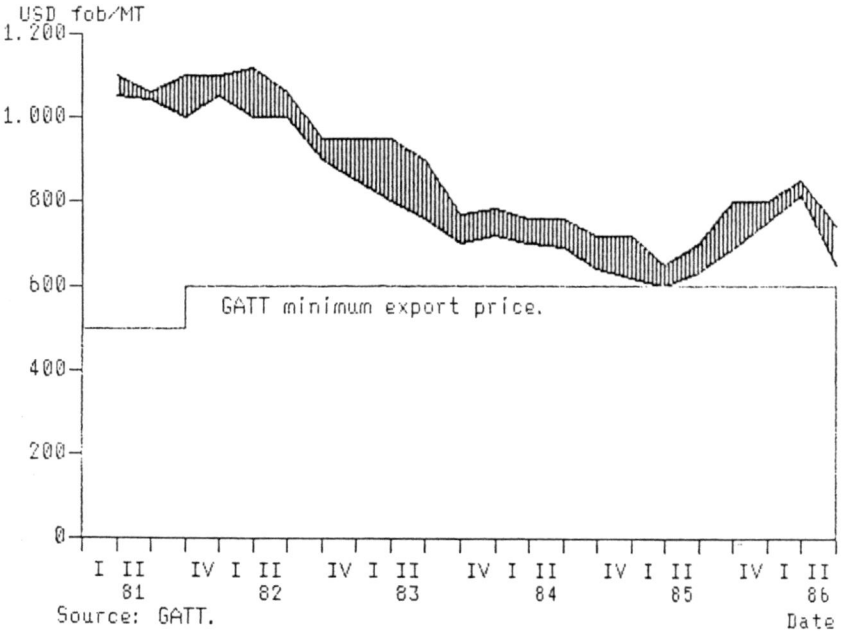

Fig. 5

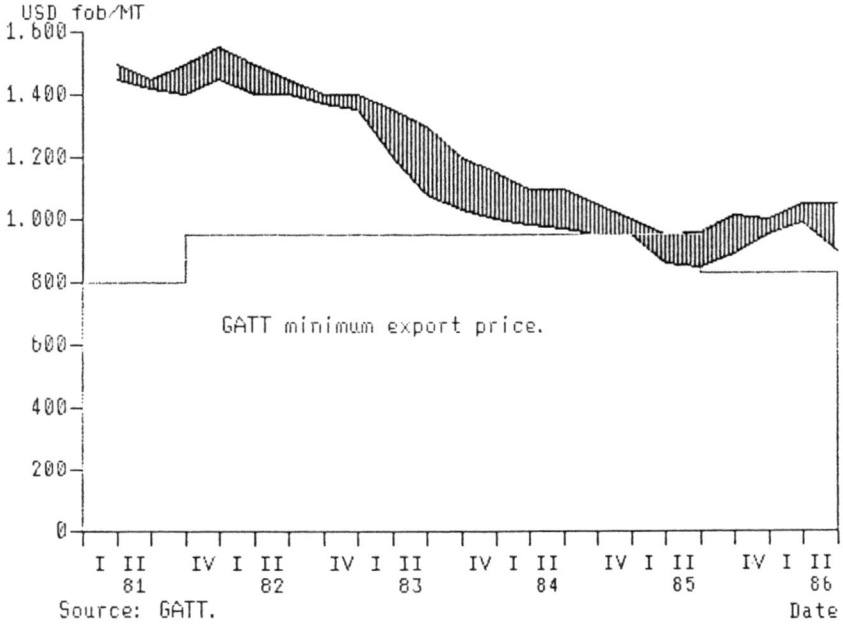

International prices.
Whole milk powder.

Source: GATT.

Fig. 6

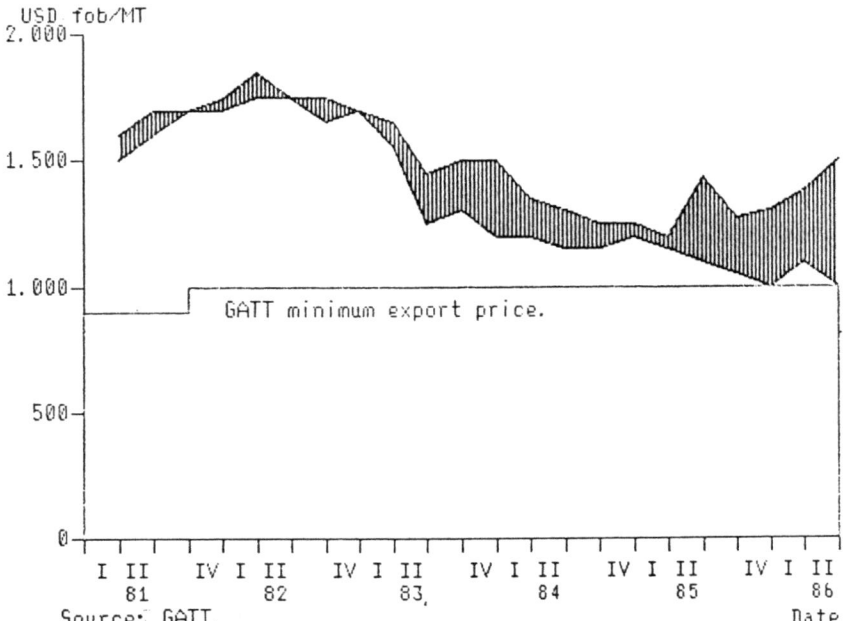

International prices.
Cheddar Cheese.

Source: GATT.

Fig. 7 Denmark, Exports of cheese.

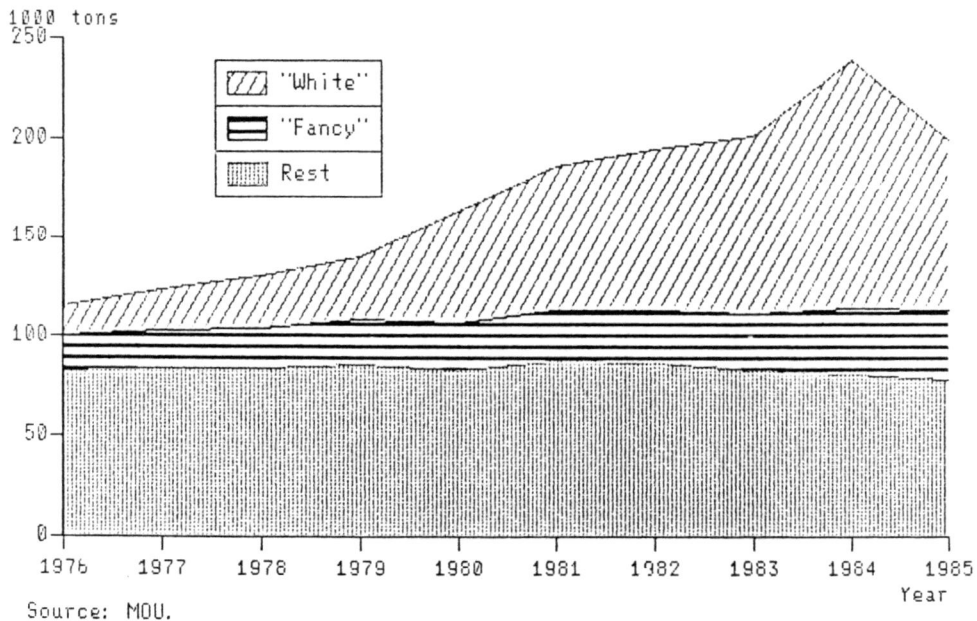

Source: MOU.

Fig. 8 Denmark, Exports of cheese.

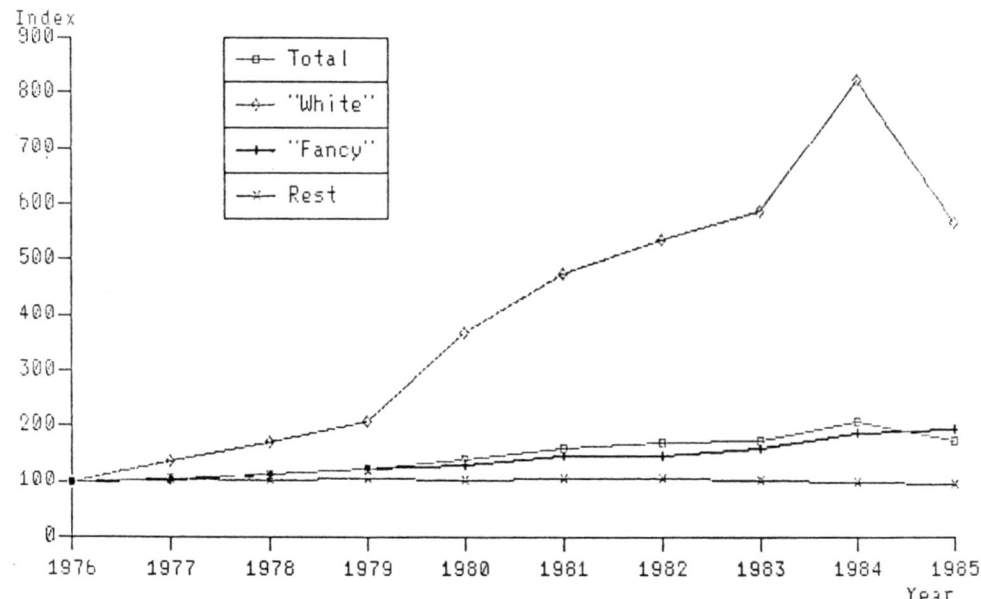

Fig. 9   Denmark, Production of cheese.

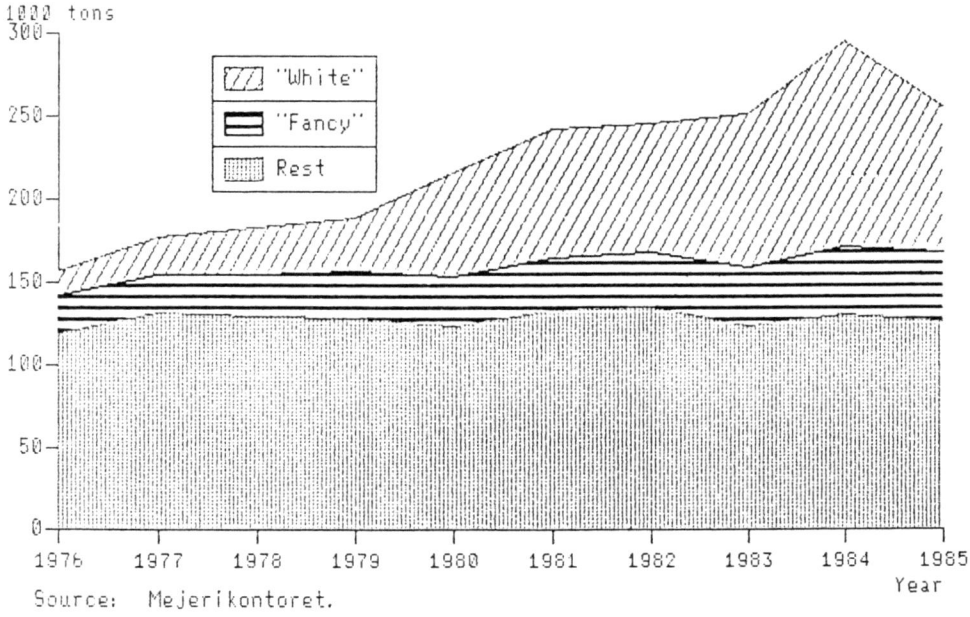

Source: Mejerikontoret.

Fig. 10   Denmark, Production of cheese.

DEMAND TRENDS IN THE LAST DECADE AND CHALLENGES TO BE MET TO THE YEAR 2000 - FOR PRESERVED MILK PRODUCTS

Dr K. J. Kirkpatrick
New Zealand Dairy Board
P.O. Box 417
Wellington
New Zealand

In this talk today I would like first of all to cover briefly the trends of the past 10 years or so in the world production and trade of preserved milk. By preserved milk I mean milkpowders, evaporated and condensed milk, and milk proteins. I am also including Ultra High Temperature treated milk in this category because while certainly a liquid product, it is nonetheless a preserved milk.

Following, then, a resume of the trends of the last decade, I will move on to some speculations about future challenges for the preserved milk industry. Here I will focus primarily on three major areas of challenge and opportunity for preserved milks which I have selected out of what are of course numerous significant challenges for us to meet in the next 14 years.

I would like to say at the outset that the greatest challenge we face in the coming years for all milk and milk products must be that of bringing supply into line with demand. If a better market balance can be achieved, the supply pressure from current surpluses would ease, and a price recovery could be achieved. The issue of oversupply is a thorny one which (fortunately) I have not been asked to talk about. I merely make these comments in passing as the supply problem must provide the backdrop for any discussion about challenges in the future for the milk industry as a whole.

In terms of volume, skimmilk powder forms the largest part of the international export market for preserved milk. As most skimmlk powder is imported by developing countries, the development of these markets and the evolution of their economies including access to foreign exchange has determined the size and nature of the trade in skimmilk powder.

During the late 1960's and in the 1970's the dependence of the developing countries on evaporated milk imports was gradually eroded by the development of recombining industries and the importation of skimmilk powder for recombining. Since the mid-1970's the demand for skimmilk powder has risen significantly in the oil exporting countries. World skimmilk powder production rose from 3.7 million tonnes in 1975 to 4.2 million tonnes in 1981, and further to just under 5 million tonnes in 1983 before easing back to 4.5 million tonnes in 1984. The EEC is the largest producer and although it consumes 1.5 million tonnes, 1.2 of this is for calf or animal feeds.

Between 1980 and 1982 the commercial export market for skimmilk powder contracted by 20% from 783,000 tonnes to 625,000 tonnes, while in the same period the quantity of skimmilk powder exported as food aid expanded from 194,000 tonnes to 250,000 tonnes.

In the past three years the commercial world trade volume of skimmilk powder has remained relatively static. The volume of trade has risen slightly as prices have decreased, and with low fat prices, this has favoured milk reconstitution and recombination of milk products. Meanwhile food aid exports of skimmilk powder have continued to expand, to 356.4 thousand tonnes in 1984.

In response to the decline in demand for evaporated and condensed milk during the 1970's, world trade in these products continued to show a gentle overall decline into the 1980's, with total exports slipping below 1 million tonnes in 1983 and recovering slightly in 1984. Exports from the EEC, the major exporter of condensed milk, dropped from 824,000 tonnes for the average for 1980 and 1981 and 720,000 tonnes average for 1983 and 1984. The decreased demand for evaporated milk is evident both in the developing countries and in the developed countries. The decline in demand for evaporated milk is broadly due to the improved technology of instant wholemilk powder and more recently to the availability of UHT whole milk, both of which products combine the advantages of long shelf life at ambient temperatures but are closer in taste and function to fresh whole milk.

In the late 1970's world trade in wholemilk powder more than doubled - growing from 275,000 tonnes a year in 1975 to over 700,000 tonnes in 1980. But in the early 1980's the import market for wholemilk powder plateaued and in 1983 and 1984 dipped to around 590,000 tonnes per annum. The expansion of import demand for wholemilk powder in the late 1970's was due to the development of technically improved products both in terms of flavour and being more "instant" when recombining the milk at home. This coincided with the growth in consumer incomes, especially in the oil exporting countries, leading to increased demand for wholemilk powder particularly in consumer packs.

Continued growth in demand for wholemilk powder will depend on price competitiveness with recombined, fresh, and UHT milk, and also on the functional properties of the product. Consumers will seek a powder which will instantly combine with water, whatever the water temperature, and immediately produce a product indistinguishable from fresh milk. If this can be truly achieved wholemilk powder will have the possibility of eroding cartoned fresh milk, especially in advanced industrial countries.

World production of casein and caseinates has grown steadily from over 100,000 tonnes in 1975 to 180,000 tonnes in 1980 and to 250,000 tonnes in 1984. Currently around 60% of total production is traded internationally and the current international market totals over 200,000 tonnes. This market has been growing at around 5% per annum over the last 10 years.

The marked growth in casein consumption over the past decade has

been made possible by the sophisticated technological development and
refinement of milk proteins during this period. The main use of
casein was formerly in the industrial sector, but with the emphasis on
its high nutritional value and the binding and other functional
qualities of milk proteins, the primary focus for these materials has
now moved to the food and feedstuffs sector. Nevertheless there
remain some important industrial, technical uses of casein.

Looking to the future challenges in the preserved milk area,
three major prospects come to my mind which I will cover briefly in
the remainder of my address. Firstly, the importance of East-Asia
where there is rapid economic progress; secondly the potential of UHT
products; and thirdly the technical achievements required to ensure a
further growth in the market for milk proteins.

There will of course be many other challenges along the way -
such as the problem of over-supply, which I have already mentioned;
another is the need for the further development of a truly instant
wholemilk powder to suit consumer tastes and the desire for
convenience in the home; the future of the skimmilk powder
recombining industry will be challenged by new methods and products.
I believe there is a strong future for skimmilk powder as an
ingredient in recombined milk in the rapidly growing markets of the
Middle East and North Africa where demand should continue to expand
despite the temporary drop in oil incomes and especially as prosperity
spreads through the population. There will be some development of
wholemilk powder for recombining (quite distinct from the consumer
product) and of other milk ingredients to further improve the
qualities of the recombined product.

I mentioned the importance and the potential of East Asia as a
specific challenge over the next decade or so.

The increasingly industrialised countries of North-East Asia
(Taiwan, South Korea, China, Malaysia, Singapore and Thailand) will
present us with a growth potential in terms of the sophisticated use
of preserved milk products. As the economies of especially Taiwan,
South Korea and China develop, the consumers and customers in these
nations will seek highly specialised forms of preserved milk,
including increasingly the pure proteins area.

The economic face of East Asia is changing. Strong economic
growth in this region will continue and the importance of the region
in economic terms will certainly become a challenge for the marketing
of preserved milk products.

Consumption of dairy products is generally low, relatively
speaking, in these Asian countries, but there has been growth in many
of them over recent years. For example, in Japan skimmilk powder
consumption per capita has increased from 957gm in 1975 to 1,508gm in
1984. Drinking milk consumption in Japan has increased in the last
nine years from 3.2 million tonnes per annum to 4.3 million tonnes per
annum.

In the rapidly developing nations of East Asia, patterns of food
consumption are thus also beginning to change, and this change is
heralded by a growing appreciation of the nutritional qualities of
milk. While the taste of many processed dairy products (such as

cheese) is not immediately suited to the palates of these peoples, the taste for and appreciation of fresh wholemilk or its preserved counterparts is not such a barrier to overcome. With increasing urbanisation and rising incomes, the patterns of consumer spending will no doubt become closer to those of consumers in the Western world. I see potential there therefore, for growth in milk consumption in general, and especially in the milk proteins area.

Another field of development which represents a challenge to us is that of UHT treatment for liquid milk, cream and other dairy products. UHT treatment is a fine example of the invaluable contribution which food scientists and technologists have made to our industry. The challenge in this area in order to achieve growth in the sophisticated consumer market, will be firstly to improve still further the resemblance of UHT products to their fresh counterparts, and secondly to market these preserved products carefully to capture their qualities of convenience. Outside sophisticated consumer markets, however, I believe that UHT products also have definite growth opportunities, despite the inherent disadvantages associated with costs of packaging and transport. Certainly the extra transport costs of UHT products compared, for example, with milk powders are a disincentive, but the convenience factor of the UHT product combined with its closeness in taste and function to fresh products warrants close examination of opportunities. In this context we must also be aware that the spread of electricity and hence refrigeration in developing nations will influence the relative progress of the different forms of consumer preserved milk products.

The last area that I would like to discuss and which represents a considerable opportunity over the next decade and a half, is the field of milk proteins - these being caseins, caseinates, whey proteins and other minor components of milk. This is an area of business in which I am closely involved personally.

Notwithstanding the continuing controversy regarding animal fats and cholesterol which surrounds milkfats, there is universal recognition of the high nutritional quality of milk proteins, their value and importance in the diet, even at relatively low usage levels. Furthermore, milk proteins have a great range of functional properties that they can impart to processed foods generally. The technological progress of the last two decades will continue, and market opportunities will grow, particularly with the spread of processed convenience forms of indigenous foods throughout the world. While the United States market has been a leader in the development of processed convenience foods, the same basic principles applied to local foods will undoubtedly present opportunities for functional milk proteins through Asia, Europe and the rest of the world.

Increasing recognition is being gained by the special health related and potentially therapeutic properties of minor milk protein components. The further fractionation of milk into its components to be used individually or reassembled into speciality nutritional food products is expected to gain significant commercial momentum over the next decades. Opportunities would include foods for newborn animals, and human beings with special dietary requirements.

While technological opportunity will be a major driving force, it must be recognised that a number of existing areas of use for milk proteins are essentially price related. Against this, the relatively low price prevailing for basic milk protein products, while encouraging consumption, tends to reduce the commitment to expensive ongoing research and development needed to exploit the technical opportunities that are seen to abound. There is evident need for sufficient technologists who are equally at home in the technologies of milk product manufacture and in the complexities of subsequent use in complex food and other use systems.

The uses for casein where there is significant potential to expand demand provided the appropriate product and market development is accomplished include the following; cheese analogues, specialised nutritional foods, bakery ingredients, convenience foods, animal and pet foods, plastics and specialised technological and industrial uses.

To sum up - it is evident that the current trends for preserved milk products are seen as likely to continue for the remainder of this century (and perhaps beyond). The relative progress of individual products as well as the category as a whole will be affected by many influences not specifically mentioned but most particularly the demographic changes in advanced and developing countries (ageing populations, economic development and changes in consumer preference such as life style, nutrition preferences, convenience foods development).

It is very easy for those of us deeply involved in our dairy industries to think in product terms and in terms of the technological factors that restrict and permit progress. However it is really consumer and customer preferences that determine progress in the long run. We must be constantly attentive to their needs if the developments outlined in this talk are to be realised.

UNITED STATES DAIRY FARMER PROMOTION

E. Hoy McConnell, II
Managing Director/CEO
D'Arcy Masius Benton & Bowles, Inc.
200 East Randolph Drive; Chicago, Illinois 60601 U.S.A.

ABSTRACT. Total milk production in the U.S. has been increasing faster than consumer demand -- resulting in a growing surplus problem. To counteract this situation, U.S. dairy farmers are undertaking an aggressive and multi-faceted effort backed by a $120 million non-brand advertising campaign to stimulate increased sales. Four key programs will be analyzed, including fluid milk, cheese, ice cream and an all-product campaign for dairy calcium.

1.   U.S. DAIRY PROMOTION BACKGROUND

In addition to having more dairy product advertising accounts in more countries around the world than any other advertising agency, D'Arcy Masius Benton & Bowles has had the privilege of being the leading advertising agency for non-brand dairy promotion programs in the United States for over 12 years. Because of this, we have been closely involved with important changes which have taken place over the last two years in dairy farmer-funded promotion in the U.S. and the significant effect these changes have had in the marketplace.

Three years ago, the dairy situation in the U.S. was not a very pretty picture:
- Surplus production was a big problem.
- Fluid milk sales were in a slow but steady decline.
- Cheese sales had begun to plateau.
- Butter and ice cream sales were flat.

It all added up to a very difficult marketing situation for dairy farmers as well as dairy product processors.

Today, things are different. Milk sales are on the upswing, cheese sales are accelerating, ice cream sales are revitalized, and even butter has begun to show some positive growth.

What accounts for such a change in marketing circumstances?

It didn't just happen. An aggressive non-brand promotion program funded by dairy farmers helped make it happen.

To begin with, because of the unfavorable marketing situation facing the U.S. dairy industry, the dairy farmer cooperative organizations knew they could not afford to continue to do "business as usual" and maintain a good price for milk. A major new industry initiative was needed. And, this could only be accomplished through national legislation.

After much debate and consideration, new dairy legislation was enacted by the U.S. Congress in the Fall of 1983. This legislation was significant in that, among other things, it:
- Changed the structure of dairy farmer promotion from a fractionalized, regionally-oriented system to a unified, nationally-based program.
- Doubled the funding for dairy farmer promotion programs, increasing consumer advertising dollars for non-brand promotion from $60 million to over $120 million.

As a result, in the Fall of 1984, a significant new thrust in U.S. dairy product advertising began. The new promotion structure and the increased levels of funding gave dairy products the chance to be much more competitive in the marketplace...provided, of course, the advertising was strong, on target and effective.

The focus here is the advertising actually developed since late 1984 for cheese, ice cream, dairy calcium and fluid milk.

2.   DAIRY PRODUCT PROMOTION PROGRAMS

2.1 **Cheese**

Cheese is the second most important dairy product in the United States. It receives one-third of the total advertising budget and is one of the most important product priorities to the U.S. dairy farmer:
- Thirty per cent of all milk produced is used to make cheese.
- Until just recently, cheese was the only major U.S. dairy product showing any sales vitality, even though that growth was slowing down.

Importantly, however, cheese has a very positive image with consumers because it's seen as:
- Convenient
- Versatile - it can be eaten as a snack, light meal, or even as a main dish
- Interesting - it offers many different tastes and textures
- A good complement to other foods
- Healthy
- Fashionable for entertaining

There is another very important consideration for industry-based cheese advertising. Unlike almost all other U.S. dairy products, there are strong brand advertising programs for cheese, totalling over $75 million per year.

With an already positive image of cheese, the real challenge of the non-brand program is to stimulate the total market by getting consumers interested in having cheese more often by trying new types and new ways to eat cheese.

The advertising developed to do this glorifies cheese as a product. It's a celebration of cheese in advertising.

To accomplish this, the songs used to deliver the cheese-selling message are songs familiar to most people in the U.S. -- indeed, to people everywhere. The words are changed to make them songs about cheese. The "Cheese, Glorious Cheese" lyrics are set to the <u>Oliver</u> tune of "Food, Glorious Food."

Cole Porter's Broadway show-stopping song, "It's the Tops," is the music behind visuals of cheese being sprinkled or poured over fresh, healthful vegetables as a topping. The familiar Mexican "Cielito Lindo" plays to lyrics of "Aye, Aye, Aye" as a positive appeal to use a variety of cheeses with ethnic foods. Hank William's country-music ballad, "Hey Good Lookin', What'Ya Got Cookin'," translates to a woman being asked, "How 'Bout Cookin' Something Up With Cheese?"

To accomplish a somewhat different job -- convincing mothers that real cheese is far superior to imitation cheese -- an irresistable baby girl, happily munching a slice of cheese, smiles through the commercial to George and Ira Gershwin's "You Can't Take That Away From Me."

In each case, the advertising strategy is to invite consumers to use their imaginations in devising new ways to serve and use cheese. The appeal to do so involves both mouth-watering cheese photography and memorable music.

2.2 Ice Cream

One of the new non-brand promotion programs created as a result of the 1983 dairy legislation is for ice cream. And, we've found, ice cream not only is a treat to eat, it's a treat to advertise.

Consumer attitudes about ice cream are consistently positive. It is seen as seen as fun, refreshing, great tasting, and especially good to eat.

Since most of the ice cream brands advertise mainly during the summer, the national dairy farmer advertising campaign in the U.S. is planned to coincide with this important summer selling season.

The non-brand advertising developed by DMB&B for ice cream sums up both the spirit and seasonality of the product with a strong call-to-action to "Taste Summer" with ice cream. Colorful, happy, cool and refreshing best describe the consumer appeals in these ads.

The new ice cream advertising program is also an excellent example of how cooperation between dairy farmers and processors can magnify promotion impact in the marketplace.

2.3 Calcium

The calcium content of all dairy products is another new advertising program developed since the 1983 legislation. It has been created by one of our respected competitors, McCann-Erickson.

Today, calcium is <u>the</u> major health topic in this nation. It's difficult to pick up any women's magazine without finding some article about problems of calcium deficiency and the need for increased calcium intake as a preventative to brittle bone disease (osteoporosis).

The calcium content of dairy products represents a special promotion opportunity because milk and milk products account for over 70% of the calcium in our diets.

However, reminding women that dairy products are an excellent source of calcium is simply not enough. New calcium/dairy product advertising communicates, in a somewhat unexpected manner, that calcium intake relates to looking good and feeling good, in addition to maintaining good bone health. This advertising features strikingly attractive women of all ages, and relates the importance of calcium throughout one's life to the high levels found in dairy products. The theme of this advertising is "Calcium the Way Nature Intended."

2.4 Fluid Milk

Fluid milk is _the_ most important dairy product to U.S. dairy farmers:
- Fluid milk represents the largest volume use of all milk produced.
- Dairy farmers receive a higher price for fluid milk than for milk used to make other dairy products.

Furthermore, and very importantly, there is virtually no brand advertising for milk, so dairy farmer milk advertising is the only major consumer support the product receives.

Milk's primary competition in the U.S. is soft drinks. And, it's very formidable competition because soft drinks advertisers spend vast sums -- currently over $400 million a year in media advertising alone.

The key consumer target for DMB&B's milk advertising is young adults, age 18-34. This segment not only accounts for about 25% of total consumption but, importantly, represents the age group when milk consumption declines most rapidly because of the variety of alternative beverages available -- soft drinks, beer, wine, coffee.

On the plus side, people in the U.S. universally see milk as a product that is good for them -- healthful and nourishing. However, compared to soft drinks, milk is also seen as bland and old-fashioned -- simply not as socially acceptable as soft drinks. To U.S. consumers, milk is just not very contemporary.

The job, then, for milk advertising is to reposition milk... to change its image so that it is viewed as a more contemporary beverage...a beverage that is more personally relevant and one that fits into the lifestyle of the young adult target audience.

We believe we've found the way to do that by capitalizing on the growing and broad-based interest in the U.S. on health and fitness. We link milk's nutritional benefits to vitality, being fit and feeling good. Thereby, we show how milk is a natural part of today's healthy way of living.

The "Milk...America's Health Kick" advertising is full of energy. There is an upbeat, energetic, fresh and friendly attitude. From the beginning, these commercials have portrayed active people of all ages drinking milk, looking good, and obviously feeling good.

3. EXPENDITURES AND EFFECTIVENESS OF U.S. DAIRY PRODUCT PROMOTION

Based on the depth and variety of the advertising I've discussed, I hope you can begin to appreciate the strength of the new dairy farmer-funded promotion programs in the United States. And, this discussion has not included any of the butter advertising or additional sales promotion and retail merchandising programs that are also a part of the total effort.

Table 1, below, summarizes the combined national and local annual advertising investment being made in the U.S. to support dairy products:

Table 1

U.S. DAIRY PRODUCT ADVERTISING EXPENDITURES
PLANNED FOR 1987

Product	Advertising $ ($ Millions)
FLUID MILK	$ 45
CHEESE	40
BUTTER	15
CALCIUM	15
ICE CREAM	5
Total	$120 Million

As Table 2, below, indicates the sales increases these dairy advertising programs have helped produce are truly impressive:

Table 2

U.S. DAIRY PRODUCT SALES
1984-1986

	1985 vs. 1984	January-June 1986 vs. 1985
Total Dairy Products	+3.6%	+4.2%
Fluid Milk	+1.5	+1.2
Cheese	+4.1	+8.2
Ice Cream	+5.6 (summer)	
Butter	+2.2	+2.7

To put these sales gains in some perspective, I can tell you that if these increases continue at the same pace for the rest of this year, then 1985 and 1986 will go down in history as the two strongest back-to-back years for dairy sales in the U.S. since before W.W. II.

I would note in closing that even with the success of the new non-brand dairy advertising programs in the U.S., surplus production continues to be a problem. So I'm <u>not</u> trying to suggest to you that advertising and promotion is the total solution to the challenge of balancing supply and demand.

Nevertheless, with a strong, effective non-brand promotion program in place, one that has already demonstrated positive results, there is optimism in the U.S. dairy industry today. And, it's supported by dairy sales forecasts that indicate the future continues to look rosey.

At the same time, on the supply front, new programs are being undertaken to reduce production. In other words, the industry is taking the action necessary to build a strong, balanced dairy market in the U.S., addressing both the supply and demand sides of the dairy equation.

It's a very exciting time to be part of dairy marketing in the United States. To see advertising produce tangible results. To see increasing sales of products that are important to the health of America. To be proud of what these products represent to the health and well-being of consumers.

MARKET AND MARKETING OF DAIRY PRODUCTS
Do advertising, sales promotion and nutrition education pay?

Eva Schmekel
Swedish Dairies Association
Box 24
101 20   STOCKHOLM
Sweden

ABSTRACT. It is difficult to give a straight answer in terms of cash. To achieve any impact on the market, the dairy industry must earn the trust and confidence if the consumer and the retail trade which can only be earned on the basis of a thorough understanding of the products. The consumer with only a vague concept of the product will prove an easy prey on the propaganda of competitors and the mainstream of public debate.

To get his message across to the consumer in a modern supermarket, the manufacturer needs to ensure that the volume and quality of his promotional material is on a par with that of his competitors.

A continuous flow of information on the importance of dairy products to a balanced diet must be made available to physicians, teachers, journalists and other moulders of public opinion. These critical groups need to be kept supplied with information and induced to adopt a positive attitude towards the products as their opinion on nutritional matters is the one that finds most frequent expression in the media and elsewhere.

Do advertising, sales promotion and nutrition education pay?

My answer to that question is definitely - yes! What remains is to provide evidence that all the billions of dollars spent on advertising, promotion and other marketing activities actually result in devidens for producers. However, the question is not that simple. The model that is really economically feasible has not yet been described. Therefore, it is necessary to choose from the knowledge that is available and combine it with one's own experience when trying to answer this question.

But before I go into the various aspects of assessment, I would like to strongly emphasize a hypothesis:

In order to be successful on the market, a company and its products must have the confidence of its consumers, the trade and opinion-builders. It is necessary that those persons have considerable knowledge of the products and that the company's own profile be serious and responsible. Otherwise consumers can easily fall victim to competitive propaganda in the public debate.

Measuring of effects, belief and knowledge

As I mentioned earlier, it is difficult to measure the economic results of marketing activities. The final goal is, of course, to achieve sales, but how is one to know which activities out of the entire marketing mixture are most effective: It is advertising, sales promotion, actual sales work, prices, product characteristics, PR activities or simple occasional debates in mass media that determine the results of a campaign?

In Sweden, where milk comsumption is high and stable, it is not possible to directly prove that a traditional milk campaign based on advertisments, sales promotion or shop campaigns played any role at all the short run. On the other hand, effects with regard to butter are immediately noticeable.

The marketing of milk has always had a long-term goal, namely that of influencing consumer knowledge and attitudes in a positive direction. Then it is only natural that direct sales effects of one single promotion activity cannot be noted immediately.

It is probable that an analysis of the value of marketing must be made both from a long and short-term perspective.

Let us study some assessment instruments that are available for milk and dairy products.

Many marketing institutes spend a great deal of time and money investigating immediate effects of various marketing activities. They study the effects of price changes, attention value of advertisements, etc. However, when it comes to measuring effects it is no longer so easy.

Let's take a closer look at some factors, more or less measurable.

1. Special prices or price reductions.
2. Coupon offers.
3. Orders of brochures.
4. Has knowledge increased?
5. Value of nutrition information.

1. Special prices

Special prices is nowadays an important means of promotion in Sweden and many other places.

Some statistics from Sweden indicate the scope:

1.1 Consumers buy more than 40 per cent of their needs at special prices on, among other commodities, coffee, ice-cream, margarine, ketchup, biscuits, detergents etc.

1.2 Two out to three women constantly look for special prices in advertisement.

Retail marketing is completely dominated by price offers.

Manufacturers contribute greatly to that marketing. At least half - and often more - of a 10 per cent reduction on a commodity is paid by the producer. However, a successful campaign always results in returns in the form of increased sales.

But is it worthwile in the long run?

Example: During one week recently, butter was sold at special prices: minus 10 per cent in the large retail chains. Consumers were informed of that fact partly through the shops' weekly ads and partly in the shops themselves.

During that period, sales increased five-fold. Sales then soon returned to the earlier level. But the long-term effect remains. Spread out over one year, increase in sales will perhaps reach one per cent.

In Sweden the particular commodity group yellow fat is often marketed at special prices. Practically every week it is possible to buy some kind of cooking fat at a special price in one of the three big retail chains. It thus appears that the market would not undergo any changes were the system of using special prices to disappear. However, apart from the immediate sales effect the system provides the possibility for a quick introduction on the market of new products.

With regard to the introduction and succes of Bregott, butter/soya-oil blend produced by the dairies it is interesting to note that a new product like this would hardly have met with the same acceptance on the market without the special-price system. In addition, Bregott is a high-priced product which is also bought by price-conscious consumers during periods of special price offers.

2. Coupon offers,

which is another form of price reduction, have also become popular during the past few years. Through the media of mail, advertising newspapers, direct advertising or a coupon on packages, consumers are given an opportunity to turn in coupons in their shops.

Coupons have always been extremely popular. In the USA, according to my sources, all of 80 per cent of the households use coupons in some form or another. Many actively seek coupons and there are even coupon clubs that help their members use and exchange coupons.

Example: In Sweden we recently tested the use of Bregott lid as a coupon. "Send in the lid of one package and receive 10 crowns in return". 100 000 lids were sent in, which was a great success.

Using this model, the consumer gives no opportunity to store the reduced-price commodity at home. Neither is there any risk that the consumer will get the idea that the product is of lower quality just because it is being sold at a lower price like when special prices are used.

3. Orders

Dairy industries all over the world have a comprehensive recipe series for households. Test kitchens create recipes and advertising agencies present the delicacies in tempting brochures, ads and display material. It is natural that dairies use recipes for selling their products. Butter, cheese, cream and milk are basic ingredients that are almost always available in households.

The object is to get housewives (or house husbands) to use them by giving tips in the recipes about how dairy products play an important role in getting good results.

From my Swedish examples, I would like to choose the marketing of cream.

The dairies annually carry out two big campaigns for cream. In addition to ads with various themes, each campaign includes a recipe brochure which is distributed to the shops via the dairies' sales personnel. Each brochure is printed in 3 million copies. This means that in theory each household in Sweden receives one brochure a year.

There are interesting statistics that confirm the value of this kind of cream promotion. Since the middle of the 70s, when active work with recipes was started, cream sales have increased by 13 million kilograms. At the same time, the use of cream in households has changed.

Ten years ago, 68 per cent of all cream was used for desserts and baked goods, 26 per cent for cooking. Today these figures are 35 per cent for cooking, while the share used in desserts and the like has decreased to 57 per cent.

4. Has knowledge increased?

As mentioned earlier, good knowledge of dairy products is important for continued market success.

This is particularly true of milk. Recently it was determined that not less than 82 per cent of Swedes consider milk to be healthful. It is, of course, of great importance that this percentage be retained. For natural reasons, marketing efforts are mainly of a long-term nature and it could be of interest to analyze long-term milk promotion that increases knowledge.

In the middle of the 1970-ties confidence to milk was high. At that time, the dairies had already over a period of ten years concentrated their message on information regarding its nutritional value for various consumer categories such as children, athletic youth, the elderly, nursing mothers, etc.

During the end of the 70s, marketing was concentrated on the naturalness of milk, its purity and other sensitive arguments. A considerable decrease in knowledge was then noted.

This was the time for dairies to once again assume the role of educater, through campaign messages that stated: Milk provides 8 out of 10 of the important nutrients. This led to increased knowledge of the advantages of milk.

Then came the time to refine the message. In a series of ads, information about the nutritional value of milk and its importance for well-being was coupled together with wild animals and the conceptions people have about their special characteristics in a humoristic way.

Example: Like this one: Milk is rich in vitamin A, good for darksight and reproduction - so if you are going out tonight, drink milk.

The results were two-sided. The consumers loved the beautiful pictures and but only half of the consumers were aware of the fact that the message was milk. But the attitude that milk is healthful and an important food was strengthened in this more intelligent group.

5. Nutrition education. Does it pay?

It is suitable to consider this question of my headline separately, since the target groups involved here are influenced by other media.

Nutrition experts, journalists and teachers are not influenced by special prices or recipe brochures.

Activities in this area are mainly aimed at creating a positive attitude among important target groups within food and nutrition in order to influence them to present dairy products in a positive way in their external information channels.

When approaching the group nutrition experts, it is important that the dairies have an open and positive dailogue. Those experts can even be given a certain influence when it comes to assortment policies.
I give you an example:
In August, milk with 1 per cent fat was replaced by milk having 1,5 per cent fat.

The introduction of milk with 1,5 per cent fat was preceded by not only market surveys and taste tests but also by detailed consultations with nutrition experts. They backed up our hypothesis that 1,5 per cent milk would replace a certain percentage of the milk with higher fat content, resulting in a decrease in total fat consumption through milk.

Already now there are marked indications of such a development. We find it very encouraging that nutrition experts also in the periodical mass media debate regarding the dangers of fat have given moral support to the dairy industry when the fat content of a milk type was discussed.

Such a PR effect of nutrition education really pays.

Conclusion

All dairy products are in various ways vulnerable to competition. In marketing and information, their unique advantages must always be brought forth. And so I revert to the subject of knowledge. Well-informed consumers, knowledgeable retailers and trustworthy cooperation with nutrition experts lay the foundation for success.

May I then conclude with a question reated to my knowledge hypothesis.

If the consumer doesn't have the firm opinion that milk is good for him, most other sales arguments are of little value.

Because why should he choose milk instead of other beverages if the only determining factors are the attractiveness of competitors' prices, taste, packaging or availability?

DISCUSSION

S.Y. Ali (Pakistan) : Is it possible to find alternative uses for butter?
B.A. Joyce (Canada) : Alternative uses will not provide better return than for instance recycling into animal feed.

Name unknown (South Africa) : What is the reaction of manufacturers of brand articles on the new non-brand promotion programmes in the U.S.?
E. Hoy McConnel (USA) : The reaction is very positive. Generic campaigns have the intention to enlarge the market for dairy products. Brand and non-brand campaigns support each other.

L.W.J. Hurd (Canada) : Has the study at Cornell University regarding brand versus generic advertising already yielded any results?
E. Hoy McConnel (USA) : Generic advertisement has added an impact on the market. The upward consumption trends as mentioned in my paper were the results of the complete marketing package.

P. Oosterhoff (Canada) : I never saw a study which indicates the point where advertising has no longer any influence. This kind of generic promotion has no negative side effects.

Name unknown (Australia) : Where are the figures mentioned in Mr. Hall's paper from?
R. Hall (UK) : The figures were obtained from the Australian Dairy Corporation. The fall in per capita consumption in the USA would – unlike in the EEC – be more than offset by population growth. However, long term forecasts are not very accurate.

Name unknown (Fed.Rep. Germany) : Can you give an explanation for the rather large differences in the per capita consumption of the various countries? Is the importance of dairy substitutes not much greater than suggested in the paper of Mr. Hall?
R. Hall (UK) : I have no explanation for the differences in the per capita consumption. With respect to substitutes it should be taken into account that in the USA the naturalness of dairy products is stressed and that many other countries will assume a similar attitude in the coming years.

Name and country unknown : Would it not be recommendable to start international cooperation in the nutrition/health field?
C. Chevalier (Canada) : Cooperation on a global basis is an outright must.

Seminar III: Market and Marketing of Dairy Products

Session 4: Global Aspects

Chairman: J. Empson (U.K.)
Secretary: Drs. P. J. Poot (The Netherlands)

The changing role of government in dairy policy

Milk supply management programmes, their strengths and weaknesses

THE CHANGING ROLE OF GOVERNMENT IN DAIRY POLICY

G. Haydock
Organisaton for Economic Co-operation and Development*
2, rue André Pascal
75775 Paris Cédex 16
France

ABSTRACT. Current tougher efforts to restrain milk production are seen as a continuation of past insufficiently effective measures to deal with surpluses. Generally speaking current policies will still not eliminate surpluses, though some countries may be near to achieving balance. The multilateral approach to solving trade problems has not so far been successful but a forthcoming OECD study suggests that the necessary adjustments may not be as severe as most governments fear.

1. When I was asked to prepare this paper at a time when my days at the OECD were coming to an end I supposed I was being asked to say how different things were now from 17 years ago when I first joined the OECD Secretariat. My first reaction was to think that the discussions we have been having recently were not much different from those I first encountered and further reflection on the question has not changed my view. Does this mean that the preoccupations of governments have not changed over the period? Except perhaps in one important country I think they have not changed, or at least not changed very much, though the rapid growth of milk surpluses has made governments take more stringent measures.

2. Three years ago the OECD published a study on what was called "positive adjustment policies" in the dairy sector i.e., it was looking at the extent that policies contributed to bringing about a better balance between supply and demand, with a more efficient and more dynamic utilisation of production resources. I mention this study because it identified what was called "the awesome array of roles" that governments expect the dairy sector to play. Not only has it to provide a regular, assured and adequate supply of essential but perishable milk and milk-based foodstuffs, but in many countries it also has to provide an adequate supply of beef. In addition in most

---

* The present paper reflects only the views of the author and does not involve the responsibility of the OECD.

countries, and particularly those in Europe, it is expected to fulfill wider social objectives, including the prevention of urban drift and rural depopulation, the maintenance of rural employment opportunities and certain other regional and environmental considerations. The study recognises that other agricultural sectors are also called upon to play a role in meeting these objectives, but emphasises that the impact on dairying is generally greater than elsewhere because in most countries it is characterised by relatively small farm units, by a fairly high labour input, mainly family, and usually by a very high proportion of farms engaged in this activity.

3. The study points out that it is the conflict between disparate, and sometimes incompatible, economic, social and strategic objectives that is at the heart of the adjustment problems in the dairy sector in most OECD member countries. Though the report did not say so in so many words, it is my belief that governments almost always give priority to the social -- even humanitarian -- objective of supporting dairy farmers' incomes whenever there is a clash between this and the economic objective of preventing the waste of resources entailed by surpluses or even the financial objective of observing budgetary constraints. Rightly or wrongly, the latter objective generally weighs more heavily with governments than optimising the use of resources.

Australia and New Zealand

4. Let me make myself clear. I am not condemning governments for acting in this way, nor am I saying that they have not done anything to try to reduce surpluses. I am simply saying that though governments have taken measures they have almost never been sufficiently severe to reduce milk supplies to the level of meeting domestic demand and exports that do not need the use of subsidies of some sort, because of the effect this would have on their dairy farmers' incomes.

5. The case of traditional exporting countries comes immediately to mind as a special case, particularly Australia and New Zealand. They are almost alone of the major dairying countries not to have a support buying mechanism of some sort. Apart from the Supplementary Minimum Price scheme the New Zealand producer had no direct help from his government (and this only once, in 1978/79), though like everywhere else there was indirect help (particularly cheap fertilisers). Even this scheme has now been removed. In the absence of government support the industry has done what it could to help itself, most importantly by stabilising revenues through a system of reserves and borrowings from the New Zealand Reserve Board, and if it was helped in its borrowings by a very low rate of interest it paid for this by receiving an equally low rate of interest on its deposits. More recently the government has had to give a guarantee so that the Board can continue to borrow, as it now has to do so on commercial terms. Given this history of virtual non-intervention by the New Zealand government it is no surprise that the first efforts to constrain

production have been taken by the Dairy Board and not the government. I refer first to the moratorium on new entrants to milk production and secondly to a scheme to pay farmers to produce less. This is expected to reduce production by 1-2 per cent in 1986/87.

6. The Australian government intervenes more actively than that of New Zealand, without however going so far as to have a support-buying system. Australian policy for many years has been to encourage the dairy industry to adapt to the realities of the world market situation, and Australia is the only example of an OECD country where production is now significantly lower than it was at the beginning of the 1970s. Over the years Australian policy was to provide money to encourage farmers to move out of milk production or to become more efficient. This policy has largely come to an end, but the other major policy measure, the price pooling arrangements under which low returns from exports are offset by higher domestic prices, have been adjusted several times to make both dairy farmers and the manufacturing industry more aware of the realities of the world market. Milk production has nevertheless ceased to decline in recent years, but it is the hope of the Australian government that the new arrangements just introduced as from the 1986/87 season will lead to a significant outflow of milk producers and some reduction in milk production. But it is fairly typical of the cautious approach of many OECD governments in not wishing to hurt their producers that the new arrangements will be gradually phased in over a three-year period, despite several years of warning about the realities of the market and a year-long discussion of the new measures before they were adopted.

North America

7. Canada may be thought of as an exceptional country where for quite a long period the milk supply has been kept close to the level of commercial demand, at least for butterfat, by government policy. The quota system for manufacturing milk makes only small provision for exports of products containing butterfat, though the pattern of demand is such that some surpluses of non-fat solids are generated. But the arrangements include a price formula that substantially compensates producers for price increases, and I understand that the discretionary element in the formula that could be used to limit price increases has never been used. Moreover, a Federal subsidy is given, whose only merit is that it is left unchanged from year to year and thus declines in real terms. But given the very limited scope for competition from imports, one would have thought that if manufacturing milk supplies were largely at the market level producers could be expected to get all their returns from the market, particularly as quotas are transferable. Of course it has to be remembered that one-third of Canada's milk goes for liquid consumption and earns higher prices for those with quotas. All this -- as well as the volume of criticism from Canadian economists -- suggests that the Canadian government still accords a high priority to maintaining the incomes of dairy farmers, though it was one of the first to make a major impact on its surpluses.

8. After Canada, let us go across the border to the United States. I had the United States in mind when in my introductory remarks I said there was perhaps one important exception to my general view that the preoccupations of OECD governments had not changed very much in recent years. I think a case can be made for arguing that the current efforts of the United States to reduce milk production are so much tougher than in the past that they can be regarded as a new policy. The herd buy-out programme was introduced with little warning and it means that a certain number of producers are abandoning milk production for five years and, more important, their cows are out for good. The most recent forecast I have seen showed production turning downwards in the third quarter of this year, and it will be lower for 1987 as a whole than for 1986. I will discuss the programme's longer term effects on production in a minute.

9. Before doing so I should like to review United States policy in the last few years. The previous major action was the diversion programme, but this was operative for only a short period (from January 1984 to March 1985) and its effect on production (a fall of 3 per cent) was equally short. This diversion programme -- like the current buy-out programme -- was accompanied by provisions linking the actual support price of manufacturing milk to the volume of support purchases: the higher the purchases the lower the price. This principle was included in the Food and Agriculture Act of 1981 for application in 1982/83, while in practice the basic support price was first left unchanged and then successively reduced.

10. Tribute has to be paid where it is due. The United States government had stood up to the Congress and was not merely sending verbal signals that the expansion in milk production should be checked; it was no longer increasing support prices and was even reducing them. It should be remembered that as late as 1977 the Congress had raised the minimum at which the support price could be fixed from 75 to 80 per cent of parity and required the support price to be adjusted (upwards of course) twice a year instead of once. But despite the new climate production continued to advance in the early 1980s, and it was not until the diversion programme was enacted that production turned downwards, but only for 15 months.

11. Thus until the spring of 1986 the United States had taken no action -- either price reduction or administrative constraint -- severe enough to reduce production to market requirements. The question is: has it done so now through the herd buy-out scheme? Tougher though the scheme is than previous measures, it is difficult to give an answer and this is why I said earlier there was "perhaps" one major exception to the general rule that the preoccupations of governments have not changed very much. Forecasts presented by the USDA this summer to the OECD Group on Dairy Products show that although the number of dairy cows is expected to continue to fall for several years milk production will begin to turn upwards in 1988, under the remorseless influence of rising yields per cow. Moreover,

these are expected to make an upward jump around the turn of the
decade as bovine growth hormone comes into commercial use.
Nevertheless USDA thinking is that by the beginning of the next decade
dairy supplies will be more or less in balance with commercial demand,
thanks to a rather abrupt reversal of the consumption trend which some
people expect to continue. But if demand does not increase as
forecast, and supplies remain excessive, it is the current intention
to take steps to introduce further measures of a set-aside or herd
buy-out nature. But there is no legislative provision for doing so
and, so far as I am aware, there is no public political commitment to
do so. Nevertheless, let us give credit for good intentions and say
that the United States may be on the way to achieving a satisfactory
domestic supply/demand balance, though without leaving any significant
market for other dairy exporting countries. It may be noted in
conclusion that there is no talk about quotas as a possible solution;
they seem to be contrary to the basic free market American approach,
though outsiders may be pardoned for paying more attention to the
frequent departures in practice in agricultural policy from this
approach.

Japan

12. Japan is not often talked about in the dairy context.
Historically there is no tradition of consumption of manufactured
dairy produce in Japan and nearly two-thirds of Japanese milk still
goes for liquid consumption. Nevertheless both production and
consumption have been increasing, the former more quickly than the
latter. I will not go into all the details of government policy on
dairying, but the producer price of milk is high in Japan, with the
government operating a guaranteed price for a standard quantity. Over
a lengthy period the standard quality was kept unchanged, though more
recently it has been raised. The guaranteed price was also kept
stable and though some very modest increases were subsequently made
there was even a reduction of 2.8 per cent in the fiscal year April
1986 to March 1987. This resulted primarily from a decline in feed
costs as a result of the high value of the yen, but the change was
intended to discourage the over-supply of milk. These government
measures have been supplemented by a producer-operated quota sheme,
but in the Japanese context I think we can be sure that the system has
at least the government's blessing, and this remark also applies to
the recent reductions in the quota made because surpluses had
developed, though these were small by international standards.
Imports are controlled as part of the overall dairy policy, but they
are reasonably significant for cheese and skim milk powder and the
government has preferred to try to curtail production rather than
eliminate imports. This, too, may perhaps be regarded as a change in
the government's approach to the dairy sector.

Europe

13. Turning now to Europe, there has been a long history of attempts
to constrain the growth of milk production. Sooner or later all have

turned to quotas after years of insufficient action. I do not wish to enter into judgement on the efficiency of different methods of restraining production -- that is the task of another speaker this afternoon -- but I do wish to make the point that the recent adoption of quotas, not only by the Community but also by Nordic countries, does not represent a new role for governments in dairy policy. It is yet another attempt to reduce milk supplies without seriously hurting farmers; there was certainly no aim of bringing about an immediate supply/demand balance.

14. Looking first at Austria, this country introduced a quota scheme in 1978 in recognition that the "crisis coin" or co-responsibility levy on milk producers and an incentive payment to keep milk on the farm were not enough. The effect on producers was in theory a sharply increased deterrent to extra production, the levy changing from about 15 per cent of the average producer price to 40 per cent of the price of milk in excess of the producer's quota. The quotas allocated corresponded to about 120 per cent of domestic consumption of butterfat and a rather higher level of solids-not-fat, and the impact of the levy on over-quota production was reduced because in practice the levy was only applied if the national quota was exceeded by more than 22 per cent. Moreover, the quotas of producers who delivered in excess of their quotas were raised in subsequent years. Though this last provision is no longer applied, the system remains essentially unmodified.

15. Switzerland, too, began to apply quotas a year before Austria, again after a long history of attempts to curb production by applying a levy and taking ad hoc measures such as a slaughter scheme from 1968 to 1970. Initially quotas were conceived as an emergency measure but the system is still in force, though many adjustments have been made. Some of the initial severity of the scheme was blunted by applying penalties only when the quota for all the members of a co-operative was exceeded, instead of being applied to producers individually. For a temporary period the highest mountain areas were exempt from the scheme, but production shot up in these areas and they had to be brought back into the scheme. An interesting feature is the fact that the quota cannot exceed a certain limit per hectare so as to discourage intensive production. Consideration has been given to a system of transferability of quotas, but so far this has not been adopted. Switzerland has long been an exporter of cheese and the national quota has been fixed accordingly; though production tends to be higher than the federal authorities would like and requires significant support from them the Swiss quota may be considered to be broadly in line with market requirements.

16. Norway introduced a voluntary quota scheme as long ago as 1977, called at the time a "bonus arrangement" or "voluntary two-price system." Dairy farmers who reduced deliveries received a higher price than those who did not. So far as I am aware this was Norway's first attempt to reduce production, though this country also pursued a policy of reducing the use of concentrated feeds by raising their

price and making subsidies available for the production of roughage.
However, this gentle approach did not prevent the growth of supply
pressures and a more orthodox scheme was introduced in 1983. This in
turn had to be tightened in 1985.

17. Finland is well known for its inventiveness in devising measures
to curtail all forms of agricultural production, including several
schemes for the slaughter of dairy cattle, limitations on the size of
dairy herds, declining real prices, incentives to retain milk on the
farm, and so on. But these did not suffice, and like Norway Finland
introduced a voluntary quota scheme in 1981. Again like Norway this
did not solve the problem and a compulsory scheme was introduced at
the beginning of 1985. Sweden was the last of the Nordic countries to
apply quotas. This it did in the middle of 1985 after a long period
of relying on more traditional methods of supply constraint.

18. The history of the EEC has been similar to that of other European
countries. It goes back a long way. Who now remembers that the first
cow slaughter scheme was implemented from 1969 to 1970 and a beef
conversion scheme from 1969 to 1971? Under these two schemes about
500 000 cows were removed from dairy production. They were part of a
wide ranging programme aimed at reducing the farm population generally
through aids to the retirement of older farmers and assistance for
younger ones to find other employment, the removal of marginal land
from agricultural production and general structural improvement. The
proposals to reduce the dairy herd envisaged a four year programme
with cow numbers falling from 22 to 19 million, but became overtaken
by events as milk supplies tightened and shortages began to appear.
Who remembers shortages nowadays?

19. Though the shortage was short-lived it inevitably influenced
attitudes for some years to come. After all, dairying depends heavily
on weather-sensitive pasture, and though it does not seem necessary
nowadays to make contingency plans for shortages, governments were
perhaps forgiveably slow to recognise that times had changed.
Characteristically, proposals of the Commission for putting a levy on
milk supplies if butter stocks exceeded 300 000 tonnes and related
measures for 1974/75 were rejected by the Council. However, as well
as pursuing a policy of declining real prices, the Community again
took direct measures in 1977: a further joint programme for conversion
to beef production or the non-marketing of milk, this usually
entailing the slaughter of the cows.

20. Though considerable attention was given over the years to
increasing consumption, the supply/demand situation continued to
worsen. It may or may not be a coincidence but the final stages of
the discussions within the OECD on the report on positive adjustment
policies were immediately followed by Community proposals for a
super-levy, known to the rest of the world as a quota scheme.
Implemented on 1st April 1984 for a five year period, the scheme
resulted in a fall in deliveries of over 4 per cent in the first year,
which may seem very large. Indeed it was large, except in comparison

with the magnitude of the problem. Moreover the fact that most
countries operated quotas on the basis of dairies rather than
individual producers blunted the effect of the super-levy provisions
and explains at least in part the fact that the reduction in quotas
for the second year was largely ignored by dairy farmers. Recognising
the cut in deliveries that had been achieved was not sufficient the
governments of the member states have agreed to cut quotas further
-- but not at once, desperate as the Community's stock situation had
become. This provides an outstanding example of the preferences of
governments for giving priority to farmers' income over economic
factors. It is also shown by the general practice of redistributing
quotas of producers who abandoned production, rather than taking the
opportunity to reduce the level of deliveries, which is, after all the
object of the exercise.

Trade

21. The growth of production in OECD countries first had the effect
of making those countries which were importers self-sufficient
-- thereby reducing the possibilities for exports to OECD countries
and still more increasing the competition on other markets. Most OECD
countries have barriers of some sort to imports, though let us not
forget the substantial concession on butter made to New Zealand by the
EEC or the significant imports of cheese made by the United States and
by Japan. Even before the surplus situation became as bad as it is
now countries had protective barriers against imports because of the
much lower prices for the relatively small quantities that enter
international trade than those prevailing in the main consuming
countries.

22. Naturally this situation did not make for good relations between
importers and exporters. The problem was recognised during the Tokyo
Round and out of it was born the International Dairy Arrangement and
the corresponding Council and protocol committees. It was probably
the hope of the exporting countries that the minimum prices for
exports fixed under this Arrangement, in continuation of those
previously fixed within the GATT or the OECD, would gradually be
raised to remunerative levels, as a result of the policy measures
taken by importing (or potentially importing) countries reflecting the
discussions in the Council. However these hopes were doomed to
disappointment. As we have seen governments have hesitated to take
sufficiently severe measures to get supply and demand into balance.
At first minimum prices were raised but they soon came down again
under the weight of accumulating surpluses. The problem then became
to ensure that the minimum prices were respected, particularly as the
Community took an opportunity to rid itself of some of its stocks of
butter so old as to be unsaleable at normal prices -- a situation
which the framers of the Arrangement had not foreseen -- and
technically broke the Arrangement. Relations between competing
exporters were already strained and this development led the United
States and Austria to leave the Arrangement and other exporting
countries have threatened to do the same. But the spirit of

co-operation is not completely dead. New Zealand, through its Dairy
Board, is suggesting a multilateral effort to dispose of a significant
part of surplus butter stocks, a suggestion reminiscent of its taking
over a substantial quantity of surplus butter from the United States
some years earlier in the interests of orderly marketing. Reducing
the current level of stocks would be a major achievement, but the real
test of international co-operation would be the second stage: ensuring
that the stocks did not build up again.

23. It so happens that a piece of work coming to fruition in the OECD
may help countries to make the necessary effort. A study has been
carried out on the consequences of reducing the level of support for
all major commodities by all OECD countries at the same time. Support
has been calculated in a wide sense, and it is perhaps no surprise
that the dairy sector has been found to be the most heavily
supported. Moreover the volume of milk put on the market by countries
not supporting or not significantly supporting their dairy sector is
relatively small and cannot be dramatically increased. So as
countries begin to reduce their support for dairying and demand for
imports increases the effect on the world price -- as given by the
econometric model devised to carry out this study -- is very marked,
greater than for any other major product. Thus the impact on
individual dairy industries in importing countries would not be very
great. This multilateral approach is shown to be much more effective
than individual countries acting separately. This is not the place to
give a detailed account of these results and in any event the study
has yet to be presented and accepted by the OECD Council.
Nevertheless I may express the personal hope that the study will
encourage governments to reappraise their policies and to join
together to tackle the problem effectively.

THE CHANGING ROLE OF GOVERNMENT IN DAIRY POLICY

T. O'Dwyer
Director Animal Products
Commission of the European Communities
Rue de la Loi 200
1049 Brussels, Belgium

ABSTRACT. National milk policies must be viewed in the context of agricultural policies in general, characterized by support and income maintenance on the one hand, and trade policy/international coexistence, on the other. Problems have increased as technological development stepped up the productivity in milk production when economic/financial forces put a break on consumption. The bill for controlling structural surpluses is heavy and only cures symptoms. Therefore, new policy orientations are required. Basically, policies reflect differences in producer structures, different degrees of organization, industrialization, international trade, etc. A necessary coordination of policies will require time and difficult negotiations to end up with final, binding results. With regulation of production follows pressure for compensating policies in the structural and social fields as well as for trade. Planning and carrying out of such far-reaching changes of policy orientation represents important elements of the change of Government intervention in the milk sector. It also includes some type of sharing of costs of storage and non-marketing of surpluses as well as active policy cooperation at international level.

1. INTRODUCTION (1)

As the title of this afternoon's session indicates, a change is taking place in the governments role concerning dairy policies. As it has happened before, changes of aims and targets in agricultural policies, and dairy policies being no exception, continue as normal ingredients of economic history in both developed and developing countries. Thus, "agricultural policies" are key words in the understanding these changes, and "trade policy" is definitely an element with rapidly increasing demand for changes in traditional governmental attitudes. Here again, dairy products represent no
----
(1) I would like to thank my colleague, Mr. Bjornstad, of the Dairy Division of the Commission, for his work in drafting this paper.

exception! Within this framework of national agricultural policies
with their characteristics of support and income maintenance and the
necessity of international coexistence, the milk problems have not
become easier to solve, as technological development has stepped up
the degree of productivity. At the same time, economic and financial
forces have put a brake on the development of global consumption of
milk. In short, national dairy policies are no longer limited to
questions of support and/or income maintenance. They have become
increasingly concerned with the structural imbalances. The main
focus has become control of stocks and at the other extreme paying
producers not to produce! The bill is heavy. An important part of
dairy policy formulation is thus finding ways of achieving the old
policy objectives at reasonable cost. This, Mr. Chairman, is the
background for the changing role of Government in Dairy Policy, and I
shall try briefly to outline some of its main implications.

2. BASIC SUPPORT POLICIES

From time to time individuals, institutions and/or organisations feel
tempted to analyse and compare dairy policies around the world. They
normally conclude, that these policies are very different and
comparisons almost impossible without confining the analysis to some
few elements only, and of a general and accepted nature (e.g. the
prices obtained for the milk produced, measures for import protection
and export support). I shall not elaborate on this analysis.
However, by referring to the European Community Market organization I
may illustrate that an adaptation to "one" dairy policy will always
force participating countries to amend original, national principles
in the area of prices and trade. The EEC-example has probably shown
that any internationally organized cooperation, even as in the
Community case when based on a common Treaty and general commitments
to a defined market policy, is going to require time and difficult
negotiations to obtain final, binding results. However, this is not
a reason for avoiding cooperation. But national reluctance is
understandable when taking into account the differences in basic
producer structures, degree of organization and industrialization,
international trade and so on. In this respect the role of the ECU
and the discipline and procedures to be followed in monetary matters
have certainly limited the scope for national divergencies in the
Community.

Against this background we must see the policies dictated by the
market developments, e.g. decrease in real terms of support prices or
the installation of production-limiting measures (quotas, special
levies, etc.). Such policies are mainly installed to cure the
symptoms and not the basic elements of the disorder. Therefore,
there is an increasing pressure for compensating policies, in
particular in the social and structural fields. Community policies
in these fields are not yet as developed as the market mechanisms in
Common Agricultural Policy. This state of affairs, which probably
can be recognized also in non-EEC countries, is at the root of the

on-going discussion on new perspectives for agricultural policy : a
transfer of land from milk production to other non-surplus
productions, a "temporary" transfer of several millions of dairy cows
into frozen beef, and - last but not least - a transfer of milk
producers to other activities or to pension schemes. This can only
be handled with responsibility if integrated in longer term policies
for trade, structure and the social sector.

Hence, milk policy is no longer an independent activity, and
total budgetary availabilities normally fix a limit to the practical
possibilities.

In the planning and carrying-out of these essential and
far-reaching changes one sees also the changing role of Government in
dairy policy. If the traditional price support policies and related
trade policies tend to widen the imbalance between supply and demand
for dairy products, nationally and internationally, the ship is off
course. But easy solutions are not available, and economic laws
alone do not solve the problems in weak structures with few or no
economic alternatives. Thus, governmental and public involvement is
necessary, at least to ensure a framework for the structural
development in the future.

3. CHANGE OF ORIENTATION

In the Community and in most countries of the Western World the
question seems no longer to be if, but rather to be how policies
should be changed. The financial burdens of agricultural policies,
of which dairy policies play an increasingly important role, are
weighing heavily on almost all public purses. Under the pressure of
increasing market imbalance, the expenses for internal and external
market disposal of dairy products have dominated support policies.
As a consequence the purely non-commercial parts of the markets have
increased and now account for nearly 50% of the international
bulk-trade of dairy products, i.e. butter and SMP. Through the
accumulation of stocks, and with sales of stored products partially
substituting for fresh products, the relative cost-benefit
development of the market policies is becoming increasingly difficult
to support. As international market prices have reached a level
giving no satisfactory remuneration in the very best structures,
there is cause for concern in all countries involved with production
and trade of dairy products. Furthermore, if support policies
intended for the less developed production structures in reality
over-favour the top producers, the contribution of these policies to
the increasing market imbalance is evident.

It is said that the role of the administration of dairy policies
has changed from that of managing markets to that of managing
surpluses. The truth is that the concentration on
production-limiting methods on the one hand and on non-commercial
disposal measures on the other, not only contributes to higher
budgetary expenses in the short run; it also makes it more difficult
to maintain a decent living for small and medium sized milk

producers, as it makes it more difficult for consumers to accept the payment of dairy products at full cost-price levels. This double set of problems shows that "market balance" is not a final aim if it cannot be obtained through balanced measures, creating a background for economy as well as welfare.

In the Community the basic structure of milk production is characterized by more than 50% of the dairy farmers with less than 10 cows, varying from 12% only in the UK, with France, Germany and Ireland in the range of 35-45%, when Italy and Greece are at the top end with more than 80% and 95%, respectively, of their producers having less than 10 cows.

Globally these 50% of the dairy herds have only 13% of the Community dairy cows and produce less than 10% of the milk. This repartition indicates that an important number of milk producers have legitimate claims on the dairy policy, without being able to hope for reasonable earnings based on a market orientated policy alone. This is why social and structural policy aspects have got to be considered. This consideration must include such elements as income supplements, regional development programmes and general early retirement systems, to mention but a few examples.

The EEC Commission has already this Spring presented to the Council of Ministers a package of proposals (1) to reinforce actual legislation and to create new possibilities for reorientation, reconversion or extensification of agricultural production. The following examples can be mentioned : early retirement systems, woodland development, investment incentives for reorientation of production for young farmers, income aids in less favoured areas, etc. These measures are not exclusively directed towards milk production but cannot avoid having a particular impact in this sector.

4. MARKET FRAGILITY

The adaptation of milk production to sales possibilities is evidently a time-consuming process involving a series of elements in addition to quota-measures and assisted sales of dairy products.

Actual stocks of surplus products amount to nearly 2 years of international trade with dairy products (in milk equivalent). Whereas international trade varies in importance between countries, it nevertheless amounts to 8-9% of the global milk supply in OECD, USSR and Eastern Europe. Hence, the pressure on the market is obvious and is accentuated by the time factor involved in policy amendments on a global scene. Since 1984 the Community has made real efforts in not only avoiding annual increases in milk output but in actually decreasing output. Measures recently introduced will further strengthen this tendency. However, decreases in milk output in the Community have not been matched by similar tendencies in other important producing areas in the world. Unfortunately also, especially this year, the reduced milk output is not reflected in

---
(1) COM(86) 199 final/2 (31.7.1986)

stock levels within the Community.

To consider the international market for dairy products as an automatic outlet for surplus production would be a fatal misconception. Economic factors (e.g. international prices for oil and raw materials from developing countries, debt burdens of individual countries and unstable currencies) as well as limited financial power for food aid purposes, limit the scope for market expansion in the short/medium term. Governments have therefore a particular responsibility for participating in some type of sharing of costs to avoid a market break-down, i.e. costs of storage and non-marketing of surpluses. The cooperation on international pricing for dairy products as in operation in the GATT arrangements cannot fulfil its aims without wider participation and reinforced coordination of surplus disposal policies. If any such measures are to be effective on the external market scene, they must be seen against the background of domestic policies aimed at output control.

Even those countries giving only a marginal contribution to the world markets - very often those with a high cost production - are carrying their part of the responsibility. The impact of world market conditions on national dairy policies is not as limited as the proportion of production could indicate. This Congress is being held at a crucial moment when the trade-powers are seeking a basis for negotiating the future world markets for agricultural products. Recent developments indicate that cooperation rather than "war" may characterize the future situation. However, only those countries conducting a multipurpose dairy policy including social considerations and supply security, as well as a competitive price policy, contribute to the mutual, economic and longer-term growth which is essential for further market developments.

These orientations have been under consideration for some time. However, against the fragility of the present imbalanced market situation, they may now represent a more valid approach for governments in their efforts to adapt national and international dairy policies to the new situation.

5. SUMMARY AND CONCLUSIONS

National milk policies must be viewed in the context of agricultural policies in general, characterized by support and income maintenance on the one hand, and trade policy/international coexistence, on the other. Problems have increased as technological development stepped up the productivity in milk production when economic/financial forces put a break on consumption. The bill for controlling structural surpluses is heavy and only cures symptoms. Therefore, new policy orientations are required. Basically, policies reflect differences in producer structures, different degrees of organization, industrialization, international trade, etc. A necessary coordination of policies will require time and difficult negotiations to end up with final, binding results. With regulation of production follows pressure for compensating policies in the structural and social fields as well as for trade. Planning and carrying out of such far-reaching changes of policy orientation represents important elements of the change of Government intervention in the milk sector. It also includes some type of sharing of costs of storage and non-marketing of surpluses as well as active policy cooperation at international level.

TRENDS IN WORLD TRADE IN DAIRY PRODUCTS AND FUTURE PROSPECTS

W. Krostitz
Meat and Dairy Specialist
Commodities and Trade Division
UN Food and Agriculture Organization
Via delle Terme di Caracalla
00100 Rome, Italy

ABSTRACT. Although the expansion of world milk production has slowed since the 1960s supplies have increasingly exceeded effective demand. This led to an accumulation of large stocks of dairy products, high government expenditure on milk price support and generally depressed prices in international trade. During the 1970s, trade increased faster than production, but the proportion of world milk output which entered international trade reached only 5 percent at the beginning of the current decade and has started decreasing more recently.

In international trade in dairy products the principle of comparative advantage has largely ceased to be practised. Low-cost producing and exporting countries, both developed and developing, have lost market shares. The larger part of supplies now consists of heavily subsidized exports from western Europe and North America where milk prices have been supported at relatively high levels and have virtually lost their market-clearing function. In contrast, milk production lagged behind demand in the developing countries as well as the USSR and eastern Europe. Especially the petroleum exporting developing countries and the USSR expanded their imports of dairy products between the early 1970s and the early 1980s. However, the outlook is for rising self-sufficiency in the main deficit regions. Hence, although supplies in exporting countries will probably remain ample, international trade in dairy products may stagnate or even contract in the next several years.

1. REVIEW OF TRENDS AND PATTERNS OF INTERNATIONAL TRADE IN DAIRY PRODUCTS

1.1 Production and Consumption

On average during the past 15 years world milk production has increased at a rate of 1.7 percent a year. i.e. slightly less than population. Compared with the 1960s the expansion of milk production has slowed down in the developed regions and accelerated in the developing countries. Nevertheless, dairying is still very much concentrated in the developed regions, with the EEC, the USSR and the United States producing well over half of the world output. The developing countries, with about three quarters of world population, account for less than one quarter of the world's milk output. Asia is by far the largest producer in the developing regions and has recently also experienced the fastest growth in milk output. The developing countries' share in world production of major milk products is also small (Tables I and II).

TABLE I: WORLD DAIRY SITUATION AT A GLANCE

	Milk Production	Net [1] Imports	Availability [1] [2] Total	Per caput
	(........ million tons ....)			kg
WORLD TOTAL				
1972-74	414.0	...	414.0	106
1982-83	486.5	...	486.5	105
1984-85	503.6	...	503.6	105
Developing countries				
1972-74	82.7	7.8	90.6	32
1982-83	112.2	16.5	128.7	37
1984-85	119.2	18.0	137.2	38
Developed countries				
1972-74	331.2	-7.8	323.4	292
1982-83	374.3	-16.5	357.8	301
1984-85	384.4	-18.0	366.4	303

1/ Milk and milk products in milk equivalent.
2/ Including non-food use and waste.

The shift in production growth from developed to developing countries partly reflects developments in demand. In the developed countries, especially in North America, western Europe and Oceania food consumption of milk and milk products has been stagnating since the 1960s while feed use has become uneconomic. By contrast, in the developing countries demand has been strong, owing to rises in population and incomes, accompanied by urbanization and the associated change in food consumption habits. Although levels of per caput intake are low, the consumption of milk and milk products has a long tradition in Latin America, large parts of Africa and western and southern Asia. However, demand has more recently also begun rising fast in the urbanized areas of East Asia. Notwithstanding accelerated growth in domestic output, demand outpaced production of milk in developing countries in most of the period under review.

1.2 Trade

Although until the early 1980s international trade in dairy products rose faster than world production (Table III), exports (excluding intra-EEC trade) in terms of milk equivalent accounted for just 5 percent of world milk output in the first half of the current decade. Up to the early 1970s, trade had occurred mainly in the form of butter, cheese, skim milk powder, casein and condensed and evaporated milk. However, over the past decade, while trade in these commodities has risen (Tables IV and V), it is trade in whole milk powder, milk-based infant foods and other dairy specialities which has experienced the most impressive growth.

Owing to the divergent trends in demand and production in developed and developing regions, not only the volume but also the direction of trade flows has changed considerably. During the first two decades after the Second World War, most of international trade in dairy products had been among developed countries, with the low-cost producing countries of Oceania as the main exporters and several countries in western Europe and later on Japan as the main importers. North America changed from a net exporter in the 1960s to a net import position by the mid-1970s but has again become a sizeable net exporter in recent years. Western Europe moved from a net import to a net export position. Until the 1960s the twelve countries of the present European Community had constituted the world's largest net import area. However, since the 1970s the European Community has been by far the largest net exporter of dairy products, accounting for over half of world exports in the early 1980s. Oceania, the leading export region in the past has hardly been able to maintain its sales abroad in absolute terms and by the first half of the 1980s its share in world exports had decreased to between a quarter and one-third. Among the developed market economies only Japan still has a sizeable net import demand.

In contrast, the centrally planned developed countries which, up to the 1960s, had been a net exporting region, have become a net importing region. In the first half of the 1980s the USSR has been the world's largest importing country. Most of the USSR dairy products imports have been butter. Among the developing countries, the few traditional exporters such as Argentina, Uruguay, Nicaragua, Zimbabwe and Kenya, have lost ground in international markets, with some of them becoming temporary or even permanent importers. Overall, the developing countries have been the biggest group of net importing countries in international dairy trade since the early 1970s (Table III).

Up to the 1960s, developing countries had played a major role only as importers of condensed and evaporated milk and to a lesser extent of milk powder. However, with the establishment of local recombining industries, especially in South-East Asia, imports of condensed and evaporated milk were largely replaced by purchases of skim milk powder and butter oil (anhydrous milk fat). In the second half of the 1970s, when incomes in the developing regions, particularly in the petroleum-exporting countries, rose rapidly, imports of condensed and evaporated milk increased again, accompanied by an even faster rise in imports of other dairy products. While imports of skim milk powder and butter oil for the recombining of liquid, condensed and evaporated milk and various other milk products continued to increase, the petroleum exporters also became important new outlets for cheese, butter, whole milk powder and partially skimmed milk powder. In recent years whole milk powder and milk-based infant foods have actually been the biggest item on the dairy import account of the developing countries, notwithstanding national and international campaigns to promote breast feeding.

TABLE II: PRODUCTION OF MAJOR MILK PRODUCTS (THOUSAND TONS)

	1972-74 average	1980-81 average	1983-85 [2] average
BUTTER			
World total	6 303	6 936	7 657
Developed countries	5 033	5 396	6 004
Developing countries	1 270	1 539	1 653
CHEESE [1]			
World total	8 771	11 432	12 470
Developed countries	7 562	9 979	10 921
Developing countries	1 210	1 453	1 549
MILK POWDER (all types)			
World total	5 733	7 369	7 962
Developed countries	5 423	6 902	7 439
Developing countries	310	467	523
CONDENSED AND EVAPORATED MILK			
World total	4 352	4 663	4 595
Developed countries	3 642	3 701	3 609
Developing countries	709	962	986

[1] Including cottage cheese and curd. [2] Provisional.

TABLE III: INTERNATIONAL TRADE IN MILK AND MILK PRODUCTS [1]
(MILLION TONS OF MILK EQUIVALENT [2])

	1972-74 average	1980-81 average	1983-85 [3] average
EXPORTS			
World total	17	26	25
EEC	8	15	12
Oceania	5	5	6
North America	1	2	4
IMPORTS			
World total	17	26	25
Developing countries	8	18	19
USSR and eastern Europe	1	3	3
Japan	1	1	1

[1] Excluding EEC intra-trade. [2] Dairy products converted into milk equivalent on the basis of total milk-solids content. [3] Provisional

TABLE IV: EXPORTS OF MAJOR MILK PRODUCTS (THOUSAND TONS) 1/

	1972-74 average	1980-81 average	1983-85 3/ average
BUTTER			
World total	489	915	800
EEC	110	500	347
Other western Europe	37	30	47
Oceania	239	236	241
North America	8	27	44
Argentina	8	-	3
CHEESE			
World total	500	735	864
EEC	170	333	427
Other western Europe	124	176	171
Oceania	118	132	145
Argentina	6	3	6
MILK POWDER			
World total	1 070	1 793	1 850
EEC	500	1 070	775
Other western Europe	67	79	112
Oceania	318	349	391
North America	152	218	396
Argentina	9	6	7
CONDENSED AND EVAPORATED MILK			
World total	560	824	748
EEC	450	574	528
North America	22	130	107
CASEIN			
EEC 2/	2	25	44
Oceania	51	64	77
Argentina	7	1	1

1/ Excluding EEC intra-trade. 2/ Net export. 3/ Provisional.

Total dairy product imports of developing countries doubled between the early 1970s and the early 1980s, reaching over 18 million tons of milk equivalent, or about two thirds of world imports at that time. Within this group the petroleum-exporting developing countries raised their imports more than three times to over 8 million tons of milk equivalent. Owing to economic recession, exacerbated by the effects of sharply lower petroleum prices on export revenues, commercial imports of the developing countries have fallen since the early 1980s though they still absorb the largest part of world exports.

With the exception of the Philippines, Singapore, Thailand and Cuba, countries with large milk recombining industries, all leading commercial importers of dairy products in the developing regions have been petroleum exporters in recent years. The most important among them have been Mexico, Venezuela, Nigeria, Algeria, Libya, Saudi Arabia, Kuwait, Iraq, Iran, Malaysia and Indonesia. Lower-income countries have also imported sizeable amounts of milk products, though to a considerable extent as food aid, with India, Egypt and Pakistan being the largest recipients. In fact, apart from rising populations and incomes, the increase in dairy product imports of developing countries reflects ample availabilities of milk products offered by the developed market economies at heavily subsidized prices or free altogether. Several East European countries and, even more so, the USSR have also taken advantage of the opportunities arising from the surplus problems of certain OECD countries.

TABLE V: IMPORTS OF MAJOR MILK PRODUCTS (THOUSAND TONS) 1/

	1972-74 average	1980-81 average	1983-85 2/ average
BUTTER			
World total	500	891	815
Developing countries	188	460	461
of which: OPEC	46	165	184
USSR	82	232	226
CHEESE			
World total	490	710	860
Developing countries	90	292	380
of which: OPEC	30	181	233
North America	129	129	154
Japan	40	73	78
MILK POWDER			
World total	1 060	1 805	1 860
Developing countries	729	1 529	1 581
of which: OPEC	112	420	444
USSR	24	73	58
Japan	73	102	108
CONDENSED AND EVAPORATED MILK			
World total	565	814	752
Developing countries	441	785	720
of which: OPEC	160	421	398
CASEIN			
United States	50	63	90
Japan	17	21	23

1/ Excluding EEC intra-trade. 2/ Provisional.

1.3 National Policies and their Implications for International Trade and Prices

The structural changes in international dairy trade partly reflect different trends in economic growth between developed and developing countries. However, national dairy policies have also had important international repercussions. In many developing countries the effects of rising incomes on demand for milk and milk products were reinforced by policies aimed at low consumer prices.

Except in several oil exporting countries which were able to afford consumer and/or producer subsidies on a large scale, such price policies, generally coupled with relatively liberal import policies, tended to discourage the development of local milk production. In fact its comparative advantage has been considered to be very low in a situation of almost chronically depressed prices in international trade. This applies not only to the lending policies of national and multilateral development banks but also to the investment policies of transnational companies, which have been playing a major role in international dairy trade. Although such companies have made considerable investments in milk processing and distribution in developing countries their interest in the development of local milk production has lessened in view of the ample availabilities of cheap raw materials in international markets.

Policies aimed at low consumer prices still prevail in large parts of Latin America, Africa and South-East Asia. However, over the past 10 to 15 years a growing number of countries in North-East, South and West Asia as well as North Africa and to a lesser extent Latin America and sub-Saharan Africa have given high priority to the development of domestic dairying. Several oil exporting countries, notably in the Near East and, until recently Venezuela, while continuing a rather liberal import policy, have subsidized local milk production. Others have controlled imports, both commercial and concessional, in such a way that disincentive effects on domestic dairying were avoided. Prominent examples of rapid milk production development during the past decade include, in particular, India, China, the Republic of Korea, Saudi Arabia, Cuba and Venezuela.

Governments in eastern Europe and the USSR, though regularly adjusting producer prices, have also pursued policies of stable consumer prices. In the USSR and the German Democratic Republic retail prices of major milk products have been kept virtually unchanged for a quarter of a century. The subsidies required to implement these policies rose substantially. More recently a number of eastern European countries have begun to adjust consumer prices to the cost of production and marketing.

Conversely, in the developed market economies of the northern hemisphere, policies have tended to raise the degree of self-sufficiency in milk and milk products, especially in western Europe. While in most continental European countries market protection in the agricultural sector has a long tradition 1/, protection of western European dairy markets has been substantially increased since the 1960s. As discussed in earlier FAO documents 2/, the reasons for protective policies are many and include the desire to achieve certain levels of national self-sufficiency and to secure stability in domestic markets, balance of

---

1/ See Tracy, M.: "Agriculture in western Europe, Challenge and Response, 1880-1980", Granada Publishing Ltd., London, Toronto, Sydney, New York, 1982, and Priebe, H. "Die subventionierte Unvernunft", Siedler-Verlag, Berlin, 1985.

2/ "Protectionism in the Livestock Sector", document CCP: ME 80/4, FAO, Rome, 1980 and "International Trade in Dairy Products - Review, Prospects and Issues", document CCP: 85/16, FAO, Rome, 1985.

payments or employment considerations and, in particular, the objective of enabling farmers to participate in general income developments. In the dairy sector, social considerations have played a specially important part in the formulation and implementation of support policies because, at least until recently, milk production has been primarily on small and medium-sized family farms and the proportion of farms which engage in dairying has been high.

In the EEC, the world's largest milk producing area, a common dairy policy was established in 1968. While variable import levies and export refunds are the main elements regulating external trade in milk products, there are also provisions for support (intervention) purchases of butter, skim milk powder and, to a lesser extent, cheese in domestic markets. The geographical coverage of the EEC's common dairy policy was widened in 1973 when the United Kingdom, formerly the world's leading importer of dairy products, as well as Denmark and Ireland became members, and in 1981 when Greece joined the Community. From 1986 Spain, which for many years has already practised a support policy similar to that of the EEC, and Portugal, will gradually apply the Community's common agricultural policy.

Switzerland, Austria, Norway, Sweden and Finland have also regulated external trade by a variety of measures and supported milk producer prices at relatively high levels. Support prices have also been high in Japan which restricts imports of dairy products, with the main exceptions of cheese and casein. In Japan restriction of imports is coupled with provisions for the stabilization of domestic butter, milk powder and condensed milk prices and with deficiency payments on manufacturing milk. The United States milk market is supported by provisions to remove surplus butter, cheese and skim milk powder from the market. Except for casein and cheese, the United States has normally allowed only insignificant imports of dairy products and been a net exporter of milk products overall since the late 1970s. Canada combines a support system for butter and skim milk powder with deficiency payments on manufacturing milk. The governments of New Zealand and Australia have supported milk prices at considerably lower levels than countries in the northern hemisphere. However, as in North America and some western European countries, the average price to milk producers in Oceania is raised above the levels of manufacturing milk prices by systems aimed at achieving relatively high returns from the liquid milk market. 1/

In Japan demand for milk and milk products, which has been stimulated by economic growth and changing food consumption habits, and milk ion have risen by and large at the same pace during the period under review. However self-sufficiency ratios increased in North America and western Europe where demand for milk and milk products as a whole has been rather slack since the 1960s. Although dairying has gradually become more specialized and benefited by rapid technical progress, governments have hardly adjusted support price policies. There has thus been an increase in prices of milk and milk products relative to other

---

1/ For more details on dairy policies and structural changes of dairy industries in the developed market economies see: "Positive adjustment policies in the dairy sector", OECD, Paris, 1983.

agricultural commodities, especially in countries where markets for
vegetable fats and protein foods as well as animal feeds have been left
relatively free. Throughout the developed market economies, butter
consumption in particular has been adversely affected by an increase in
its price relative to vegetable fats. At the same time, milk and milk
products, unless subsidized, have become uneconomic as livestock feed
whereas the relationship between milk producer prices and concentrate
feed prices has changed in favour of dairy farmers.

In this situation of opposite trends of demand and production,
western European and North American countries first restricted imports.
During the past decade imports of butter, milk powder and condensed milk
of northern hemisphere developed market economies have been severely
limited, mainly by non-tariff barriers such as quantitative restrictions
and variable levies. Cheese imports have been restricted to a lesser
extent while those of casein and lactose have remained largely free.
When output eventually exceeded effective demand, surpluses were disposed of at subsidized prices in both domestic and international
markets.

The main features of surplus disposal on domestic markets in
western Europe have been subsidized feed use of liquid and dried skim
milk and special sales, at reduced prices, of butter to consumers, the
food processing industry and, most recently, the compound feeds industry. In the EEC, subsidized feed use of liquid and dried skim milk has
been of the order of 2 million tons of skim milk powder equivalent in
recent years, accompanied by sizeable subsidization of skim milk for the
manufacture of casein, most of which is exported. Other western European
countries, notably Austria and Finland, have also channelled significant
amounts of subsidized milk products into the feeds sector. At the same
time the EEC, Austria, Switzerland and the Scandinavian countries have
heavily subsidized exports. In the United States domestic and external
food donation programmes have absorbed the bulk of surpluses. Canada
has disposed of its surpluses mainly in international markets, partly as
food aid, although milk producers have contributed substantially to the
financing of export losses.

All in all, subsidized disposals of milk and milk products by
governments of the developed market economies have been around 40
million tons of milk equivalent annually in recent years, with the EEC
and the United States accounting for most of this. In the United
States, on average during fiscal years 1983/84 to 1985/86, dairy
products at an equivalent of 5 million tons of milk were removed from
the market, i.e. about 8 percent of all milk marketed by farmers. In
the EEC of Ten, although milk collection by dairy plants decreased from
the peak of 104 million tons in 1983 (the year prior to the imposition
of quotas) to 100 million tons by 1985, sales of milk and milk products
at the full market price have averaged less than 70 million tons of milk
equivalent in recent years.

As the cost of surplus disposal increased considerably, governments
in the northern hemisphere began taking action to curb milk production.
Measures included dairy cow slaughter premiums, co-responsibility levies
on the milk price, levies on concentrate feeds and - probably the most
drastic step - limitation of the price guarantee to specified quotas of

milk marketed by individual farmers. Such systems of milk marketing quotas have been applied in Canada since the late 1960s and were subsequently introduced in some smaller western European countries and, in 1984/85, in the EEC. Only in the United States, and to a lesser extent Japan, have governments lowered the nominal support price with a view to curbing output and stimulating demand. This has been accompanied by measures to reduce the dairy cattle population.

In absolute terms, expenditure on dairy price support has reached particularly large levels in the EEC and the United States but, in relation to milk marketings, expenditure in Austria, Finland, Norway, Sweden and Switzerland has also been substantial. In Canada, however, the proportion of subsidies to total dairy farm incomes has decreased over the past decade. Including subsidies in Japan and Oceania but excluding aids by national governments in the European Community (which also appear to have risen), expenditure on dairy price support in the developed market economies is estimated to have reached over US$ 8 billion per year in the first half of the 1980s. In addition to these direct costs, the administration of increasingly complex dairy policies results in further expenses. Finally, the production, processing and storage of milk and milk products over and above market needs has led to a large misallocation of resources and, in particular, a waste of nutrients.

For commodities with narrow international markets such as those for dairy products, changes in the demand/supply position of individual countries, due to economic or political factors or weather, have characteristically resulted in a high degree of instability in international prices. During the past 15 years or so, the combined effect of slack demand, rapid technical progress and protectionist policies, notably in the developed market economies of the northern hemisphere, has accentuated this instability about a generally depressed level of prices in international dairy trade. For a short period in the early 1970s (at the time of the world food and feed crisis) prices in international dairy trade rose to about the levels of domestic support prices in North America and western Europe. Following a sharp decrease around the mid-1970s, international prices increased towards the turn of the decade, mainly in response to rapidly rising import demand in the petroleum-exporting developing countries. However they remained well below the levels of domestic support prices in western Europe and North America. Since the early 1980s international prices have again decreased sharply.

During the period under review, low-cost, efficient producers in exporting countries, both developed and developing, have been gradually excluded from their traditional markets which are now being supplied from domestic high-cost production. At the same time low-cost producers have lost part of third markets to high-cost producers whose governments subsidize the sale of surplus products abroad. In recent years, approximately three-quarters of total dairy exports has been subsidized. Confronted by this situation, Oceanian low-cost producers have hardly been able to maintain export volumes despite a growing international market, while most traditional suppliers among developing countries have had to withdraw from export marketing. Also, during the mid-1980s returns to

milk producers in Oceania, notably New Zealand, have fallen much more
than in the northern hemisphere.

1.4 Food Aid

Since the 1950s donations of milk products as food aid have been a
particular feature of surplus disposal programmes of several developed
market economy countries. However, as food aid has been by and large an
accidental by-product of dairy support policies, shipments have fluctua-
ted considerably over the year. Until the 1960s most food aid in dairy
products came from the United States. Between the late 1960s and the
early 1980s the EEC was by far the largest donor, but in most recent
years the EEC has reduced food aid in dairy products in favour of grains
while the United States has again made substantial shipments in the
1980s. Certain quantities have also been provided by Canada, some
smaller western European countries and occasionally Australia, New
Zealand and Japan. Skim milk powder and butter oil have been the
principal items, but some countries have also made available cheese,
condensed milk and whole milk powder. On average during the first half
of the 1980s food aid in dairy products was over 3 million tons of milk
equivalent per year, more than one tenth of total world dairy products
exports. However, the food aid component accounted for nearly one third
of dairy imports of those developing countries which do not export
petroleum.

1.5 International Market and Price Stabilization Arrangements

Since the mid-1950s, under the FAO Principles of Surplus Disposal, the
Consultative Sub-Committee on Surplus Disposal (CSD) has monitored food
aid transactions in agricultural commodities, including dairy products,
with a view to minimizing possible harmful effects on commercial trade
and domestic production of recipient countries. In 1963, under the
auspices of OECD, an arrangement concerning minimum export prices for
whole milk powder was established, followed in the early 1970s by similar arrangements concerning skim milk powder and butter oil (anhydrous
milk fat) under GATT. During the Tokyo Round of the Multilateral Trade
Negotiations the International Dairy Arrangement (IDA) was established.
   The IDA, which came into operation in 1980, has essentially
formalized, and extended to additional products, the previous OECD and
GATT arrangements. Under the IDA, minimum export prices have been set
for whole, skim and butter milk powder, butter and butter oil (anhydrous
milk fat) and certain cheeses. In the early 1980s, actual prices in
international trade were generally well above IDA minimum levels, but
more recently an increasing part of transactions, especially in butter,
has been below these levels. Up to 1984 there were 18 participants but
in 1985 Austria and the United States withdrew from the Arrangement.

2. MEDIUM TERM OUTLOOK

During the remainder of the 1980s world demand for milk and milk products is expected to grow at a relatively slow pace, with most of the

rise likely to occur in the developing countries. With economic growth projected to be most pronounced in Asia, this continent is likely to see the fastest expansion in the consumption of milk and milk products. However, overall in the current decade the rise in consumption in the developing countries will probably be less rapid than in the 1970s.

In the developed regions there seems to be scope for increased demand in eastern and southern Europe, the USSR and in Japan. In North America, Oceania and North West Europe population growth is slow, and income elasticities of demand for a number of dairy products have for some time been around zero or even negative. While per caput consumption of cheese and fermented milk products could increase further, especially in North America, that of butter, liquid milk and condensed and evaporated milk may stagnate or continue to decline. If relative prices of milk and milk products are not lowered, butter will continue to face strong competition from vegetable fats and competition between milk and vegetable proteins will grow.

Rising domestic demand will continue to stimulate milk production, at least in parts of the developing regions. Notably in Asia the effects of economic growth will probably be supported by high priority on dairying in agricultural development plans. In this continent growth in milk output may well keep pace with the rise in demand. Production and consumption are projected to grow in particular in the world's most populous nations, China and India. In India, milk output is projected to reach 65 million tons by 2000, compared with less than 40 million tons at present. The Chinese production target for 2000 is 30 million tons, ten times the present level. Efforts to raise milk output have also been undertaken in the Near East and other petroleum exporting developing countries. However, these countries should remain substantial importers of dairy products. In Latin America, improved economic conditions would probably also stimulate dairy production. However, whether Latin America will make fuller use of its relatively good potential for dairying depends largely on government price and import policies. On the whole in the developing regions, the degree of self-sufficiency which fell sharply in the 1970s, may increase somewhat.

In eastern Europe and the USSR where milk yields are still relatively low, rising cow numbers and yields could result in production growth exceeding that of demand, especially if consumer prices are adjusted towards the cost of production and marketing. Japan foresees continued expansion of both production and consumption. While Oceania has a great potential to raise milk output, the realization of this potential depends on an improvement in conditions in international trade.

In most countries of western Europe and in Canada milk marketings are now governed by quota systems. While such systems have halted the expansion of milk supplies, quotas in most countries still allow farmers to deliver considerably more milk than markets can absorb at the guaranteed price. In the EEC where milk quotas will be reduced further in the next few years, the excess of supply over effective demand should decrease though adjustment is likely to be slow. In spite of heavy penalization of above-quota deliveries, milk collection in the EEC not only continues to exceed the level of the global quota, but there is also a remarkable increase in the milk-solids, especially butter-fat, content of milk marketed by farmers. Generally, in western Europe and Canada, governments have tended to grant higher milk prices to farmers in compensation for the restriction on their marketings whereas prices of grains and other crops are under downward pressure. Hence, relative consumer prices of milk and milk products have been rising. In the United States, a combination of measures to curb production and increase demand could reduce excess output more quickly. However, in view of rapid technological progress and low feed prices, additional measures would appear to be necessary if full equilibrium between supply and remunerative outlets were to be achieved in the US dairy market.

Hence, for some time to come, supplies in exporting countries will probably remain ample. At the same time, self-sufficiency in major deficit regions tends to rise. In this situation, international trade in dairy products may stagnate or even shrink in the next several years, with little prospect for a major recovery in prices.

MILK SUPPLY MANAGEMENT PROGRAMMES: THEIR STRENGTHS AND WEAKNESSES

Professor Dr. Friedrich Hülsemeyer
Federal Dairy Research Centre, Institute for Business Administration and Market Research, Kiel, FRG
Postbox 6069
D-2300 Kiel 14

ABSTRACT
Adjustment of milk production to the demand is the essential consequence of the growing disproportion between relative trends of the market and technological progress.
   The action margin of agricultural market policy ranges between a stronger emphasis on the function of prices as a means of market equilibration - an approach which is considered problematic with respect to social and regional policy - and an administrative production planning which is questionable from an organizational and structural policy standpoint.
   On the background of a further escalating surplus situation on the butter and skimmed milk powder market the European milk market policy is forced to act. The result, however - a very soon further reduction of the price supported quota by at least 10 to 15 % or a restricted obligation to support the EEC price level - cannot be predicted reliably.
   What is certain is that a price cut will be the more probable solution, if policy due to its inability to arrive at a consensus will turn out to be too weak in the longer term to meet such a development by an adequate strategy.

MAIN TEXT
Agricultural policy - in fact, who does it still offer an advantage to?

According to Dale A. HATHAWAY (1) the agricultural problem exists "for laymen in the form of surpluses whose increasing quantities and costs are regularly reported;
   for agricultural scientists in the form of unsatisfactory allocation of production factors;
   for farmers mainly in the form of low and varying income despite hard work, careful management and often high capital investment;
   for members of parliament in the form of a hole of milliards in the budget;
   for politicians in the form of a trap promising to an increa-

sing extent a premature end of their political career, namely when they are trapped between unsatisfied farmers and furious taxpayers - with but little hope to satisfy one of both parties, let alone both".

The errors of the past

A better description can hardly be given when assessing also the current situation of the European milk market policy - more analytically perhaps to do better justice to the specific conditions.

Dairy farming is the most important agricultural branch of production in the Community: About two fifths of all farms covering approx 40 % of the agricultural area are dealing with milk production which constitutes nearly one fifth of the value of agricultural end production.

In considering further the strong spatial concentration of dairy husbandry in areas which scarcely allow alternatives of production it becomes abundantly clear how much importance must be attached to the development of this market for the economic situation of farming.

These facts may provide an explanation why the common E.E.C. agricultural policy has, for such a long time, persistently put aside overall economic necessities in favour of incomes policy aspects.

Indeed, as early as 1969, that means only one year after the Common Regulation on Milk and Milk Products came into force, the European Parliament took the view "that on the background of the alarming development of surplus stocks of butter and skimmed milk powder the system of the common milk market policy had to be reviewed in order to solve existing problems" (2).

Since then, nearly half a generation of professors of rural economy and economics has predicted the financial collaps not only, but mainly of this market organization.

So, one may ask what was this scepsis based on?

In the field of external protection functioning of the common milk market regulation has, from the beginning, been impaired by the imbalance between supported milk fat and milk protein prices, on the one hand, and plant fats and proteins imported mainly free of tax, on the other hand.

In terms of prices these two imbalances discriminate both: butter consumption and usage of skimmed milk and skimmed milk powder in feeding.

Above all, however, they are responsible for a favourable price relation oilmeal/cake: milk which is the most important incentive for intensifying milk production.

On the other hand, it is also true that the foreign trading regulation on oilseed, oils and oilmeal/cake is subject to international agreements within the framework of the General Agreement on Tariffs and Trade (GATT) and can, therefore, not be corrected autonomously by the Community. Not less irrealistic is a deconsolidation of the GATT obligations, because the Community is hardly able to offer compensations and scaring of the most important exporter, the USA, would be unthinkable both from the economic and the political viewpoint.

On this background a problematic decision was taken by the Council of Ministers: for reasons of income support the price target for milk aimed at on the home market in the form of the target price and hedged by a good 90 % by the intervention prices for butter and skimmed milk powder was set above the equilibrium price, i. e. above the level that would have settled down in the case of 100 per cent self-sufficiency in the EEC, whilst renouncing a regulation of production at the same time.

Hence, the intervention measures taken on the milk market have been allocated a function that was in several respects contrary to the prevailing market conditions (3).

Nevertheless, this view has almost never been taken seriously, because the belief in the possibilities of expansion of the spontaneous and supported domestic and external demand and - associated therewith - in the financial capabilities of the milk market organization have been almost unshakeable.

Now, however, it is pointless to reflect how much easier an adaptation of milk production to the demand would have been at an earlier time, unless these reflections will contribute to initiate the necessary change from neutralization to causation policy on other problematic agricultural markets at a much earlier date.

The scope of action of agricultural policy

The urgent but at least in the EEC very lately answered question relating to the starting-points of production restriction cannot be dealt with without asking about the priority function of the milk price:

According to the economic theory the price assumes a double task: on the one hand, to bring about a balance between supply and demand and, on the other hand, to achieve sufficient income of the competitive suppliers for consumption and investment purposes.
If now the real situation mainly on the European milk market shows that under the prevailing general structural setting the price is apparently not able to assume this double role, it appears appropriate to conclude that one of these functions has to be transferred to another instrument. This actually means:

The priority of the market balance function of the price, which means a lower classification of the milk price in the general agricultural price level, forces a compensation of the ensuing individual income losses by speeding up structural changes at the stage of production, unless the government compensates either permanently or at least transitorily for these income losses by direct income transfers.

On the other hand, adherence to the income function of the milk price calls for ensuring the market balance by supply regulating measures.

Market versus control

An answer to the advantage of a change in system in the one or the other direction must be based, above all, on the prevailing general setting (4):

Because of the strong dependency on farm land and high capital intensity of dairy farming the volume of milk production is, over the short term, almost completely inflexible in terms of prices.

Undoubtedly, the reaction to changes in prices is becoming more intense in the course of time; here, however, where falling prices are the focus of interest it will be less intense compared with increasing prices, because grassland farmers have at best limited production alternatives at their disposal.

If, in addition, the general economic setting, mainly employment aspects, unduly restrict the farmers' adaptability, a policy of falling prices cannot lead to any considerable restriction in terms of quantity either in the medium-range; but it will necessarily lead to a substistence economy for many dairy farmers.

In the longer term, however, there is no doubt about the effectiveness of such an adjustment strategy, but then with the inevitable consequence that the previous extensive objective of the agricultural policy will change to an increasing extent into a spatial concentration of production in locations where the most favourable natural and economic preconditions for production are given. However, these will not be identical with the greater part of locations which are currently reserved for dairy farming.

Here, the political dimension of this subject-matter becomes clearly recognizable: That means the goal conflict between a development that is exclusively oriented according to economic criteria and other (agricultural) political purposes, e. g. maintenance of the highest possible number of farmers, broad allocation of property, preservation of land cultivated by man and infrastructure of rural regions.

These competing goals have one thing in common (5): "They exhibit an non-economic character. Hence, one connot decide upon them according to economic points of view. This does not mean, economy has nothing to do with them, but the decision on whether these reasons must be considered or not must be taken before making any economic consideration" and this the more rapidly and compulsory the more clearly it can be foreseen that one of the bases of the market policy followed so far, namely the guaranteed unlimited demand associated with minimum price guarantee threatens to fail as a result of cost-explosion.

So if "the resolving of the free-market sense" alone is a doubtful regulative in terms of social and regional development policy in order to realize the necessary adaptation of milk production to the demand, then the emphasis of prime importance laid on the market balance function of the producer price could be combined with general payment of direct income transfers as a compensating instrument ensuing from the government's social obligations.

In fact, such a political alternative of producer price reduction and compensation of income shares achieved so far via the market by direct income transfers is, as far as it aims at product neutralization and not at product stimulation, cheaper on surplus markets from the aggregate economic viewpoint compared with the same income effect achieved via price policy with the consequence of increasing consumer prices and, further, increasing expenses for inferior utilization of additional production stimulated by a price increase.

Further, one may well expect that besides the supply reducing effect resulting from dropping producer prices the demand stimulating effect produced by correspondingly reduced consumer prices will ease on the market to certain extent.

Nevertheless, there is a reason for the assumption that to the extent to which an increasingly larger proportion of agricultural income will be allotted to direct transfers these transfers would be, at least to the same extent as the expenses incurred by the current market regulations, the focus of public criticism and this all the more so because full compensation of income losses resulting from producer price reduction to the balance level by transfer payments financed by public authorities would be, under the prevailing conditions, necessarily associated with a noticeable increase in current government expenses for the organization of the milk market in favour of income support.

This scepticism, that questions not at all practicability of sociopolitically oriented income transfers, became concrete - first on the level of farmers' lobby - in the alternative conception that adherence to the income function of the price to an extent to which the market balance is thereby infringed with the consequence of follow-up costs which are high and, what is more, ineffective in terms of income, should be supplemented by a quota system regulating milk supply.

Application of this idea to practical milk market policy cannot surprise anybody seriously:

a) Agriculture as a whole must actually have an income motivated interest in maximizing the consumers' expenses for food, of course with the consequence of efficient marketing.

Under the given conditions of saturation demand this is only realizable by an increase in product prices. This, however, urgently presupposes a restriction of supply.

b) From the European point of view the decision on the quota system was consequent in that, it has actually been the only possibility of efficiently restricting milk production over the short term.

c) Further, the solution has also suggested itself inasmuch as nearly all countries of the Western World concerned with surplus problems meet these difficulties with quota controls.

Such a practice is based on the theoretically exact fact that in the case of not very price sensitive demanded goods and functioning foreign trade protection quota setting would be an effective instrument of governmental price support policy - provided administratively imposed restriction of production is controllable and adequate.

Indispensable keeping of quotas requires a clear canalization of product sales (bottleneck principle) which is certainly an exception under the prevailing conditions on the agricultural markets.

As regards milk, however, one may assume that the main proportion of the yield can be processed economically only in dairies. So, the technical difficulties of quota setting on this market can be managed - the same is true for sugar-beet production. However, a quota system is associated with substantial problems also on these markets:

On the one hand, they concern the reproach to produce a more or less structure-preserving effect, either directly and more far-reaching by not marketable quotas, which deprive the farmer along with production planning also of the possibility of cost minimization, or indirectly because of the price of saleable quotas, which burdens the growth-oriented farms with additional production costs.

On the other hand, imposed restriciton of factor input in the limited line of production leads to increased engagement in alternative lines of production as a result of the rather inflexible employment of land and labor in agriculture. Hence, restriction of production of one agricultural product finds itself confronted with increased production of other agricultural products. So, total production is not restricted. As a result, the agricultural income is altogether not raised essentially. But price and income safeguarding for quota products is faced with price and income reductions for increasingly produced non-quota products - a fact that from the viewpoint of the encouraged farmers casts doubt upon any temporal limitation of a quota system and makes, in addidition, any more restrictive adaptation of production to the demand extremely difficult.

To this brief analysis of instrumental effects the question must be added whether a political practice which is directed towards protection of national advantages is actually able to manage a quota system appropriately:

At least in the European Community the Council of Ministers has, absolutely informed about the market situation - despite the experience gained with the sugarmarket policy - neither been able to reduce the production volume sufficiently to the actual sales potential, nor politics have been so consequently to combat problematic trends of the quota system relevant in terms of quantity - in the form of fat regulation and regional quota balance.

Hence, on the background of a further escalating surplus situation on the butter and skimmed milk powder market the European milk market policy is again forced to act. The result, however - a very soon further reduction of the price supported quota by at least 10 to 15 % or a restricted obligation to support the EEC price level - cannot be predicted reliably.

What is certain is that a price cut will be the more probable solution, if policy due to its inability to arrive at a consensus will turn out to be too weak in the longer term to meet such a development by an adequate strategy.

Literature:

(1) D. A. HATHAWAY, Government and Agriculture, Economic Policy in a Democratic Society, New York 1963, p. 81.

(2) <u>Amtsblatt der Europäischen Gemeinschaften</u> Brussels, Jg. 12 (1969), No C41, p. 21.

(3) R. PLATE, 'Agrarmarktpolitik'. Volume 1: 'Grundlagen'. München, Basel, Wien 1975, p. 153. - The same author, 'Agrarmarktpolitik'. Volume 2: 'Die Agrarmärkte Deutschlands und der EWG.' München, Basel, Wien 1970, p. 334 and next pages.

(4) R. PLATE, Volume 1, a.a.O., p. 82.

(5) G. WEINSCHENCK, 'Was von Bonn und Brüssel zu erwarten ist - Möglichkeiten und Grenzen einer Neuorientierung der Agrarpolitik'. In F. HÜLSEMEYER, F. KUHLMANN, G. WEINSCHENCK, R. E. WOLFFRAM; 'Agrarmarktsituation der 80er Jahre aus wissenschaftlicher Sicht' (Archiv der DLG, **66**), Frankfurt am Main 1980, p. 60.

MILK SUPPLY MANAGEMENT PROGRAMMES, THEIR STRENGTHS AND WEAKNESSES

G. Syrrist
Agricultural University of Norway
Department of Dairy and Food Industries
P.O. Box 36
1432 Ås-NLH
Norway

ABSTRACT. The strengths and weaknesses of milk supply management programmes can hardly be evaluated on an absolute and general basis. One meaningful appraisal of their appropriateness would, in the author's opinion, be to register to what extent such a programme fulfils political goals in a given situation in a given country. The situation in Norway in the beginning of the 1980-ies may serve as an illustration. It is argued that instead of studying how different supply management systems influence the conditions and structure in the milk production, one should pose the problem in the opposite way and ask: Which milk supply management systems are in conformance with the primary targets for the economy and the secondary targets for the milk production? This way of stating the problem may add to the mutual understanding necessary in our international organization.

1. A DEFINITION

In this contribution a milk supply management programme is a set of regulations enacted by a country or a group of countries in order to affect the milk production on the farm level in a favorable direction.

This definition covers actions taken by an individual country as well as a group of countries, and it comprises international agreements obtaining legal force after ratification.

It includes both decisions made to decrease the total supply, to reduce its growth rate or to redistribute a chosen supply volume among milk producers, e.g. a quota system, and it also includes stimuli to initiate, foster and protect a national or regional milk production.

Last but not least, it implies that there exists a political basis for the evaluation. It means that the definition is applicable for conditions in socialist as well as in capitalist countries.

In our international meeting I think we ought to choose a so wide definition if we will make the discussion interesting for the majority of the nations represented.

2. MY APPROACH

How to contribute to the discussion of our topic on this basis?

My point will be that instead of asking how different milk supply management systems influence the conditions in the primary milk production, we have to pose the problem in the opposite way and ask: Which milk supply management systems are in conformance with the primary targets for the economy and the secondary targets for milk production? I shall illustrate how this can be done by means of an example.

Being in the Netherlands, let us remind ourselves that the great economist from this country, Jan Tinbergen, in the early fifties demonstrated the need for as many policy variables or instruments as there are policy ends or targets. Henri Theil, another outstanding dutch economist, extended this approach by replacing the targets with a welfare function to be optimized and by letting the instruments enter this function and reflecting the costs associated with the use of different instruments.

These brilliant attempts for their time in macroeconomic theory to formalize the complicated nature and structure of political decision-making, nevertheless have shortcomings in the description of the intricacy of political life to day.

Maybe we could talk about a hierarchy of targets or primary targets, secondary targets and so on. At the same time secondary targets are instruments to fulfil the primary targets, the tertiary targets belong to the instruments employed to realize the secondary etc. It is my impression that economic growth, full employment, moderate inflation and possibly a few others like individual freedom and unpolluted environment in most economies are rated as primary targets, whereas agricultural policy and its subsidiary, milk production, have to accept a role as instruments or as secondary targets.

Moreover, modern technology has led to a predominance of large scale enterprises which together with unions and other interest organizations has given the industrial society traits of corporatism. As a consequence the economic mechanisms have become even more complex than in earlier decades.

I postulate that different countries may have different welfare indicators and that certainly the milk production is of greater importance in some economies than in others. Accepting this, the only way of giving our topic a realistic exposition is to discuss the programmes individually and in relation to the conditions and political targets in each case.

I will use the situation in Norway as an example (2). This is done because Norway is a small country with a surveyable political structure. After the excellent work of group C19 (1) there is not much to be added in the description. My angle will therefore be a different one.

Going back to my definition of a milk supply management system it would have been interesting if similar studies were made for countries with other agricultural and political conditions, e.g. New Zealand, India and Poland. Both the available time and my insufficient insight prohibit such an extention.

3. MILK PRODUCTION AS INSTRUMENT IN THE NORWEGIAN ECONOMY

Figure 1 displays at the top some primary targets which are directly influenced by agriculture. In the middle are given the secondary targets for the milk production, and at the bottom you find the central elements in our quota system. The figure is organized in such a way that you can follow the main connections between the levels vertically. If there is a priority between the elements on the same level, it decreases from the left to the right. It is necessary to give some further comments to the figure because of its condensed form.

To the upper row:

NUTRITIONAL BASIS: Agriculture shall contribute to assure the nutritional basis for the population through efficient resource utilization in food production.

REGIONAL DEVELOPMENT: The production capacity shall be made use of in such a way that agriculture together with other sectors foster the regional development of the rural society.

ECOLOGICAL BALANCE: Agriculture shall participate in the efforts to obtain a balanced ecological system. The production shall take place with due care of the environment.

ECONOMIC GROWTH: The income and working conditions in agriculture must give foundation for development in the business in such a way that the abovementioned targets can be attained.

To the second row:

TOTAL MILK PRODUCTION: The domestic market shall be furnished with milk and dairy products.

REGIONAL MILK PRODUCTION: The location of milk production shall be in compliance with the targets for regional development and efficient resource utilizations.

ROUGHAGE AND SILAGE FEEDING: The feeding of roughage and silage shall be stimulated through information and economic incentives.

INCOME MEASURE: The income per man-year on an efficiently run farm shall correspond to the average annual wage of an industrial worker.

4. THE TWO-PRICE QUOTA SYSTEM

The majority of the milk producers are, together with other agricultural producers, organized in two farmers' unions, which negotiate on behalf of their members with the political authorities about the income measures for agriculture.

In the negotiations the farmers' unions had very early to accept that losses due to a surplus of milk above the volume necessary to furnish the domestic market, should be carried by the milk producers. It caused discussions of regulation systems as soon as the overproduction reached worrying quantities after the World War II. Thus, in the 1960-ies a two-price system was elaborated, but not brought about, because the balance in the market was restored without intervention.

From 1975 better economic conditions stimulated the milk production, whereas a slight decrease in milk consumption was experienced. Hence,

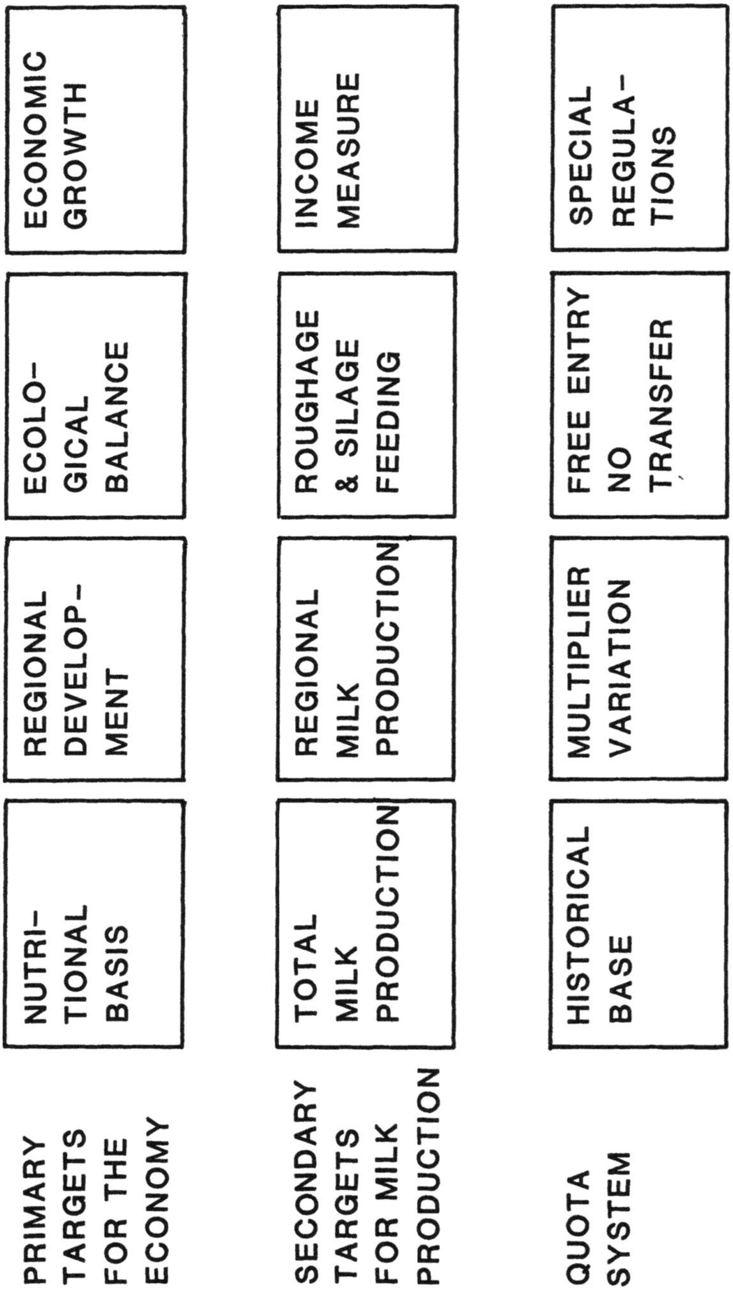

PRIMARY TARGETS FOR THE ECONOMY	NUTRI- TIONAL BASIS	REGIONAL DEVELOP- MENT	ECOLO- GICAL BALANCE	ECONOMIC GROWTH
SECONDARY TARGETS FOR MILK PRODUCTION	TOTAL MILK PRODUCTION	REGIONAL MILK PRODUCTION	ROUGHAGE & SILAGE FEEDING	INCOME MEASURE
QUOTA SYSTEM	HISTORICAL BASE	MULTIPLIER VARIATION	FREE ENTRY NO TRANSFER	SPECIAL REGULA- TIONS

Figure 1. Primary and secondary targets and central features in the quota system.

the situation very soon called for action. To begin with a bonus system was preferred, but it disclosed after a few years negative side effects. As a consequence we went over to the quota from January 1983.

So to the bottom row, where you see the system we have chosen within the admissible alternatives.

HISTORICAL BASE: The quota is allotted to each dairy farm. The deliveries in the last three years form the basis for the calculation. Some amendments to this rule have been made in order to eliminate the possibility of building up future quota basis through present overproduction.

MULTIPLIER VARIATION: The product of the base and a so-called multiplier gives the quota. The multiplier varies regionally, and for small producers and for producers granted special regulations a higher multiplier is used than for the others.

Any quantity above the annual quota receives a milk price of 60 øre per liter. The regular price is more than 4 times higher.

FREE ENTRY-NO TRANSFER: A quota can not be sold or transferred, but a newcomer may apply for and will get a quota at any time. In such cases the quota is based on the feed resources on the farm and restricted to a maximum of 86 000 litres per year or 2 man-years per holding.

There is no incentive to encourage the farmers to quit milk production. Nevertheless, 3 to 4 per cents leave the business every year.

SPECIAL REGULATION: For not causing drastic change in the short run, four categories of dairy farms are granted special regulations. They are shown in Figure 2. A few additional remarks may be in place. When there has been

a) a recent and significant investment in expansion of the farm buildings or
b) a change in ownership or restart of the milk production or
c) a cultivation of new land or a long term lease of additional acreage or
d) a fire or other damage causing temporary cut-backs in the base period,

it may be applied for such regulations. The additional base given to category a and b is scaled down over a period of seven years. For category c the period is three years.

5. EVALUATION

Firstly, we may conclude that it has been possible from a situation with increasing surpluses to bring the production down and in line with the demand. So unless we experience a drastic drop in the milk consumption, the system could be abolished. It will, however, be in operation also in the coming years.

Secondly, the system underpins the political targets. It is also in conformance with the basic principle in the dairy co-operative which states that the membership and the compulsory delivery obligation follow the farm, not the farmer.

Thirdly, the system has been introduced and operated without severe set-backs in the solidarity and loyalty necessary for the dairy co-ope-

ratives to function properly. One reason for this is undoubtedly that
the system was thorougly discussed on beforehand and implemented with
sufficient flexibility to avoid drastic reductions in volume and changes
in economic conditions. Thus, all producers have been urged to seek
assistance in filling out application forms for special regulations and
2/3 of the producers utilized this opportunity. About 50 per cents of
the holdings attained such treatment from the beginning.

BASIS FOR SPECIAL REGULATIONS	SCALING DOWN PERIOD, YEARS
a. INVESTMENTS	7
b. OWNERSHIP	7
c. CULTIVATION	3
d. DAMAGES	

Figure 2. Basis for special regulations and period for their withdrawal.

This smooth start had its price, and it was the other half of the members who carried that burden through a reduction of their multiplier. It gave rise to heated discussions between producers from different regions and categories. However, since the special regulations applied to very limited time spans, the exceptional cases were largely accepted.

This extensive use of special treatments has also inflicted a tremendous amount of work on the administration of the co-operative organizations and some official agencies, but after the first avalanche of applications had been dealt with, the situation became easier to manage.

6. SUMMING UP

Now, why does a gay from a Lilliput in the dairy world bother you with these details on the quota system in Norway? The goal has been to illustrate by means of this example how every milk supply management system should be tailored to fit the political targets in the society in question. Such a system must be of an ad hoc-nature, adapted to place, time and wishes. I will argue, that starting with the primary targets for a given economy and successively narrowing down the class of admissible supply management systems by taking into account the implications following from secondary and further targets is an approach which can add to the mutual understanding needed in an international forum like IDF.

7. REFERENCES:

1. 'Quota Control on the Milk Supplies and Supply Management'.
   Report of Group C19 to Annual Sessions in The Hague (Netherlands), September 1986, IDF C-DOC 1986.

2. Tømte, E.D. and Sand C.: 'The Norwegian Milk Quota System'.
   Twelfth European Seminar of Agricultural Economists, Helsinki, Finland, May 26.-29. 1986.

SUMMARY OF DISCUSSION

During the discussion it was emphasized that the increase of the milk price is much lower than the increase of the income during the last years. Therefore it seems not understandable to say that milk is becoming a luxury good?
W. Krostitz (Italy) answered by explaining that we must compare the increase of the milk price with those of other consumable goods such as vegetable oils and that in this respect milk and milk products are becoming luxury goods.

Further it was stressed that in the USA the situation for the milk farmer is very bad. It is the opinion that the buying regulations for milk for supporting the price can not hold up.

Seminar IV: Modern Methods of Analysis of Milk and Milk Products

Chairmen: Dr. Ir. J. Koops (The Netherlands)
Prof. Dr. W. Heeschen (Fed. Rep. Germany)
F. Harding (U.K.)
Secretary: Dr. V. Palo (Czechoslovakia)

MODERN METHODS OF ANALYSIS OF MILK AND MILK PRODUCTS: RAPID
DETERMINATION OF MAIN COMPONENTS

R.J. Brown
Department of Nutrition and Food Sciences
Utah State University
Logan, Utah 84322-8700
U.S.A.

ABSTRACT. Moisture, fat, protein and lactose are the milk components of most importance to those who test dairy products. Newly available testing techniques which are fast, robust and affordable coupled with improved capacity for collecting, processing and storing data are making assays for these components possible in routine and on-line applications where they were previously impractical. The same advances in instrumentation are improving accuracy and precision and making calibrations better.

1. THE CLIMATE FOR MILK AND DAIRY PRODUCT TESTING

Milk and dairy product testing for moisture, fat, protein and lactose has come a long way since the methods now used as standards for calibration of rapid instruments were used for every test. This progress has been made by the combined efforts of and cooperation between those in the dairy industry using the tests and those in the industry that has manufactured and sold the ever improving generations of instruments.

1.1. The users of testing equipment

Those who buy milk testing instruments are interested in testing milk and dairy products. They are not concerned with whether the instrument uses Babcock, Gerber, dye binding, Kjeldàhl, refractive index, light scattering, IR, NIR, FTIR, ATR-FMIR or any other procedure. They normally do not want to know details of how the instrument works or how it is calibrated. They want an instrument and instructions for running it that are simple enough to be used by anyone they hire to work in their laboratory. Many different, and sometimes conflicting, factors affect acceptability of milk testing instruments. Some of these are price, reliability, configuration (automatic, on-line, manual, etc.), precision, accuracy, adaptability to present laboratory procedures, speed of operation, testimonials and rumors from those using various instruments and other information.

1.2. The manufacturers of testing equipment

People who make and sell instruments have different needs than those who buy them. They must make a reasonable profit while maintaining a substantial investment in development of new products. Sales of each generation of instruments must finance research to produce the next generation, or there will not be a next generation. None of these things are of immediate concern to those who buy and use the instruments. Nor is simply making a better product a guarantee of success in the instrument business.

2. ADAPTATION OF LABORATORY ASSAYS TO DAIRY USES

Most analytical methods that have been adapted from their original laboratory configurations for application in industrial settings are based on very simple, usually old, technology (Table 1). This applies to instruments used for testing of milk and dairy products. The requirement for reliability in the environment found in dairy plants (or even in dairy laboratories) coupled with cost considerations has limited the choices available. Those methods applied in industrial settings represent the state of the art in instrumentation 30 years ago (Hirschfeld, et al., 1984). Current state of the art methods have not been applied in the dairy industry outside of research laboratories. This condition does not exist without reason.

2.1. Requirements of Laboratory Instruments

All laboratory testing procedures must be both accurate and precise. Where concentrations of substances being tested are low, there is a critical requirement for sensitivity. Selectivity is an additional necessity if the samples being tested are, like milk and dairy products, mixtures or solutions. The only way around this is to separate the component in question from all others before measurement. Such purifications are subject to errors and add time and expense.

2.2. Additional Requirements of Industrial Instruments

When laboratory methods are moved into industrial settings, such as dairy plants or dairy plant laboratories, all of the constraints of laboratory tests go with them. But, there is now much more strain on all of these factors. (Accuracy is no less important when a farmer's income depends on it than in a research laboratory.) Accuracy and precision are much harder to maintain under these new conditions and the people called upon to maintain them are less capable of doing so. Sensitivity and selectivity of many kinds of assays decrease after many samples have been run.
   The problems encountered outside of research laboratories have additional limitations, not critical to laboratory tests.

Table 1. Industrial instrument utilization

Widespread
Temperature
Pressure
Flow
Density
Refractive index
Ultraviolet-visible colorimetry

Infrared absorption (filter)
Gas chromatography
Indicator tags
pH electrodes
Acoustic velocity
Amperometry

Occasional
Wet-chemical processor
Chemical tape systems
Specific-ion electrodes
Fluorometry
Infrared absorption (scanning)
Mass spectrometers
Liquid chromatography

Ultraviolet Spectrometry
Gamma-rayabsorption
Dielectrometry
Magnetometry
Particle counters
Viscosity

Potential
Chemiluminescence
Microwave absorption
Optical emission
X-ray fluorescence
Raman
Infrared emission
Nuclear magnetic resonance
Near-infrared reflectance

Acoustic emission
Chemresistor
Piezobalances
Acoustic attenuation
Turbulent flow
Array chromatography
Electrophoresis

(Hirschfeld, et al., 1984)

Maintaining stability of instruments is much more difficult under plant conditions. Calibrations are less frequent and usually less reliable. Calibrations against laboratory tests run by someone else in a different location using different samples are a special problem (Sjaunja, 1982). In addition to stability of measurements, the ability of instruments to continue operating in the same way day after day is challenged by plant conditions. Instruments must be much more robust to withstand forms of abuse and everyday wear to which laboratory instruments are not subjected.

And finally, instruments must be affordable or they will not be used. Purchases of instruments to be used in production facilities are not looked upon in the same way as purchases of laboratory instruments. Capital equipment costs require justification based on increased income from use of the equipment.

3. TRENDS IN INSTRUMENTATION

Within the last five years, the emphasis in development of

instruments has shifted toward intelligent instruments (Hirschfeld, 1985). Limits that had been reached in measurements are now being overcome by mathematical algorithms. Calibrations are done much more easily, and are often done by the instrument itself. Instruments are able to work much faster, and often without operator intervention. Measurements from instruments are recorded, sorted, transformed, sent to distant locations, compared with previous results, matched with results from other instruments, and used in many other ways that until very recently would have been done manually. There is little similarity between what was available five years ago and now.

3.1. The Microcomputer Revolution

Most computer professionals regarded the first personal computers as toys when they were introduced in 1978 (Crecine, 1986). Desktop computers are now available for less than $2000 that take up less than 30 cm^2 of desk space, operate at 1-2 million instructions per second, and have 1.1 megabytes of primary memory and 20 megabytes of long term memory. The specifications of these small computers exceed those of most mainframe computers in use eight years ago, without the necessity of sharing computer time with many other simultaneous users. Small computers of the next generation, the first of which were introduced this month at about the same price as their predecessors, are even more powerful.

3.2. Intelligent Instruments

Microcomputers have changed the way instruments are built, the way they are used and even the way they are thought about (Heller and Poterzone, 1983). Many more research laboratory instruments will be able to move into industrial situations because of microcomputers. These intelligent instruments will be capable of extraordinary resolution and sensitivity. They will be able, using computer intelligence, to distinguish better between true signals and unwanted noise. It will be possible to find very small signals in very noisy data. Signal processing between the sensors and the output will make these advancements invisible to the users.

3.2.1. <u>Sensors</u>. Measurement by instruments is dependent on sensors that respond to some characteristic of the sample. These sensors may be things such as pH probes, light absorbance measurements in specific areas of the spectrum, or any other measurable phenomenon linked to what is being measured. Each sensor gives a signal that is converted by some kind of calibration into the units necessary to make sense of the measurement.

Before computers, it was necessary to use a separate sensor specific for each constituent measured. Much effort went into making sure that sensors gave linear responses. These sensors could not be influenced by anything other than their specific target. When several sensors were similar, the same method could be used to convert signals from all of them into useful data.

3.2.2 <u>Computers with Sensors</u>. This has all changed with addition of computers to instruments.

> "We no longer require linearity in a sensor, but merely a monotonic response. We may use internal standardization in lieu of reproducibility, redundancy in lieu of the very difficult last increment of reliability, correction based on internal or external information in lieu of insensitivity to perturbation....a merely fair performance is enough if it is repeatable, well understood, and coupled with a high enough signal-to-noise ratio to allow the data-processing algorithms to proceed with a reasonable chance of success" (Hirschfeld, et al., 1984)

Measurements still depend on sensors, and better sensors make better measurements, but many of the shortcomings of sensors are now overcome in the data processing step. The trend is toward instruments containing a computer attached to one or more sensors.

Sensors can now be used that are not specific for only one component in a mixture if the ability to interpret their signals is provided. If each sensor responds differently to at least two components in the sample mixture, then the concentration of each component can be determined by solution of linear equations (Jurs, 1986; Warner, et al., 1977). For maximum information from an instrument, the ideal situation is a large group of nonselective sensors attached to one computer.

Even more information is available from large arrays of sensors if pattern recognition, or fingerprint, methods are used (Smith, et al., 1985). This requires the computer to "learn" the combination of responses from all the sensors caused by each possible component in a mixture. This can be done simultaneously for many components.

4. APPLICATIONS TO MILK AND DAIRY PRODUCT TESTING

Fixed filter infrared spectroscopy has been the leading method for rapidly measuring moisture, fat, protein and lactose in milk and dairy products for many years (Goulden, 1956, 1961, 1964). Several generations of new instruments have been developed, with recent models following the trend toward computerization. This has allowed improvements in stability and reliability of measurements, it has facilitated ease and accuracy of calibrations and it has made the instruments easier to use. It has not overcome the limitations imposed by the small number of sensors (filter bands) available (Grappin, 1984; Sjaunja and Anderson, 1985).

Although fixed filter infrared absorption still has desirable features, other procedures which allow for use of many more sensors should now be adopted for dairy testing. Methods are available capable of very quickly scanning over a whole series of sensors, thus allowing much more powerful data processing methods to be used.

Any number or combination of these readings can then be used to measure any component. Handling of this large amount of data is no longer a problem with computers built into all new instruments.

5. CONCLUSIONS

Testing techniques which are fast, robust and affordable coupled with improved capacity for collecting, processing and storing data can make tests for milk components even better than they are now. The amount of information available from instruments with many nonspecific sensors will allow us to do a better job of measuring the components we now measure and to measure other components as they become important. Development of such instruments should be (and probably is) going on right now.

REFERENCES

Crecine, John P., Science 231, 935 (1986).

Goulden, J.D.S., J. Sci. Food Agric. September 7, 609 (1956).

Goulden, J.D.S., Nature 191, 905 (1961).

Goulden, J.D.S., J. Dairy Res. 31, 273 (1964).

Grappin, R., Challenges to contemporary Dairy Analytical Techniques (The Royal Society of Chemistry, London, 1984).

Heller, S. and R. Poterzone, Eds., Computer Applications in Chemistry (Elsevier, New York, 1983).

Hirschfeld, T., Science 230, 286 (1985).

Hirschfeld, T., J.B. Callis, and B.R. Kowalski, Science 226, 312 (1984).

Jurs, P.C., Science 232, 1219 (1986).

Sjaunja, L.-O., Studies on milk analyses of individual cow milk samples. Thesis, Swedish University of Agricultural Sciences (1982).

Sjaunja, L.-O., and I. Andersson, Acta Agric. Scand. 35, 345 (1985).

Smith, A.B., A.M. Belcher, G. Epple, P.C. Jurs and B. Lavine, Science 228, 175 (1985).

Warner, I.M., E.R. Davidson and G.D. Christian, Anal. Chem. 49, 2155 (1977).

DETERMINATION AND EVALUATION OF HYGIENIC QUALITY

Frank O'Connor
An Foras Taluntais
Moorepark Research Centre
Fermoy
Co. Cork

1.   ABSTRACT

The fundamental aim of the dairy processor is to secure a
raw material which will not limit his options in processing
or jeopardise the quality of the finished product. To
achieve this, critical quality parameters must be clearly
understood and defined and the necessary control procedures
rigidly applied all along the route from the cow to the
product. The operation of a comprehensive control
programme which takes full advantage of modern analytical
methodology is the key to successful milk quality control
in today's industry. The application of these methods is
discussed in detail in the paper.

2.   INTRODUCTION

Milk production systems have altered dramatically over the
past few decades with the streamlining of the milking
operation and the use of refrigerated milk tanks to store
milk on the farm for 48 to 72 hours before collection.
Hence, milk may not be processed for three to four days
after production thereby placing major emphasis on quality
control at each stage of production, storage and assembly.

Although top quality products can be made from milk from
modern hygienic production and assembly systems experience
has taught the processor that to maintain the required
quality, encouragement of good producers with bonus
payments and elimination of poor supplies from the bulk is
necessary.

To measure quality, and reward achievement methods must be
available which will accurately differentiate between good
and poor quality supplies.

Since there is no one test which will give a measure of

all the parameters a number are usually used. The
combination of tests should then present an accurate
picture of the quality of the milk from the points of view
of overall bacterial content, udder health status and the
presence of contaminants or inhibitory substances.
Accordingly, most milk quality grading schemes include a
combination of test methods to evaluate milk supplies
although the specific tests used and the importance
attached to the test results differ widely throughout the
dairying world. In this paper I shall examine the most
common and appropriate methods employed in the dairy
industry at the various stages of milk production up to the
point of process.

3. TEST METHODS

Testing methods for milk are many and can be conveniently
grouped depending on the information required under three
headings namely (a) rejection, (b) quality control and (c)
quality payment. A suitable test accurately measures the
required parameter within ranges specified, should be
simple to operate and gives a rapid result at an acceptable
cost. The speed of an assay can be the most important
factor in choosing a method for accepting or rejecting a
milk supply while at laboratory level the accuracy of a
method may be the determining factor.

Test methods which indirectly measure a particular
parameter have to be calibrated against a standard method.
In my paper I will give an indication of strength of the
relationship between indirect and standard methods.

3.1 Rejection Tests

An ideal rejection test carried out at the point of milk
collection should be simple, give an immediate result and
should be capable of differentiating between acceptable and
unacceptable milk quality supplies. If the ideal rejection
test was available it would limit the need for any routine
quality testing as all unacceptable quality would have been
eliminated from the supply at the time of collection.
However, up to the present time we do not have such a
rejection test, rather we employ tests which are geared to
detect grossly abnormal or contaminated milk.

3.1.1 At milk production

At farm level the producer who decides if some milk should
be added to his bulk supply on the bases of e.g. udder
health or antibiotic residues, is operating his

own rejection system. To check udder health the producer may use the California Mastitis Test (CMT). This test gives a rough indication of the somatic cell count of the udder and hence of udder health status. The test requires just one reagent which is mixed with an equal quantity of milk on a simple paddle device and the result is easy to read. However, it must be remembered that milk from cows in very early or late lactation may have high somatic cell counts without any inflammation of mammary glands hence some caution is necessary in interpretation of results. This is a very useful and simple test which all producers should be geared to carry out on an ongoing basis.

Antibiotic residues in milk from treated cows can render very large supplies unfit for human consumption or unsuitable for processing. If recommended procedures are followed when using such products there should be no difficulty with residues. However, accidents will happen and to guard against the consequences of such occurrences, antibiotic residues in milk can be determined using test kits suitable for use at farm level e.g. Delvo Test, Penzym Test.

The Delvo test for use at farm level is similar to that used in many milk control laboratories with results becoming available after two and a half hours. While at laboratory level the procedure is relatively simple it may present some problems for untrained operators.

The Penzym test can detect beta-lactam antibiotic residues in milk. The method is simple to use as all reagents have been dried together and presented in tablet form with the enzyme already in tubes. The sensitivity claimed for the method is in the range .009 to .017 IU penicillin per ml with results available in about 20 minutes. The method will not detect antibiotics other than those of the penicillin family. The cost of a test by both the Delvo and Penzym is high relative to cost of large scale antibiotic testing methods in control laboratories. However, if considered against the potential losses which their use can prevent, the cost may not be a major disincentive.

3.1.2   At Pick Up

Where milk is picked up at farm level it is the tanker driver's decision to accept or reject the supply. If he accepts the supply he takes a sample which will be analysed at the laboratory for quality payment purposes. For non refrigerated milk a quick check may be made on doubtful quality supplies by measurement of pH or acidity.

This is a very simple test but only grossly contaminated supplies will be identified. Reliable bacteriological rejection tests are not yet available for refrigerated bulk tank supplies although the temperature of milk in tanks may be monitored and temperature limits set for rejection. The search for a rapid and effective rejection test at the point of reception is now a high research priority as a breakthrough in this area could lead to a drastic reduction in routine laboratory testing programmes.

3.2　Quality Control at the Factory

Ideally tanker supplies arriving at the factory should not contain reject quality milk as this should have been eliminated at the pick-up point. However, mistakes or accidents can happen at production level resulting in reject quality milk occasionally being delivered to the factory. Hence, for effective quality control tanker milk supplies should be monitored on an ongoing basis. In particular the presence of inhibitors and the general bacteriological status of the supply should be examined. The general criterion for tests used for this purpose is that they give a rapid result.

3.2.1　Inhibitory substances

For a rapid assay for antibiotics, tests such as Charm, Penzym and Spot tests can be used. The Charm test is a radiometric assay based on the specific binding of antibiotics to enzyme sites on microorganisms. By varying the reagents used in the test a range of antibiotics including penicillin, streptomycin, novobiocin, tetracycline, sulfamethazine, erythromycin and chloroamphenicol can be detected at levels similar and in some cases better than those obtained using more familiar methods. A well equipped kit is provided by the manufacturer and trained technicians can easily carry out the assay and report results in less than 15 minutes.

The Spot test is based on an agglutination antigen/antibody reaction. The test will detect only beta-lactam products e.g. penicillin G, cephapirin and cloxacill'in at quoted levels similar to those obtained by the biological methods (e.g. $<.005$ IU penicillin). The method is simple to carry out and results are available within ten minutes.

The Penzym has already been discussed in the previous Section.

While all of these rapid methods are in use for screening tanker milk supplies, the Charm test does have an advantage over the others in the range of antibiotics which can be detected. The cost of these rapid tests are high in comparison with standard test methods but as previously pointed out the cost has to be weighed against possible savings which can accrue from rapid detection of contaminated loads.

3.2.2 Bacterial Content

The quickest way to estimate the microbial content of milk is by direct microscopic examination of a stained smear of the sample. The method has rarely been used on a large scale to classify milk samples due to the tedious nature of counting under the microscope. In recent times a new approach to the concept has proved more successful and has resulted in the development of one semi-automated method (DEFT) and a second fully automated method (Bactoscan). The basis of both of these methods has been to separate the bacteria from the milk sample and then stain and examine the film under the microscope.

3.2.2.1 Direct Epifluorescent Filter Technique (DEFT)

In this method the bacteria are separated from the treated milk sample on a membrane filter which is then stained with a fluorescent dye and counted using epifluorescent microscopy. The precision of the method is satisfactory and correlation coefficients of better than 0.9 have been obtained with Standard Plate Colony Count procedure. Operator fatigue, a feature of manual microscopic counting of samples has been overcome by automating the counting using an image analyser while microscopic field selection has been greatly facilitated by the use of electronically operated microscope stages. Results using the method are available within 30 minutes and up to thirty samples can be counted in an hour. The method is still relatively laborious and skilled operators are required if consistent results are to be obtained. Using the method an assessment of the flora of the supply is also possible. The cost of the method is much more expensive than that of the colony count procedure.

3.2.2.2 Bactoscan

In the Bactoscan method, which is totally automated, the bacteria are separated from the treated milk samples using a gradient centrifugation technique. The bacteria in the gradient solution are then stained with a fluorescent dye and counted on a revolving disc under a microscope using a continuous flow technique. A sample passes through the system in about seven minutes and the capacity of the instrument is about 70 samples per hour. While correlation coefficients with Standard Plate Counts of around

0.80-0.84 are normally obtained, the variation around a predicted plate count can be quite high. As a test for screening tanker milk supplies at the point of intake the Bactoscan method combines speed of assay with a high level of accuracy. Although simple to operate on a day to day basis, a highly skilled technical back-up is essential to ensure smooth and efficient performance. In summary, this instrument, although expensive to purchase, offers convenience and relative speed of assay at a competitive price.

3.2.2.3 ATP Method

This assay gives an indirect measure of numbers by estimating the adenosine triphosphate (ATP) content of the sample after first degrading the background ATP of the milk sample. Results compare favourably with plate counts when numbers are high ($10^6 - 10^7$ per ml) but less so at lower levels. The method is simple to operate and results can be determined in about ten minutes. Despite the speed of assay the method has not found widespread favour.

3.2.2.4 Impedance Measurement

Changes in electrical resistance (impedance) of a medium in which microorganisms are growing, can be used to estimate the numbers of bacteria in a sample. Equipment, simple to operate has been available for some time past to measure impedance but because bacteria must grow for some time before they can effect impedance changes the method does not give a result as fast as in the preceeding methods. However poor quality milk with bacterial plate counts 10^6 can be detected by the system within 1-2 hours and hence the method has application in a platform situation.

3.3 Quality Payment

When testing milk samples for quality payment purposes there are a wide spectrum of analyses which may be used. Typically routine testing includes the content and nature of bacteria, the presence and levels of inhibitory substances, the somatic cell count as an indicator of herd udder health and added water. In some instances parameters such as free fatty acid levels, flavour, etc. may be included. In this paper however, I shall confine my review to those tests most commonly used throughout the dairy industry.

3.3.1 Bacteriological Quality

3.3.1.1 Plate Counts

Bacterial count estimates (based on colony forming units) of producers supplies for payment purposes are usually carried out using the automated loop/automated counting system which has a high throughput capacity (approx. 200 samples/hr) but a minimum of 2 day delay before results are known. For smaller laboratories the spiral plating system may be used to good effect although caution in instrument use and plate reading is essential if reliable data is to be acquired.

In recent times we have used an automated diluter successfully in preparing dilutions directly in the plate. The method is simple to use and relatively operator proof. The precision and accuracy of the method is excellent with no carry over between samples and any one or more dilutions may be used. A trained operator may achieve a throughput of up to 100 samples per hour in single dilutions. The diluter is cheap in comparison with automated loop system, is inexpensive to operate and has the advantage of flexibility for making a number of dilutions.

3.3.1.2 Bactoscan

The Bactoscan method due to its degree of automation and greater compatability with other automated methods and speed of data reporting is now gaining widespread interest throughout the larger routine dairying laboratories and may well supercede plate counting methods. The absence of a perfect relationship with the Plate Count Method does not necessarily invalidate the Bactoscan method as applied to quality grading of milk supplies. In fact our studies and experience elsewhere have shown that when used to separate milk supplies into broad quality categories the Bactoscan method compares quite favourably with plate count procedures although its sensitivity in differentiating between low count supplies is, as yet, somewhat limited.

3.3.1.3 DEFT

The DEFT method is excellent for estimating bacterial counts on milk samples for payment purposes when numbers of samples are small. The method offers great scope for further automation to increase the speed of sample throughput and reduce operator strain.

3.3.1.4 Impedance

Impedance measurement may also be used in milk quality payment systems. We have found that while there is some variation around predicted counts nevertheless producer samples may be divided into selected grades with a reasonable degree of certainty. However as milk quality improves impedance detection times increase. Hence the time until results become available increases which may considerably reduce the capacity of the impedance system with consequent increase in costs.

3.3.1.5 Specific Groups of Bacteria

The determination of specific groups of bacteria in milk samples e.g. psychrotrophs, thermodurics, sporeformers, coliform etc. is still carried out by the traditional methods but impedance measurements using suitable media show distinct promise for the future.

Likewise a Limlus Lysate assay for the determination of gram-negative bacteria in milk samples is a method which is gaining popularity in recent times. The test is a measure of the amount of lipopolysaccharides produced by gram negative bacteria. Good correlations have been found between test titres and gram negative bacterial counts of milk samples and results are available after an hour. The material cost of the test is quite expensive and skilled operators are required for its operation.

3.3.2 Inhibitory Substances

For routine testing for inhibitory substances the tendency is to use methods which will detect as wide a variety of antibiotics as possible and this invariably means using a biological method. There are a number of microorganisms now available which are sensitive to a range of antibiotics but the two most commonly used cultures are probably B. stearothermophilus (Disk assay - Delvo method) and Str. thermophilus (acidification methods). Semi-automated systems are now available for some of these methods which give high sample throughput. While the sensitivity of these organisms to penicillins in general is excellent these are less sensitive to some of the other antibiotics used in mastitis treatment. The material cost of testing using semi-automated biological methods is far less than using the modern rapid methods and it is doubtful if the rapid methods currently available will displace the current large scale biological methods unless some major breakthrough occurs. Expansion of the antigen/antibody agglutination type test for products other than beta-lactans in the future offers great possibilities for future testing in this areas.

3.3.3 Somatic Cell Counts

Somatic cell counts in supplier milk samples are generally estimated using Fossomatic or Coulter counter. In the Fossomatic instrument which is fully automated with a capacity of 180 samples per hour a sample of the milk is mixed with buffer and dye and microscopically examined under ultra violet light. In the Coulter system which may also be automated results are obtained on the basis of particle size. Results obtained by these methods are satisfactory provided instruments are kept properly calibrated. while determination of somatic cell counts on a monthly basis should provide adequate information on herd udder health status.

4 CONCLUSION

Milk testing whether at farm or laboratory level is expensive and laborious. However, some testing is absolutely essential to help ascertain and improve or maintain desired quality status. In my opinion many quality payment schemes are overelaborate and seem to have lost sight of their primary objective i.e. to maintain a high quality raw material for the processor. It could be argued that this objective could be easily attained with greater cost effectiveness by a reduced testing load, greater selectivity in the choice and application of tests and in particular a more effective response in applying the test data. For example it makes little economic sense to operate a highly elaborate routine testing programme while failing to take the necessary action against persistent defaulters the quality of who's supply is likely to exercise the predominant influence on the quality of the total supply. Equally the importance of good tanker hygiene and operating practice must not be overlooked and appropriate measures adopted where deficiencies are evident.

In conclusion routine milk testing should have clear objectives, be highly selective and cost effective and provide the catalyst for immediate and effective action at a practical level.

DISCUSSION

D.B. Stewart (UK) : Is it necessary to obtain the lowest possible bacterial counts rather than to obtain the best milk quality for milk products. Those two goals are not necessarily identical.
F. O'Connor (Eire) : This has not been my experience.

F. Pollack (Israel) : Did you include *Clostridia* in the thermoduric bacteria?
F. O'Connor (Eire) : Yes.

P.B. Waddy (Canada) : How do you consider oxidation problems and how do you correct for it at farm level?
F. O'Connor (Eire) : It is a quality problem to be corrected by good farm management practices.

W.H. Heeschen (Fed.Rep. Germany) : What methods can be used for milk bacterial counts lower than 100 000 per ml?
F. O'Connor (Eire) : Colony counts and - after improving the Bactoscan- this method.

R. Bossuyt (Belgium) : How do we have to evaluate simplified methods for bacteriological testing? One of the biggest problems being the lack of accuracy, precision and repeatability of the reference method (Standard plate count). May it be that some "simplified" methods give us more and more precise information on bacteriological quality than the "reference" methods?
F. O'Connor (Eire) : While modern methods may be more precise than the SPC it is normal to use SPC as the standard reference method. For quality payment purposes the precision is not as important as the ability of the method to grade milk supplies into suitable categories.

F. Harding (UK) : Frequency of sampling and sample handling errors exceed the testing errors experienced with modern instruments. Therefore, are we not receiving diminishing return when focussing further attention on testing?
R.J. Brown (USA) : The results of a list cannot be better than the sample deserves. We are improving precision, which is already good enough. We should work on variables that will improve accuracy.

M. van Boekel (the Netherlands) : In discussing the calibration of instruments we should realize that reference methods are also inaccurate. So it will be very difficult to judge whether future self-calibrating instruments will be correct or not.
R.J. Brown (USA) : I agree. Emphasis should now be placed on accounting for variables that affect readings, such as fat saturation levels.

J. Shields (UK) : I think that instrumentation is moving towards microprocessors and non-selective sensors.
R.J. Brown (USA) : I agree.

Seminar V: New Methods of Concentrating and Drying

Chairman: Prof. Dr. B. Hallström (Sweden)
Secretary: Dr. K. Masters (Denmark)

REVERSE OSMOSIS: ITS TECHNICAL, TECHNOLOGICAL,

ECONOMICAL AND LEGAL ACHIEVEMENTS AND LIMITATIONS

Bernard S. Horton
Horton International, Inc.
Cambridge, Mass. U.S.A.

I. INTRODUCTION

After accepting the suggestion that I give this lecture on reverse osmosis (RO), I was then assigned what at first seemed a long and complicated title. However, upon reflection I decided to use that title, with some arbitrary definitions of its contents, as a means of organizing the information to be presented.

In addition, to help put the status of RO in perspective, I have given the subject a human frame of reference in terms of its age. With the indulgence of my audience, I considered the development of RO for the dairy industry before the appearance of the first full-scale plant in 1969 to be part of the gestation period of the process. Then I have taken RO's life in the dairy industry to have begun in 1969 and have broken its life since then into three periods - infancy, childhood, and teenage (teens). Most of the information presented here will be keyed to these three periods. A forward view of RO's more mature life will then be given.

II. GENERAL STATUS

Before going into the specific issues of the four areas of achievement and limitation included in the title of this lecture, a quick review of three indicators of the maturity of RO will serve as a background. These indicators are shown in Table I.

RO, having been perceived as a "heatless" concentration process for foodstuffs, was reputed as offering great energy cost savings when compared to the evaporation of whey. In the U.S.A. there was also the idea that RO would make the hauling of whey from small cheese plants to centralized whey processing plants an attractive economic proposition which also had the clear benefit of alleviating the serious and growing whey disposal problem.

RO commenced its real life in the southern end of the South Island of New Zealand by acting to expand the capacity of a very old evaporator in a Cheddar cheese whey condensery. The relative energy costs for water removal weighed heavily in RO's favor for this role. In the U.S.A., at the time, it became apparent that RO could greatly reduce the cost of

Table I

REVERSE OSMOSIS

in the

DAIRY INDUSTRY

	Infancy 1969-1972	Childhood 1972-1981	Teenage 1981-1986
Cumulative Number of Plants Sold	10	90	210
Cumulative Installed Membrane Area, m^2	400	13,000 (NIZO 1979)	41,000 (NIZO 1982) 65,000 (1985)
Primary or New Applications:	Evaporator Expander Preconc'n for Roller Drying Hauling from Small Cheese Plants (2X)	Evaporator Expander Hauling from Small Cheese Plants (2X) UF Permeate Conc'n Milk Conc'n	Higher Conc'n (25% TS) before Hauling Conc'n for On-site UF

producing roller dried whey ("popcorn") by preconcentrating it from 6% total solids (TS) to 10 - 12% TS, thereby removing up to one-half the water by a process many times more efficient than roller drying in terms of the energy required. Unfortunately, but not surprisingly, there were a number of problems with the first membranes and equipment during RO's infancy and the process did not enjoy its expected rapid growth.

During what I call the childhood of RO's life, 1972-1981, not even the "oil crisis" of 1973-1974 really accelerated the adoption of the process, mostly because it was not yet considered reliable enough and, in the U.S.A., wasn't ready for use in a U. S. Department of Agriculture (USDA) graded plant. It was the oil crisis of 1978-1979 which finally provided the impetus for rapid growth. This can be seen from Table I which cites two surveys by NIZO of RO membrane area installed in the dairy industry. Thus, most of the 90 plants sold by the end of RO's childhood came between 1978 and 1981.

To follow the progress of RO with respect to its uses in the dairy industry during this period, the role as an evaporator expander was most popular in the U.S.A. and the hauling of two-fold (2X) concentrated whey from smaller cheese plants began to spread. This first example of concentrating whey ultrafiltration (UF) permeate appeared in 1972, in the U.S.A. In 1974 in France the first commercial use of RO for concentrating milk began, in this case for the production of yoghurt; this general application area has not proliferated.

It is obvious from Table I that RO has enjoyed its greatest growth in the past five years, its teenage year (teens). More particularly this growth occured in 1981 through 1984 before improvements in evaporation to be discussed later made for a serious competitive situation and before some saturation of the practical market for RO was realized on both sides of the Atlantic Ocean (given the limits on the maximum solids level it could reach).

The point on solids level also will be discussed later, but is noted here because one of the improvements in RO which was important during its early teens was the adoption of a design of the system which enabled one to concentrate whey routinely from 6% TS to 25% TS (or slightly higher with some wheys). This 4X concentration has been of more interest in Europe than in the U.S.A.

In 1981 in the U.S.A., a large-scale RO plant for concentrating whey at a cheese factory prior to on-site UF was installed as the most economical process scheme for making whey protein concentrates; this idea has not had much further acceptance.

The surge in the use of RO of whey which had started in 1978-1979 in the U.S.A. was slowed by the recession of 1982 and some market saturation, while in Europe the first real surge, which started in 1981, was over by 1984-1985 as the market for hauling whey from small cheese factories, especially in France, was well penetrated. Of the 210 plants noted in

Table I as being sold to date, nearly 100 were sold in the U.S.A. This represents a faily respectable penetration of the market in terms of the percentage of total number of cheese factories approximately 750 in 1985.

III. TECHNICAL ACHIEVEMENTS AND LIMITATIONS

For purposes of this lecture's title, "technical" has been taken to refer to the membrane itself. Table II presents some highlights in the history of that part of the RO process which does the actual work in separating water from whey, milk, and UF permeates.

The two membrane materials noted are cellulose acetate and derivates (CA), and what are known as thin-film composites (TFC). The latter have a coating, which is the working RO membrane, formed on a porous support such as a UF membrane that has large pores.

Despite the shortcomings of CA with respect to its ability to withstand high pH's, a temperature of at least $50^0 C$ and exposure to practical chlorine levels during sanitation (disinfection), it is evident that the benefits of RO caused those concentrating whey to live with these shortcomings. A breakthrough in two of these properties appeared at the beginning of RO's teenage years with the adoption of TFC membranes from the water desalination application area into the concentration of whey and UF permeates. As can be seen from Figure 1, graciously provided by one of the two membrane equipment suppliers making this type of TFC membrane under license from a U.S. manufacturer, the pH and temperature limitations are less restrictive than those for CA membranes.

While this TFC membrane has no chlorine tolerance at all, its advantages have spurred on its acceptance to the point where it has been fitted in approximately one-half of the RO installations in the dairy industry. This is the "now approx. 50/50" annotation in Table II.

Table II states that TFC has much better retention of low molecular weight (MW) organic compounds. Figure 2 shows this more dramatically. This property results in a better quality of RO permeates, making disposal or reuse easier. In particular the TFC membrane referred to here has a much better retention of lactic acid than does a CA membrane.

Figure 3 shows another advantage of this TFC membrane, the ability to operate at higher pressures than CA membranes, thus offering the possibility of enjoying higher flux rates and lower plant costs.

IV. TECHNOLOGICAL ACHIEVEMENTS AND LIMITATIONS

Just as "technical" has been taken to refer to the RO membrane itself in organizing this lecture, "technological" in the title has been taken to refer to those features of the process which convert the membrane into a practical system. Four of these features are included in Table III.

With respect to packaging the RO membrane in an element or module, this is a subject which raises much controversy. Efficiency and reliability of performance of a RO system and its costs are at stake. What is most interesting to note is that, in both concept and details of realization, radically different configurations have been made to work. The spiral-wound configuration, offering the potential of being lowest in cost beca of the relatively low unit cost of the membrane elements and the systems was thought to be prone to plugging and fouling because of the internal

Table II

"TECHNICAL" FEATURES

of

REVERSE OSMOSIS

in the

DAIRY INDUSTRY

	Infancy 1969-1972	Childhood 1972-1981	Teenage 1981-1986
Materials	CA	CA	CA now approx. TFC 50/50
Operating pH limits	3 to ~7	3 to ~7	CA: 3 - ~7 TFC: 3 - 11
Maximum Operating Temp at Optimum pH, °C	40	40	CA: 40 TFC: 60
Practical Chlorine Resistance	Very Slight	Very Slight	CA: Very Slight TFC: None
Retention of Low MW Organics	Acceptable	Acceptable	CA: Acceptable TFC: Much Better

Figure 1.   Comparison Between the Resistance of a CA and a TFC Membrane to pH and Temperature

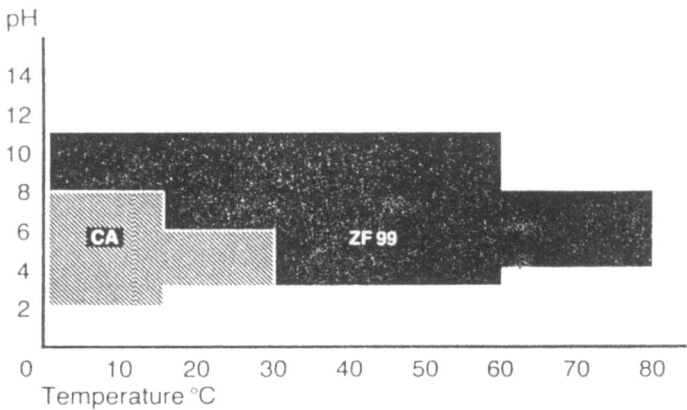

Source: P.C.I.

Figure 2.   Comparison Between the Permeate Qualities of a CA and a TFC Membrane Operating on Sweet and Acid Wheys

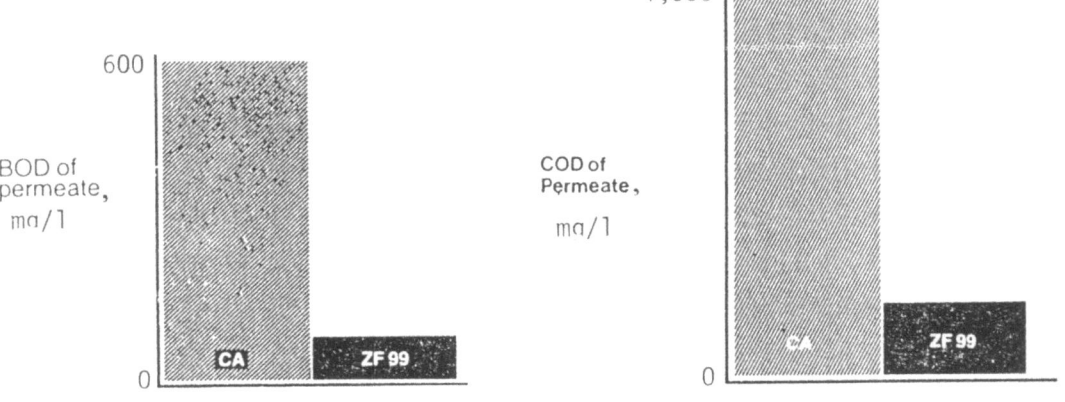

(Data for concentration of sweet whey from 6% to 28%TS)

(Data for concentration of lactic acid whey from 6% to 12%TS)

Source: P.C.I.

Figure 3. Comparison Between the Capabilities of a CA and a TFC Membrane for Operating at Different Pressure and Temperature Combinations

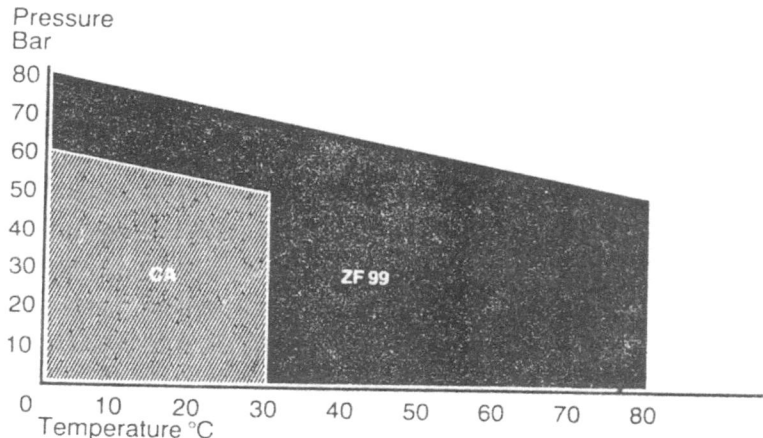

Source: P.C.I.

Table III

"TECHNOLOGICAL" FEATURES

of

REVERSE OSMOSIS

in the

DAIRY INDUSTRY

	Infancy 1969-1972	Childhood 1972-1981	Teenage 1981-1986
Membrane Configurations in Commercial Use	Tubular Flat Plate	Tubular Flat Plate	Tubular Flat Plate Spiral Wound
Systems Designs	Once-through, One-stage	Once-through, One-stage	Stages-in-series; Once-through, One-stage
Pretreatments	None	None	CO_2 to control pH so $CaPO_4$ won't ppt
Processing Milk for Cheese	---	---	Shearing problem noted and solved

design of the elements. However, it has been shown to work satisfactorily and both whey UF permeate and whole whey are concentrated in spiral-wound RO plants. The first commercial ones probably did appear before 1981, but the entry in Table III under RO's teenage period is a more realistic assessment of the timing.

Through RO's childhood, the design of systems to process whey continuously did not require the same arrangement of a series of stages as did UF because relatively low concentration factors were involved (2X - 3X for RO and 5X - 20X for UF) and the membrane elements could be arranged to provide diminished internal volume as removal of water progressed (thus maintaining velocity to some degree). Therefore, a once-through, one-stage design was used. However, the combination of a desire to attain higher concentrations (4X - 5X) and some problems with fouling in the last membrane elements at the end of the once-through flow path led to the use of the stages-in-series arrangement (once a good high pressure recirculation pump was found for use at each stage). As a historical note, during the pilot development of the first whey UF permeate RO plant in 1970-1972, a three-stage design was used to attain 25 - 28% TS, approximately a 5X concentration; each stage was once-through and the pressure was let down after the first and second stages.

Until 1981 or 1982, whey was concentrated by RO without any pretreatment other than separation (of fat) or clarification, pasteurization and, for some membranes configurations, filtration. One supplier working on the problem of membrane fouling by calcium phosphate in whey then found that careful control of the pH in the range below 6 could avoid precipitation. This pH control was accomplished by injecting carbon dioxide into the whey, where it remains just long enough to do its job before the concentrate is used or further processed. This interesting technological achievement is applied mainly in Europe where most of the RO plants concentrating whey to the higher solids levels are found.

While RO is not yet in commercial use for cheesemaking, a fact to be discussed below, it has received fairly extensive attention on the pilot scale for the manufacture of both Cheddar and cottage cheese. The technological problem found, especially with Cheddar cheese, was that the valve where the concentrate exited the RO system caused enough shearing to change the condition of the fat globules; inferior cheese resulted. This problem was solved by using a tapered tube rather than a valve to let the pressure down gently. In fact, this shearing problem is not peculiar to RO. There is a counterpart in UF where the extensive exposure to shearing during recirculation of milk in each stage of a plant, but particularly in the latter stages where the milk is more concentrated, has been shown to result in troubles with attaining the proper texture of the UF cheese.

V. ECONOMICAL ACHIEVEMENTS AND LIMITATIONS

Table IV summarizes some economic subjects of importance to RO and its acceptance and growth in the dairy industry.

The key economic subject, of course, is cost. RO was brought into the dairy industry as having advantages over evaporation in both capital and operating costs. This remained true through RO's childhood, especially toward its end when sharp jumps in fuel prices made the electrically driven membrane process look very attractive indeed. The great potential which might have then been seen for RO in its teenage years because of this cost benefit did not, however, materialize. The suppliers of evaporators

Table IV

"ECONOMICAL" FEATURES

of

REVERSE OSMOSIS

in the

DAIRY INDUSTRY

	Infancy 1969-1972	Childhood 1972-1981	Teenage 1981-1986
Costs versus Evaporation	Lower	Much lower especially after 1978	MVR and drop in fuel costs change picture
Maximum Practical Concentration, %TS	12-18	12-18	25-30
Product Quality with Whey	"No heat" process	Not a major issue	?
Benefits with Milk	---	Increase solids for yoghurt	Increase yield ∼1% for Cheddar

responded to the energy cost issue mainly by developing mechanical vapor recompression (MVR) designs which put the evaporation process largely on an electrically driven basis. These MVR designs have been steadily improve so that compact versions with lower cost compressors can make evaporation competitive with RO at the more dilute end of the concentration range, say 6% TS to 12% TS (2X). In addition to the breakthrough offered by MVR, the more common thermal vapor recompression (TVR) evaporation process was improved with multiple-effect designs demonstrating much higher water removal efficiencies than in the past. As a further help to evaporation and a block to RO, recent drops in fuel oil costs have changed the entire energy cost picutre, at least temporarily.

Even when RO enjoyed its greatest cost advantage, it was not able practical to concentrate whey beyond 18% TS. This might be considered a technological limitation, but it is really an economic one because it kept RO from competing with evaporation when wanting to attain 40% TS to 50% TS. Even when RO became routinely viable in producing whey concentrates with 25% TS 30% TS it still meant that evaporation was required. This limitation remai and will be discussed at the end of this lecture when a perspective on PO's future is given.

Product quality was a claimed advantage of RO in its earliest days, but it is difficult to say that this really has been translated into an ecoomic virtue in food products employing whey concentrates. Table IV shows a question mark for this subject during the past five years because there have once again been claims that RO is more "gentle" with whey than is evaporation and because no public evidence is on hand to cite here.

Not giving up on the concentration of milk by RO, Table IV includes comments on the benefits which have been realized to date. Beside the one commercial plant making yoghurt mentioned earilier, the economics of which do not seem compelling, there has been a preference for using UF to get higher yield increases, at least when UF also can be employed in the same dairy to make ymer or other cultured products. This yield increase benefit is the key in cheesemaking and, in the author's opinion, puts RO in a less favorable position than it should enjoy. As Table IV indicates, RO has been demonstrated to offer approximately a 1% yield increase when making Cheddar cheese. This increase comes from an increase in all the solids rather than just the whey protein concentrated by UF. Further, a concentration factor of only 1.25X - 1.5X gives this result, there is little issue of attaining normal quality cheese, and payback of the RO plant can be rapid. Nevertheless, the dairy world focuses on using UF to get much higher yields. At least one aspect of the choice between RO and UF for making those cheeses with standards of identity should favor RO, as will be seen in the next section.

VI. LEGAL ACHIEVEMENTS AND LIMITATIONS

RO has had to work its way through a variety of governmental and related regulatory issues to gain acceptance for use in the dairy industry, especially in the U.S.A., whose requirements can be used to discuss the "legal" status of the process.

The U.S. government's Food and Drug Administration (FDA) has prime responsibility for materials in contact with foodstuffs. Membranes must be made of a material already approved for direct contact or must be shown by the manufacturer not to introduce any harmful substances under the conditions of the application. Happily for RO in its infancy, CA was a material already looked upon with favor in many other uses. When TFC membranes were developed they had a much tougher time winning approval and each of the two suppliers making the membrane now in use on whey separately had to submit evidence that harmful compounds, such as the solvents used in manufacturing the membrane, were not being extracted.

In contrast to the U.S. government approval of materials in contact with foodstuffs, there is no direct approval of the equipment used in food plants. Rather, the USDA grades food plants on a voluntary basis requested by the plant and indicates whether or not the equipment design meets certain guidelines for being sanitary. If the equipment has shortcomings, the USDA will define these. It is up to the equipment supplier to deal directly with the USDA in having details of a RO system reviewed for their suitability. The USDA has been quite helpful over the years in guiding membrane equipment manufacturers on the specific needs for attaining sanitary design. While initial work commenced during RO's infancy, it is fair to say that much of the effort took place during RO's childhood so that it became a fairly routine procedure by RO's teenage years.

The USDA's influence with respect to sanitary design is backed up by an industry-government cooperation which approves sanitary design standards for various pieces of equipment written by committees of the manufacturers themselves. The resulting 3-A Sanitary Design Standards are published. Suppliers whose equipment meets the standards are given an identity number to indicate approval. A committee on membrane process equipment was formed in the mid-1970's and the first draft of a design standard was released in 1977. As stated in Table V, a standard has yet to appear. Among the many reasons for this are the time needed to convince sanitarians that the structure below the surface of the porous membrane itself did not constrain cleaning and disinfection, the practical need not to prevent one or more membrane configurations from gaining acceptance, the problem with accepting internal dimensions and radii within membrane elements that run against those in more traditional equipment, etc.

If RO was being pushed more strenuously for processing milk for cheesemaking, the pressure to have 3-A Sanitary Design Standards would be greater. As it is, no one has asked the FDA to give approval of process for manufacturing cheeses with Standards of Identity. Based on full-scale pilot work in making Cheddar cheese, the fact that there is a precedent for concentrating milk (by evaporation) and the fact that no significant change in composition or organoleptic properties of the cheese occur, one would expect that RO could gain the FDA's approval fairly readily. UF, the process currently preferred by industry, has been having trouble winning over the FDA.

Table V

"LEGAL" FEATURES

of

REVERSE OSMOSIS

in the

DAIRY INDUSTRY

	Infancy 1969-1972	Childhood 1972-1981	Teenage 1981-1986
FDA-type Approval of Membrane	CA easily approved	---	TFC: OK after long efforts
Approval for Use in USDA Graded Plants	---	Elements and systems changed as needed	No Longer a major issue
3-A Sanitary Design Standards	---	First draft attempted ca 1977	Still no standard
Processing Milk for Cheese, etc.	---	Plant in France for yoghurt ca 1973-1974	Full-scale pilot plant in U.S.A. shows OK for Cheddar. No gov't approval yet.

VII. THE FUTURE OUTLOOK FOR RO

Now that review of the topics included in the title of this lecture has been fulfilled, it is appropriate to see how well RO's life is progressing as it passes into adulthood. Are some of the limitations of the process being overcome? Will there be a larger role for the process in the dairy industry or has it gone as far as can be expected?

Table VI lists three responses to these questions.

A. RO MEMBRANES WITH CHLORINE RESISTANCE

Two membrane suppliers have recently announced the development of membranes, in the spiral-wound configuration, that are claimed to have sufficient resistance to chlorine to make daily disinfection by inexpensive hypochlorite compounds practical for the first time. Since these membranes were developed for desalination of waters with low residual chlorine levels, extensive evaluation in a dairy is required before this potential cost-saving breakthrough can be realized. The test data on continuous exposure make it hopeful that daily sanitizing with 50 - 100 ppm free chlorine may be possible.

B. ONCE THROUGH PROCESSING TO HIGH SOLIDS

A commercial RO product exists which might be put to use in concentrating whey or milk to 40% TS - 50% TS, thus overcoming the limitation which still makes RO fall short of the capabilities of evaporation. Table VII lists some features of this product. The key is its basic construction from porous stainless steel tubing whose diameter can be stepped down to provide one long path of diminishing volume. Since nothing specific is known about the product's performance on whey or milk, it is much too early to tell whether this version of RO will help open more opportunity for the process in the dairy industry.

C. "LOOSE" REVERSE OSMOSIS

During RO's infancy, CA membranes were developed which had only approximately 50% retention of sodium chloride while still exhibiting 98% or higher retention of lactose. The author was involved in their evaluation for the preconcentration and partial demineralization of whey before electrodialysis. Results of the evaluation were published by the New Zealand Dairy Research Institute in 1972. Unfortunately, concerns over the amount of BOD in the permeate were paramount at the time and RO hardware was too immature to see the idea through.

Today, TFC membranes exist which have much lower retention of monovalent salts. They exhibit some ion exchange properties, so extensive testing with different wheys and milk must be performed before specific selectivity data can be published. Table VIII does list some estimates of selectivities with whole sweet whey.

Table VI

LATE TEENAGE TO EARLY ADULT DEVELOPMENTS
IN REVERSE OSMOSIS FOR THE DAIRY INDUSTRY

- The emergence of membranes with practical resistance to chlorine

- A possible breakthrough in once-through processing to high solids

- Commercial realization of the 15-year old idea of "loose" RO

Table VII

INORGANIC/CERAMIC/MINERAL MEMBRANE
ON
POROUS STAINLESS STEEL TUBING

- Once-through processing using tapered flow path.

- Very high temp. capability, wide pH capability -- cleaning and sanitizing should be excellent (sterilize if necessary).

- Oxide other than zirconium being used for food processing, so FDA approval may be easier.

- Maximum concentration for dairy unknown.

- Costs for dairy processing not known.

∴ too early to tell

Table VIII

APPROXIMATE RETENTION VALUES

FOR A

"LOOSE" REVERSE OSMOSIS MEMBRANE

IN A

SPIRAL WOUND ULTRA-OSMOSIS^{TM} SYSTEM

Whole Sweet Whey

Na^+ : 2%

K^+ : 2%

Ca^{++} : ~80% (Due to binding to protein?)

Mg^{++} : 10%

Cl^- : <10%

$PO_4^=$: 70%

Lactose : 98%

Lactic Acid : may be 0%, but verification needed, especially as function of pH

Table IX

APPLICATIONS FOR
"LOOSE" REVERSE OSMOSIS

- Converting "salt whey" to normal whey while solving a disposal problem.

- Concentrating and partially demineralizing permeate prior to further processing into lactose and lactose derivatives.

- Partial demineralization of lactose mother liquor (delactosed whey).

- Preconcentration and partial demineralization of whey for electrodialysis.

- Treating brine solutions for re-use.

Obviously, the potential for partially demineralizing while concentrating becomes exciting. For the sweet whey case a straightforward 3X concentration (two-thirds of the water removed) is said to give 60% demineralization. For the treating of "salt whey" from Cheddar cheese manufacture, the ability to get 90% demineralization with the use of diafiltration has been developed in two commercial systems operating in the U.S.A. This enables the whey processor to convert the salt whey to normal whey and, in many cases, to solve a disposal problem.

Besides the "salt whey" application, "loose" RO (or nanofiltration or Ultra-Osmosis^{TM} as this new variation is being called), there are a number of other possibilities. These are listed in Table IX and indicate that RO should have a bright future as it matures further in its life in the dairy industry.

MULTISTAGE EVAPORATION AND WATER VAPOUR RECOMPRESSION WITH SPECIAL EMPHASIS ON HIGH DRY MATTER CONTENT, PRODUCT LOSSES, CLEANING AND ENERGY SAVINGS

H. G. Kessler
Technische Universität München
Institute of Dairy Science and Food Process Engineering
D-8050 Freising-Weihenstephan
West Germany

ABSTRACT. With regard to energy saving measures the recent developments in evaporation are mechanical recompression and processing with low temperature differences. Consequently longer residence times of the product in a plant and higher evaporation temperatures are required. But this can increase deposit formation and shorten running times. Evaporation experiments with whey and ultrafiltration-permeates will show influences on overall heat transfer coefficients and possibilities of reducing deposits, increasing running times and efficiency of evaporation plants.

1. INTRODUCTION

Concentration of milk and whey by evaporation in a vacuum was started as early as the middle of last century. For a period of 50 years batch evaporation was used for this purpose in which the energy requirement - in relation to the amount of water evaporated - was more than 100 %. Fig. 1 depicts the development in evaporation techniques.

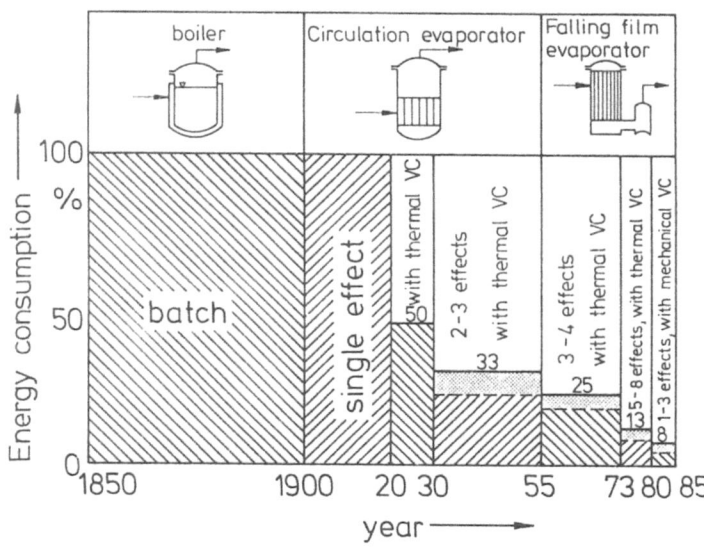

Fig. 1:
Development of evaporation plants

Milk – The vital force, 545–558.
© 1987 *by D. Reidel Publishing Company.*

It is evident that the introduction of the circulation evaporator ushered in a second phase which lasted for another half a century. However, during this period there appeared already evidence of technological progress in the form of thermal vapour recompression and the use of multieffect processing techniques which reduced energy requirements to less than 50 %. The disadvantages of circulation evaporation, namely the longer residence times and the increase in browning and deposit formation were considerably reduced by the introduction, in 1955, of the falling film evaporator. At the same time, but especially after 1973, energy requirements were reduced to less than 15 % adding more effects to the evaporator. Finally, the increase in energy prices which happened once again led to the introduction, in 1979/80, of mechanical vapour recompression whereby the energy requirement was lowered to less than 10 %.

The most recent developments are the change from the expensive radial compressors to simple fans. Because pressure or temperature differences of no more than 4 - 5 K can be achieved with these machines, only single effect plants with subdivided evaporators are practicable.

2. ENERGY ASPECTS

In order to make the energy requirements of the various processes comparable, in this contribution the electrical and heat energy consumed in each process was converted to primary energy and calculated as the energy required for the removal of 1 kg of water. For that purpose the electrical energy consumed was multiplied by a factor of 3 and the heat energy by a factor of 1.2.

Because vapour recompression contributes so much to energy saving, its function and the energy requirements are illustrated in enthalpy / entropy diagrams.

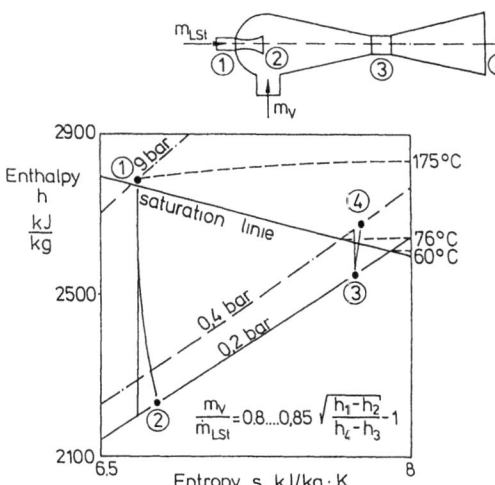

Fig. 2a Steam-jet vapour recompressor and h-s-diagram

Fig. 2a illustrates the operation of the steam jet vapour compressor and the amount of vapour that can be sucked in by it.

The ratio of the mass of vapour to that of the live steam can be calculated from the differences in enthalpy arising from the expansion of live steam and the compression. As this ratio is 1 when the temperature of the saturated steam is raised by 15 °C, a smaller temperature difference of e.g. 3 °C will make it possible to suck in and compress more than 3 kg of vapour with 1 kg of live steam (Fig. 2b).

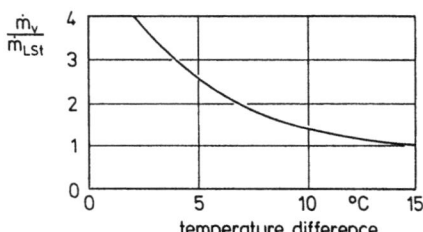

Fig. 2b: Ratio $\frac{\dot{m}_{vapour}}{\dot{m}_{Live-steam}}$ and temperature increase by thermal recompression

Measures such as the reduction in temperature differences (3 to 4 °C) and the increase in the number of effects (up to 8) increase the cost of the plant considerably but make it possible to reduce the energy consumption to about 10 % of the steam used, which for the whole plant corresponds to about 400 kJ of primary energy per kg of water removed.

Fig. 3 is a schematic representation of the hs-diagram for a triple effect evaporation with a radial vapour compressor.

The compressor sucks in vapour at 59 °C from the last effect and compresses it to a pressure which corresponds to a saturation temperature of 71 °C. The resulting increase in temperature to 125 °C is reduced to 71 °C by the injection of condensate. This heat as well as the heat of the condensate and of the concentrate are used for the preheating of the product. The introduction of radial compressors made it possible to lower the primary energy consumption to about 200 kJ/kg of water removed.

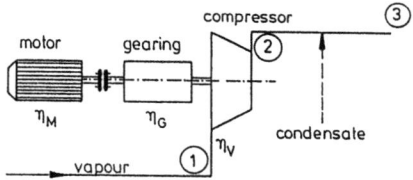

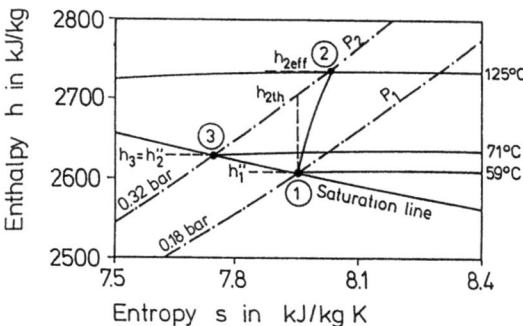

Single effect evaporators in which the vapour is compressed by simple fans have been in operation since 1983. By these the saturation temperature is increased by 4 - 5 °C.

The energy consumption can again be obtained from the enthalpy/entropy diagram of Fig. 4.

For plants with smaller evaporation capacities of up to 5 t/h the isentropic efficiency $\eta_{isentrop}$ is about 70 to 75 % and for plants with capacities of up to 30 t/h the isentropic efficiency is about 85 %. In the example in Fig. 4 a value of 80 % was chosen. η_M at 93 % includes frictional losses at shafts and clutches, losses due to variable drives, frequency regulators, control panels etc.

When the calculation of the energy consumption for a temperature increase from 60 to 64 °C takes into account the energy consumed by every part

Power of the el. motor:

$$P_{el} = \frac{\dot{m}_v(h_{2th} - h_1'')}{\eta_V \cdot \eta_G \cdot \eta_M}$$

Fig. 3: Radial vapour recompressor and h-s-diagram

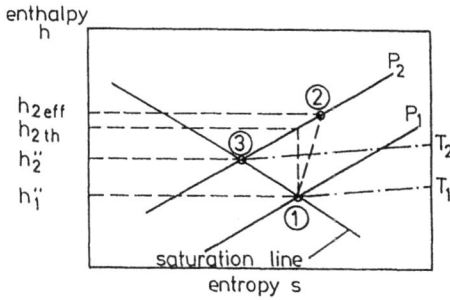

Power of el. motor

$$P_{el} = \frac{m_v (h_{2th} - h_1'')}{\eta_M \cdot \eta_{isentr}}$$

$\eta_M \approx 0.93$
$\eta_{isentr} \approx 0.80$

with $T_1 = 60°C$

Temp. diff. °C	$h_{2th} - h_1''$ kJ/kg	primary energy kJ/kg
4	28.5	115
5	35.7	144

Fig. 4: Mechanical recompression by fan

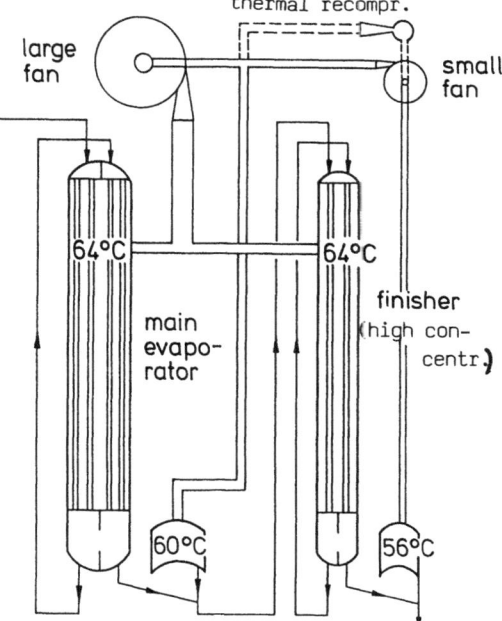

Fig. 5: Double effect evaporator with 2-fan-mechanical recompression

of the plant (circulating pumps, vacuum pumps, steam consumption on starting up) then the energy required is composed of the following: Fan drive ~ 38 kJ/kg, other electrical drives ~18 kJ/kg, steam heat ~ 25 kJ/kg. This adds up to a total primary energy requirement of ~ 198 kJ/kg of water removed. In principle all mechanical compressors could be driven by gas or fuel motors whose exhaust heat could then be used for heating. This would reduce the primary energy requirement considerably.

Fig. 5 is a schematic representation of a double effect evaporator with two fans which is suitable for higher concentrations. The bulk of the liquid is evaporated in the first effect in which the temperature difference between the heating steam and the evaporating product is 4 °C. This temperature difference can be increased in the second effect by means of the small fan to compensate for the increase in boiling point at high concentrations. Alternatively, using a thermal vapour compressor instead of the small fan will increase the temperature difference even further.

However, no further energy saving is possible with this kind of double effect evaporation. A further reduction in the specific energy requirement can be obtained by using reverse osmosis for pre-concentration prior to evaporation. About 50 % to 60 % of the total water to be extracted can be removed economically by reverse osmosis. It is not advisable to use reverse osmosis to achieve higher concentrations, for three reasons:
1. the rate of permeation becomes very low,
2. the specific energy requirement increases considerably,
3. the passage of soluble constituents through the mem-

brane increases markedly. The biological oxygen demand becomes too great.

concentration process	spec. electrical energy kJ/kg	spec. steam energy kJ/kg	spec. primary energy kJ/kg
freeze concentr.	55...75	-	- 600...800
triple effect with therm. recompr.	18	736	-1000
multiple effect with therm. recompr.	30	245	- 400
triple effect with mech. recompr.	59	25	- 200
single effect with fan	56	25	- 200
reverse osmosis	15...30	-	- 45...90

Table I: Concentration processes and energy consumption in kJ per kg water removal

Table I shows a comparison of the primary energy consumption of a number of concentration processes.
The table shows that further developments in freeze concentration have made the process nowadays more economical, as far as primary energy consumption is concerned, than the triple effect evaporation with thermal vapour recompression which has been widely used for such a long time. On the other hand it is evident that the energy consumption of evaporation processes cannot be less than 200 kJ/kg - cleaning and ancillary operations included. Greater energy savings can only be obtained by combining reverse osmosis with evaporation. Such a combination might make a primary energy consumption of 100 to 150 kJ/kg achievable.

3. THE RELATION OF OVERALL HEAT TRANSFER TO DEPOSIT FORMATION

As the energy utilization is increased by these innovations, the area required for the plant also increases. As a result of the developments in evaporation technology the well known phenomenon of deposit formation has become a problem. The manufacturers of a modern plant with mechanical vapour recompression and large evaporating surfaces would like to choose high temperatures in order to keep the compressor as small as possible. But high temperatures combined with high residence times mean increased deposit formation and increased energy consumption.
  To study the factors affecting deposit formation with a view of discovering measures for counteracting it, a inve-

stigation with a vacuum falling film pilot plant (Fig. 6) has been carried out at our Institute (Fiedler, J. Diss. TU München, 1985). The main part of the plant was a 2 m long evaporation tube with an internal diameter of 40 mm. The chief characteristic of the arrangement was that the tube or tubes could easily be exchanged for analysis of the deposit and that the amount of heating steam condensing on the tube was measured as mass flow rate. This made it possible to determine the overall heat transfer coefficient k continuously.

To start with, the effects of the evaporation temperature and of the solids concentration on the k_0 value of the still clean 2 m long tube were determined. This is shown in Fig. 7 for whey.

The values obtained were higher than those found in practice as revealed by measurements on industrial evaporators (Table II).

It was discovered from these that the k_0 values were lowest in the second effect. The only possible explanation of that is a decreased

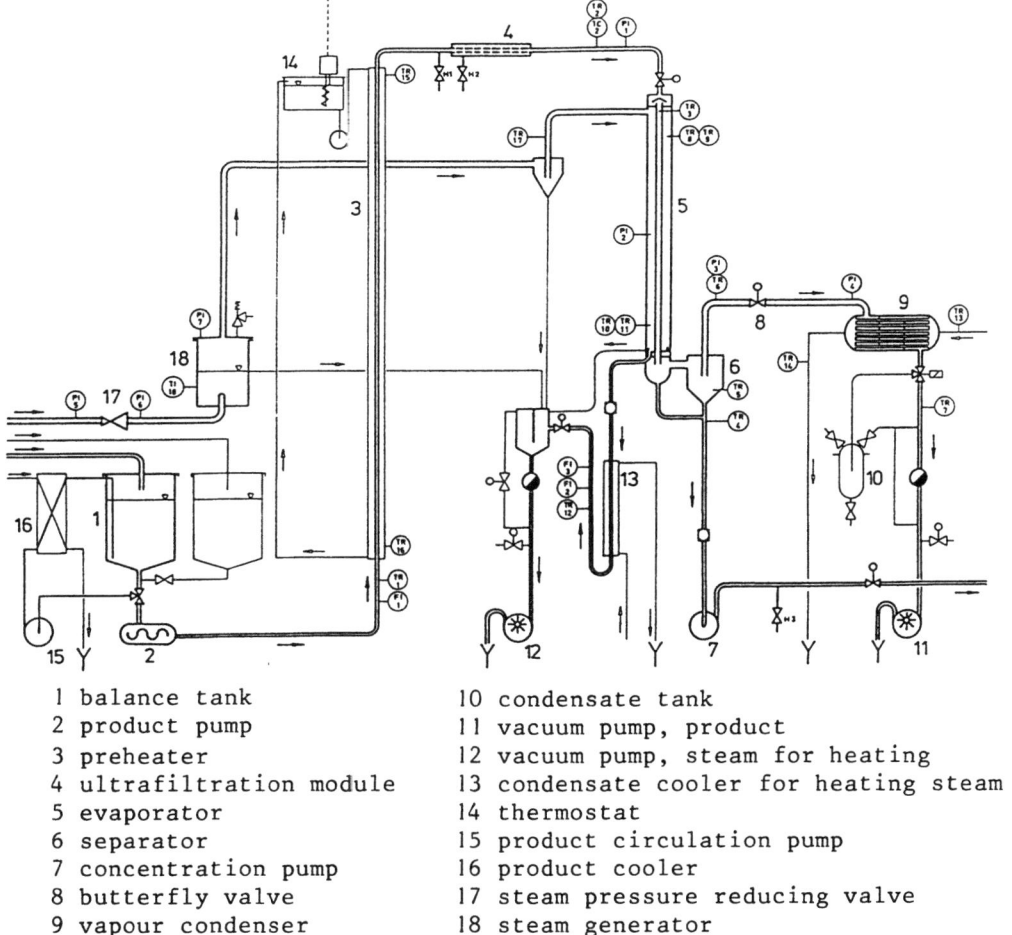

1 balance tank	10 condensate tank
2 product pump	11 vacuum pump, product
3 preheater	12 vacuum pump, steam for heating
4 ultrafiltration module	13 condensate cooler for heating steam
5 evaporator	14 thermostat
6 separator	15 product circulation pump
7 concentration pump	16 product cooler
8 butterfly valve	17 steam pressure reducing valve
9 vapour condenser	18 steam generator

Fig. 6: Pilot plant of an one-tube-evaporator

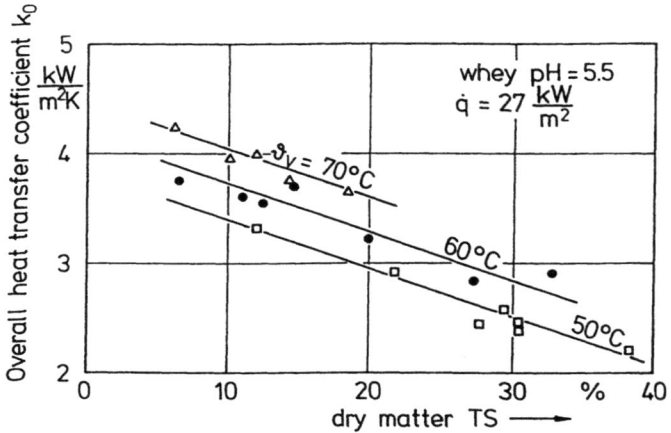

Fig. 7: Overall heat transfer coefficient k_0 at the beginning of a run

rate of heat transfer due to the presence of inert gas. This is produced in the first effect by evaporation and reaches the second effect together with the heating steam. Experiments on the single tube pilot plant designed to test this hypothesis confirmed the marked effect of inert gas on the k_0 value (Fig. 8). Only traces of air expressed by the ratio of air pressure to the total pressure caused a steep decrease.

effect	concentration in %		k_0
	from	to	W/m²K
I	5.7	8	2 300
II	8.0	13	1 500
III	13.0	31.5	1 800

Table II: Overall heat transfer coefficient k_0 of a 3 effect evaporator for whey

Another factor accounting for the differences in k_0 values obtained in experiments and in practice is the length of the tubes (e.g. 8 m) used in industrial plants. This is because the heat transfer in condensation is inversely proportional to the 4th root of the tube length.

The following graphs show a number of results on deposit formation in an evaporator tube and its effect on heat transfer and the thermal resistance of the deposit.

Figs. 9 show for low concentration whey that deposit formation falls with decreasing evaporation temperatures. If long running times are required then evaporating temperatures should range from about 58 to 60 °C. Figs. 9a and b differ in the composition of the deposit. At lower pH values the proportion of salts in the deposit is considerably smaller than at the higher pH value.

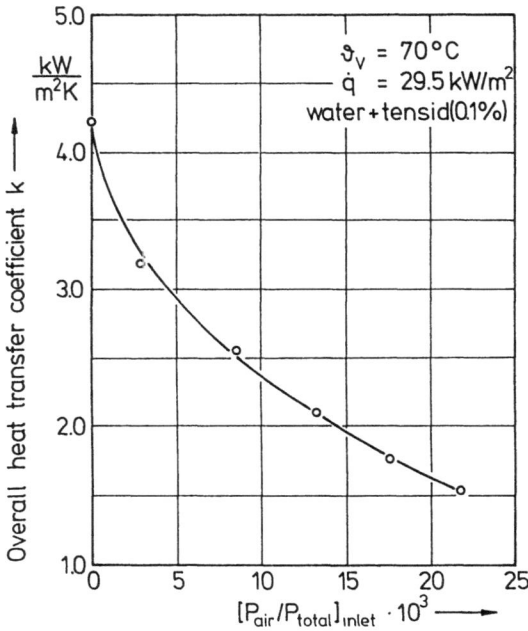

Fig. 8: Decrease of k due to aircontent in the steam for heating

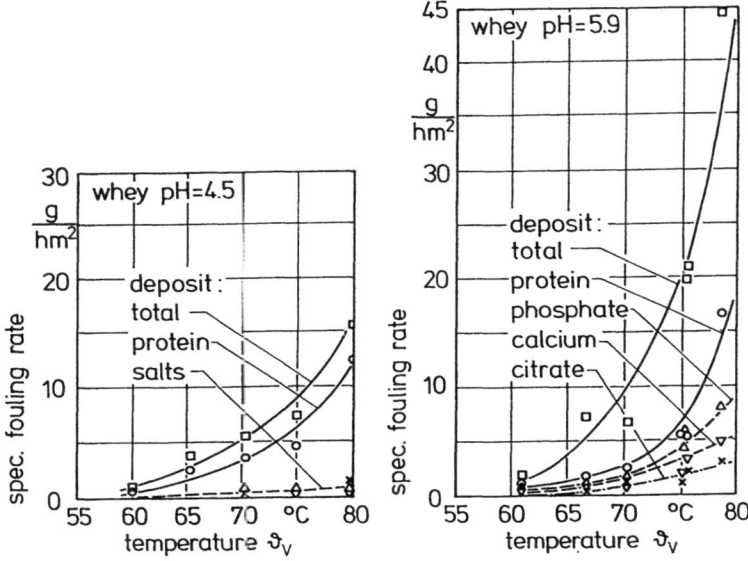

Fig. 9: Fouling rates during evaporation of whey

It is well known that evaporator tubes carry the heaviest deposit at the inlet. We therefore carried out a number of experiments with acid whey permeate from ultrafiltration to study the effect of tube length on deposit formation at different values of heat transferred per unit surface area ($\dot{q}$) which was kept constant during each run (Fig. 10). It was indeed shown that the specific rate of deposit formation in the upper part of the tube increased with $\dot{q}$, particularly at high values of $\dot{q}$. The initial temperature differences ($T_H - T_V$), relative to the mean $\dot{q}$ values were 4.8, 7, 10 and 13.4 K.

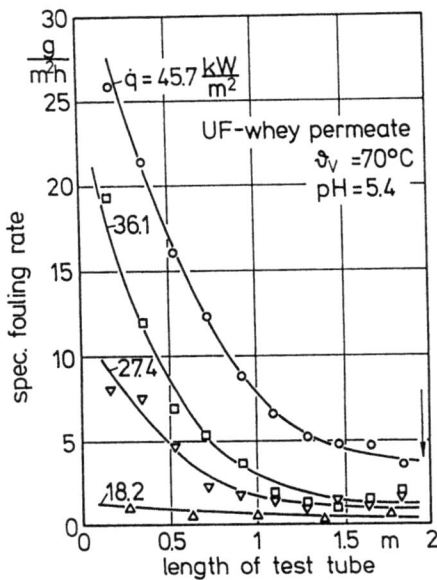

Experiments with wheys of different concentrations showed that the decrease in the relative overall heat transfer coefficient k/k_0 becomes greater as the solids concentration of the whey rises (Fig. 11). With whey of a solids concentration of 27.4 %, the k_0 value of 3,000 W/m^{2}K had fallen after 8 hours of operation to about 40 % of its initial value, i.e. to 1,200 W/m^{2}K.

The fall in k_0 is caused by the deposit whose thickness increases with time t. The change in the value of the overall heat transfer coefficient k with time is described by the following equation:

Fig. 10: Spec. fouling rate in dependence on the length of the test tube and on the heat transferred per unit surface

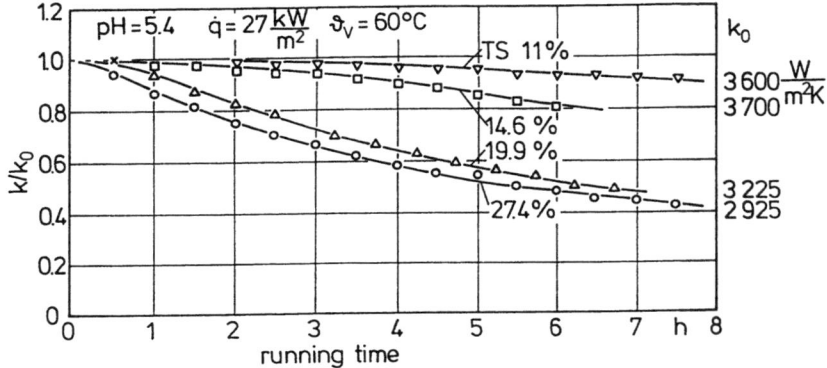

Fig.11: Ratio of the overall heat transfer coefficient during evaporation of whey concentrates

$$\frac{1}{k} = \frac{1}{k_0} + (\frac{s}{\lambda})_d$$

where λ is the heat conductivity of the deposit (d). The term $(\frac{s}{\lambda})_d$ represents the thermal resistance R_d of the deposit so that:

$$R_d = \frac{1}{k} - \frac{1}{k_0} = (\frac{s}{\lambda})_d$$

The results of Fig. 11 were used to plot R_d against running time in Fig. 12. It is seen that R_d increases linearly with time. After a certain induction period, which differs from experiment to experiment, the rate of product deposition increases linearly with time. This rate rises as the solids content of the product increases.

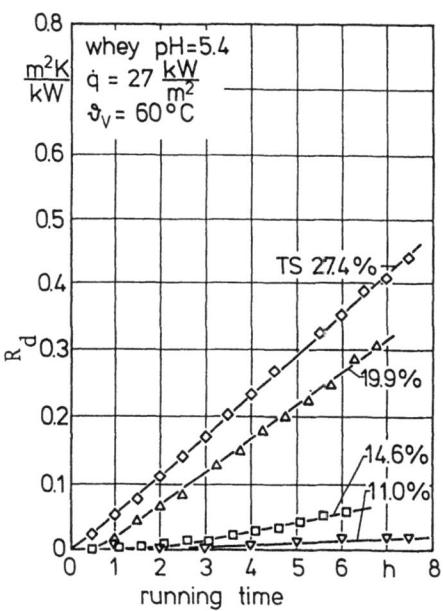

Fig.12: Thermal resistance of the deposit during evaporation of whey concentrates

As it is the milk salts which increasingly determine deposit formation as the whey concentration increases and because deposit formation is strongly influenced by the heat $\dot{q}$ transferred per unit surface area, experiments were carried out with whey of 30 % solids concentration to study this (Fig. 13).

The slopes of the straight lines in Fig. 13 demonstrate the great influence of $\dot{q}$ on deposit formation. Although the evaporating temperature was only 50°C, when the $\dot{q}$ value was 38.38 kW/m² (which corresponds at k_0 = 2,210 W/m²K to an initial temperature difference of $\Delta T_0 = T_H - T_V$ = 18 K) the amount of deposit formed after 6.5 hours was so great that the total heat transfer coefficient k_0 was reduced to 30 % of its initial value i.e. from 2,100 to 700 W/m²K (Fig. 14).

For wheys of low solids content the rate of deposit formation had been found to be strongly dependent on temperature but this was not the case, at least in the region of lower temperatures, for highly concentrated wheys (Fig. 15).

Except for the plot at 65°C, all other lines show that there are equal increases in the thermal resistance of the deposits and therefore equal increases in deposit mass with running time. The only differences were those in the induction phase. Because the deposits from concentrated wheys consists mostly of salts, the concentrates were subjected to a pretreatment and subsequently evaporated at T_V = 50°C. The purpose of the pretreatment was to promote crystallisation of the salts and thus to prevent supersaturation. It consisted of raising the temperature during concentration above the evaporation

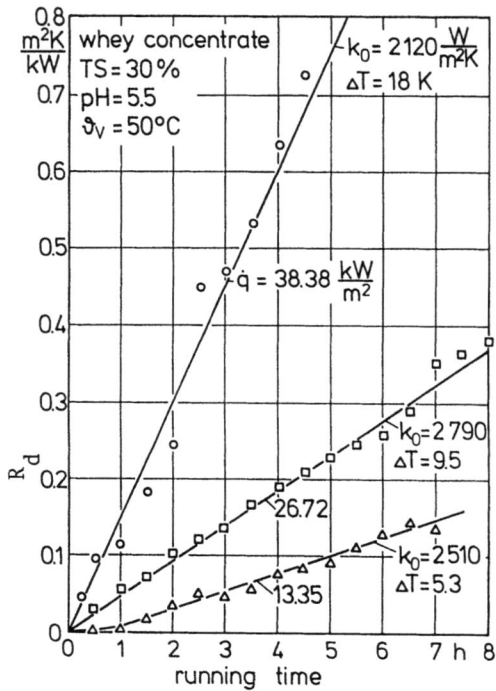

Fig. 13:
Thermal resistance of the deposit during evaporation of whey concentrate in dependence on the heat transferred per unit surface

temperature ($\Delta T = T - T_V$) by variable amounts and also of holding the concentrates hot for various periods of time. The excess temperature was about 1.5 K in each case. In addition, experiments were done involving heating at a temperature difference of 8.3 K, all other conditions remaining constant. A further test run was done in which the whey concentrate was pretreated by holding it in the balance tank at 80°C for 40 minutes.

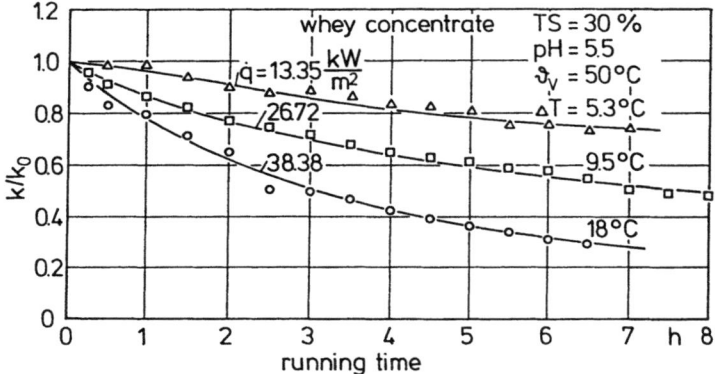

Fig. 14: Ratio of the overall heat transfer coefficient during evaporation of whey concentrate in dependence on heat transferred per unit surface

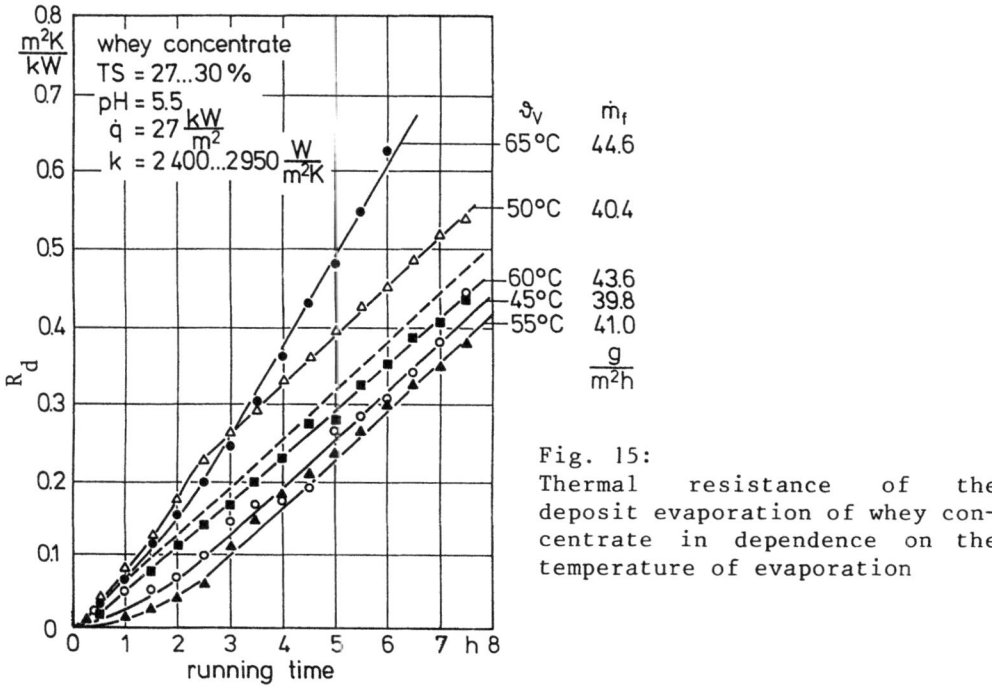

Fig. 15:
Thermal resistance of the deposit evaporation of whey concentrate in dependence on the temperature of evaporation

Fig. 16:
Thermal resistance of the deposit during evaporation of whey concentrate in dependence on pretreatment

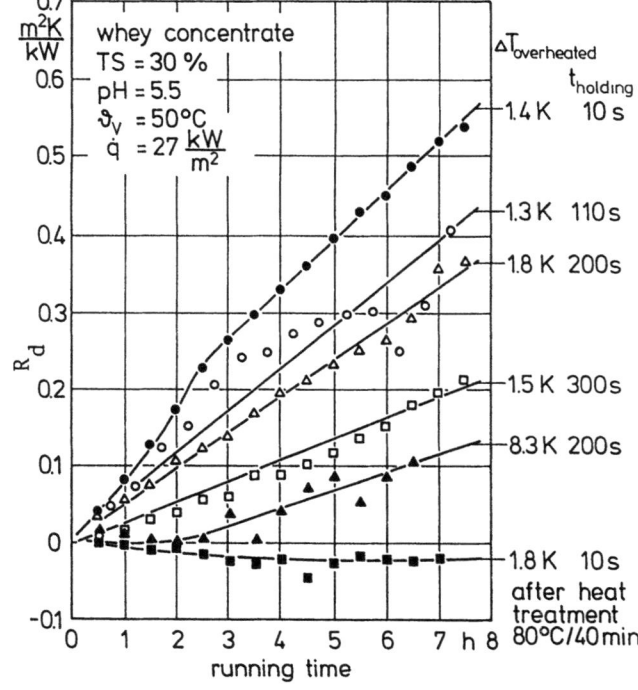

Fig. 16 shows the changes in the thermal resistance of the deposit during evaporation. The effect of the pretreatment of the product on the deposit is evident. It affects not only the thermal resistance of the deposit but also its specific mass.

The negative value for the thermal resistance in the experiment with the pretreated concentrate (80°C/40 min) is due to the improvement in heat transfer ($k > k_0$). This may be explained by an increase in the number of nuclei on the surface of the tube which initiates and promotes the production of vapour bubbles.

4. CONCLUSIONS

The results of the experiments lead to the following conclusions:
* To reduce the formation of deposits of salts the pH values should be as low as possible.
* Supersaturation should be reduced by appropriate treatment of the product to be evaporated. The principle of superheating and holding treatment can be applied between the effects of multieffect plants. Although this treatment is limited to tubes of a certain length because the product becomes more concentrated as the tube length increases, it will at least achieve a deposit free evaporation head.
* The heat transferred per unit surface area should not be more than about 18 kW/m^2.
* With high total solids concentrates the effect of temperature on deposit formation decreases while, at the same time, its rate increases. This is because the deposit consists increasingly of salts. When the rate of deposition of salts is high, proteins become included in the deposit at the same time even at low temperatures. Because of the high proportion of salts in the deposit, measures such as heating, holding and the lowering of the heat transferred per unit area of the heating surface will all be effective in reducing deposit formation.
* Analyses have shown that the composition of the deposit is strongly influenced by the pH value of the product. The higher the pH value and the concentration the greater will be the proportion of salts in the deposits. This will have a great influence on the choice of cleaning methods. Thus it was observed that the usual cleaning sequence of alkali/acid was ineffective when the deposit contained a high proportion of salts. In such cases the sequence "acid" has to be followed by an additional alkali run to get the plant clean. The reason for that is that the acid is necessary to dissolve the very stable salt structures in the deposit and only then can the alkali remove the remaining proteins. This is specially true for deposits from the evaporation of highly concentrated products. It might therefore be advantageous to use a special cleaning program for the last effects of a plant.
* Because protein denaturation is the cause of deposit formation by proteins, it should be possible to minimise it (at least for low concentration products) by choosing an evaporation temperature

$\vartheta_V \approx 60°C$. This is especially applicable to sweet whey which has a pH value above the isoelectric point of ß-lactoglobulin (5,3).

The project deposit formation during evaporation was sponsored by Ministry of Economics via Arbeitsgemeinschaft industrieller Forschungsvereinigungen and Forschungskreis der Ernährungsindustrie.

NEW DRYING TECHNIQUES FOR IMPROVED PROCESSING AND FOR A WIDER
PRODUCT VERSATILITY

W B Sanderson and A G Baucke
New Zealand Dairy Research Institute
Palmerston North
NEW ZEALAND

ABSTRACT. Recent developments in spray drying include the supply of
high capacity driers incorporating integral fluid-beds and the
increasing use of nozzle atomization.
   New systems employing multiple drying stages have allowed versatility in the manufacture of product types with characteristics not previously attainable.
   Integrated air heating and heat recovery systems are being introduced. More plants are being supplied with fire and explosion control systems and bag filters for environmental and economic benefits.

1. INTRODUCTION

The past five years of increasing costs and decreasing returns for standard products has resulted in a considerable change in emphasis in the types of spray drying facilities being offered. The need for increased production efficiency, a reduction in energy consumption and a greater flexibility in the types of products that can be spray dried, has dramatically influenced the designs of spray driers and associated equipment. Of particular note is the apparent consensus of opinion as to the best means of achieving these results since there is now a distinct similarity in the design of driers and options being offered by the major drier manufacturers (8, 11, 17).
   The development of driers for standard products has been much less spectacular than that for speciality applications. In the main, emphasis has been placed on reduction in capital costs, ease of operation and more effective cleaning systems, by attention to the detail of design.
   The choice of atomization systems is still a vexing question for designer and customer alike. There seems to be a consensus that, other constraints such as bulk density being set to one side, centrifugal disk atomization has the advantage from the point of view of the maximum feed concentration achievable and thus increased energy efficiency.
   Where it is necessary to maximise the bulk density of the powder, which would possibly be of interest to a company planning to focus on the manufacture of skim milk powder for the commodity trade, or to

improve powder appearance and flow characteristics, the nozzle atomization system still appears to win out. At least for the time being, the best quality instant wholemilk powder is produced with nozzle atomizers and many of the manufacturers producing infant foods appear to prefer nozzle atomization. The design of a drier capable of optimal operation with each of the two atomizing options does not seem to have been attained.

2. POWDER COLLECTION

Improvements in powder collection efficiency have been sought (12), but, in Europe and the United States of America the emission standards are so low that in most cases, bag filters or wet scrubbers are necessary (6, 22, 23). To the majority of plant owners, the economics of the operation is of lesser importance than is absolute compliance with the environmental constraints. In other words, the value of the recovered product is frequently insufficient to pay for the cost of the equipment needed to capture it. It is likely that the situation in many other countries will develop in the same fashion as it has in Europe and the USA.

3. PLANT CAPACITIES

Increasing consolidation of milk supply to single sites has made possible the realization of the economic advantages of large-scale operations. The capacity of spray driers has increased accordingly from a maximum size of about four to five tonnes of evaporation per hour only five years ago to eight tonnes per hour today. In the main, these larger driers are operating on standard products such as skim milk powder; however, some large driers (six tonnes per hour and greater) have been installed for the production of instant wholemilk powders.

The physical dimensions of some of these high capacity plants have been kept to manageable proportions by technological innovations such as the introduction of integral fluidized beds (IFB) and multiple stage drying. These developments will be outlined in more detail later. Thus, modern high capacity plants may well require even less building space than many of the plants previously installed.

4. PRODUCT VERSATILITY

There has been a small but steady fall in demand for the products generally considered to be part of the bulk commodity trade, for example skim milk powder. The profitability of such production has been eroded by surpluses and competition for markets.

The drop in demand for standard products has stimulated interest in driers with lower operating costs for traditional products and, as well, in driers capable of handling products with a higher value (20). Most of these products are more difficult to dry and the quality attributes required in them are more difficult to attain.

Examples of these high value/high difficulty products include:

high fat (70%) milk powders
instant wholemilk powder
40% fat whey powders
50% fat/mother liquor blends
whey permeates
ice cream powders (high fat/high sugar)
hydrolysed whey products
high acid products (fermented milks)

5. ENERGY CONSUMPTION

The energy crisis of the 1970's has also played a part in shaping the design of spray driers (7, 9, 16) and in the use of heat recovery systems. Various forms of air-to-air heat exchangers which include plate, tube and heat pipes have been installed (1, 15, 19) to raise the ambient inlet air temperature, and reductions in energy consumption of 20% or more have been achieved.

Plant purchasers have become more conscious of the low thermal efficiency of the old style single and two-stage spray driers. Plants of these types were generally limited to inlet and outlet temperatures in the range of 160-180°C and 70-105°C respectively. Improvements in design allowing higher inlet air temperatures and lower exhaust air temperatures has now evolved.

6. INTEGRAL FLUIDIZED BEDS

The drive for better efficiency and product diversity was limited because, with the designs available at that time, the minimum temperatures and maximum attainable moisture content in the product from the primary drying zone was constrained by the difficulty in transferring the moist, sticky powder from the drying chamber to the next (secondary) stage of drying. This problem was overcome with the development, initially by Niro Atomizer, of the "integrated fluid-bed", a static fluid bed built into the base of the primary drier cone (2, 4, 17, 18). Since the product did not come into contact with hot metal surfaces, higher moistures and thus lower outlet air temperatures could be employed. Today, most major manufacturers of spray driers (Niro Atomizer, APV Anhydro, Stork and MKT) have developed the integral fluidized bed and offer this as one of the options in their plant design. Integral fluidized beds can also be fitted to older conical based spray drying plants (17, 18). They offer increased throughput and efficiency. Typically, increases in inlet air temperatures of 20 to 40°C have been possible on modified plants without detrimental effects on the quality of the products and with consequential increases in throughput of 50% or more.

The high moisture product (10-18%) falls, without mechanical handling onto the fluidized bed of powder particles which have a relatively low moisture content. The bed which typically has a depth of some 400 mm, but which is adjusted according to the type of product being manufactured, moves the product around the bed annulus by means of the angled air entry through the bed plate until it is discharged, at lower moisture content, through a rotary valve. The residence time in the bed can be from 10 to 20 minutes and during this time the particles are dried, either to their final moisture content, or to some intermediate moisture level if an external third stage of drying is incorporated. An external system is normally installed for powder cooling prior to sifting.

7. THREE-STAGE DRYING

The use of an external fluid bed drier extends the two-stage drying concept to three-stage drying. These driers offer substantial energy savings over that achieved by the traditional single and two-stage units primarily due to the more efficient use of the inlet drying air and to the greater efficiency of fluid bed driers over primary spray drying chambers (3, 21). Inlet air temperatures of 240°C and outlet air temperatures of less than 70°C have been claimed leading to upwards of 50% savings in energy in the drying operation (21).

The range of products which can be processed by such driers is considerably broader than that which can be dried readily with the now traditional two-stage drier. However, there are limitations on the maximum fat content which can be handled without problems and it is also claimed by some of the manufacturers of such driers that, even with fines recycling, there is a limit to the degree of agglomeration that can be achieved, particularly if a disk atomizer is used.

To overcome these problems, further developments, again by Niro Atomizer but also more recently by APV Anhydro (8, 17, 18) have led to a new design in the basic primary drying chamber of the three-stage drying operation. The Niro Multi-Stage Drier (MSD) uses a basic cone shaped chamber with exhaust air outlets in the roof. High pressure nozzles are employed for atomization. The drier is fitted with an integral fluidized bed, an external vibrating fluidized bed drier/cooler and a special, low velocity scalping cyclone. The nozzles are located in the central inlet air throat at the top of the drier and, on some versions, the position of the nozzles can be altered to allow adjustment of the degree of spray interference, and thus the amount of agglomeration.

Because of the arrangement of the drier exhaust ducts positioned in the roof of the drier adjacent to the inlet and because of the upwards air velocity in this region, a considerable proportion of the powder (40-70%) is extracted along with the exhaust air. Some of this powder can collide with the stream of fresh concentrate droplets issuing from the nozzles and agglomeration can occur at this point. The powder which is entrained in the exhaust air is separated by a scalping cyclone. The very low rotating velocity in this cyclone minimizes the "g" forces acting upon the powder and this helps to minimize any build-up of deposits inside it, even when drying high-fat products. Because the scalping cyclone is inherently inefficient as a collector, it is necessary to incorporate a second, high efficiency cyclone to capture residual powder in the exhaust air.

The large quantity of "fines" removed in this way are returned to the integral fluid bed by means of a vibrating tube conveyor. Depending on the drying conditions at this point, further agglomeration may take place.

Since agglomeration is an integral part of the drying operation in the MSD drier, the products from this type of drier tend to have a relatively low bulk density when compared with that from the more traditional driers. The flow properties of the products from this drier are generally superior to that produced in other driers and this is particularly noticeable with high fat powders.

The APV Anhydro Spray Bed drier also employs a cone shaped primary drying chamber with exhaust air discharge at the top of the chamber (8). The drier is fitted with an integral fluidized bed and may also have an external fluid bed drier/cooler. Although low velocity scalping cyclones are not generally employed, fines may be returned to either the atomizing zone, the integral fluidized bed or to the external fluid bed. The drier is offered with either pressure nozzle or centrifugal disc atomization. A low level nozzle, mounted directly above the integrated fluid bed, can also be installed. The use of the low level nozzle is claimed to increase the levels of agglomeration and reduce the amount of fines carried over to the cyclones. For some applications this system may also be used as a form of codrying in which an addition component can be dried with the product. The powders produced on these driers are free flowing with average particle sizes of between 250-500 microns being possible depending on the type of product being manufactured and the operating conditions employed.

Extreme degrees of agglomeration may not be of immediate interest for many purchasers of new drying plants. However, with the growing interest in the manufacture of convenience consumer products there does appear to be an interest in the development of highly agglomerated products. A super instant wholemilk powder which, in the marketing concept could be similar to the highly agglomerated instant coffee, may well offer opportunities. Whether such a product could be made with the attractive appearance and flow properties, while at the same time maintaining all of the other desirable functional properties of a good instant wholemilk powder, has, however, yet to be realized.

8. DRY INGREDIENT INCORPORATION

The fines recirculation systems can be used for incorporation of dried ingredients such as sugars, starches etc, rather than dissolving them in the concentrate. While this process may well improve the ability to process the final product it does not produce a totally homogeneous powder but for many applications this may not be a disadvantage. A novel application for the fines recirculation system, is its use for 'upgrading' standard powders to an agglomerated product. Previously dried standard powder is metered into the fines return system to the integral fluidized bed or to the atomizing zone, at the same time as freshly produced concentrate is fed to the spray nozzles. The dried powder is agglomerated with the atomized product and higher moisture powder in the integral fluidized bed. It has been claimed that addition rates of up to 30% can be tolerated without any significant detrimental effect on the properties of the powder although the capacity of the external fluidized bed and the powder conveying system tends to limit the amount of dry powder which can be added in this manner.

9. STATIC FLUIDIZED BED COOLING

A recent development by APV Anhydro has been the successful operation of an externally mounted non-vibrating fluidized cooling bed. The bed has been used in the manufacture of skim milk powder and also successfully tested on wholemilk powder. The advantage of a lighter, less complex fluidized bed is its lower cost. However, in cases where more difficult cohesive products are to be handled it is unlikely to be as suitable as the more traditional vibrating fluidized bed.

10. TALL-FORM DRIERS

Until recently the availability of tall-form driers to the dairy industry has been limited to one or two drier manufacturers. These driers are typically tall, with small diameter chambers employing nozzle atomization. The exhaust air ducting was usually through a bustle at the top of the conical section and powder was discharged through the bottom of the cone directly onto a vibrating fluidized bed secondary drier/cooler. Fines recycling to the atomizing zone over the nozzles was often employed to produce high quality agglomerated powders. This type of drier would appear to be one of the preferred options for the manufacture of infant foods and instant wholemilk powders.

Recently APV Anhydro has taken the integral fluidized bed and incorporated this design feature into a tall-form drier either as a cooler or as a secondary drier in which case an external fluidized bed cooler would be used. The drier is basically a straight sided cylinder with minimal conical transition to the integral fluidized bed, further reducing the possibility of particle build-up on the walls. The drier is available with high capacities and is claimed to be low in capital costs and building requirements yet highly versatile in its product capabilities.

Stork, which has supplied traditional and tall-form type driers to the dairy industry in the past have added a new range of driers in which the exhaust air is ducted from the roof of the drier, similar to that offered by Niro Atomizer and APV Anhydro. In one option, the drying chamber design, however, is somewhat between a traditional low profile and tall-form drier and is referred to as a "wide-bodied" drier (diameter approximately equal to the length of the straight side of the upper chamber). It employs pressure nozzle atomization and can be equipped with external fluidized bed secondary drying and fines recycling for agglomeration.

involves capturing the partially dried powder on a moving perforated belt (10). The bed of powder builds up as a porous agglomerated mat which is slowly transported through the secondary drying stage and the cooling stage. The drier has a very low profile which reduces building costs.

There are a number of installations, particularly in Europe and the United States, which are used for the production of high fat, high sugar or whey based products for which the design of the Filtermat would appear to be particularly suitable. Specifically this is related to the fact that only a very small proportion of the product passes through the porous bed to the cyclones. This means that thermoplastic products and products with high fat contents such as cheese powders, whey/fat blends, butter powders and the like, which, if dried in conventional driers, will eventually build up deposits in the cyclones, can be readily handled by the Filtermat.

12. FIRES AND EXPLOSIONS

As all operators and managers of milk powder plants know, the risk of a fire or explosion in a spray drying plant and ancillary equipment is fairly high (24), particularly in the manufacture of high fat powders. The spontaneous ignition of milk powders has been well documented (5). In a recent document prepared by the International Dairy Federation (13), based on a paper prepared for the Dutch dairy industry (14), factors which may lead to a fire or explosion, such as powder build-up in the primary drying chamber, and recommendations to minimize such risks, are presented. Justifiably the emphasis is placed on good operator training and management to avoid situations which may lead to fires. However, drier manufacturers have also been addressing the issue and have begun offering a variety of monitoring and relief systems. TV monitors can be used to supplement physical checking by operators. High exhaust air temperatures, carbon monoxide (25), burnt particles and even sparks and flames can be detected and used as an early warning against fires. Explosion relief ducting from the drier to the exterior of the building is becoming more common in new installations and automatic fire extinguishing systems as well as explosion supression devices can also be installed although the cost justification for the installation of explosion supression would require close examination.

13. CONCLUSIONS

The equipment which is now available for use in the spray drying of milk and milk-based products has been the subject of considerable innovation and development, particularly during the past five years.

For an industry which intends to continue a programme of diversification of its product range, there is a wide choice of equipment offering. Options include; type of atomization, chamber configuration and aerodynamics, integral fluidized beds, external vibrating and non-vibrating fluidized beds, air sweeps, fines return systems, co-drying, dry ingredient incorporation and alternative powder recovery and energy recovery systems.

The most cost effective system for one product may not necessarily be the same for another. With the wide variety of options available today it is even more important that the plant chosen be designed for the specific product requirements of the purchaser.

14. REFERENCES

1. Anon. (1982) : <u>Technique Laitiere</u> **No. 968** 84.

2. Anon. (1983) : <u>Nordisk Mejeriindustri.</u> **10** (2) 87-88.

3. Anon. (1983) : <u>Food Engineering.</u> **55** (10) 103.

4. Anon. (1984) : <u>Technique Laitiere</u> **No. 987** 25-27.

5. Beever, P. (1984) : <u>Journal of the Society of Dairy Technology.</u> **37** (2) 68-71.

6. Burykin, A.I., Kharitonov, V.D., Kuz'min, V.M., & Fadeeva, L. Ya. (1983) : <u>Molochnaya Promyshlennost.</u> **No. 4** 14-18, 47.

7. Darlington, R. (1982) : <u>Journal of the Society of Dairy Technology.</u> **35** (3) 82-86.

8. Dickinson, P.W. (1986) : <u>APV-Anhydro A/S.</u> Bulletin H-28 Eng.

9. Gronlund, M. (1984) : <u>Institution of Chemical Engineers Symposium Series.</u> **No. 84** 397-408.

10. Hansen, P.S. (1982) : In <u>Proceedings of the Third International Drying Symposium.</u> **Vol. 2** 288-289. Drying Research Ltd, Wolverhampton, UK.

11. Hayashi, H. (1984) : <u>Japanese Journal of Dairy and Food Science.</u> **33** (6) A167-A171.

12. Horikawa, M., Saito, T. & Fukushima, M. (1985) : <u>Reports of Research Laboratory, Snow Brand Milk Products Co.</u> **No. 81** 1-14. Sappora, Japan.

13. International Dairy Federation, (1986) : <u>Report B-Doc 128</u> to Annual Sessions, The Hague.

14. Jansen, L.A. (1984) : <u>Voedingsm iddelentechnologie.</u> **17** (20) 74-77.

15. Kessler, H.G. (1982) : Molkereitechnik **53** 58-61.

16. Khazhinskii, Yu. N., Kharitonov, V.D., Bazikou, V.I., Zelenin, V.M., Granovskii, V. Ya., and Kuznetsov, P.V. (1982) : <u>Molochnaya Promyshlennost'.</u> **No. 6** 35-38.

17. Pisecky, J. (1983) : Dairy Industries International. **48** (4) 13-15, 17.

18. Pisecky, J. (1985) : Journal of the Society of Dairy Technology. **38** (2) 60-64.

19. Richardt, K. (1980) In Drying '80 II. Proceedings of the Second International Symposium. 379-386. Hemisphere Publishing Corp. Washington, USA.

20. Saito, Z. (1984) : Japanese Journal of Dairy and Food Science. **33** (6) A185-A193.

21. Sorensen, N. Skytte, and Kyllesbech, H. (1986) : Presentation to Danish Technical Days, Moscow. APV Anhydro A/S (K-41 Sov/Eng).

22. Swientek, R.J. (1983) : Food Processing, USA. **44** (2) 74-75.

23. USA, International Association of Milk, Food & Environmental Sanitarians; USA Public Health Service; USA, Dairy Industry Committee. (1982) : Dairy and Food Sanitation. **2** (6) 253-256.

24. Willmann, N. (1982) : Nordeuropaeisk Mejeri-Tidsskrift. **48** (6) 219-220.

25. Zockoll, C. (1985) : Deutsche Molkerei-Zeitung. **106** (33) 1085-1090.

SUMMARY OF DISCUSSION

B.S. Horton (USA) was asked about the performance of "loose" reverse osmosis (RO) membranes, especially to what extent they demineralize whey. A typical demineralization rate of 60 % was given for a three-fold concentration situation. Shearing problems in RO units were also brought up, and it appears that these problems no longer occur in present dairy designs. The use of carbon dioxide to reduce fouling in RO membrane systems was discussed. It was recognized that some suppliers recommend this approach, but results on actual performance are not available in the literature.

H.G. Kessler (Fed.Rep. Germany) was asked whether any results were available at solids contents above the 30 % level used as basis in his paper on fouling aspects. Could the trends at 30 % be extrapolated to 55/60 % solid conditions? No clear answer could be given, as no tests had been carried out at such high solids contents. Reference to pretreatment up to 80°C and 40 minutes was stated in the paper without reference to quality. It was asked whether there was any quality influence, but no data could be given as this aspect was not studied. The possibility to further improve thermal efficiency in falling film evaporators was discussed. The author felt that although further development work will take place, limits are being reached. Perhaps future overall energy savings in concentration can be achieved by membrane filtration/evaporation combinations. A comment from the audience suggested, however, further thermal savings are possible by a better understanding of the aspects of minimizing fouling. The nucleate boiling phenomenon that contributes to this is an important area for future research in improved evaporator design.

W.B. Sanderson (New Zealand) was asked about the dangers of integrated fluid bed spray dryers producing over-agglomerated, compact particles that dissolve slowly. Although this new design concept is in an early development stage, results already indicate that the problems of slowly dissolving particles are less serious than with powders from conventional two stage drying. This was due to the lower processing temperatures and major agglomeration taking place in the integrated fluid bed. The question of a relation between feed concentration/feed quality on dried powder quality was raised. There is a clear relation, although difficult to generalize. For example, increasing feed solids can increase powder quality by improving bulk density, free flowability, etc., but at certain solids contents atomization becomes no longer complete and powder quality the deteriorates. This is why 46 % appears upper limit for pressure nozzle atomization use, while rotary atomizers are preferred for handling higher feed concentrations. The influence of drying/dryer design on powder properties was discussed in relation to caseinates. It was concluded that dryer design, operating temperatures, particle residence time all play an important role, and greater flexibility of modern spray dryer designs is giving the opportunity to produce better, more soluble powders.

Seminar VI: Conversion of Feedstuffs in the Ruminant Under Different Conditions

Chairman: Prof. Dr. Ir. A. J. H. van Es (The Netherlands)
Secretary: Prof. Dr. Ir. S. Tamminga (The Netherlands)

INTAKE AND COMPOSITION OF TROPICAL FEEDS

A.J.H. van Es and S. Tamminga
Institute for Livestock Feeding and Nutrition Research
P.O. Box 160, 8200 AD Lelystad and
Agricultural University Wageningen
P.O. Box 88, 6700 AB Wageningen
The Netherlands

ABSTRACT. Basic feed resources in tropical developing countries are grasses (from fallow land, rangeland and other uncropped areas) and cereal crop residues. Intake and digestibility are limited by high cell wall and low protein. Potential protein supplements consist of oil seed cakes and legume forages and fodder trees. Factors that limit nutritive value of these feed resources are discussed in relationship to appropriate methods of laboratory analysis and development of feeding systems for tropical dairy production.

1. INTRODUCTION

Animal husbandry in most developing countries differs considerably from that in developed countries. Nearly always production level - milk yield, growth rate - is lower, often far lower. Most farms are small, those that are large are usually run extensively. In many cases livestock is kept for more than one purpose, e.g. cattle not only for milk or meat production but also for work. Even cattle dung is in many densely populated areas an important product as fuel for cooking. Especially in regions with long humid periods and communal grazing cattle and buffaloes suffer from gastrointestinal parasites. Also exoparasites and specific diseases may lower production levels. Local breeds or crossbreds with, say, one third exotic - north american or european - blood for increasing production level, have a higher disease resistance than imported "exotic" animals.

In some parts of the year or nearly always transport of feeds, products and animals is difficult and expensive; so is getting help from extension officers and veterinarians.
The warm climate may negatively affect both the animals, especially the productive ones which produce more heat, and the forages which lignify sooner. In general feeds are at times scarce and usually of low quality.

Namely the small farmer is reluctant to accept new, modern methods of farming as his level of farming education is low and his financial state is such that he cannot buy expensive equipment or feed to overcome possible failures of these innovations. Often his government shows not much interest in him (although this is improving slowly) and extension services are usually small or lacking. Finally sociological factors may prevent changes in conventional animal husbandry.

Thus it will be clear that direct, unadapted transfer of "exotic" technology aiming at higher milk yield or growth rate from developed to developing countries will not be very successful. There are many examples that such activities turned even out to be complete failures. Instead, innovations should only be introduced after thorough study of the local circumstances and then first on a small scale. Usually, lack of continuous supply of suitable feed is the most important limiting factor. Cross breeding, better health and other care will only result in higher production levels if the amount and/or quality of the feed is improved at the same time. Unfortunately, as already mentioned, feed supply is usually low and of low quality and this situation cannot easily be improved. The only solutions appear to be the use of local feeds, crop residues and by-products, which were not or seldom fed sofar, improvement of pastures and, especially, fitting the diet better to the requirements of the ruminants, taking advantage of our better understanding of microbial fermentation in the forestomachs (Preston and Leng, 1986). Using imported feeds from developed countries instead of local feeds will on the long run become too expensive; moreover, usually the developing countries cannot afford to spend money for such a purpose.

2. RUMEN FERMENTATION AND VOLUNTARY FEED INTAKE

2.1. Factors controlling feed intake

Animal production depends on an adequate supply of metabolic nutrients to the animal's intermediary metabolism. Factors governing this are voluntary feed intake and nutrient density in the ingested feed. If voluntary intake is too low, maintenance takes up a large proportion of the available nutrients (in exceptional cases even more than becomes available from ingested feed), resulting in poor or even negative production and a poor efficiency of feed conversion, a situation quite common in many tropical countries.

Voluntary intake in ruminants is controlled by a number of factors, associated with the animal, its feed or its environment. Important characteristics under tropical conditions in this respect are genetic potential and physiological status of the animal, taste and physico-chemical properties of the feed and temperature and humidity as environmental factors.

2.2. The digestive system in ruminants

The digestive system of ruminants contains two distinct parts, the forestomachs (reticulorumen + omasum) and the lower digestive tract (abomasum + intestine). An important control factor for voluntary feed intake is believed to be the capacity of the reticulorumen to digest (size reduction, degradation and passage)ingested feed. This capacity depends on the amount of feed the rumen can hold, the rate of particle size reduction, the rate of degradation and the rate of passage to the lower tract.

In the reticulorumen the ingested feed is subjected to particle size reduction, microbial fermentation and passage to the lower gut. As a result, a large proportion of the potentially degradable feed is converted to microbial cells and microbial waste products such as volatile fatty acids (VFA), fermentation gases (methane and carbon dioxide) and heat.

Microbial cells are an important source of protein (essential amino acids), minerals (P, S) and vitamins (B-vitamins) for the animal, whereas the VFA (largely absorbed from the stomachs) are a valuable source of energy.

Small particles, together with microbial cells, either adhered on feed particles or dispersed in the fluid are then passed on to the lower tract and the digestible part solubilised and absorbed.

2.3. Variation in holding capacity of the reticulorumen

Rumen size ranges from 13-18 % of the animal's live weight (Van Soest, 1982). Dry matter content in rumen ingesta varies between 7 and 17 %. Rumen capacity to hold dry matter therefore varies between 1 and 3 % of body weight and depends on animal breed, physiological status of the animal, energy content of the diet and the presence of sufficient microbial growth factors.

Less improved breeds of sheep showed a greater ability to adapt their rumen capacity to poor quality roughages than improved breeds (Weyreter and Von Engelhardt, 1984). Similar suggestions have been made for cattle (Chesson and Ørskov, 1984). Native tropical breeds seem to have a higher rumen capacity than breeds from Europe or the United States. Lactating cows appeared to have a higher rumen capacity than non lactating ones (Tulloh, 1966), whereas pregnant sheep had a lower rumen capacity than non pregnant (Forbes, 1986). Increasing the digestible organic matter of the diet from 520 to 760 g/kg dry matter (dm) decreased the rumen holding capacity from 40 to 18 g dm/kg digesta free body weight (Weston, 1984). This does suggest that with better quality feeds rumen pool size is not only determined by its capacity to hold material, but that metabolic control is more important. Supplementing a poor quality diet of chopped oaten hay with a continuous infusion of urea increased rumen capacity to hold organic matter with 20 % (Egan and Doyle, 1985).

2.4. Particle size reduction

The process of partical size reduction is as yet poorly understood (Ulyatt et al., 1986). One reason is differences in techniques in particle size analysis (see P.M. Kennedy, 1984). The main contributing factor to particle size reduction is chewing, both during eating and during rumination. The contribution of microbial fermentation seems restricted to weakening the forage cell wall structure sothat breakdown

by chewing during rumination is facilitated (Ulyatt et al., 1986).
Based on the fact that usually a large proportion of feed particles
present in the reticulorumen is small enough to be passed on to the
lower tract, it was concluded that particle size reduction is not the
rate limiting step in the capacity of the reticulorumen to digest feed.

2.5. Rate of degradation

Microbial activity in the reticulorumen and as a result (rate of) degradation of feed depends on properties of the feed as well as on the rumen environment.
  Maturing of the feed increases the encrustation of the cell wall components with lignin and/or silica. As a result, the proportion of the feed which is potentially not digestible increases and the rate of degradation of the potentially digestible part decreases, particularly of lignocellulose (Tamminga and Van Vuuren, 1986). In tropical forages this encrustation is often higher than in forages grown in temporate regions, reason why tropical forages are usually degraded comparatively slowly in the rumen.
  Factors in the ruminal environment which are important are the presence of sufficient rumen degradable N, S and P and more specific microbial growth factors such as peptides, iso-acids, amino acids and possibly B-vitamins (Tamminga, 1986). Next to these growth factors rumen pH is important, particularly for the activity of rumen protozoa, cellulolytic bacteria and methanogenic bacteria. For a good ruminal degradation pH should not fall below 6.0.

2.6. Rate of passage to the lower tract

Rate of passage can be looked upon as a probability of a feed particle to leave the reticulorumen. Amongst factors which may prevent this are the size of the feed particles, their functional specific weight and their probability to get entrapped in the math of long material (Faichney, 1986). It was found (Pienaar and Roux, 1984; Egan and Doyle, 1985) that potentially digestible material stayed longer in the rumen than potentially undigestible material. The reason for this is thought to be a difference in functional specific weight (Hooper and Welch, 1985) between particles which are or are not being degraded, caused by gas and/or air spaces in or on the degrading material and differences in hydration or ion exchange.

2.7. Consequences under tropical conditions

The main factors preventing an efficient microbial fermentation in ruminants in tropical countries are a deficiency of microbial growth factors (N, S, P, iso-acids, peptides, vitamins), a too large proportion of potentially undegradable material in tropical feeds and a too slow rate of degradation of the potentially degradable part.
  Improvements are possible by supplementing such feeds with the necessary growth factors and treatment of the cell wall rich crop residues (alkali- or steam-treatment). Offering the animals enough feed

to enable selection of the better parts, may also have a positive effect on intake and animal production (Zemmelink, 1980). More practical applications will be discussed in later sections.

3. SUPPLY AND UTILISATION OF NUTRIENTS

3.1. Supply and nature of end products of fermentation

Energy containing nutrients resulting from rumen fermentation and intestinal digestion can be divided in ketogenic, aminogenic and glucogenic nutrients. End products of rumen fermentation which are of use to the host animal are the digestible part of microbial cells (mainly aminogenic), the volatile fatty acids (ketogenic and glucogenic) and the digestible part of feed components escaping degradation in the forestomachs (ketogenic, aminogenic, glucogenic).

The total amount of nutrients becoming available for the host animal depends on the amount of feed ingested, the proportion of the feed which is potentially degradable, the rate at which nutrients are extracted from the degradable part and the residence time of the feed in the digestive tract.

For roughages rate of degradation is often positively related to its proportion of potentially digestible organic matter. However a better quality feed not only provides the animal with more nutrients per kg ingested feed, the animal will also ingest more feed. Increasing feed intake usually reduces the transit time in the digestive tract, both in the rumen and in the lower tract. As a result, slightly less nutrients are being extracted from the feed. For each level of maintenance the reduction varies between 10 and 50 g/kg organic matter ingested. To some extent this reduction is compensated for by a smaller loss of energy in methane and in urine.

With high quality diets some 25 % of the digested energy is present as protein (Tamminga, 1982). With poor quality diets this may be reduced to less than 20 % (Egan and Doyle, 1985) and there are some indications that this may also impair voluntary feed intake under tropical conditions. Depending on the production level the required ratio varies between 17 (maintenance) and 26 % (high level of milk production) (Tamminga, 1982).

The ratio aminogenic energy/total energy can be manipulated (Preston and Leng, 1986) either by changing the conditions in the rumen (rumen degradable N, S or P, iso-acids, easily available energy) or by the inclusion of less degradable protein (e.g. fishmeal) in the diet.

Between 15 and 30 % of the total energy digested becomes available as glucogenic energy, stored in propionic acid, starch escaping degradation in the rumen or in glucogenic amino acids. Manipulation of this ratio is possible (Sutton, 1985). Means are including easily degradable ingredients in the diet which enhances the proportion of propionate in ruminal VFA or by feeding ingredients with high levels of slowly degradable starch (e.g. rice bran)., sothat a significant proportion can escape degradation in the rumen, be digested in the small intestine and the resulting glucose absorbed.

In non producing animals between 20 and 25 % of the digestible energy ingested appears to be needed as glucogenic energy. In producing animals this ratio is likely to be considerably higher (Preston and Leng, 1986).

3.2. Utilisation of nutrients

After being absorbed, nutrients are utilised for maintenance, reproduction or the production of draught power, meat (growth), wool or milk. The highest priority is for reproduction and maintenance. If feed intake is too low to provide enough nutrients for this the result may be a loss of energy. Once the requirements for reproduction and maintenance have been met the animal will produce. Depending on its physiological status, its genetic potential, the level of feeding and the nature of the absorbed nutrients, nutrients will be directed towards the mammary gland or to the body. This nutrient partitioning process is controlled by the animal's endocrine system (Bauman and Currie, 1980). This control is not restricted to a partitioning between mammary gland and body. Within the body a partitioning takes place between protein and fat.

In dairy cows a high ratio of aminogenic to total energy seems to direct nutrients towards the mammary gland rather than to the body; under such conditions nutrients may even be liberated from body reserves and directed towards the mammary gland (Oldham, 1984). A high ratio of glucogenic to total energy on the other hand causes nutrients to be directed to the body rather than to the mammary gland. This may result in a reduced milk fat content (Sutton, 1985). High levels of long chain fatty acids or fats in the absorbed nutrients seem to have a similar effect. The reason for this is as yet rather unclear.

4. DIETS TO BE USED FOR MILK PRODUCTION BY RUMINANTS

In view of sufficient supply with nutrients at the tissue level the animal's intake and digestion should be as little impaired as possible, furthermore the absorbed nutrients should meet what is required for maintenance and milk yield and, in draught animals, for work with respect to total energy and its partition in keto-, amino- and glucogenic energy, protein, minerals and vitamins. Voluntary intake, digestibility and absorption of energy, amino acids and glucose precursors all depend highly on the extent and kind of microbial fermentation in the forestomachs as was explained earlier. This fermentation is retarded by low contents of degradable N and energy of the diet and in some cases also by low S and P. Moreover, intake of much easily fermentable material may lead to temporary low pH of the rumen fluid resulting in considerable reduction in rate of plant cell wall degradation by the microbes. Even when microbial fermentation is optimal, nutrient supply at very high potential production level may not meet all the animal's demands, usually resulting in lower actual production. In this case using small amounts of feeds in the diet with protein, starch and/or fat that is not or only partially degraded in the forestomachs by the

microbes, might be beneficial. However, the non degradable part of
such feeds should be still reasonably digestible by the animal's own
digestive enzymes in true stomach and small intestine, otherwise there
is no benefit from it.
Another advantage of using such slowly degrading feeds is that they do
not lead to sudden drops in pH after high intake and that they may
give a steady though low supply of nutrients and growth factors for the
microbes in the reticulorumen.

Optimal use of the separate feedstuffs clearly only will be made
by the animal when these are fed as part of a diet which meets its requirements and it is not stressed by disease or heat, receives sufficient and good water and has not to walk much.

5. AVAILABLE FEEDS IN DEVELOPING COUNTRIES AND THEIR PROPERTIES

5.1. Available feeds

Feeds in developing countries are pasture and savannah, residues from
crops (rice straw, stovers of maize, millet, sorghum), parts of trees,
sugar cane or parts of it, banana or parts of it, some tubers (manioc,
sweet potato), by-products from industries producing human food (brans
from rice, wheat, other cereals; molasses, bagasse; soya-, groundnut-,
cottonseed- and coconut cakes; citruspulp, pine apple wastes; etc.).
For various reasons - low keeping quality, high transport costs, necessity to obtain foreign currency - often much of these by-products
is not used as animal feed in the countries where they were produced.
For example, Sansoucy (1986) mentions that two third of the oilseeds
is grown in developing countries whereas these use themselves only one
quarter of the oilcakes although such feed protein is very beneficial
when added to the main feed which is usually low in protein. However,
in many cases urea, a chemical that to some extent can be used as a
N feed for ruminants and also for improving the digestibility and voluntary intake of straws, is available, often at a low price because
of subsidies.

5.2. Feeds, diets and feeding value

In developed countries the feeding values of the various feedstuffs
for ruminants are mainly determined by their content for digestible
energy (DE) and for digestible crude protein (DCP). For composing
diets, DE values of the various feeds can be considered to be additive,
whereas for conventional diets this is for DCP values approximately
so too. It occurs very seldom that microbial fermentation of such diets
in the forestomachs is impaired because of shortage of N, easily fermentable energy, S or P, a consequence of the better quality of the
forages - straws and very mature forages are seldom fed - and of the
greater variety of feeds used in the concentrate mixture. Moreover
from the diets mostly enough ketogenic, aminogenic and glucogenic
energy is absorbed for moderate to high milk yields. Together this
explains the additivity of DE and often also DCP values of the separate

feeds when used for a mixed diet.

In developing countries most forages are low in N and fat and have a low digestibility and for composing diets only few feeds are used. Therefore due attention should be paid to compose with the often very limited means available, a diet which impairs microbial fermentation in the forestomachs as little as possible and provides sufficient absorbable nutrients. In such circumstances feeding values of the feeds when fed alone are usually lower, often considerably so, than when fed in a well-composed diet: there exists no or little additivity of feeding values of feeds.

5.3. Feed analysis

Feed analysis in developing countries should give that information on the feeds which is needed for composing optimal diets and not only give estimates of DE and DCP as in developed countries. The nylon bag technique, suspending samples for various duration in the rumen of ruminants, informs on contents of easily fermentable matter and N as well as on extent and rate of degradation of the other components. The in vitro digestibility technique using rumen fluid or a mixture of cellulases, provided some samples of similar kind as the feeds under test with known in vivo digestibility are included in each run, gives an estimate of the digestibility of organic matter under fairly favourable fermentation conditions. In vitro digestion with pepsin may tell how much of the undegraded part of a feed's protein will be digested in the ruminant's stomach and small intestine. Data on easily fermentable components (protein, non-protein-nitrogen; sugar, starch) may also be obtained from analysis of a hot-water extract. For wet feeds knowledge of dry matter content is needed, for feeds contaminated with soil or sand total and HCl-insoluble ash content. Depending on whether deficiencies for such elements are to be expected or not, analysis for Ca, P, S, etc. may be needed. For feeds with more than 5 % fat analysis of the fat content may be useful as fat of good quality can be a good energy source. Extracting the fat with hexane or petroleumether boiling between 40º-60ºC might be better than with diethylether as it gives a purer fat and has a lower fire risk. Some information on the degree of saturation of the fat's fatty acids is useful when the diet contains 5 % or more of it as unsaturated fatty acids still more than saturated ones may harm microbial forestomach fermentation.

The lack of additivity of feeding values of feeds in developing countries is the reason why feed analysis has an objective which differs so markedly from that in most developed countries.

5.4. Properties of some feeds in developing countries

In warm climates grasses mature rapidly, resulting in highly lignified dry matter of low digestibility with low contents of N, fat, easily fermentable carbohydrates and sugar. Legumes do so too but to a smaller extent (Minson, 1980) and provide more Ca and P. Preservation of young material of high digestibility and higher protein and fat content in the rainy season often is unsuccessful or too expensive. In some cases grass is collected from road sides which are common property which also interferes with preservation. Due to scattered or stemmy growth of herbage grazing requires more physical work of the animals.

Straws of rice, wheat, other cereals and stovers of maize, sorghum and millet have very low N and fat contents; their digestibility is variable and usually moderate to low. Still in well-composed diets or after treatment with alkali - preferably urea as NaOH or NH_3 is usually too expensive and may give health risk for man - they can provide much energy to ruminants.

Foliage of legume trees because of their tannin content may be useful as slow release protein. In some parts of East-Asia unfortunately growth of Leucaena suffers very much from an insect.

Sugar cane yields very high amounts of organic matter per unit of area half of which is (too) easily fermentable whereas the other half is hardly degradable; all is low in N. Cane molasses have a high palatability and provide excellent fermentable carbohydrates and some S, Ca and K; however both properties may easily lead to acidosis in the forestomachs. Fed as block or liquid lick they are eaten more slowly and can be used as carrier for additives (urea, minerals). Cane bagasse has a low feeding value.

Rejected banana fruit in the green state when the carbohydrate is still present as starch is a good cattle feed although very low in N. Banana leaves and stems have a high water content.

Tubers like manioc and sweet potatoes have high contents of easily fermentable carbohydrates but hardly any N and fat. Manioc leaves (for instance after drying) may have potential as a source of N and minerals.

Quite a variety of feeds are by-products from food technology. These may differ considerably in digestibility, protein, carbohydrate, fat, lignin and ash content. The degradability of their proteins may depend on tannin content and heat treatment. As to carbohydrates starch in rice and maize millings are less easily degraded by rumen microbes than starch of wheat or, of course, sugars. In oil cakes fat content is higher after the expeller than the solvent process.

6. REFERENCES

Bauman, D.E. and W.B. Currie, 1980. 'Partitioning of nutrients during pregnancy and lactation: A review of mechanisms involving homeastasis and homeorhesis'. *J. Dairy Sci.*, 63: 1514-1529.

Chesson, A. and E.R. Ørskov, 1984. 'Microbial degradation in the digestive tract'. In: *Straw and other fibrous by-products as feed* (F. Sundstol and E. Owen, eds.), Elsevier, Amsterdam, pp. 305-339.

Egan, J.K. and P.T. Doyle, 1985. 'Effect of intraruminal infusion of urea on the response in voluntary food intake by sheep'. *Aust. J. Agric. Res.*, 36: 483-495.

Faichney, G.J., 1986. 'The kinetics of particulate matter in the rumen'. In: *Control of Digestion and Metabolism in Ruminants* (L.P. Milligan, W.L. Grovum and A. Dobson, eds.), Reston Books, Prentice Hall, New Jersey, USA, pp. 173-195.

Forbes, J.M., 1986. *The voluntary food intake of farm animals*, Butterworth, London, pp. 1-206.

Hooper, A.P. and J.G. Welch, 1985. 'Effects of particle size and forage composition on functional specific gravity'. *J. Dairy Sci.*, 68: 1181-1188.

Kennedy, P.M., 1984. *Techniques in particle size analysis of feed and digesta in ruminants*. Can. Soc. Anim. Sci. Occ. publ. no. 1.

Minson, D.J., 1980. 'Nutritional differences between tropical and temperate pastures'. In: *Grazing animals* (F.H.W. Morley, ed.), Elsevier, Amsterdam, pp. 103-157.

Oldham, J.D., 1984. 'Protein-energy interrelationships in dairy cows'. *J. Dairy Sci.*, 67: 1090-1114.

Pienaar, J.P. and C.Z. Roux, 1984. 'Differential rates for the outflow of fermentable and non-fermentable organic matter from the rumen'. In: *Techniques in particle size analysis of feed and digesta in ruminants* (P.M. Kennedy, ed.), Can. Soc. Anim. Prod. Occ. publ. no. 1. pp. 175.

Preston, T.R. and R.A. Leng, 1986. *Matching Livestock Production systems to available resources*. ILCA, Addis Ababa, Ethiopia.

Sansoucy, R., 1986. 'The zootechnical aspects of feeding livestock on the basis of non-conventional feed resources in developing countries'. In: *Expert consultation on animal feed manufacturing*, Barneveld College, Barneveld (Neth.), III: 1-11.

Sutton, J.D., 1985. 'Digestion and absorption of energy substrates in the lactating cow'. *J. Dairy Sci.*, 68: 3376-3393.

Tamminga, S., 1982. 'Energy-protein relationships in ruminant feeding: similarities and differences between rumen fermentation and post-ruminal utilisation'. In: *Protein contribution of feedstuffs for ruminants* (E.L. Miller, I.H. Pike and A.J.H. van Es, eds.), Butterworths, London, pp. 4-17.

Tamminga, S., 1986. 'Prospects for supplementation of cell wall rich crop residues in tropical countries'. In: *Rice straw and related feeds in ruminant rations* (M.N.M. Ibrahim and J.B. Schiere, eds.). Straw utilisation Project, Kandy, Sri Lanka, pp. 208-217.

Tamminga, S. and A.M. van Vuuren, 1986. 'Formation and utilisation of end products of lignocellulose degradation in ruminants'. *Anim. Feed Sci. Technol.* (in press).

Tulloh, N.M., 1966. 'Physical studies on the alimentary tract of grazing cattle. IV. Dimensions of the tract in lactating and non-lactating cows'. *N.Z. J. Agric. Res.*, 9: 999-1008.

Ulyatt, M.J., D.W. Dellow, A. John, C.S.W. Reid and G.C. Waghorn, 1986. 'Contribution of chewing during eating and rumination to the clearance of digesta from the reticulorumen'. In: *Control of Digestion and Metabolism in Ruminants* (L.P. Milligan, W.L. Grovum and A. Dobson, eds.), Reston Book, Prentice Hall, New Jersey, USA, pp. 498-515.

Van Soest, P.J., 1982. *Nutritional Ecology of the Ruminant*. O & E Books, Corvallis, OR, USA.

Weyreter, H. and W. von Engelhardt, 1984. 'Adaptation of Heidschnuken, Merino and Blackhead sheep to a fibrous roughage diet of poor quality'. *Can.J. Anim. Sci.*, 64 (suppl.): 152-153.

Weston, R.H., 1984. 'Rumen digesta load in relation to voluntary feed consumption and rumination in roughage-fed young sheep'. *Can. J. Anim. Sci.*, 64 (Suppl.): 324-325.

Zemmelink, G., 1980. *Effect of selective consumption on voluntary intake and digestibility of tropical forages.* Thesis Agric. University, Wageningen.

SUPPLEMENTATION OF TROPICAL FEEDS

T.R. Preston
Convenio Interinstitucional para la Produccion
Pecuaria en el Valle del Rio Cauca (CIPAV),
Apartado Aereo 7482, Cali, COLOMBIA

ABSTRACT. The challenge for the nutritionist entrusted with the task of developing feeding systems in developing countries is not "how to maximise rates of productivity per animal". Rather, the aim should be to optimise the use of the available resources within the framework of "matching the production system to the available resources".
   The approach should be based on: (i) an understanding of the needs for the actual nutrients that are required in the physiological and metabolic activities associated with the different functions that the animals perform in successive stages of the production cycle; (ii) a survey of the available resources and their characteristics in terms of their capacity/constraints to provide the required nutrients; (iii) the manipulation/supplementation of the basic feed in order to maximise rumen function and then to balance the products of fermentation, in accordance with animal needs and always measuring response in economic rather than biological terms.
   Attention is drawn to the advantages in the tropics of using carbohydrate-rich feeds derived from sugar cane because of its high rate of biomass production and efficiency of solar energy use; and of legume trees as sources of both micronutrientes for rumen microorganisms as well as of bypass nutrients for the animal; and to the opportunities for upgrading fibrous residues by chemical pre-digestion.

1. INTRODUCTION

The feed resources available for ruminant feeding in developing countries are markedly different from those commonly used for this purpose in the industrialised countries. Cereal grains are the staple of the human diet, and are too expensive or simply are not available for aninal feeding. Other low-fibre feeds (eg: agroindustrial byproducts such as cereal offals) are preferentially reserved for monogastric animals. The predominant feed resources for ruminants are those not suitable for the previous two categories; they are generally

TABLE I
Tropical feeds as sources of fermentable organic matter

Resource	Rumen degrad.	N cont. in DM	Glucogen. VFA ratio (%)	Pre-digest. (??) (%)	Suppl. (??) [C3/(C2 + C4)]
Pastures					
Wet season	50-60	1-2	<0.2	No	??
Dry season	<45	<1	<0.2	No	Yes
Cut forages	45-55	<1.5	<0.2	*	Yes
Sugar cane					
Whole plant	60-65	<1	0.2-0.25	*	Yes
Tops	55-60	<1	<0.2	*	Yes
Bagasse (factory)	<30	0	<0.2	Yes	Yes
Bagasse (trapiche)	50	<1	<0.2	Yes	Yes
Cereal straw	40-50	<1	<0.2	Yes	Yes
Husks (oilseeds)	<40	<1	<0.2	Yes	Yes
Pulps					
Coffee	<50	<1	<0.2	Yes	Yes
Citrus/pineapple	60-70	<1	0.2-0.3	No	Yes

rich in cell wall material and low in protein (Table I). The main consequence is that rumen fermentation must be the predominant mode of digestion and therefore transactions in this organ are the determinants of efficient use of the available feeds.

Appreciation of this situation was the reason for the proposal that "in developing countries" animal production strategy should aim to "match livestock systems with available resources" rather than to "maximise rates of productivity" (Preston and Leng 1986).

The suggested strategy comprises the following steps:

* Understanding of the physiological and metabolic needs of the ruminant for a specific function (eg: growth, lactation, work, reproduction)

* A survey of available feed and biological resources including their classification in terms of their capacity (and/or limitations) to supply the required nutrients.

* Development of feeding systems based on the principle of maximising rumen function and "balancing" absorbed nutrients, by

manipulation and/or supplementation of the basal CHO-rich basic feed resource.

2. THE NEEDS OF THE RUMEN AND OF THE ANIMAL

2.1 The rumen

The first step is to maximise the intake of fermentable carbohydrate then to optimise the rate at which this is fermented, in a way which will favour microbial growth rather than VFA production. Essential needs are for fermentable N as ammonia, the levels of which in rumen fluid should exceed 200 mg/litre in order to optimise fibre digestion (Krebs and Leng 198-). Micronutrients (amino acids, peptides, branched-chain VFA, vitamins and minerals) are also needed. It also appears to be an advantage to have all or some of these nutrients intimately associated with long (>10mm?) fibres which are the preferred niches of bacteria.

2.2 The animal

Synthesis of tissues and of milk requires a supply of amino acids and glucose (and/or its precursors) in addition to the 2-carbon units derived from the volatile fatty acids; part of the needs for glucose (for oxidation to NADPH) are more appropriately met by long chain fatty acids (LCFA). Glucose is a necessary fuel for the brain and central nervous and reproductive systems and a preferred fuel, along with the AGCL, for muscular activity (Leng 1985).

The amounts of amino acids and glucose precursors produced in the rumen (from microbes and propionic acid) are usually inadequate when fibrous feeds are fermented and supplementation is then required with feeds providing these nutrients in a form which will enable them to "bypass" or "escape" the rumen fermentation. The balance of supplementary nutrients required will depend on the needs for a particular productive function (eg: lactation, work, reproduction etc) and the supply emanating from the rumen fermentation.

3. SOURCES OF NUTRIENTS

The major tropical feeds rich in fermentable carbohydrate (Table I) are described in terms of their limitations as sources of nutrients for an efficient rumen fermentation (potential degradability, supply of fermentable N, micro-nutrients and "long fibres") and with respect to the final balance of nutrients available for absorption. Particular issues with most of these feed resources are their low N content and the low propionate production in the rumen when fermented. Table II summarises the main sources of the supplements required for the rumen and for the animal.

Two feed resources merit special attention for tropical regions. Sugar cane is already known for its high rate of production (biomass/ha/yr) and efficiency (biomass energy/solar energy). Recent

research shows that this capability can be put to effective use for
the production of feed - for both monogastric and ruminant animals -
and fuel for on- and off-farm use, in intensive integrated farming
systems (Preston 1980, 1986 unpublished data). N-fixing perennial
legume trees (eg: Gliricidia, Erythrina and Leucaena spp) also have
high yield capacity and are proving to be excellent sources of
micronutrients for rumen microorganisms and of bypass nutrients for
the animal. Their potential to improve the climatic and soil
environment is also being increasingly recognised.

TABLE II
Tropical feeds to supplement the rumen and/or the animal

SUPPLEMENTS FOR THE RUMEN:

Fermentable N
  Urea, ammonia, poultry litter

Micronutrients
  Foliages from forage trees and food crops and shrubs
  (Leucaena, gliricidia, erythrina; cassava, sweet potato;
  canavalia)
  Grasses and herbaceous legumes

Long fibres
  Stoloniferous grasses
  Forage crops, sugar cane and foliage from food crops

SUPPLEMENTS FOR THE ANIMAL

Protein (amino acids)
  Leguminous trees and shrubs (Leucaena, gliricidia,
  erythrina, canavalia)
  Milling offals (from rice, maize, wheat)
  Cakes and meals from oilseed processing (cotton,
  groundnut, palm kernal, sunflower, sesame)
  Meals from animal slaughter (fish, meat and bone, blood)

Glucose (and/or precursors)
  Milling offals
  Grains (Maize> rice> sorghum)

Long chain fatty acids
  Polishings, brans (wheat/rice)
  Calcium soaps
  Oilseed cakes/meals

---

4. DEVELOPMENT OF FEEDING SYSTEMS

4.1 Pastures and green forages

A distinction must be made between green (wet season) and dry (dry season) pastures and forages. The rumen ecosystem provided by greee pastures is rarely limiting and responses are only likely to be obtained from bypass nutrients, especially where lacating animals are concerned. "Energy" is usually limiting, but this problem is best resolved by stimulating intake of the basic feed (by balancing nutrients) rather than supplementing with easily fermentable carbohydrate. The data in Table III illustrate these principles.

The foliage from the legume tree Gliricidia sepium is a low-cost source of bypass protein. It improved dramatically the growth rate of steers fed a basal diet of a tropical grass (Table IV). The foliage from the legume tree Erythrina poeppigiana was equally effective in stimulating milk yield in goats fed the same basal diet (Table V).

TABLE III
Supplementation of setaria grass for milking cows (Mapoon et al 1977)

Supplement	Persistency of milk yield (%)	Change in livweight (kg/d)
None	70	-0.7
Molasses/urea (600 g/litre milk)	80	-0.6
Cereal grain and offals/oilcakes (500 g/litre milk)	100	0.0
Groundnut cake (200 g/litre milk)	98	+0.2

TABLE IV
Supplementation of King grass with gliricidia foliage for growing cattle (ICA Tulipana, unpublished data)

Gliricidia foliage (% of diet, fresh basis)	Weight gain (g/d)
0	170
25	370
50	380

TABLE V

Milk production of goats given a basal diet of King grass, reject bananas and leaves from the "Poro" tree (Erythrina poeppigiana)(Esnaola and Rios 1986)

Erythrina leaves in diet (% of LW, DM basis)	Milk yield (g/d)	
	High potential	Moderate potential
0	390	280
0.5	700	500
1.0	820	520
1.5	950	640

When pastures are dry, the first constraint is at the level of the rumen and sources of fermentable-N and micronutrients are the first priority. After these have been provided, additional improvements in animal performance can be obtained with a source of bypass nutrients (Table VI).

TABLE VI

Supplementation of simulated dry pasture (Spear grass hay) with fermentable N and S, micronutrients (alfalfa) and or bypass protein (cottonseed cake) for pregnant beef cows (Lindsay et al 1982)

	No supp	Urea/S	Urea/S + alfalfa	Urea/S + alfalfa + bypass prot.
Intake of hay (kg/d)	4.2	6.2	6.7	8.1
Calf birth weight (kg)	22	31	32	32
Weight change of cows (kg/d)	-0.82	-0.31	0.41	0.75

4.2 Cereal straws

As for all feed resources, the need is to balance the nutrients needed first for the rumen fermentation and then for the animal. However, even when this is done, the low fermentability of the carbohydrate in the straw is a serious constraint to the level of animal performance that can be achieved. Partial saponification of the lignin:carbohydrate linkages, and/or partial hydrolysis of the hemicellulose fraction of the cell wall, can be achieved by treatment with alkalis and acids and is usually a first step in the development of commercial feeding systems based on straws and similar fibrous crop residues.

TABLE VII
Ammoniated (urea-ensiled) rice straw and gliricidia forage for lactating buffaloes (Perdok et al 1982)

Straw	Untreated		Treated	
Gliricidia forage (kg/d)	0.0	6	0.0	6
Cow data:				
DM intake (kg/d/100 kg LW)	2.8	2.8	3.7	4.0
Liveweight change (g/d)	-90	+60	+60	+130
Milk yield (kg/d)	2.2	2.6	3.0	3.4
Cows milking >84d (%)	60	90	100	90
Calf data:				
Liveweight gain (g/d)	165	265	295	345
Suckled milk intake (kg/d)	0.95	1.03	1.03	1.15

TABLE VIII
Supplementation of ammoniated straw with leucanea leaves and/or rice polishings for weaned calves (Forero O and Preston T R unpublished data)

	Fresh leucaena (kg/d)	
	0	2.0
Rice polishings (g/d)	Liveweight gain (g/d)	
0	180	580
250	300	590
500	590	600

The data in Tables VII and VIII, which show the effects on animal productivity of supplementing ammoniated straws with micronutrients and/or bypass nutrients, demonstrate that rumen function and the final balance of nutrients are limiting factors on fibrous residues whose potential rumen degradability has been improved by chemical treatment.

4.3 Sugar cane (the whole plant and/or the tops)

With 50% soluble sugars in the dry matter, availability of fermentable carbohydrate should not be a problem with sugar cane. Nevertheless, the fibre "load" associated with the sugars is a constraint and there are apparent advantages to the animal when the degradability of this fibre is increased by alkali treatment.

Supplementation plays a key role in the development of animal production systems based on sugar cane (Table VII). Rumen function on whole sugar cane is seriously impaired by the low N content which results in inadequate ammonia levels in the rumen. Micronutrients also appear to be lacking (note the response to the highly fermentable sweet potato tops); and there is a significant response to the balancing of nutrients for the animal by providing rice polishings, the nutrients in which (starch, protein and lipids) largely bypass the rumen fermentation (Elliott et al 1978a,b).

TABLE IX
Effect of fermentable-N (urea) level, a source of rumen micronutrients (sweet potato foliage) and or bypass nutrients (cottonseed cake) on performance of steers fed a basal diet of derinded sugar cane (Meyreles et al 1979)

Supplement	Liveweight change (g/d)	
	20 g urea/ kg DM	60 g urea/ kg/DM
None	+30	-50
Sweet potato forage	350	550
Cottonseed cake (CSM)	350	450
SW foliage + CSM	700	1000

4.4 Sisal pulp

The residue after extracting the fibre from the leaf of the cactus Agave forcroydes is a pulp (dry mattercontent about 20%) which has a digestibility of about 55%, contains less than 1% of N and is extremely imbalanced in minerals. The non-fibre fraction is mainly in the form of soluble sugars and ferments rapidly to lactic acid (Harrison 1984). Supplementation with the limiting mineral elements and with urea is of little benefit and animals lose weight on this ration (Table VI). Supplements of green forage bring about significant improvements in the rumen ecosystem (Gutierrez and Elliott 1984) leading to increased voluntary intake. When complemented with bypass nutrients the sisal pulp becomes a production ration and mineral supplementation yields a further response in animal performance (Table X).

TABLE X
Lamb growth rates (or losses) on ensiled sisal pulp with different supplements (Rodriguez et al 1982)

Supplements	Liveweight change (g/d)	
	Without minerals	With minerals
None	-40	-30
Foliage from fodder tree*	35	45
Soybean meal (SBM)	50	65
Foliage and SBM	90	125

*Brosimum alicastrum

5. CONCLUSIONS

The experimental data reviewed in this paper show that supplementation strategies which aim to improve the rumen ecosystem and/or the balance of absorbed nutrients can bring about significant improvements in animal performance on the typical feeds available in tropical countries.

In such situations, the successful development of ruminant feeding systems requires an understanding of the nutrient needs of the animal according to its productive state; knowledge of the likely deficiencies in the digestion end products when the basal feed is fermented in the rumen; and appropriate manipulation/supplementation practices designed first to optimise rumen function and then to provide the balance of nutrients required by the animal.

The manipulation of the basal feed resource and the supplementation should be achieved whenever possible by means of resources available, or capable of being produced, on the farm. Purchased inputs shpuld be minimised and animal response to dietary manipulation/supplementation should be assessed by economic rather than biological criteria.

6. REFERENCES

Elliott R, Ferreiro H M, Priego A and Preston T R 1978a Rice polishings as a supplement in sugar cane diets: the quantities of starch (glucose polymers) entering the proximal duodenum Tropical Animal Production 3:30-35

Elliott R, Ferreiro H M, Priego A and Preston T R 1978b Estimate of the quantity of feed protein escaping degradation in the rumen of steers fed chopped sugar cane, molasses/urea supplemented with varying quantities of rice polishings Tropical Animal Production 3:36-39

Esnaola M A and Rios C 1986 Hojas de 'Poro' (Ethythrina poeppigiana) como suplemento proteico para cabras lactantes Tropical Animal Production (in press)

Gutierrez E and Elliott R 1984 Interaccion digestiva de la pulpa de henequen (Agave fourcroydes) y el pasto estrella de Africa (Cynodon plectostachyus) In: Alternativas y valor nutritivo de algunos recursos alimenticios destinados a produccion animal Informe provisional No 16 Fundacion Internacional para la Ciencia: Stockholm pp229-246

Harrison D G 1984 Sisal by-products as feed for ruminants se World Animal Review 49:25-31

Krebs G and Leng R A 1984 The effect of supplementation with molasses/urea blocks on ruminal digestion Animal Production in Australia 15:704

Leng R A 1985 Muscle metabolism and nutrition in working animals In: Proceedings ACIAR Workshop on 'Draught Animal Power for Production James Cook University:Townsville (In press)

Lindsay J A, Mason G W J and Toleman M A 1982 Supplementation of pregnant cows with protected proteins when fed tropical forage diets Proceedings Australian Society of Animal Production 14:67-78

Mapoon L K, Delaitre C and Preston T R 1977 The value for milk production of supplements of mixtures of final molasses, bagasse pith and urea, with and without combinations of maize and groundnut cake Tropical Animal Production 2:148-150

Meyreles L, Rowe J B, and Preston T R 1979 The effect on the performance of fattening bulls of supplementing a basal diet of derinded sugar cane stalk with urea, sweet potato forage and cottonseed meal Tropical Animal Production 4:255-262

Perdok A B, Thamotaaram M, Blum J J, Van DenBorn H and Van Velun C 1982 Practical experiences with urea ensiled straw in Sri Lanka In: Maximum Livestock Production from Minimum Land (Editors: T R Preston, C Davis, F Dolberg, M Haque and M Saadullah) Bangladesh Agricultural University and BARC:Dacca

Preston T R and Leng R A 1986 Matching Livestock Systems with Available Feed Resources International Livestock Centre for Africa:Addis Ababa

Preston T R 1980 A model for converting biomass (sugar cane) in animal feed and fuel In: Animal Production Systems for the Tropics Publication No:8 International Foundation for Science:Stockholm

Rodriguez A 1983 Supplementation of ensiled henequen (sisal) pulp for sheep Tropical Animal Production 8:74-75

FEED CONVERSION AND NUTRIENT PARTITIONING

K.Rohr and H.J.Oslage
Institute of Animal Nutrition,
Agricultural Research Center (FAL)
Bundesallee 50, D-3300 Braunschweig
Federal Republic of Germany

ABSTRACT. There are various constraints to ruminant production in the tropics and subtropics. Heat stress is known to affect feed intake, efficiency of energy utilization and fertility. The resulting decrease in productive and reproductive performance is associated with alterations of the endocrine status. Nutritional strategies to alleviate thermal stress essentially aim at a lower heat increment of the ration. In most developing countries, however, production levels are more dependent on nutrient availability than on the climatic conditions per se. The principle diet components are rapidly maturing forages, crop residues and byproducts. High-energy concentrates are scarce and often unbalanced in composition. Providing the lactating animal with sufficient amounts of energy, glucose precursors and available protein becomes rather difficult under these circumstances. Moreover, most tropical diets are deficient in certain macrominerals (especially phosphorus) and trace elements. Measures to improve the supply position of the ruminant animal are discussed in some detail in this paper.

1. INTRODUCTION

There are enormous disparities in milk production between different regions of the world. While the self-sufficiency rate in the industrialized countries exceeds 100 %, milk is still in short supply in most of the Third World countries.

During the last decades, increases in milk production per cow have been an effective means of improving the profitability of dairy farming in the developed countries. Even with the introduction of quota systems, the high-yielding cow will be indispensible. From an economic point of view, the "dilution" of maintenance requirements with increasing production level is the most important factor. Raising the yield per recorded cow (305-d lactation) in the Federal Republic of Germany from 3840 kg FCM in 1959 to 5360 kg FCM in 1983 reduced the maintenance portion of the net energy (NE) intake from 46 % to 38 %. Such a progress in productive performance has been achieved both by the use of animals of high genetic merit and by an improved nutrient supply. Adequate feed intake from a balanced diet is provided by supplementing high-quality roughages with cereals and high-protein concentrates which are normally available at reasonable prices.

Animal performance in the developing countries of the tropics and subtropics is largely affected by the adverse climatic conditions and by the shortage of high-quality feedstuffs. High production levels are therefore diffcult to obtain and many of the feeding and management systems of the industrialized world have turned out to be inappropriate for these areas. The object of this paper is to describe the constraints to ruminant production in the tropics and subtropics and to indicate means for major improvements through the application of relevant research findings.

2. THERMAL STRESS

2.1 Effects of thermal stress on animal performance

Thermal stress is caused by those factors that hamper heat transfer from the animal to its environment. High ambient temperature is usually the primary cause of thermal stress, although other factors such as intensive radiation, high humidity and low wind velocity can intensify the stress (Beede and Collier, 1986).

A reduction in voluntary feed intake with a concomitant drop in milk production or growth rate is a well-known response to increasing heat load. In dairy cattle, dry matter (DM) intake begins to decline at temperatures above about 25° C (Ragsdale et al., 1949, 1951). The role of fluctuating temperatures, however, must be taken into account. The depressing effect of high daytime temperatures can partly be counteracted by nighttime lows (Mc Dowell et al., 1976). The environmental temperature at which feed consumption begins to decline is dependent on ration composition. The greater the portion and the poorer the quality of roughage, the greater and the more rapid will be the reduction in intake with increasing temperatures (NRC, 1981; Beede and Collier, 1986). Decreases in roughage intake may contribute to decreased volatile fatty acid (VFA) production in the rumen and may alter the acetate:propionate ratio (Niles et al., 1980).

In general, ration digestibility is slightly increased during heat stress (NRC, 1981). Even with equal DM intakes, heat-stressed cattle showed some superiority to those kept at lower temperatures (Warren et al., 1974; Lippke, 1975). Therefore, the increase in digestibility cannot solely be attributed to lower rates of intake. There is some evidence that a lower rate of passage (i.e. a prolonged residence time in the rumen) may be the main causative factor (Warren et al., 1974; Schneider et al., 1984).

Absorption of nutrients is likely to be retarded during heat stress. This hypothesis is based on experiments with sheep which show a decrease in gastrointestinal blood flow during exposure to hot environments (von Engelhardt and Hales, 1977).

Our knowledge of the effect of high ambient temperatures on nutrient utilization is still incomplete. There is no doubt, however, that thermal stress reduces the efficiency of converting feed energy units into production energy units. In experiments with heat-stressed dairy cows, milk energy decreased about twice as much as digestible energy intake (Mc Dowell et al., 1969). Evidently, with a body temperature higher than normal additional energy is needed to accelerate heat dissipation and to support increased water and electrolyte turnover. According to the NRC (1981), maintenance requirements of a dairy cow (600 kg LW, 27 kg milk/d) increase by 20 % when the ambient temperature is elevated from 15 - 20° C to 35° C.

Metabolic health problems that have been discussed in relation to heat stress comprise respiratory alkalosis, ketosis and rumen acidosis (Dale and Brody, 1954; Niles et al., 1980; Collier et al., 1982a). The question, however, arises whether these metabolic disorders are of major relevance at production levels that can be achieved in the developing countries. Thermal stress per se does not appear to influence the milk somatic cell content of healthy cows (Paape et al., 1973; Collier et al., 1981).

All facts indicate that extreme environments have an impact on reproductive efficiency. Thermal stress has been shown to reduce the duration and intensity of estrus (Hall et al., 1959; Gangwar et al., 1965). Conception rates and fertility are known to decrease in subtropical and tropical climates (Roman-Ponce et al., 1977; Ingraham et al., 1974). The days around estrus and mating are the most critical period for fertility, especially in terms of embryonic losses (Gwazdauskas, 1985). Thermal stress during late pregnancy reduces uterine blood flow, placenta weight and fetal growth (Roman-Ponce et al., 1978; Brown et al., 1977). Reduced birth weights have been reported for lambs and calves (Brown et al., 1977; Collier et al., 1982b).

Responses in productive and reproductive performance to heat stress are associated with alterations of the endocrine status. Thus, chronic exposure to heat depresses thyroid activity in cattle (Yousef et al., 1967). As far as the pituitary is concerned, thermal stress causes a decrease in secretion rate and plasma concentration of growth hormone with a concomitant increase in antidiuretic hormone (Mitra et al., 1972; El-Nouty et al., 1980). Adrenal function is effected as follows: temporary increase in cortisol level (Christison and Johnson, 1972), significant decrease in plasma aldosterone (El-Nouty et al., 1980), high and sustained plasma levels of catecholamines (Alvarez and Johnson, 1973). As to reproductive hormones, plasma prolactin levels are elevated during acute thermal stress (Wettemann and Tucker, 1974). Experimental results concerning alterations in luteinizing hormone and progesterone have not been consistent (Gwazdauskas, 1985).

2.2 Allevation of thermal stress

Three main strategies have been proposed for reducing the effects of heat stress: a) choice of animals with more efficient thermoregulation, b) physical modification of the environment and c) improved feeding systems.

Selection and breeding strategies cannot be discussed here, the reader is therefore refered to corresponding review articles (Finch,

1986; Mc Dowell, 1982). Measures of physical protection are also beyond the scope of this paper. It may, however, be pointed out that provision with shades as well as evaporative cooling are effective means of improving animal performance (Roman-Ponce et al., 1977; Ingraham et al., 1979; Stott and Wiersma, 1974).

With regard to improved <u>nutritional strategies</u>, the selection of feedstuffs with lower heat increment has been suggested. This can be accomplished by increasing the concentrate portion of the diet. American workers (Stott and Moody, 1960; Rainey et al., 1967) found that low roughage rations increased FCM yield and decreased body temperature. It must, however, be emphasized that concentrates are in short supply in most developing countries; a corresponding adjustment in diet composition is therefore irrelevant. Additonal dietary fat may be another means to reduce total body heat load and to improve the efficiency of energy utilization. As heat-stressed cows showed only minor responses to dietary lipids during short-term experiments (Moody et al., 1967), more research is needed with regard to this topic.

With higher crude protein (CP) in diets for heat-stressed cows (20.8 vs. 14.3 % CP in DM), DM intake and FCM yield were shown to increase by 11 % and 4.3 %, respectively (Hassan and Roussel, 1975). A simultaneous increase in blood glucose possibly indicated an improvement in tissue availability of glucose. However, in view of the complex N transactions in the rumen and the pronounced protein: energy interrelationships, no general conclusion can be drawn from this experiment.

Maintenance of homeostasis in hot environments may call for a higher supply with certain minerals, especially potassium and sodium. K and Na in excess of the NRC (1978) recommendations increased the milk yield of heat-stressed Holstein cows significantly (Mallonee et al., 1985; Schneider et al., 1986). It stands to reason that water supply becomes increasingly important with increasing temperatures. Water needs during heat stress as compared to thermoneutrality have been calculated to rise 1.2 - 2-fold (Beede and Collier, 1986).

3. MILK PRODUCTION AS DEPENDENT ON AVAILABLE NUTRIENTS

Production levels in the tropics and subtropics often are more affected by a shortage in high-quality feedstuffs and by an imbalance of the diet than by the hot environment per se. In addition, adverse effects of parasites on food intake, efficiency of energy utilization and protein supply pose serious problems to the livestock industry (Dargie, 1980).

The principle sources of animal feed are forages (predominantly natural pastures), crop residues (mainly straw from wheat and rice) and a number of byproducts. Tropical forages reveal a more rapid maturation than temperate fodder plants. They are higher in lignin and total cell wall constituents and lower in nitrogen, all contributing to lower digestibility, slower rate of passage and reduced intake. High-energy supplements are usually in short supply in the developing countries. Moreover, most concentrates are low in lipids and contain little starch. Providing the lactating animal with sufficient amounts of energy, glucogenic compounds and absorbable amino acids often becomes difficult under these circumstances. A lack in certain macrominerals and trace elements has also to be encountered.

A comprehensive discussion of mineral supply goes beyond the limits of this paper. It should, however, be emphasized that lack of phosphorus is the most common deficiency, especially in grazing areas. Other macroelements (calcium, magnesium, sodium and sulfur) and quite a number of microelements (copper, cobalt, iodine, selenium and zinc) are also in short supply in many regions of Latin America, Africa and Asia (Mc Dowell et al., 1983). In specific areas, toxic concentrations have been reported for copper, fluorine, manganese, molybdenum and selenium. Production responses to carefully directed mineral supplementation have been remarkable (Mc Dowell et al., 1983). Oral supplementation of minerals is more economic than the use of mineral-containing fertilizers.

3.1 Energy intake

Similar to temperate forages, intake of tropical grasses increases with increasing digestibility (Minson, 1980). In the humid tropics, energy supply may therefore be improved by utilizing pastures in a young vegetative phase. Such a procedure, however, presupposes higher stocking rates which in turn will eventually limit production because of fatigue or overgrazing.

With more mature forages, intake mostly is limited because of an insufficient supply of rumen-degradable nitrogen (RDN) to the rumen microbes. Urea as a source of RDN increased forage DM intake of steers and cows by 50 - 60 % (Mullins et al., 1984; Lindsay et al., 1982). Due to the improved nutrient supply, increases in calf birth weight (Lindsay et al., 1982) and milk yield (Rohr, 1962) were observed. Supplying a certain amount of rumen-undegradable protein (UDP) in addition to RDN has been proven to be even more effective: increments in roughage intake of pregnant and lactating cows of around 80 - 90 % have been reported (Lindsay and Loxton, 1981; Lindsay et al., 1982). Forage intake of crossbred and zebu heifers increased by 15 %, when Pennisetum purpureum in the Sao Paulo region was supplemented with a mineral mixture (Oliveira et al., 1982).

With most tropical forages, milk production above 5 - 8 kg/cow/day calls for supplemental energy. As far as the type of supplement is concerned, Australian workers (Cowan and Davison, 1978) found similar responses in milk yield to molasses (3 kg/day) and maize (2.4 kg/day). Up to 3.6 kg of molasses/cow/day gave increases of milk production up to 2.5 kg (Chopping et al., 1970). With a lower protein supply, poorer responses to molasses may be expected (Preston and Leng, 1986). A reduction in forage intake due to increasing amounts of molasses (Dixon, 1984) can be explained by the effect of soluble carbohydrates on rumen pH and on cellulolytic activity.

In a restricted grazing system, similar responses in terms of feed intake, milk yield and calf growth rate were obtained (Gill et al., 1981) when ad libitum molasses (with 2.5 % urea) was replaced by sugar cane juice (0.8 % urea). The low rumen degradability of sugar cane precludes its inclusion in the diet where grazing is freely available

(Gill et al., 1981).

In several countries in Asia, crop residues (mainly straws) are the predominant basal feedstuffs. With an insufficient amount of concentrates, energy supply hardly exceeds maintenance requirements. Under these circumstances, rather small additions of high-quality green forage will lead to improved animal performance. Milk yield of Surti buffaloes increased by 0.4 kg/day (15 %) when fodder legumes (<u>Glyricidia maculata</u>) were added to a diet based on rice straw (Perdok et al., 1982). There is strong evidence that high-quality green forage through its content in easily-fermentable cellulose increases the number of cellulolytic microbes floating free in the rumen fluid (Preston and Leng, 1986). Part of the effect of additional green forage has been suggested to be due to stimulation of fungal growth. There is general agreement that fungi act as initiators of fermentative breakdown of cell wall components. At present, however, it is impossible to pass a judgement on the significance of the fungi. The development of large numbers of sporangia on fibre may not indicate a substantial role as digesters of forages (Windham and Akin, 1984).

Various treatments have been proposed to improve the digestibility and voluntary intake of lignocellulosic materials. Among these, the application of ammonia or urea appears to be the most appropriate in the developing countries. Urea treatment of rice straw increased DM intake and milk yield in lactating Gir cattle by about 40 % (Perdok et al., 1982). In Sahiwal heifers, the use of treated straw increased daily liveweight gains from 70 to 350 g (Perdok et al., 1982). A future alternative to urea could be the treatment with animal urine.

3.2 Provision of glucogenic compounds

In the ruminant animal, only minor portions of the digestible carbohydrates escape fermentation in the forestomachs. In view of its considerable glucose needs, the lactating animal has therefore to rely on gluconeogenesis (mainly from propionate). Many tropical diets, however, reveal a low glucogenic potential. This can be attributed to a low production of propionate in the rumen relative to acetate (mature forages) or butyrate (molasses-based diets). With such diets, provision of starchy ingredients which are relatively resistant to ruminal fermentation may considerably improve glucose availability. "By-pass" characteristics have been reported for maize, sorghum and rice starch (Waldo, 1973; Rowe et al., 1979). Nevertheless, strategies for supplying "bypass" starch to dairy cows in the tropical and subtropical countries have still to be developoed. Grazing cows in midlactation, which received a rice-based supplement, showed only minor responses in milk yield and tended to partition energy towards tissue deposition (Throckmorton and Leng, 1984). The increase in liveweight gain was associated with increased insulin secretion. A more positive effect might be expected in early lactating cows. Where absorbable amino acids are in short supply, "bypass" starch should only be fed in combination with protected protein. The simple reason for this is that microbial protein synthesis is reduced with less fermentable starch (Lebzien et al.,1983).

An increase in glucose availability might also be achieved by manipulating rumen fermentation towards higher propionate production. In this connection, the application of additives such as polyether ionophores must be taken into consideration. Ionophore-induced reductions in milk fat content, that have been observed with high-concentrate feeding (van Beukelen et al., 1984), are unlikely to occur with the diets in question. Egyptian workers reported significant increases in actual milk yield and fat corrected milk due to monensin (Mohsen et al., 1981/1984). Cows fed antibiotic tended to have earlier uterine involution and post-partum ovulation and shorter service period. Monensin and Lasalocid were effective in accelerating puberty of heifers (Mosely et al., 1982) and bulls (Neuendorff et al., 1982).

Small amounts of poultry litter have also been suggested as means of increasing propionate production (Preston and Leng, 1986). The effect of this waste material, however, appears to be diet-dependent: increases in propionate were reported for diets based on molasses (Fernandez and Hughes-Jones, 1981) but not for rations consisting of lucerne hay and some concentrates (Cañeque and Galvez, 1984).

3.3 Amino acid supply

A shortage in absorbable amino acids may result from an insufficient amount of microbial protein (MP) synthesized and/or from a high degradability of feed protein in the rumen. Efficiency of microbial net synthesis (g MP/unit of carbohydrates fermented) is generally assumed to be low with most tropical diets, although experimental evidence for this assumption is inconclusive. A reduced flow of microbial N to the intestines has been associated with high populations of protozoa in the forestomachs. Protozoa are preferentially retained in the rumen (Weller and Pilgrim, 1974); they are characterized by a high maintenance energy requirement and by intensive N recycling within the rumen. Defaunation resulted in higher protein flow to the duodenum and increased body-weight gain and wool growth in sheep (Bird and Leng, 1985) and body-weight gain in cattle (Bird and Leng, 1978). To the authors' knowledge, no experiments with dairy cows have yet been reported.

When taking the negative influence of a high population of protozoa for granted, one has to think about methods of defaunation. Apart of the well-known effect of a number of chemicals (detergents), one might speculate on the effiency of certain dietary lipids. Small additions of free linseed oil or coconut oil have been shown to increase duodenal flow of total N and bacterial N in sheep significantly (Knight et al., 1978; Ikwuegbu and Sutton, 1982). In some cases, linseed oil virtually eliminated protozoa from the rumen. Dutch experiments with dairy cows also indicate a depressing effect of beef tallow on numbers of protozoa (Tamminga et al., 1983). The positive effect, however, may partly be counteracted by a lower fibre digestibility (Ikwuegbu and Sutton, 1982).

Increasing the amount of UDP would be the alternative approach for improving amino acid supply. This should preferably be achieved by selecting local protein feeds with low degradabilities. Tropical legume forages rich in tannins appear to be quite effective in increasing UDP absorption from the intestines (Barry and Manley, 1984). However, with some negative effect on fibre digestion, tannin-rich plants should contribute at most 25 % to ration dry matter (Preston and Leng, 1986). Among concentrate ingredients, cottonseed cake has been proven to be rather resistent to ruminal fermentation (Parra et al., 1984). Fish meal (Preston and Leng, 1986) and artificially protected oil meals (Kaufmann and Lüpping, 1982) are certainly more effective in increasing UDP flow to the duodenum. However, the range of applications of these ingredients will largely be dependent on their cheapness and on the level of production.

The authors wish to thank Dr. Schafft, Institute of Animal Production of the Technical University Berlin, for stimulating discussions.

4. REFERENCES

Alvarez,M.B. and Johnson,H.D. (1973). J.Dairy Sci. 56, 189.
Barry,T.N. and Manley,T.R. (1984). Br.J.Nutr. 51, 493.
Beede,D.K. and Collier,R.J. (1986). J.Anim.Sci. 62, 543.
Bird,S.H. and Leng,R.A. (1978). Br.J.Nutr. 40, 163.
Bird,S.H. and Leng,R.A. (1985). In: Biotechnology and Recombinant DNA Technology in the Animal Production Industries in Australia. Reviews in Rural Science 6, pp. 109-117 (R.A.Leng, J.S.F.Barker, D.Adams and K.Hutchinson, editors). Armidale: University of New England.
Brown,D.E., Harrison,R.C., Hinds,F.C., Lewis,J.A. and Wallace,M.H. (1977). J.Anim.Sci. 44, 442.
Cañeque,V. and Galvez,J.F. (1984). Anales de Instituto National de Investigaciones Agrarias, Ganadera No. 19, 23.
Chopping,G.D., Deans,H.D., Sibbick,R., Thurbon,P.N. and Stoko,J. (1970). Proc.Aust.Soc.Anim.Prod. 11, 481.
Christison,G.I. and Johnson,H.D. (1972). J.Anim.Sci. 35, 1005.
Collier,R.J., Eley,R., Sharma,A.K. and Pereira,R.M. (1981). J.Dairy Sci. 64, 844.
Collier,R.J., Beede,D.K., Thatcher,W.W., Israel,L.A. and Wilcox,C.J. (1982a). J.Dairy Sci. 65, 2213.
Collier,R.J., Doelger,S.E., Head,H.H., Thatcher,W.W. and Wilcox,C.J. (1982b). J.Anim.Sci. 54, 309.
Cowan,R.T. and Davison,T.M. (1978). Aust.J.Exp.Agric.Anim.Husb. 18, 12.
Dale,H.E. and Brody,S. (1954). Missouri Agr.Exp.Sta.Res.Bul. 562.
Dargie,J.D. (1980). In: Digestive Physiology and Metabolism in Ruminants, pp. 349-371 (Y.Ruckebusch and P.Thivend, editors). Lancaster: MTP Press Limited.
Dixon,R.M. (1984). Trop.Anim.Prod. 9, 30.
El-Nouty,F.D., Elbanna,I.M., Davis,T.P. and Johnson,H.D. (1980). J.Appl.Physiol. 48, 249.
Fernandez,A. and Hughes-Jones,M. (1981). Trop.Anim.Prod. 6, 360.
Finch,V.A. (1986). J.Anim.Sci. 62, 531.
Gangwar,P.C:, Branton,C. and Evans,D.L. (1965). J.Dairy Sci. 48, 222.
Gill,M., Berry,S., Vasquez,O. and Preston,T.R. (1981). Trop.Anim. Prod. 6, 127.

Gwazdauskas,F.C. (1985). J.Dairy Sci. **68**, 1568.
Hall,H.J., Branton,C. and Stone,E.J. (1959). J.Dairy Sci. **42**, 1086.
Hassan,A.A. and Roussel,J.D. (1975). J.Agric.Sci., Camb. **85**, 409.
Ikwuegbu,O.A. and Sutton,J.D. (1982). Br.J.Nutr. **48**, 365.
Ingraham,R.H., Gilette,D.D. and Wagner,W.C. (1974). J.Dairy Sci. **57**, 476.
Ingraham,R.H., Stanley,R.W. and Wagner,W.C. (1979). Amer.J.Vet.Res. **40**, 1792.
Kaufmann,W. and Lüpping,W. (1982). In: Protected Proteins and Protected Amino Acids for Ruminants, pp. 36-74. (E.L.Miller, I.H.Pike and A.J.H. van Es, editors). London: Butterworths.
Knight,R., Sutton,J.D., McAllen,A.B. and Smith,R.H. (1978). Proc.Nutr.Soc. **37**, 14A.
Lebzien,P., Rohr,K. and Schafft,H. (1983). Landbauf.Völkenrode **33**, 57.
Lindsay,J.A. and Loxton,I.D. (1981). In: Recent Advances in Animal Nutrition in Australia, p. 1 A (D.J.Farrel, editor). Armidale: University of New England.
Lindsay,J.A., Mason,G.W.J. and Toleman,M.A. (1982). Proc.Austr.Soc.Anim.Prod. **14**, 67.
Lippke,H. (1975). J.Dairy Sci. **58**, 1860.
Mallonee,P.G., Beede,D.K., Collier,R.J. and Wilcox,C.J. (1985). J.Dairy Sci. **68**, 1479.
McDowell,L.R., Conrad,J.H. and Ellis,G.L.(1983). Feedstuffs **55** (38) 31.
McDowell,R.E. (1982). Southern Coop.Ser.Bul. **259**.
McDowell,R.E., Moody,E.G., van Soest,P.J., Lehmann,R.P. and Ford,G.L. (1969). J.Dairy Sci. **52**, 188.
McDowell,R.E., Hoover,N.W. and Camoens,J.K. (1976). J.Dairy Sci.**59**,965.
Minson,D.J. (1980). In: Grazing Animals, pp. 103-157 (F.H.W.Morley, editor). Amsterdam: Elsevier.
Mitra,R., Christison,G.I. and Johnson,H.D. (1972). J.Anim.Sci. **34**, 776.
Mohsen,M.K., El-Keraby,F. and El-Safty,M.S. (1981, publ. 1984). Agric.Res.Rev. **59**, 15.
Moody,E.G., van Soest,P.J., McDowell,R.E. and Ford,G.L. (1967). J.Dairy Sci. **50**, 1909.
Moseley,W.M., McCartor,M.M. and Randel,R.D. (1982). In: Nutritional Influences on Reproductive Performance of Replacement Heifers, Techn.Bul. **82**, Texas A & M University.
Mullins,T.J., Lindsay,J.A., Kempton,J.T. and Toleman,M.A. (1984). Proc.Austr.Soc.Anim.Prod. **15**, 487.
Neuendorff,D.A., Rutter,L.M., Peterson,L.A. and Randel,R.D. (1982). In: Nutritional Influences on Reproductive Performance of Replacement Heifers, Techn.Bul. No. **82**, Texas A & M University.
Niles,M.A., Collier,R.J. and Croom,W.J. (1980). J.Anim.Sci. **50**, (Suppl. 1), 152.
NRC, National Research Council (1978). Nutrient Requirements of Dairy Cattle, 5th Revised Edition. Washington DC: National Academy of Sciences.
NRC, National Research Council (1981).Effect of Environment on Nutrient Requirements of Domestic Animals. Washington D.C.: National Academy Press.
Oliveira De,M.E., Veiga,J.S.M., Nogueira Filho,J.C.M., Rocha, U.F. and

Veiga,M.C.V. (1982). Revista Facult.Medic.Veterin.Zootech.Univ.Sao
    Paulo **19**, 177.
Paape,M.J., Schultz,W.D., Miller,R.H. and Smith,J.W. (1973).
    J.Dairy Sci. **56**, 84.
Parra,A., Combellas,J. and Dixon,R. (1984). Trop.Anim.Prod. **9**, 196.
Perdok,H.B., Thamodevam,M., Blom,J.J., Van den Born,H. and van Valuw,G.
    (1982). Proc.Sem.Maximimum Livest.Product. from Minimum Land.
    Joydevpur, Bangladesh, 15-18 Febr. 1982.
Preston,T.R. and Leng,R.A. (1986). Matching Livestock Production
    Systems to Available Resources (Pretesting Edition). Addis Ababa:
    ILCA.
Ragsdale,A.C., Worstell,D.M., Thompson,H.J. and Brody,S. (1949).
    Missouri Agr.Exp.Sta.Res.Bul. No. **449**.
Ragsdale,A.C., Thompson,H.J., Worstell,D.M. and Brody,S. (1951).
    Missouri Agr.Exp.Sta.Res.Bul. No. **471**.
Rainey,J., Johnston,J.E. and Frye,J.B. (1967). J.Dairy Sci. **50**, 966
Rohr,K. (1962). Kieler Milchwirtsch. Forsch.ber. **14**, 353.
Roman-Ponce,H., Thatcher,W.W., Buffington,D.E., Wilcox,C.J. and van
    Horn,H.H. (1977). J.Dairy Sci. **60**, 424.
Roman-Ponce,H., Thatcher,W.W., Caton,D., Barron,D.H. and Wilcox,J.C.
    (1978). J.Anim.Sci. **46**, 175.
Rowe,J.B., Ravelo,G., Bordas,F. and Preston,T.R. (1979). Trop.Anim.
    Prod. **4**, 241.
Schneider,P.L., Beede,D.K., Hirchert,E.M. and Wilcox,C.J. (1984).
    J.Dairy Sci. **67**, (Suppl. 1), 120.
Schneider,P.L., Beede,D.K. and Wilcox,C.J. (1986). J.Dairy Sci. **67**,2546
Stott,G.H. and Moody,E.G. (1960). J.Dairy Sci. **43**, 871.
Stott,G.H. and Wiersma,F. (1974). In: Proc.Int.Livestock Environ.Symp.,
    p. 88. St.Joseph (Mi): Amer.Soc.Agr.Eng.
Tamminga,S., van Vuuren,A.M., van der Koelen,C.J., Khattab,H.M. and
    van Gils,L.G. (1983). Neth.J.Agric.Sci. **31**, 249.
Throckmorton,J.C. and Leng,R.A. (1984). Proc.Aust.Soc.Anim.Prod. **15**,
    628.
van Beukelen,P., van Lingen,A.F.V., Peters,M.E., Wensing,T. and
    Breuking,H.J. (1984). Zentralblatt für Veterinärmedizin A **31**, 350.
von Engelhardt,W. and Hales,J.R.S. (1977). Amer.J.Physiol. **232**, E 53.
Waldo,D.R. (1973). J.Anim.Sci. **37**, 1062.
Warren,W.P., Martz,F.A., Asay,K.H., Hilderbrand,E.S., Payne,C.G. and
    Vogt,J.R. (1974). J.Anim.Sci. **39**, 93.
Weller,R.A. and Pilgrim,A.F. (1974). Br.J.Nutr. **32**, 341.
Wettemann,R.P. and Tucker,H.A. (1974). Proc.Soc.Exp.Biol.Med. **146**, 908
Windham,W.R. and Akin,D.E. (1984). Appl.Environm.Microbiol. **48**, 473
Yousef,M.K., Kibler,H.H. and Johnson,H.D. (1967). J.Anim.Sci. **26**, 142.

SUMMARY OF DISCUSSION

Most of the discussion emphasized the role of rumen fermentation. In this context the role of rumen protozoa was discussed. It was suggested that their role in animals fed with poor quality diets may be wasteful rather than useful. Defaunation may have potential to improve the protein supply to the host animal. This opinion was questioned because of the role protozoa are thought to play in the degradation of cell wall components. Fats or fatty acids were mentioned as possible agents to control or reduce the numbers of protozoa in the rumen.

The use of protected proteins and ways to achieve a high degree of protection were also discussed. Protein feeds with a high degree of natural resistance against degradation (coconut cake, palm kernel cake, rice polishings) and this approach seems preferable to the use of chemical protection with, for instance, formaldehyde.

Supplementation of crop residues with legume tree leaves (*Gliricidia, Leucaena*) was also discussed. Part of the unfavourable response may be due to the presence of tanning or anti-protozoal agents in such supplements. Because of the danger of diseases or insects, it was recommended not to concentrate on only one species of legume tree, but always to use a mixture.

Questions were asked concerning the treating of crop residues with ammonia, because of the danger of the formation of 4-methyl imidazole, causing the animals to go crazy. This phenomenon has only been observed if temperature is high and if soluble sugars are present in significant quantities; a combination which seems rather rare with the treatment of cell wall-rich crop residues.

Finally it was stated that the approach towards feeding animals in developing countries has changed in recent years. More emphasis is on the maximum utilisation of local breeds of animals and local feed resources. There was a general agreement in this respect. The general feeling was also that more knowledge is needed on tropical feeds. Emphasis should be on feeds which can be utilised by ruminants or on a mixed utilisation as fuel and as feed for non-ruminants and ruminants.

Seminar VII: Fermented Milks

Session 1

Chairman: Dr. F. M. Driessen (The Netherlands)
Secretary: Dr. J. A. Kurmann (Switzerland)

DIFFÉRENTS LAITS FERMENTÉS, LEUR SIGNIFICATION ET LEUR CARACTÉRISTIQUE.

Ewa LIPIŃSKA
Central Union of Dairy Cooperatives
Hoża 66/68
00-682 Warszawa
Poland

1. CONDITIONS GÉNÉRALES DE LA PRODUCTION ET LA SIGNIFICATION DES LAITS FERMENTÉS

Sur de grandes étendues de la terre les laits fermentés étaient pendant des siècles et sont encore de nos jours préparés au foyer ou artisanalement, en appliquant le procédé le plus simple d'inoculation: par une petite quantité de lait fermenté antérieurement. Par suite de l'industrialisation intense advenue dans certains pays au cours du XIX-ème et XX-ème siècle le mode de production des laits fermentés a fortement changé. On y assiste à une augmentation permanente de volume de lait transformé, au perfectionnement continu des moyens techniques mis en oeuvre.

Parmi les nombreux aspects de la signification des laits fermentés certains paraissent particulièrement importants; ceci concerne la différence entre leur rôle dans la nutrition humaine dans les pays en voie de développement et dans les pays industrialisés. Dans les premiers, ou des couches importantes de la population souffrent de sous-alimentation, les laits fermentés constituent une source accessible parfois unique, de protéines animales, qui se prête également à la préparation des préserves pour les périodes moins favorables. Dans les pays riches, industrialisés, où la nourriture est abondante, les laits fermentés fonctionnent comme aliments sains, agréables au goût, bien conditionnés, de longue conservation, s'accordant avec le nouveau style de vie. Les valeurs diététiques et même thérapeutiques des laits fermentés constituent un autre aspect important de leur signification. Ces dernières qualités, reconnues empiriquement depuis des époques tres éloignées /22, 30/ ont été popularisées dans les pays occidentaux dès le début du siècle par les théories de Metchnikoff sur le rôle bénéfique pour la longévité des aliments riches en lactobacillus /28/. Notons enfin que pour l'industrie laitière les laits fermentés constituent un débouché dont l'importance grandit continuellement.

Les caractères organoleptiques des laits fermentés dépendent en partie de la qualité du lait, de sa teneur différenciée en extrait sec et en matière grasse , des procédés technologiques: d'homogénisation, de thermisation, de brassage, de refroidissement, de conditionnement final et de la fermentation même. On agit aussi sur la composition des levains, en y introduisant des diverses souches bactériennes pour améliorer l'arôme, la viscosité ou la valeur physiologique des laits fermentés /4, 18, 26, 42/. L'idée heureuse des industriels suisses d'allier les qualités des laits fermentés à celles des fruits a conduit à son tour à l'introduction dans les laits fermentés de nombreux additifs: fruits, sirops de fruits, édulcorants, substances fortifiant l'arôme des fruits, produits colorants, stabilisants et emulsifiants, ainsi que des conservants - d'habitude, mais pas toujours, entraînés par les concentrés de fruits /35, 40/.

2. CARACTÉRISTIQUE DES LAITS FERMENTÉS

Les laits fermentés sont produits dans différents états physiques, répondant aux besoins multiples de la nutrition moderne: comme produits gélifiés, fermes et brassés; naturels ou additionnés de fruits, de sirops, de légumes; comme boissons, pâtes, mousses, congelés et comme produit séchés au soleil, par atomisation ou par lyophilisation /Figure 1/. Ces produits si divers, difficiles à classifier seront caractérisés par groupes, selon le citère de la microflore essentielle des levains utilisés pour leur production /Tableau 1/. Seulement certains des laits fermentés seront brièvement décrits. Ce choix de critère qui a déjà servi à des classifications précédentes /3, 18, 20, 22/, est pleinement justifié par le rôle de la microflore; elle constitue la la clef de voûte des caractères distincts des laits fermentés. Grâce à l'environnement qui pendant des siècles sélectionnait dans le lait dans les régions à climat chaud et tropical une microflore thermophile, et dans les régions à climat modéré et froid une microflore mésophile, nous avons actuellement des laits fermentés aussi différents que le yaourt, les laits scandinaves, le kéfir.

2.1. Laits fermentés traditionnels à microflore essentielle termophile

Ces laits dont la microflore essentielle est composée de souches Streptococcus thermophilus et Lactobacillus bulgaricus portent différents noms régionaux, particulièrement dans les pays du Proche-Orient, d'Europe Méridionale, ainsi que dans certaines parties de l'Afrique et de l'Asie. Ils sont produits avec du lait à composition variée, provenant de différents animaux: brebis, chèvre, bufflesse, vache, mais aussi: jument, chamelle, ânesse, yak et zébu. En plus de la microflore essentielle, ces laits fermentés peuvent contenir d'autres bactéries lactiques, homo- et hétèrofermentatives, des bactéries acétiques, des levures et des moisissures. Ces deux facteurs, notamment les différences de composition du lait et la diversité de la microflore rendent les qualités gustatives des laits traditionnels plus riches que celles des mêmes produits confectionnés industriellement avec du lait de vache pasteurisé et uniquement des levains sélectionnés

/18, 22/.

De nombreux laits fermentés traditionnels ont été l'objet d'études approfondies /6, 18, 30, 40, 43/. Ceci concerne: Kiselo mijako,Yaourt /Bulgarie, Turquie/, Dahi /Inde/, Gioddu /Sardaigne/, Katyk /Arménie/, Leben, Laban /Iraq, Egypte, Liban/, Prostokvaša joužnaja /URSS/, Roba, Rob /Soudan, Iraq/, Tahro /Hongrie/, Tiaouri /Grèce/ Riaženka, Varenets /URSS/; des produits concentres: Brano /Bulgarie/ Grusovina /Yougoslavie/, Hooslanka, Žentica /Ukraine/, Labneh, Lebneh/ Liban/, Leben Zeer /Egypte/; des produits dilués: Eyran /Turquie/, Doogh /Iran/; des produits desséchés: Kashk, Kaskg /Iran/, Jub-Jub /Liban, Labneh Anabaris /Proche-Orient/ et d'autres qu'il n'est pas possible d'énumérer ici. L'importance économique des différents laits traditionnels est très inégale.

2.1.1. <u>Le yaourt</u>. Dans la grande famille des laits fermentés le yaourt est devenu le favori de l'époque moderne. Les caractères favorables à un tel développement résident dans le goût et l'arôme rafraichissants, la stabilité de la symbiose de la microflore essentielle de ce lait: S. thermophilus et L. bulgaricus; le caractère du gel, qui se prête à la production des yaouts brassés et permet d'en confectionner des produits en différents états physiques, ainsi que de s'associer à des additifs très variés /Figure 2/. Environ 10-20% des yaourts sont consommés comme produits naturels, le reste avec des additifs divers, dont il a déjà été question. La production du yaourt, y compris le conditionnement final se laisse automatiser /9/. En appliquant les procédés ultrapropres il est possible d'atteindre 20-30 jours de conservation du produit /31/.

A partir du yaourt on confectionne les produits le plus diversifiés: le yaourt à boire - populaire dans certains pays d'Europe; le yaourt congelé - en sorbets, ou soumis au foisonnement avec des sirops et du sucre - populaire au Canada et aux Etats Unis. Les concentrés de yaourt /pâtes/ sont utilisés pour la confection de sauces et desserts: crèmes, flans, puddings et cocktails. Le yaourt séché - par atomisation ou lyophilisation - sert à confectionner des laits fermentés instantanés, d'autres aliments /pain, biscuits, confiserie, "chips"/ et à des fins thérapeutiques /22, 35, 41/. Le yaourt carbonaté - saturé des CO_2 acquiert ainsi de nouvelles valeurs gustatives. Sa conservation est également prolongée /5/. Les formules diététiques et médicamenteuses des yaourts seront présentées ultérieurement.

2.1.2. <u>Laits fermentés à microflore d'origine intestinale</u>. Certaines souches de bactéries d'origine intestinale: Lactobacillus acidophilus, Bifidobacterium, ainsi que Propionibacterium et Lactobacillus casei sont utilisées seules ou ensemble avec des levains traditionnels pour la production des laits fermentés ayant des qualités à la fois gustatives et thérapeutiques. Il est possible de sélectionner des bactéries intestinales possédant de bonnes aptitudes technologiques tant pour l'intensité de l'acidification que pour le caractère du caillé et les qualités gustatives /27, 36/. Par ailleurs, on sélectionne pour ces laits, de même que pour les laits maternisés fermentés, des souches qui forment l'acide lactique à configuration optique L/+/ et des substances antibiotiques /18, 22, 36, 39/. Il y a même des essais de se servir de souches à génotypes déterminés /37/. La présence de bactéries d'origine intestinale rapproche la microflore des laits fermentés industriellement et celle des produits naturels.

Les laits fermentés avec la microflore d'origine intestinale et ses préparations lyophilisées servent souvent avec succès dans la prévention et le traitement des troubles gastriques ainsi que comme adjuvant pendant l'application prolongée des antibiotiques et des remèdes anticancéreux /18, 36, 41/. On les utilise aussi comme probiotiques dans la prévention et le traitement des diarrhées de jeunes animaux:porcelets, veaux et autres /2, 10, 19/. Il faut néanmoins remarquer que les effets constatés empiriquement n'ont pas encore obtenu une pleine justification scientifique; certaines recherches mènent à des résultats contradictoires, p. ex. en ce qui concerne l'influence de ces laits sur le taux de cholestérol dans le sang /11, 29/. Il règne l'opinion que les connaissances dans ce domaine n'en sont qu'à leurs débuts /25/. Les laits fermentés avec L. acidophilus sont connus depuis environ 60 ans, fermentés avec les bifidobactéries - depuis 30 ans environ. Ces laits éveillaient de sérieuses réticences, mais depuis quelques années ils gagnent lentement l'acceptation /13/.

2.1.2.1. Laits fermentés par les monocultures. Les monocultures de L. acidophilus et L. casei servent au confectionnement de laits fermentés et de produits dérivés. On compte parmi eux le lait à l'acidophile, connu en premier aux Etats-Unis - peu savoureux, car l'acide lactique formé au cours de la fermentation est accompagné de quantités minimes de substances arômatiques. Pour remédier à ce défaut il a été créé, également aux Etats-Unis, le lait doux à l'acidophile - lait pasteurisé additionné de 10^6 germes /cm^3 environ de L. acidophilus. Bien que privé de produits du métabolisme bactérien, ce lait garde en partie le caractère avantageux d'une culture vivante, de faciliter la digestion du lactose /12/. Par des techniques différentes on produit industriellement des laits concentrés /pâtes/ à l'acidophile: Pastolact, Rimacide, Acidofilnaja pasta et autres /18, 22, 33/.

Yacoult est un lait additionné d'extrait de Chlorella, fermenté ensuite avec L. casei /Shirota/, connu au Japon. Avec S. faecalis sousesp. liquefaciens est confectionné le lait fermenté Dubu. On connait aussi un lait fermenté par les bifidobactéries seules /17, 18, 36/.

2.1.2.2. Laits fermentés à microflore mixte contenant des bactéries intestinales. Ces laits à noms divers contiennent des bactéries sélectionnées intestinales en plus de la microflore des produits traditionnels. On compte parmi eux: Aco-yaourt /yaourt et L. acidophilus/; Lunebest /yaourt, L. acidophilus, B. bifidum/; un yaourt avec des bifidobactéries naturellement résistantes à certains antibiotiques /26, 27/. Les laits fermentés de type BAT et relatés: Bioghourt /L. acidophilus, S. thermophilus, parfois aussi L. bulgaricus/; Biogarde, Bifigourt /L. acidophilus, S. thermophilus, B. bifidum/ sont connus en RFA, mais aussi dans d'autres pays; p. ex. au Japon le lait Mil-Mil /B. bifidum, B. breve, L. acidophilus/; au Danemark, le lait Cultura /L. acidophilus, B. bifidum/; au Chili, le Progourt /fermenté par des

streptocoques mésophiles et additionné de cultures de L. acidophilus
et B. bifidum/ /13, 36/. Biokys, lait de type BAP /B. bifidum, L. acidophilus, Pediococcus acidilactici/ et Elvit, enrichi biologiquement
en vit. B_{12} et acide folique /L. acidophilus, levain à kéfir, Propionibacterium freudenreichii sous-esp. shermanii/ sont connus en Tchécoslovaquie /4, 36/. Un babeurre Idéal /streptocoques mésophiles
et L. acidophilus/; Acidofilin /L. acidophilus, streptocoques mésophiles, levain à kéfir/; Acidofilnaja prostokvaša /L. acidophilus, levain
a kéfir/ et enfin un lait à l'acidophile et aux levures, ayant des propriétés antibiotiques sont produits en Union Soviétique /18, 39/.

2.1.2.3. <u>Laits maternisés contenant de la microflore intestinale.</u>
Ces laits à composition chimique changée pour les rendre semblables
au lait de femme se trouvent à mi-chemin entre les laits fermentés
de consommation générale et des laits fermentés médicamenteux
quand ils contiennent la microflore appropriée: les bifidobactéries
et les lactobacillus. Les nombreuses recherches concernant l'effet
des laits maternisés à germes vivants sur les enfants malades/troubles intestinaux, allergies/ et sains, d'âges différents, même sur les
enfants nés avant terme, ont souvent permis de constater des résultats bénéfiques de leur consommation /16, 33, 36/.

On compte parmi les laits maternisés ceux destinés aux nourrissons jusqu'à 4-6 mois: Bifidoline et un lait fermenté à pH 5, 9 /Bifidobacterium/; Maljutka /L. acidophilus/; Vita /L. bulgaricus/ et pour
les enfants de plus de 6 mois: Malyš /L. acidophilus/, Vitalact Kislomoločnyi /L. acidophilus, L. lactis, levain à kéfir/; Napitok Dietski
/yaourt de lait maternisé/ /16, 33, 36, 39/. On connait aussi les laits
maternisés fermentés, ensuite lyophilisés /Vita - L. bulgaricus/ ou
séchés par atomisation /Femilact -Bifidobacterium, L. acidophilus/
ainsi que les laits maternisés séchés, additionnés de cultures lyophilisées de Bifidobacterium ou de L. acidophilus, p. ex. le lait maternisé
Töpfer et autres, produits en petites quantités pour les buts thérapeutiques dans de nombreux pays /8, 16, 36/.

2.2. Laits fermentés traditionnels à microflore essentielle mésophile
homo - et héterofermentative

Ce groupe comprend des laits fermentés aux propriétés organoleptiques diverses. Quand le lait est fermenté avec des bactéries
mésophiles /Streptococcus lactis sous-esp. lactis, Streptococcus lactis sous-esp. cremoris, Streptococcus lactis sous-esp. diacetylactis
et Leuconostoc/ le facteur majeur de l'arôme est le diacétyle. Les
levains contenant en plus des levures procurent aux laits fermentés
un goût levuré et piquant, dû à la présence de l'alcool et du gaz carbonique.

2.2.1. <u>Laits à microflore essentielle homofermentive</u>

2.2.1.1. <u>Laits fermentés filants.</u> Ces produits traditionnels des pays
scandinaves portent des noms locaux différents: Langmölk, Taettmjölk,
Langfil, Piimä, Viili. Les streptocoques contenus dans les levains

produisent des exoglycoprotéines, substances mucilagineuses, qui confèrent une consistance filante à ces laits. Le levain du lait finlandais Viili contient en plus le Geotrichum candidum /22,30/.

2.2.1.2. Laits fermentés à microflore mésophile non filante. Ces laits sont produits dans plusieurs pays d'Europe, particulièrement dans l'Est. On compte parmi eux: Aerin, Lactorol, Lubitelskij, Prostokwaśa obyknoviennaja, Ruskij, Sauermilch, le babeurre artificiel obtenu par fermentation du lait écrémé et les crèmes acides à faible teneur en matière grasse. Pour améliorer les propriétés rhéologiques de ces produits on les additionne de souches filantes S. thermophilus /18,30,39/. En employant des techniques d'égouttage, d'emprésurage et d'ultrafiltration on confectionne industriellement avec les mêmes levains des laits fermentés concentrés: Ymer au Danemark, Lactofil, produit similaire suédois, Skyr islandais, Polkrem polonais et aussi des produits traditionnels: lait de cave /Kellermilch, Kaeldermaelk/, lait stocké /Lagermilch/, lait d'automne /Herbstmilch/ - ces trois derniers en voie de disparition.

2.2.1.3. Babeurres acides. Ce groupe comprend le babeurre naturel, séparé par barattage d'une crème acide et le babeurre séparé d'une crème non acide, soumis ensuite à la fermentation. La consommation du babeurre très populaire aux Etats-Unis diminue ces dernières années /22/.

2.2.2. Laits à fermentation mixte, lactique et alcoolique

2.2.2.1. Le kéfir. Le produit traditionnel, confectionné en petites quantités est préparé avec les grains à kéfir, constitués par une symbiose de bactéries homo- et hétérofermentatives,/S.lactissous-esp. lactis, S.lactis sous-esp. cremoris,Lactobacillus caucasiens,Leuconostoc, parfois aussi Acétobacter/de levure comme Torula kéfir - liés par un polysaccharide, le Kefiran /30/. Dans la production industrielle du kéfir on utilise comme levain le lait fermenté séparé après la fermentation des grains à kéfir. Le goût et l'arôme d'un tel kéfir sont moins prononcés. De nouvelles technologies de production du kéfir, partant de levains lyophilisés, ont été élaborées /21,32/.

2.2.2.2. Le koumiss. Le produit traditionnel est confectionné en minimes quantités en URSS avec du lait de jument, plus riche que le lait de vache en albumines, peptones, acides aminés et immunoglobulines. Il sert dans le traitement de la tuberculose. La microflore essentielle du koumiss est constituée de levures qui fermentent le lactose: Toroula koumiss et Saccharomyces lactis, et de Lactobacillus orenburgii, L. bulgaricus, L. acidophilus. Le produit fermenté contient entre autres 1,75 - 2,5 % d'alcool et du CO_2 /30/. Industriellement le koumiss est produit à partir du lait de vache à faible teneur en matière grasse, additionné de protéines du lactoserum /18/. Certaines recherches préconisent la production du koumiss d'un lait de vache ultrafiltré, fermenté d'abord par une culture à yaourt, ensuite par la levure boulangère /34/. Kurunga, produit du lait de vache et Airan, du lait de brebis, similaires au koumiss, sont consommés en Mongolie.

2.4. Autres types de laits fermentés

2.4.1. Additionnés de blé, légumes et des concentrés de lactosérum. Il est de première importance de mettre à la portée des populations à faible revenu des aliments peu coûteux, de haute valeur nutritionnelle, contenant, en partie au moins, des protéines animales dont le manque hante non seulement le présent, mais aussi l'avenir de beaucoup d'habitants de notre globe.

Cette fonction d'accroissement de la quantité de nourriture contenant des protéines animales, et en même temps, d'augmentation de leur valeur biologique au delà de la somme arithmétique des protéines, peut être accomplie par les laits fermentés, additionnés de plantes ou de leurs extraits riches en protéines /soja, fève, cacahouette, lupin et autres/ ainsi que du blé /1, 7/. Mentionnons ici les protéines animales accessibles dernièrement en quantités importantes: les concentrés de lactosérum, déjà utilisés, et ceux de l'avenir, préparés à l'aide de nouvelles technologies de séparation, soit soumis à des modifications dirigées /14, 24/.

2.4.2. Laits fermentés médicamenteux. Ces laits constituent parfois une aide précieuse dans la prévention et le traitement de différentes maladies. Il existe des formules très variées de ces laits, préparés à base de yaourt, de lait à l'acidophile, de kéfir et d'autres. On compte parmi eux des laits fermentés - sans cholestérol /à matière grasse remplacée par des huiles à acides gras polyinsaturés/; pour les régimes "basses calories" /à 12 % d'extrait sec et 0,2 % de matière grasse/; pour les diabétiques /contenant 0,3-0,6 % de lactose/: pour les allergiques /avec du lait de soja/; pour traiter la constipation /enrichis de fibres diététiques/; pour diètes intensives, entre autres pour les sportifs /à protéines hydrolysées / et du type "Balcan" /fortifiés en protéines, matière grasse, magnésium et acide orotique/ enrichis en vitamines, en calcium et autres /35, 36, 40/.

. . . .

A la fin quelques mots sur les produits rapproches qui ne sont pas des laits fermentés à part entière ou ne sont pas considérés comme tels. Le yaourt pasteurisé - se conserve en température ambiante pendant un temps prolongé. Il est d'habitude moins cher, bien qu'aussi savoureux que le yaourt à germes vivants. Mais il ne procure pas les avantages d'ingestion des bactéries lactiques vivantes. Cet argument, habilement utilisé par la publicité a une portée vraiment remarquable: en Italie, à Milan, il a fait diminuer la consommation du yaourt pasteurisé de 59 % en 1975 à 39 % en 1984 /23/. Signalons ici le pouvoir extraordinaire du yaourt à germes vivants de faciliter la digestion du lactose /12/. Ce caractère a une importance particulière pour les régions de malnutrition, souvent accompagnée de l'intolérance au lactose /38/. Les laits partiellement fermentés, ensuite acidifiés chimiquement et les laits gélifiés et acidifiés par les acides organiques et les enzymes, peuvent être attachés aussi à ce groupe. Pour des raisons diététiques et leur prix bas ces produits gagnent des consommateurs dans certains pays /15/.

3. STATISTIQUES ET CONCLUSIONS

Les extraits des donnees statistiques publiees par la FIL-IDF pour les années 1979, 1983 présentent la consommation de différents laits fermentés en kg par habitant dans les pays-membres de cette organisation /Tableau 2/.

Ces données permettent de constater que la consommation des laits fermentés dans l'année 1983 comparée à celle de 1979, dans les pays qui ont fournis des réponses comparables a augmenté respectivement:
- pour les yaourts, en 20 pays sur 23 répondants,
- pour les autres laits fermentés, en 8 pays sur 11 répondants,
- pour les babeurres, en 6 pays sur 12 répondants.

Les informations recueillies en 1986 de certains Comités nationaux de la FIL-IDF comparées à celles publiées en 1981 /25/ démontrent que la consommation du kéfir, du koumiss et des laits fermentés contenant la microflore intestinale est suivie par ces Comités dans un nombre très restreint de pays, qui a quand même augmenté en 5 années de 4 à 11. La consommation par habitant et par an est relativement importante, allant de 1 à 6 kg env. aux Pays-Bas, en Finlande, Hongrie, Suède, Tchécoslovaquie et URSS; de 0,5 kg env. au Danemark, en Pologne, RFA et en Suisse.

· · · ·   ·   · · · ·

Les données présentées laissent croire que les trends d'augmentation de la consommation des laits fermentés différents peuvent se maintenir dans l'avenir. Dans ce même sens doivent agir les contributions des techniques du génie génétique et de la fermentation à l'aide des bactéries incluses qui peuvent permettre de créer des laits fermentés peu couteux et aux caractères très avantageux. On peut aussi supposer qu'à mesure qu'augmentera la conscience de la dégradation du milieu écologique les préoccupations de santé deviendront de plus en plus importantes et l'attrait des laits fermentés, grâce à leur image de produits diététiques et, dans une certaine mesure thérapeutiques, ne fera qu'augmenter.

SESSION 1

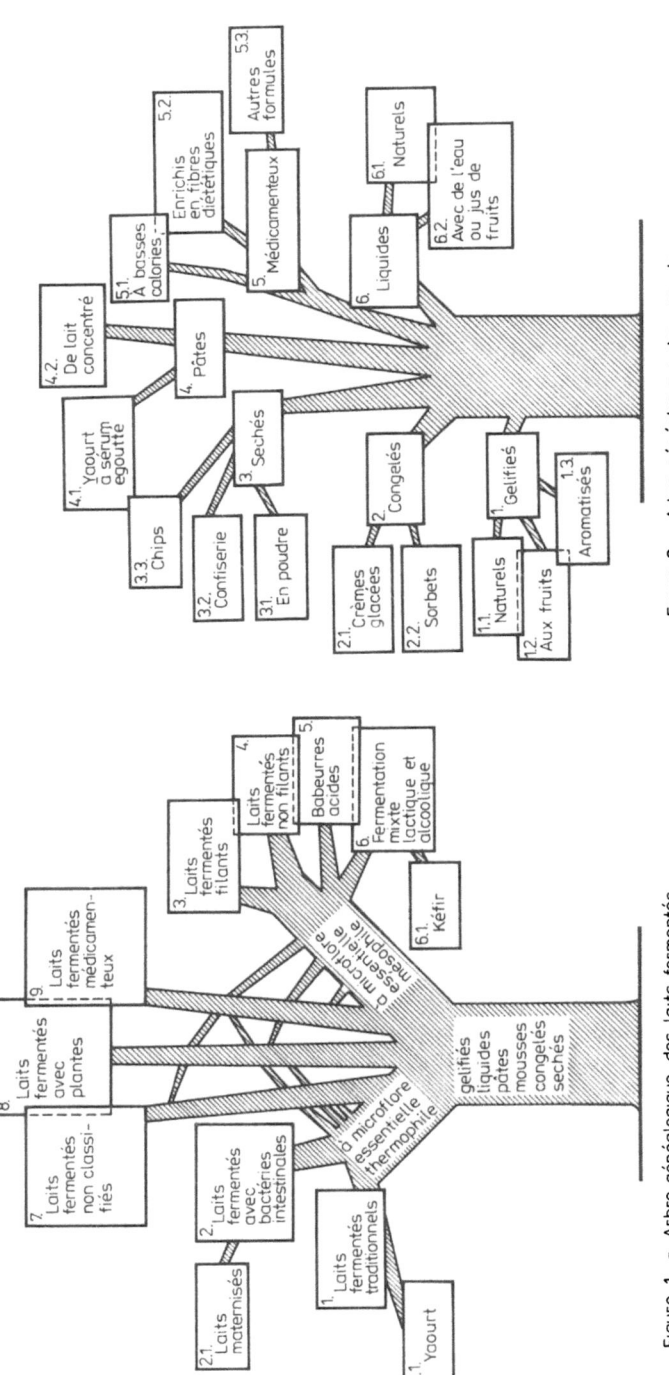

Figure 1 – Arbre généalogique des laits fermentés

Figure 2 – Arbre généalogique des yaourts

Tableau I - Essai de classification des laits fermentés
           traditionnels et nouveaux[1]

1. Laits fermentés à microflore essentielle lactique, thermophile et homofermentative; temperature d'incubation de 3o /35-40/ 45°C
1.1. Yaourt et laits similaires traditionnels en différents états physiques: gélifiés, liquides, congelés, concentrés, /pâtes/, mousses séchés
1.2. Laits fermentés à microflore essentielle intestinale, fermentation acide, sans production appréciable de gaz et d'alcool
   1.2.1 Levains constitués de monocultures
   1.2.2 Levains constitués de microflore mixte
   1.2.3 Laits maternisés
2. Laits fermentés à microflore essentielle mésophile, homo- et hétérofermentative; temperature d'incubation 10 /15-20/ 30°C
2.1. Fermentation lactique à microflore essentielle homofermentative
   2.1.1. Laits fermentés filants /scandinaves/
   2.1.2. Laits fermentés non filants, préparés avec des cultures de beurrerie, y compris les babeurres artificiels et crèmes acides à teneur variée en matière grasse
   2.1.3. Babeurres acides: classique et nouveaux
2.2. Fermentation mixte, lactique et alcoolique
   2.2.1. Kéfir à grains et à levains /sans grains/
   2.2.2. Koumiss et produits similaires
3. Autres types de laits fermentés
3.1. Additionnés de blé, légumes et concentrés de lactoserum
3.2. Laits médicamenteux
3.3. Laits fermentés non classifiés

---

[1] Selon Koroleva et Kondratenko /1978/, Oberman /1985/, Rašič et Kurmann /1978/ /1983/, Tamine et Robinson /1985/

Tableau II - La consommation des laits fermentés et des babeurres en kg par habitant dans les pays-membres de la FIL-IDF dans les années 1979, 1983 1/

	Année de recensement/consommation en kg					
	1979			1983		
	yaourt	autres laits fermentés	ba-beurres	yaourt	autres laits fermentés	ba-beurre
Autriche	5,3	2,2	2,1	6,2	2,4	2,2
Australie	1,7	-	-	2,2		
Belgique	4,8	-	2,8	4,9 O		2,8
Brésil	0,5	0,2	-			
Canada	1,7	-	0,7	1,8		0,6
Chili	1,0	-	1,8	1,8	-	
Danemark	9,2	7,8	9,4	9,6	8,4	9,8
Espagne	0	0	-	6,8 O		
Finlande	8,0	27,0	3,9	8,8 R	29,1	2,4
France	8,8 R	0	-	11,8 R		
Grande Bretagne	2,5	-	-	2,5		
Hongrie				0,9	1,1	0,04
Inde	3,5	-	-	4,0		14,1
Irlande	1,8	-	0	2,6 O		4,3 O
Islande	5,4	-	-	5,0	15,6	
Israël	5,8	10,2	-	7,3	8,9	
Italie	1,2	-	-			
Japon	1,0	1,7	-	2,7	4,8	
Luxembourg	5,1	-	-	5,7		2,1 O
Norvège	2,0	7,6	-	3,3	12,1	
Pays-Bas	16,2	-	9,3	17,7		9,9
Pologne	0,2	0,8	1,5	0,2	0,5	0,9
RFA	6,6	1,3	2,4	7,3	1,3	2,5
Suède	3,9	19,4	0,2	4,4	21,0	0,1
Suisse	13,5	-	1,0	15,0		1,0
Tchécoslovaquie	1,6	2,3	3,0	1,9	2,8	4,3
URSS	-	6,5	-	7,0		
USA	1,2	-	2,0	1,4		1,9

Légende:
    O - Indication qui n est pas séparee
    R - Indication révisée
    - - Pas d'indication
    1/ - Compilation FIL-IDF /1981, 1985/

BIBLIOGRAPHIE

1. Abou-Donia S. A. /1984/ N. Z. J. Dairy Sci. Technol. 19 /1/ 7-18
2. Aldrovandi V., Ballarini G., Galeffi F., Monetti P. G. /1984/ Documenti Veterinari 5/1/ 51-54
3. Camus A. /1964/ Inventaire des laits fermentés actuellement connus dans les pays adhérents à la FIL-IDF Ann. Bull. 9-16 FIL-IDF
4. Černa J. /1984/ Průh. Potravin 35 /4/ 192-195
5. Choi H. S., Kosikowski F. V. /1985/ J. Dairy Sci. 68 /3/ 613-619
6. Dagher S., Ali A. /1985/ J. Food Protect. 48 /4/ 300-302
7. Deka D. D., Rajos R. B., Path G. R. /1984/ Egyptian J. Dairy Sci. 12 /2/ 291-299, selon DSA 47, 4677
8. Dedicová L., Drobohlav J. /1984/ 'Le lait maternisé fermenté en poudre pour l'alimentation des nourrissons - "Femilact"' FIL-IDF Doc. 179, VI-VII
9. Driessen F. M., Ubbels J., Stadthuders J. /1977/ Biotechnology Bioengeneering 19 /6/ 841-851
10. Ervolder T. M., Gutkov A. V., Gutkov S. A., Dušenin N. V., Trubnikov N. K. /1984/ Moloč. Prom. /8/ 18-20
11. Gilliland S. E., Nelson C. R., Maxwell C. /1985/ Appl. Environ. Microbiol. 49 /2/ 377-381
12. Gilliland S. E. /1985/ Cult. Dairy Prod. J. 20 /2/ 28-33
13. Hansen R. /1985/ North European Dairy J. 51 /3/ 79-83
14. Jelšik A., Prepoková J. /1984/ Zborn. Prac Vysk. Ustavn. Mljekarsk. Ziline 8 111-117
15. J. P. G. /1984/ Revue Lait. fr. nr 431; 55, 56, 58, 65-67
16. Kondratenko M. S. /1984/ Prům. Potravin. 35 /2/ 73-76
17. Kook H. K., Gee H. K., Young H. K. /1983/ Korean J. Dairy Sci. 5 /3/ 205-211
18. Koroleva N. S., Kondratenko M. S. /1978/ Simbiotičeskije zakwaski termofilnyh bakterii v proisvodstvie kislomoločnyh produktov, Piščevaja Promyšlennost, Moskwa; Tehnika, Sofia
19. Kotowski K. /1985/ Medycyna weter. 41 /3/ 134-137
20. Kosikowski F. V. /1966/ /1977/ Cheese and Fermented Milks. Edwards Brothers Inc. Ann. Arbor. Michigan., selon /22/
21. Kramkowska A., Kornacki K., Bauman B., Fesnak D. /1982/, XXI Congr. Intern. Lait., vol. 1, 304-305
22. Kurmann J. /1984/ 'La production des laits fermentes dans le monde.' FIL-IDF, Doc. 179, 12, 24
23. Leali L. /1983/, Scienza Tech. Latt. -casear. 34 /5/ 374-383
24. Les composants du laits. Nouvelles technologies, nouveaux marchés. Journées d'études. Interprofession Laitière /1984/ Paris
25. 'Les produits laitiers fermentés dans l'alimentation humaine' /1983/ FIL-IDF, Doc. 159, 3, 15, 29
26. Lipińska E. /1979/ Acta Alim. Polonica XXIX /4/ 359-363
27. Lipińska E., Kosikowska M., Jakubczyk E., Lipniewska D., Mamczarek B. /1979/ Roczn. Inst. Przem. Mlecz. XXI /1/62/ 35-51
28. Metchnikoff E. /1908/ The Prolongation of Life. G. P. Putnam's Sons, New York, 1st Ed.

29. Massey L. K. /1984/ J. Dairy Sci. 67 /2/ 255-262
30. Oberman H. /1985/ Microbiology of Fermented Foods. Vol. 1, 167-195. Elsvier Applied Science Publishers. London, New York
31. Odet J. /1984/ 'Le conditionnement des laits fermentés.' FIL-IDF Doc. 179, 123
32. Pettersson H. E. /1984/ Freeze-dried concentrated starter for kéfir, FIL-IDF Doc. 179, XVI-XVII
33. Piščevaja i biologičeskaja cennost' moločnyh produktov detskovo pitanja. VNIIMI, Red. P. F. Krašenin, Agropromizdat /1985/ Moskwa
34. Puhan Z., Vogt O. /1984/ Milchw. Ber. Wolfpassing u. Rotholz, nr 78, 7-14
35. Rašič J. L., Kurmann J. /1978/ Yoghurt, Scientific Grounds, Technology, Manufacture and Preparations. Technical Dairy Publishing House, Copenhagen
36. Rašič J. L., Kurmann J. /1983/ Bifidobacteria and their Role, Birkhäuser Verlag, Basel, Boston, Stuttgart
37. Sarra P. G., Dellaglio F. /1984/ Microbiologica 7 331-339
38. Sareen D., Saxena S. /1982/ Pediat. Clin. India 17 /2/ 21-25, selon DSA 46, 2944
39. Semenihina V. F. /1984/ 'Les autres laits fermentés.' FIL-IDF Doc. 179, 113, 114
40. Tamine A. Y., Robinson R. K. /1985/ Yoghurt, Science and Technology, Pergamon Press, Oxford
41. Tomoda T., Nakano Y. /1984/ Bull. Osaka Med. School 30 /1/ 14-18
42. Warchoł-Drewek Z., Roczniak B. /1982/ XXI Congr. Intern. Lait. Vol. 1, 309
43. Warsy J. D. /1982/ J. Agric. Res. 20 /4/ 211-218

Characteristics of Cultures used for the Manufacture of Fermented Milk Products.

S. E. Gilliland, Ph.D
Animal Science Department
Oklahoma State University
Stillwater, Oklahoma 74078-0425
U.S.A.

ABSTRACT. The ability to grow in milk and produce lactic acid from the lactose is perhaps the most important characteristic shared by the starter bacteria. Yet variations in characteristics among species and/or strains of this group of bacteria result in a wide variety of cultured milk products. Close taxonomic similarities among some species within genera often make correct identification difficult. New methodologies such as DNA homology tests have proven to be helpful in identifying some species. Many of these bacteria possess characteristics enabling them to aid in controlling undesirable microorganisms. Some also can have beneficial roles in nutrition and/or health.

1. INTRODUCTION

While it is recognized that yeasts and/or molds may be involved in some cultures used in the manufacture of fermented milk products, this paper will focus on the lactic acid bacteria involved. The bacterial species most often present in cultures used for the fermentation of milk products are listed in Table 1. Within each species listed, many variants are encountered, which in some cases makes exact identification difficult. Yet the very fact that there are variant cultures or strains occurring in each species permits the production of a wide variety of cultured milk products having unique properties. The names of the organisms used in Table 1 may or may not be the ones with which you are currently familiar. They are the ones used in Bergey's Manual of Determinative Bacteriology (3). New and improved techniques of studying identity characteristics have revealed many new relationships among various traditional species. This has resulted in the renaming of some species.
    Characterization of the starter bacteria will be approached from basically three standpoints: (1) Those characteristics which make them useful in the manufacture of fermented milk products (2) Identity characteristics. (3) Characteristics making them useful beyond the culturing of milk.

2. CHARACTERISTICS MAKING CULTURES USEFUL

The bacteria listed in Table 1 all have some similar characteristics which make them important in the manufacture of fermented milk products. The general functions of starter cultures used in this regard are to produce lactic acid, acetic acid, aromatic flavor components, and carbon dioxide (31). Additionally possibilities relative to changes in the product, include lipolytic and proteolytic activities. These types of enzymatic activities can produce a wide variety of desirable changes and in some cases undesirable ones. Of these general functions, the production of acid from lactose is generally considered as the main one.

While the organisms listed in table one, each have the ability to hydrolyze lactose, the mechanisms whereby the hydrolysis occurs varies. Some species possess β-galactosidase (β-gal) as a primary enzyme for hydrolyzing lactose. These organisms include Lactobacillus bulgaricus, L. lactis, Streptococcus thermophilus, and L. acidophilus (14, 18, 31, 36). Streptococcus lactis, S. cremoris, and L. casei rely on β-D-phosphogalactoside galactohydrolase (β-P-gal) as the primary means of hydrolyzing lactose (14, 31, 36).

In the case of β-gal, lactose is hydrolyzed to glucose and galactose. The glucose is further metabolized to lactic acid via the Embden Meyerhof pathway for glycolysis and galactose is converted to glucose-6-phosphate which also is further metabolized via the Embden Meyerhof pathway in the homofermentative species (14, 31, 36). However, the conversion of galactose to glucose-6-phosphate requires several additional enzymes and thus it may be more efficient to ferment the glucose rather than the galactose moiety of lactose. This may explain why some species of starter culture bacteria accumulate free galactose in milk culture (14). In heterofermentative species the hexoses are metabolized via the hexose monophosphate (pentose phosphate) pathway (4, 14, 36) which involves the removal of carbon dioxide from the hexose and a 3/2 cleavage of the resulting pentose with the 3 carbon unit being converted into lactic acid and the 2 carbon unit being converted to acetaldehyde or ethanol.

For organisms in which the lactose is hydrolyzed by β-P-gal, the glucose is further metabolized via the Embden Meyerhof or the hexose monophosphate pathway. Galactose-6-phosphate resulting from hydrolysis of lactose-6-phosphate is metabolized via the Tagatose pathway (36).

Since lactic acid is a major end product of each pathway, these pathways probably constitute the most important metabolic pathways in bacteria involved in the production of fermented dairy products. The level of enzyme activity for hydrolyzing lactose by these organisms can limit the rapidity of acid production by the organism (19, 36).

The major volatile flavor component produced by the starter culture bacteria is diacetyl. The leuconostocs (primarily L. cremoris) and S. lactis subsp. diacetylactis are the primary ones involved. While most work relative to the mechanisms of diacetyl production has been done with S. lactis subsp. diacetylactis it is speculated that citrate metabolism is the same in both genera of bacteria (4, 40). The major difference between the two genera with regard to diacetyl production appears to be that the leuconostocs do not produce diacetyl above pH 5 whereas the streptococci initiate production at a higher pH. While there is still some debate over the exact intermediates involved, citrate is metabolized to yield carbon dioxide, diacetyl and acetoin (4, 40).

Virtually all of the bacteria involved in starter cultures are very complex with regard to nutritional requirements. Milk provides all of the nutrients needed for growth by this group of organisms although perhaps not in optimum concentrations or in the most readily available form thus limiting growth by some. However, since variations occur among strains of the individual species it is possible to select ones with adequate activity for producing fermented milk products. While the ability to grow in milk is essential for use of most of the cultures, there are cases where growth is not essential. An example of this would be in the use of leuconostocs to produce flavor in pre-acidified milk products (15). If these organisms are produced properly so that they contain the necessary complement of enzymes they are able to produce diacetyl in the acidified milk without growing.

In addition to problems associated with the fastidious nature of this group of microorganisms, they share some other common characteristics which can present problems. One of these is their sensitivity to antibiotics. Generally speaking, all species of bacteria used as starter cultures for dairy products are sensitive to antibiotics. The primary problem arises when antibiotic residues are present in the raw milk supplies used for manufacture of cultured products.

Another factor which can create tremendous problems is the action of bacteriophage. While most of the attention relative to bacteriophages has in the past been directed toward the streptococci there are bacteriophage specific for each of the species of bacteria which are encountered in cultured milk products. Fortunately there is variation among strains of each species with regard to sensitivity to specific bacteriophages. Thus it is possible to alternate the use of strains having different bacteriophage specificity in order to help prevent this problem. With the rapidly developing field of genetic manipulation it may be possible to develop improved strains having resistance to bacteriophage (27).

Some characteristics and/or desirable activities of starter bacteria are controlled by plasmids. Examples of these include lactose hydrolysis (12, 32), proteolytic activity (12, 32), and phage resistance (27) for S. lactis and/or S. cremoris. Citrate utilization by S. lactis subsp. diacetylactis and L. cremoris is also under plasmid control (25). Antibiotic resistance and the ability to produce bacteriocins also are plasmid controlled (12, 32). If a plasmid(s) which controls one or more of the desirable characteristics is lost from the culture during routine handling of the culture or through mistreatment of the culture, then the desirable characteristic may be lost. However, the fact that some of these desirable characteristics are under plasmid control provides the opportunity to develop and improve cultures through genetic manipulation involving plasmid transfer technology.

3.0. IDENTITY CHARACTERISTICS

Characteristics used for identifying bacteria involved in dairy fermentations are important because improper identification can lead to misconceptions concerning a culture's use and/or performance. This is more important in some cases than in others. It is especially important when a culture is expected to perform a specific function when other similar or closely related species cannot. Historically the identification of these bacteria has been based on morphological, physiological and phenotypic traits such as the ability to ferment various carbohydrates and/or their action on other substrates. A number of previous publications have summarized the identification traits of the lactobacilli (9, 14, 35, 38), the leuconostocs (4, 7, 9), and the streptococci (5, 9, 36). The lactobacilli are Gram positive rods which occur in pairs or chains that are catalase negative. They obtain their energy from fermentation and are facultatively anaerobic. In the past identity characteristics which permitted differentiation of species within this genus involved primarily differences in ability to ferment various carbohydrates. The leuconostocs are Gram positive cocci occurring in pairs or chains which are catalase negative and facultatively anaerobic. They obtain their energy from fermentation, however, most will not produce sufficient acid to coagulate milk. Identification characteristics separating species within this genus have generally included optimum growth temperatures and action on various carbohydrates. The streptococci are Gram positive cocci occurring in pairs or in chains which are catalase negative, facultatively anaerobic and obtain energy from fermentation. Additional identification characteristics separating the important species in this genus have included growth temperature, diacetyl production, and action on various sugars.

Because it is often difficult to distinguish between these organisms based on morphology, fermentation patterns, and other physiological characteristics, other comparisons may be done including characteristics of cellular components. Some of these appear to be useful while others do not. The percentages of guanine plus cytosine in the DNA of the bacteria can be a useful identity characteristic. However, those species within a genus having similar phenotypic characteristics usually have similar guanine plus cytosine percentages. On the other hand, those having similar guanine plus cytosine percentages may not be closely related based on phenotypic characteristics.

Serological properties have been used primarily to determine specific strains within a given genus and/or given species (7, 38). Perhaps the most widely recognized serological grouping of starter bacteria is the Lancefield grouping of streptococci (36). However, several species occur in each group. As an example in which serological typing had been applied to a given species, L. acidophilus has been reported to involve at least four serological types (39). These four serological types appear to be related to four different biotypes of L. acidophilus isolated from the intestinal tract of different animal species (33). However, the relationships between serological types and host animals were not clear cut.

The electrophoretic mobilities and/or serological reactions of selected enzymes from starter bacteria provides yet another useful tool. Lactate dehydrogenase, glucose-6-phosphate dehydrogenase, and glyceraldehyde-3-phosphate dehydrogenase are enzymes which have been used in this capacity (9, 24, 38). These techniques are useful in differentiating species, as well as strains within a given species (11, 24).

Perhaps one of the best techniques for confirming the identification of bacteria is DNA-DNA homology. With this technique some phenotypic groups of bacteria (individual species) may be further divided or some of those which have been considered to be of different phenotypic characteristic (different species) may be shown to be more closely related. As examples within the genus Leuconostoc, L. mesenteroides, L. dextranicum, L. cremoris have all been shown to be very closely related based on DNA-DNA homology studies (8, 23). S. lactis, S. cremoris, and S. diacetylactis have been shown to comprise one DNA homology group, thus these organisms have been designated as S. lactis, S. lactis subsp. cremoris, and S. lactis subsp. diacetylactis (10).

Within the species L. acidophilus at least six DNA homology groups have been reported leading to the suggestion that this species should be further divided into additional species (24). There appears to be some relationship between these six different DNA homology groups with antigenic groups reported earlier (39).

In addition to involving critical techniques which may be difficult, the final decision concerning confirmation of identity based DNA-DNA homology tests may not be clear cut. There does not appear to be a clear understanding of the percentage of the homology with the reference strain required for conclusive identification of an organism. There still appears to be a certain amount of subjective evaluation involved in making these decisions.

4.0. USEFUL CHARACTERISTICS OTHER THAN CULTURING MILK

The starter bacteria possess beneficial characteristics beyond those necessary for the manufacture of fermented milk. The acid(s) produced by the starter culture bacteria and hydrogen peroxide produced by some exert inhibitory action toward both spoilage and pathogenic bacteria. Additionally many of the starter culture bacteria produce antibiotic or antibiotic-like substances. These characteristics have focused considerable attention on their role in food preservation (13). Some of the antibiotic like materials produced are bacteriocins which are active against closely related strains or species. While these generally do not have a broad spectrum of activity they likely are important in enabling one strain(s) to become well established and predominate in the presence of other starter culture bacteria.

Some cultured or culture containing milk products have nutrition and health benefits beyond those which can be derived from milk alone.

A considerable proportion of the world's population suffers from a condition referred to as "lactose malabsorption". Consumption of some cultured or culture containing milk products can improve lactose digestion in such individuals. Those starter culture bacteria possessing the enzyme β-galactosidase appear to be most important in this respect. Yogurt containing viable starter bacteria as well as acidophilus milk have been shown to be beneficial for persons unable to adequately digest lactose (16, 26, 28). Some studies have shown that yogurt is more digestible than the milk from which it is manufactured (1). Other studies have reported increased growth in rats fed cultured yogurt compared to those fed uncultured milk (22). Both yogurt and acidophilus milk have been reported to contain activities which are detrimental to the formation of certain types of cancer (6, 20, 37). Some fermented milk products have been shown beneficial in relation to reducing serum cholesterol levels (17, 21, 29). Selected strains of L. acidophilus possess the ability to assimilate cholesterol (17). Using pigs on a high cholesterol diet we have shown that feeding one of these selected strains of L. acidophilus resulted significantly lower levels of serum cholesterol than in those not receiving the culture.

There currently is much interest in the relationship of certain of these bacteria particularly the lactobacilli, (primarily L. acidophilus and L. casei) in relation to human health and nutrition. An additional culture receiving attention is Bifidobacterium bifidus, formerly named L. bifidus (2, 30). This organism is considerably different than the lactobacilli normally encountered in dairy fermentations in that it is anaerobic and heterofermentative. They also form branched cells when grown in certain media (34). In studies involving any of these bacteria it is very important that a number of desirable characteristics be considered. It is anticipated that in the future, much additional research will be done concerning the role of cultured or culture containing milk products in nutrition and health.

REFERENCES

1. Breslaw, E. S. and D. H. Kleyn. 1973. J. Food Sci. 38:1016.
2. Brown, C. D. and P. M. Townsley. 1970. Can. Inst. Food Technol. J. 3:121.
3. Buchanan, R. E. and N. E. Gibbons. eds. 1974. Bergey's Manual of Determinative Bacteriology. 8th ed. Williams and Wilkins, Baltimore, Maryland.
4. Cogan, T. M. 1985. p 25. in Gilliland, S. E. (ed). Bacterial Starter Cultures for Foods. CRC Press, Inc., Boca Raton, Florida
5. Deibel, R. H. and H. W. Seeley. 1974. p. 490. in Buchanan, R. E. and N. E. Gibbons (eds) Bergey's Manual of Determinative Bacteriology. 8th ed. Williams and Wilkins, Baltimore, Maryland.
6. Esser, P., C. Lund, and J. Clemmesen. 1984. Milchwissen. 38:257.
7. Garvie, E. I. 1974. p 510.. in Buchanan, R. E. and N. E. Gibbons (eds) Bergey's Manual of Determinative Bacteriology. 8th ed. Williams and Wilkins. Baltimore, Maryland.
8. Garvie, E. I. 1974. Int. J. Syst. Bacteriol. 26:116.

9. Garvie, E. I. 1984. p 35. in Davies, F. L. and B. a. Law (eds). Advances in the Microbiology and Biochemistry of Cheese and Fermented Milk. Elsevier Applied Science Publishers, Ltd., London, England.
10. Garvie, E. I. and J. A. E. Farrow. 1982. Int. J. Syst. Bacteriol. 32:453.
11. Gasser, F. 1970. J. Gen. Microbiol. 62:223.
12. Gasson, M. J. and F. L. Davies. 1984. p 99. in Davies, F. L. and B. A. Law (eds). Advances in the Microbiology and Biochemistry of Cheese and Fermented Milk. Elsevier Applied Science Publishers Ltd., London, England.
3. Gilliland, S. E. 1985. p. 175. in Gilliland, S. E. (ed). Bacterial Starter Cultures for Foods. CRC Press, Inc., Boca Raton, Florida.
14. Gilliland, S. E. 1985. p 41. in Gilliland, S. E. (ed). Bacterial Starter Cultures for Foods. CRC Press, Inc., Boca Raton, Florida.
15. Gilliland. S. E., E. D. Anna, and M. L. Speck. 1970. Appl. Microbiol. 19:890.
16. Gilliland, S. E. and H. S. Kim. 1984. J. Dairy Sci. 67:1.
17. Gilliland, S. E., C. R. Nelson, and C. Maxwell. 1985. Appl. Environ. Microbiol. 49:377.
18. Gilliland, S. E. and J. W. Nielsen. 1986. Abs. Annual Meeting of American Soc. Microbiol. Abs. No. P-16, p 278.
19. Gilliland, S. E., M. L. Speck, and J. R. Woodard, Jr. 1972. Appl. Microbiol. 23:21.
20. Goldin, B. R. and S. L. Gorbach. 1984. Am. J. Clin. Nutr. 39:756.
21. Grunewald, K. K. 1982. J. Food Sci. 47:2078.
22. Hargrove, R. E. and J. Alford. 1978. J. Dairy Sci. 61:11.
23. Hontebeyrie, M. and F. Gasser. 1977. Int. J. Syst. Bacteriol. 27:9.
24. Johnson, J. L., C. F. Phelps, C. S. Cummins, J. London, and F. Gasser. 1980. Int. J. Syst. Bacteriol. 30:53.
25. Kempler, G. M. and L. L. McKay. 1981. J. Dairy Sci. 64:1527.
26. Kim, H. S. and S. E. Gilliland. 1983. J. Dairy Sci. 66:959.
27. Klaenhammer, T. R. and R. B. Sanozky. 1985. J. Gen. Microbiol. 131;1531.
28. Kolars, J. C., M. D. Levitt, M. Aouji, and D. A. Savaiano. 1984. New Eng. J. Med. 310:1.
29. Mann, G. V. and A. Spoerry. 1974. Am. J. Clin. Nutr. 27:464.
30. Marshall, V. M., W. M. Cole, and L. A. Mabbitt. 1982. J. Society Dairy Technol. 35:143.
31. Marshall, V. M. E., and B. A. Law. 1984. p 67. in Davies, F. L. and B. A. Law (eds). Advances in the Microbiology and Biochemistry of Cheese and Fermented Milk. Elsevier Applied Science Publishers Ltd., London, England.
32. McKay, L. L. 1985. p. 159. in Gilliland, S. E. (ed). Bacterial Starter Cultures for Foods. CRC Press, Inc., Boca Raton, Florida.
33. Mitsuoka, T. 1969. Zentralblatt fur Bakteriol. Parasit. Infektion. und Hygiene (Abt. I Orig) 210:32.
34. Paupard, J. A., I. Husain, and R. F. Norris. 1973. Bacteriol. Rev. 37:136.

35. Rogosa, M. 1974. p 576. in Buchanan, R. E. and N. E. Gibbons (eds). Bergey's Manual of Determinative Bacteriology. Williams and Wilkins. Baltimore, Maryland.
36. Sandine, W. E. 1985. p 5. in Gilliland, S. E. (ed) Bacterial Starter Cultures for Foods. CRC Press, Inc., Boca Raton, Florida.
37. Shahani, K. M., B. A. Friend, and P. J. Bailey. 1983. J. Food Protection. 46:385.
38. Sharpe, M. E. 1981. p 1653. in Starr, M. P., H. Stolp, H. G. Truper, A. Balows, and H. G. Schlegel (eds). The Prokaryotes. A Handbook on Habitats, Isolation, and Identification of Bacteria. Vol II. Springer-Verlag. New York.
39. Shimohashi, H. and M. Mutai. 1977. J. Gen. Microbiol. 103:337.
40. Stadhouders, J. 1974. Milchwissenschaft. 29:329.

Table 1. Bacterial species most often involved in fermented milk products.

Lactobacillus acidophilus	Leuconostoc dextranicum
Lactobacillus bulgaricus	Leuconostoc lactis
Lactobacillus casei	
Lactobacillus helveticus	Streptococcus cremoris
Lactobacillus lactis	Streptococcus lactis
	Streptococcus lactis subsp. diacetylactis
Leuconostoc cremoris	Streptococcus thermophilus

INTERRELATIONS BETWEEN STRAINS IN STARTERS FOR CULTURED DAIRY PRODUCTS

V.F.Semenikhina
All-Union Dairy Research Institute (VNIMI)
Moscow
USSR

ABSTRACT. In the USSR a great amount of cultured dairy products is manufactured. For their production only multistrained starters are used. Studies on the interrelations between strains in starters allow to develop completely new ones with improved properties. For manufacture of fermented milks the next combinations of cultures are used: mesophilic lactic streptococci; mesophilic and thermophilic lactic streptococci; thermophilic lactic streptococci and lactobacilli. The microflora of these starters is selected by steps with strain combinations control of acid producing ability and organoleptic characteristics in pasteurized milk; strain combinations selected are simultaneously cultivated. Instead of individual strains cultivation it would be applied in further works the cultivation of starters. They are periodically tested on their basic properties which are of interest to fermented milks production, those being on the level or even better as compared to properties of strains composing these starters.

In the USSR a wide range of cultured dairy products is manufactured, their production being constantly increased. Among them large quantities account for kefir, curds, cultured cream, ryajenka, varenez. Such a wide popularity of these products in the USSR may be explained by their high dietetic and therapeutic properties.
   The peculiarity of cultured dairy products as compared to all other food products is that they contain a great number of viable cells, mainly lactic acid bacteria, - up to billion per gram, as well as valuable nutritional substances. These bacteria and their metabolites determine the individual properties of one or another product. Cultured dairy products possess a high antibiotic activity towards undesirable intestine microflora. Protein in

these products acquires a finely flocculent form. As a result of this, its digestibility is increased and transformation of casein to paracasein decreases its allergic action. Therefore, in the USSR cultured dairy products are successfully used for infant nutrition /4,10/.

For the last years low fat products gained a wide popularity thanks to their high tasteful qualities and physicians' recommendations. Moreover, in the USSR a large amount of national cultured dairy products is manufactured.

Studies on the interrelations between cultures composing starters allow to improve properties of traditional products and to develop completely new starters with improved characteristics intended for developing new cultured dairy products.

Starters used should ensure preparing the products with required consuming properties as well as guarantee technological processes to be stabilized.

In our country for manufacture of the traditional dairy products (curds, cultured cream, etc.) starters from mesophilic lactic streptococci are preferred. This microorganism group is the most sensitive to unfavourable conditions. Therefore, for developing starters consisting of mesophilic lactic streptococci the last ones are selected according to their resistance to polyvalent phages, their rotation in starters, strains with homogeneous cell population being selected /1/.

Starters from thermophilic lactic streptococci and bacilli are not widely used in our country for yogurt production as in West Europa. In our country this type of starters is used for the manufacture of matsoni, ryajenka, prostokvasha metchnikovskaya, etc.

Starters consisting of thermophilic and mesophilic lactic streptococci are of certain interest. Owing to thermophilic streptococci presence they are hardly attacked by bacteriophage and are more resistant towards seasonal changes of milk quality. When developing together with mesophilic lactic streptococci, thermophilic lactic streptococci receive the possibility to multiply at much lower temperatures than optimal ones for this species. These starters are resistant to bacteriophages, produce coagulum quicker than individual cultures composing them and ensure moderate acidity. Using these starters one can receive coagulum with a good water holding capacity. Therefore, these starters are used for cultured dairy products manufacture by stirred method as well as for production of cultured cream and curds. In our country only multistrained starters are used for cultured dairy products manufacture.

When developing multistrained starters it is necessary to achieve a stable equilibrium of strains for ensuring the constant quality of starters and products. Only in this case a complete advantage of multistrain starters is realized.

Multistrain starters in which a strain equilibrium is achieved as a result of a great number of passages are selected.

It should be taken into account that the nature of strain interrelations is more dependent on medium composition on which microorganisms are incubated than on the correlation between microorganisms and other factors.

It is rather difficult to determine interrelations between microorganisms in such a composite medium as milk. In different conditions one and the same microorganism may influence another one as a stimulator or as a inhibitor.

Therefore, the correct choice of methods for studying interrelations between microorganisms is of great importance /2,3/.

To determine the interrelations nature dense nutrient media are used (agar block method, bacto-streep method, etc.) for estimating growth inhibition or growth stimulation zones of one microorganism by metabolites of another one.

Cells of microorganisms tested and substances produced by them are concentrated by these methods on dense nutrient media. In such liquid medium as milk cell concentration of microorganisms interrelated will be uniform and considerably lower than on dense nutrient media.

For studying interrelations between microorganisms in starters it is more advisable to use the method of serial dilution of filtrates of microbe culture liquid in liquid nutrient media.

Determination of interrelations between microorganisms composing the combination (starter) is one of the main problems at their selection.

Works connected with selection of microorganisms for multi-species multistrain starters is the most difficult and laborious. Selection of microflora of such starters is carried out by steps with strain combination testing the acid production activity and organoleptic characteristics in pasteurized milk; in combinations selected simultaneous cultivation of strains is conducted.

This method is used for selection of starter microflora for cultured dairy products.

Interrelations between mesophilic streptococci

Starters from mesophilic lactic streptococci are composed of S.lactis, S.cremoris and S.acetoinicus. Because of higher sensitivity of S.cremoris to biologically defective milk starters for curds and cultured cream manufacture are composed only of S.lactis and S.acetoinicus.

Pearce /1970/ showed a possibility of achieving symbiosis consisting of S.lactis and S.cremoris for preparing two-strain starters for cheese; Bannikova L.A. /1966/ showed such a possibility for preparing multi-strain starters for cultured dairy products. In this case acid production increase in combinations up to 30-50 % was achieved as compared to pure cultures. It was supposed that stimulation was conditioned by adenine formation by one of the strains. However, in tests with addition of adenine to lactic streptococci cultures /Pearce, 1970/ acid production was not stimulated. Apparently, this effect might be explained by that between microorganisms the complicated process of metabolism took place.

Antagonistic interrelations between lactic bacteria are caused, apparently, by releasing specific antibiotic substances. The ability of S.lactis to produce the antibiotic substance - nisin - was determined by Whitehead and Cox in 1933 for the first time. Its polypeptide nature was also proved. Nisines show an antibiotic effect on all streptococci, lactobacilli, clostridia, spore-forming bacteria, etc.

Insufficiently careful account of intraspecific antagonism presence in lactic streptococci may lead to transformation of multi-strain starters to single-strain ones.

Therefore, starters from mesophilic lactic streptococci are compiled by several steps.

At first a starter base consisting of 3-4 strains of S.lactis is selected and their combination is tested. Those bases are selected which coagulate milk at the level of the most active strain or faster and which are possess with good organoleptic characteristics.

Then S.cremoris is added to this base and prepared mix is tested according to combination and organoleptic characteristics.

The combination of S.acetoinicus (aroma producing streptococci) is composed as a base. This combination consists of 2-3 strains. Strains combination is tested by milk coagulation period of time by each individual strain and by organoleptic properties.

Those combinations are selected which coagulate milk at the level of the most active strain or faster.

The final stage is combination of the base with the aroma producing streptococci selected.

Strains combination composing starters are tested according to pasteurized milk coagulation time as compared to its coagulation time by the base and according to organoleptic characteristics.

Starters selected are tested for the absence of antagonistic strains and for their stability. For this purpose starters subculture on sterilized skim milk (incubation temperature + 25°C; storage temperature 3-5°C) is carried out in 15-20 days. Afterwards starters are tested on pasteurized milk noting coagulation time, organoleptic properties, diacetyl and acetoin formation, carbon dioxide and volatile fatty acids production; starters retaining their original properties are selected.

Studies on the starter microflora selection showed that strains with similar acid production capacity were combined better /1/.

According to the method described starters coagulating pasteurized milk in 4,0 hours possessed good organoleptic characteristics, accumulated a great amount of aromatic compounds and volatile acids. These starters consisting of S.lactis and S.acetoinicus appeared to be stable and were selected.

Interrelations between lactic-acid bacteria with different temperature optimum of growth

In view of the fact that thermophilic streptococcus and lactic-acid streptococci are more stable and less exposed to bacteriophage as compared to mesophilic lactic-acid streptococci, the development of starters consisting of these cultures combinations is of great interest.

Studies carried out on the simultaneous cultivation of lactic-acid bacteria with different temperature optimum of growth showed that their physiological properties could be changed. Thermophilic lactic-acid streptococci acquire an ability to grow rather well in the presence of mesophilic ones at temperatures which are considerably lower than temperature optimum of their growth /7,8,9/.

Thermophilic lactic-acid streptococci are growing rather well at 30°C in combined culture with mesophilic streptococci. Only occasionally individual mesophilic streptococci displace the thermophilic ones from the combined culture. Therefore, before composing combinations it is necessary to test interrelations between individual cultures of thermophilic and mesophilic streptococci.

Combinations of thermophilic streptococci with S. lactis at 30°C coagulated milk 1 hr quicker than pure cultures of mesophilic streptococci; at 40°C 1,5-2 hr quicker than pure cultures of thermophilic streptococci.

Combinations of thermophilic streptococci with S. lactis var.acetoinicus at 30°C coagulated milk considerably quicker than pure cultures of aroma producing streptococci; at 40°C they coagulated milk for such a period of time as thermophilic streptococci.

Therefore, the use of thermophilic and mesophilic lactic-acid streptococci allowed to develop starters with improved properties. It was determined /6,7/ that thermophilic lactic-acid streptococcus ensured preparing rather viscous coagulum restoring its structure after mechanical action and lactic-acid streptococci increase water holding capacity of coagulum.

The influence of different cultivation temperatures on starters properties was studied. Cultivating these starters at 37°C (as compared to 32°C) leads to decreasing coagulum viscosity and to increasing the destruction rate. Moreover, coagulum ability to syneresis is increased. It may be explained by that at higher temperature more favorable conditions are created for thermophilic streptococcus growth. The quantity of mesophilic lactic-acid streptococcus that influenced on the coagulum density is decreased and water holding capacity of milk gel is diminished.

At cultivation temperature of 30°C the coagulum strength was decreased and its reparative capacity became worse. It may be explained by poor growth of thermophilic streptococcus at 30°C. The structure of this type was characterized by its low reological properties and low ability to repairing.

At cultivation temperature of 32°C the conditions for growth of both thermophilic and mesophilic lactic-acid streptococci were created that ensured the formation of milk gel with the stable structure. The structure of this type was characterized by higher reological properties as well as by repairing and water holding capacity.

While studying the different combinations such as S.thermophilus + S.lactis; S.thermophilus + S.lactis var. acetoinicus; S.thermophilus + S.lactis var. acetoinicus + S.lactis it was determined that the most pronounced combination with above-mentioned properties was noticed in case of S.thermophilus + S.lactis. Moreover it was stated that the viscosity and strengthening properties of acid coagulum formed by thermophilic and mesophilic lactic-acid streptococci depended on their chains length. Strains of lactic-acid bacteria formed chains may be considered as important additional structure elements of acid coagulum that influence a product consistency. Therefore, for preparing cultured dairy products with the required consistency it is possible to carry out the directional selection of strains of lactic-acid streptococci on the nature of bacterial cells location. Such starters are successfully used for the manufacture of low-fat cultured dairy products with different flavourings. The use of these starters for preparing curds and cultured cream allowed to accelerate the acid formation process and to improve the finished product consistency /5/.

Interrelations between thermophilic lactic-acid streptococci and lactobacilli

Starters consisting of thermophilic lactic-acid streptococcus and L.bulgaricus are widely used for the manufacture of yogurt, Bulgarian sour milk, etc. /2,II/.

Works on the development of symbiotic starters consisting of above-mentioned microorganisms are carried out in Bulgaria as well as in other countries.

Studies on the great number of combinations of thermophilic streptococcus and L.bulgaricus showed that there was no clear correlation between acid-producing capacity of pure cultures and their combinations. Apparently it may be explained by not the total result of acid accumulation of two simultaneously developing microorganisms but by more fine mechanism of their mutual action on the nature of multiplication and vital activity. It was determined that filtrates of culture liquid of some strains of thermophilic streptococcus had stimulating influence on L.bulgaricus while others acted as inhibitors.

When L.bulgaricus is combined with inhibiting its growth streptococci, milk coagulation process is slowed down.

Therefore, it is advisable to select strains of L. bulgaricus and thermophilic streptococcus preliminary, the number of bacilli in combined culture with streptococcus being determined in 3 hr after inoculation (their number should not be less 2-3 in microscopic field) and milk coagulation time being measured (2 hr 30 min - 3 hr 20 min). In this case thermophilic streptococcus filtrates must stimulate the growth of L.bulgaricus in dilution range from 1:4 to 1:16. Afterwards symbiotic combinations are tested on acetaldehyde formation. Its quantity should not be less than that one formed by pure culture of L. bulgaricus including into this starter. Organoleptic assessment of such starter showed it had a clearly pronounced aroma. After that combinations are tested on their antibiotic activity. If starter filtrate inhibits the growth of E.coli in dilution of 1:2 and slows down its development in dilution of 1:16, this starter posesses higher antibiotic activity.

For developing symbiotic starters studies on the ability of L.bulgaricus and S.thermophilus to long keeping in culture at re-inoculation are of great importance. The method for studying this ability was suggested by Bulgarian investigators /Nikolov, 1966/.

The main point of this method is that combined cultures of L.bulgaricus and thermophilic streptococcus are daily re-inoculated in milk for 15 days followed by testing this starter microscopic pattern. Nowadays more durable cultivation (up to 6 months) of combinations of L. bulgaricus and thermophilic streptococcus is carried out in Bulgaria. As a result of this the starters which are stable in production conditions at different seasons were developed.

Starter symbiosis stability is tested by daily re-inoculation into sterile skim milk for 15 days (1 % of starter at 43°C). Starter is considered to be stable and symbiotic if it coagulates milk during 15 re-inoculations for 2,5-3,5 hr and if from 5 to 15 rods are visible in microscopic field with favourable growth of streptococcus.

If necessary, the proportion of thermophilic streptococci and L.bulgaricus may be controlled by the volume of starter added and by holding it at room temperature before cooling.

If lacbacilli prevail in the starter, its quantity may be decreased to 0,5 % and immediately after coagulum formation it is cooled.

If streptococci prevail in the starter, its quantity should be increased up to 1,5-2 % and after coagulum formation the starter is held additionally for 1-2 h at room temperature and then cooled.

In further works starters cultivation would be applied instead of individual strains. Starters are tested **periodically** on their production-effective properties which should be at the level or even better than those ones of strains including into these starters.

References

1. Bannikova L.A. "Selection of lactic acid bacteria and their use in dairy industry". "Piscevaia promyshlennost", 1975, 256 s.
2. Koroleva N.S. "Technical microbiology of wholemilk products", "Piscevaia promyshlennost", 1975, 270 s.
3. Koroleva N.S.,Semenihina V.F. "Sanitary microbiology of milk and milk products", "Piscevaia promyshlennost",1980, 250 s.
4. Semenihina V.F. "Development of new dietetic cultured milk products". Trudy conferencii "Novaia technologiia i tehnika izgotovleniia kislomolochnyh produktov",ChSSR, 1982, s 11-16
5. Koroleva N.S.,Bannikova L.A.,Semenihina V.F.,Caregradskaia I.V. "Preparation and usage of starters for quarg and sour cream production". "Express-informaciia celnomolochnoi promyshlennosti", 1982, N 8,s 27-30
6. Semenihina V.F.,Zadoiana S.B.,Bannikova L.A.,Mytnik L.G. "Usage of S.thermophilus in the dairy industry". Tezisy dokladov Vsesoiuznoi konferencii "Primenenie termofilnyh mikroorganizmov v promyshlennosti",Moskva,1983,s 72-73
7. Koroleva N.S.,Lozoveckaia V.T. "Role of bacteria in acidic coagulate structure formation". Trudy VNIMI "Sovershenstvovanie metodov selekcii i prigotovleniia zakvasok v molochnoi promyshlennosti", 1984,s 8-9
8. Koroleva N.S.,Piatnicina I.N.,Lozoveckaia V.T. "Selec-

tion of starters for cultured milk drinks manufacture". "Molochnaia promyshlennost", 1984, N 7, s 19-21
9. Semenihina V.F. "Other fermented dairy products". International symposium "Fermented dairy products", Avignon, France, 1984
10. Semenihina V.F.,Grudzinskaia E.E.,Sedunova N.V. ""Improvement of cultured dairy products biological value". Trudy konferencii "Problemy i puti racionalnogo ispolzovania moloka", 1986, 48 s
11. Bulgarian cultured milk (M.Stefanova Kondratenko, P. Vasilev Gruev, A.Filipov Andreev etc.), Zemizdat, Sofia, 1985, 285 s.
12. Pearch L.E. "Growth stimulatory between lactis streptococci". XVIII Internationa Dairy Congress, Sydney, 1970, V, p. 109

NUTRITIONAL ASPECTS OF FERMENTED MILK PRODUCTS

M.I. Gurr
Nutritional Consultant
Milk Marketing Board
Thames Ditton
Surrey KT7 OEL,
UK

ABSTRACT. Fermented products are a palatable and economical source of a wide range of nutrients. The nutrient composition is similar to that in milk, but concentrations of vitamins are in general a little lower, with the possible exception of folic acid. Concentrations of lactic acid, galactose, free amino acids and fatty acids are increased as a result of the fermentation. Lactose intolerant individuals tolerate lactose when it is consumed in yoghurt better than when it is taken in equivalent quantity in milk. The mechanism of this effect has not been clearly established. Yoghurt organisms do not colonize the gut, but the ingestion of live organisms in cultured products influences the numbers, types and enzymic activities of organisms in the gut in ways that may be beneficial to health.

INTRODUCTION

Because of the breadth of the subject and the limited time available, I intend to be quite selective. Much of my own work and indeed research in Western Europe generally has been on yoghurt and most of my remarks will relate to this product, although I shall make reference to other cultured products where appropriate. As to subject matter, I will give a very brief account of the nutritive value of fermented milks compared with natural milk and then concentrate on two main aspects: the digestion of lactose and the possible benefits of cultured products for lactose intolerant individuals and finally the influence of ingested cultured products on the gastrointestinal microflora and the possible health benefits this may confer.

NUTRITIVE VALUE

The products I am going to discuss are based on cow's milk and therefore we can expect their nutrient content to be broadly similar to that of the milk from which they were made. The composition will be modified by (1) changes in milk constituents brought about during the fermentation by the action of the microorganisms upon them. (2) the addition of nutrients and other chemical substances supplied by the microorganisms during the fermentation. (3) the presence of the microorganisms themselves and their associated enzymes. (4) materials added in manufacture.

Energy

The chief sources of energy in milk are fat and lactose.

The energy value of yoghurt is very similar to that of the milk from which it is made. However, when the solids-not-fat are increased in the basic yoghurt mix then, on a weight for weight basis, yoghurt may provide the consumer with a higher intake of protein, carbohydrate, calcium and certain B group vitamins than milk (Robinson 1977).

It has frequently been claimed that the fat is more digestible in cultured products than in milk because a certain degree of 'predigestion' has taken place (eg Wasserfall, 1972). Shahani and Chandan (1979) cited the fact that lactic cultures possess lipase activity as evidence that the lipids in cultured products are partially degraded. Their assays for lipase activity, however, were done with cultured organisms as the source of lipase and tributyrin emulsions as substrates and did not prove that bacterial lipases acted on lipids in the cultured products. Other authors have shown that starter bacteria have limited ability to hydrolyse fat (Stadhouders and Veringa, 1973). The presence of free fatty acids in yoghurt (Rasic & Kurmann, 1978) should not, theoretically, aid digestion and absorption of fat and could in fact give products of poor organoleptic properties (Tuckey & Stadhouders, 1968). More recently, Schaafsma (personal communication) could find no significant differences in milk fat digestibility in vivo in rats fed milk, yoghurt or pasteurized yoghurt. It seems unlikely that major claims can be made on the basis of better milk fat digestibility. Most adult human beings digest milk fat extremely efficiently. It remains a possibility that cultured products could be beneficial to individuals who are unable easily to digest fat and it would be worthwhile testing this experimentally.

The trend in many countries is towards the production of low fat yoghurts which are sometimes promoted as beneficial to those seeking a low energy diet. There is no doubt that such products could form a valuable and palatable part of an energy-controlled diet but claims are valid only for products that have not had their energy value increased by the addition of substantial amounts of sucrose as a sweetener.

Lactose and its metabolic products

During the fermentation of milk, the microorganisms generally use lactose as a substrate, converting it into lactic acid. As a result, the lactose concentration in yoghurt is lower than that in unfermented milk, provided that no supplementation with skim milk powder was made during manufacture. The extent to which this is beneficial to those people who are unable to digest lactose will be discussed in a later section of the paper.

Whereas fresh milk contains a negligible quantity of lactic acid, the fermentation process results in the conversion of some lactose into lactic acid (Rasic & Kirmann, 1978). Lactic acid may be beneficial by (a) acting as a preservative for the product (b) contributing a mildly sour and refreshing taste (c) influencing the physical properties of the casein curd to promote digestibility (d) improving the utilization of calcium and other minerals and (e) inhibiting the growth of potentially harmful bacteria in the gut. Its energy value is 15kJ/g compared with 16 kJ/g for lactose. Lactic acid occurs in yoghurt as two isomers: L(+) and D(-) lactic acid. The D(-) iosomer is metabolized only very slowly in man compared with L(+) lactic acid (Kielwein & Daun, 1979) and if taken in excess can lead to metabolic disturbances. The World Health Organization recommended not more than 100 mg D(-) lactic acid per kg body weight should be consumed daily, although Giesecke and Stangassinger (1977) claim that only 60 mg/kg/d can be metabolized by man. A large part of the D(-) lactic acid intake is metabolized by the liver or is excreted in the urine by adults. D(-) lactic acid normally represents about 40-50% of total lactic acid in yoghurt. Only when the diet is extremely unbalanced will the D(-) isomer be increased enough to lead to disturbances and none have so far been reported.

Because of the breakdown of lactose during fermentation, the concentration of galactose is higher in cultured products than in unfermented milk. Galactose is normally absorbed very rapidly from the gut and metabolized to glucose in the tissues. A rare inborn error of metabolism, galactosaemia, in which the patient is unable to convert galactose into glucose, resulting in excessively high blood galactose concentrations is associated among other symptoms, with cataracts (Segal, 1978). Richter and Duke (1970) observed that rats fed an exclusive diet of yoghurt developed cataracts, an effect they attributed to the galactose content of the yoghurt. However, it should be emphasized that the extrapolation of these animal results is invalid because, in the first place, no human beings eat diets consisting entirely of yoghurt and secondly, rats have a limited ability to convert galactose into glucose, whereas apart from rare cases of galactosaemia, human beings have an abundance of the enzyme needed to metabolize galactose. It can certainly be excluded, therefore, that normal consumption of yoghurt will be associated with cataracts as has sometimes been suggested.

Protein and amino acids

The total amino acid content and composition of yoghurt does not differ substantially from that of the milk from which it was made, but the free amino acid content is higher due to proteolytic action of the microorganisms (Shahani & Chandan, 1979; Rasic & Kurmann, 1978, Miller et al, 1964).

The protein quality of milk is already very high and in our laboratory the biological value of yoghurt, as measured by a rat growth assay was not improved significantly above that of milk (Hewitt & Bancroft, 1985). Most healthy human beings digest proteins very

efficiently and it is unlikely that the 'predigestion' of part of the
protein in cultured products and the finer coagulation of the curd
(Renner, 1983) will result in improved digestibility for them.
However, these properties may be helpful for those whose ability to
digest protein is already impaired due to illness and this possibility
should be verified experimentally in double blind trails.

Vitamins

Breed, diet, climate, geographical location, stage of lactation and
other factors can influence the vitamin content of cow's milk (Gregory,
1967; Scott & Bishop, 1986) which in turn will affect the vitamin
content of the cultured product. The amounts of the various vitamins
in the milk base from which the cultured products are made will also be
influenced to different extents by the heat treatment it receives in
the preparative stages of manufacture. More significant will be the
influence of the microbial innoculum (Hartman & Dryden, 1974). While
many lactic bacteria require B vitamins for growth, several cultures
are capable of synthesizing certain vitamins. It is therefore
impossible to quote 'typical' values for vitamin content of cultured
products, although an indication of changes that can occur after heat
treatment, fermentation and storage is given by IDF (1983).
Unfortunately most vitamins are generally present in lower
concentrations in yoghurt than in milk with the exception in some cases
of folates (IDF, 1983; Hewitt & Bancroft, 1985). Deficiency of this
vitamin can occur even in affluent societies and the higher folate
content of yoghurt might be a nutritional advantage, although it would
have to be shown that the folates in question have biological activity
for man.

Minerals

Fermentation has little effect on the mineral content of milk and like
milk, yoghurt is an excellent source of essential minerals,
particularly calcium, phosphorus, magnesium and zinc. The nutritive
value of a food depends not only on its nutrient composition but on the
bioavailability of those nutrients, namely the proportion that can be
absorbed and utilized by the body. Lactose improves the absorption of
calcium and other minerals (Schaafsma, 1983) and it is important to ask
whether the decrease in lactose concentration that occurs during
fermentation is associated with lower mineral bioavailability. In
recent experiments with rats, Schaafsma (personal communication) showed
that reduction in lactose concentration either by fermentation or
treatment of milk with lactase resulted in a somewhat lower
bioavailability of calcium and other minerals and a small decrease in
bone mineral content. This is consistent with other results in rats
(Dupuis et al, 1985) and lactase deficient human subjects (Smith et al,
1985) showing that calcium from yoghurt is not better utilized than
that from milk. Nevertheless, it should be emphasized that the
differences are small and that the bioavailability of calcium from all
dairy foods is very much higher than calcium from plant sources.

Many cultured products sold on the supermarket shelf today are not simply the fermented equivalent of milk. They may be fortified by the addition of skim milk powders, caseinates, ultrafiltered concentrates, fruit pulp, stabilizers, flavours and colourings, many of which will modify the nutritive value by increasing the concentration of proteins, sugars, polysaccharides and other nutrients (Deeth & Tamime, 1981).

LACTOSE DIGESTION AND LACTOSE INTOLERANCE

To be efficiently absorbed from the gut, lactose must be digested into its constituent sugars, glucose and galactose, by the enzyme lactase.

Lactose is a normal constituent of human milk as well as cow's milk and in babies of all races the enzyme lactase is present in the gut to digest the milk lactose. In most of the world's races this enzyme is lost during the first and second decade of life and only peoples of Northern European origin, their overseas descendents and some isolated African and Indian tribes maintain a high intestinal lactase activity throughout life. It is generally believed that this change is genetically programmed and that the amount or activity of the enzyme is not influenced by lactose in the diet. People who have a low activity of intestinal lactase may develop gastrointestinal symptoms upon the ingestion of lactose, which may include diarrhoea, flatulence and abdominal pain caused by bacterial fermentation of undigested lactose in the colon and the resultant generation of gases. Different individuals tolerate different amounts of lactose: many in whom lactase activity is low or absent are quite able to tolerate modest amounts of dairy products taken as components of a regular mixed diet; others become quite ill with quite small amounts of milk products.

Anecdotal evidence suggests that yoghurt is better tolerated than milk by lactase deficient people and if this were true then many would find it unnecessary to reject dairy products which are the source of many essential nutrients. The advantage of yoghurt has been attributed either to its lower lactose content, or the lactase activity of Lactobacillus bulgaricus and Streptococcus thermophilus which survive passage through the stomach and might contribute to lactose digestion in the small intestine (Kilara & Shahani, 1976). As explained earlier, not all yoghurts have a lower lactose concentration than milk, so caution must be exercised in recommending cultured products as 'low lactose products'.

Recently, several reports on the tolerance of yoghurt compared with milk by lactase deficient human subjects, have provided more rigorous scientific evidence for the benefits of the cultured product (Gilliland & Kim, 1984; Kolars et al, 1984; Savaiano et al, 1984). For example, Kolars et al (1984) provided 10 healthy lactose intolerant subjects the following test meals: lactose (20 g in 400 ml water); milk (400 ml containing 18 g lactose); yoghurt (440 g containing 18 g lactose or 270 g containing 11 g lactose). Although the larger amount of yoghurt matched the amount of lactose provided by 400 ml milk, the smaller amount of yoghurt was designed to provide about the same fat and protein load as 400 ml of milk. Samples of the subjects' breath were taken every hour for eight hours after consuming the test meal and the concentration of hydrogen in the breath gases was measured. The principle of this technique is that hydrogen is produced by fermentation of unabsorbed carbohydrate substrates (eg lactose) reaching the colon. A proportion of hydrogen is absorbed and is excreted in the breath, the rate of breath hydrogen excretion being roughly proportional to the amount of carbohydrate reaching the colon. Thus when lactose is readily digested and absorbed in the small intestine, little goes on to reach the colon and little hydrogen is produced. Therefore, individuals producing large amounts of hydrogen are those who have digested lactose very poorly; low concentrations of hydrogen indicate good digestion.

As indicated in Figure 1, the amount of hydrogen expelled after ingestion of yoghurt by these 10 lactase deficient subjects was only one-third of that expelled after taking milk despite the fact that the lactose content was the same. Diarrhoea or flatulence was reported by 80% of those drinking milk but only by 20% of those eating yoghurt. The authors measured the lactase activity in samples of duodenal juice aspirated from three of the subjects. Significant activity was measured 20 minutes after ingestion of yoghurt. In two patients this activity then returned to its former value over the next 40 to 60 minutes whereas in the third patient the level of activity continued to rise for one hour and then declined. The authors concluded that the microbial lactase activity in the ingested yoghurt was responsible for the improved digestibility of lactose from this food.

Although this study is one of the most convincing published to date, it has two major drawbacks. The number of subjects studied was extremely small. Ten subjects only were studied but in only 3 was lactase activity measured directly and in two of these three the activity was short lived: within an hour it had returned nearly to baseline levels. The most important criticism was that it was not conducted in a double-blind fashion, in which neither subjects nor experimenters knew which diet they were receiving. It cannot therefore be excluded that there was a 'placebo effect' such that patients were expecting to react better to yoghurt and actually did so.

Savaiano et al (1984) and Gilliland and Kim (1984) demonstrated that the consumption of heated yoghurt, by lactase deficient subjects resulted in the production of more breath hydrogen than from a meal of unheated yoghurt, indicating greater digestibility of lactose from unheated than heated yoghurt. The bacterial lactase activity was much lower in the heated product and the results were interpreted as indicating digestion of lactose by the bacterial enzyme. The hydrolytic activity was higher in the presence of bile salts (Gilliland and Kim, 1984) perhaps as a result of disruption of the bacterial cells and a release of enzyme activity.

Other recent studies throw some doubt on whether the lactase activity of yoghurt bacteria contributes to lactose digestion in the gut. Because of the difficulty of making direct measurements of lactose digestion activity in human beings, many research workers have turned to the adult laboratory rat as a model for human lactose intolerance, since the young rat quickly loses its intestinal lactase

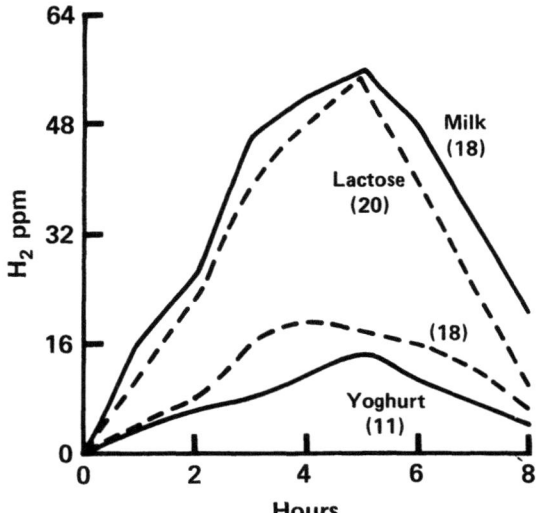

Figure 1

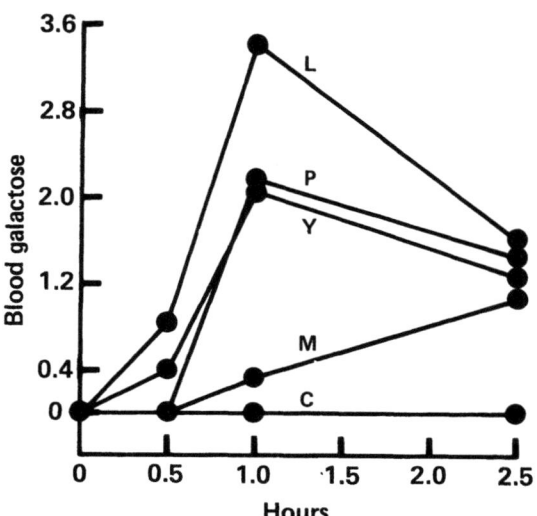

Figure 2

after weaning on to solid food. Schaafsma et al (unpublished results: personal communication) fed five groups of rats, for six weeks, diets based on milk (M, n = 12), lactase-treated milk (L, n = 12), yoghurt (Y, n = 12), pasteurized yoghurt (P, n = 12) or a commercial rat diet (C, n = 6). Since galactose released from lactose by digestion in the small intestine is rapidly absorbed, and is present to a limited extent in foods other than those based on milk, its appearance in the blood following a meal that contains lactose can provide a measure of lactose digestion. Schaafsma and colleagues found that, using blood galactose appearance as a measure of lactose digestion the lactase treated milk demonstrated the highest digestibility, while the lactose in yoghurt and pasteurized yoghurt also showed significant digestibility. Milk lactose was poorly digested while no galactose was detected in the blood following a meal of commercial diet which did not contain galactose (Figure 2). It should be noted, however, that these blood galactose profiles reflect not only galactose released by digestion of lactose in the gut but also galactose present in the yoghurt and pasteurized yoghurt as a result of the hydrolysis of lactose during the fermentation. It is likely that a large part of the rise in blood galactose concentration can be accounted for by absorption of this free galactose in the diet. The release of some galactose from unfermented milk which does not contain free galactose (see Figure 2) provides evidence that part of the blood galactose appearance is due to hydrolytic activity in the gut.

They demonstrated that pasteurization had inactivated the bacterial lactase which was present in abundance in the unpasteurized yoghurt. Since apparent lactose digestibility was no less from pasteurized yoghurt than from non-pasteurized yoghurt, we can rule out the possibility that bacterial lactase in the cultured product contributes significantly to lactose digestion in the gut.

Garvie and her colleagues (Garvie et al, 1984) fed rats a yoghurt diet and compared the activities of enzymes which hydrolyse lactose in the gut contents and the mucosal cells with the activities in control groups fed the base milk from which the yoghurt was made or on a standard rat chow diet. There were no changes in lactase activity in the gut mucosa suggesting that the yoghurt in the diet does not stimulate the animal's inate ability to hydrolyse lactose, although there were large increases in lactose-hydrolysing activity in the gut contents due to the presence of the bacterial enzyme (Figure 3). This is consistent with the general opinion that dietary lactose does not induce lactase activity in the gut mucosa (eg see De Groot & Hoogendoorn, 1957). However, the literature is not clear on this matter since Besnier et al (1983) found that mucosal lactase of mice could be stimulated by yoghurt feeding. There is also some evidence that intestinal lactase can be stimulated in man (Buts et al 1986).

In summary, it is not yet possible to offer a satisfactory explanation of why lactose digestion (as measured by appearance of lactose in blood or decreased concentrations of breath H_2) appears to be more efficient when the sugar is ingested as yoghurt than in the form of unfermented milk. Despite some discrepancies in the literature, the results presented here do not substantiate the

YOGHURT FEEDING AND LACTOSE-HYDROLYSING ACTIVITY — RAT

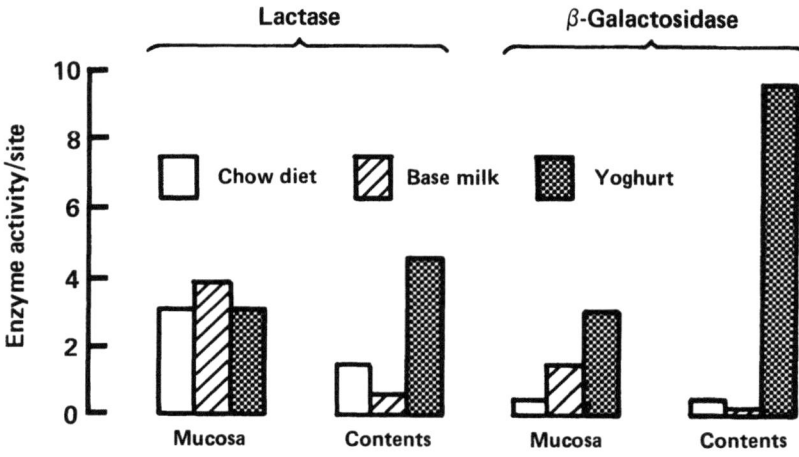

Figure 3

Table 1 : Survival of yoghurt bacteria in the gut (Garvie et al 1984)

	\multicolumn{2}{c}{small intestine}	\multicolumn{2}{c}{large intestine}
	Lacto-bacilli	Strepto-cocci	Lacto-bacilli	Strepto-cocci
Continuous yoghurt from 7d	8.0	7.2	6.1	7.5
Yoghurt discontinued for (3 hours	5.6	5.0	6.9	8.1
(24 hours	< 3	3.8	< 3	6.8
Chow (7d): dose of yoghurt at start	< 2	< 2	< 2	< 2

Table 2 : Comparison of the effect of yoghurt and acidified base milk on the viable counts[a] of coliform bacteria and lactobacilli and the pH of gastric contents of 14 day-old pigs (Ratcliffe et al 1986)

		Base milk		Yoghurt		Acidified base milk	
Stomach	Coliforms	7.25	(0.816)	3.16***	(0.320)	4.40**	(1.726)
	Lactobacilli	7.10	(1.216)	7.99*	(0.419)	5.04**	(1.221)
	pH	4.6	(0.67)	4.0*	(0.53)	4.2	(0.42)
Duodenum	Coliforms	6.53	(0.782)	3.96***	(0.994)	4.07***	(1.156)
	Lactobacilli	6.72	(1.001)	8.02**	(0.587)	4.64**	(1.246)
	pH	5.8[b]	(0.32)	5.8[b]	(0.77)	5.7[b]	(0.22)
Colon	Coliforms	9.17	(0.707)	7.69**[c]	(0.997)	8.62	(0.792)
	Lactobacilli	6.86	(1.647)	8.13*[c]	(0.483)	6.65	(1.797)
	pH	7.1[b]	(0.13)	7.2[b]	(0.30)	6.9[b]	(0.14)

[a] $\log_{10}$ colony forming units per g wet weight

Values are means of 10 observations with SD in parentheses except [b] based on observations from 4 pigs only

* P<0.05, ** P<0.01, *** P<0.001 significance of difference from base milk, [c] yoghurt also different from acidified base milk P<0.05, 9 df.

Table 3 : Coliforms and Lactobacilli in the digestive tract of baby rats fed a host-specific fermented milk for the first 3 days of life (Cole and Fuller 1984)

$\log_{10}$ cfu/g wet weight

Age (days)	Site	Coliforms		Lactobacilli	
		Control	Dosed with L5	Control	Dosed with L5
5	Stomach	6.01	3.69*	7.57	6.67
	Small intestine	4.10	2.35*	5.55	5.21
	Colon	5.48	3.97	6.51	6.19
10	Stomach	6.45	5.57	7.14	7.03
	Small intestine	4.67	4.05	5.01	5.63
	Colon	7.46	6.50	6.46	6.59

* P<0.05

hypothesis that the digestion is brought about by the microbial enzyme in the product, or by stimulation of the gut enzyme. Nor can the benefits of yoghurt be attributed necessarily to an intrinsically lower content of lactose in the food. The following hypothesis may be worth pursuing: That cultured products, because of their acidity and the finer dispersion of protein in the stomach, retard the emptying of the gastric contents into the duodenum. Any lactose hydrolysing activity present whether from indigenous or bacterial origin would have longer in contact with the substrate and digestion would be more efficient even when the specific activity of the enzyme was low.

INFLUENCE OF CULTURED PRODUCTS ON THE GUT MICROFLORA AND POSSIBLE THERAPEUTIC BENEFITS

Diet may influence the microflora in a number of ways. Firstly, it might supply substrates for bacterial growth. Secondly, diet may influence various aspects of gut physiology such as acid and bile secretion and motility in ways that preferentially inhibit or encourage certain microbial populations. Thirdly, components of the diet may have anti-bacterial properties. Fourthly, organisms introduced into the diet in fermented products may exert an influence on the existing ecology encouraging beneficial populations to flourish and suppressing harmful ones. They may prevent colonization of a habitat by competing more effectively than an invading strain for essential nutrients or sites of adhesion or making the local environment unfavourable for the growth of the invader by producing antibacterial substances. It is the last two aspects that we are mainly concerned with in considering the effect of cultured products on the gut microflora.

Survival of ingested organisms from cultured products in the gut

In making the proposition that the ingestion of live organisms in cultured products may enhance the beneficial effects of the indigenous flora, it has been implied that they must be capable of surviving the acid conditions of the stomach and passing down the gastrointestinal tract in large numbers and filling an ecological niche. Although several authors have reported that many lactic acid bacteria do survive passage through the gut, others have not found sustained counts of these organisms after feeding foods containing live bacteria (International Dairy Federation 1983). The author and his colleagues have used gnotobiotic rats to try to answer this question and fed them yoghurt in several ways (Table 1). Germ-free rats on a standard chow diet did not become colonized with Lactobacillus bulgaricus or Streptococcus thermophilus after feeding a small amount of yoghurt (Garvie et al 1984). When yoghurt was fed continuously the yoghurt bacteria survived in the gut in large numbers but when feeding of the cultured product was discontinued, and the rats were reintroduced to the standard chow diet, yoghurt bacteria were still detectable throughout the gut after 3 hours but thereafter the numbers declined (Garvie et al 1984). Thus yoghurt bacteria survive while feeding is continued, but do not colonize and are rapidly eliminated.

Influence of ingested organisms from cultured products on the balance of the gut microflora

Eight two-day old piglets were fed yoghurt and compared with eight piglets given the base milk from which the yoghurt was made (Ratcliffe et al 1986). Because the piglets tended to prefer the yoghurt, it was necessary to pair-feed them to equalize food intakes. Yoghurt had a strong depressing effect on the coliform bacteria in the stomach and duodenum and a weaker effect in the colon (Table 2). The lactobacillus count was increased throughout the gut. In other experiments the coliform depression in the colon and the lactobacillus exhancement in the duodenum failed to reach significance and this illustrates a major problem of research on the intestinal microflora that there may be huge between-animal variations even in inbred strains of animals. The pH value was reduced in the stomach but not in the small intestine or colon. To try to distinguish between the effects of live bacteria and the associated acid content of the cultured product animals were fed the base milk acidified with lactic acid to the pH of yoghurt. Coliforms were suppressed but the viable counts of coliforms in these pigs were always slightly higher than those in yoghurt fed pigs. The count of lactobacilli was decreased in the stomach and duodenum by acidified base milk. It seems that the bacteriological changes in the stomach and duodenum are due, at least in part, to lactic acid but the changes in the colon are not and may be due to upgrowth of indigenous lactobacilli and their inhibition of E.coli.

Techniques to improve the chances of survival of ingested bacteria and their impact on the indigenous flora

Ingested microorganisms are likely to have a more beneficial effect if, as well as surviving passage through the stomach, they can adhere to epithelial surfaces and colonize the gut. Organisms most likely to do this are those that are indigenous to the species that is consuming the cultured product. Therefore we have developed a range of cultured products in which the milk has been fermented with "host-specific" organisms (Marshall et al 1982a, 1982b; Gurr et al 1984). It was our hypothesis that the organisms in these products, when fed to the appropriate animal, will be more able to colonize the gut and will therefore be more successful in aiding indigenous organisms to combat invading pathogens. Baby rats were dosed with 0.1 ml of either water or with milk fermented with Lactobacillus salivarius L5 isolated from rat gut and which was shown to adhere to stomach squamous cells and small intestinal brush border. They were dosed once each day for the first three days after birth (Cole and Fuller, 1984). At five days, the coliform count was reduced in all regions of the gut and the differences were significant in the stomach and small intestine. After 10 days, coliform counts were still lower in the treated animals but not significantly so (Table 3). The treatment did not have a significant effect on lactobacilli in the gut. Although these experiments do not present conclusive evidence for colonization of L5 in the gut of these young animals, it seems unlikely that the significant depression of E.coli would have occurred without colonization by the L5 strain. The loss of effect by 10 days of age may be due to changes in the naturally acquired indigenous microflora and the establishment of strains that compete with and eliminate L5.

In the experiment with piglets described earlier, feeding with a product fermented with a strain of Lactobacillus reuteri isolated from pig intestinal tract, produced effects on the microflora that were not different from these obtained by feeding conventional yoghurt. The L.reuteri strain attached to stomach squamous cells and was potentially a good colonizer of the intestine. The lack of effect may be related to the age of the piglets at dosing. After two days on the sow, the piglets had already acquired an indigenous lactobacillus flora (Barrow et al 1977) which is difficult to displace. Dosing at birth may be more effective as was demonstrated with rats.

These experiments do not prove that simply because the numbers of E.coli are reduced in perfectly healthy animals, there will be an improvement in health in animals or human beings that are suffering from an E.coli infection. The beneficial effects of cultured products on gastrointestinal ailments, although often quoted, still lack rigorous proof (IDF, 1983). More research is needed, with statistically validated double blind trials, preferably using host-specific strains of organisms.

Influence of ingested organisms from cultured products on the metabolic activities of the gut flora

Human diets supplemented with Lactobacillus acidophilus significantly depressed the activity of β-glucuronidase (Goldin et al 1980), an enzyme that may have significance in relation to the incidence of colon tumours (Reddy et al 1973). It has been shown that feeding yoghurt decreased the β-glucuronidase activity in the chicken caecum (Cole et al 1984). More recently, Fuller and his colleagues have developed a model to study dietary effects on human flora which involves associating germ-free rats with a human intestinal microflora. After weaning at three weeks, the rats were fed a control rat chow diet or a diet incorporating 35% by weight of butter. In some experiments yoghurt was included in the drinking water at a concentration of 40% (v/v). In one experiment butter induced a 2.5 fold increase in β-glucuronidase activity which was reduced by 50% by feeding yoghurt. These results were highly significant but have not been reproduced in other experiments illustrating yet again the extreme variability in the gut microflora and the difficulties in observing significant dietary effects.

I conclude that the ingestion of diets containing live organisms can have significant effects on the indigenous flora. Interpretation is complicated by the extreme variability of the flora and factors such as the age of the animal and the ability of the ingested strain to colonize the gut must be considered. Effects of the live organisms per se can be distinguished from those of the lactic acid present in the cultured product. Techniques such as the feeding of host-specific strains and the use of models in which germ-free animals are associated with human flora may yield further information.

Many factors contribute to the quality of foods in the human diet: palatability, availability at an affordable price and the provision of a wide range of important nutrients. Cultured dairy products possess all these qualities and in addition, evidence for the therapeutic benefits of the live bacteria they contain is no longer anecdotal but receiving well established scientific support.

REFERENCES

Barrow, P.A., Fuller R. and Newport, M.J. (1977) Infection and Immunity, 18, 586-595.
Besnier, M.O., Bourlioux, P., Fourniat, G., Ducluzeau, R. and Aumaitre, A. (1983) Ann. Microbial., 134A, 219-230.
Buts, J.P., Bernasconi, P., van Crainest, M.P., Meldaque, P. and Demeier, R. (1986) Pediatric Research, 20, 495-498.
Deeth, H.C. and Tamime, A.Y. (1981) J. Food Protection, 44, 78-86.
De Groot, A.P. and Hoogendoorn, P. (1957) Netherlands Milk and Dairy Journal, 11, 290-303.
Dupuis, I., Gambier, G. and Fournier, P. (1985) Sciences des Aliments, 5, 559-585.
Garvie, E.I., Cole, C.B., Fuller R. and Hewitt, D. (1984) J. Appl. Bacteriol., 56, 237-245.
Giesecke, D. and Stangassinger, M. (1977) Ernahrungsumschaw, 24, 363-364.
Gililand, S.E. and Kim, H.S. (1984) J. Dairy Science, 67, 1-6.
Goldin, B.R., Swenson, L., Dwyer, J., Sexton, M. and Gorbach, S.L. (1980) J. National Cancer Institute, 64, 255-261.
Gregory, M.E. (1967) J. Dairy Res., 34, 169-181.
Gurr, M.I., Marshall, V.M.E. and Fuller R. (1984) IDF Document 179, pp 54-59.
Hartman, A.M. and Dryden, L.P. (1974) In: "Fundamentals of dairy chemistry" 2nd Ed. pp 325-401. Eds B.H. Webb, A.H. Johnson & J.A. Arnold.
Hewitt, D. and Bancroft, H.J. (1985) J. Dairy Res., 53, 197-207.
International Dairy Federation (1983) Document 159, Cultured dairy foods in human nutrition.
Kielwein, G. and Daun, U. (1979) Deut. Molkerei Ztg., 101, 290-293.
Kilara, A. and Shahani, K.M. (1976) J. Dairy Science, 59, 2031-2035.
Kolars, J.C., Levitt, M.D., Aouji M. and Savainano, D.A. (1984) New Engl. J. Med., 310, 1-3.
Marshall, V.M.E., Cole, W.M. and Mabbitt, L.A. (1982a) J. Soc. Dairy Technol., 35, 143-144.
Marshall, V.M.E., Cole, W.M. and Vega, J.R. (1982b) J. Dairy Res., 49, 665-670.
Miller, I., Martin, H. and Kandler, O. (1964) Milchwissenschaft, 19, 18-25.

Rasic, J.L. and Kirmann, J.A. (1978) Yoghurt: Scientific grounds, technology, manufacture and preparation. Tech. Dairy Publ. House, Copenhagen.

Ratcliffe, B., Cole, C.B., Fuller, R. and Newport M.J. (1986) Food Microbiology, 3 (in press).

Reddy, B.S. and Wynder, E.L. (1973) J. National Cancer Institute, 50, 1437-1442.

Renner, E. (1983) Milk and Dairy Products in Human Nutrition. W-GmbH, Volkswirtschaftlicher Verlag, München.

Richter, C.P. and Duke, J.R. (1970) Science, 168, 1372-1374.

Robinson, R.K. (1977) British Nutrition Foundation Bulletin, 21, 191-197.

Savaiano, D.A., Abou El Anouar, A., Smith, D.E. and Levitt, M.D. (1984) Amer. J. Clin. Nutrition, 40, 1219-1223.

Schaafsma, G. (1983) IDF Document 166, pp 19-32.

Schaafsma, G., Derekx, P., Dekkr, P.R. and de Waard, H. (1986) (Personal communication).

Scott, K.J. and Bishop. D.R. (1986) J. Soc. Dairy Tech., 39, 32-35.

Segal, S. (1978) In: "The metabolic basis of inherited disease". Eds Stanbury, J.B., Wyngaarden, J.B. and Fredrickson, D.S. 4th Ed., McGraw-Hill.

Shahani, K.M. and Chandan, R.C. (1979) J. Dairy Science, 62, 1685-1694.

Smith, T.M., Kolars, J.C., Savaiano, D.A. and Levitt, M.B. (1985) J. Clin. Nutrition, 42, 1197-1200.

Stadhouders, J. and Veringa, H.A. (1973) Netherlands Milk and Dairy Journal, 27, 77.

Tuckey, S.L. and Stadhouders, J. (1968) Netherlands Milk and Dairy Journal, 21, 158-165.

Wasserfall, F. (1972) Ernährungsumschau, 19, 155-158.

ACKNOWLEDGEMENTS

I thank my colleagues C.B. Cole, R. Fuller, D. Hewitt, M.J. Newport, B. Ratcliffe who were associated with the experimental work reported in this paper. I also acknowledge support of this work by the Agricultural and Food Research Council and the European Economic Community. It is a pleasure also to acknowledge fruitful discussion with Dr G. Schaafsma (NIZO, Ede, The Netherlands) and to express my thanks to him for permission to reproduce Figure 2. Dr D. Levitt and his collegues and the editors of The New England Journal of Medicine are also thanked for permission to reproduce Figure 1.

DISCUSSION

G. Schaafsma (the Netherlands) : Is folic acid synthesized by the yoghurt biologically available for humans?
M.I. Gurr (UK) : The folic acid has been determined by microbiological assay. It is still to be shown that the folic acid synthesized by the culture can be utilized by the human being.

B. Bianchi-Sálvadori (Italy) : 1. our research has shown adhesion of *L. bulgaricus* to the intestine wall in mice and multiplication of *L. acidophilus* and *B. bifidus* in infants. Thus it seems not necessary to add these organisms to fermented milks.
2. Can you give more details about the experimental rat model with human flora.
M.I. Gurr (UK) : 1. There are many different strains and types of *L. bulgaricus* . We have never demonstrated that these organisms from yoghurt can adhere to the epithelial surface, but this does not exclude that it could occur with some strain.
2. Dr. Fuller used a rat model as a vehicle to carry a human flora. This was done by infecting a germ-free rat with human faecal flora. This flora remained stable for nearly two years. We do not think that the organisms adhere to the epithelial surface but when studying the effects of diet, the changes are those expected of a human flora.

V.T. Meriläinen (Finland) : Does anyone of you know whether fermented milks might have any unbenefical effect? In our country high sugar, acid yogurts have been critisized to cause caries.
M.I. Gurr (UK) : I do not know of any well documented scientific papers that demonstrate a cariogenic effect of cultured products. However, sucrose is well known to be cariogenic and therefore there could be a problem with products that have high sugar levels.
S.E. Gilliland (USA) : There are some not convincing papers suggesting an adverse effect. Certain cheeses are beneficial in helping to prevent caries.

J. Stadhouders (the Netherlands) : Does *L. acidophilus* when grown in milk produce characteristics compounds other than those produced by the yoghurt bacteria?
S.E. Gilliland (USA) : There is possibility that fermentation products produced by *L. acidophilus* could have an influence. Most of our studies have involved the use of non-fermented milk to which cells of *L. acidophilus* are added. Thus we do not include their fermentation products in the milk. We still observe some beneficial effects.

G. Schaafsma (the Netherlands) : Are you sure that the phenomenon that lactase-deficient individuals can tolerate yoghurt better than milk, is attributable to the lactase activity of the culture and not to other factors like gastric emptying rate?
S.E. Gilliland (USA) : We observed that when the lactase activity of the culture was destroyed by heating, the effect on the tolerance was lost.
G. Schaafsma (the Netherlands) : Can you then explain why lactase deficient people demonstrate lower symptoms of intestinal discomfort and have less hydrogen in their breath after the ingestion of pasteurized yogurt as compared to milk as published recently.
S.E. Gilliland (USA) : I have to make two comments:
1. It is very important that the amount of lactose ingested is controlled, as we have done in our studies.
2. The observation of symptoms of gastro-intestinal discomfort is very influenced by phychological factors.

F.M. Driessen (the Netherlands) : According to US investigations *L. acidiphilus* and *B. bifidus* rapidly decrease during the shelf-life of fermented milks. Can you comment on that?
E. Lipinska (Poland) : In Poland the shelf-life of these products is 48 or 72 hours. There are no problems with a decrease of *L. acidophilus* and *B. bifidus*. These bacteria may still have high viable numbers after cold storage for 2 weeks.

S. Negri (Italy) : 1. What is the difference between antibiotics and bacteriocins?
2. Can you give more information on anticarcinogenic activity of Lactobacilli? Is it a direct or an indirect activity?
S.E. Gilliland (USA) : 1. Antibiotics have a larger spectrum of activity; bacteriocins have a much more restricted spectrum being active against only those strains or species closely related to the producer organism.
2. Some have been reported to be indirect, others direct. More research is needed.

C. Barth (Fed.Rep. Germany) : I would like to make a comment concerning your suggestion that metabolites of lactobacilli might inhibit the host's cholesterol synthesis. When infusing acetate or propionate to the pig large intestine we did not observe any change of the serum cholesterol level. Although we did not measure cholesterol synthesis, I would not give much credit to the mechanism suggested above.

Seminar VII: Fermented Milks

Session 2

Chairman: Prof. Dr. Z. Puhan (Switzerland)
Secretary: Dr. J. Auclair (France)

YOGHURT AND CULTURED BUTTERMILK

V. T. Meriläinen
Valio Finnish Co-operative Dairies' Association
P.O. Box 390
00101 Helsinki
Finland

ABSTRACT. Fermented milks, especially yoghurt, have increased their popularity in almost every country during the latest years. Their organoleptic quality and properties are determined by the starters used in the production, processing techniques and the ingredients added after fermentation in purpose to flavour the product. Production costs of fermented milks are affected by starter handling techniques in the dairy, the treatment of milk in the process, fermentation techniques, the treatment of milk in the product after fermentation and the packaging. All these aspects - their current situation and their possible future development - are discussed in the presentation.

1. INTRODUCTION

Fermented milks, especially yoghurt, have enjoyed increasing interest and popularity in most of the countries where they are actively marketed. They have been the subject of many seminars and symposia in recent years (IDF Bulletin Documents 179/1984), and of numerous good reviews and books (RASIC and KUKMANN 1978, TAMIME and ROBINSON 1983, OBERMAN 1985, BOTTAZZI 1985). Lately their nutritional and therapeutical value has received more and more attention. The dairy industry in many countries is also eager to use these arguments in advertising its products. In addition, fermented milks have other advantages over beverage milk (Table I).

Yoghurt and cultured buttermilk are the best known and studied fermented milks. The present product profile and use of yoghurt differ from those of buttermilk. Yoghurt can be called a product of the future and younger people with a modern image. It is sold in a variety of forms, combined with a variety of other ingredients (Table II), and its production has increased in all countries (Table III). Cultured buttermilk can be called a product of the past with a slightly oldfashioned image. It is traditionally produced mainly in Scandinavia and the northern areas of continental Europe, and it has decreasing consumption.

The difference between yoghurt and buttermilk lies in the starter. which gives each its characteristic flavour, structure and viscosity.

Yoghurt is traditionally produced with thermophilic <u>Streptococcus thermophilus</u> and <u>Lactobacillus bulgaricus</u> mixed starters and for cultured buttermilk multiple mixed strain mesophilic BD or B-starters are used. The selection of starters provides possibilities to improve existing fermented milks and create new types of product. Also the flavouring of both products with sweetened fruit and berry pulp, juices and cereals has become an essential part of their manufacturing process and a subject for active future development. (KAMMERLEHNER 1985). It is interesting that the consumption patterns in different countries depend on flavouring. In the Netherlands most of the yoghurt is purchased unflavoured in large packages, whereas in Switzerland, France, Germany and many Scandinavian countries most is consumed ready-flavoured in small packages.

The organoleptic properties and quality of all fermented products depend not only on their composition (fat, protein, lactose, minerals, total dry solids and added ingredients) but also on the stages of the manufacturing process, such as the mechanical treatment which the raw milk receives, heating, homogenization, fermentation, breaking the coagulum, cooling, packaging, and storage. (KLUPSCH 1986). The manufacturing process is only the hardware of the production, while the starter perhaps is the third and most important biological factor controlling the properties and quality of yoghurt and cultured milk in everyday production. Naturally these three the most important factors are closely linked to each other and from the standpoint of product properties and quality they should be studied as a whole.

2. THE MANUFACTURING PROCESS

The manufacturing process of yoghurt and cultured buttermilk can be divided into six main steps or phases. In each we can mention the salient features of the machinery and equipment available and the operating parameters (Table IV).

2.1. Preparation of the milk

The preparation of the milk involves the standardization of its fat and dry matter content. Also a preliminary addition of sucrose for flavoured, sweetened products can be made at this stage. The adjustment of the dry matter content is very important in yoghurt production because it has a considerable effect on the rheological properties of the finished product. Generally, the dry matter content is increased by 1,5 - 2,5 % in the SNF content. Numerous methods of dry matter standardization are available (Table V). The choice is affected by investment costs, the scale of production, plant running costs and product quality considerations. Also the legislation in some countries determines the methods that can be used. It should be remembered that by starter selection, especially by the use of ropy strains, the need to increase the dry matter content can be reduced or almost totally removed. This matter will be discussed in the section on starters.

In commercial practice the milk for fermented milks manufacture is heated batchwise at 80 - 85°C for 30 min. In many plants, however, milk is heated in a continuous process in a closed system at 90 - 95°C for 5 min or more. The considerably improved hygienic conditions in plants with modern CIP cleaning systems and the natural antimicrobial properties of fermented milks provide grounds for the consideration of an interesting alternative way to improve the waterbinding properties of milk protein, that is cold renneting the milk. The partial cleavage of paracasein, and the resulting reaction with calcium improves the water-binding and allows the use HTST-pasteurization (75°C/20 sec). This leads to a considerable saving of energy costs in the processing. (NORDLUND 1984). Also in a recent study, the application of UHT-treatment (138°C/3-6 sec) in the manufacture of yoghurt has given promising results. (SCHMIDT et al 1985).

In the manufacture of fermented milk the air content of the milk should be as low as possible. The known protocooperation among the yoghurt bacteria requires the oxygen content of milk to be reduced below 4,0 mg/l. (DRIESSEN 1981). In the manufacture of Scandinavian type of cultured buttermilk de-aeration of the milk improves the keeping quality, especially wheying off is reduced. Its effect on viscosity is only slight. (BITSCH and GARDHAGE 1977, LEPORANTA and VÄLIAHO 1985). The de-aeration of milk is generally performed in one-stage evaporators.

The homogenization of the milk is done at a pressure of 20 MPa (200 kg/cm^2) and a temperature of 55°C at least. (GALESLOOT 1958, LEPORANTA and VÄLIAHO 1984). Homogenization at these conditions affects not only the fat but also the casein micelles, restructuring them and improving their water-binding capacity. Homogenization improves the viscosity and the keeping quality of the products.

2.2. Fermentation

Standardized multiple mixed strain starters are used for the fermentation in the manufacture of yoghurt and cultured buttermilk. The milk is cooled to a fermentation temperature of 30 - 43°C for yoghurt and 21 - 24°C for cultured buttermilk. The exact fermentation temperature is matched to the specific properties of the starter in use, because temperature determines the starter growth and thus affects the rheological and flavour properties of the product. The use of lower temperatures in conjunction with ropy starters gives better viscosity. Depending on the fermentation temperature and the scale of production tanks of capacity 5000 - 12000 l are generally used.

Breaking the coagulum as well as the cooling method are very important steps affecting the rheological properties of the final product. Naturally both operations are adjusted in accordance with the type of starter. After cooling the temperature of the fermented milk is generally 18 - 20°C in yoghurt production and 3 - 8°C in cultured buttermilk production. Tubular coolers or plate heat exchangers with 4 - 7 mm between the plates are most often used for the cooling in continuous production lines. If a high viscous product is wanted the temperature after cooling should be higher to allow for cold setting; in this case more efficient cooling in the cold storage is needed to avoid the development of excess acidity. (KESSLER and BÄURLE 1980).

2.3. Production line

The components of the production line such as fermentors, valves, pipelines, fittings and the installations can affect the final quality of fermented milks, especially their rheological properties. STEENBERG (1973) showed that mechanical stress during different phases causes loss of viscosity. The capacities of the pumps and the pipes should be matched, so that large pressure differences are avoided. Pumps in the fermented milks production lines should be positive pumps. The whole installation after the fermentation should be as simple as possible, and also natural hydrostatic pressure should be used.

2.4. Added ingredients

In the manufacture of flavoured products the extra ingredients are usually added after the fermentation stage. Their composition and quality affect the sugar content, flavour, viscosity, and also keeping quality of the final product. The ingredients vary from unsweetened fruit purees to high-sugar fruit preparations (60 - 65°Brix), which may contain added stabilizers and preservatives (sorbic or benzoic acid or their salts). Their flavour is often enchanced with natural, nature identical or fully artifical flavours. Obviously the components of added ingredients play a role not only in flavouring but also, through carry-over, as other functional food additives. For example stabilizers in fruit preparations affect the viscosity and keeping quality. Some common stabilizers of the fruit preparations have marked effect on the mouthfeel. (MERILÄINEN et al 1985).
Today, one major trend in dairy product development is to use as little food additives as possible in fermented milks or avoid additives altogether. In some European markets products are advertised intensively as being free of stabilizers and preservatives. (UNTERHOLZNER 1985). The cost of a flavouring ingredient in relation the prize of the product is high. In this way, future developments in the preparation of fruit ingredients may require more changes also in the processing of flavoured fermented milks. As the scale of production of the dairy grows fruit processing in the dairy plant itself may become cost-effective.

2.5. Packaging

Fermented milks can be packaged in a wide variety of containers, ranging from 1,25 dl plastic cups to 1 l cartons. Glass bottles or cups have increased in popularity in some countries. This is a good example how extra value can be added to a traditional dairy product. The rheological properties of the product can be destroyed if the rate of shear of the packaging machines is high. The latest development in packaging machines is the use of low-velocity piston type filling cylinders. After filling the individual packages are closed with multi-pack covers and delivered to cold storage. Type of package, filling temperature and cooling rate in cold storage must all receive careful consideration

2.6 Storage

Freshly-packed, freshly-stored fermented milk undergoes cooling until attains temperature equilibrium in cold storage. Temperature changes affect important quality criteria: acidity and viscosity. Good viscosity and stability in yoghurt are achieved, when it cooled as much as possible in the cup without excess acidity development. In some cases it can be cooled from even $30^{\circ}C$ down to $6-7^{\circ}C$ during at least 12 hrs. For this purpose cooling must be very rapid to avoid excess acidity, and efficient cooling tunnels must be used. Of course the choice of cooling method and temperature at the different stages of manufacture depend on cost estimates and the desired product quality.

In particular we should remember the most important factors affecting the viscosity of the product; the dry matter content of the milk, the starter, the type of fruit preparation and the cooling method.

3. STARTERS

Starters are the biological factor in the manufacture of yoghurt and cultured buttermilk. Their importance in product properties and quality is at least as great as the hardware, that is the process itself. It is well known that they create the flavour through the carbohydrate metabolism, but we are becoming more and more interested in their role in the determination of rheological properties, and keeping quality.

Nowadays this is the case for both cultured buttermilk and yoghurt alike. In Sweden and in Finland we have high viscous cultured buttermilk products, for example långfil and viili, made with ropy mesophilic starters, which may be in their physical character liquid, semi-liquid or semi-solid. Dutch yoghurt is also well known for its ropy character.

Speaking generally the questions concerning product quality changes in product characteristics which can be affected by starter selection, must not be made, unless we are prepared to accept the risks of unknown consumer reaction.

3.1 Starter technology

In recent years, the preparation of starters in the dairy for manufacture of fermented milks has changed also drastically. Modern frozen or freeze-dried concentrated starters for bulk starter and direct use are replacing the traditional techniques based on maintenance of mother cultures, which make use of liquid or freeze-dried starters from the starter supplier. If a dairy plant wants to produce its own quality products and be independent of external suppliers of starters it can become self-sufficient in the traditional sense. According to a recent study in Germany, surprisingly the modern starter systems for cheese production are no more economical than the traditional systems. (BETZ and KLEIN 1986).

3.2 Starters in the formation flavour

The compounds which contribute to the flavour of yoghurt may be divided into non-volatile acids, volatile acids, carbonyl compounds and miscellanous compounds. (TAMIME and ROBINSON 1983). It has been shown that the typical aroma and flavour of plain yoghurt is directly associated with its carbonyl compounds, mainly acetaldehyde. (ROBINSON et al 1977, TAMIME 1977).

In our investigations it was a surprise to find that yoghurts with the same acetaldehyde level, but different viscosity had different tastes: plain ropy yoghurt was evaluated flavourless and plain yoghurt with normal viscosity (non-ropy) was evaluated as having a normal flavour. (HUTTU-HILTUNEN et al 1986). This demonstrates that not only the content of the flavour component, but also the mouthfeel and the flavour release are important in sensory flavour evaluation of yoghurt. In parallel with acetaldehyde in yoghurt diacetyl is the most important flavour component in cultured buttermilk. It was elucidated by JÖNSSON and PETTERSON (1977) that daicetyl is formed from citrate by non-enzymatic oxidativedecarboxylation of -acetolactate and enzymatically from activated acetaldehyde and acetyl Co-A. Diacetyl is reduced to acetoin by the diacetyl reductase present in lactic acid bacteria or contaminating flora.

The manufacturing process of cultured buttermilk should be such that it stabilizes the diacetyl content by preventing natural reduction for example by as rapid cooling as possible or by promoting the oxidative decarboxylation of α-acetolactate, f.ex. by aeration. This, however, is easier said than done, because this kind of operations affect at same time the viscosity and the resistance to wheying off.

3.3. Starters and viscosity

In the foregoing paragrafes frequent reference has been made to the slimy or ropy character of both yoghurt and cultured buttermilk starters. If one uses such a starter the slime or polysaccharide produced by starter organisms acts a food stabilizer, preventing syneresis and graininess and providing a product with natural thickness.

Yoghurt microstructure studies (KALAB et al 1983) and analysis on slime composition of S. thermophilus, S. cremoris and L.bulgaricus (SCHELLHAASS 1985) have learned us to understand more about the starter bacteria function in rheology of fermented milks. Surface polymers appear to form interconnected nets between the bacteria cells. Some of the polymers become detached from the cell surface and affiliated with the hydrated casein micelles in fermented milk. The monosaccharide composition was found to be galactose and glucose in the ratio of 2:1. NMR-studies suggested that first two glucosyl residues are β-linked and the third is α-linkage. Slime production is strongly favoured by the use of lower fermentation temperatures.

In many studies the mucous polymer has been shown to contain also protein as well. For instance MACURA and TOWNSLEY (1984) found the slime of mesophilic strains to contain 47 % protein concluding that the material is probably glycoprotein. In practice ropy starters, especially the mesophilic ones, show culture instability. In many cases this is due to bacteriophages (SAXELIN et al 1983) like wise slime formation may be coded by unstable plasmids, as recently shown by VEDAMUTHU and NEVILLE (1986).

In the manufacture of fermented milks ropy starters arouse increasing interest, because of their potential for considerable savings in production costs through the use of lower dry matter contents. Our results show that yoghurt of almost the normal organoleptic quality can be produced with 1 % lower dry matter content (reduction from 12,5 % to 11,5 % with 2,5 % fat in milk) with ropy starter than with the traditional starter. (HUTTU-HILTUNEN et al 1986). In many countries the yoghurt type is standardized and consumers are used to it, so that changes in ropiness should be made very carefully.

4. FUTURE TRENDS IN THE MANUFATURE OF YOGHURT AND CULTURED BUTTERMILK

We are living in a changing world and our consumers have an invariable appetite for innovation. Yoghurt and cultured buttermilk are no exception. This means that in the future our traditional products will have to become much more versatile for instance in respect to flavour, texture, shelf-life, different added ingredients, special health effects, special enjoyment etc. Some of our production lines have to be as flexible as possible.

In the manufacture of major bulk products the change will include prefermentation or continuous processing in order to reduce the production costs. These aspects have already been studied in the Netherlands. (DRIESSEN et al 1977). The reduction of dry matter content of yoghurt will be achieved by the more widespread usage of ropy starters. Also it will be possible to produce certain types of fermented milk with single strain starters, as suggested by MARSHALL and MABBIT (1980), MARSHALL et al (1982).

Traditional yoghurt and cultured buttermilk technology form good base on which to add the use of new starter strains in order to modify product profile or to develop totally new product types. Examples are already existing on Lactobacillus acidophilus and bifidobacterium strains. In this way flavour and physical properties can easily be changed. (HUNGER 1985). The dairy industry should also remember that soy milk production is increasing and that yoghurt and cultured buttermilk fermentations can be carried out in soy milk bases. (PINTHONG et al 1980, MCBEIDE and STETHCMEYER 1981). We are well aware of the value and the difference of soy milk and how it differs from cow's milk. Obviously the dairy industry should not rush into soy based production. It is, however, worth keeping eyes open in that direction, because it is better to start competitive production than let it be monopolized by some other industry.

Table I Characteristics of fermented milks compared to ordinary plain milk

Easier to produce
Better keeping quality
Different flavour
Different acidity
Better nutritive value
Better digestability
Possible therapeutical effects

Table II Classification of yoghurts

Product type	Physical state	Addition of sugar and fruit
Yoghurt	Liquid, gelatinous	+/-
Concentrated/condensed yoghurt	Gelatinous, pasty	+/-
Soft/hard frozen yoghurt	Solid	+
Dried yoghurt	Powder	+

Table III Consumption of yoghurt and cultured buttermilk (kg per capita) in some European countries in 1984
(Data from REINHARDT and IDF Group B38)

	YOGHURT			CULTURED BUTTERMILK
	Unflavoured	Flavoured	Total	
Netherlands	13,0	4,5	17,5	8,8
Switzerland	1,4	14,4	15,8	- 2)
Finland 1)	6,9	8,1	15,0	24,6
France	7,1	4,8	11,9	0,7
Denmark	1,1	9,0	10,1	-
Germany	2,3	6,5	8,8	2,0
Austria	2,7	4,5	7,2	-
Belgium	2,9	2,6	5,5	-
Sweden	1,3	3,6	4,9	2,2
Great-Britain	0,2	2,5	2,7	-
Italy	0,4	1,6	2,0	-

1) Consumption of viili is included
2) Data not available

Table IV The main stages in the manufacture of fermented milks

PREPARATION OF MILK

1. Standardization
   fat
   dry matter
   (sugar)
2. Heat treatment
   Homogenization
   De-aeration

FERMENTATION

1. Starter
2. Incubation
3. Cooling

PACKAGING

1. Type of package
2. After-treatment
3. Delivery

STORAGE

1. Temperature
2. Time

PRODUCTION LINE

1. Fermentors
2. Valves
3. Pumps
4. Pipes
5. Installation

ADDITIONAL INGREDIENTS

1. Composition
2. Quality

Table V   Possible methods of standardization of milk dry matter for the manufacture of yoghurt

PROCESS/ADDITIONS	EFFECT
Evaporation	Increase all constituents
Ultrafiltered whole milk	Increase protein and fat
Ultrafiltered skim milk	Increase protein
Concentrated (RO) milk	Increase all constituents
Whole milk powder	Increase all constituents
Skim milk powder	Increase SNF
Caseinate	Increase casein
Evaporated whole milk	Increase all constituents
Evaporated skim milk	Increase SNF
Whey protein concentrate	Increase protein

REFERENCES:

BETZ, J. and KLEIN, K. (1986). Dtsche Milcwirtschaft 37, 791.
BITSCH, Z. and GARDHAGE, L. (1977). Nord Mejeri-ind. 4, 38.
BOTTAZZI, V. (1985), In 'Biotechnology, vol 5' Ed. Rehm, H-L. and Reed, G., Verlag Chemie, p. 315.
DRIESSEN, F.M., UBBELS, J. and STADHOUDERS, J. (1977_9). Biotech. Bioeng. XIX; 841.
DRIESSEN, F.M., UBBELS, J. and STADHOUDERS, J. (1977_6). Biotech. Bioeng. XIX; 841.
DRIESSEN, F.M. (1981), In 'Mixed culture fermentations', Ed.Bushell, M.E. and Slater, J.H., Spec. publ. Soc. Gen. Microbiol. Acad. Press p.99.
GALESLOOT, T.E. and HASSING, F. (1973). NIZO-Mededeling 7, 15.
HUNGER, W. (1985). Dtsche Molkerei-Ztn. 106,826.
HUTTU-HILTUNEN, E., MERILÄINEN, V. and VÄLIAHO, A. (1986). Unpublished results.
IDF Bulletin, Document 179 (1984). International Dairy Federation.
JÖNSSON, H. and PETTERSON, H-E. (1977). Milchwissenschaft 32, 587.
KAMMERLEHNER, J. (1985). Dtsche Molkerei-Ztn. 106, 868.
KACAB, M. ALLAN-WOJTAS, P. and PHILLIPS-TODD, B.E. (1983). Food Microstructure 2, 51.
KESSLER, H.G. and BAURLE, H.W. (1980).Molk-Ztn. Welt der Milch 34,1652.
KLUPSCH, H.J. (1986). Dtsche Molk-Ztn. 107, 590.
LEPORANTA, K. and VÄLIAHO. A. (1984). Karjantalous 67, 36.
LEPORANTA, K. and VÄLIAHO. A. (1985). Karjantuote 68, 17.
MACURA, B. and TOWNSLEY, P.M. (1984). J. Dairy Sci 67, 735.
MARSHALL, V.M.E., COLL, W.M. and MABBIT, L.A. (1982). J. Dairy Res. 49, 147.
MARSHALL, V.M.E. and MABBIT, L.A. (1980). J. Soc. Dairy Technol.33,129.
MERILÄINEN, V., OLKINUORA, L. and VÄLIAHO. A. (1985). Unpublished results.
NORDLUND, J. (1984). Karjantuote 67, 6.
OBERMAN, H. (1985), In 'Microbiology of Fermented Foods vol 2' Ed. Wood, B.J.B., Elsevin Applied Science Publishers, p. 167.
PINTHONG, R., MACRAE, R. and ROTHWELL, J. (1980). J.Fd.Technol. 15,647.
RASIC, J.L. & KUKMANN, J.A. (1978). Yoghurt. Technical Dairy Publishing House.
REINHARDT, W. (1986). Dtsche Milchwirtschaft 37, 446.
ROBINSON, R.K., TAMIME, A.Y. and CHUBB, L.W. (1977). The Milk Industry 79, 4.
SÄXELIN, M-L., NURMIAHO-LASSILA, E-L., MERILÄINEN, V. and FORSEN, R. (1983). Poster in 'Symp. in Lactic Acid Bacteria in Foods, The Netherlands.
SCHELLHAASS, S.M. (1983). PGD Thesis. Univ. Minnesota.
SCHMIDT, R.H., VARGAS, M.M., SCHMIDT. K.L., & JEZESKI, J.J. (1985). J. Food Process and Preserv. 9, 235.
STEENBERGEN, A.E. (1973). NIZO-Mededeling nr 7, 34.
TAMIME, A.Y. (1977). Dairy Ind. Int. 42, 7.
TAMIME, A.Y. and ROBINSON, R.K. (1983). 'Yoghurt', Pergamon Press.
UNTERHOLZNER, O. (1985). Dtsche Molkerei-Ztn. 106, 14.
VEDAMUTHU, E.R. and NEVILLE, J.M. (1986). Appl. Envirom. Microbiol. 51, 677.

OTHER PRODUCTS

J. Lj. Rašić
Food Research Institute
Bul. AVNOJ 1
21000 Novi Sad
Yugoslavia

ABSTRACT. The manufacturing technology of kefir and kumiss and fermented milk products containing intestinal strains of lactobacilli and bifidobacteria is briefly described. Characteristics of some commercial types of baby foods incorporating bifidobacteria or other lactic acid bacteria are outlined in tabulated form. New types of fermented beverages are mentioned.

1. INTRODUCTION

Specific fermented milk products include traditional products obtained by fermenting milk with lactic acid bacteria and yeasts, and products made by using intestinal strains of lactobacilli and bifidobacteria.
 This paper outlines briefly the manufacturing technology of some known fermented milk products.

2. TRADITIONAL FERMENTED MILKS

2.1. Kefir

It is produced in USSR, Poland, W.Germany and other countries. Traditional manufacture of kefir involves the use of grains which resemble cauliflower florets (1-2 mm to 3-6 cm or more in diameter). They contain a complex microflora in an immobilised system composed of a polysaccharide of microbial origin, although some denatured milk protein may be associated with the polysaccharide matrix. The microflora includes mesophilic lactic streptococci, leuconostocs, lactobacilli, yeasts (fermenting and non-fermenting lactose) and often acetic acid bacteria. The microbial composition of kefir grains may show variations depending on the origin of grains and the methods of their cultivation (17,30,21,29). Kefir culture and production culture (A and B) are prepared as follows (17).

A. - Skimmilk (pasteurised at 95°C/10-15 min; tempering)
   - Grains added to milk (ratio 1:20 to 1:50)
   - Incubation (18-22°C/ca. 24 h; stirring after 15-16 h and 22 h)
   - Grains sieved out ---- adding to new lot of milk
   - Kefir culture

B. - Skimmilk (pasteurised at 95°C/10-15 min; tempering)
   - Inoculation with 1-3% of kefir culture
   - Fermentation (20-23°C/ca. 15-16 h)
   - Ripening, refrigeration

The use of kefir culture in making kefir gives more typical flavor of the product than does the use of production culture, but adding small amounts of kefir culture to the production culture improves flavor of kefir (17).

Kefir is usually made as a stirred product from the pasteurised (90-95°C/3-5 min) homogenised whole milk. The manufacturing procedure is shown in Table I.

TABLE I. Setting Conditions for Kefir

Type of starter	Bulk start.	Incubation/Process
Grains for kefir culture	2-3% (2-5%)	22-25°C/10-12 h until 0.8% lactic acid is produced; then ripening at 14-16°C/ca.12 h; cooling
Lyophilised conc. (a) for bulk start.milk	1%	22°C/18-22 h; followed by cooling and stirring

(a) Contains homofermentative lactic streptococci, citric acid fermenting streptococci, lactobacilli, yeasts (ratio ca. 75:24:0.5:0.1). Bulk starter cultivated at 22°C for 20-22 h (pH 4.4-4.6). Reference: 35

The product has sour cream-like consistency, specific flavor and contains ca. 1.0% lactic acid, ca. 0.1% ethanol and small amounts of carbon dioxide. Other types include lowfat or skimmed kefir with not less than 11% SNF, kefir with up to 6% fat, fruit kefir, baby kefir and freeze-dried kefir.

Difficulties in preparing a good quality culture have led to the development of single-use starters consisting of pure microorganisms usually isolated from the grains. The variable microbial composition of starters, sometimes containing nearly only streptococci resulted in making kefir of varying composition and organoleptic properties (17,49,24,14). A new method of preparing a kefir culture for use as an intermediate culture inoculum (20) consists of mixing grains with neutralised kefir in a mixer, then freeze-drying and subsequent mixing with a dried yeast preparation. Reported the quality of kefir produced is similar to that obtained by using the grains.

Another method of preparing a lyophilised concentrated culture for use as a bulk starter inoculum (35) consists of separate cultivation, concentration and freeze-drying of kefir microorganisms with subsequent mixing at a desirable ratio (Table I). As reported starter activity and kefir quality are uniform, including possibilities to vary the starter composition for various demands. Possibly the use of starters consisting of pure microorganisms in making kefir may lead to the development of a variety of products with a kefir-like flavor.

2.2. Kumiss

It is traditionally obtained by lactic and alcoholic fermentations of mare milk with a mixed culture of L.bulgaricus and lactose-fermenting yeasts (3,17,18). Kumiss is produced in limited quantities mainly in USSR for therapeutic purposes. Mare milk contains more lactose than cows milk, but less fat, protein and ash; and its protein contains less casein than cows milk but more whey protein. Hence in fermented mare milk protein precipitates without changing the texture of product. The manufacturing procedure of kumiss is shown in Table II.

TABLE II. Setting Conditions for Kumiss

Culture organ.	Medium	Start.	Incubation/Process
L.bulgaricus Lactose-fermenting yeasts	Mare milk (past.)	ca.30%	26-28°C/2-4 h; in the beginning stirring for 1/2 to 1 h, then at intervals; bottling and ripening at 4-6°C/1-3 days
L.bulgaricus L.acidophilus Sacch.lactis	Cows milk and whey (past.)	10%	26-28°C/several hrs; then stirring, aeration and cooling to 16-18°C; stirring at intervals, then bottling and ripening at 0-4°C/1-3 days

The product is a frothy drink of liquid uniform consistency and specific flavor. The contents of lactic acid and ethanol respectively range from 0.54-0.7% and 0.7-1% in weak kumiss (1 day ripening) to 0.9-1% and 1.7-2.5% in strong kumiss (3 days ripening).

Shortage of mare milk has led to the development of kumiss from cows milk whose technology is based on the modification of cows milk and the use of a mixed culture (18,36,4,46) (Table II). Two new types of starter have been suggested to improve the organoleptic properties and shelf life of kumiss. The first consists of S.lactis, L. bulgaricus, Sacch.lactis and kumiss yeasts, and the second of S.lactis, L.bulgaricus, Torula lactis and Acetobacter aceti (47). Possibly further advances in technology, including modification of cows milk may occur with the aim to simulate kumiss from mare milk.

3. FERMENTED MILK PRODUCTS CONTAINING HUMAN INTESTINAL STRAINS OF LACTOBACILLI AND BIFIDOBACTERIA

3.1. Products Containing Lactobacilli

Acidophilus products are the most common products containing lactobacilli. While each kind of product requires specific procedures all have some commonality in the selection and treatment of milk. A high quality milk is standardised for composition with subsequent homogenisation and heat treatment. Table III shows some pure acidophilus products.

TABLE III. Setting Conditions for Pure Acidophilus Products

Culture organisms	Bulk start.(%)	Incubation/Cooling
- Acidophilus milk -		
L. acidophilus	1-2 (2-5)	37-38°C/18-24 h
Acid-fast strains	3-5	about 40°C/several hrs
- Acidophilus paste -		
L. acidophilus	3-5	about 40°C/several hrs whey separation follows

Acidophilus milk devised in 1922, USA (16) is obtained by lactic acid fermentation of milk with a culture of L.acidophilus. Slow growth of acidophilus bacteria is promoted by adding stimulating nutrients to the milk or by using acid-fast strains (37,25,18). Fresh acidophilus milk contains about 1% DL-lactic acid and 100-1000 million/ml viable L.acidophilus, but lacks flavor. The viscosity of stirred products is often improved by using "supplementary" mucogenic strains of L.acidophilus. A variety of acidophilus milk includes low-calorie products, flavored products, vegetable oil fortified products or additionally protein/vitamin fortified products (4,1). Acidophilus milk is produced in many countries in limited quantities and the consumption is often linked with its suggested therapeutic values (7,41,45). Acidophilus paste, produced mainly in USSR, is a concentrated acidophilus milk which has pleasant taste when fruit flavored. It contains 80% moisture (skimmed product),1.6-1.8% lactic acid and 20-30 billion/ml L.acidophilus (18,4). The unappetising flavor of acidophilus milk has led to the development of the so-called combined products (Table IV).

Acidophilus yoghurt, produced mainly in Western countries, is a product of characteristic flavor and reportedly contains 10-50 million/ml L.acidophilus in addition to large number of yoghurt organisms (12,37). Acidophilus buttermilk, made in Scandinavia and some other countries, usually contains 10% L.acidophilus and has pleasant flavor (23,48). Acidophilus natural buttermilk (originating from USSR) is a product of agreable flavor, prepared from sweet buttermilk standardised to 1% fat (4). Products made by new combinations of starter organisms include: a) a mild acid product, called Bioghurt, produced mainly in W. Germany (31); b) Acidophilin, a product of specific flavor, made in USSR; c) Acidophilus-yeast milk (originating from USSR), a product of sparkling yeasty-like flavor; made with L.acidophilus and lactose-fermenting yeasts (18) or sucrose-fermenting yeasts. In the latter case the milk is fortified with 2-3% of sucrose.

TABLE IV. Setting Conditions for Combined Acidophilus Products

Culture organisms	Bulk start.(%)	Incubation/Cooling
— Acidophilus yoghurt —		
L. acidophilus	1	
L. bulgaricus	1	42-43°C/3-4 h
S. thermophilus	2	
— Acidophilus buttermilk —		
L. acidophilus	1-2 (2-5)	37-38°C/16-24 h
Cream culture	1-2	21-24°C/14-18 h
organisms (a)		followed by mixing
— Acidophilus natural buttermilk —		
L. acidophilus (30%)		
S. diacetilactis (70%)	3-5	30-32°C/9-10 h
— Bioghurt —		
L. acidophilus		
S. thermophilus	6	42°C/ca. 4 h
— Acidophilin —		
L. acidophilus (1/3)		
S. lactis (1/3)	5-8	32-35°C/6-8 h
Kefir culture (1/3)		
— Acidophilus-yeast milk —		
L. acidophilus		30-33°C/4-6 h
Yeasts	3-5	followed by
		10-17°C/6-10 h
(a) S.lactis, S.cremoris, Leuc.cremoris, S.lactis subsp. diacetilactis		

A fermented drink called Yakult (made in Japan and some Far East countries) is prepared from reconstituted skimmilk powder, sterilised and fermented with L.casei at 37°C, then cooled and mixed with the flavored syrup of sucrose and glucose; subsequently it is diluted with sterilised water and bottled (34). The product contains hundreds million/ml L.casei and about 1.2% protein, 0.1% fat and 16.5% carbohydrates.

3.2. Fermented Milks Containing Bifidobacteria

The first commercial process for making fermented milks containing bifidobacteria has been suggested in 1968 (43), followed by a variety of products developed. Specific procedures for each kind of product are shown in Table V, while selection and treatment of milk are similar as for acidophilus products.

TABLE V. Setting Conditions for Products Containing Bifidobacteria

Culture organisms	Bulk start.(%)	Incubation/Cooling
	- Bifidus milk -	
B. bifidum	approx. 10	37-42°C/until coagulat.
	- Bifighurt -	
B. bifidum	6	42°C/ca. 4 h
S. thermophilus		
	- Biogarde -	
B. bifidum		
L. acidophilus	6	42°C/ca. 4 h
S. thermophilus		
	- Biokys -	
B. bifidum		
L. acidophilus	2-5	30-31°C/until coagulat.
P. acidilactici		
	- Cultura -	
B. bifidum	Direct set	37°C/ca. 16 h
L. acidophilus	as directed	
	- Mil-Mil - (a)	
	- Yoghurt with bifidobacteria -	
B.bifidum/longum	5	40-42°C/3-4 h approx.
Yoghurt culture (b)		
	- Yoghurt with bifidobacteria and L.acidophilus -	
B.bifidum/longum	Direct set	
L. acidophilus	as directed	40-42°C/3-5 h
Yoghurt culture	0.1-1.0	
(a) Culture organisms: B.bifidum, B.breve, L.acidophilus		
(b) S.thermophilus, L.bulgaricus		

Bifidus milk as the original bifidus product is obtained by fermenting milk with B.bifidum, sometimes B.longum. Slow growth of bifidobacteria may be improved by using a large inoculum during culturing and/or by adding stimulating nutrients to the milk (26,38,5,19). The product has a pH of 4.3-4.7, a typical flavor, slightly spicy and contains 100-1000 million/ml bifidobacteria whose numbers decline by 2 logs during cold storage for 1-2 weeks.

Additional lactic acid bacteria as S.thermophilus or

Pediococcus acidilactici have been used to help acidification in the products called Bifighurt, Biogarde and Biokys. The first two (made in W.Germany and other countries) are mild acid products containing respectively B.bifidum, and L.acidophilus and B.bifidum in addition to S.thermophilus. Bulk starter milk contains 1.5% yeast extract (15). The third (made in Czechoslovakia) is prepared from milk standardised to 15% TS using a mixed culture of B.bifidum, L.acidophilus and P.acidilactici (ratio 1:0.1:1) (13).

New products containing bifidus and acidophilus bacteria, called Cultura and Cultura drink (made in Denmark), have been developed. The first is made by fermenting protein-enriched whole milk with highly concentrated cultures of B.bifidum and L.acidophilus. It has firm body, specific flavor and a shelf life of at least 20 days; and contains hundreds million/ml each of B.bifidum and L.acidophilus (10). A fermented drink called Mil-Mil (made in Japan) is a similar product colored with carrot juice and sweetened with glucose or fructose (33) (Table V).

The use of bifidobacteria as such or together with L.acidophilus in making "special yoghurts" (32,37) resulted in products of specific flavor (made in W.Germany, and other countries). They are usually prepared by simultaneously fermenting milk with cultures of yoghurt and B.bifidum or B.longum or additionally L.acidophilus. Possibly buttermilk may also be used as a carrier of bifidus and acidophilus bacteria.

4. FERMENTED BABY FOODS

Bifidobacteria are the predominant intestinal organisms of breast-fed infants; and the potentially beneficial functions of these bacteria in the gut have led to their use in baby foods (9,39). B.bifidum is the species most often used as such or together with its growth-promoting substances (e.g. lactulose, purified pigs gastric mucin, hydrolysed casein) to modify the gut flora of artificially-fed infants and to protect against enteric infections or side-effects of antibiotic therapy (28,42,38).

Bifidobacteria were used the first in 1948/49 in the making of baby foods (27) but commercial production has begun much later. Table VI shows some types of products. Other types of baby foods are shown in Table VII.

5. OTHER FERMENTED BEVERAGES

In addition to drinks mentioned, there are others made from milk, natural buttermilk or whey. They are usually flavored, and in the case of milk often diluted with water or fruit juice. Fermented drinks made from milk or sweet buttermilk incorporate either: a) L.acidophilus or together with bifidobacteria or L.casei (38,10,50); b) L.acidophilus combined with a cream culture organisms (4,48,22); c) combined mesophilic and thermophilic lactic streptococci containing mucogenic strains (17,44); d) other cultures

TABLE VI. Baby Foods Containing Bifidobacteria

Product-name	Culture bacteria	Medium heat-treated	Referen.
a.Lactana-B,dried	B.bifidum	Milk formula	28
b.Bifiline,liquid	Bifidobacteria	Milk formula	19,44
c.Femilact,dried	B.bifidum L.acidophilus P.acidilactici	Cream 12% fat and other components	6

Contains: a. lactulose and B.bifidum; b. ca. 0.6% lactic acid and large number of bifidobacteria; formula fermented with 5% of a culture at 37°C/8-10 h; c. reconstituted, 0.25% lactic acid and large number of culture bacteria; cream fermented with 2-5% a culture at 30°C, cooled and mixed with other components, homogenised and dried.

TABLE VII. Baby Foods with Other Lactic Acid Bacteria

Product-name	Culture bacteria	Medium heat-treated	Referen.
a.Eledon,dried	S.lactis	Half-skimmed milk formula	2
b.Pelargon,dried	S.lactis	Whole milk formula	8
c.Malutka,liquid	L.acidophilus	Milk formula	19,44

Contains: a. L(+)-lactic acid and S.lactis; formula fermented, homogenised, dried; may contain added carbohydrates; for babies suffering from digestive disorders; b. L(+)-lactic acid and S.lactis; formula fermented and mixed with heat-treated carbohydrates, homogenised, dried; for babies after 2 months of age; c. ca. 0.6% lactic acid and large number of L.acidophilus; formula fermented with 1-2% a sel. culture at 37°C/5-6 h; for babies up to 3 months of age.

Fermented drinks made from whole or deproteinised whey incorporate either lactic acid bacteria as such or together with yeasts,e.g. a yoghurt flavored drink containing L.acidophilus and L.casei, acidophilus-yeast drink, whey kwas or pure yeasts,e.g. whey champagne, milk kwas from whey, etc. (11,4).

6. REFERENCES

1. Alm,L. (1982) Dissertation, Stockholm
2. Ballarin,O.(1971) FAO/WHO/UNICEF, Pag.Doc. 1.14/19

3. Berlin,P.J.(1962) IDF Annual Bull., 4
4. Bogdanova,G.J.,Bogdanova,E.A.(1974) New Whole Milk Products of Improved Quality /Ru/ Pishch.Prom.Moscow
5. Collins,E.B.,Hall,B.J.(1984) J.Dairy Sci. 67, 1376
6. Dedicova,L.,Drbohlav,J.(1984) IDF Bull.Doc.179,Pos.VI
7. Gilliland,S.E.(1979) J. Food Prot. 42, 164
8. Grieder,H.R.(1969) Praxis, 58, 1236
9. Gurr,M.Y.,Marshall,V.,Fuller,R.(1984) IDF Bull.Doc.179
10. Hansen,R.(1985) North European Dairy J. 3, 79
11. Holsinger,V.,Posati,L.,DeVilbiss,E.(1974)J.Dairy Sci. 57, 849
12. Hull,R.R.,Roberts,A.V.,Mayes,J.J.(1984) Austr.J.Dairy Technology, Dec. p. 164
13. Hylmar,H.J.(1978) Prumysl Potravin, 29, 99
14. Klupsch,H.J.(1985) North European Dairy J. 1, 14
15. Klupsch,H.J.(1968)Sauermilcherzeugnisse.Th Mann,Hild.
16. Kopeloff,N.(1926)Lactobacillus Acidophilus.Bail.Tind. and Cox, London
17. Koroleva,N.S.(1975) Technical Microbiology of Whole Milk Products /Ru/ Pishch. Prom.Moscow
18. Koroleva,N.S.(1982) 21st Int.Dairy Cong. 2, 146
19. Koroleva,N.S. et al.(1982) Mol.Prom. 6, 17
20. Kramkovska,A. et al.(1982) 21st Int.Dairy Cong.1, 304
21. Kunath,R.,Kandler,O.(1983) in Lactic Acid Bacteria in Foods, p.53. The Netherlands Soc.Microbiology
22. Kurmann,J.A.(1984) IDF Bull. Doc. 179,8
23. Lang,F.,Lang,A. (1978) Milk Ind. 80, 22
24. Mann,E. (1984) Revue Lait.Francaise,No.429 (reprint)
25. Marshall,V.M.,Cole,W.M.,Vega,Y.(1982) J.Dairy Res.49, 665
26. Marshall,V.M.,Cole,W.M.,Mabbitt,L.A.(1982) J.Soc.Dairy Technology, 35, 143
27. Mayer,J.B. (1948) Z. Kinderheilk. 65, 319
28. Mayer,J.B. (1966) Mschr. Kinderheilk. 114, 67
29. Merilainen,V.F.(1984) IDF Bull.Doc. 179,89
30. Molska,J. et al. (1982) 21st Int.Dairy Cong.1, 305
31. Mülhens,K. (1967) Medizin und Ernährung, 8, 11
32. Mülhens,K.,Stamer,H.(1969) Milchwissenschaft, 24, 25
33. N.N. (1978) Packaging, 49, 37
34. N.N. (1979) Yakult. Yakult Honsha Co.Ltd. Tokyo
35. Pettersson,H.E.(1984) IDF Bull.Doc.179,Pos. XVI
36. Puhan,Z.,Gallmann,P.(1980)North European Dairy J.8,220
37. Rašić,J.Lj.,Kurmann,J.A.(1978) Yoghurt. Technical Dairy Publ. House, Copenhagen
38. Rašić,J.Lj.,Kurmann,J.A.(1983) Bifidobacteria and Their Role. Birkhäuser Verlag, Basel,Boston,Stuttgart
39. Rašić,J.Lj. (1983) North European Dairy J. 4, 80
40. Rašić,J.Lj. (1984) IDF Bull. Doc. 179
41. Sandine,W.E. (1979) J. Food Prot. 42, 259
42. Schnegans,E. et al. (1966) Sem. Hop.Paris, 42, 457
43. Schuler-Malyoth,R. et al.(1968) Milchwiss.,23, 554

44. Semenikhina, V.F. (1984) IDF Bull. Doc. 179, 120
45. Shahani, K.M., Ayebo, A.D. (1980) American J. Clinical Nutrition, 33, 2448
46. Shamgin, V.K. et al. (1978) 20th Int. Dairy Cong. Brief Communications, p. 266
47. Shigaeva, M.K., Ospanova, M.S. (1982) 21st Int. Dairy Cong. 1, 308
48. Skovdal, A.B., Wedc, E. (1978) Dansk Vet. Tidsskr. 61, 105
49. Teply, M. (1972) Zborn.ref.konf.kisl.vyr.Vratna dol. 16.
50. Nahaisi, M.H., Robinson, R.K. (1985) Dairy Industr. International, 50 (12)

FERMENTED MILK PRODUCTS IN DEVELOPING COUNTRIES WITH EMPHASIS ON
THOSE PRODUCED FROM EWE'S AND GOAT'S MILK

C.H.Kehagias
School of Food Technology and Nutrition
Technological Educational Institute of Athens
Saint Spyridon Street
12210 Egaleo, Athens, Greece

ABSTRACT. The significance of fermented milk products from ewe's and goat's milk for developing countries is discussed. The technology and the organoleptic characteristics of traditional set and drained type yoghurt, which are produced in Greece from ewe's and goat's milk are described. Yoghurt from ewe's milk has specific flavour and smoother and firmer texture than yoghurt from cow's and goat's milk. Ewe's milk is considered irreplaceable for production of high quality yoghurt. Yoghurt from goat's milk is not offered to the market, while among the rural people the preparation of the drained type is favourable. Because of the increased demand of yoghurt during last years and the seasonal production of ewe's milk many consumers got also used to yoghurt from cow's milk. It is expected that fermented products from ewe's and goat's milk will continue to play in the near future the same role as they played in the past from nutritional and social point of view. However, there is need to take measures to protect the quality of these products by legislation and international agreements and to give more emphasis on studying their specific characteristics.

1. INTRODUCTION

The developing countries, with almost three quarters of the global population produce only 23% of its milk. This quantity of milk represents an output per capita ten times less than that of the developed economies. About 90% of all milk produced in the world is cow's milk. Buffaloe's milk accounts for 6% while the rest is ewe's and goat's milk. Nearly 85% of cow's milk is produced in developed countries while on the other hand developing countries account for about half of the world's ewe's milk and almost three fourths of its goat's milk. North Africa and the Middle East have the smallest number of large ruminants but they have 32% of the total number of ewes and goats. It is also interesting to notice that goats and ewes in many countries cannot be replaced by large ruminants in their ability to use limited resources in difficult climatic conditions. In these countries, although total milk consumption is low the percentage of fermented milk products is relatively high. Table 1 gives the contribution of goat's and ewe's milk in total milk production in some selected countries. It can be seen

TABLE I. Production and percent contribution of goat's and ewe's milk to total milk production in selected countries (FAO Production Yearbook 1984).

Country	Ewe's		Goat's	
	1000 MT	%	1000 MT	%
Algeria	180	20.7	160	18.4
Bangladesh	17	1.0	517	31.7
Denmark	0	0	0	0
Ethiopia	61	8.0	95	12.5
France	1060	3.0	475	1.4
Greece	596	34.7	420	24.5
Guinea Bis.	28	24.6	79	69.3
Italy	647	5.6	125	2.1
Iran	705	26.9	223	8.5
Mexico	136	8.9	415	27.0
Niger	13	5.4	124	51.2
Somalia	100	18.1	290	52.6
Spain	255	3.6	370	5.2
Sudan	136	8.9	415	27.0
Turkey	1300	22.4	625	10.8
U.K.	0	0	0	0
U.S.A.	0	0	0	0
U.S.S.R.	90	0.1	330	0.4

that goat's and ewe's milk has a special interest for some developing countries and gives support to marginal human populations. On the other hand, the contribution of goat's and ewe's milk to the total milk production of most developed countries ranges from zero to negligible. Taking into consideration the importance that goat's and ewe's milk has in many developing economies, in this paper emphasis will be given to the technology and characteristics of the fermented milk products produced from these two kinds of milk. In Greece, nearly 60% of its total milk is produced from ewes and goats and the experience we have gained with the fermented milk products from these two species will also be presented here.

2. ADVANTAGES OF FERMENTATION OF EWE'S AND GOAT'S MILK

Milk from ewes and goats in most countries is still produced under primitive conditions. Animals are milked by hands, milk is not cooled and remains at high environmental temperatures above 30°C for a few hours. These problems are intensified by the small quantities of milk produced per animal and per farmer in isolated areas where transportation of milk is very difficult. Milk produced under such conditions should be consumed soon or preserved by fermentation. Therefore, the production of fermented products from ewe's and goat's milk is very essential from social, economical and nutritional point of view for many rural people.

Ewe's and goat's milk has specific flavour which is found by many people objectional. It is worthwhile noticing that this undesirable flavour is not observed after fermentation. Another characteristic of the fermented products from ewe's and goat's milk is their traditional character. Fermented products from these two kinds of milk vary from one area to the other because their production is influenced by many factors related to the tradition of the specific area.

3. TRADITIONAL FERMENTED MILK PRODUCTS AND TRENDS IN TECHNOLOGY

Among the fermented milk products produced from ewe's and goat's milk in Greece, yoghurt is the most important. Fermented milk products other than yoghurt are rarely produced or offered to the market from ewe's and goat's milk. Two types of yoghurt are very popular now, one is the so-called traditional (set type) while the other is the drained type (stragisto or sakkoulas). Although yoghurt from cow's milk has lower price than the traditional type produced from ewe's milk, the last product is produced in large quantities not only by rural people but also by many family units and large dairy plants. In fact ewe's milk is considered as irreplaceable for the production of traditional type of yoghurt. In contrast to that, the traditional type from goat's milk is not commercially available, it is prepared to a small only scale by rural people that don't have in their possession ewe's milk. The drained type yoghurt that is offered to the market was traditionally prepared only from ewe's and by rural people also from goat's milk. The demand for the drained type yoghurt increased very much during last years and since ewe's or goat's milk is not available in large quantities all the year, some dairy plants prepare a similar product by mixing cow's with ewe's milk or by using cow's milk only. The drained type became very popular during the last years, and it is also used for the preparation of a very popular appetizer known as Tzatziki (mixture of drained type of yoghurt with cucumber, vinegar, olive oil, garlic and spices). The emulsifying properties of the drained type yoghurt from ewe's milk for the preparation of tzatziki are appreciated very much by housekeepers.

The manufacturing procedure for the production of both, the traditional (set type) and the drained type yoghurt from ewe's milk is briefly outlined. <u>Traditional type:</u> Milk should be standardized, because yoghurt should have a minimum fat content 6.6% and is not homogenized, since the formation of a thick white skin on the surface is a desirable characteristic. Milk is batch heat treated at 80-85°C for 20-30 min and is left for slow cooling down to incubation temperature (42-47°C) in open containers (from clay or plastic) to facilitate skin formation.

Special care is also taken when the inoculum is added not to disturb the skin. During the last years, some big dairy plants produce also a set type yoghurt (not traditional) from homogenized ewe's milk with continuous heat treatment. Small dairy plants use their own cultures and most of them say that they don't face problems with them. Large dairy plants use commercialy available cultures. The special fermentation conditions for ewe's milk have not been studied. However, it is known that ewe's milk coagulates sooner as compared to cow's and goat's milk. The properties of ewe's and goat's milk and the distinct fermented products that are prepared from them should stimulate research on the non commercial cultures, in order to select those that give the most desirable characteristics to the products. <u>Drained type</u>: The manufacture of this, involves the preparation of set type yoghurt from ewe's milk in wooden containers or vats. After coagulation yoghurt is cooled down and the gel is broken. The yoghurt is transferred to cloth bags which are left in a cool room for draining. The bags are emptied in a mixing vat where the gel is worked. After working the containers are filled. When the drained yoghurt is packed in bags, it is called Sakkoulas. Large quantities of yoghurt are also packed in wooden barrels and are sold to the consumers by weight. The final product should have a minimum fat content 8%. In modern dairy plants serum might be separated by ultrafiltration or after coagulation by centrifugation. In such large units the product might be standardized by adding cream after separation of the serum. In all cases milk is coagulated by a mixed yoghurt culture.

4. ORGANOLEPTIC CHARACTERISTICS

Yoghurt from ewe's and goat's milk has specific flavour and aroma which is distinguished from that of cow's milk. It is known, that flavour and aroma is affected by lactic acid and the volatile compounds, that are formed during fermentation, among which acetaldehyde is the most important. Figure 1 shows the acid development during fermentation of ewe's, goat's and cow's milk by a mixed yoghurt culture. The highest acidity was obtained with ewe's milk and the lowest with cow's milk. When cultures ferment milk for shorter time for yoghurt preparation, the pH values are similar for the three kinds of milk. However, the acidity that is developed during storage of yoghurt from goat's milk seems to be more perceptible to consumers compared with that from ewe's and cow's milk (Kehagias, 1986). The type of milk that is used for the manufacture of yoghurt affects also the formation of the volatile compounds. Acetaldehyde content of yoghurt from goat's milk is lower than that from cow's and ewe's milk (Abrahamsen et al.,1978; Kehagias, 1986). The relatively flat flavour that sometimes is observed in the yoghurt from goat's milk might be related to the low acetaldehyde content. It has been found that quantitative differences in volatile acids exist between the yoghurt prepared from cow's, ewe's and goat's milk (Rasic and Kurman, 1978) and as it is known volatile acids play a balancing role in the flavour of yoghurt.

The organoleptic characteristics of yoghurt are also affected to great extend by the physical properties of the yoghurt gel. The higher acceptability of yoghurt from ewe's milk and the lower of yoghurt from goat's milk could be rather attributed to the textural characte-

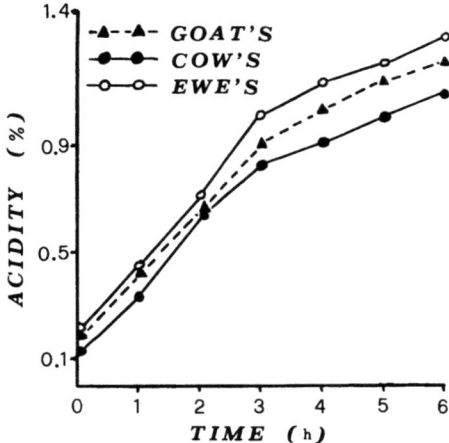

Figure 1. Acidity in mixed yoghurt cultures

ristics. Yoghurt from ewe's milk has smoother and firmer texture than yoghurt from cow's and goat's milk. It has been found that yoghurt from ewe's milk has the highest penetration force and viscosity and the lowest volume of separated serum as compared with yoghurt from cow's and goat's milk (Kehagias et al., 1985). The lowest penetration force and viscosity and the highest volume of separated serum was exhibited by yoghurt from goat's milk of Saanen breed, while yoghurt from indigenous goat breeds gave higher or lower values than yoghurt from cow's milk respectively. Taking into consideration the composition of milk from these species, we might assume that the firmer consistency of yoghurt from ewe's milk could be related to the higher concentration of the various constituents in this milk. However, by selecting breeds of goats with high protein and fat content and increasing the total solids by concentration up to 25%, it was not possible to get the high values of penetration force obtained with yoghurt from ewe's milk, with much lower total solids content (Figure 2). These results indicate that the gross compositional data are not sufficient to explain the superiority of yoghurt from ewe's milk in its textural characteristics. It has been found that the viscosity and the stability of emulsions were higher with sodium caseinates from ewe's milk as compared with those from cow's and goat's milk (Kehagias and Dalles 1981). The functionality of the proteins from ewe's milk has to be studied in relation to the physical properties of yoghurt.

5. NUTRITIONAL VALUE

It is known that the compositional data of yoghurt reveal only one part of the nutritional value. Taking this into consideration, we must recognize that in most of the cases goat's and ewe's milk fermented products can replace those of cow's milk in the diet of people suffering from allergy to cow's milk. As it was expected, the viable counts

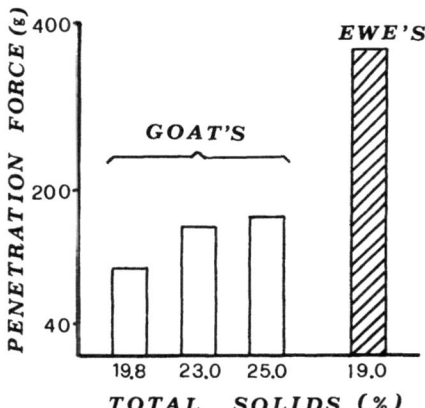

Figure 2. Penentation force of yoghurt from ewe's, goat's and concentrated goat's milk

of yoghurt from goat's and ewe's milk were similar (Abrahamsen and Holmen 1981, Kehagias and Dalles 1984). However, biochemical characteristics of yoghurt microorganisms growing in goat's and ewe's milk might be different from those growing on cow's milk. There are also indications that yoghurt from ewe's milk has higher lactase (Kehagias and Dalles 1984) and lower lactose content than yoghurt from cow's milk (Dalles and Kehagias 1984). The biological value of the proteins in yoghurt has also been compared, and it was found that the proteins of goat's milk had higher biological value than the proteins from cow's milk (Rasic and Kurman 1978). However, it is necessary to emphasize that there are enormously large gaps in our knowledge on the nutritional and physiological properties of the fermented products from ewe's and goat's milk.

6. THE YOGHURT MARKET

Production and consumption of yoghurt has increased, in Greece, during the last years. Thirty thousands tonnes of yoghurt were produced in 1969 from 379 units, 36000 tonnes in 1978 from 185 units and 43000 tonnes in 1983 from 37 units (GNSS 1986). These data show that although the production has increased the number of the units producing yoghurt is decreasing. In the above data is not included the production of yoghurt by very small household units which are quite common in Greece. We should also take into consideration that 44% of ewe's and goat's milk is utilized by the farmers and a part of it is used for yoghurt production. Forty years ago, all yoghurt that was offered to the market was produced from ewe's milk. The increased demand for yoghurt during last years forced large dairy plants to show interest and produce also yoghurt from cow's milk. A survey that was conducted recently (Athanasatos et al.,1985) revealed that 87% of Greek households buy yoghurt and among yoghurt consumers the highest preference was for traditional yoghurt from ewe's milk followed by the drained type yoghurt and yoghurt from cow's milk. According to the survey, most of the consumers prefer traditional yoghurt from ewe's milk and the drained type because they find them superior in taste. Yoghurt from cow's milk is preferred by the consumers who want to consume less fatty products. Retail prices for traditional yoghurt from ewe's milk are 70% higher than the prices of yoghurt from cow's milk. Under the existing price structure, the drained type yoghurt from ewe's and cow's milk has the same prices.

7. FUTURE NEEDS

Fermented products from ewes and goats milk played in the past a very significant role, securing food and money for rural people of many developing countries. Certainly for some non-industrialized countries these products will play equally important role in the future. In Greece, yoghurt from ewe's milk has still a very good market. However, the future doesn't look bright, optimism alone is not enough. Young people, don't show much interest in raising ewe's and goat's in isolated mountainous areas, under the current economical and social conditions. It is easier for them to work in a non dairy farm, in the industry e.t.c. Certainly, similar situations and problems also exist in other countries producing goat's and ewe's milk. It is difficult to predict what will happen in the future. However, it is necessary to take special measures to give stimuli to the people to continue goats'and ewes' farming. As it was already mentioned, fermented products other than cheeses are not commercially available for Greek consumers from goat's milk. This is of importance since large quantities of goat's milk remain in the farms and are not properly utilized especially when the small cheese plants which collect ewe's and goat's milk are closing down at the end of ewe's lactation period, when goat's milk production is at its peak. We should study carefully the market and find the key factors that influenced consumer acceptability, especially of the fermented products from ewe's and goat's milk. These products have special characteristics and this should be taken into consideration for developing products. New products should be developed while the traditional ones should be protected and improved. The textural problems that we have with set type yoghurt from goat's milk should be overcome by improving the technology or by developing new products. Fermented products with low fat content should also be developed from ewe's milk. Until now, the traditional products from ewe's milk were preferred by people with low income. As the cost of the products will be increasing, the quality should be improved in order to attract more demanding consumers. For manufacturing high quality products, the quality of raw milk should be improved by refrigeration and better transportation conditions. It is also essential to assure the genuineness of these products by developing methods for identification of milk from other species and protect the traditionality of them by national legislation and international agreements.

8. REFERENCES

Abrahamsen, R.K., Svensen, A. and Tufto, G.N.1978. "Some bacteriological and biochemical activities during the incubation of yoghurt from goat's and cow's milk" 20th International Dairy Congress, Paris, E: 828-829.

Abrahamsen, R.K., and Holmen, T.B.1981. "Goat's milk yoghurt made from non-homogenized and homogenized milks, concentrated by different methods". J.of Dairy Res., 48:457-463

Athanasatos, D.,Michalopoulos,G., Zeugaridis, S., Kamenidis,C.,Patsis P., and Stamatiou, A. 1985. "The Greek market of milk and milk products". Special Edition by the Agricultural College of Athens.

Dalles, T.,and Kehagias, C.,1984. "Chemical composition of commercial types of yoghurt and changes during cold storage of yoghurt from sheep's milk". Bulletin of Greek National Dairy Committee. 1:32-41

Greek National Statistical Service, 1986. Personal communication.

FAO,1984. "Production yearbook". 44

Kehagias,C.,and Dalles,T.,1981. "Functional properties of sodium caseinates derived from ewe's, goat's and cow's milk. Milchwissenschaft, 36:29-31.

Kehagias,C.H., and Dalles, T.N.1984. "Bacteriological and biochemical characteristics of various types of yoghurt made from sheep's and cow's milk".J. of Food Protection, 47:760-761

Kehagias,C.,Komiotis,A., Koulouris,S., Koroni,E., and Kazazis,J.,1985. "Physico-chemical properties of set type yoghurt made from cow's,ewe's and goat's milk" IDF seminar on production and Utilization of Ewe and Goat milk, Athens (In Press).

Kehagias,C.H.,1986."Comparative studies on the yoghurt from ewe's, goat's and cow's milk" Unpublished results.

Rasic,J., and Kurman, J.A.,1978. "Yoghurt". Technical Publishing House. Jyllingevej 39, Copenhagen, Denmark.

YOGHURT - LEGAL ASPECTS

F. Winkelmann
Senior Officer, Meat and Dairy Service
Animal Production and Health Division
Food and Agriculture Organization of the UN
00100 Rome
Italy

ABSTRACT. In the field of food legislation, FAO and WHO, provided a forum for international action to protect the consumers and facilitate food trade by establishing the Codex Alimentarius Commission (CAC). Standards for milk products are elaborated by a Government Expert Committee on Milk and Milk Products, a subsidiary body of the CAC. This Committee established standards for yoghurt and flavoured yoghurts comprising product definitions, compositional requirements including provisions for food additives and labelling requirements.

The tasks assigned to FAO by its Member Nations include raising levels of nutrition and securing improvements in the efficiency of production and distribution of all food and agricultural products, with particular reference to bettering the condition of rural populations.
In the field of food legislation, FAO, in cooperation with WHO, arranged for international action to protect the consumer and to facilitate trade in foodstuffs by removal of non-tariff obstacles to this trade caused by differing national food legislations. In 1962, the two agencies established the FAO/WHO Food Standards Programme and set up the Codex Alimentarius Commission (CAC) which held its first session in 1963.
CAC took over and continued on a world-wide level the work of the Codex Alimentarius Europaeus which had been set up in 1958 jointly by the International Commission on Agricultural Industries and the Permanent Bureaux of Analytical Chemistry. The supreme rule of the European Food Codex was based on the two legal principles of health and honesty in food production and trade; it read as follows:

> "Supreme law in honest food trade is the well-being of the consumer, his protection against damage to health and his protection against misguidance and fraud. All economic and technical considerations are subordinated to this supreme law."

Although all modern national food law systems are said to be based on these principles of health and honesty, food laws diverge to such an extent that a former German food law official illustrated this situation in his "Credo of World Food Laws" as follows:

> "Food laws applicable not only in Europe but also in the whole world are a hodgepodge of archaic patch-work regulations far behind the times, the technology of food, and the needs of consumers."

This statement was made twenty years ago. In the meantime the standards work of the Codex Alimentarius has grown significantly (approximately 200 standards have been submitted to Governments for acceptance or will be submitted for acceptance soon), thus paving the way towards securing international agreements on the following legal aspects of foods which constitute the main components of the majority of food standards already developed or under consideration by FAO and WHO, namely composition and quality, additive, contaminant, pesticide residue, hygiene, analytical and labelling aspects. After having reached agreement on the standards governments are invited to accept them in certain specified ways. FAO and WHO not being supranational but international organizations have no power to enforce food legislation neither on international nor on national level. Governments, however, are free to proceed in accordance with their national and constitutional procedures of implementation.

With regard to the European Common Market it is interesting to note the "Gentlemen's Agreement" made by the EEC members as a group to accept Codex Standards as appropriate.

THE CODE OF PRINCIPLES CONCERNING MILK AND MILK PRODUCTS AND ASSOCIATED STANDARDS

On the international level FAO provided a forum for worldwide action in the dairy field by establishing a Committee of Government Experts on the Use of Designations, Definitions and Standards for Milk and Milk Products which later became a subsidiary body of the Codex Alimentarius Commission.

The work commenced in 1958 when the Committee of Government Experts at their first session agreed upon a preliminary text of Code of Principles concerning Milk and Milk Products which was based on a draft prepared by the International Dairy Federation (IDF).

This Code provides a framework of definitions and designations for "milk", "milk products", "composite products" and "other products". The means by which the objectives of the Code of Principles are to be achieved are summarized in the Preamble of the Code and read as follows:

"The purpose of this Code of Principles is to protect the consumer of milk and milk products and to assist the dairy industry on both the national and international levels by: Ensuring the precise use of the term "milk" and the terms used for the different milk products: Avoiding confusion arising from the mixing of milk and/or milk products with non-milk fats and/or non-milk proteins: Prohibiting the use of misleading names and information for products which are not milk or milk products and which might thereby be confused with milk or milk products: and Establishing (a) definitions and designations: (b) minimum standards of composition: and (c) standard methods of sampling and analysis for milk and milk products".

In short, the Code in following the above mentioned supreme rule aims at consumer protection and fair practices in the trade of milk and milk products.

In 1969 (during its 12th Session) the Joint FAO/WHO Committee of Government Experts on the Code of Principles concerning Milk and Milk Products (hereafter referred to as the "Milk Committee" decided to elaborate a standard for yoghurt. With this decision the Milk Committee embarked on a task which turned out to be one of the most difficult to complete satisfactorily. It took in fact 5 years from the first draft which the Secretariat prepared in 1970, to the adoption in 1975 of the two resulting Standards No. A-11(a) for yoghurt (yogurt) and sweetened yoghurt (sweetened yogurt) and No. A-11(b) for flavoured yoghurt (yogurt) and products heat treated after fermentation.

As it would be impossible to attempt to refer to all the individual national legislations concerning (the manufacture of and trade in) yoghurt the following major problem areas were selected for consideration with regard to international action undertaken, namely:

(i)   designations, definitions, labelling;
(ii)   compositional and labelling requirements;
(iii)   food additives and pesticide residues; protection of consumers' health.

DESIGNATIONS, DEFINITIONS AND LABELLING THE CODE OF PRINCIPLES AND THE (NEW) CODEX STANDARD FOR FOOD LABELLING

Yoghurt and sweetened yoghurt are covered by the provisions of Article 2 "Milk Products" and flavoured yoghurt by the provisions of Article 3 "Composite Products".

The following short summary is intended to recall major points of the designations for milk, milk products, composite products and imitation products.

Articles 1 and 2 state that the term "milk" shall mean exclusively the normal mammary secretion obtained from one or more milkings without addition or extraction and that the terms used to designate "milk products" shall only be employed for products exclusively derived from milk as just defined. The terms used for milk products may be employed when substances necessary for the manufacturing process are added, provided that these substances are not intended to take the place in part, or whole, of any milk constituent.

Both the term "milk" and the terms used to designate milk products can be used in association with words to designate the type, grade, origin, intended use of the milk or milk product or to describe physical treatments or modifications in composition, provided that the modification is restricted to an addition and/or withdrawal of natural milk constituents. The same provision is also valid for "composite products" which are covered by Article 3, i.e. products of which no part takes the place of any milk constituent and of which milk or a milk product is an essential part. Article 3 further requests that, if composite products are designated in terms which are suggestive of milk or milk products or the dairy industry, the label indicates the milk or milk products used as well as the other essential constituents. Article 3 allows the use of the term milk and the terms employed for milk products together with a word or words to designate composite products such as flavoured yoghurt.

In conclusion, it would appear that the most important feature of the above definitions of milk products and composite products is that prohibiting the replacement of any milk constituent.

Milk substitute or imitation products, i.e. products in which milk solids are wholly or partly replaced with non-milk ingredients are covered by Article 4. The Milk Committee by its Decision No. 6 has stipulated that such imitation products
1. shall meet the essential compositional requirements of the milk or corresponding milk product, apart from the nature of the constituents being replaced;
2. shall not contain additions other than those used in the corresponding milk product, except for harmless additives technologically necessary for the replacement and optional additional nutrients as appropriate;
3. shall be produced under hygienic conditions;
4. shall conform to the hygienic quality standards and to such maximum levels of contaminants normally

applicable to the corresponding milk product;
5. shall be labelled in accordance with:
   (i) Article 4 of the Code of Principles and shall preferably be designated according to Article 4.2(b) rather than Article 4.2(a) and furthermore, if a distinct name is used, it should conform with Article 4.1 and should not be suggestive of a milk product;
   (ii) the "Codex General Standard for the Labelling of Prepackaged Foods" (CODEX STAN 1-1981);
   (iii) the appropriate sections of the standard for the corresponding milk product in other respects.

Note: In formulating the "Decision No. 6" recognition is given to the fact that Article 4.1(a) which makes provision for the use of the word "imitation" in front of the name of the product is now deprecated by the International Dairy Federation (IDF).

DEFINITION

The FAO/WHO Standards No. 11(a) and 11(b) define yoghurt and flavoured yoghurt respectively as "the coagulated milk product obtained by lactic acid fermentation through the action of Lactobacillus bulgaricus and Streptococcus thermophilus from milk and milk products...". This definition which was used in the first draft of the standard 1/ prepared by the Secretariat of the Milk Committee in 1969/70 remained unchanged.

The Milk Committee considered proposals to delete the reference to specific micro-organisms and suggested the inclusion of other species such as Lactobacillus acidophilus and agreed that Lactobacillus bulgaricus and Streptococcus thermophilus were essential for the production of yoghurt and should remain in the definition, but that the use of other suitable lactic acid producing cultures should not be excluded. These cultures were listed as "optional additions".

While international agreement was reached rather quickly on these proposals, the Government Experts of the Milk Committee had to consider during the following four consecutive Sessions a proposal to include in the definition a requirement that the micro-organisms in the

---

1/ The first drafts covered both yoghurt and flavoured yoghurt by means of one single standard.

final product must be viable and abundant in order to distinguish between "conventional" and heat treated yoghurts.

DESIGNATIONS AND LABELLING

At its (14th) Session in September 1971 the Milk Committee had agreed that the future standard on yoghurt should also cover heat-treated yoghurts. However, a clear distinction was to be made on the label between "conventional" yoghurt and a post-fermentation heat-treated product and governments were asked to make proposals regarding the terminology for these products. One year later (at its 15th Session) the Committee concluded that heat-treated yoghurt should not be designated "yoghurt" even with a qualifying term (such as "heat treated") mainly because of the view that it was essential for any product called "yoghurt" to contain specific bacteria in viable form and in abundance and that it was improper to use the term "yoghurt" to denote any product that did not correspond to the standard product traditionally sold under this name.

At its (16th) Session in 1973 the Milk Committee in order to avoid further delays decided to split the draft standard for yoghurt into one for plain yoghurt and a standard for flavoured and heat-treated yoghurt. As regards the standard for "plain" yoghurt (No. A-11(a)) the Milk Committee had agreed that the micro-organisms in this product had to be "viable and abundant" and submitted it to Governments for acceptance in 1975. The standard for flavoured yoghurt (No. A-11(b) could only be finalized one year later when the Milk Committee adopted a revision of the standard based on the agreement that the product heat treated after fermentation was a wholesome product and that the difficulty in dealing with the two products (heat-treated and not) was predominantly related to labelling. The designations of products heat treated after fermentation were left to governments which were requested to notify the specific names exclusively provided in their national regulations.

It is interesting to note in this context the following opinion voiced by a commercial firm on this subject:

The firm asserted that correct labelling was the issue to be addressed by the government authorities concerned in their proposed yoghurt standard, but not whether yoghurt must contain live cultures when sold to the consumer.

The firm submitted results of a consumer survey demonstrating that consumers buy yoghurt for its good taste and nutrition and consider a pasteurized yoghurt product to be yoghurt.

Consumers who had eaten yoghurt at least twice in the past 60 days were surveyed. The research included three samples of 200 respondents each in two districts, and 500 respondents nationally.

The survey results showed that pasteurized yoghurt was satisfactory to 69% of the consumers surveyed, that 68% responded there is "no difference" between live cultured and pasteurized yoghurt, and 16% said they would not eat a live cultured product.

The firm contended that proper labelling will allow consumers to purchase the product they desire whether live cultured or pasteurized. (Food Chemical News P.26 Aug 6, 1979).

COMPOSITIONAL AND LABELLING REQUIREMENTS

Perhaps the most interesting compositional aspect for yoghurt from the legal point of view - in addition to the replacement of milk constituents dealt with before - is the use of reconstituted and/or recombined milk for the production of yoghurt.

The Milk Committee covered this aspect by its Decision No. 5 which provides definitions for reconstituted and for recombined milk products and stipulates that standards adopted under the Code of Principles should apply to products covered by these standards whether made from milk, reconstituted or recombined milk or by reconstitution or recombining milk constituents, unless the standards provide otherwise.

The Milk Committee decided that yoghurt made from recombined or reconstituted milk should be labelled accordingly.

The origin of Decision No. 5 dates back to 1961 when the Government of Malaysia requested the Milk Committee to permit the use of the designations "evaporated milk" and "condensed milk" (Standards No. 3 and 4 respectively) for products made by recombination techniques.

The compositional and labelling requirements of the two yoghurt standards distinguish between "yoghurt"(for yoghurt with not less than 3.0% m/m milk fat content) "partly skimmed yoghurt" or any other suitable qualifying description (for yoghurt with less than 3.0 but more than 0.5% m/m milk fat) accompanied by a milk fat statement in multiples of 0.5%, e.g. 1.0%, 2.0% etc., and "skimmed yoghurt" or any other suitable qualifying description (for yoghurt with less than 0.5% m/m milk fat).

Further, it is required to declare
- a complete list of ingredients
- the net contents
- the name and address of the manufacturer, packer, distributor, importer, exporter or vendor
- the country of origin

- the date of production, sell-by date or minimum durability date
- the lot and producing factory (in clear or code).

These labelling requirements which are based on the Codex General Standard for the Labelling of Prepackaged Foods underwent some amendments in accordance with the new edition of that labelling standard. Proposals for revised labelling provisions were considered by the Milk Committee at its 21st Session in June 1986.

FOOD ADDITIVES

The replacement of the <u>general</u> wording used for permitted additives in the original text of FAO/WHO milk product standards i.e. "harmless substances necessary for the manufacturing process" by <u>definitive</u> lists of food additives with maximum permitted quantities reflects the general trend for establishing clear-cut provisions for the use of food additives. The Joint FAO/WHO Expert Committee on Food Additives (JECFA) has published procedures for the testing of food additives to establish their safety for use.

The comprehensive animal studies to be carried out to establish the safety for use of food additives include:

(1) Acute toxicity studies with at least three species of animals one of which should be a non-rodent, both sexes to be used in at least one species, the test material to be administered orally and parenterally: the observation period should be 2-4 weeks.

(2) Short-term toxicity studies with at least two species of animals including a rodent and a non-rodent, feeding on a sufficient number of levels, ensuring that at least one level has no effect and that doses are included that produce definite toxic effects, if this is possible: the test span should be 90 days for the rat and 10 per cent of the life-span of other animals (1 year for the dog).

(3) Long-term toxicity studies including carcinogenicity, with two species of animals (both sexes) throughout their life-span.

(4) Studies of the metabolic and biochemical activity and examination of enzymatic processes which might be affected, the effect of additives on the nutritive value of the diet and of the possibility of formation of toxic substances during processing, storage and household use.

General Principles for the use of Food Additives have been published by the Codex Alimentarius Commission.

They read as follows:
- (a) All food additives, whether actually in use or being proposed for use, should have been or should be subjected to appropriate toxicological testing and evaluation. This evaluation should take into account among other things, any cumulative, synergistic or potentiating effects of their use.
- (b) Only those food additives should be endorsed which so far as can be judged by the evidence presently available, present no hazard to the health of the consumer at the levels of use proposed.
- (c) All food additives should be kept under continuous observation and should be re-evaluated whenever necessary in the light of changing conditions of use and new scientific information.
- (d) Food additives should at all times conform with an approved specification, e.g. the Specifications of Identity and Purity recommended by the Codex Alimentarius Commission.
- (e) The use of food additives is justified only where they serve one or more of the purposes set out from (a) to (d) and only where these purposes cannot be achieved by other means which are economically and technologically practicable and do not present a hazard to the health of the consumer:
    - (i) to preserve the nutritional quality of the food: an intentional reduction in the nutritional quality of a food would be justified in the circumstances dealt with in sub-paragraph (b) and also in other circumstances where the food does not constitute a significant item in a normal diet:
    - (ii) to provide necessary ingredients or constituents for food manufactured for groups of consumers having special dietary needs:
    - (iii) to provide aids in manufacture, processing, preparation, treatment.

The technological justification for using certain additives in milk products is provided by the Milk Committee. The additives and the maximum levels suggested by the Milk Committee are subject to endorsement by the Codex Committee on Food Additives (CCFA). The work of JECFA is twofold: the WHO Group is responsible for evaluating the toxicological data and setting the ADI (Acceptable Daily Intake), while the FAO Group is concerned with establishing specifications for the identity and purity of the additives, as well as methods of analysis to check the various criteria of identity and purity.

The Milk Committee has restricted the use of food additives to flavoured yoghurts. Standard No. A-11(b) (Standard for flavoured yoghurt (yogurt) and Products Heat Treated after Fermentation) has provision for flavours (natural, nature identical and artificial), number of stabilizers and certain food colours and preservatives which come exclusively from flavouring substances as a result of carry over.

PESTICIDE RESIDUES AND PROTECTION OF CONSUMER'S HEALTH

The Codex Alimentarius Commission (CAC) which gives high priority to tackling international difficulties in health and trade encountered in the use of pesticides convened in 1966 the Codex Committee on Pesticide Residues (CCPR). The CCPR advises the CAC regularly on all matters relating to pesticide residues in food and in animal feed. It is specifically responsible

(a) to establish maximum limits for pesticide residues in specific food items or in groups of food; (MRL);

(b) to establish maximum limits for pesticide residues in certain animal feeding stuffs moving in international trade where this is justified for reasons of protection of human health;

(c) to prepare priority lists of pesticides for evaluation by the Joint FAO/WHO Meeting on Pesticide Residues (JMPR);

(d) to consider methods of sampling and analysis for the determination of pesticide residues in food and feed;

(e) to consider other matters in relation to the safety of food and feed containing pesticide residues, and

(f) to establish maximum limits for environmental and industrial contaminants showing chemical or other similarity to pesticides, in specific food items or groups of food.

The CCPR relies on data supplied by governments and on the evaluation and on the recommendations of the JMPR which is composed of specialists acting in an individual capacity, not representing governments. The JMPR establishes Acceptable Daily Intakes (ADI) for individual pesticides on the basis of toxicological data and recommends MRLs on the basis of appropriate residue data which reflect registered or approved usage of the pesticide in accordance with "good agricultural practice" (GAP).

The CCPR examines the findings of the Joint FAO/WHO expert meetings, in the light of government comments and finally presents the MRLs to the CAC for adoption and publication in the Codex Alimentarius for acceptance by governments.

The Acceptable Daily Intake (ADI) of a chemical is that which can be taken daily in the diet over a lifetime without appreciable risk to the health of the consumer. Good Agricultural Practice in the Use of Pesticides (GAP) is the officially recommended or authorized usage of pesticides under practical conditions at any stage of production, storage, transport, distribution and processing of food, agricultural commodities, and animal feed bearing in mind the variations in requirements within and between regions, which takes into account the minimum quantities necessary to achieve adequate control, applied in a manner so as to leave a residue which is the smallest amount practicable and which is toxicologically acceptable. The Maximum Residue Limits (MRL) are expressed in milligrammes of pesticide per kilogramme of the food. MRLs in milk and milk products have been proposed for the following pesticides:

Milk and Milk Products	Pesticide	Codex MRLs
Milk	aldrin and dieldrin	0.06 CXL
Milk	bromophos-ethyl	0.008 CXL
Milk	carbaryl	0.1(*) CXL
Milk	carbophenothion	0.004 CXL
Milk	chlordimeform	CXL
Milk	chlorfenvinphos	0.008 CXL
Milk	chlormequat	0.1(*) CXL
Milk	chlorobenzilate	0.05(*) CXL
Milk	chlorpyrifos	0.01(*) CXL (on a fat basis)
Milk	crufomate	0.05 CXL
Milk	diazinon	0.02 CXL
Milk	dichlorvos	0.02 CXL
Milk	dioxathion	0.008 CXL
Milk	diquat	0.01 (*) CXL
Milk	endrin	0.0008 CXL
Milk	ethion	0.02 CXL
Milk	fenchlorphos	0.08 CXL
Milk	fenitrothion	0.02(*) CXL
Milk	heptachlor	0.006 CXL
Milk	methidathion	0.008 CXL
Milk	monocrotophos	0.002(*) CXL
Milk	paraquat	0.01(*) CXL
Milk	thiabendazole	0.1(*) CXL
Milk	trichlorfon	0.05 CXL
Milk	cyhexatin	0.05(*) CXL

Milk and Milk Products	Pesticide	Codex MRLs
Milk	propoxur	0.05 (*) CXL
Milk	pirimiphos-methyl	0.05 (*) CXL
Milk	chlorpyrifos-methyl	0.01 (*) CXL
Milk	carbofuran	0.05 (*) CXL
Milk	methamidophos	0.01 (*) CXL
Milk	pirimicarb	0.05 (*) CXL
Milk	fenbutatin oxide	0.02 (*) CXL
Milk	DDT	0.05 (*) CXL
Milk	lindane	0.01 CXL
Milk Products	carbaryl	0.1 (*) CXL
Milk Products	chlormequat	0.1 (*) CXL
Milk Products	chlorpyrifos	0.01 (*) CXL (on a fat basis)
Milk Products	monocrotophos	0.02 (*) CXL
Milk Products	cyhexatin	0.05 (*) CXL

(*) Limit of Determination

At its 21st Session held in June 1986 the Milk Committee stressed the need to pay attention to health considerations (e.g. contamination by toxic residues) relating to milk and milk products which are consumed by vulnerable groups of population. The Milk Committee noted especially that the recent accident involving a nuclear power plant had demonstrated the lack of an agreed approach to dealing with scientific and health aspects of radio active contamination of milk. The Milk Committee recommended that FAO, WHO and Codex should give consideration to dealing with the contamination of food with radio nuclides.

The Executive Committee of the Codex Alimentarius Commission has recommended at its 33rd Session the establishment of a new Codex Committee on Environmental Contaminants which would concern itself with radio nuclides and other contaminants capable of creating problems for international trade in food.

Selected References
Joint FAO/WHO Committee of Government Experts on the Code of Principles concerning Milk and Milk Products
- Reports of the Thirteenth to Twentieth Sessions CX5/70 13th to CX5/70 - 20th
- Code of Principles etc. 8th edition, CAC/Vol. XVI, Ed. 1 FAO/WHO 1984, Joint FAO/WHO Food Standards Programme, Codex Alimentarius Commission.
- Procedural Manual, 6th Edition (in print) FAO/WHO
  F. Winkelmann "Legal Aspects - Milk Products of the Future", Society of Dairy Technology, United Kingdom 1974

DISCUSSION

D. Makriniotis (Greece) : 1. A post fermentation heat-treated yoghurt, reinoculated with a yoghurt culture may be considered as a typical yoghurt.
2. What is the policy with respect to preservatives carried over by added flavours?
F. Winkelmann (Italy) : In both questions the national governments have to decide.

Z. Puhan (Switzerland) : What is the role of acetic acid bacteria in kefir?
J.Lj. Rasic (Yugoslavia) : Acetic acid bacteria contribute to the viscosity of kefir and partly to its flavour. Also they contribute to the maintenance of a symbiotic relation between the various kefir micro-organisms.

J. Rossi (Italy) : How can CO_2 formation be controlled when kefir is manufactured with starters instead of with grains?
J.Lj. Rasic (Yugoslavia) : Acetic acid bacteria contribute to the viscosity of kefir and partly to its flavour. Also they contribute to the maintenance of a symbiotic relation between the various kefir microorganisms.

J. Rossi (Italy) : How can CO_2 formation be controlled when Kefir is manufactured with starters instead of with grains?
J.Lj. Rasic (Yugoslavia) : It is possible to produce kefir by using pure micro-organisms: lactic streptococci (e.g. *S. lactis*), aroma producing streptococci (e.g. *S. lactis* , subsp. *diacetilactis*) or *Leuconostic cremoris* , *L. casei* and lactose fermenting yeasts (e.g. *Saccharomyces lactis*). Possibly, cultured milk with lactic streptococci and aroma-producing streptococci or *Leuconostic cremoris* can be mixed with cultured milk with lactobacilli and yeasts, then packed, cold-stored for 24 hours and distributed.

F.M. Driessen (the Netherlands) : What is your opinion about the production of acetic aldehyde in so-called monostrain yoghurt, which is milk fermented with *S. thermophilus* alone?
V.T. Merilaïnen (Finland) : As far as I know, normal amounts of acetic aldehyde compared to yoghurt are produced, provided hydrolyzed casein is added.

F.M. Driessen (the Netherlands) : Do you have any preference for packaging fermented milks in air tight or air permeable coated cartons? There are arguments in favour for either.
V.T. Meriläinen (Finland) : This is a very difficult question. The answer is related to the type of product. Especially oxidative off-flavours shall be avoided.

CONCLUSIONS from session 1 and session 2 by Dr. F.M. Driessen (the Netherlands) and Prof.Dr. Z. Puhan (Switzerland)

1. Nowadays a large variety of fermented milks is produced because of their good organoleptic properties. In the near future the health properties will become more important.

2. At present microorganisms for the production of fermented milks are selected for growth in milk. In the near future selection will also be based upon interaction with gastro-intestinal microflora and metabolic properties.

3. The influence on the immune system and the depression of enzymes involved in the development of cancers are new areas of scientific interest, which need further elucidation.

4. In the future genetic engineering of microorganisms and new technological possibilities will allow to create new fermented milks in which the favourable properties are emphasized.

5. New products should be developed while the traditional ones should be maintained and in some cases improved.

6. A need exists for a general standard for fermented milks as well as standards for products other than yoghurt.

Seminar VIII: Milk as a Source of Ingredients for the Food Industry

Session 1

Chairman: Dr. Ir. J. M. G. Lankveld (The Netherlands)
Secretary: Prof. Dr. E. H. Reimerdes (Fed. Rep. Germany)

Milk powder

Lactose and lactose derivatives

THE USE OF MILK POWDER IN FOOD PRODUCTS

B. K. Mortensen
The Danish Government Research Institute For Dairy Industry
Roskildevej 56
DK-3400 Hillerød
Denmark

1. ABSTRACT

A wide range of milk powders is available to the food industry with various levels of fat and with specified properties and composition. The subject discussed in the present report includes only ordinary milk powders without fractionation and modification of specific components. The use of different types of milk powder in meat products, convenience foods, beverages, dietetic products and pet foods is covered, and examples are given within the different areas of application.

2. INTRODUCTION

Milk is regarded as a nutritionally very valid food commodity and is probably the best documented food stuff in existence. It has therefore been long recognized that milk and dairy products are very useful additives for incorporation into other foods. Two major reasons could be mentioned. First of all addition of dairy components increases the nutritional value of non-dairy foods, and secondly it might improve important functional properties like flavour, mouth-feel, consistency, stability etc. of the final product.

The main topic of this seminar, Milk as a source of ingredients for the food industry, opens a very broad angle towards application of the different components in milk. Some of them will be discussed in more details later in the programme, and I shall therefore cover only the use of milk powders like whole milk powder, skim-milk powder, buttermilk powder and whey powder without fractionation or modification of specific components.

3. MILK POWDERS

3.1. Whole milk powder

Whole milk powder is easier to use in food formulations, easier and less expensive to transport and easier to store than fluid milk, but it has due to its sensitivity towards fat-oxidation a limited storage stability. The development of the very distinct oxidation flavour can be delayed by packaging of the whole milk powder under nitrogen or carbon dioxide. Addition of antioxydants to the powder is another possibility.

Although the mentioned flavour problems to some extent have restricted the use of whole milk powder, large quantities are being used in the food industry. The main application of the whole milk powder in non-dairy foods seems to be in confectionery products like candy and chocolate, in different cakes, cookies and biscuits and in convenience foods like soups.

3.2. Skim-milk powder

Skim-milk powder is the most used dairy ingredient in the food industry. It has the same advantages as whole milk powder but at the same time a much higher storage stability.

Different types of skim-milk powder are marketed. High heat skim-milk powder given a strong heat treatment during the manufacturing process is used in the bakery industry, and the high water absorption ability of the high heat product is also advantageous in confectionery products and in comminuted meat products. Medium heat powder has a less distinct cooked flavour and is used in products such as ice cream where both flavour and water binding are important parameters. Low heat powder is preferred in e.g. beverages due to its excellent flavour and solubility properties. If it is important that the product is easily dispersed and solubilized use of instantized skim-milk powder may be preferred.

Generally speaking skim-milk powder has been and still is used in large quantity in the food industry, but over a wide range of applications it is now facing competition from e.g. caseinates and soy proteins.

3.3. Buttermilk powder

Dried buttermilk contains a higher concentration of fat than skim-milk powder, and much of this fat is phospholipids which provide excellent emulsifying and whipping properties. Except from this specific difference dried buttermilk is very similar in composition to skim-milk powder, and the two are frequently used interchangeably. It should be mentioned, however, that the relatively high content of unsaturated phospholipids makes the product less suitable for long term storage.

3.4. Whey powder

The interest for whey powder has in recent years shifted towards de-lactosed whey powder and whey protein concentrates. Whey powder is, however, still used in meat products, in bakery products and in confectionery. The bakery industry is without doubt the principal user of whey solids.

4. APPLICATION AREAS

In the following is given a review of recent published findings on the use in non-dairy foods of the different powders already mentioned with specific focus on examples of application in meat products, convenience foods, beverages, dietetic products and pet foods, as confectionery and bakery products are discussed in the following papers.

4.1. Meat products

When an animal is processed the primary cuts are first selected as steak, fillet, ham etc. The larger muscles left are then processed into coarsely comminuted products like meat balls, hamburgers, bratwurst etc., while the trimmings from the carcass and parts not considered for separate processing serve as basic material for finely comminuted products like luncheon meat, frankfurters, liver spread etc.

It is important that meat products are stable during processing through steps like e.g. cooking, smoking, drying, cooling, frying, slicing and mechanical handling which means that there should be absolutely no separation of free fat or water from the products.

The myofibrillar proteins are the main stabilizers in meat, but the content of these proteins in finely comminuted meat products is often low at the same time as the amount of free fat could be rather high. A main technical problem in processing comminuted meat products is therefore binding of water and fat. The fat needs to be emulsified, and it is also of importance that the product sets upon heating so that the water is bound and the fat particles in the meat are included in a homogeneous product.

In the manufacture of processed meat products functional properties like emulsifying capacity, emulsion stability, water binding and cohesion of particles are therefore extremely important. To improve these properties of the minced meat, non-meat binding agents are often added. It might be that only low amounts of non-meat ingredients are added in order not to change the authentic character of the product, but if the additive has a neutral character and a sufficiently low price it could also be regarded as a meat extender replacing lean meat. The extension limit is in such cases often set by insufficient structure of the final product.

Skim-milk powder is widely used as a neutral filler with a good water binding effect in comminuted products like meat balls, breakfast sausages and liver products where the addition of skim-milk powder reduces the cooking losses. A limitation for use of skim-milk powder might be the content of lactose which can give a Maillard reaction during heat treatment for example in canned wieners where it is very unsuitable that the product undergoes a browning reaction during processing. In other cases where the browning reactions contribute to the special flavour of the cooked product the lactose plays a positive role. Calcium-ions might negatively influence the capacity of meat to bind water. It might therefore be advantageous to use skim-milk powder with a reduced calcium content.

Whey powder can be used instead of skim-milk powder. It is cheaper and gives a good consistency of the product due to the gel forming properties of the whey proteins, but the high content of lactose sets a limit for the use of the product.

The beneficial effect of adding skim-milk, buttermilk or whey powder to finely comminuted processed meat products have been reported for many years. From recent reports a few examples could be mentioned:

- In Yugoslavia it has been shown that addition of 1.5 to 4 % skim-milk powder to products like liver patés, mortadella, frankfurters and chopped tinned meats improves flavour and consistency. Especially the emulsification of fat was improved (1).

- It has also been stated that addition of 3.5 % calcium-reduced skim-milk powder to Braunschweiger sausages greatly enhanced flavour and stability of sausages cooked at high temperatures (2).

- From Canada it has been reported that 1.4 % whey powder could be utilized as the sole fermentable carbohydrate source in dry sausages if appropriate starter cultures are selected (3).

- 2.5 to 15 % skim-milk powder added to chicken sausages was shown in India to give the best flavour and the most acceptable product as compared to additions of semolina, maize, starch or pasteurized egg white. No significant variation in appearance and tenderness was revealed (4).

- In production of breaded broiler drumsticks it has been shown that the protein source used in the pre-breading mix affects the binding of the breading to the poultry skin. Skim-milk powder gave a better adhesion than soya or whey protein (5).

- The nutritive value of sausages in terms of protein/fat ratio, protein digestibility and tryptophan/hydroxy-proline ratio can be improved by adding a preparation consisting mainly of a ripened buttermilk concentrate with addition of blood and collagen-rich material. The best results were achieved by addition of 10 to 15 % of the preparation (6).

- Skim-milk powder can be added to minced fish to produce fish pies and fish sausages. A mix containing approximately 5 % skim-milk powder in combination with other additives was proved advantageous (7, 8).

4.2. Convenience foods

Convenience foods are foods that require a minimum of preparation from the consumer, i.e. they can be foods that require only addition of a single ingredient (water or milk) followed by cooking. They can also be foods that require no processing at all like ready-to-eat salad dressings or puddings etc., or foods that only require heating.

In a variety of convenience foods dairy ingredients exhibit a number of useful functions as flavour enhancers, whiteners, emulsifiers, thickeners, whipping agents etc., and furthermore dairy ingredients in many cases provide the image of old-fashioned quality of the final product that many consumers are seeking.

Generally speaking milk powders add a pleasant, enhanced and balanced flavour and a milky white colour to convenience foods which to many consumers associate with richness.

The present trend, argued by the changing lifestyle of the consumer towards safe, wholesome, easy-to-prepare foods, has led to a rapid development of convenience foods. Only a few examples from a very broad application range can be mentioned here, but obviously there are great possibilities for utilizing dairy ingredients in convenience foods.

- In dehydrated soup mixes skim-milk powder in combination with other dairy ingredients can be used as a whitener and flavouring agent. Buttermilk powder in addition to acting as a whitener imparts richness to the mix. Skim-milk powder can also be used in canned soups and in frozen soups for the same reasons.

- As in dry soups skim-milk powder can be utilized as a whitener and flavouring agent in dry sauce mixes. Buttermilk powder also imparts richness to these products, and furthermore acidified buttermilk solids can be used as an acidulant in dry sauces. It has also been experienced that whey proteins stabilize sauces as they are less prone to cook onto the walls of the processing equipment.

- The main dairy ingredient in salad dressings is sour cream, preferably supplied in dried form due to the wish for simplification of the processing. Skim-milk and buttermilk powder can be used for enhance of flavour and as a whitener. These powders in addition act as emulsifiers and thickeners.

- Important convenient foods are the broad variety of cereal products. Milk powders, both whole milk powder and skim-milk powder, can be used in breakfast cereals, e.g. in muesli, to eliminate the need for adding of milk. A German patent describes a product ground and combined with rolled oats, dry yeast and wheat bran. The blend is then mixed with quarg, yoghurt or dried milk, and other foodstuffs such as fruit, cocoa, chocolate and sugar are added (9). For instant cereal gruels, e.g. for infant feeding, milk powders are also used as ingredients due to the positive effect on the biological value and on the digestibility of the product (10). An example from the USSR is groats of high nutritive value achieved by preparing a mix of 73 % rice, 15 % wheat and 10 % skim-milk powder meant for use in infant feed, and a special product for sportsmen containing 90 % oats and 10 % skim-milk powder (11).

- From Italy is reported the use of skim-milk powder and whey protein concentrate in the formulation and production of pasta. The best products were obtained when the traditional ingredients including maize, germ flour, pea flour, pea protein concentrate and soya protein isolate were fortified with the mentioned dairy ingredients (12).

- Also the developing world has shown an interest in utilizing milk powders in the production of nutritious convenience foods. An instant mung-bean soup has been developed in the Philippines by adding skim-milk powder to mungo grits (13). Another example is Vadas, a common Indian fried fast-food dish based on 30 % soya beans and 70 % pulse-mix fortified with up to 10 % skim-milk powder. The fortification increased the fat absorption during frying and the nutrition value of the final product mainly due to an increased content of methionine and cystine (14).

4.3. Beverages

Beverages cover the whole range from ready-to-drink products requiring no preparation at all, over concentrated liquids requiring dilution, to powders requiring reconstitution.

Among dairy ingredients the beverage industry is mainly interested in the protein components of milk since it is of major importance to the flavour of the products. Protein may present a number of technical problems in the manufacturing process e.g. foaming, and specially designed milk protein preparations are therefore now finding their way into the market and will most probably eventually replace most of the traditional milk powders in the beverage industry.

In a number of beverages, typically dry mixed products containing appreciable amounts of milk components, milk powders serve as the main ingredient. Other dry ingredients are typically sucrose or other sugars, flavour, colour and stabilizers, perhaps also an acidifier. In other beverages milk components are only minor ingredients.

A few examples of utilization of dairy ingredients in beverages could be mentioned:

- Recipes for dairy based beverages are reported from Canada. An example is a dry beverage mix containing 50.01 % whole milk powder, 43.72 % sucrose, 3.50 % citric acid, 3.50 % carboxymethyl cellulose as a stabilizer and small amounts of flavouring and colouring agents. The mix is reconstituted with milk or water to provide an acidified milk drink without any curdling (15).

- From Australia is reported on a drink containing milk, skim-milk powder, malt, whole egg powder and wheat germ but no artificial ingredients like stabilizers and flavour agents. Other variants of the same drink contain carob or honey (16).

- In acidic beverages, typical citrus flavoured juices, however, traditional milk powders are not very useful. Instead whey proteins can be incorporated due to their stability at a low pH. An example from USA is a protein enriched citrus soft drink incorporating with a commercial citrus soft drink formulation, a protein-fortifier consisting of a mixture of whey protein concentrate with high-heat skim-milk powder. The fortifier is added at a rate sufficient to provide at least 0.1 % protein in the final drink (17).

- Chocolate milk powder is a typical example of products with a high content of milk powder. Many products also include whey solids to replace non-fat milk solids, among other reasons because lactose enhances natural chocolate flavour.

- In products like coffee and tea milk and milk powder have traditionally served as whiteners. However, non-dairy whiteners now dominate this market.

4.4. Dietetic products

Dietetic products are strictly speaking foods that serve a dietetic purpose by virtue of increasing or decreasing the content of certain nutrients or physiologically active substances or changing the ratio between them in the diet.

They serve a specific purpose in the case of e.g. illness, deficiency, functional disorder, allergies or overweight. They may also be of importance during pregnancy or breast-feeding or as products designed for infants and young children and for the elderly. It is obvious that the distinction between normal food and diet products is often vague.

In most cases traditional milk powders contain too broad a range of nutrients to be used unmodified as a basic component in dietetic foods. Skim-milk powder for example contains lactose, and therefore should be avoided in diets for diabetics. Nor should unmodified milk powders be used directly in cases of lactose intolerance. In this case lactose-hydrolysed milks or lactose-reduced milks can be used. Skim-milk powder, however, can be utilized in diets for fat resorption disorders, hypocholesterinaemia and constipation or in case of treatment of gluten-sensitive enteropathy.

The use of milk proteins, caseines and whey proteins in food formulations will be discussed later today so let it just briefly be mentioned here that milk powder can be used in milk based infant foods, follow-up formulas like milk or cereal based foods for bottle feeding and in cereal based infant foods.

In the production of slimming meals or meal replacers skim-milk powder can play an important role due to the high biological value of the protein. Caseinates are, however, a cheaper source of milk protein and are no doubt more widely used.

Foods meant for elderly or for sportsmen etc. is an area where milk powders have found some use. These products, though not strictly dietetic foods, therefore deserve mentioning. Athletes, especially those engaged in muscle sports and body building, need a high daily supplement of protein, and numerous products have been developed to meet this demand.

One interesting aspect of dietetic foods worth mentioning is the increasing interest for incorporation of milk cultures containing Lactobacillus acidophilus and Bifidobacterium bifidum claimed to have a positive influence on the digestion and on the gastro-intestinal flora.

A few examples of milk based products within the area of dietetic products could be mentioned:
- Bran products are widely used in diet food, both for their low calorific value and for their ability to combat constipation. Milk products have proved to be valuable additives to bran products added before the mix is extruded into low calorie, protein enrich, easy to eat snack products.

- Products containing both milk and honey always seem to have been particularly attractive to man. A German patent describes the manufacture of a buttermilk-honey preparation in which dried buttermilk and honey are mixed to yield a sweet-sour fruity tasting granulate which can be packed for direct use or pressed into tablets (18).

- From USSR is reported the development of a cultured milk health-food containing 25 % co-precipitated milk proteins, 20 % skim-milk powder, 5 % sugar and 50 % of a Lactobacillus acidophilus base plus small amounts of vitamins. The acidophilus base contains about 2.7 millions viable acidophilus cells per ml. The product has to be reconstituted in water and is particularly intended for children (19).

- Also from USSR is described a protein-fruit product specially intended for sportsmen. The product consists of a mix of dried milk, casein and soya protein, oat broth, sucrose, chocolate and a number of salts, minerals and vitamins plus an acidophilus culture. The product is freeze-dried, comminuted and mixed with a fruit puree. The product has to be reconstituted before consumption (20).

4.5. Pet foods

Although this seminar concerns primarily human food a few remarks should be included on the utilization of milk powders in animal feed and particularly in pet foods.

The major use for powder is in calf feeds, particularly vealers, but low levels of skim-milk powder have also been utilized in pet foods until the price became too high. It is however, to a limited extent still being used in specialized mainly therapeutical pet diets, as such products normally can be sold at higher prices. It should also be possible, at least from the technical point of view, to utilize texturised skim-milk powder as a meat replacer in pet food, but again the question is whether the price is competitive.

Whey powder has to some extent become an ingredient utilized by the pet food industry mainly to boost the nutritional value of the products but also in order to improve the keeping quality of moisture products due to the impact on the water activity of the product.

Aside from the specialized pet diet products milk powders could be used in more fancy pet food products. A few examples can be mentioned:

- An US patent describes a frozen simulated ice cream product made from a blend containing 40 % dry matter. The base mix contains 13 % whey powder, 20 % de-lactosed whey powder and 30 % skim-milk powder plus fat and vitamin and mineral additives (21).

- A British patent describes a dry pet food consisting of milk, milk by-products, cereals, starches etc. which on reconstitution with water yields a palatable wet meat-like textured food. The product is fortified with vitamins and minerals. Several of the tested formulas contained milk products, e.g. fat-filled milk powder in amounts up to 20 % of food for cats and up to 10 % of feed for dogs. The difference is due to lactose intolerance in dogs (22).

5. FINAL REMARKS

A close look into the different areas of utilization defined at the beginning of this subject has revealed that the situation is not very encouraging. Many examples of application of milk powders have been reported, but none of them is likely to make the world stock of powder disappear.
In spite of the surplus situation the price of bulk milk powder is so high that most other protein sources are competitive, and dairy ingredients do not seem to have much of a chance in the unspecified, cheap end of the bulk market.
One of the conclusions of the IDF seminar on Dairy ingredients in foods held four years ago in Luxemburg was:

> "The dairy industry must intensify its efforts to identify the dairy ingredient needs of the food industry and support research and development needed to supply the full potential of the market for dairy ingredients."

This is still valid today. The only way of opening up for utilization of expensive dairy ingredients in composed foods is in my opinion specialization and design of tailor-made products with specific and refined functional properties. To do that it is extremely important to establish a close cooperation between research, experimental development and marketing.

6. REFERENCES

1. Roseg, D., Kampus, A., Petrak, T., Jelic, A., Hraste, A.: Dried skim-milk as a constituent of sausage and tinned meats. Dairy Sci. Abstr., 44, 98. (1982).

2. Chyr, C.Y., Sebranek, J.G., Walker, H.W.: Processing factors that influence the sensory quality of braunschweiger. J. Food Sci., 45, 1136. (1980).

3. Paradis, D.C., Mungal, M.: Whey utilization in fermented sausage. Evaluation of chemical, sensory and physical characteristics. Can. Inst. Food Sci. Technol. J., 17, 44. (1984).

4. Panda, P.C., India Poultry Gazette, 64, 48. (1980).

5. Suderman, D.R., Wiker, J., Cunningham, F:E: Factors affecting adhesion of coating to poultry skin. J. Food Sci., 46, 1010. (1981).

6. Schnackel, W., Dantschev, S.A.: Anwendung eines kombinierten Eiweisspräparates bei der Bruhwurstproduktion. Fleisch, 37, 158. (1983).

7. Loseva, A.V., Teplitskaya, A.M.: Cooked frozen Alaska pollack pie. Dairy Sci. Abstr., 42, 21. (1980).

8. Didenko, A.P., Goroshko, T.N.: Effect of addition of flavour compounds and binders on the structural, mechanical and sensory properties of cooked/frozen fish sausage. Dairy Sci. Abstr., 42, 21. (1980).

9. Hodapp, F.: The use of malt grain residues in combination with other biological raw materials, such as oats, dry yeast, wheat bran, in quarg or dried milk for producing a breakfast cereal. Dairy Sci. Abstr., 44, 312. (1982).

10. Blattna, J., Dostalova, J., Eisenberger, B., Holasova, M., Pickova, J., Simova, J.: Nutritive value and stability of instant cereal gruels for infant feeding. Dairy Sci. Abstr., 46, 119. (1984).

11. Lopatinskii, S., Zenkova, A., Pavlova, N.: Groats of high nutritive values. Dairy Sci. Abstr., 42, 904. (1980).

12. Pagani, A., Resmini, P., Dalbon, G.: Formulation and production of pasta using non-conventional raw materials. Dairy Sci. Abstr., 44, 22. (1982).

13. Payumo, E.M., Castillo, E.S.: Preparation and storage qualities of instant mung bean soup. Dairy Sci. Abstr., 42, 575. (1980).

14. Gulati, T., Chopra, A.K., Bhat, C.M.: Effect of supplementation of sesame and skim-milk powder on the nutritional quality of vadas. Dairy Sci. Abstr., 44, 49. (1982).

15. Sirett, R.R., Eskritt, J.D., Deriatka, E.J.: Dry beverage mix composition and process. Dairy Sci. Abstr., 44, 308. (1982).

16. Mann, E.J.: Dairy beverages. Dairy Industries International. 48, 13, (7). (1983).

17. Scibelli, G.E.: Fortification of soft drinks with protein. Dairy Sci. Abstr., 42, 904. (1980).

18. Mann, E.J.: Dairy ingredients in dietetic products. Dairy Industries International. 47, 11, (10). (1982).

19. Tolstenko, L.A., Krasheninin, P.F., Shamanova, G.P., Stolyarova, A.V., Ladodo, K.S., Barashneva, S.M.: Cultured milk health-food product. Dairy Sci. Abstr., 45, 645. (1983).

20. Polyachenko, N.S., Belova, N.M., Laricheva, K.A., Yalovaya, N.I., Gudochkova, V.M.: Protein fruit food product. Dairy Sci. Abstr., 46, 667. (1984).

21. Mann, E.J.: Pet foods. IDF Bulletin. Document 147, 87. (1982).

22. Burnett, G.S.: Dry pet foods. Dairy Sci. Abstr., 44, 401. (1982).

THE USE OF MILK POWDERS IN CONFECTIONARY AND BAKERY PRODUCTS

Rory A.M. Delaney
Research and Development Department
Frito-Lay, Inc.
900 N. Loop 12
Irving, TX 75061

Nitin Desai
Research and Development Department
Frito-Lay, Inc.
900 N. Loop 12
Irving, TX 75061

Virginia Holsinger
U.S.D.A.
Eastern Regional Laboratory
600 East Mermaid Lane
Philadelphia, PA 19118

ABSTRACT. Milk powders have been used in the confectionary and bakery industries for many years, largely on an empirical basis, because of their contributions to texture, color, and flavor of the products. Incorporation of milk powders can significantly improve the nutritional value of baked and confectionary goods as well as improve their processability.
  This review covers the following aspects of milk powders: production, manufacture, composition, and microstructure of milk powders, functional properties of milk powder components, influence of processing on milk powders, and applications of milk powders in confectionary and bakery products.

1. INTRODUCTION

Milk powders consisting of dry whole milk powder (DWM), nonfat dry milk (NDM), and buttermilk powder (BM) are derived from the dehydration of milk and milk fractions. Milk powders constitute an important part of the food proteins produced on an industrial scale for food applications. Food proteins were first produced due to the necessity to preserve foods, handle wastes, and other economical considerations; they were not produced, per se, for their functional properties. New food products can be fabricated entirely from industrially produced ingredients such as milk or soy proteins, shortenings, and polysaccharides including starch. Milk powder and other food ingredients are frequently added to foods to enhance the structure of the food and the functional properties related to processing and stability of the food product. Milk powders have been traditionally added to confectionary and bakery products as a replacement for fluid milk in the original recipes to improve the organoleptic properties - color, flavor, and texture, as well as the nutritive value of the product. This article reviews the following aspects of milk powders: production, manufacture, composition, and microstructure of milk powders, functional properties of milk powder components, influence of processing on milk powders, and applications of milk powders in confectionary and bakery products.

2. MILK POWDER

2.1 Production and Utilization of Milk Powder

Table I contains data on milk powder production in the USA. If present trends continue, the outlook for 1986 is for a 1% increase in the production of milk powders.

The confectionary and bakery industries have traditionally utilized milk powders as a replacement for fluid milk in the recipes. Typical usage of milk powders in a variety of confectionary and bakery products is shown in Table II. There has been a steady decline in the use of milk powders in the bakery industry in recent years due to replacement of milk powder by other, less expensive sources of food proteins, principally soy proteins and whey protein concentrates. This change has been made possible due to our understanding of the functional role of proteins in foods; thus replacement of expensive or fluctuating sources of proteins, such as milk powders, is possible without sacrificing product quality. The confectionary industry utilizes a large portion of the WMP produced for milk chocolate and candy manufacture. This situation could change if suitable replacements for WMP, for example whey protein concentrate or casein, can be successfully utilized. (Lim, 1980; Dodson, et al. 1984)

2.2 Manufacture of Milk Powders

The manufacture of milk powders involves the concentration and dehydration of milk/skim milk/buttermilk and is most commonly accomplished by spray-drying and, in some cases, by drum drying. Powders obtained by drum drying have very poor reconstitution properties and thus have limited usage. The spray-drying process is described by Masters (1976) and the manufacture of milk powders has been treated by Pallansch (1970), Jensen (1975), and Pisecky (1978), In Figure 1, the process for the manufacture of milk powders is outlined. The physical properties of the powders, e.g. particle size, porosity, bulk density, flowability and wettability, are largely determined by the process and processing conditions (DeVilder et al.

1976; Pisecky, 1978).

Milk crumb, a mixture of dried milk, sugar, and cocoa, is used in the UK for making milk chocolates and coatings. Milk crumb is manufactured by adding sugar to milk at 80°C to obtain an 80% solution. The mix is concentrated to 88% solids and cocoa mass is added to the mix. Sugar is allowed to crystallize (particle size not exceeding 35 microns) in mixers and the mass is moulded in trays, dried to 1-2% moisture. The product is then crushed and milled to obtain milk crumb (Lim, 1980). Milk crumb has good keeping qualities and confectionary products made from milk crumb have better texture and flavor (Edwards, 1984).

2.3 Composition of Milk Powders

Data on the typical composition of milk powders are given in Table III. NDM powder is classified according to the heat treatment given to milk prior to dehydration and is based on the amount of undenatured whey protein in the powder (Table IV). Therefore, when producing NDM powders of certain heat specifications, it is necessary to know the whey protein content in the skim milk and the heat stability of the whey proteins to determine the amount of heat treatment that must be given to milk prior to dehydration to produce NDM powders according to the heat classification (Sørensen et al. 1978). The Harland and Ashworth method, based on turbidity measurements, is used to determine the amount of undenatured whey protein in heat treated milk powders (Kuramoto, et al. 1959). HPLC and DSC methods hold promise as rapid methods for the determination of heat denaturation of proteins in milk products.

2.4 Influence of Processing on Milk Powder Components

There are several process induced changes in the chemical and physiochemical properties of milk powder components. Morr (1984) has summarized these changes for milk proteins (Table V). The manufacture of milk powder involves heating the milk during forewarming, concentration, and spray-drying (Figure 1). Therefore, heat induced changes during manufacture are of significance. Kinsella (1984), Kilara and Sharkasi (1986) have reviewed the functional properties and the influence of heat on milk proteins. Casein micelles are relatively heat stable, whereas the whey proteins, β-lactoglobulin and α-lactalbumin, denature rapidly on heating at 70°C for 30 minutes. Though the casein micelles are heat stable, several large aggregates of the micelles can be observed during electron microscopy of reconstituted NDM. The forewarming stage is the most critical stage during the manufacture of milk powders and largely determines the end use of the spray-dried powder.

The interactions between casein and whey proteins has been extensively studied and it has been theorized that κ-casein, β-lactogobulin and α-lactalbumin interact to form a thermally induced complex. Heat treatment of skim milk normally results in the formation of aggregates and loss of solubility of the protein. In the manufacture of whole milk powder, forewarming to a relatively high temperature is necessary to ensure the inactivation of lipases. Normally, extremely high heat treatment of milk is avoided since protein-sugar interactions leading to browning and protein insolubility can occur. These conditions are generally avoided during the manufacture of milk powders.

The microstructure of milk powders has been examined by scanning electron microscopy (SEM) (Figure 2). Spray-dried NDM particles are spherical in shape and are wrinkled on the surface. The wrinkles on the surface of the particles are caused by the casein in the milk which contracts on the surface of the particles during spray-drying (Buma and Henstra, 1971). These particles also contain large vacuoles which determine the bulk density of the powders.

Graf and Bauer (1976) used electron microscopy to determine the effect of processing and drying of milk on casein micelles. The micrographs in Figure 3 indicate that the processing and drying of milk causes only a slight change in the casein micelle structure. A denser arrangement of micelles can be observed in reconstituted milk when compared to fresh milk. These studies reveal that when milk powder is reconstituted the caseins still exist as micelles.

Lactose in spray-dried milk normally is in the amorphous state. Microstructure studies of milk powders reveal that "instantizing" or agglomeration of the powder leads to the conversion of lactose from a glass form to a more wettable crystalline form. The most common form of lactose observed in NDM powders is the typical tomahawk and prism crystals ranging from 60 to 170 mm in length (Kalab, 1979).

In DWM, fat is present as surface fat on powder particles or in pools and can be easily extracted by solvents. As a result of homogenization, fat is encapsulated by protein and is present as a fat/protein complex. This form of fat is not easily extracted by solvents and may add to the stability of the DWM to lipid oxidation (Buma, 1971; Walstra, and Ooertwijn, 1981).

3. FUNCTIONAL PROPERTIES OF MILK POWDER COMPONENTS IN CONFECTIONARY AND BAKERY PRODUCTS

3.1 Components of Milk

In Figure 4, the breakdown of milk into its components is shown. When milk or milk powder is added to a recipe or a food product, these components impart some characteristic functional, organoleptic and nutritional property to the food.

3.2 Milk Proteins

Casein constitutes approximately 80% of the milk proteins and is composed of α, β, κ, and γ caseins. The milk serum proteins (whey proteins) are composed of α-lactalbumin, β-lactoglobulin, immunoglobulins, proteose peptone, etc.. The proteins of casein readily interact and exist primarily in the form of micelles.

The micelle contains not only casein but also non-protein components such as calcium and phosphorous. Numerous models for the casein micelle have been proposed (Thompson and Farrell, 1974). Among these, the subunit model originally proposed by Morr (1967) and later modified by several researchers is the most widely accepted model. Recently, a subunit model with a "hairy" layer, or a hairy micelle model, has been proposed for the casein micelle (Walstra, et al. 1981; Holt, 1985).

The whey proteins of milk exist as monomers and are also capable of forming larger complexes by means of -S-S- bonds. Casein is an assembly of proteins which does not have higher structures and can be considered as a protein which is denatured in its natural state. Therefore, casein is considered to be relatively heat stable under normal processing conditions. Of the milk proteins, β-lactoglobulin readily undergoes heat denaturation at 55°C or above. Unfolding of the molecule occurs and the -SH groups are exposed. At 70°C or above, irreversible complexes are formed by means of -S-S- bonds. The whey proteins form a complex on heating and the presence of casein accelerates the interaction. The formation of a complex wherein κ-casein, β-lactoglobulin and α-lacalbumin are involved is conceivable.

In recent years the functional properties of food proteins have been extensively studied and many excellent reviews and books have been written on the subject (Pour-El, 1979; Cherry, 1981; Kinsella, 1982).

Kinsella (1970, 1971, and 1984) and Lim (1980) have reviewed the functional properties of milk powders in bakery and confectionary products. There are many important functional properties for proteins in food systems, but in confectionary and bakery products only a few of these properties are of significance since these products are water restricted, low moisture systems (Table VI).

In confectionary products viscosity, gelation, emulsification and fat binding properties are important, whereas in bakery products water binding, emulsification and foaming properties are of significance. The precise role of milk proteins in confectionary and bakery products is not well understood since food systems are highly complex (Figure 5) and functional properties studied in simple model systems may not be applicable to the food system due to interactions. Harper (1984) has suggested a model food system for evaluating protein functionality where the system is designed to study a single or several functional properties of the protein.

Milk protein is essential for the development of texture, flavor, and color of confectionary products such as toffees, caramels, fudges, milk chocolate, marshmallows, and nougats. Sugar, fat, and milk are the basic ingredients of confectionary products. The order of addition and processing conditions determine the textural properties of the products. Milk proteins aid in the emulsification and mixing of the ingredients and influence the viscosity of the matrix. Heating the mix during cooking and subsequent processing results in the unfolding of proteins and formation of fibers by means of -S-S- bonds (Kinsella, 1970). The result is a viscoelastic network which imparts structure and texture to the product. Microstructure studies indicate differences in the structure of confections prepared from DWM and milk crumb, and these differences are related to the association of protein with sugar and cocoa particles (Heathcock, 1985).

The firm chewy texture of several confections are related to the water binding properties of casein. Products such as toffees exhibit "cold-flow" phenomenon when the milk casein is replaced by whey protein (Dodson, et al. 1984). The texture of caramels and toffee is attributed to the emulsifying and water binding properties of milk proteins.

In bakery products, the addition of milk powders increases the amount of water to be added in mixing the dough. The addition of 6% NDM increases the water absorption capacity of dough by approximately 6%. High heat NDM (WPNI $\leq$ 1.6 mg/gram) is used in bakery products, particularly for bread making. Low heat NDM reduces the extensibility of dough and bread with poor loaf volume is obtained. The volume-depressant factor is related to the milk proteins of NDM and the formation of κ-casein and whey protein complex due to heat treatment reduces the effect of the volume-depressant factor (Guy, 1970). These effects may also be related to the -S-S- bond breakage in the dough protein structure whereby weak doughs with reduced carbon dioxide holding properties are produced. Breadmaking by continuous processes generally utilizes NDM at levels below 2% since higher levels produce weak doughs and bread with poor volume. In general, the addition of NDM to bread improves the texture (crumb softness), flavor, and shelf-life of the products (Dubois and Dreese, 1984). The soft texture and increased shelf-life of bread and cookies are attributed to the water binding properties of milk proteins. In cakes, NDM aids in better foam structure and texture.

3.3 Milk Lipids

The lipid content of milk powders is shown in Table III. Milk fat is typically composed of triglycerides (97%), mono- and diglycerides (1-2%), and phospholipids (1-2%). WDM contains 27-28% fat and its role in confectionary products can be characterized as: (1) impart structure and texture to the product, (2) the surface active ingredients present in fat (mono- and diglycerides and phospholipids) affect the viscosity and emulsifying properties of the mix during processing, (3) impart flavor to the product. Milk fat is compatible with cocoa butter and DWM can be added to confectionary products without altering the properties of fat. In confectionary making the oxidation of lipids by lipases and heat is encouraged to develop important flavor components. Oxidation of fat, leading to the development of the typical undesirable rancid flavor in milk powders, determines the shelf-life of the powders. The aroma compounds of milk and dairy products have been reviewed by Badings and Neeter (1980). The role of flour lipids in baking has been reviewed by Chung (1986); the significance of the role of lipids added as a milk powder constituent has not been elucidated. The surface active properties of lipid constituents in milk powder may be of some significance during baking. BM contains large amounts of surface active membrane material which would affect the viscosity and emulsifying properties of the dough or mix during processing. BM is added to bakery products to improve the flavor and shelf-life of baked products.

3.4 Lactose

Milk powders are high in lactose (35-50%) which impart important organoleptic and textural characteristics to the product. Lactose is present in milk powders in both amorphous and crystalline states and these determine the wetting properties of milk powders. Lactose plays a predominant role in imparting a characteristic color and flavor to confectionary and bakery products. Browning reactions occurring between free amino acids and lactose are well known in food products as non-enzymatic browning or Maillard reactions which produce the color and flavor of these products (Hodge and Osman, 1976). Another mechanism whereby milk sugars impart color and flavor to products is by dehydration reactions leading to yellow-brown pigments and flavor components (Kinsella, 1970). In confectionary products, lactose imparts graininess/grittiness to the products and is a desirable textural characteristic in hard candies. Addition of milk powders to bakery products improves the crust color due to browning reactions of lactose.

3.5 Minerals and Salts

The mineral and salt content of milk powders is shown in Table VII. Dairy powders are an excellent source of calcium and phosphorus and significantly improve the nutritional value of confectionary and bakery products. The effect of calcium and other divalent cations is well known in the gelling of several polysaccharides like alginates and carrageenan. Such an effect has not been demonstrated for the polysaccharides present in wheat. Studies indicate that the mineral content of NDM added to dough prolongs the fermentation time of the doughs due to the buffering action of the milk powder and decreases the amylase activity (Pyler, 1973).

4. CONCLUSION

The constituents of milk powder components play important functional roles in confectionary and bakery products. The selection and continued use of milk powder as an ingredient in these products will be largely determined by price, process functionality, and contribution to nutritive value.

5. ACKNOWLEDGEMENTS

The assistance of Ms. Elizabeth Ott in preparing this manuscript is gratefully acknowledged.

FIGURE 1
PROCESS FOR THE MANUFACTURE OF MILK POWDERS

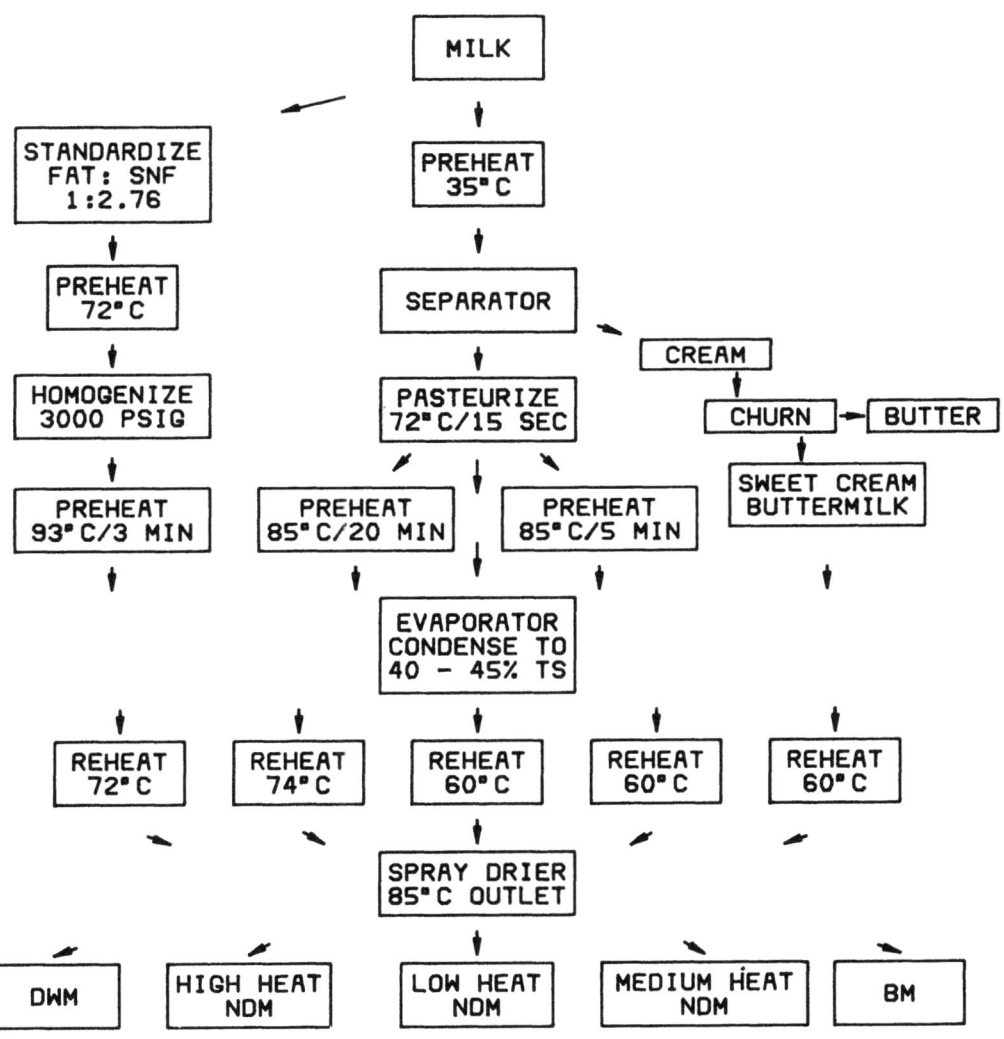

FIGURE 2
SCANNING ELECTRON MICROGRAPHS OF SPRAY-DRIED MILK POWDERS

Note the smooth and wrinkled surfaces of the particles. In the lower micrographs, the interior of the particles is visiable and contain trapped milk globules. (Reprinted with permission of SEM Inc., USA.)

FIGURE 3

(a) Casein micelles in fresh cream. The subunits are relatively loosely packed

(b) Casein micelles of a reconstituted unhomogenized milk powder. Note the denser packing of the subunits compared to 3(a).

(Reprinted with permission of Marcel Dekker, Inc., New York.)

FIGURE 4
SCHEMATIC BREAKDOWN OF MILK INTO ITS COMPONENTS

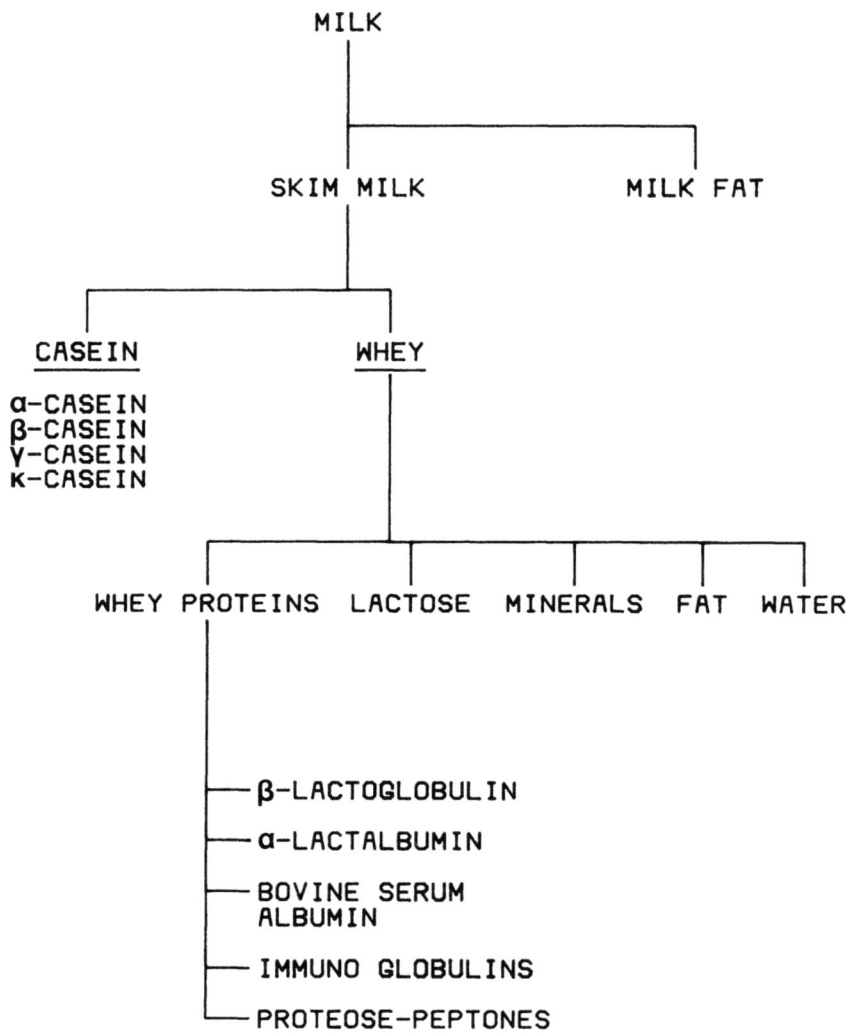

TABLE I
TOTAL MILK POWDER PRODUCTION IN THE USA[a]

Milk Powder Type	Year					
	1980	1981	1982	1983	1984	1985
Nonfat dry milk (NDM)	1,160.7	1,314.7	1,400.2	1,499.9	1,160.7	1,390.0
Dry whole milk (DWM)	82.7	92.7	102.2	111.2	119.6	118.9
Buttermilk (BM)	---	---	---	---	---	36.9

[a]Values in millions of pounds. Data from USDA Reports and American Dry Milk Institute

FIGURE 5

Figure showing the complex interactions of food ingredients in a simple food system. (Adapted from Harper, 1984.)

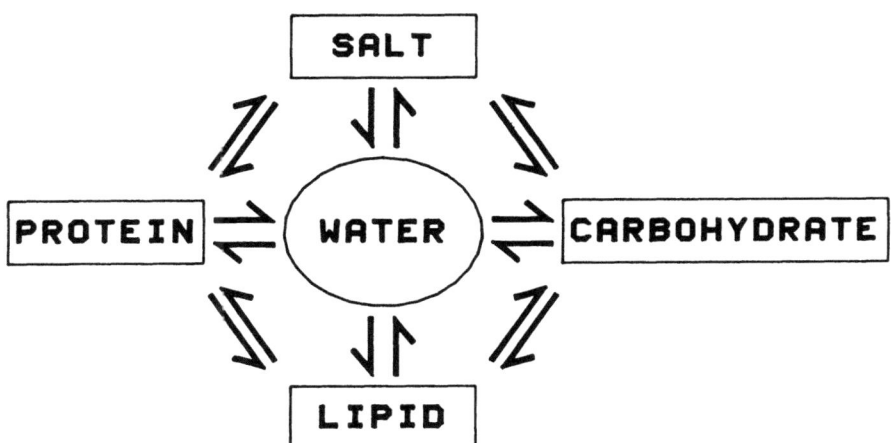

TABLE II
BAKERY AND CONFECTIONERY PRODUCTS CONTAINING MILK POWDERS

	Approximate Amount Added (%)		
Food Commodity	DWM	NDM	BM
Bread	+	1-6	+
Rolls	+	1-6	+
Cakes	1-10	5-10	0-2
Cookies	0-5	2-10	0-2
Crackers	-	2-5	-
Doughnuts	0-3	2-8	-
Danish pastry	0-2	3-15	+
Pretzels	-	3-10	-
Biscuits	0-5	5-8	2
Pies	0-10	4-8	+
Cake mixes (dry)	0-10	10-20	+
Pancake mixes	0-5	7-12	+
Waffles	-	10-15	-
Pizza dough	-	-	-
Macaroni	-	12-25	-
Batters (meats)	1-2	0-5	-
Icings & frostings	0-20	5-30	+
Chocolate	0-30	5-20	+
Fudge, fondants	0-15	3-20	+

Adapted from Kinsella (1971).

TABLE III
COMPOSITION OF VARIOUS TYPES OF MILK POWDERS, 100 GRAMS[a]

Component (%)	DWM	NDM Instant	NDM Regular	BM
Water	2.0	3.0	4.0	2.8
Protein	25.4	35.9	35.8	34.3
Fat	27.5	0.8	0.7	5.3
Carbohydrate	38.2	52.3	51.6	50.0
Ash	5.9	8.0	7.9	7.6

[a]From USDA Agriculture Handbook No. 8 (Watt and Merrill, 1975).

TABLE IV
CLASSIFICATION OF NDM BASED ON UNDENATURED WHEY PROTEIN INDEX (WPNI)[a]

Classification	WPNI/Gram Powder
High heat powder	$\leq$ 1.5 mg.
Medium heat powder	1.51 - 5.99 mg.
Low heat powder	$\geq$ 6.0 mg.

[a]American Dry Milk Institute, <u>Bulletin No. 916</u>.

TABLE V
PROCESSING INDUCED CHANGES IN CHEMICAL AND PHYSIOCHEMICAL PROPERTIES OF MILK PROTEINS[a]

Heating	Interaction of whey proteins with caseins
	Denaturation of whey proteins
	Aggregation of whey proteins
	Maillard browning reaction between proteins and lactose
	Activation of whey protein sulfhydryl groups
Acid treatment	Solubilize colloidal calcium phosphate from casein micelles
	Lower heat stability of milk proteins
	Precipitate casein and denatured whey proteins
Addition of calcium ions	Alter casein micelle structure and composition
	Lower heat stability
	Promote milk protein aggregation when heated
Rennet treatment	Hydrolyze κ-casein to release glycomacropeptide
	Lower zeta potential on casein micelle
	Coagulate modified casein micelles

[a]Morr (1984). Reprinted from <u>Journal of Food Technology</u>, with permission

TABLE VI
FUNCTIONAL PROPERTIES OF PROTEINS IN CONFECTIONARY AND BAKERY PRODUCTS

Major Functional Properties	Minor Functional Properties
Water binding	Foam stabilization
Viscosity (rheological influence)	Water holding capacity
Emulsification	Fat binding
Gelation	Flavor

TABLE VII
APPROXIMATE LEVELS OF MINERAL CONSTITUENTS IN DRY MILKS[a]

Salt Constituent	DWM	NDM	BM
	(mg per 100g)		
Major Minerals			
Calcium	909.0	1293.0	1248.0
Sodium	405.0	526.0	507.0
Potassium	1130.0	1725.0	1606.0
Phosphorus	709.0	1005.0	970.0
Chlorine	1200.0	820.0	1100.0
Magnesium	100.0	70.0	95.0
Sulfur	85.0	63.0	80.0
Trace Elements			
Zinc	1.1	1.5	1.4
Iron	0.6	0.5	0.6
Copper	0.06	0.07	0.07
Manganese	0.04	0.06	0.06
Cobalt	0.002	0.002	0.002
Molybdenum	0.05	0.07	0.07
Iodine	0.34	0.46	0.44
Bromine	2.6	3.5	3.3
Flourine	0.11	0.15	0.13
Organic Salts			
Citrates (as citric acid)	1500	2000	1900
Lactates (as lactic acid)	40	50	50

[a]Adapted from USDA Handbook No. 8 (Watt and Merrill, 1978) and Pedraja (1965).

6. REFERENCES

Badings, H. T., and Neeter, R. 1980. 'Recent Advances in the Study of Aroma Compounds of Milk and Dairy Products.' Netherlands Milk Dairy Journal **34** (1), 9.

Buma, T. J. 1971. 'Free-Fat in Spray-Dried Whole Milk. 8. The Relation Between Free-Fat Content and Particle Porosity of Spray-Dried Whole Milk.' Netherlands Milk Dairy Journal **25**, 123.

Buma, T. J., and Henstra, S. 1971. 'Particle Structure of Spray-Dried Caseinate and Spray-Dried Lactose as Observed by Scanning Electron Microscopy.' Netherlands Milk Dairy Journal **25** (9), 278.

Cherry, J. P. 1981. Protein Functionality in Foods. Washington, D.C.: American Chemical Society

Chung, O. K. 1986. 'Lipid Protein Interactions in Wheat Flour, Dough, Gluten, and Protein Fractions.' Cereal Foods World **31** (3), 242.

DeVilder, J., Martens, R. as, Naudts, M. 1979. 'The Influence of the Day Matter Content, the Homogenization and the Heating of Contrate on Physical Characteristics of Whole Milk Powder.' Milchwissenschaft **34** (2), 78.

Dodson, A. G., Beacham, J., Wright, S. J. C., and Lewis D. F. 1984. 'Role of Milk proteins in Toffee Manufacture. Part 1. Milk Powders, Condensed Milk and Wheys.' Leatherhead Food R.A. Research Report No. 491.

Dubois, D. K., and Dreese, P. 1984. In Dairy Products for the Cereal Industry. Edited by J. L. Vetter. St. Paul, MN: AACC.

Edwards, W. P. 1984. 'Uses for Dairy Ingredients in Confectionary.' Journal of the Society of Dairy Technologists **37** (4), 122.

Graf, E., and Bauer, H. 1976. In Food Emulsions. Edited by S. Friberg. Marcel Dekker.

Guy, E. J. 1970. In Byproducts From Milk. Edited by B. H. Webb and E. O. Whittier. AVI.

Harper, W. J. 1984. 'Model Food System Approaches for Evaluating Whey Protein Functionality.' Journal of Dairy Science **67**, 2745.

Heathcock, J. F. 1985. 'Characterization of Milk Proteins in Confectionary Products.' Food Microstructure **4**, 17.

Hodge, J. E., and Osman, E. M. 1976. In Food Chemistry. Edited by E. O. Fennema. Marcel Dekker.

Holt, C. 1985. 'The Size Distribution of Bovine Casein Micelles: A Review.' Food Microstructure **4**, 1.

Jensen, J. D. 1975. 'Some Recent Advances in Agglomerating, Instantizing, and Spray Drying.' Food Technology **29** (6), 60.

Kalab, M. 1979. 'Scanning Electron Microscopy of Dairy Products: An Overview.' Scanning Electron Microscopy III, 261.

Kilara, A., and Sharkasi, T. Y. 1986. 'Effects of Temperature on Food Proteins and Its Implications on Functional Properties.' CRC Critical Reviews Food Science Nutrition **23** (4), 323.

Kinsella, J. E. 1970. 'Functional Chemistry of Milk Products in Candy and Chocolate Manufacture.' Manufacture of Confections **50** (10), 45.

Kinsella, J. E. 1971. 'The Chemistry of Dairy Powders with Reference to Baking.' Advanced Food Research **19**, 147.

Kinsella, J. E. 1982. In Food Proteins. Edited by P. F. Fox and J. J. Cowden. Applied Science Publications.

Kinsella, J. E. 1984. 'Milk Proteins: Physiochemical and Functional Properties.' CRC Critical Reviews Food Science Nutrition **21** (3), 197.

Kuramoto, S., Jenness, R., Coulter, S. T., and Choi, R. P. 1959. 'Standardization of Harland-Ashworth Test for Whey Protein Nitrogen.' Journal of Dairy Science **42** (1), 28.

Lim, D. M. 1980. 'Functional Properties of Food Proteins with Particular Reference to Confectionary Products.' Leatherhead Food RA Research Report No. 120.

Masters, K. 1979. Spray Drying Handbook. Halstead Press.

Morr, C. V. 1967. 'Effect of Oxalate and Urea Upon Ultracentrifugation Properties of Raw and Heated Skim Milk Casein Micelles.' Journal of Dairy Science **50** (11), 1744.

Morr, C.V. 1984. "Production and Use of Milk Proteins in Food.' Food Technology **38** (7), 39.

Pallansch, M. J. 1970. In By-Products From Milk. Edited by B. H. Webb and E. O. Whittier, AVI.

Pisecky, J. 1978. 'Bulk Density of Milk Powders.' Dairy Industry International **43** (2), 4.

Pisecky, J. 1978. 'Instant Whole Milk Powder.' Dairy Industry

International **43** (8), 5.

Pour-El, A. 1979. *Functionality and Protein Structure*. Washington, D.C.: American Chemical Society.

Pyler, E. J. 1973. *Baking Science and Technology, Vol. 1*. Chicago: Siebel Publishing Company.

Sørensen, H. I., Krag, J., Pisecky, J., and Westergaard, V. 1978. *Analytical Methods for Dry Milk Products*. Denmark: A/S Niro Atomizer.

Thompson, M.P. and Farrell, H. M. 1973. 'The Casein Micelle – The Forces Contributing to Its Integrity.' *Netherlands Milk Dairy Journal* **27** (2/3), 220.

Walstra, P. and Oortwijn, H. 1982. 'The Membranes of Recombined Fat Globules, III. Mode of Formation.' *Netherlands Milk Dairy Journal* **36**, (2), 103.

Walstra, P. W., Bloomfield, V. A., Wei, G. J., and Jenness, R. 1981. 'Effect of Chymosin Action on the Hydrodynamic Diameter of Casein Micelles.' *Biochemistry Biophysics Acta* **669**, 258.

THE USE OF LACTOSE IN FOOD PRODUCTS

J.G. Zadow
CSIRO Division of Food Research
Dairy Research Laboratory
P.O. Box 20, Highett, Victoria 3190
Australia

ABSTRACT. The paper is a review article on the use of lactose in a range of food products. The structure and physical and chemical properties of lactose are discussed and compared with other sugars. Production of lactose by both batch and continuous process are summarized. The utilization of lactose in infant foods, confectionery, meat, packed goods and biological media are discussed. Literature on the role of lactose in calcium absorption and its effect on the intestinal flora is considered. Increased utilization of lactose will probably result from developments in infant food and confectionery together with its use as a feedstock for fermented products and lactose hydrolysis.

1. GENERAL

The effective utilization of lactose remains one of the dairy industry's most intransigent problems. Until recently, the major source of lactose has been whey, a byproduct from the manufacture of cheese and casein. However, development of ultrafiltration (UF) technology for the processing of milk and whey has resulted in increased production of permeate with high lactose content. This trend is certain to continue with UF technology being applied more widely to cheese making and to the production of whey protein concentrates (WPC). The increased volume of permeate produced has generally been aimed at better utilization of the protein fraction, which attracts the highest economic return, largely ignoring the protein-free permeate. Potential returns from permeate are significantly lower than for whey, and applications within the food industry are more restricted.
    The importance of developments in UF applications on the resulting production of permeate has been outlined by Teixeira, Johnson and Zall (1983a). These authors suggested that whilst the potential market for WPC in the USA is about ten times the 15,000 tonnes used in 1981, by 1986 the demand for lactose would have risen by about 25%. They strongly recommended development of new applications for the utilization of lactose permeate.

Current world cheese whey solids manufacture is about 4,500,000 t per annum, containing over 3,000,000 t of lactose. Current commercial utilization of lactose is only about 200,000 t per year. Much of the remaining lactose is utilized as whey - in animal food, in spray-dried whey, or disposed of on the land or in sewerage. Few statistics are available for the direct utilization of lactose by individual countries, or on a world-wide basis. However, it is likely that major uses are in infant foods and the pharmaceutical industry, where the ability of lactose to be moulded into tablets and pills is important.

It is generally recognized that world demand for lactose is inelastic, and any significant increase in production would result in a sharp price reduction. For this reason, the dairy industry has tended to utilize lactose as a raw material for the manufacture of more valuable derivatives. However, this approach often requires considerable capital, the assessment of alternative technologies, and the development of new markets.

In spite of this, it is clear that there are useful outlets for lactose within the food industry, and that additional applications are being developed. Some of these have been discussed in review articles by Teixeira, Johnson and Zall (1983b); Coton, Poynton and Ryder (1982) and Mann (1984).

Most uses of lactose by the food industry rely on its particular physiological characteristics in comparison with other sugars. For example, lactose is a useful carrier for flavour and colours, leading to its use in applications such as seasonings and baked goods. Confectioners use lactose to obtain desirable properties in their products, relying on lactose to alter the crystallization characteristics of other sugars. The reducing nature of lactose, coupled with the fact that it is not fermented by many common yeasts means that it also offers unique properties to the baking and brewing industries. Addition of lactose will increase the browning of bread crust which is often highly desirable, and, since it is not fermented, other functional properties conferred by the lactose will not be lost during manufacture. Lactose is also unaffected by beer yeasts, and may be used as a means of altering the organoleptic quality of beer.

Lactose is also used as a substrate in the production of penicillin, as seed material in the manufacture of concentrated and condensed milks, and as a raw material for the production of specialty chemicals and in some fermentations.

2. PROPERTIES OF LACTOSE

2.1. Structure

For all practical purposes, the sole source of lactose is in the milk of mammals. Other sources are rare, but it is also found as a component in the polysaccharides of some plants. It is a disaccharide yielding D-glucose and D-galactose on hydrolysis. The two monosaccharides are linked through the aldehyde group of the D-galactose moiety; thus the aldehydic portion of lactose is attached to the glucose moiety.

Lactose exists in two isomeric forms (anomers), alpha and beta, differing only in the configuration of the substituents on the number one carbon atom of the glucose residue. The solubility of these two forms is significantly different. The solubility of the alpha form is about 7g/100g. When lactose dissolves, mutarotation occurs, yielding a solution containing about 63% beta-lactose. On concentration, some alpha-lactose will precipitate, and further mutarotation will occur, with conversion of soluble beta-lactose to alpha-lactose. As crystallization proceeds, this process continues, yielding a product mainly composed of alpha-lactose monohydrate. The product obtained will therefore depend on the rate of two competing equilibria - the rate of conversion of soluble beta-lactose to soluble alpha-lactose, and the conversion of soluble alpha-lactose to crystalline alpha-lactose monohydrate crystals.

Alpha-lactose crystallizes as a hydrate, whereas beta-lactose contains no water of crystallization. When lactose solutions are dried rapidly, there may be insufficient time for crystallization of the alpha-lactose to alpha-lactose hydrate. The dry lactose is then in a form similar to that present in the liquid. A number of studies have confirmed that lactose in rapidly dried dairy products is a mixture of beta-lactose, alpha-lactose mono-hydrate and amorphous alpha-lactose. Neither beta-lactose nor alpha-lactose mono-hydrate are hygroscopic. However, anhydrous alpha-lactose is highly hygroscopic and absorbs water from the air, forming the hydrate which occupies more volume than the anhydrous form. This is the cause of the caking and lumping observed in many dried dairy products.

Manufacturing processes need to take this into account to avoid problems. Normal procedures for the manufacture of "non-hygroscopic" dairy products involve the crystallization of much of the lactose prior to drying. This can be achieved by holding the concentrate under fixed conditions to allow alpha-lactose hydrate crystals to form. Alternatively, techniques similar to "instantising" can be employed, whereby the surface of the product is humidified, or the particles partially dried to permit crystallization of the lactose before final drying.

2.2. Solubility and sweetness

Lactose is neither as sweet nor as soluble as sucrose, which substantially restricts its application in the food industry as an alternative sweetener. The relative sweetness and solubility characteristics of lactose, glucose, galactose and sucrose are shown in Table 1 (Pazur, 1970; Shah and Nickerson, 1978). As relative sweetness varies with concentration, the figures in Table 1 are meant only as a guide. The sweetness of lactose increases with concentration more rapidly than does the sweetness of sucrose, although there appears to be little difference in the effect of concentration on the sweetness of lactose, glucose or galactose. Beta-lactose is sweeter than alpha-lactose, but at low concentrations is not significantly sweeter than the equilibrium mixture.

Lactose crystallization in frozen foods may lead to undesirable calcium-protein interactions and instability (Muir, 1985). Freeze-thaw stability may be improved by a number of options, including lactose hydrolysis.

3. MANUFACTURE OF LACTOSE

Principles for the manufacture of lactose were outlined by Zirm (1895) and Aufsberg (1910), and the principles they describe remain relevant today. In general, batch production of lactose involves protein removal, perhaps by liming, heat treatment and filtration, followed by concentration of the mother liquor, refiltration, further concentration, induction of crystallization then crystal removal by centrifugation. About 50% recovery of lactose is achieved, and the mother liquor may be dried and sold as delactosed whey powder. It should be noted that permeate from ultrafiltration of milk or whey is an ideal raw material for this process, as the protein has already been removed.

For crystallization to occur, the mother liquor must be supersaturated with alpha-lactose. The rate of crystallization is important to the economics of the process, as is crystal size. However, supersaturated solutions of lactose do not nucleate readily. Beta-lactose strongly retards the growth of the alpha-hydrate crystals as do minerals, acids and vitamins (Michaels and Krevald, 1966). More recently, Visser (1980) has reported that pharmaceutical-grade lactose contains an acidic substance which crystallizes with the lactose and retards the rate of crystal growth. Ion exchange of the lactose solution removes this contaminant, yielding a non-ionic lactose which crystallizes more rapidly than does the pharmaceutical grade. Whey and permeate probably contain similar growth retarding compounds, which contribute to the practical problems of lactose crystallization. The value of demineralization in crystallization processes may lie in the removal of these compounds.

Crystallization is thus a time-consuming and complex process. The response of supersaturated solutions to seeding is uncertain, and it is extremely difficult to predict yields or average crystal size. Griffiths, Paramo and Merson (1982) examined crystal size distribution from a continuous mixed-suspension - mixed-product withdrawal crystallizer operating at a steady state. Growth and nucleation rates in this system at 30°C were studied. Rates of crystal growth were somewhat lower than would be expected from the literature at low supersaturations.

In batch processing for lactose production, steam is injected directly into the whey, with the addition of calcium·chloride, although Kwon and Nickerson (1978) have advocated use of magnesium chloride. The result of such treatment is the precipitation and aggregation of whey proteins, which are removed by filtration. The deproteinized whey is concentrated, and the lactose then crystallized. After separation by basket-type centrifuges, edible-grade lactose is obtained. Further crystallization and clarification results in pharmaceutical grade (Evans and Young, 1982). The two byproducts of the process, the mother liquor and filtercake, are also sold after drying for feed purposes. Details of a batch process for the manufacture of lactose have been described by Visser, Van den Bos and Ferguson (1986) (Figure 1), and Beunseu and Credoz (1982).

A continuous process for the manufacture of lactose could offer considerable economic benefits and be more appropriate for use by small scale manufacturers. Whilst decanting centrifuges have allowed the development of semi-continuous processes, a truly continuous commercial process is yet to be developed. Studies by Muller (1979) and Thurlby and Sitnai (1976) clearly indicate that the process must result in the growth of existing crystals rather than the formation of new ones. MacBean (1979) and Hobman (1984) have reviewed developments in lactose crystallization, and Polyanskii (1979) has examined the kinetics of crystallization from various whey syrups.

More recently, Quickert and Bernhard (1982) have outlined a method for lactose recovery based on the use of the Group II metal chlorides and sodium hydroxide as precipitating agents. Maximum recoveries were 25% for strontium, 55% for magnesium and 97% for calcium. Lactose recovery with metal hydroxides seems to involve complex formation, and is not strictly an absorption process.

4. ANALYSIS OF LACTOSE

Many methods are available for the estimation of lactose. These include the classical Munson-Walker method involving formation of cuprous oxide, the chloramine-T method (involving back titration of liberated iodine), polarimetry and the picric acid method, involving measurement of a coloured product at 520 nm. HPLC and enzymatic systems are widely utilized. Zarb and Hourigan (1979a, 1979b) have described a simple effective means for lactose determination based on cryoscopy. Using this method, buffered lactase is added to samples, and the difference in freezing point between the hydrolyzed and unhydrolyzed samples determined. This difference is directly proportional to the lactose concentration over the range 1 to 5%. The method does not require removal of protein before lactose estimation, and is specific regardless of the presence of other sugars. Variation in freezing point of starting materials is also no barrier to the process. It appears to be particularly valuable as it is rapid and only requires a cryoscope, which is commonly used in dairy factories.

Piklad, Umanskii and Gudkov (1981) have described methods for the determination of proteins and lipids in lactose, using spectrophotometric methods. Protein contents of 1.6% for crude lactose, and 0.7% for food grade lactose were reported. There was no evidence of protein in refined lactose.

5. UTILIZATION OF LACTOSE

5.1. Infant foods

Lactose is the only source of carbohydrate in mammalian milk, and is a major contributor to the energy requirements of infancy. Lactose is hydrolyzed slowly in the intestines, resulting in a steady energy supply and a comparatively constant blood glucose level between feedings.

Replacement of lactose by glucose, for example, would require a greater response from the insulin system, risking over-secretion of insulin, and consequently low blood-sugar levels. It is believed that lactose assists in the development of a favourable environment in the intestine, resulting in the development of lactic acid flora which may inhibit the growth of pathogenic flora (refer Section 6.3). Lactose also has a useful role in calcium absorption (refer Section 6.2).

Table 2 outlines some differences between human and cow milk (Visser, Van den Bos and Ferguson, 1986). Because of the different lactose contents, it is common practice to fortify infant foods based on cow milk with lactose. There is much literature available on the formulation of infant foods, including useful reviews by Mann (1977), Mathur and Shahani (1979) and Ulrich (1976). Lucas and Barr (1985) have outlined the preparation of a food for premature infants based on WPC lactose, maltodextrin, vegetable oils, glycerol mono-stearate, lecithin and vitamin concentrate. The process involves clarification, pasteurization, homogenization and heating. The food contains more than 160 mg/100mL riboflavin, and may be of benefit to infants receiving phototherapy.

5.2. Confectionery

The confectionery industry is a major user of lactose (Spurgeon, 1976; Riedel and Hansen, 1979; Mann, 1982; Estelmann, 1984). A recent report has outlined the use of lactose in fondant at Meggle Milchindustrie (Anon, 1984). The incorporation of lactose into fondant (usually comprising sucrose, glucose syrup and water), and the use of microfine lactose in fondant fillings were examined. Manufacture of a lactose-containing fondant was achieved by bringing the lactose and sucrose into solution and crystallizing them together. As an alternative, microfine lactose could be incorporated into prepared fondant. Inclusion of lactose in fondants results in control and reduction of sweetness, intensification of whiteness, and economies in operation. Meggle Milchindustrie (Anon, 1985) have also reported a lactose-containing special ingredient for toffee and fudge.

The sugar coating of cores (e.g. chocolate buttons or hazelnuts) using a mixture of sucrose and lactose has been described by Boesig (1981). The ratio of sucrose to lactose examined covered the range 90:10 to 50:50. Lactose suppresses sucrose crystallization, allowing the coating to be effected at lower temperatures and reducing the sweetness of the coating.

5.3. Meat

There is comparatively little information concerning the application of lactose in meat products, which is an area of much potential. Pinel (1981) has considered various possibilities in a review article. Lactose-based products have been suggested as being particularly suitable for addition to comminuted meat products as a flavour enhancing and binding agent. Scharner et al (1981) have described a lactose-containing product recommended for addition to fermented (salami-type) sausage formulations as a carbohydrate source for the starter culture.

5.4. Baked goods

The reducing nature of lactose, coupled with the fact that it is not fermented by baker's yeast, offers opportunities to the industry. Browning of the crust is increased, which is often highly desirable. Luksas (1984) has reviewed the applications of lactose-based products in the baking industry. Harper, Rogers and Hoseney (1983) examined the use of whey-based products in breadmaking, and concluded that the loaf depression associated with use of some whey-based products was not related to lactose concentration.

5.5. Cultures and biological materials

Recently, lactose fractions derived from whey UF permeate have been recommended as basic ingredients for a wide range of bacterial and fungal culture media (IGI Biotechnology Inc, 1984). Harju et al (1983) have considered the use of milk-based powders in the drying of starter bacteria.

5.6. Miscellaneous

Recently, lactose has been suggested for use in coffee creamers (Moran and Halstead, 1981), dietetic agents (Kowalsky and Scheer, 1981) and edible gel products (LeGrand and Paul, 1981). Other applications include use in food release agents (Kanebo Foods Ltd, 1981), ketchups and sauces (Dordevic et al, 1981), and a water-miscible starch-based product (Gasser and Badertscher, 1981).
    Jelen and Chan (1981) reported the firming of vegetables by addition of lactose. Blanched carrots, green beans and peas were retorted at 121°C in 2% sodium chloride brine containing 0 to 15% lactose. After 37 and 68 days, hardness of the vegetables was evaluated. Increasing lactose content correlated significantly with average hardness of peas and beans, and carrots to a lesser extent. All samples from brines containing more than 8% lactose showed higher average hardness than those containing less or no lactose. The increase was noticeable to an untrained panel.
    The use of lactose in dipeptide sweetener formulations has been suggested by Eisenstadt (1981) to give a product approaching the natural sweetness of glucose, and minimal addition of dipeptide sweetener.

6. NUTRITION

6.1. Lactose malabsorption

Lactose malabsorption and its implications for the development of lactose-hydrolyzed products will be discussed by others at this Symposium, and has recently been reviewed by Hourigan (1984).

6.2. Calcium absorption

Research over many years has indicated that dietary lactose assists in the absorption of calcium (Allen, 1982; Schaafsma, 1983; Renner, 1983). The enhancement of calcium absorption by lactose is partly due to increased passive diffusion (Allen, 1982), but beyond that, the mechanism is uncertain. The effect is thought to be due not to the lactose itself, but rather to its metabolic byproduct, lactic acid, as consumption of sour milk also improves calcium absorption. It has been suggested that the mechanism involves a decrease in pH of the intestinal tract from fermentation, resulting in increased calcium solubility. Soluble complexes of calcium and lactose may also contribute to the mechanism. Other work has not supported these suggestions, however. Experiments with rats showed that the effect was not due to lowering of pH by fermentation, nor to stimulation of intestinal metabolism by lactose (Wasserman and Lengemann, 1960). The suggestion that lactose forms a complex with calcium, thereby increasing absorption (Charley and Saltman, 1963), is not supported as the necessary molecular structure does not exist in lactose (Angyal, 1974). It is clear however that milk remains the most concentrated and available form of dietary calcium.

6.3. Intestinal flora

Lactose is virtually unhydrolyzed in the stomach, and little is absorbed in the upper section of the large intestine. In the next portion of the intestine it is cleaved by the enzyme lactase, and the resulting mono-saccharides provide a useful substrate for the body's flora. The lactic acid produced results in acid conditions which inhibit the growth of many putrefactive bacteria, thereby encouraging their replacement with acid-producing flora.

7. CONCLUSIONS

The utilization of lactose per se will remain difficult. Its applicability is limited by its low sweetness and low solubility, with little to offer in the way of specific advantages. Increased utilization of lactose will probably result from developments in infant foods, confectionery and meats. Other applications are unlikely to be of major world-wide significance. It is more likely that the greatest use for lactose will remain as a raw material for further processing, such as the feedstock for fermented products or in lactose-hydrolyzed products.

REFERENCES

Allen, L.H. (1982). Am.J.Clin.Nutr., 35, 783.
Angyal, S.J. (1974). Tetrahedron, 39, 1695.
Anon (1984). Confect.Manuf.Market., 21(2), 13.
Anon (1985). Confect.Manuf.Market., 22(2), 2.

Aufsberg, T. (1910). Chemiker Zeitung, 24, 481.
Beunseu, P. and Credoz, P. (1982). Fr.Pat. FR2 493 679 A1.
Boesig, W. (1981). Ger.Fed.Rep.Pat.Applic., 2 936 040.
Charley, P. and Saltman, P. (1963). Science, 139, 1205.
Coton, S.G., Poynton, T.R. and Ryder, D. (1982). Bull.Int.Dairy Fed., 147, 23.
Dordevic, J., Misic, D., Petrovic, D. and Macej, O. (1981). Mliekarstvo., 31, 3.
Eisenstadt, M.E. (1981). U.S. Pat. 4 254 154.
Estelmann, H. (1984). Dtsch.Milchwirtsch., 35, 567.
Evans, J.W. and Young, G.C. (1982). U.S. Pat. 4 316 749.
Gasser, R.J. and Badertscher, E. (1981). U.K. Pat.Appl. 2 066 643 A.
Griffiths, R.C., Paramo, G. and Merson, R.L. (1982). Symp.Ser.Am.Inst.Chem.Eng., 78(2218), 118.
Harju, M., Mattila, L., Heikonen, M. and Linko, P. (1983). Kemia-kemi, 10, 963.
Harper, K.A., Rogers, D.E. and Hoseney, R.C. (1983). J.Food.Proc.Pres., 7, 213.
Hobman, P.G. (1984). J.Dairy Sci., 67, 2630.
Hourigan, J.A. (1984). Aust.J.Dairy Technol., 39, 114.
IGI Biotechnology Inc. (1984). Int.Pat.Applic. WO 84 01 104 A1.
Jelen, P. and Chan, C.-S. (1981). J.Food Sci., 46, 1618.
Kanebo Foods Ltd. (1981). Jap.Exam.Pat. 5 609 090.
Kowalsky, H. and Scheer, H. (1981). U.K. Pat.Appl. 2 050 142 A.
Kwon, S.Y. and Nickerson, T.A. (1978). U.S. Pat. 4 399 164.
LeGrand, C.G.G.R. and Paul, R.A.E.C. (1981). U.S. Pat., 4 251 562.
Lucas, A. and Barr, R.I. (1985). U.K. Pat.Applic. GB 2142 518 A.
Luksas, A. (1984) in Dairy Products for the Cereal Processing Industry, ed. J.L. Vetter, St.Paul, Minn., Am.Assoc.Cer.Chem., 167.
MacBean, R.D. (1979). N.Z.J.Dairy Sci.Technol., 14, 113.
Mann, E.J. (1977). Dairy Ind.Int., 42(7), 26.
Mann, E.J. (1982). Dairy Ind.Int., 47(11), 11.
Mann, E.J. (1984). Dairy Ind.Int., 49(6); 49(7), 11.
Mathur, B.N. and Shahani, K.M. (1979). J.Dairy Sci., 62, 99.
Michaels, A.S. and Krevald, A. (1966). Neth.Milk Dairy J., 20, 163.
Moran, D.P.J. and Halstead, P.W. (1981). U.S. Pat. 4 305 964.
Muir, D.D. (1985). Dairy Ind.Int., 50(9), 19.
Muller, L.L. (1979). N.Z.J.Dairy Sci.Technol., 14, 119.
Pazur, J.H. (1970) in Carbohydrates: Chemistry and Biochemistry, ed. W.Pigman, D. Horton and A. Herp. 2nd ed. N.Y., Academic IIA, 69.
Pinel, M. (1981). Tech.Lait., (952), 61.
Piklad, N.G., Umanskii, M.S. and Gudkov, A.V. (1981) in Sbornik Nauthnykh Trudev.Novoe v Tekhnike i Tekhnologii Pereabotki Molochnoi Syvorotki, ed. G.G. Shiler. Uglich, Alt.Fil.Vses.Nauchn.Inst.Maslod.Syrod.Prom., 88.
Polyanskii, K.K. (1979). Izv.Vysssh.Uchebn.Zaved.Pishch.Tekhnol., (5), 99.
Quickert, S.C. and Bernhard, R.A. (1982). J.Food Sci., 47, 1705.
Renner, E. (1983). Milk and Dairy Products in Human Nutrition. Munich, W-GmbH, Volkswirtsch. Verlag, 450pp.
Riedel, D.-L. and Hansen, R. (1979). Lebensm.Ind., 26, 311.

Schaafsma, G. (1983). Bull.Int.Dairy Fed.Doc. (166), 19.
Scharner, E., Schiffner, E., Huttner, G. and Hepp, H.U. (1981). Ger.Dem.Rep.Pat. DD 152 716.
Shah, N.O. and Nickerson, T.A. (1978). J.Food Sci., 483, 1081.
Spurgeon, K.R. (1976). Cult.Dairy Prod.J., 11(4), 8.
Teixeira, A.A., Johnson, D.E. and Zall, R.R. (1983a). Food Eng., 55(5), 106.
Teixeira, A.A., Johnson, D.E. and Zall, R.R. (1983b). ASAE Publ.Am.Soc.Agric.Eng., (9-83), 78.
Thurlby, J.A. and Sitnai, O. (1976). Schweiz.Milchwirtsch.Forsch., 5, 99.
Ulrich, W. (1976). Schweiz.Milchwirtsch.Forsch., 5, 99.
Visser, R.A. (1980). Neth.Milk Dairy J., 34, 255.
Visser, R.A., Bos, M.J. van den and Ferguson, W.P. (1986). Bull.Int.Dairy Fed. (in press).
Wasserman, R.H. and Lengemann, F.W. (1960). J.Nutr., 70, 377.
Zarb, J.M. and Hourigan, J.A. (1979a). N.Z.J.Dairy Sci.Technol., 14, 171.
Zarb, J.M. and Hourigan, J.A. (1979b). Aust.J.Dairy Technol., 334, 184.
Zirm, G. (1895). Milch-Ztg., 24, 481.

TABLE 1
Relative sweetness and solubility of sucrose, lactose and some monosaccharides

	Relative Sweetness	Solubility (g/100g solution)		
		10°C	30°C	50°C
Sucrose	100(a)	66(b)	69(b)	73(b)
Lactose	16(a)	13(b)	20(b)	30(b)
D-Galactose	32(a)	28(b)	36(b)	47(b)
D-Glucose	74(a)	40(b)	54(b)	70(b)
D-Fructose	173(a)	...	82	87

(a) Pazur (1970)
(b) Shah and Nickerson (1978)

TABLE 2
Some attributes of human and cow milk

		Human	Cow
Solids (5)	- Total	12.4	12.4
	- Lactose	7.2	4.6
Calorific value (kJ/100mL)	- Total	263.7	272.1
	- Lactose	121.4	77.4
Osmotic Pressure (mOsm/L)	- Total	239.0	221.0
	- Lactose	210.0	134.0

Adapted from Visser, van den Bos and Ferguson (1986)

CAPTION TO FIGURE

Figure 1. Production of lactose and delactosed whey powder (Visser, van den Bos and Ferguson, 1986)

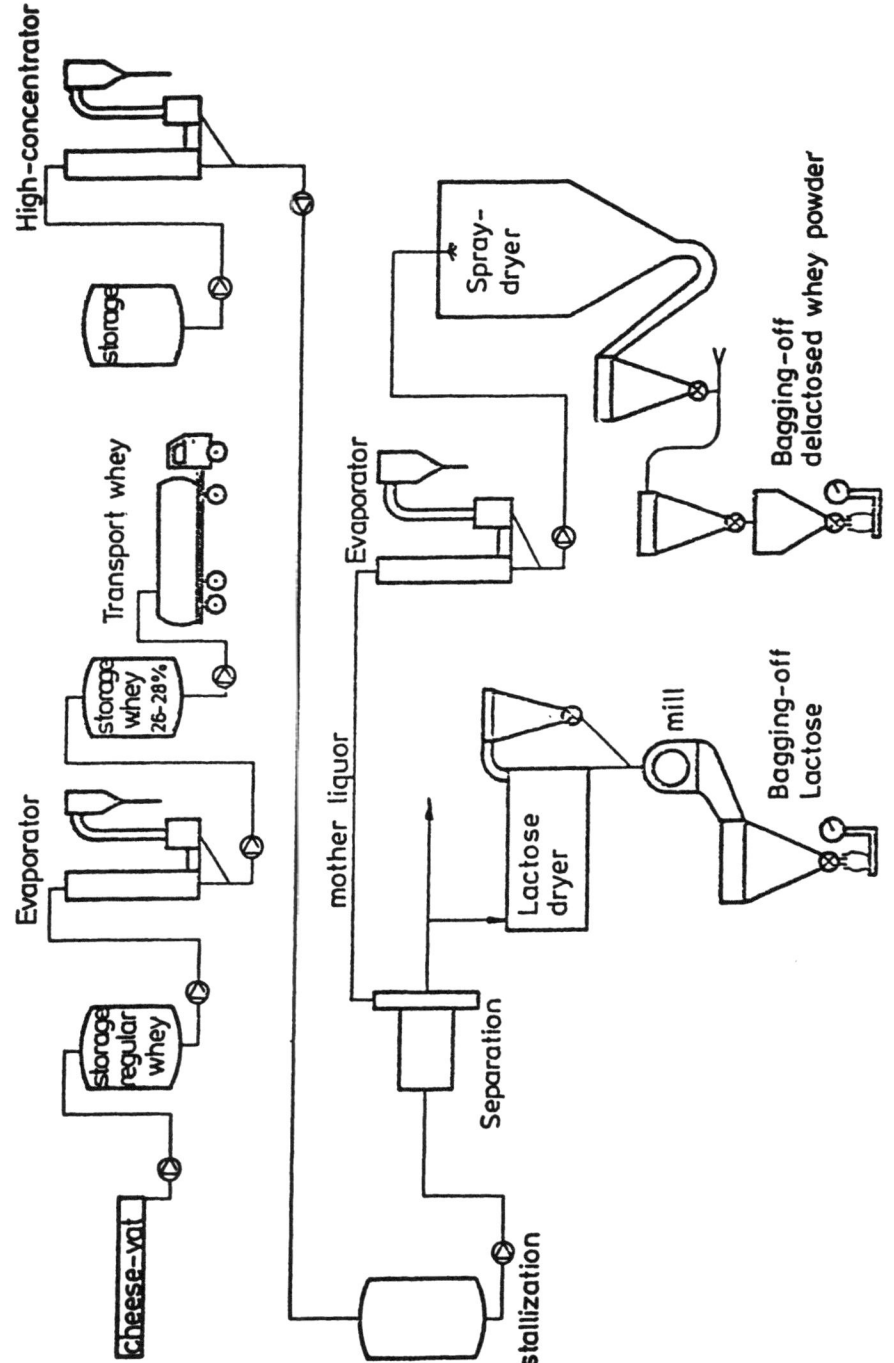

Figure 1.

DERIVATIVES OF LACTOSE AND THEIR APPLICATIONS IN FOOD PRODUCTS

Leslie A.W.THELWALL
Tate & Lyle Group Research and Development
P.O. Box 68
Reading RG6 2BX
U.K.

ABSTRACT

The properties of lactose derive from its unique chemical structure. Methods of modifying the structure of this disaccharide are discussed in terms of the use of protecting groups, with special reference to selective ester and isopropylidene acetal formation. The use and potential application of lactose derivatives are highlighted.

INTRODUCTION

Lactose[1,2] (Fig. 1,(1)) is a ß, 1-4 linked heterodisaccharide, comprising of galactose and glucose. The glycosidic linkage involves the anomeric position of galactose, and therefore the reducing properties of lactose arise from the glucose moiety (which is glycosidically-linked at C-4).

Lactose

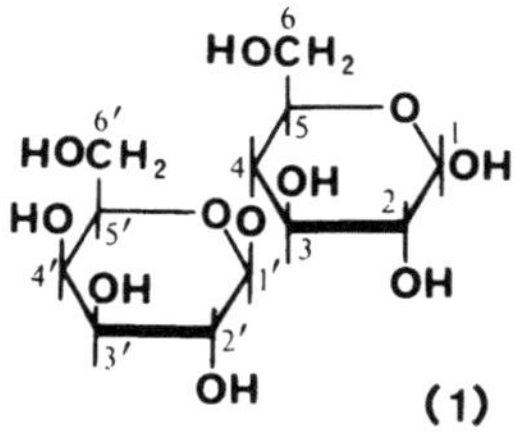

Fig. 1

(1)

Lactose contains two primary and six secondary hydroxyl groups.The anomeric hydroxyl group at C-1 is part of a hemiacetal system and consequently exhibits different chemical properties from the remaining seven. On dissolution in water, the hemiacetal ring can open, and reform to give the alternative anomeric configuration (i.e. α or ß). This equilibration is accompanied by an optical rotational change known as mutarotation. For sugars that effectively give only two species at equilibrium (e.g. glucose, mannose), a first-order kinetic law is observed.

Milk – The vital force, 749–758.

The properties of lactose derive from the unique combination of
functional groups and conformational structure. Minor alterations
in the chemical structure, such as a change in configuration of one
of the hydroxyl groups can lead to significant differences in
physical and sensory properties. For example, cellobiose[3] (Fig.
2,(2)) differs from lactose at only the hydroxyl group at C-4',
which is in the equatorial position (i.e it is a C-4' epimer). The
properties of this sugar, such as solubility, crystal habit, melting
point, sweetness etc., are quite different, however, from those of
lactose.This is also the case for maltose[4] (3), a constituent of
starch and glycogen, which differs from cellobiose by only the type
of glycosidic linkage(α vs ß).

Fig. 2 Common Disaccharides — Lactose, Cellobiose (2), Maltose (3)

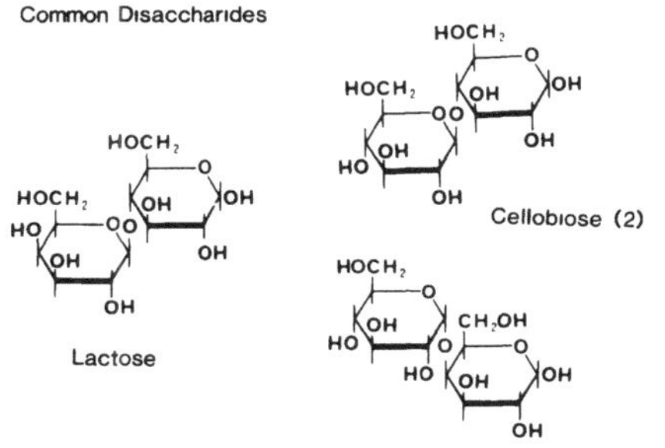

CHEMICAL MODIFICATION OF LACTOSE

In order to change the functionality at a specific position in the
lactose molecule, it is necessary to know how to direct a reaction
so that only the site of interest is modified. This can be achieved
by careful choice of reagents, solvents and reaction conditions.
The general strategy used to chemically modify carbohydrates is to
either direct reaction at the required position, or by using
protecting groups, to block all other sites prior to reaction.
The importance of selective reaction in carbohydrate chemistry is
well recognised. Selectivity is a consequence of the relative
reactivity of the hydroxyl groups (or their derivatives) towards the
reagent used[5]. The folowing examples serve to demonstrate the
application of selective reaction in the chemical modification of
lactose. Regioselective acylation of lactose has been studied by a
number of groups[6-8]. Hough and co-workers[7] determined the order
of benzoylation of the hydroxyl groups in methyl ß-lactoside (Fig.
3,(4)) (using benzoyl chloride in pyridine) to be 6'>3'>6>2>2',4'>
3. By exploiting the low reactivity of the 3-OH towards
benzoylation, the 3-epimer of lactose could be readily prepared[9].
Selective hexabenzoylation of methyl ß-lactoside gave the
2,6,2',3',4',6'-hexabenzoate(3-OH free)(5) in 33% yield. The 3-OH
group was mesylated(using methanesulphonyl chloride in pyridine) in
high yield, and then displaced by the benzoate anion. As
nucleophilic substitution of the sulphonyloxy group by the benzoate
anion occurs with inversion of configuration at C-3, the resultant

sugar is converted into the 3-epimer of lactose, 4-O-ß-D-galactopyranosyl-D-allose (6) (after conventional removal of the protecting groups).

Fig. 3 Synthesis of "Allolactose"

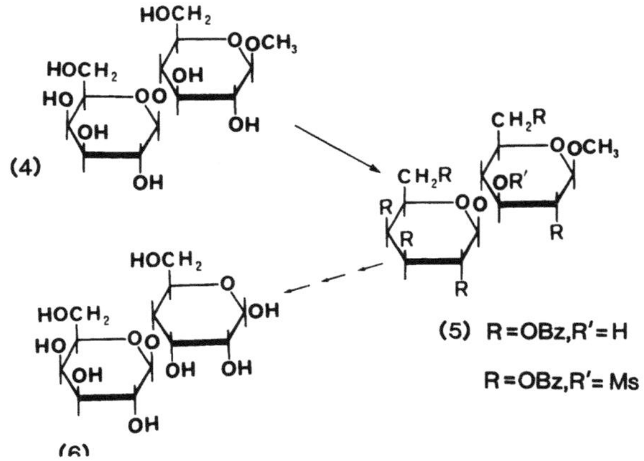

(4)

(5) R=OBz, R'=H
   R=OBz, R'=Ms

(6)

The transformation of lactose to its 2',3'-di-epimer has also been achieved[10] using selective benzoylation of a partially protected lactose derivative. By employing various molar proportions of benzoyl chloride in pyridine at -20 C, Tejima and co-workers[10] established the order of reactivity of the secondary hydroxyl groups in 1,6-anhydro-4',6'-O-benzylidene-ß-lactose (Fig. 4,(7)) to be 3'>2 3>2'. Selective tribenzoylation was then carried out on (7) to give a 41% yield of the 2,3,3'-tribenzoate(2-OH free) (8), which upon mesylation afforded the 2'-mesylate (9). Conversion of the 2'-mesylate to the talo-epoxide (10) was achieved using 1.1 molar equivalents of sodium methoxide in boiling methanol. Treatment of

Fig. 4   Synthesis of "Idolactose"

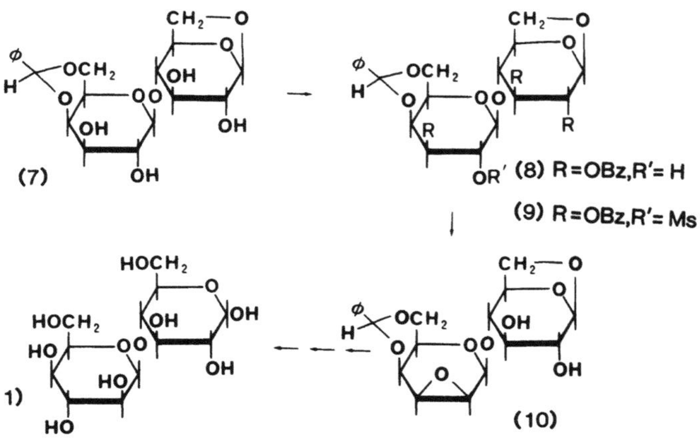

(7)   (8) R=OBz, R'=H
                                (9) R=OBz, R'=Ms

(11)   (10)

this epoxide with aqueous potassium hydroxide solution caused stereospecific cleavage of the anhydro ring to give the trans diaxial product (in which the C-2' and C-3' substituents are in the axial positions; cf. in lactose, C-2' and C-3' substituents are both equatorially displaced). Removal of the protecting groups afforded, 4-O-ß-D-idopyranosyl-D-glucose (11).

Another common method of selective protection of the hydroxyl groups in lactose is that of acetalation. Pairs of hydroxyl groups are protected by formation of a cyclic acetal. These derivatives are readily formed by reaction of a diol system with an aldehyde or ketone in the presence of an acid catalyst (Fig. 5). Isopropylidene acetals are particularly valuable as protecting groups, as they are simple to introduce, stable to most neutral and basic reaction conditions, and readily removed under mild conditions. In addition, their formation does not introduce any new chiral centres (cf. benzylidene acetals) to the molecule, and therefore eliminates the possibility of creating diastereoisomers. Recent developments in acetal formation use 2-methoxypropene and acetal exchange reactions to produce novel multisubstituted acetal derivatives of lactose. The number, position and size of the acetal rings that are formed are highly dependant on the choice of reagent used and reaction conditions employed. When lactose was treated[11] at ambient temperature with 2,2-dimethoxypropane in N,N-dimethylformamide and containing p-toluenesulphonic acid for 3h, the major product of the reaction was the 4',6'-acetal (Fig. 6, (12)). By performing the reaction at higher temperatures and for shorter periods (80-85 C for 45 minutes), the 3',4'-acetal (13) was isolated as the major product[12]. Treatment of lactose with 2-methoxypropene produces a diacetal[13] (14), in which one of the isopropylidene acetals bridges the monosaccharide units, forming a nine-membered ring system. Higher acetal formation occurs in lactose when heated at reflux temperature in 2,2-dimethoxypropane containing p-toluenesulphonic acid[14]. The crystalline tri-O-isopropylidene-lactose dimethyl acetal (15) that is obtained has only two hydroxyl groups free (2- and 6'-OH's), and is a valuable precursor to a variety of products. These four derivatives of lactose allow easy access to a number of sites for further chemical modification, which would not be readily available by other routes.

Fig. 5   Acetal Formation

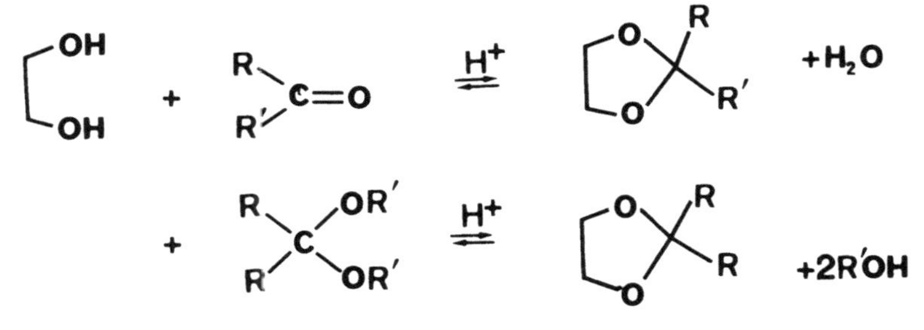

DERIVATIVES AND THEIR APPLICATION IN FOODS

Lactitol[15-17] (4-0-ß-D-galactopyranosyl-D-glucitol) (Fig. 7,(16)) is obtained by hydrogenating lactose. Hydrogenation is typically carried out under high temperature and pressure with a catalyst present. Raney nickel is a commonly used catalyst for the hydrogenation of a sugar. An alternative method is to use sodium borohydride as the hydrogenating agent[16]. The boron can be removed as volatile methyl borate under reduced pressure. Yields of 90% of lactitol can be achieved by these methods. It has been claimed that lactitol may be used as a sweetening agent, which is non-crystallisable, highly soluble, capable of retaining moisture, can confer stability on flavourings and colorants and has no food value _per se_. Lactitol has good textural properties, comparable to those of sorbitol, but without the disadvantage of a mouth-cooling effect. Flavour release in boiled sweets, however, is poor compared to sucrose. The loss of the aldehyde function on hydrogenation of lactose, imparts a greater stability of the molecule towards extreme heat and pH conditions. Lactitol offers many interesting uses for the food industry, especially in the area of dietetic and low-caloric foods. Many products can benefit from the replacement (partial and total) of sucrose by lactitol (thereby reducing the products caloric value), such as jams and marmalades, chocolate and hard candy. Due to its low hygroscopicity, lactitol can be used in baked products for diabetics, especially in crisp and biscuit-like products.

Isopropylidene acetals of Lactose

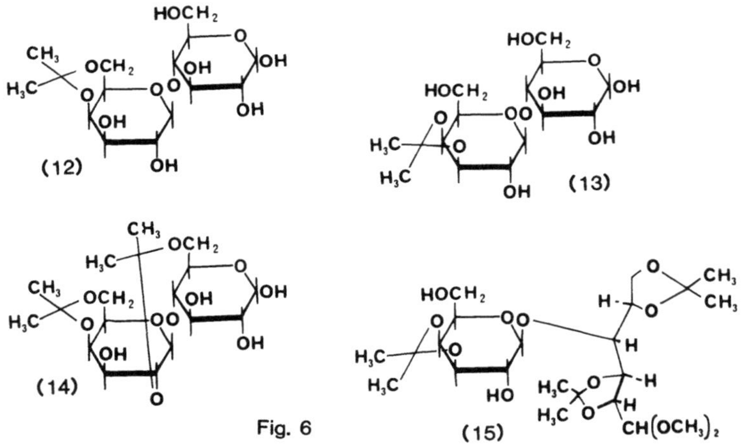

Fig. 6

Lactulose (4-0-ß-D-galactopyranosyl-D-fructose) (17) can be prepared from lactose by alkaline epimerisation. The first synthesis was achieved by Hudson and co-workers[18], who used a Lobry de Bruyn-Alberda van Ekenstein transformation of lactose in calcium hydroxide solution. More recent preparations employ the use of boric acid and base catalysts, which give yields[19] of lactulose of over 90%. This technique has been applied directly to sweet cheese whey ultrafiltrate to produce good yields of the sugar[20]. Lactulose may also be obtained by the Amadori rearrangement of p-tolyl-N-lactosylamine[21] or by reduction of lactosone(4-O-ß-D-galactosyl-D-arabinohexosone) using zinc dust in acetic acid[22]. These latter methods result in low yields of lactulose and are impracticable as a means of synthesis. Lactulose is formed in heated milks. Bernhart et al[23] reported that heat-sterilised liquid milk formulae used for infant feeding contained lactulose representing 1-5% of the total carbohydrate, whereas those formulae prepared from spray-dried powders contained little if any lactulose. When milks for infant feeding were heated, before feeding, lactulose was produced in amounts that increased with heating time. The mechanism by which lactulose is formed in heated milks is thought to be alkaline epimerisation and catalysed by amino acids. It is claimed[24] that the presence of lactulose in infant foods promotes the development of Bifidobacterium bifidum in the intestinal flora, mimicking the flora present in healthy breast-fed infants.

Parrish et al[25] have suggested that because of the greater sweetness and solubility of lactulose compared to lactose, it could be used as a partial replacement for sucrose in food applications. The sweetness of lactulose solutions was evaluated over a concentration of 5-35% (w/w), and was found to be 48-62% that of sucrose. Ingestation of large quantities (150g/day) of the sugar can, however, lead to flatulence and diarrhoea, and these are limiting factors in an otherwise ideal sugar for food use.

Lactobionic acid[26] (4-O-ß-D-galactopyranosyl-D-gluconic acid) is obtained by the oxidation of lactose. Its calcium salt, calcium lactobionate is a white, odourless, free-flowing powder. The salt is freely soluble in water and has a bland taste. Under neutral to mildly alkaline conditions (pH range 6.0-10.0), the calcium lactobionate solution is resistant to hydrolysis or degradative changes. Under mildly acidic conditions (pH range 3.5-4.0), especially on longer contact with the acid and at elevated temperatures, slow hydrolytic cleavage of the galactoside bond may occur.

Calcium lactobionate can be used as a firming agent to dry pudding mixes. It also accelerates and improves the gelling process in the reconstituted pudding. In addition, the calcium lactobionate produces an easily dispersing, non-hygroscopic, non-caking, dry pudding mix.

Sugar acids, because of their non-toxicity, have been suggested for use as sequesterants[27] in food processing. They should be effective for the prevention of calcium, magnesium and iron precipitation in dilute lye solutions prepared with hard water for use in peeling fruits and vegetables prior to canning.

The drift towards natural food additives is a continuing trend. For colouring foods brown or dark brown, there has been no suitable way other than using caramels and/or a mixture of several synthetic colorants such as those derived from tar. Oura et al[28] have devised a method of producing a brown colouring matter for foods, using natural substances. They utilised shea nut meal, a by-product of the manufacture of shea butter from shea nuts. This by-product has few outlets, other than its use as an animal feed. When heated in the presence of an amino acid (e.g. glutamic acid, leucine, valine) and a reducing sugar such as lactose, a brown colouring matter for foods is produced. It is thought that the unpleasant odours and taste associated with heated shea nut meal are significantly reduced by the inclusion of lactose and an amino acid. It was suggested that the flavour produced by the amino-carbonyl reaction of the amino acid and sugar may mask the off-odours and bitter taste. The authors claim that the colouring matter can be used in colouring confectionery, ice-creams, soft drinks, liquors, processed livestock products, processed marine products and artificial protein products(e.g. fibrous soybean protein products etc.).

The attachment of fatty acid groups to disaccharides has produced compounds which have surface active properties. Reducing sugars suffer the disadvantage of having a reactive aldehyde or ketone group, which can take part in adverse side-reactions. This precludes the direct use of lactose in the esterification process, but not as a protected derivative, or in its reduced form, lactitol. Fatty acid monoesters of lactitol exhibit physical properties similar to those of sucrose analogues. Lactitol palmitates are reported to have application as emulsifiers in foods[17]. The higher esters may prove to be useful low caloric fat substitutes for various fat-containing foods. Properties such as surface and interfacial tension, emulsification and wetting properties(amongst others) are likely to be determined by the number, position and nature of substituent esters. Mono- and di-palmitates of lactose have been synthesised from partially protected derivatives, and were reported[29] to have a degree of surface activity. Myristoyl esters of methyl ß-lactoside have also been reported[30], although the properties of these esters were not described. Sugar esters are excellent emulsifiers and an extensive range of products benefit from their inclusion, such as fats and margarines, solid flavour concentrates and frozen desserts.

Certain phosphorus-containing compounds of lactose[31] have found application as stabilising agents for edible oil- and water-containing emulsions, particularly saltless margarines, in order to suppress spattering during the frying process. Lactose combines with phosphatidyl-ethanolamine compounds, via an Amadori rearrangement reaction, to give products (Fig. 8,(19)) which are very effective anti-spattering agents in water/oil emulsion spreads, and even in emulsions where sodium chloride is absent or present in very low concentrations(where other phosphatides are less active). In the initial stage of the frying process, margarines separate essentially into an aqueous phase and an oil phase. When the temperature exceeds the boiling point of water, spattering ensues since the water evaporates vigorously entraining some of the oil phase. Although pure lactose is the preferred source as the reducing sugar, milk solids such as skimmilk powder and whey powder in a spray-dried form are suitable alternatives. The emulsions can be produced by dispersing the Amadori rearrangement products either in the aqueous or in the fatty phase in a proportion ranging from 0.05-5 wt%, based on the total composition.

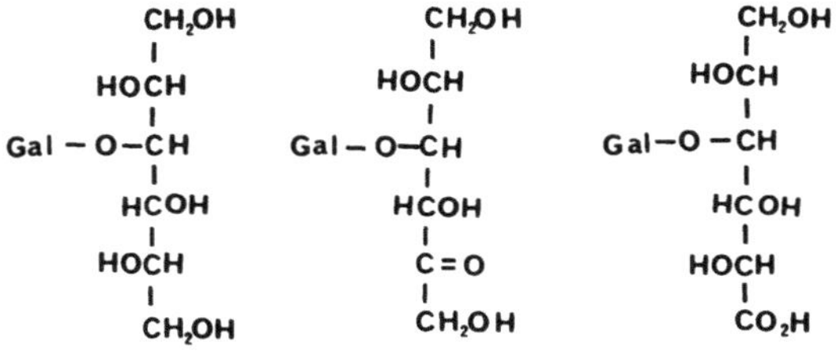

```
        CH₂OH              CH₂OH              CH₂OH
         |                  |                  |
        HOCH               HOCH               HOCH
         |                  |                  |
Gal – O – CH        Gal – O – CH        Gal – O – CH
         |                  |                  |
        HCOH               HCOH               HCOH
         |                  |                  |
        HOCH               C = O              HOCH
         |                  |                  |
        CH₂OH              CH₂OH              CO₂H

   (16) Lactitol      (17) Lactulose    (18) Lactobionic acid
```

Fig. 7

The percentage of derivatives of lactose currently used in food products is low. This is perhaps not surprising, as relatively few are screened for such purposes. To rely on serendipity appears the order of the day, as the prediction of the properties of derivatives is difficult. For example, chlorinated sugars display a range of properties depending on the nature and location of the chlorine atoms within the molecule. Certain monochlorinated disaccharides are potential anti-fertility drugs[32], whereas higher substituted derivatives of sucrose can produce compounds that are intensely sweet[33]. Chlorinated derivatives of lactose and lactulose are, however, reported to be bitter[34].

Emulsifying agents

Amadori rearrangement product

from phosphatidyl-ethanolamine and lactose

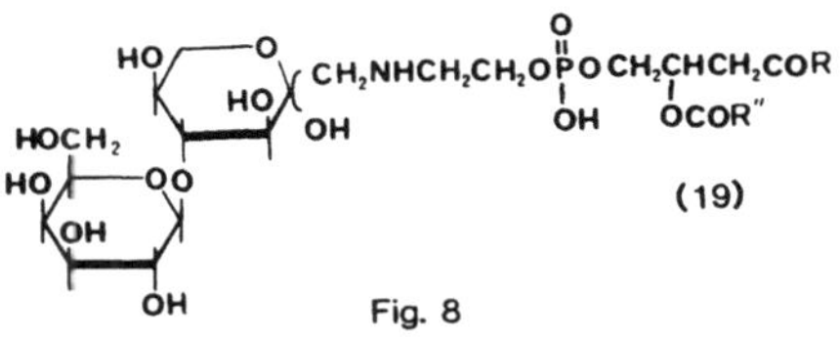

(19)

Fig. 8

Clearly, if lactose is to be any more than a by-product of the cheese manufacturing industry, investment in the form of co-operative research between manufacturing companies and food research institutes is needed.

References

1. L.A.W. Thelwall, Developments in Food Carbohydrate - 2, (Ed. C.K. Lee) 275, Applied Science Publishers Ltd., London, 1980.

2. L.A.W. Thelwall, Developments in Dairy Chemistry - 3, (Ed. P.F. Fox) 35, Elsevier Applied Science Publishers, London, 1985.

3. R.G. Edwards, Developments in Food Carbohydrate - 2, (Ed. C.K. Lee) 229, Applied Science Publishers Ltd., London, 1980.

4. E. Tarelli, Developments in Food Carbohydrate - 2, (Ed. C.K. Lee) 187, Applied Science Publishers Ltd., London, 1980.

5. A.H. Haines, Advances in Carbohydr. Chem. Biochem., 33, 11, 1976.

6. R.S. Bhatt, L. Hough and A.C. Richardson, Carbohydr. Res., 32, C4, 1974.

7. R.S. Bhatt, L. Hough and A.C. Richardson, J. Chem. Soc., Perkin 1, 2001, 1977.

8. T. Ogawa and M. Matsui, Tetrahedron, 37, 2363, 1981.

9. R.S. Bhatt, L. Hough and A.C. Richardson, Carbohydr. Res., 51, 272, 1976.

10. T. Chiba and S. Tejima, Chem. Pharm. Bull., 1049, 1977.

11. H.H. Baer and S.A. Abbas, Carbohydr. Res., 77, 117, 1979.

12. H.H. Baer and S.A. Abbas, Carbohydr. Res., 84, 53, 1980.

13. E. Fanton, J. Gelas and D. Horton, Chem. Commun., 21, 1980.

14. L. Hough, A.C. Richardson and L.A.W. Thelwall, Carbohydr. Res., 75, C-11, 1979.

15. K. Hayashibara and K. Sugimoto, U.S. Patent, 3,973,050, 1976.

16. T. Saijonmaa, M. Heikonen, M. Kreula and P. Linko, Milchwissenschaft, 33, (12), 733, 1978.

17. J.A. van Velthuijsen, J. Agric. Food Chem., **27**, (4), 680, 1979.

18. E.M. Montgomery and C.S. Hudson, J. Amer. Chem. Soc., **52**, 2102, 1930.

19. K.B. Hicks and F.W. Parrish, Carbohydr. Res., **82**, 393, 1980.

20. K.B. Hicks, D.L. Raupp and P.W. Smith, J. Agric. Food Chem., **32**, 288, 1984.

21. S Adachi, J. Agric. Chem. Soc. Jap., **31**, 97, 1957.

22. S. Adachi and S. Patton, J. Dairy Sci., **44**, 1375, 1961.

23. F.W. Bernhart, E.D. Gagliardi, R.M. Tomarelli and R.C. Stribley, J. Dairy Sci., **48**, 399, 1965.

24. A. Mendez and A. Olano, Dairy Sci. Abst. **41**, 531, 1979.

25. F.W. Parrish, F.B. Talley, K.D. Rosss, J. Clark and J.G. Phillips, J. Food Sci., **44**, 813, 1979.

26. H.S. Isbell and H.L. Frush, J. Res. Natn. Bur. Stand., **6**, 1151, 1931.

27. C.L. Mehltretter, B.H. Alexander and C.E. Rist, Ind. Eng. Chem., **45**, (12), 2782, 1953.

28. M. Oura, H. Tsumura and H. Kubota, U.S. Patent, 4,229,483, 1980.

29. L.A.W. Thelwall, L. Hough and A.C. Richardson, U.S. Patent., 4,284,763, 1981.

30. R.M. Munavu, B. Nasseri-Noori and H.H. Szmant, Carbohydr. Res., **125**, 253, 1984.

31. K.H. Todt and W.A.M. Castenmiller, Eur. Pat. Applic., 0 096 439, 1983.

32. G.M.H. Waites, W.C.L. Ford, R. Khan, and H.F. Jones, Brit. Patent., 1,595,941, 1981.

33. L. Hough and S.P. Phadnis, Nature, **263**, 800, 1976.

34. R.L. Olsen, PhD Thesis, Purdue University, U.S.A., 1983.

SUMMARY OF DISCUSSION

The dominant subject concerned the future development of new milk-based raw materials and the relevant parameters, especially composition and physico-chemical treatment. Everybody is asking for tailor-made products based on milk for the food industry. The discussion showed that for the food industry tailor-made milk-based ingredients, processing parameters and processing costs are closely related and have to be carefully considered.

The second subject concerned lactose, its derivatives and lactose hydrolysis. The major interest regarded the different properties of lactose and its different forms as well as the application of hydrolyzed lactose. The legal status of hydrolyzed lactose are of interest for marketing purposes. The epimeric products such as lactulose and epilactose are of interest for the pharmaceutical industry. To what extent new derivates can be used in the food industry is still questioned. More has to be investigated, concerning the properties, including digestibility, emulsifying activity, toxicity, sweetness and colour formation of the derivatives.

Seminar VIII: Milk as a Source of Ingredients for the Food Industry

Session 2

Chairman: Dr. J. G. Zadow (Australia)
Secretary: Dr. Ir. P. J. J. M. van Mil (The Netherlands)

Milk proteins

THE USE OF MILK PROTEINS IN FOOD FORMULATIONS

C. V. Morr
Department of Food Science
Clemson University
Clemson, SC 29634-0371
USA

ABSTRACT. Large quantities of commercially manufactured milk protein products, including caseins, caseinates, whey protein concentrates, total milk proteinates, co-precipitates, and lactalbumins are being used by the food industry as functional and nutritional ingredients in a wide range of formulated food products. Caseins and caseinates are produced from skim milk by rennet or acid treatments; whey protein concentrates are produced from cheese whey by ultrafiltration, ion exchange adsorption or insoluble ion complex formation; and total milk proteinate, co-precipitate and lactalbumin are produced by a combination of pH and/or calcium ion adjustment and heat treatment of skim milk or whey. These milk protein products offer special advantages over other protein sources for safety, nutritional quality, flavor, color, protein concentration, availability, cost, compatibility with processing conditions and with other ingredients in the formulation. Important functional properties of these milk protein products include fat emulsification, whipping and foaming, viscosity, stabilization, water binding, hydration and gelation.

1. INTRODUCTION

Milk protein products, e.g., caseins, caseinates, whey protein concentrates and others, are important in that they provide additional economic returns to the dairy processors, result in improved utilization of milk proteins, and their manufacture aids in alleviating the tremendous waste product disposal problem facing the dairy industry. These milk protein products compete well with alternate food protein products for many applications and meet the following general requirements of food ingredients: (a) free of toxic microorganisms and chemical residues, (b) contain minimal concentrations of off-flavor and off-odor compounds that detract and limit their acceptance in food product formulations, (c) contain high concentrations of protein, (d) are highly digestible and contain high concentrations of essential amino acids with a high degree of bioavailability, (e) are compatible with other food ingredients and processing conditions, and (f) are readily available at competitive prices.

2. MANUFACTURE OF MILK PROTEIN PRODUCTS

Reliable data are not available for world milk protein production and utilization. However, it is estimated that about 220 thousand tonnes of caseins and caseinates are produced annually in the world (Morr 1984 & 1985). The large quantities of whey produced annually as major byproduct of the cheese industry represents a potential source for at least 650 thousand tonnes of high quality whey protein. Large amounts of these protein products are involved in international trade that is controlled by export-import quotas, tariffs and compositional standards. Utilization of milk proteins is strongly influenced by government regulations concerning formulated food product standards of identity. The industry has the capacity for producing considerably greater amounts of milk protein products, but such increased production can only be realized if the demand for their use by the food industry can be enhanced.

A number of different milk protein products are manufactured from skim milk and cheese whey (Morr 1982, 1984 & 1985). These products include caseins, caseinates, total milk proteinates, co-precipitates, whey protein concentrates and lactalbumins. Milk protein products range in protein concentration from about 35 to 95%, depending upon the fractionation process employed (Table 1).

Casein is precipitated by acidification of skim milk or coagulated by treating skim milk with rennin. The resulting casein curd is washed and purified to remove residual lactose and milk minerals and processed to produce acid casein or rennet casein. The curd may also be solubilized with one of several alkalies and dried as the corresponding caseinate. Co-precipitates and total milk proteinates are combinations of caseins and whey proteins that are produced by appropriately adjusting the pH and heating skim milk. Whey protein concentrates are produced by ultrafiltration or ion exchange adsorption to recover the whey proteins from whey. Lactalbumins are produced by heating whey and the addition of acid to precipitate the denatured proteins. Most of the whey protein concentrate is produced by ultrafiltration, but one major cheese manufacturing company in the U.S. has begun manufacturing 95% whey protein concentrate by the ion exchange adsorption process. More recently there has been several reports in the literature of processes for fractionating the proteins from whey protein concentrate into β-Lg and α-La enriched products that offer additional opportunities for improving milk protein utilization (Slack et al., 1986).

3. FUNCTIONAL PROPERTIES OF MILK PROTEIN PRODUCTS

The inherent physicochemical and functional properties of casein and whey proteins are modified during their fractionation and processing for the manufacture of milk protein products. Thus, they represent a wide spectrum of solubilities and functional properties that are useful

Table I. Composition of milk protein products[a]

Protein products	Source	Treatment	Protein	Mineral	Lactose	Milkfat
Acid casein	Skim milk	Acid	95	2	0.2	1.5
Rennet casein	Skim milk	Rennet	89	7-8	-	1.5
Caseinates	Skim milk	Acid/alkali	94	4	0.2	1.5
Co-precipitates	Skim milk	Heat/Acid	90-94	4-5	1.5	1.5
Total milk proteinates[b]	Skim milk	Alkali/Heat/Acid	95	4-5	0.1	1.0
Whey protein concentrate	Whey	Ultrafiltration	35	2-20	40-60	6-8
Whey protein concentrate	Whey	Ultrafiltration	75	3-5	8-10	6-8
Whey protein concentrate[c]	Whey	Ion Exchange	95	3	-	0.5
Lactalbumins	Whey	Heat/Acid	86	1-2	3-5	3-4

[a] Morr 1984 & 1985
[b] Anonymous 1984
[c] Palmer 1982

for formulating different food products. For example, sodium and
potassium caseinates are highly soluble at alkaline and neutral pH and
at low calcium concentrations. Rennet casein is virtually insoluble in
the presence of even low calcium ion concentrations due to its
enzymatically modified conformational state. Co-precipitates and
lactalbumin, which are subjected to high heat conditions during their
manufacture, are virtually insoluble at acidic and neutral pH
conditions due to their highly denatured and complexed state. Total
milk proteinates are highly soluble at neutral pH conditions even
though they too have undergone sustantial amounts of denaturation and
protein-protein interaction during their manufacture. Whey protein
concentrates manufactured with mild temperature and pH conditions
retain a high degree of protein solubility and functionality. However,
whey protein concentrates manufactured under more drastic processing
conditions are also subject to a lack of solubility and functionality.

The solubility and functionality of these milk protein products
are further influenced by the types and concentrations of the other
ingredients that compose the formulated food products. The severity of
heating conditions used in manufacturing the formulated food product
is also important for determining the solubility and functionality of
milk protein products.

The key functional properties of milk protein products that are
important in controlling their functionality include water binding,
hydration, fat emulsification, viscosity, stailiization, gelation,
foam expansion, stabilization and texturization. One or more of each
of the above functional properties are critically important for the
successful manufacture of specific formulated food products. For
example, successful manufacture of meringue and other confectionery and
bakery products requires that the milk protein provide excellent
hydration, solubility, viscosity, foam expansion and stability and
texturization properties that are relatively competitive with those of
alternate food protein products.

Milk protein products are used as key functional ingredients in
numerous food products including dairy foods, frozen desserts, ice
cream novelties, breakfast bars, breakfast cereals, pasta products,
bakery products, meringues, whipped toppings, low calorie margarine,
coffee creamers, cheese analogs, fruit juices, UHT products, cultured
dairy products and others. In addition, large amounts of casein-whey
and casein-soy blends are manufactured for used in the bakery and
confectionery industries.

4. UTILIZATION OF MILK PROTEIN PRODUCTS IN THE UNITED STATES

The U.S. uses about 35 thousand tonnes of casein and caseinate annually
in human food product formulations. Of this amount, about 55% is used
for making cheese analogs, 16% for coffee creamers and 15% for bakery
products (Table II). The food industry uses about 40% of the total whey
supply in human food products and only about 6% of the total whey
supply is used for manufacturing whey protein concentrate. The industry
uses about 35 thousand tonnes of liquid whey solids, 217 thousand

Table II. Utilization of casein and caseinate in the U.S.[a]

Products	Amount (Thousand tonnes)	(%)
Cheese analogs	19.0	55
Coffee creamers	5.6	16
Bakery products	5.2	15
Dessert toppings	3.1	9
Miscellaneous	1.7	5
	34.6	100

[a] A Survey of Utilization and Production Trends. 1985. Whey Products Institute. Chicago, IL

tonnes of dry whey solids and 17.2 thousand tonnes of whey protein concentrate in human food product formulations. About 50% of the total 37 thousand tonnes of whey protein concentrate manufactured in the U.S. is used for human food product formulation. Of the total whey protein concentrate manufactured, the food industry uses about 56% in dairy products, 20% in protein blends and the balance in infant foods, bakery products, institutional food products, soft drinks, dietary products and confectionery products (Table III).

Table III. Utilization of whey protein concentrates in the U.S.[a]

Products	Amount (Thousand tonnes)	(%)
Dairy	9.6	56
Blends	3.5	20
Infant foods	0.5	3
Bakery	0.6	3.7
Institutions	0.7	4.2
Soft drinks & dietary	0.5	2.6
Prepared mixes	0.05	-
Confectionery	0.2	1.0
Meat	0.05	-
Other	1.5	8.9
	17.2	100

[a] A Survey of Utilization and Production Trends. 1985. Whey Products Institute. Chicago, IL

5. REFERENCES

1. Anonymous 1984. Product Bulletin PB#1100/1145.0. New Zealand Milk Products, Inc. Petaluma, CA, USA.
2. Morr, C.V. 1982. 'Functional properties of milk proteins and their use as food ingredients,' In <u>Developments in Dairy Chemistry -1</u>, P.F. Fox, ed. Applied Science Publishers. New York.

3. Morr, C.V. 1984. 'Production and use of milk proteins in food,' Food Technol. 38(7): 39-42, 44 & 46-48.
4. Morr, C.V. 1985. 'Manufacture, functionality and utilization of milk protein products,' In Milk Proteins '84, T.E. Galesloot & B.J. Tinbergen, ed. PUDOC, Wageningen, The Netherlands.
5. Palmer, D.E. 1982. 'Recovery of protein from food factory wastes by ion exchange,' In Food Proteins, P.F. Fox & J.J. Condon, ed. Applied Science Publishers, New York.
6. Slack, A.W., Amundson, C.H. & Hill, C.G. 1986. 'Production of enriched β-lactoglobulin and α-lactalbumin whey protein fractions J. Food Proc. & Pres. 10: 19.

THE USE OF CASEINATES IN FOODS

S.S.Gulayev-Zaitsev
The Ukrainian Meat and Dairy
Research Institute
252105, Kiev, M.Raskovoy, 4a
USSR

ABSTRACT. World production of casein and caseinates redoubled during the last decade and amounts approximately to 240,000 metric tons per year at present. The development of novel production methods for caseinates with high biological value and oriented properties furthered the progress in the field.
The paper features the scientific and technological Progress achieved in the use of caseinates for the manufacturing of meats, sausages, bakery, macaroni, pastry and fat emulsions, low-fat products, cheese substitutes and other foods.
The scope of application for caseinates is constantly widening, particularly by their use in special purpose blends together with other dairy or vegetable proteins, blood plasma and carbohydrates.
In future it looks promising to extencify researches aimed at the studiing of interconnections between the structural peculiarities of the casein system sub-units and its functional properties.
The experience gained in caseinates application for food manufacturing shows that the efficiency of their functional properties depends largely on the chemical composition and the nature of the component distribution in the product structure, and therefore the problems of daire proteins interaction with fats, carbohydrates, minerals and other components are to be elaborated on the molecular level.

I. INTRODUCTION

World casein and caseinate production redoubled during the last decade and amounts approximately to 240,000 metric tons per year at present. Major casein and caseinates manufacturers are New Zealand, France, Poland, Ireland, FRG, the Netherlands, USSR and USA. FRG, USSR, Italy and Japan

are the biggest consumers.
The progress in the field was stimulated by development of the new production technologies of caseinates with high biological value and oriented properties.
At present the development of caseinates manufacturing technology features two major trends: to improve and to boost the efficiency of existing technologies and to manufacture modified and novel types of caseinates having special compounds and properties designated for their oriented application /1-8/. As a result the caseinates with improved composition capable to withstand a broad pH range were created. They possess sophisticated technological and functional properties, i.e. Wettability, dispersancy, solubility, viscosity, water-binding, fat-emalsifying and foam-forming capabilities. The bulk of caseinates is produced in powder form, while the minority of them are granulated. Nevertheless there exist technologies for caseinates manufacturing in wet form. We gained some experience in an efficient use of caseinates by blending them with other components, i.c. with proteins of animal and vegetable origin. The biological value of caseinates is corroborated by their fulvalue amino-acid composition, easy liability to the proteolytie enzymes in the alimentary canal and by their high assimilation. All these unique features render caseinates to be indispencible components of many food products. Economical expediency of the caseinates can be attributed to their relatively low cost /9/. Now the problems connected with the functional properties of dairy proteins and their application in food products is often discussed in special publications, at the IDF sessions, seminars, congresses / 6, 10-19/.
The present paper features the progress in science and technology for the various uses of caseinates in the food industry.

2. DISCUSSION

2.1. Meat products. As early as 10 years ago up to 90 % of caseinates produced were used for meat products. Recently their alternate uses expanded and covered other fields of the food industry changing this ratio. Nevertheless the meat industry consumes now well over the half of world caseinate production. The wide use valuable dairy proteins in the form of sodium caseinate for meat products depends on its good water-binding and emulsitying capacity, high protein concentration and low sugar and Calcium cation percentage. The efficiency of caseinate application is also assured by high solubility and simple processing; optimum concentrations yield products with high organoleptic specifications.
The broad-scale application of caseinates for the meat in-

dustry is carried out in the USA, FRG, USSR, france and Poland. The existing meat product assortment with caseinates includes scores of items: cooked and semismoked sausages, canned meats, pastes, ravioli, chops, steaks, and a large number of meat substitutes as well as products for traditional and dietetic nutrition. Sodium caseinate is introduced into sausage recipes either as powder or as a protein/fat emulsion due to its high emulsifying capacity. Accordingli to this parameter sodium caseinate surpasses coprecipitates, whole milk powder and a number of vegetable proteins. Due to its high surface activity sodium caseinate forms adsorption films on the fat/water interface that withstand the heat processing. Fat emulcification rate with caseinates rises with pH and the optimum emulsion stability is achieved at pH 7,0 /19/.
Being a component in the sausage stuffing sodium caseinate reveals its functional properties depending on the composition and particularly on the total protein proportion /10/. The use of functionally active and composition stable proteins permits to reduce many negative factors/age, breed, fatness of the cattle, thermal conditions of meat, etc/influencing the quality of the staffing and the endproduct. Sodium caseinate concentration in sausages usually does not exceed 2-3 % which is an optimum for emulsifying capacity and a limit that could not be augmented without the risk of degrading the organoleptic characteristics. Meat industries accumulated some experience in using sodium caseinate combined with other protein components, i.e. milk powder, whey or soya-been proteins, coprecipitates, casecitates, albumins, isolates, blood plasma and starch, in a number of cases the total functional characteristics in these compositions surpars the ones of individual components.
Soviet researchers /6/ studied the possibilities of using sodium caseinate in a combination with soya-been protein, blood plasma, protein stabilizers and starch for sausage manufacturing. As a result they developed the recipies and processes for cooked sausages.
An assortment of canned meats, canned meat stuffings and pastes with the use of sodium caseinate also for children's and athletes nutrition has been developed in the USA, the USSR, FRG, Cuba, etc. The use of sodium caseinate in these products prevents from water separation during the sterilization. Apart from caseinates or casecitates these receipts could also include starch and soya-been isolate.
The use of sodium caseinate in manufacturing of comminuted meat products like chops, ravioli, beef steakes is an efficient treud in Europe and North America.
According to the technology developed in the USSR /20-22/ these products could also include soya-been proteins and blood plasma.

Meat substitutes based on texturized proteins form an important group of products. Manufacturing of texturized proteins demands casein and caseinates. Procedures for texturized proteins manufacturing and technologies for their application in varions meat products are developed in the USSR / 23/.

2.2. Bakery and macaroni products. The use of high-quality dairy proteins in bakery yilded a group of food products with oriented composition and properties, low-fat ones included. The balanced nutrition theory considers the reducing of carbohydrate content in these products by increasing of protein concentration to be an important objective which is in progress in many countries.
At present they developed modified caseinate compositions and special protein concentrations containing caseinates for bakery and macaroni industry. Caseinate replaces to 20 % of flower in bread, macaroni and biscuit dough, which could highten the protein content, for instance, in bread-up to 20 %. The flavour of this bread is retained, with biological value increasing. Calcium or sodium caseinates blended with skim milk powder are included into flower substitutes developed in the USA. The dietetic and baby products: rolls, macaroni, bisqucts etc, manufactured with the use of caseinates in the USSR feature high consuming qualities /24, 25/.

2.3. Pastry. The use of caseinates in this group of products is stimulated by the nessessity of reducing of carbohydrates (mainly suger) and increasing of protein components. Further, it is reasonably to replace skim milk powder in pastry by caseinate/whey blends from the economie viewpoint.
The pastry dough that contains caseinates possess better stability, structure, colour and reological characteristics. Caseinates are used in pastry products at a commercial scale in Belgium /20/ increasing thus the protein content up to 20 % 50-80 % of flower is replased by calcium caseinate in special sorts of busquit for those suffering from diabetes.
The so-called wet and juicy cakes carrying calcium caseinate/whey powder blend in their receipt are very populer in the USA /26/.
These components give a rich taste, good consistency and retain water. Similar blends in concentrations of 4,5 % are used for busquit products with soft, wet, but not crisp consistency that have along shelf life.
Calcium caseinate of the DMV company (Holland) is widely used in busquit manufacturing. Due to the low sodium and carbohydrates content it could be used for many dietetic and baby products.

Soviet workers developed technologies for pastry, khalva, etc /17/ with the use of sodium caseinate. The use of caseinate in the khalva manufacturing increased its biological value by 7 % and lowered its sugar content by 4 % and resulted in better fat-binding.
American confection companies developed milk substitutes containing 8-13 % of sodium caseinate, 35-45 % of whey powder and 47 % of whey protein concentrate.
Caseinates found application in pastry decorative creams where they effct their foam-formig and stabilyzing qualities resulting in creams with homogenous consistency and a good storing stability.

2.4. Fat products. Fat products manufacturing with the use of dairy proteins is widely spread. These inglude varions fat emulsions of animal and vegetable origin incthe form of mayonnaise, sauces, sour cream, coffee whiten ners, and low-fat products like butter and margarine. Caseinates possess a valuable property to form fat emulsions that are stable to high-temperature processing, defrosting and spray-drying followed by non-coalescent solubilization /27/.
Big volumes of caseinates are used for coffee whiteners (up to 1.000 metric tons in the USA).
Many technologies and patents for the production of low-fat products like butter and margarine are developed now in a number of conntries. They contain 35-65 % of fat and 8-15 % of skim milk dry matter, mostly protein. Milk fat, its blends with vegetable or margarine oils are used as a fat base in these products. Milk and vegetable proteins, caseinates included, form the protein components. Compositions of low-fat butters and margarines include also structure stabilizers (starch, gelatin, agar-agar, etc), flavour additives (fruit, vegetables, cocoa, coffee, chicoory), emulsifiers and preserving agents.
Some products contain fat in a dispersion state with a continuons water/protein phase. In this case the main function of caseinates is to create a high aggregation stability of the fat emulsion. Further, they give a multifold increase in plasma viscosity and possibly form gel structures in the solution with stabilizers and other additives. Nevertheless, in many cases for the purpose of better similarity to butter the processing of low-fat products includes the destabilization of fat emulsion and the forming of a continions fat phase or for the products with a continions fat phase the water solution of protein components undergoes dispersioning.
Soviet reseaerchers developed the processes for new types of butter. It contains dry or condenced milk, butter-oil, milk protein concentrates produced by the membrane processing and caseinates /28-30/. These butters are produced by the processing of high-fat cream. In analyzing the effect

of protein components on the structure and consistency of low-fat butters the auther together with A.P.Beloussov found that the mechanical strength of these butters (45 % of fat, 10,7-19,0 % of non-fat solids including 5,3-10,0 % of protein) at the temperature above 16°C is higher comparatively to this parameter in butters of the traditional composition (78 % of fat).
This permitted to introduce and develop a fresh view at the physical structure of low-fat butters and at the effect produced by protein (casein, caseinates) on its formation. The important detail in this structure is the presence of the continuons water phase inter layers stabilized by the protein interphase adsorption layers of a certain mechanical strength. The latler is a result of the hydrophobic interaction between the non-polar groups of casein globules sweeped in the interfaces and of the hydrophobic interaction of those globules with the liguid fat phase. The presence of free fat acids of small concentration in the liquid fat phase contributes to the additional gain in strength of the interphase layers /31/. These film structures pierce the food block and give another influence on the total mechanical streugh of the low fat butter with the increased protein concentration side by side with the crystal structure formed by the plate triglyceride crystals in the fat phase.

2.5. Cheeze substitutes. The use of caseinates for cheeze substitutes manufacturing is rapidly growing of late. Hard cheeze analogs like Mozarella, Cheddar, Tilsit and soft paste-like cheezes and processed cheezes are popular in the USA, FRG, France, UK and other countries. These cheezes also contain non-dairy raw material: vegetable fats, flower, emulsifiers, organic acids, flavourings and colours with caseinates used as a protein component. Cheap raw material and simple technology together with high organoleptic characteristics of end products approximatins to ones of natural cheezes makes this production rather promising. Cheeze substitutes usually contain 15-35 % of dairy proteins, presented mostly by the calcium caseinate, with its functional propertiss assuring the optimum product structure. In some cases they use potassium and sodium caseinates blended with calcium caseinate.
The process for the Camembert production (FRG) with higher whey protein content presents some interest. This technology provides adding of caseinates into skim milk to form a coagulum capable of binding whey proteins.

2.6. Desserts. Whipped dairy desserts are popular in many countries. Apart from the dairy base these products include fats, proteins, carbohydrates ensuring good

physico-chemical and organoleptic characteristics. Casein and caseinates, whey proteins, gelatine and egg wnites are used as proteins of animal origin. Whipped dairy desserts are produced by coipping of the blend containing besides milk components also fruit-and-berry raw material, sweeteners, emulsifiers, foam-formers, stabilizers, flavourings and colours. Dessert whippability and stability of their structure is a result of a joined effect of foam-formers, stabilizers and emulsifiers.
Milk proteins presented in these systems are mainly foam-formers; they accelerate forming and homogenious bubble distribution of the gaseons phase and its fixation in the liquid phase.
At present they developed modified caseinates that are soluble in warm and cold water, effect their functional properties throughout a wide pH range and are stable to heat processing. These are the caseinates of the Dutch DMV company and American DMI company.
The manufacturing of whipped milk desserts develops in two trends: redy-to-use whipped products and blend powders to be whipped at home. The assortment of various whipped desserts is developped in the USA, Great Britain, FRG and in other countries: whipped natural and artificial cream, milk-based-or milk substitute-based desserts, creams, cocktails, mousses. Soviet workers made powder blends for whipped products /18/ , and also for acid milk desserts produced according to the technology developed in cooperation with the DMV company (Holland).

2.7. Milk products. All industrialized countries increase the production of low-fat products. This is achieved by lowering fatcontent and by hightening protein concentration. To enrich milk and liquid acid milk products they developed water soluble spray-dried potassium caseinates with lower phosphate and calcium caseinate concentrations not exceeding 1,2 %. For these purposes sodium caseinate or the blend of sodium, potassium and calcium caseinates is used in a number of countries.
FRG produces "branded" milk with caseinate, Czechoslovakia, Creat Britain and Bulgaria release acid milk products. Soviet workers produce buttermilk beverage and kefir and developed the process for low fat sour creams production with the use of sodium caseinate. These products feature good taste and consistency and also slower whey separation during storage.

2.8. Other fields of application. There exists a number of developing brunches in the food and the combined feed industries where caseinates are used. These are: the manufacturing of food concentrates, fruit-and-vegetable canned products, fish stuffings, liquers and soft drinks, baby foods and whole milk substitutes.

3. CONCLUSION

Caseinates found application in various foods and now it is difficult to point a food brunch working without the use of caseinates. One may forsee that the renewal of the caseinate assortment on the novel technologies base wile take place in the nearest future. This will broaden the existing spheres of their application and will create new ones.
The most important objective in the milk protein study is the research of the interaction between the structural idiosyncresies pertained to casein system sub-units and its functional properties and also the elucidation of the environmental impact on the structure and functional properties of casein micelles and caseinates. The knowledge of these regularities and the protein and caseinate structure control by means of chemical, enzymatie and physical processes can result in the developing of new caseinate production methods at the scientific background. The experience of the caseinate application in foods shows that their efficiency depends largely on the chemical composition and the component distribution in the product structure. In this respect the problems linked to the milk protein interaction with fats, carbohydrates, minerals and other components are to be further studied at a molecular level.

REFERECES

I. Segalen P. 'Actuelites dans le domaine des caseins et caseinatea'. Revue laitiere francaise,1981,No 400,p 83-89.
2. Kyle W.S.A. 'The usa of calcium seguestrants as viacosity modifiers in caseinate manufacture'.Brief Communications to the XXI Int.Dairy Congr. Moscow,1982,v.1,b$\underline{2}$,p 81-82.
3. Костин Я.И.'Молочный белок, казеин (казеинаты), копреципитаты'. В кн.: XXI Международный молочный конгресс. Материалы.-М.: Агропромиздат,1985, т.$\underline{2}$, с. I72-I82.
4. Puigserver Antoine J., Senlourminia C. et al. 'Covalent attachment of amino acids to casein. J.Chemical modification and rates of in vitro enzymatic hydrolysis of derivatives'.J.Agr.and Food Chem.,1979,v.27,No 5,p 1098-1104.
5. 'Bridel obtient le Prix du Conseil Superior des Etablissements Classes'. La technique laitiere,1984,No986,p 23-25.
6. Рогов И.А., Журавская Н.К., Жаринов А.И., Ясырева В.А., Рослова А.Р.,Куликова В.В. Современные тенденции использования белковосодержащего сырья животного и растительного происхождения при производстве мясных продуктов. - М.: ЦНИИТЭИмясомолпром СССР, 1981.- 32 с.
7. Сергеева В.Ф., Дьяченко П.Ф.'Физико-химические свойства

казеинатов щелочных и щелочноземельных металлов'. В кн.: Совершенствование методов анализа качества молока и молочных продуктов. -М.: Пищевая промышленность, 1973, с. 61-68.

8. Дьяченко П.Ф. Новое в технологии пищевого казеина и казеинатов .- М.,1971.- 43 с.

9. Jensen K., Jul M. 'Vegetable and Dairy Protein Products in Meat'. European Congress of Meat Research Workers. Review Papers. Moscow, 1977, 2, p 152-180.

10. Салаватулина Р.М. Рациональное использование сырья в колбасном производстве.-М.:Агропромиздат,1985.- 256 с.

11. Robinson B.P., Short J.I. & Marshall K.R. New Zealand Journal of Dairy Science and Technology, 1976,11,p114-126.

12. Morr C.V. New Zealand Journal of Dairy Science and Technology, 1979a, 11, p 185-194.

13. Morr C.V. Journal of Dairy Research, 1979b, 46,p 369-376.

14. Morr C.V. In: "Developments in Dairy Chemistry.-Proteins". P.F.Fox, ed., 1982, p 375-399. Applied Science Publishers Ltd., London.

15. Fox P.F. & Mulvihill D.M. Journal of Dairy Research, 1982, 49, p 679-693.

16. Fox P.F. & Mulvihill D.M. 'Physico-chemical and functional properties of caseinates'. Proceedings of International Dairy Federation Symposium. "Physico-chemical aspects of protein-rich dairy products",Helsingor(May 1983).

17. Харламова О.А. 'Молочные концентраты в производстве кондитерских изделий'. Хлебопекарная и кондитерская промышленность, 1982, № 10, с. 11-15.

18. Гроностайская Н.А., Холодова Т.А. Функциональные свойства растворимых молочнобелковых концентратов и использование их в производстве пищевых продуктов.- М.: ЦНИИТЭИмясомолпром СССР, 1977, № 4, с. 16-27.

19. Кайл В.С.А., Бэйбар И. 'Функциональные свойства белков-производных казеина'. В кн.: XXI Международный молочный конгресс. Краткие сообщения. - М., 1982, т.1,кн.2,с.52-53.

20. Колесникова Л.А., Успенская Н.Р. 'Влияние казеината натрия на качество котлет'. В кн.: XXIII Европейский конгресс научных работников мясной промышленности. - М.: Пищевая промышленность, 1980.- с.456-459.

21. Золотарева В.И., Чирикова М.М. 'Применение соевых белков и казеината натрия при производстве котлет и пельменей! В кн.: Достижения в области исследования сырья и продукции мясного производства./Под ред.А.Ф.Савченко.- М.,1981, с. 6-13.

22. Юрченко Т.И., Афанасьева Д.А., Золотарева В.И. 'Использование казеината натрия при производстве котлет и пельменей'. Мясная индустрия СССР, 1977, № 5, с.25-26.

23. Толстогузов В.Б., Дианова В.Т., Рогов И.А. и др. 'Влияние белковых волокнистых разбавителей на биологическую ценность вареной колбасы'. Мясная индустрия СССР,

1980, № 10, с. 36-39.
24. Кузьминский Р.В., Патт В.А., Синельникова Э.М.,Проблемы повышения пищевой ценности продуктов питания (достижения и перспективы) в СССР и за рубежом. - М.: ЦНИИТЭИпищепром СССР, 1982, серия 12, вып. 7, с. 23-24.
25. Кротова К.И. Увеличение выработки хлебо-булочных диетических изделий: (с применением казеината натрия). - М.: ЦНИИТЭИпищепром СССР, 1983, серия 14, вып. 7, с.15-16.
26. Информационный бюллетень ММФ, 1985, № 1 (июнь).
27. Mann E.J. 'Utilization of milk proteins'. Journal of Society of Dairy Technology, 1971, 23, 4, p 145-150.
28. Вышемирский Ф.А. Улучшение ассортимента и повышение качества сливочного масла. - М.: ЦНИИТЭИмясомолпром СССР, 1978, - 48 с.
29. Гуляев-Зайцев С.С., Добронос В.Г., Березанский М.М., Евчев И.М. 'Разработка и внедрение технологии производства новых видов сливочного масла'. В кн.: Сборник трудов интенсификации процессов маслодельного производства. - Киев,1981, с. 93-100.
30. Gulyayev-Zaitsev S.S., Dobronos V.G., Berezansky M.M. and Zhirnaya N.M. 'Low-calorie butter'. Brief Communications to the XXI Int.Dairy Congr., Moscow, 1982, v 1, b 1, p 328-329.
31. Измайлова В.Н., Ребиндер П.А. Структурообразование в белковых системах. - М.: Наука, 1974, - 270 с.

EMPIRICAL OBSERVATIONS AND THEORETICAL CONSIDERATIONS ON WHEY PROTEIN FUNCTIONALITY IN FOOD PRODUCTS

J.N. de Wit
Netherlands Institute for Dairy Research
P.O. Box 20
6710 BA EDE
The Netherlands

ABSTRACT. The notion "functionality" refers to the functional demands made on food products regarding their desired properties. Protein functionality appears to be poorly related to the well-known functional properties of whey proteins, properties which form the basis for theoretical explanations.
   In this paper basic protein knowledge is used directly to explain protein functionality in food products. Empirical observations on the water-binding capacity of whey proteins and gelatin during cooling in set yoghurt and during sterilization of meat products are discussed in terms of the effects of temperature on the structure and thermodynamic behaviour of these proteins.
   From the results of this analysis it appears that differences in amino-acid composition and sequence govern the water-binding capacity through a different folding behaviour. In particular the greater tendency of whey proteins to fold at low temperature due to hydrophobic effects, and to disperse at high temperatures as a result of disorder effects, keeps these proteins from water-binding in a matrix structure during cooling and sterilization.

INTRODUCTION

Whey proteins are well-known ingredients for improving food products, because of their high nutritional quality and their versatile functional abilities. The behaviour of whey proteins during food processing is, however, very complex and lack of adequate predictive tests limits their functional use. Many attempts have been made to solve this problem following two different working methods, as shown in Fig. 1. These are: the systematic approach of estimating functional properties in aqueous solutions (shown on the left-hand side), and the empirical approach of assessing protein functionality by using whey protein products directly in food products (shown on the right).
   The notion of functional properties is often used in relation to physico-chemical properties of proteins in aqueous solutions or in simple model systems. Characterization of these properties informs us

FUNCTIONAL CHARACTERIZATION OF WHEY PROTEINS

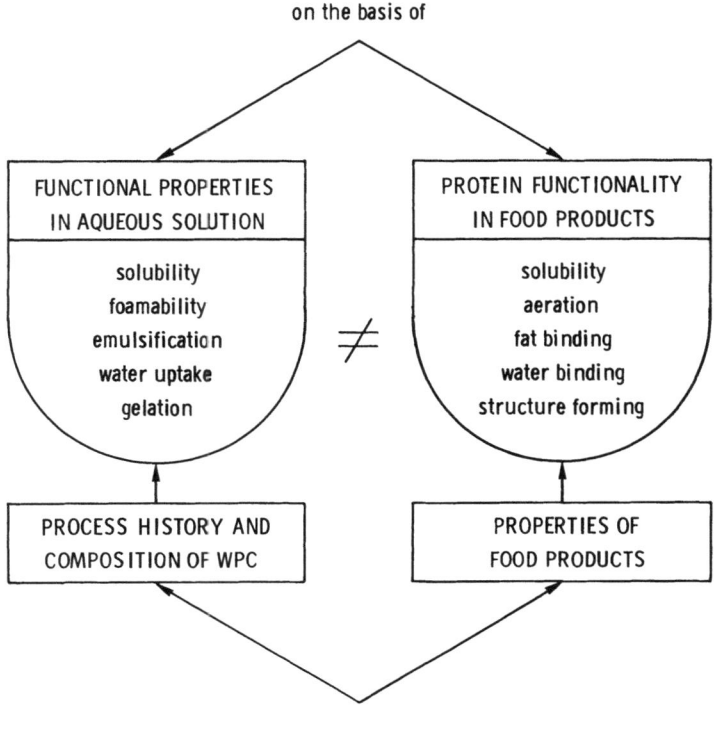

Figure 1. Functional characterization of whey proteins on the basis of functional properties in aqueous solution and protein functionality in food products.

primarily of the process history and composition of the whey protein products involved. Lack of internationally standardized characterization methods, however, limits the usefulness of this approach.

The term functionality refers to the functional demands made on food products, regarding their desired properties such as aeration, fat-binding, water-binding, structure-forming capacity, etc. These functional demands are not simply related to similar functional protein properties in aqueous solutions. This implies that the functional requirements of food products are frequently solved by trial and error through additions of arbitrarily selected food proteins before food processing. For every new product this empirical procedure has to start again, without any possible help from systematic background information.

This raises the question, whether the functionality in complex food products cannot be studied directly by a more theoretical approach of protein functionality. Examples of such an approach will be discussed now by using the water-binding capability of some quite different proteins in two representative food products. The products selected are set yoghurt, which may exhibit serious water-binding problems during cooled storage, and comminuted meat products, which often release water and fat during sterilization.

EMPIRICAL APPROACH OF PROTEIN FUNCTIONALITY IN FOOD PRODUCTS

The set yoghurts shown in Fig. 2 were prepared from (5 min, 85 °C) preheated skim-milk which was, after cooling to 45 °C, inoculated with 2.5 % ISt-starter. The figure shows the syneresis observed after 1 hour upon a storage period of 24 hours at 4 °C. (1) indicates the reference set yoghurt, and (2) this yoghurt fortified with 1 % proteins from whey protein concentrate, (3) that with 1 % Na-caseinate, and (4) with 0.5 % gelatin. The released whey (in the tubes) was slightly coloured with methylene blue for visualization.

These results show that gelatin supports the water-binding in set yoghurt much better than do whey proteins, with caseinate being in an intermediate position. It is generally assumed that gelatin

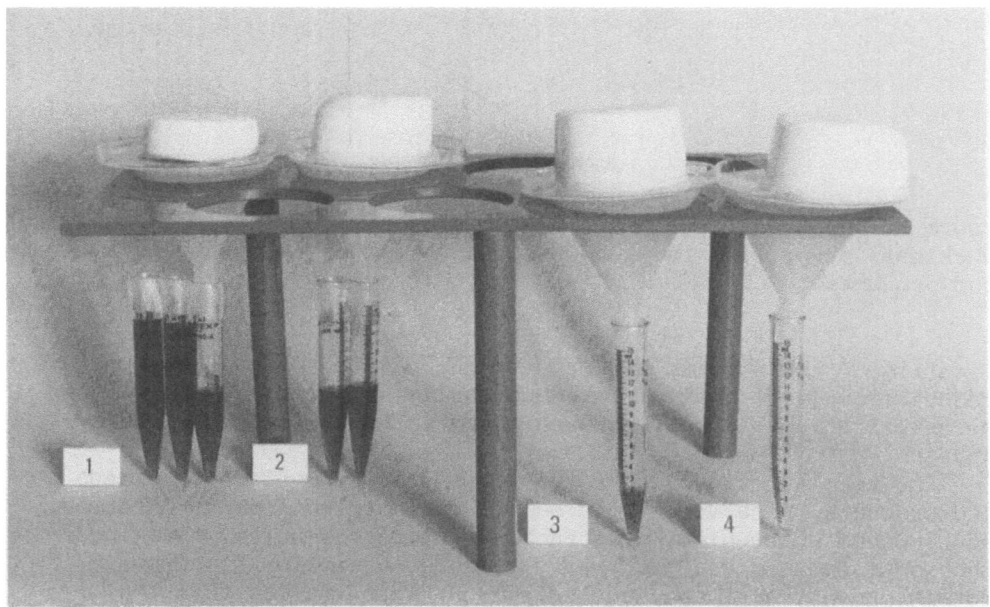

Figure 2. Set yoghurt from skim milk (1), fortified with 1 % whey proteins (2), 1 % Na-caseinate (3), and 0.5 % gelatin (4). The coloured liquid indicates the amount of syneresis after 1 hour upon a storage period of 24 hours at 4 °C.

Figure 3. Model product for Frankfurter sausage, showing fat cap and jelly separation.

immobilizes water, during cooling, within its extended network, and the important questions are now: What is the driving force for the water-binding of gelatin, and is it possible to modify the whey proteins in order to obtain these gelatin-like properties during cooling?

Another water-binding problem is observed during sterilization of comminuted meat products. Fig. 3 shows an example of a Frankfurter type product, showing fat cap and jelly separation after sterilization. This product was prepared from 42 % lean meat and contains about 40 % fat; it was, after chopping, sterilized for 30 minutes at 110 °C. This figure clearly illustrates that this product needs functional support for both fat- and water-binding. It is generally accepted (Schut, 1976) that heat-induced gel- or matrix-structures are required for both fat- and water-binding in meat products.

Heat-induced gelation is a well-known property of whey proteins, but it has been shown (De Wit, 1984) that addition of whey proteins hardly improves the water- and fat-binding properties of Frankfurter type meat products during sterilization processes (above 100 °C). Lauck (1975) observed, however, that whey protein concentrates enhance water- and fat-binding in Frankfurters after heat treatments between 70 and 80 °C. Hermansson (1975) argued that heat treatments of meat systems around 80 °C are best for the support of water-binding by whey proteins in a gel- or matrix-structure. These observations again raise the questions as to the driving forces responsible for these features, and as to the possibility of controlling them.

It is clear that answers to the above-mentioned questions are of paramount importance for an optimal use of protein functionality. To this end we need more basic information on the effect of temperature on the structure and behaviour of proteins.

THERMALLY-INDUCED STRUCTURAL CHANGES OF PROTEINS

It is well known that the molecular structure of most proteins in aqueous solution is sensitive to changes in temperature. The effect of heating and cooling on the structural changes of α-lactalbumin (α-La) and collagen (the precursor of gelatin) is only qualitatively known, and Fig. 4 illustrates schematically the most relevant changes. Alpha-lactalbumin is a small globular whey protein which, in aqueous solution, unfolds during heat treatment at the denaturation temperature (62 °C) to a completely unfolded structure (Pfeil, 1981). Upon cooling, the molecules will reversibly refold again to their native configuration (and molecular size) as shown in Fig. 4. Creighton (1985) has argued that refolding may occur within a minute or so. The reversible folding behaviour of α-La is, however, dependent on the presence of non-reduced disulphide bonds.

Tropocollagen is the basic subunit of collagen tissue (the main constituent of animal frameworks). It is a long rod-shaped molecule with a length of 280 nm, composed of three super-coiled protein chains. Tropocollagen distinguishes itself from other proteins by its high contents of glycine (30 %) and proline (about 25 %). Sequences of the type Gly-Pro-Pro (or Hypro) are common in the non-polar regions, and are responsible for the association of three peptide chains, each coiling along a left-hand three-fold axis (Ledward, 1979).

Heating of a tropocollagen solution above its transition temperature (about 40 °C) results in dissociation and unfolding of the polypeptides to random-coil structures. Subsequent cooling below the transition temperature will refold the polypeptide chains. At the protein concentrations commonly used, this may, however, partly result in the formation of network structures through intertwining of some

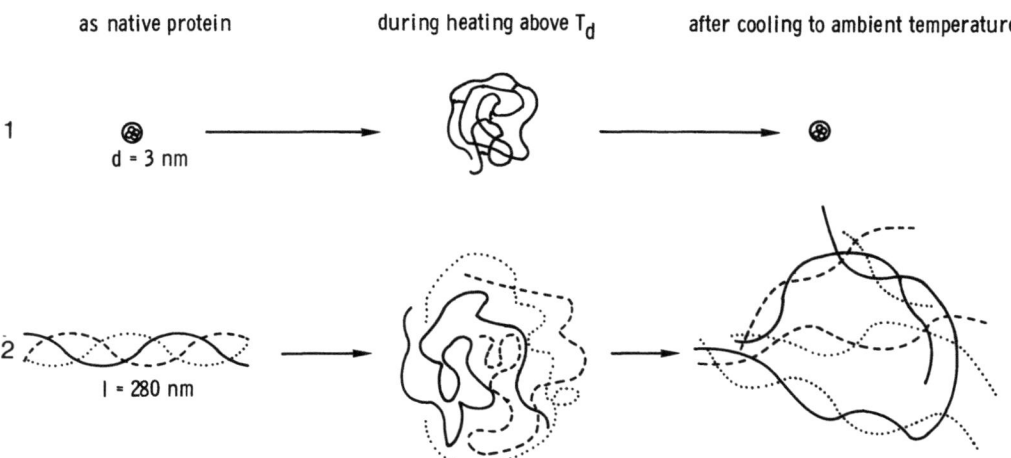

Figure 4. Schematic illustrations of structures of α-lactalbumin (1) and tropocollagen (2) before, during and after heat-treatment as indicated.

refolded polypeptide structures (Ryan, 1977), as shown in Fig. 4. Gelatin shows upon cooling a behaviour identical to that of tropocollagen (Ryan, 1977).

These completely different heat-induced changes in the structures of α-La and gelatin give a plausible explanation for their different gelation behaviour during cooling. It is clear that α-La is unable to form network structures during cooling for entrapping water as does gelatin.

These differences between α-La and gelatin, however, are extremes. Other whey proteins such as β-lactoglobulin (β-Lg) and serum albumin, or α-La in combination with these proteins, show aggregation upon thermal unfolding, and aggregation prevents a normal refolding upon cooling. These thermally-induced aggregation processes may, depending on the conditions (protein concentration, pH, etc.) result in either a gel or coarsely aggregated particles (Ledward, 1979). The thermally-induced protein gels so obtained entrap water during their formation.

However, there are exceptions: Pantaloni (1965) observed that when a β-Lg solution is heated quickly to 95 °C, it does not aggregate at all within 10 minutes. Aggregation then only starts after subsequent cooling down. Tombs (1970) reported that serum albumin solutions heated to 70 °C may form gels, but when these solutions are heated quickly to 100 °C, gelation is not observed at all. Both these features point to some systematic changes of protein behaviour near 100 °C, changes which are relevant for the ability to form matrix structures during sterilization processes of meat.

More accurate results obtained with Scanning Calorimetry recently enabled protein chemists to evaluate a more elaborated thermodynamic analysis of protein denaturation (Privalov 1979, 1982). These results are important for elucidating the driving forces involved during the heating and cooling processes of proteins in food systems. But before discussing these, we need some more information on the thermodynamics of protein denaturation.

THERMODYNAMIC APPROACH OF PROTEIN DENATURATION IN FOOD PRODUCTS

Unfolding of globular proteins is accompanied by an endothermal heat effect (heat uptake). This effect may be observed by differential scanning calorimetry (DSC), a technique by which differences in heat flows to a sample and a reference sample, are measured as a function of temperature or time. For reviews on details of DSC of globular proteins, we refer to Privalov (1974), and Wright (1982).

Fig. 5 shows a thermogram of α-La, obtained with a DuPont-990 Differential Scanning Calorimeter, at a heating rate of 5 K/min. Alpha-lactalbumin is the only whey protein that shows reversibility in its thermally-induced unfolding behaviour, and therefore this protein allows a thermodynamic analysis of the effects of heating and cooling processes. The following parameters may be derived from the DSC-thermogram in Fig. 5.

The minimum point of the curve (T_d), corresponds to

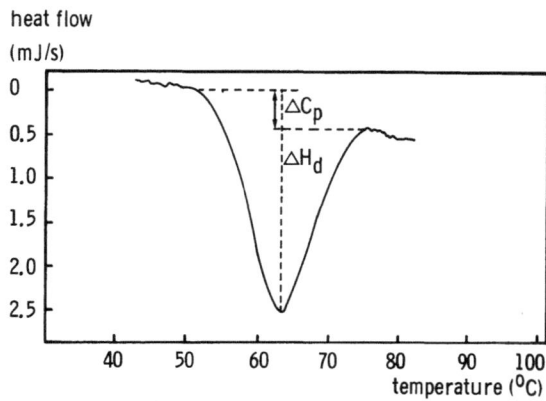

Figure 5. DSC-thermogram of α-lactalbumin, recorded at a heating rate of 5 K/min at pH 6.0.

the denaturation temperature (62 °C for α-La). At this temperature the change in Gibbs free energy (ΔG), between the folded and unfolded protein structure, is zero. ΔG, however, is the net result of two large counteracting free energy contributions: ΔH, the calorimetric-measurable enthalpy change during unfolding, derived from the peak area of the thermogram (276 kJ/mol α-La), and TΔS, which is the free energy contribution due to the order-disorder transitions of both protein side chains and solvent molecules during unfolding.

Another important thermodynamic parameter is the change in heat capacity (ΔC_p), which is the difference in heat capacity between the native and denatured state of the protein (as indicated in this thermogram). The heat capacity is generally attributed to the excess heat needed to melt ice-like (ordered) water, present around (non-polar) solutes. This quantity is, however, only measurable with very sensitive scanning calorimeters. Recently Pfeil (1981) used such a calorimeter for studying the unfolding of α-La, and reported a ΔC_p-value of 4 kJ/mol.

Now we have sufficient experimental quantities to calculate the driving forces for denaturation and renaturation of α-La as a function of temperature, using the well-known thermodynamic equations (1-5).

$$\Delta G(T) = \Delta H(T) - T\Delta S(T) \qquad (1)$$

in which:

$$\Delta H(T) = \Delta H(T_d) + \Delta C_p [T - T_d] \qquad (2)$$

and:

$$\Delta S(T) = \Delta S(T_d) - \int_{T_d}^{T} [\Delta C_p / T] \cdot dT \qquad (3)$$

moreover: $\Delta G(T_d) = 0$, giving: $\Delta S(T_d) = \Delta H(T_d)/T_d$ (4)

Substituting eqs. (2) to (4) in (1) gives:

$$\Delta G(T) = H[T_d - T]/T_d + C_p[T - T_d] - T\Delta C_p \ln[T/T_d] \quad (5)$$

Fig. 6 shows the results, calculated for the temperature range between 0 and 110 °C. The curves for ΔH (solid line) and TΔS (dashed line) illustrate the typical compensation effects of two big thermodynamic quantities of the same magnitude. The small hatched area between these curves is a measure of the stabilizing free energy (ΔG) of the native protein structure, and this is zero at the denaturation temperature (62 °C). ΔG increases at lower temperatures as a result of dominating enthalpy effects, and its (absolute) value may also be considered as the driving force for refolding or renaturation of the protein during cooling.

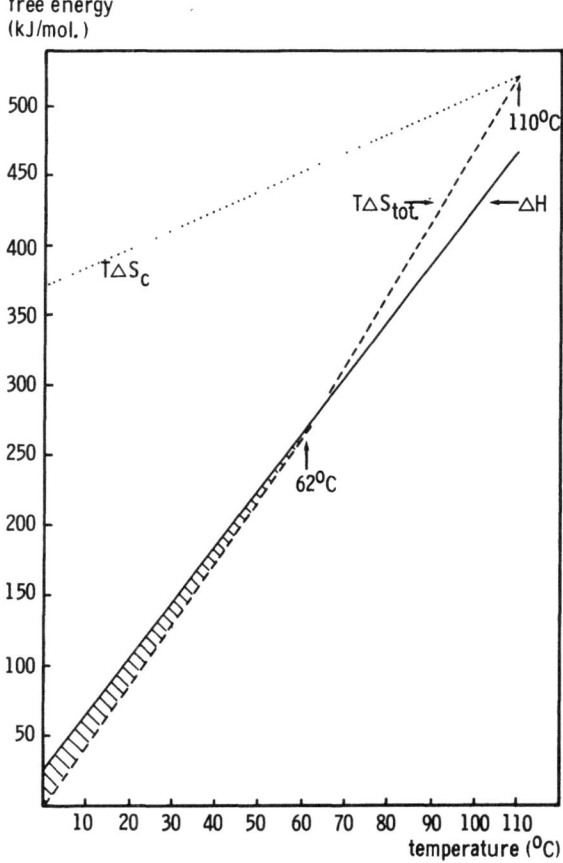

Figure 6. Temperature dependence of ΔH (—), TΔS (--), $T\Delta S_c$ (...) and ΔG(///) calculated for α-lactalbumin (see text).

When complete folding or renaturation is hampered (as observed
with other whey proteins), aggregation occurs. If we assume that the
hiding of hydrophobic groups by hydrophobic aggregation during cooling
is induced by driving forces identical to those during protein
refolding, then both phenomena have important consequences for
syneresis during cooling, as will be discussed below.

At temperatures far above the denaturation temperature the
entropic (disordering) effects dominate the enthalpic (binding)
contribution, as shown in Fig. 6 by the dashed line above 62 °C. These
entropy-driven reactions might explain why fast heat treatments up to
100 °C and higher prevent the formation of aggregates (as observed by
Pantaloni), and inhibit the formation of gels (as observed by Tombs).
Moreover, this information gives important indications why whey
proteins are more suitable for the entrapment of water and fat in a
matrix structure during heat treatments of meat products near 80 °C,
than during sterilization processes (above 100 °C).

DRIVING FORCES FOR WATER-BINDING OF PROTEINS DURING COOLING

We already discussed that the total entropy change may be split up in
an order-disorder contribution of both the protein side chains and the
water molecules. Privalov (1979) reported that at 110 °C the
contribution of water molecules to the total entropy of unfolding is
small, and that the observed entropic effects should be fully
attributed to the configurational freedom of the protein side chains
only.

If we assume as a first approximation that this conformational
entropy (ΔS_c) is temperature independent, then we may calculate its
free energy contribution also at other temperatures. The so calculated
$T\Delta S_c$ data are plotted by the dotted curve in Fig. 6. The area between
this dotted ($T\Delta S_c$)-curve and the (dashed) total entropy curve ($T\Delta S$)
then indicates the significant entropy decrease during cooling of

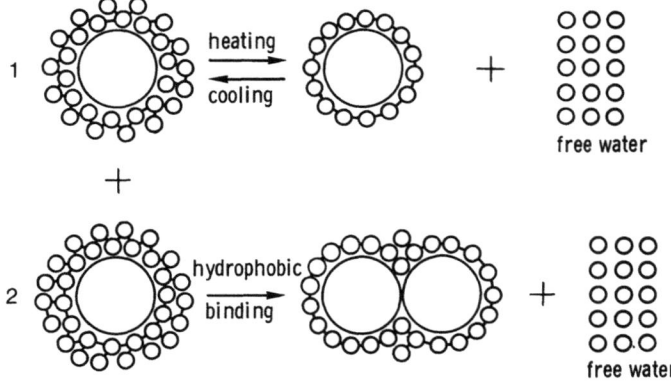

Figure 7. How water accommodates non-polar solutes of proteins after
heating and cooling (1), and after hydrophobic bonding (2).

unfolded proteins in solution, as a result of the ordering effect of the water molecules.

This ordering effect of water around charged and hydrophobic groups appears to be directly related to water-binding in the way schematically shown in Fig. 7. Lauffer (1975) measured this change as a function of both temperature (illustrated by 1 in Fig. 7) and protein aggregation (illustrated by 1 + 2) with a sensitive spring-balance technique. He weighed the mass loss caused by dialysis of the, so-called, "free water" out of a dialysis bag, and related his results to the entropy changes of the water molecules. This implies that the water-binding capacity of proteins is much higher at low temperatures, provided that the proteins do not fold or aggregate. This folding tendency is, in fact, the main difference between gelatin and whey proteins, caused by significant differences in the amino-acid sequence of these proteins.

The folding (or aggregation) tendency of whey proteins hampers an increased exposure of their non-polar groups during cooling processes. We scored a limited success by a thermal modification of whey proteins through activation of disulphide exchange reactions (NIZO, 1980). Fortification of set yoghurt with these modified proteins indeed reduced the syneresis during cooling.

CONCLUSIONS

Summarizing, very large differences in amino-acid composition and sequence between whey proteins and gelatin give rise to a strongly different folding behaviour of these proteins after heat treatments. Upon cooling, the extended chains of gelatin and tropocollagen form an intermolecular matrix with many still exposed hydrophobic groups. Both the extended chains and the exposed hydrophobic groups enhance the binding of water molecules. Whey proteins, however, will hide their hydrophobic groups by refolding or aggregation through enthalpy-driven processes, thus preventing the formation of a gel and promoting the release of water molecules (syneresis in yoghurt) during cooling.

Thermally-induced gelation of whey proteins occurs just above the denaturation (unfolding) temperature at a sufficient protein concentration. However, fast heat treatments up to 90 °C and higher may prevent the gelation by entropy-driven processes, which has consequences for the binding of fat and water in meat products. Only strong (e.g. covalent) bonds can resist the entropic energy at temperatures above 100 °C.

From this theoretical analysis, it appears that it is unlikely that whey proteins can be modified to gelatin-like properties by the usual modification procedures. A slightly improved water-binding capacity might, however, be achieved by a controlled thermal modification, resulting in conducted thiol/disulphide exchange reactions in whey proteins.

To conclude, by means of examples of yoghurt and meat products I have illustrated that a direct confrontation of empirical observations in food products with basic protein knowledge may be very informative when predicting protein functionality. For a number of products this approach is perhaps more effective than that by way of the so-called functional properties.

REFERENCES

Creighton, T.E. (1985). 'The problem of how and why proteins adopt folded conformations'. J. Phys. Chem. 89(2452).
De Wit, J.N. & E. Hontelez-Backx. (1984). 'Functional properties of whey proteins in food systems'. Voed. Midd. Technol. 17(2)19.
Feeney, R.E. and J.R. Whitaker. Ed. Advances in Chemistry, series 160. Am. Chem. Soc. Washington D.C.
Hermansson, A.M. (1975). 'Functional properties of added proteins, correlated with properties of meat systems'. J. Food Sci. 40(595).
Lauck, R.M. (1975). 'The functionality of binders in meat emulsions'. J. Food Sci. 40(736).
Lauffer, M.A. (1975). 'The role of water' Ch. 2(21) in: Entropy-driven processes in Biology. A. Kleinzeller, Ed. Springer-Verlag. Berlin.
Ledward, D.A. (1979). 'Proteins' Ch. 1(1) in: Effects of heating on foodstuffs. R.J. Priestley, Ed. Appl. Sci. Publ. Ltd., London.
NIZO (1980). 'A method of preparing modified whey proteins'. British Patent No. 2063273.
Pantaloni, D. (1965). 'Structure et changements de conformations de la β-lactoglobulin en solution'. Theses Faculty of Science, University of Paris.
Pfeil, W. (1981). 'Thermodynamics of α-lactalbumin unfolding'. Biophys. Chem. 13(181).
Privalov, P.L. (1974). 'Thermal investigations of biopolymer solutions and scanning microcalorimetry'. FEBS letters 40(140).
Privalov, P.L. (1979). 'Stability of proteins. Small globular proteins'. Advanc. Prot. Chem. 33(167).
Privalov, P.L. (1982). 'Stability of proteins. Proteins which do not present a single cooperative system'. Advanc. Prot. Chem. 35(1).
Ryan, D.S. (1977). 'Determinants of the functional properties of proteins and protein derivatives'. Ch. 4(67) in: Food proteins.
Schut, J. (1976). 'Meat emulsions' in: Food emulsions. S. Friberg Ed. Marcel Dekker. New York.
Wright, D.J. (1982). 'Application of scanning calorimetry to the study of protein behaviour in foods' Ch. 2(61) in: Development in food proteins-1. B.J.F. Hudson Ed. Applied Sci. Publ. London.

"Modification of Milk Proteins to Improve Functional
Properties and Applications"

John E. Kinsella and Dana M. Whitehead
Institute of Food Science
Cornell University, Ithaca, NY

Abstract
The posession of a range of functional properties is a critical requirement of functional food ingredients. This is of growing importance as more food products are increasingly formulated. Milk proteins display many useful functional properties but to expand their application in foods, modifications, i.e. separation/fractionation, chemical, enzymatic, or genetic techniques may improve their performance in various products. The properties required in products and the results of modification are discussed.

Physicochemical Properties of the Major Milk Proteins
The proteins of milk are responsible for many of the desirable properties of milk and cheeses and are being increasingly employed as functional ingredients in a broad range of food products [10,11,23,31,32]. In many instances, milk protein powders reflect to some degree the inherent physical properties of the major protein components i.e. caseins (85% of the total milk protein) and whey proteins [23].

Caseins are a family of heterogenous phosphoproteins which self-associate into micelles in the presence of calcium, citrate, and phosphate [23,50]. Associations among the caseins and between other components are influenced by temperature, pH, and ionic strength [11]. The physicochemical and structural attributes of the four major casein fractions, as1-, as2-, β-, and K-caseins and the two major whey proteins, β-lactoglobulin (b-Lg) and α-lactalbumin (a-La) are summarized in Table I [43,53]. Several significant structural features which are characteristic of the caseins are the content and uneven distribution of acidic (ester phosphate), nonpolar amino acids, and the high number of proline residues present along the polypeptide chain (Table I). The relationship between average hydrophobicities (Hϕave) and molecular weight is typical for globular proteins which undergo self-association [43].

The as1-casein fraction has a highly charged, highly solvated region between residues 41 and 80 containing the 8 phosphoseryl residues, and 3 strongly hydrophobic regions, 1-40, 90-110, 130-199. The as2-casein fraction which displays the greatest hydrophilicity of the casein family, has 3 charged clusters (containing the 11 phosphoseryl residues) comprised of residues 8-16, 56-61, 129-133 and a strongly hydrophobic C-terminal segment (residues 160-207). β-Casein, the most hydrophobic casein, has a negatively charged N-terminal region (residues 1-40), but the remainder of the molecule has no net charge and is rich in nonpolar amino acids. Thus, β-casein contains clearly delineated hydrophilic and hydrophobic domains. K-casein is strongly amphipathic since its C-terminal segment is strongly anionic and contains the single phosphoseryl group and the 0-3 charged polysaccharide moieties [11,23,52].

Comparison of the primary structures of the whey proteins with those of the caseins reveals several key differences which account for differences in their physicochemical and functional properties [31,43]. Whey proteins contain lower concentrations of proline and nonpolar amino acids and no ester phosphate or sugar groups, but contain higher concentrations of sulfur amino acids and a more even distribution of different amino acids along the polypeptide chain (Table I). Thus, the native whey proteins lack the amphiphilic properties of the caseins and also exhibit compact, globular conformations with substantial alpha-helical content [2,11].

Secondary structure distributions for β-casein and K-casein, b-Lg and a-La, based on data obtained by circular dichroism (CD) and optical rotatory (ORD) analysis, are summarized in Table I. The unusually high content of nonpolar amino acid and proline residues coupled with the relative lack of disulfide bonds results in little or no helical content and much more unordered structure in caseins than in whey proteins. However, recent studies have indicated that the as-caseins contain some β-structure and that a significant fraction of residues exist in β-turns [5,43].

Functional Properties of Milk Proteins

The proteins in milk i.e. caseins and whey proteins, whether collectively in milk powders or individually as caseins and whey proteins, display a wide range of functional properties [23,32] and are used im many food products because of these properties, i.e. water holding, viscosity control, coagulation, gelation, adhesion, and surface active properties (emulsions, foam formation and stabilization) (Table II) [23].

However, for many applications, the caseins and whey proteins do not possess the appropriate functional properties; hence, some degree of modification may be beneficial. Modification of properties for a particular application presupposes a knowledge of the particular physical properties required in a product and an expectation that modification will impart these desired properties. However, at present, there is inadequate information available concerning the basis of functional properties and how modification affects these in the context of designing proteins for specific applications.

Different applications generally require a different array and sequence of properties [22]. Thus in a beverage, solubility, pH tolerance, stability, storage temperature, and compatibility with flavor and other components, mouth-feel etc., are critical. In bread and bakery products, good mixing characteristics, lack of excessive moisture sorption, surface activity, compatibility with foam formation, limited browning tendency and neutral flavor are desirable qualities. In frankfurter type products, water binding, emulsifying, and gelling properties are required; while in foams the ability of proteins to form continuous, cohesive viscoelastic films is vital [23].

A thorough comprehension of protein functional properties necessitates a fundamental knowledge of protein structure, since the properties of proteins are dictated by the primary sequence of proteins [22]. Thus, the physical and structural features required for particular food product purposes demands greater clarification in order to proceed with rational modification of milk proteins by physical, chemical, enzymatic, or genetic methods. (Table III)

Modification of Milk Proteins

1. *Physical*: The functional behavior of milk proteins can be adjusted for improved performance in specific products by fractionation/enrichment or heat induced modification.

Fractionation/separation of the major components of milk is practiced commercially to prepare fractions for specific uses. Thus, in many foods, surface activity is a required functionality [23,33,43]. The unique, segregated hydrophilic-hydrophobic structures present in the caseins are generally believed to impart the high surface activity diacritic of these proteins. While the whey proteins, with several hydrophilic and hydrophobic domains in their molecular configurations, are less surface active, even when unfolded [2]. Caseins and caseinates are commonly used in food product applications where solubility, heat stability, and surface-active properties (emulsifying, foaming) are required [32,34]. Caseinates are adsorbed more quickly at interfaces than other dairy proteins and rapidly lower the surface tension. β-casein is the most effective in reducing surface tension, followed by as1-, K-casein, and then the whey proteins [12].

Heating induces physicochemical changes in whey proteins and induces complex formation with K-casein in micelles. Thus, specific 'preheat' treatments (85-100 C for 30 min) are used in the preparation of non-fat dry milk to achieve optimum properties i.e. inactivation of loaf depressing factors for breadmaking/bakery purposes [51]. This may involve oxidation of the free thiol group of b-Lg (which promotes interaction with K-casein via disulfide interchange) following its exposure during preheat treatment; and/or induce association of foam destabilizing peptides (proteose-peptones) with the milk proteins [10,23,33] thereby inactivating them.

2. Chemical: A significant portion of the research conducted on food protein modification has involved chemical derivatization of milk proteins, but little has focused on the rational modification of milk proteins with the objective of specific applications. The most common chemical derivatives are the alkali caseinates, especially sodium caseinate, which is widely used in the food industry [23,35]. Sodium caseinate has good emulsifying properties and is used in coffee whiteners, whipped toppings, etc. [9,23].

Calcium coprecipitates with varying solubility and viscosity characterisitics have been prepared [35], but the extent of their applications has been limited to date. They may be useful in semisolid and cereal food products as a rich source of calcium.

Chemical modification of proteins, in addition to altering physical properties, has also provided insights into the structure-activity behavior of food proteins. The most common chemical treatment of proteins involves derivatization of reactive amino acid side chains yielding a modified version of the native protein (Table IV). Modification of the net charge of proteins by acylation of the ϵ-amino groups of lysyl residues using acylanhydrides has been widely studied to enhance solubility, emulsifying, and foaming properties [8,21,44,46].

Acetylation of the caseins with acetic, maleic, or succinic anhydride increased solubility at pH 4.4 but reduced the nutritional value of the proteins [12] (Table IV). Succinylation of a-La resulted in increases in solubility, water absorption, fat adsorption, and emulsifying capacity with only slight losses in net protein and protein efficiency ratio [44] (Table IV). However, whippability and emulsion stability were adversely affected. The modified protein did not coagulate upon heating to 80 C. Presumably the increase in negative charges effected an increase in intramolecular repulsion causing molecular rearrangement. Acetylation of whey proteins also improved heat stability; however, succinylation generally increased emulsifying and foaming abilities to a greater extent than acetylation [3].

Improvement of the amphipathic nature of as1-casein would be expected to occur upon covalent attachment of palmitoyl residues to the lysyl residues, since the hydrophobic ligand would permit greater spatial flexibility and enable the bulky fatty acyl protein molecules to associate by lipophilic interactions [16,17]. From conformational analysis, lipophilization with palmitic acid did not cause any marked rearrangements in the secondary or tertiary structures [16]. Palmitoylation improved the ability of as1-casein to form and stabilize emulsions and increased foam stability and activity as the number of moles of palmitoyl residues attached per mole of protein increased (up to 6) [16]. Foam density decreased and foaming activity increased with an increase in ligand length [17].

Covalent attachment of carbohydrate moieties, e.g. glucose, maltose, and lactose, to the ϵ-amino group of lysyl residues of certain proteins has also been explored to improve functional properties. The non-charged, hydrophilic substituents attached to proteins would be expected to result in variations of net charge, hydrophobicity, and possibly conformation, thus creating a protein with modified surface active properties and solubility. Glycosylated caseins have increased solubility in the pH range 2.5 to 4.5. Lactosyl- and glucosyl-caseins have displayed increased viscosities of 4- to 28-fold, respectively, although foaming capacities remained constant and emulsifying properties were slightly improved [6]. The free amino groups of b-Lg were modified to varying degrees with glucosaminyl or maltosyl residues [48] and physicochemical and functional properties were assessed. Altered conformation of the modified proteins as reflected by a decrease in the helical content [47], an increased susceptibility to proteolysis [49], and by enhancement of relative fluorescence intensity [48] compared to native b-Lg, was manifested by altered surface active properties. Increases in foam stability and emulsifying activity of glycosylated b-Lg were observed (Table IV), although the introduction of substituent moieties increased net hydrophilicity, it may have impeded protein-protein interactions (via steric, hydration and perhaps osmotic effects) and resulted in the weakening of interfacial films with reduction in foam stability [13].

The carboxyl groups of proteins can be esterified by suspending the protein in an appropriate alcoholic solution (e.g. methanol, ethanol, butanol) containing HCl as an acid catalyst [28,29]. Methyl and ethyl esters of the carboxyl groups are readily formed, yielding proteins with an enhanced net positive charge at pH 7 [39], and changes in the conformation of the modified proteins as indicated by exposure of the aromatic, nonpolar residues to a more polar environment [28,29], and altered or extended functional properties of the proteins [15].

Changes in functional properties following amidation or esterification of b-Lg are summarized in Table IV [15,28,29]. Methylation and ethylation of b-Lg improved surface activity at an oil/water interface [15]. EtO-b-Lg exhibited the greater emulsion stability of the esterified proteins and was 3.7-fold more concentrated at the o/w interface compared to the native protein, but a decrease in the emulsifying activity was also noted [29]. The enhanced surface activities of the modified proteins may be due to introduction of alcohol residues into the protein, but major contributions to the surface activities result from conformational rearrangements of the proteins upon exposure to the alcohol during the reaction and, to changes in the net 'surface' charge on the proteins [39]. The enhancement of net positive charges on the reacting 'surface' of b-Lg possibly disrupted the tertiary structure and allowed the protein to facilely interact electrostatically and hydrophobically for film formation. These negatively charged food proteins may, by interactions with anionic molecules, result in potentially useful complexes for food products.

Chemical phosphorylation of proteins has been proposed as a method of improving the functional properties of whey proteins [26,27,54]. Covalent attachment (to the ε-amino acids) of 13 mol P/mol protein was achieved by reacting b-Lg with phosphorus oxychloride [55]. A solution (6% w/v) of the phosphorylated protein gelled upon dialysis against 100 mM Ca. Emulsions prepared with the phosphorylated protein derivative were more stable at pH 5-7 compared to those made with native b-Lg and viscosity of an emulsion prepared with the modified b-Lg was approximately double that of the native protein [55] (Table IV). Phosphorylation of b-Lg altered the net charge and thus the amphipathic nature of the protein and this could account for the alteration in emulsifying power.

Since caseins are naturally occuring phosphoproteins, The effect of increasing the phosphorylation state of casein would be expected to yield proteins with altered structure and functional properties [34,52]. Phosphorylation of casein with phosphorus oxychloride [26] or phosphoric acid/triethylamine [57] increased their emulsifying activities and gel-forming properties. The enhanced charge repulsion on the modified proteins may result in emulsion stabilization.

Electrostatic interactions between phosphorylated caseins, via calcium cross-bridging, are important in stabilizing the micelle structure formed by self-association of casein polymers. Dephosphorylated as- and β-caseins can still form micelles [4,56], in fact, self-aggregation is enhanced in the case of dephosphorylated -casein, indicating some electrostatic repulsion occurs in native -casein. A decrease in binding of calcium ions would be expected to occur in dephosphorylated caseins, as would destabilization of the as1-, K-casein complex and thus of the casein micelle. Dephosphorylation of the caseins decreased their thermal stability, presumably because of reduced charge repulsions in the system [10]. Thus, the presence and position of phosphoserine residues obviously contribute to the stability of the casein micelles but investigation of their influence is difficult with natural casein micelles, which contain large quantities of calcium phosphate structures [43,50,52].

3. Enzymatic: Enzymatic modification has a long tradition of use in altering milk proteins and improving functional behavior. For example, chymosin, which cleaves K-casein at the Phe-Met bond, destabilizes the casein micelle to form a coagulum [7] and is the basis of cheese manufacture.

Modification with enzymes is desirable because of specificity, safety, and the potential of enzymes to modify such a wide range of functional groups on proteins, e.g. acyl, glycosyl, phosphate, and polypeptide groups, which can be introduced under controlled conditions. In this paper, the use of enzymes to change the functional properties of isolated proteins is reviewed with emphasis on enzymatic hydrolysis and polymerization.

In general, foaming agents originating from proteins are prepared by partial hydrolysis which results in improved foaming behavior [1]; however, foam stability is generally decreased by this treatment [14] because of changes in molecular size and structure occurring during the hydrolysis reaction [18].

The relationship between nonpolar amino acid content and surface activity as it relates to foam strength and stability was explored by correlating the differences in large surface adsorption with large external hydrophobic regions of peptic hydrolyzates derived from five common food proteins [18]. Although there were exceptions, it was generally concluded that protein hydrolyzates with large surface hydrophobic regions adsorbed more readily at interfaces and rates of surface desorption were lower; however, secondary structures and the content of nonpolar amino acids in the protein hydrolyzates bore no close correlation with their respective foam stabilities [18].

Casein (as1-) is strongly adsorbed onto fat globule surfaces at the hydrophobic N-terminal region [40]. An N-terminal peptide containing 23 amino acid residues was isolated and purified from a peptic hydrolyzate of as1-casein and the emulsifying properties were compared to the parent protein [41]. The 23 amino acid residue peptide derived from as1-casein, designated as1(1-23), exhibited similar emulsifying activities compared to as1-casein in neutral pH, but the emulsion capacity of peptide as1(1-23) was approximately one half that of as1-casein. The extent of absorption of peptide as1(2-23) onto an oil surface was high [41]. The high surface adsorption and the low emulsion capacity of peptide as1(1-23) reflects a different emulsification mechanism. The low emulsion capacity may indicate that the peptide is too small to achieve the surface coverage required in emulsion systems [45].

Complete hydrolyzates are of value in special intravenous diets. In cases where injury or disease prevents digestion or absorption of ingested protein, intravenous feeding of amino acids derived from complete proteolysis of milk proteins is an important nutritional source [25].

Transglutaminase cross-links casein to yield large polymers with substantially modified rheological properties [37,38]. The enzyme crosslinks the components of casein through the ϵ-(γ-gluta-myl) lysyl groups with K-casein exhibiting less reactivity than either as1- or β-casein. The polymerization process had no adverse effects on the functional properties of the caseins other than rheological characteristics and also served to localize the components in the casein micelle structure [37].

4.   **Genetic**: Because the structure of milk proteins affects their physical behavior, modification of structure by altering the amino acid composition and/or sequence of milk proteins may provide a long term approach in redesigning these proteins for specific application. Thus genetic manipulation offers a potentially powerful tool in the design and improvement of food proteins with desirable functional properties. Recent advances in recombinant DNA technology, i.e. oligonucleotide synthesis, gene restructuring, cloning, and transgenetic methods may be used for the addition, deletion or alteration of specific amino acids in milk proteins to change certain properties, since a single substitution in the amino acid sequence can markedly alter the functional characteristics in a protein [19,42].

TABLE I

Chemical and Physical Properties of Major Milk Proteins

Properties	Caseins				Whey Proteins	
	$a_{s1}-$	$a_{s2}-$	B-	K-	B-Lg	a-la
MWr (daltons)	23 612	25 228	23 980	19 005	18 362	14 174
Tot. residues	199	207	209	169	162	123
Conc. in skim milk (g/l)	12-15	3-4	9-11	2-4	3.0	0.7
# residues phosphorylated	8	11	5	1	0	0
1/2 Cystine	0	2	0	2	5	8
CHO content	0	0	0	NANA, N-Ac-$GalNH_2$,gal	0	0
Prolyl res. per molecule	17	10	35	20	8	2
A_{280} nm (cm^2/g)	1.01	1.11	0.46	0.96	0.94	2.01
Secondary structure (%)						
α-helix	n/a	n/a	9%	23%	15%	26%
β-sheet	n/a	n/a	25%	31%	50%	14%
β-turns	n/a	n/a		24%	18%	
random coil	n/a	n/a	66%			60%
H ϕ_{ave}	4.89	4.64	5.58	5.12	5.03	4.68
pI	4.96	5.27	5.20	5.54	5.2	4.2-4.5
partial specific volume (ml/g)	0.728	0.720	0.741	0.734	0.751	0.735

TABLE II

Functional Properties Performed by Proteins in Food Systems[1]

Functional Property	Physical basis	Food System
Solubility	Protein solvation	Beverages
Water binding and absorption	Hydrogen bonding; water entrapment	Meat, Sausages, Breads, Cakes
Viscosity	Thickening; water binding; association	Soups, Gravies
Gelation	Protein matrix/network formation	Meat, Curds, Cheeses
Adhesion	Protein as adhesive agent	Meats, Sausage, Bakery, Pasta
Elasticity	Hydrophobic bonding in gluten	Meat, Bakery
Emulsification	Formation-stabilization of fat globule membrane	Bologna, Soup, Cakes
Foaming	Form stable films and entrap gas; retain lamellar water	Whipped toppings, Chiffons, Angel cakes
Fat adsorption	Binding of free fat	Meat, Sausage, Doughnuts
Flavor-binding	Adsorption, entrapment, release	Simulated foods

[1] Adapted from Kinsella, 1981. in "Food Proteins", ed. P.F. Fox and J.J. Condon, Applied Science Publishers: London; p. 52.

TABLE III

Approaches for Rational Modification of Proteins to Improve Functional Properties

Modification Process		Example
1.	Physical	- Fractionation/separation - Thermal modification
2	Chemical	- Acylation, esterification, phosphorylation, amidation, glycosylation
3.	Enzymatic	- Hydrolysis, glycosylation, cross-linking
4.	Genetic	- Recombinant DNA techniques: site-directed mutagenesis, cloning, transgenetic manipulation

TABLE IV

Changes in Functional Properties of Chemically Modified Caseins

Reactive Group	Type of Modification	Protein	Change in Functional Property	Ref
ϵ-amino (lys)	Acetylation	Casein	>solubility (pH 4.4), <nutritional value	12
	Succinylation	a-La	>solubility, >water absorption, >fat absorption, >emulsifying capacity <emulsion stability	44
		Casein	>emulsifying capacity >foaming ability	3
	Palmitoylation	Casein	>emulsifying capacity >foam activity, >solubility	17
	Glycosylation	Casein	>solubility (pH2.5-4.5), >emulsifying properties	6
		b-Lg	>viscosity, >foam stability, >emulsifying activity	46,47
Carboxyl (glu, asp)	Esterification	b-Lg	>surface activity	15
			<emulsifying activity, >emulsion stability (EtO-), >conc. at o/w interface	29
ϵ-amino (lys)	Phosphorylation	b-Lg	>gelling properties, >emulsion stability, >viscosity	55
Hydroxyl (ser, thr)		Casein	>viscosity, >water absorption, <emulsifying activity, gel-forming properties	26
			>Ca binding	57

> = increased
< = decreased

Modification of caseins to increase the content of essential amino acids [20], or of b-Lg to delete cysteine residue 121 and improve its heat stability by reducing aggregation, may now be possible.

Conclusion

Knowledge of the physical and functional properties of milk components in dairy foods and rigorous attention to these in the preparation of functional ingredients cannot be overemphasized in their importance to the dairy industry. Functional ingredients, either as milk powders or dried whey preparations, must not be viewed as marginal by-products but as premier products with a growing market potential in the food industry.

In this context, there is a need for more research to elucidate the physicochemical basis of functional behavior (i.e. structure-activity relationships) and to develop quantitative standardized methods for measuring functional properties. More importantly, to aid in the development of methods, i.e. physical, chemical, and especially enzymatic, for modification of dairy proteins in order to meet specific criteria in particular food products. There is sound rationale for the concurrent conduct of recombinant DNA research directed towards the rational modification of milk proteins for optimum performance in foods.

Acknowledgements

We gratefully acknowledge support from the following: National Dairy Board, Dairy Research Foundation, Wisconsin Milk Marketing Board, and a special award from the General Foods Foundation.

References

1. Adler-Nissen, J., Proc. Biochem., **12**,18 (1977).
2. Arai, S., Watanabe, M., Hirao, N. in "Protein Tailoring for Food and Medical Uses" (Feeney, R. E. and Whitaker, J.R., Eds.) pp. 75-94; Marcel Dekker: New York (1986).
3. Bech, A.-M., Dairy Ind. Int., **46**,25 (1981).
4. Bingham, E.W., J. Agric. Food Chem., **24**,1094 (1976).
5. Byler, D.M. and Susi, H., Biopolymers, **25**,469 (1986).
6. Canton, M.C. and Mulvihill, D.M., Irish J. Food Sci. Technol. **6**,200 (1982).
7. Creamer, L.K. and Olson, N.F., J. Food Sci., **57**,631 (1982).
8. Franzen, K.L. and Kinsella, J.E., J. Agric. Food Chem., **24**,788 (1976).
9. Fox, P.F., Int. Dairy Fed. Bulletin, No.125, p. 22 (1980).
10. Fox, P.F. in "Developments in Dairy Chemistry - 1" (Fox, P.F., Ed.) pp. 189-229; Applied Sci. Publ.: London (1982).
11. Fox, P.F. and Mulvihill, D.M., J. Dairy Res. **49**, 679 (1982).
12. Fox, P.F. and Mulvihill, D.M. in Proc. Int. Dairy Fed.-Symp. Physicochem. Aspects Dehydr. Protein-rich Milk Prod., Denmark, p.188 (1983).

13. Graham, D.E. and Phillips, M.C. in "Foams" (Akers, R.J., Ed.) pp. 237-255, Academic Press: New York (1976).
14. Halling, P.J., CRC Crit. Rev. Fd Sci. Nutr., **13**,155 (1981).
15. Halpin, M.I. and Richardson, T., J. Dairy Sci., **68**,3189 (1985).
16. Haque, Z. and Kito, M., J. Agric. Food Chem., **31**,1232 (1983).
17. Haque, Z. and Kito. M., J. Agric. Food chem., **32**,1392 (1984).
18. Horiuchi, T., Fukushima, D., Sugimoto, H., Hattori, T., Food Chem. **3**:35 (1978).
19. Jimenez-Flores, R., Kang, Y.C., Richardson, T. in "Protein Tailoring for Food and Medical Uses" (Feeney, R.E. and Whitaker, J.R., Eds.) p. 155; Marcel Dekker:New York (1986).
20. Kang, Y.C. and Richardson, T., Food Technol., Oct. p. 89 (1985).
21. Kinsella, J.E. and Shetty, J.K. in "Functionality and Protein Structure" (Pour-El, A., Ed.), ACS Symp. Ser. No. 92, pp. 37-63 (1979).
22. Kinsella, J.E. in "Food proteins" (Fox, P.F. and Condon, J.J., Eds.) pp. 51-103; Applied Science Publ.:New York (1982).
23. Kinsella, J.E., CRC Crit. Rev. Food Sci. Nutr., **21**,197 (1984).
24. Lyster, R.L.J., J. Dairy Res., **39**,279 (1972).
25. Manson, W., Int. Dairy Fed. Bulletin, No. 125, p. 60 (1980).
26. Matheis, G., Penner, M.H., Feeney, R.E., and Whitaker, J.R., J. Agric. Food chem., **31**,379 (1983).
27. Matheis, G. and Whitaker, J.R., J. Agric. Food chem., **32**,699 (1984).
28. Mattarella, N.L., Creamer, L.K., Richardson, T., J. Agric. Food Chem., **31**,968 (1983).
29. Mattarella, N.L. and Richardson, T., **31**,972 (1983).
30. McKeekin, T.L. in "Milk Proteins" (McKenzie, H.A., Ed.) vol. I, pp. 3-17; Academic Press:London (1970).
31. Morr, C.V. in "Functionality and Protein Structure" (Pour-El, A., Ed.) ACS Symp. Ser. **92**, p. ,Washington, D.C. (1979).
32. Morr, C.V. in "Developments in Dairy Chemistry - 1" (Fox, P.F., Ed.) pp. 375-399; Applied Science Publ. :London (1982).
33. Morr, C.V., in ACS Symp. Ser. **147**, p. 201, Washington, D.C. (1982).
34. Modler, H.W., J. Dairy Sci., **68**,2195 (1985).
35. Muller, L.L. in "Developments in Dairy Chemistry - 1" (Fox, P.F., Ed.) pp. 315-338; Applied Science Publ.:London (1982).
36. Nakai, S. and Li-Chan, E., J. Dairy Sci., **63**,715 (1980).
37. Nio, N., Motoki, M., Takinami, K., Agric. Biol. Chem., **49**,2283 (1985).

38. Nio, N., Motoki, M., Takinami, K., Agric. Biol. Chem., **50**,851 (1986).
39. Richardson, T., J. Dairy Sci., **68**,2753 (1985).
40. Shimizu, M., Takahashi, T., Kaminogawa, S., Yamauchi, K., J. Agric. Biol. Chem., **31**,1214 (1983).
41. Shimizu, M., Lee, S.W., Kamonogawa, S., Yamauchi, K., J. Food Sci., **49**,1117 (1984).
42. Stryer, L., Biochemistry - 2nd Ed., pp. 92-93; W.H. Freeman and Co.:New York (1981).
43. Swaisgood, H. E. in "Developments in Dairy Chemistry - 1" (Fox, P.F., Ed.) pp. 1-61; Applied Science Publ.:London (1982).
44. Thompson, L.U. and Reyes, E.S., J. Dairy Sci., **63**,715 (1980).
45. Tornberg, E., J. Food Sci., **45**,1662 (1980).
46. Waniska, R.D. and Kinsella, J.E., J. Agric. Food chem., **29**,826 (1981).
47. Waniska, R.D., Ph.D Thesis, Cornell Univ., Ithaca, NY., 1981.
48. Waniska, R.D. and Kinsella, J.E., Int. J. Peptide Prot. Res., **23**,467 (1984).
49. Waniska, R.D. and Kinsella, J.E., J. Agric. Food Chem., **32**,1042 (1984).
50. Waugh, D.F. in "Milk Proteins" (McKenzie, H.A., Ed.) vol. II, pp. 4-79, Academic Press:London (1971).
51. Webb, B.H. and Whittier, E.O., By-Products from Milk, AVI Publ., Westport, CT (1970).
52. West, D.W., J. Dairy Res., **53**,333 (1986).
53. Whitney, R.M., Brunner, J.R., Ebner, K.E., Farrell, H.M., Josephson, R.V., Morr, C.V., Swaisgood, H.E., J. Dairy Sci., **59**,785 (1982).
54. Woo, S.L., Creamer, L.K. and Richardson, T., J. aAgric. Food chem., **30**,65 (1982).
55. Woo. S.L. and Richardson, T., J. Dairy Sci., **66**,984 (1983).
56. Yoshikawa, T., Agr. Biol. Chem., **38**,2051 (1974).
57. Yoshikawa, M., Sasaki, R., Chiba, H., Agr. Biol. chem., **45**,909 (1981).

THE USE OF MILK FAT AND MILK FAT COMPONENTS IN FOOD PRODUCTS

W. Banks
The Hannah Research Institute
Ayr KA6 5HL
Scotland

ABSTRACT. The declining use of butter as a spread is accentuating the need to find alternative markets for milk fat. Most such markets demand that the physical properties of the fat conform to defined criteria. The ways in which this may be achieved by adding other fats, fractionation, inter-esterification and hydrogenation are discussed. However, no unique physical property of milk fat, either itself or in modified form, has yet been discerned, and therefore competition from cheaper vegetable fats is inevitable.

The traditional end use of milk fat has been as butter, but the steady decline over much of the Western world in the consumption of this commodity in the last 20 years has posed a considerable problem to the dairy industry. It has become necessary to develop new uses for milk fat and the search for these outlets has emphasised the fact that the dairy industry must consider itself as part of the wider food industry.
    To many consumers, the image of butter has become tarnished. All fat is under pressure from medical opinion, but saturated fats are particularly susceptible. An amalgam of hostile medical opinion, poor rheological properties at refrigeration temperature, changes in eating habits and price (relative to margarine) has resulted in the consumer having a lower regard for butter.
    Fortunately, the loss of image that has affected butter does not yet appear to have been transferred to cream. However, it is difficult to see how the sale of cream could be expanded sufficiently to counteract the decrease in butter consumption. For example, one of the success stories of the last decade has been the development of cream liqueurs. Currently, some 4 million cases of cream liqueur are produced each year in the British Isles. This production makes use of approximately 6000 tonnes of anhydrous butterfat annually; between 1982 and 1983, the decrease in butter consumption in the British Isles was equivalent to 20,000 tonnes of anhydrous butterfat. Even in the case of these liqueurs, we have obtained samples from European manufacturers in which milk fat has been entirely substituted by vegetable fat. Substitution decreases costs, but it may improve some

aspect of performance. For example, using a fully hydrogenated
vegetable oil avoids the possibility of oxidative rancidity occurring
in the liqueur.

As the image of milk fat as a superior product declines in the
eyes of the consumer, it becomes necessary to consider it as merely
another fat, which must compete in the market in terms of performance
and price. The function of the scientist and technologist is to define
milk fat in terms of its composition and structure and then to relate
these parameters to the properties of the fat.

Well over 400 fatty acids have been identified in milk fat, but
only about a dozen occur in sufficient amount to affect the physical
properties of the fat. Genetic factors dictate the spectrum of acids
present in the fat, but the relative amounts of each reflect mainly
feeding practice. The range of values that we have encountered in a
series of feeding trials are shown in Table 1. High diene and triene
contents are associated with the feeding of "protected" vegetable oils
and under other circumstances the 18:2 and 18:3 contents are close to
the minimum values. Thus whilst the total degree of unsaturation in
milk fat is not very large, there is sufficient to cause problems,
e.g. the development of oxidative rancidity in cream liqueurs.

As might be expected from its fatty acid composition, milk fat
gradually melts over a wide temperature range. The functional
properties of a fat are to a large extent dominated by the ratio of
solid/liquid at a given temperature. For example, the ease with which
a yellow fat can be spread on bread is largely determined by that
ratio. The melting "fingerprint", obtained by means of differential
scanning calorimetry, of a typical milk fat is shown in Figure 1.
Only a small proportion of the fat is molten at 7°C, hence the
difficulty in obtaining good spreadability at refrigeration
temperature. Only by imposing extreme dietary situations on the cow
can this situation be rectified. However, between 7° and 18°C, the
temperature range encountered in the home, a considerable amount of
milk fat becomes molten and the spreadability accordingly increases.
If a vegetable oil is added to butter, spreadability at refrigeration
temperature is increased. The softening point of the fat mixture is
lower than that of butter, and the amount of melting that occurs on
going from refrigeration to room temperature may make the product too
liquid. It may then be thought necessary to add a hardened
(hydrogenated), high melting fat to the mixture. Thus, by a suitable
choice of vegetable fats one may produce a range of table spreads,
based on milk fat, possessing engineered physical and rheological
properties. Certainly, this approach makes full use of the only
unique property of milk fat, namely its delicate flavour.

Physical properties may also be manipulated by the process of
fractionation. Effectively, the melting curve shown in Figure 1 is
divided into constituent portions. For example, by allowing the fat
to crystallise at 10°C, separating the crystals from the liquid,
crystallising the latter fraction at 25°C and repeating the
separation, three fractions of very different properties may be
obtained. Outlets for the fractions may be found in shortenings, or
in spreads. By mixing appropriate amounts of the first and third

fractions given in the above example, a material having good spreadability at refrigeration temperature and reasonable stand-up properties at room temperature can be obtained. In fact, this approach has been used commercially in New Zealand.

The fatty acids of milk fat are arranged in the triglycerides in a preferred manner. For example, butyric (4:0) and caproic (6:0) acids are found almost exclusively in stereospecific position sn 3 of the triglyceride. By submitting the fat to the process of inter-esterification, a random distribution of the fatty acids can be imposed. There are consequential changes in the melting fingerprint of the milk fat, but they are relatively minor in nature and do not confer any marked advantage in terms of physical properties.

Although commonly regarded as a saturated fat, milk fat does contain an appreciable quantity of oleic acid (18:1), see Table 1. A completely saturated fat results from the process of hydrogenation, which is widely used in the case of vegetable oils. A fully saturated milk fat has its melting spectrum moved to appreciably higher temperatures.

The reverse of hydrogenation, i.e. the introduction of some degree of unsaturation, is presently being investigated at the experimental stage. Unfortunately, all known methods of enzymic desaturation depend upon the presence of a free acid rather than one esterified in a triglyceride. Since any process must therefore involve hydrolysis, de-hydrogenation and esterification, it would be difficult to call the product "butterfat". Without that appellation, the desaturated product would offer few advantages over a butteroil/vegetable oil mixture.

In the case of chocolate, milk fat enjoys a legal advantage, since only a limited number of fats are specified for use in that product. Increasing the proportion of milk fat in chocolate has attracted much attention in the past decade. Unfortunately, the technical evidence shows that a significant increase in the proportion of milk fat cannot be achieved without a deleterious change in the properties of the chocolate. Even hydrogenating the milk fat does not appear to offer any hope of improvement. In this case, the limitation lies in the cocoa butter/milk fat interaction and is thus technical in nature.

This paper has sought to show that various strategies are available for altering the physical properties of milk fat to tailor it for specific end usses. However, whether modified or not, milk fat possesses no unique physical properties and must therefore compete against many other fats, which invariably have a marked advantage in terms of price. In this respect, it is ironic that in the United Kingdom the dairy farmer effectively subsidises margarine production by using extracted oil seed meal as a major feed ingredient.

TABLE 1. Fatty acid compositional (wt %) range found in milk fats

Fatty acid	Range
$4:0^a$	2.5 - 4.2
6:0	1.5 - 2.4
8:0	0.8 - 1.5
10:0	2.5 - 3.9
12:0	2.8 - 4.0
14:0	8.0 - 15.0
16:0	23.0 - 48.0
18:0	5.0 - 20.0
18:1	17.0 - 42.0
18:2	1.5 - 20.0[b]
18:3	0.5 - 15.0[b]

[a] Number of carbon atoms : number of double bonds.

[b] High diene and triene contents of milk fat result from feeding "protected" vegetable oils.

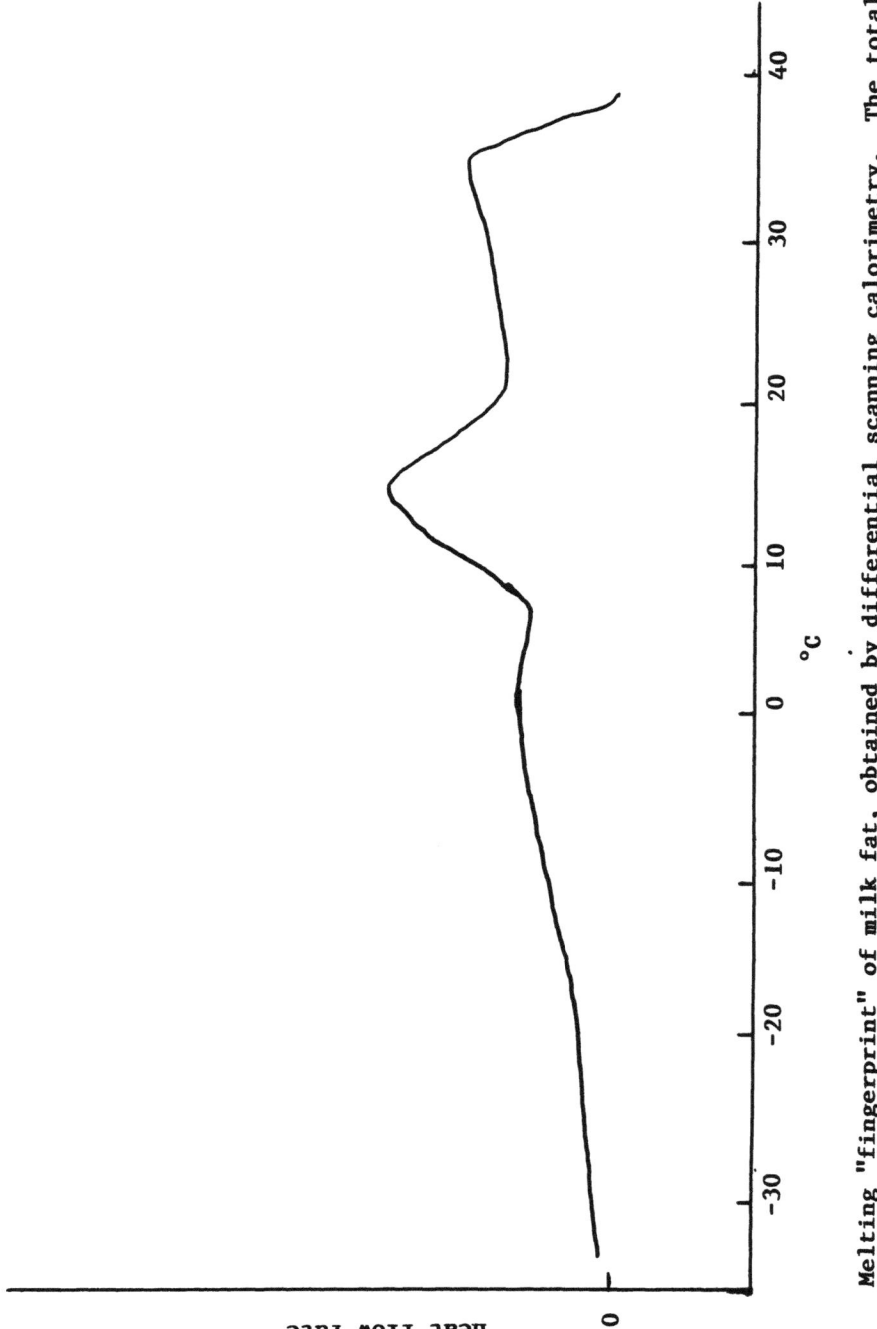

Melting "fingerprint" of milk fat, obtained by differential scanning calorimetry. The total area under the curve is proportional to the heat of melting of the fat; the ratio of the area up to a given temperature to the total area gives an estimate of the proportion of the fat that is molten at the particular temperature

SUMMARY OF DISCUSSION

The desirability of a rapid method for the determination of the water binding and the emulsifying properties of milk proteins in meats was suggested. In the discussion it became clear that no such simple test would readily be developed. It was suggested that a product test should be developed in collaboration with the meat-industry, and that data from the test should be used to devise a more rapid, empirical method in conjunction with the assessment of basic physico-chemical properties.

The methodology used for manufacturing a 95 % whey protein concentrate (WPC) containing only 0.1 % fat as outlined by C.V. Morr (USA) was discussed. The process used employed an ion-exchange absorption method for removal of protein from whey. This method was believed to result in effective separation of lipid from the protein fraction.

The factors affecting the good foaming and foam stability characteristics of egg white were discussed extensively. It was agreed that fat was a major factor reducing the foam characteristics of WPC, and efforts should be made to keep fat levels as low as possible. The type of lipid incorporated in WPC was also seen to be important. It was suggested that the main difference between some of the egg-white proteins and almost every other protein is that the former easily denature at the air-water interface, thereby forming insoluble multilayers that lend mechanical strenght to the air bubbles. Any attempt to achieve similar qualities in other proteins such as WPC must involve these denaturation properties.

Some concern was expressed regarding high levels of trans fatty acids in some samples reported by W. Banks (UK). It was suggested that these can be controlled to an acceptable level under normal conditions.

Seminar IX: Integrated Quality Control and Assurance

Session 1

Chairman: Ir. M. G. van den Berg (The Netherlands)
Secretary: B. Oterholm (Norway)

Raw milk quality, prerequisite for excellent milk and dairy products

THE QUALITY WORLD AND ITS TERMINOLOGY

M.G. van den Berg
Melkunie Holland
P.O. Box 222
3440 AE Woerden
Holland

Quality is everybody's concern; as a consumer, as a manufacturer and as a quality professional.

1. THE CONSUMER

If a product fails to satisfy expectations, we believe it to be below-standard. We show our dissatisfaction by telling our families of its shortcomings, by not buying the product for a while or by complaining to the supplier.
   Dissatisfaction with a product is shown in the consumer's unwillingness to purchase it and this is the main weapon of the consumer. When many consumers use this weapon the supplier will notice a fall in sales and consumer demand without necessarily knowing what has caused it. Thus the success of a product is definitely consumer-linked: "The consumer is always right".
   To make a complaint takes so much effort that the consumer only feels it necessary if there is something very obvious to complain about. Complaints about household goods are more evident than those about food products.
   It could be important for consumers to organize an unanimous judgement of products and to publish their findings: consumerism.
   In any case the consumer always has the last word about quality.

2. THE MANUFACTURER

A manufacturer is obviously involved with quality, but in a very different way to the consumer. Production and the products themselves must be constantly reviewed in order to increase the value, to add "quality" aiming to be successful in business on a short and long term-basis.
   This is the subject of numerous books and articles. Crosby (1), for example, in his book "Quality is free", demonstrates the role of "quality" in the success of a company and he makes "quality-

management" the centre of importance.

In Clifford and Cavanagh's book "The winning performance" (2) quality also plays an important role, but the word "quality" itself is seldom used; instead their emphasis is placed on "product value".

These McKinsey researchers come to the conclusion that a good quality policy is essential for the success of the best "midsize companies" in the U.S.A.; some quotations:
"Almost all of the winning companies compete on the basis of value, not price; they are superior in quality to the industry average".
"Successful entrepreneurs take note of consumer needs and realize that consumers know they get what they pay for".

Success is shown by the relationship between profitability and product quality of 525 midsize business, see fig. 1.

Fig. 1, from Clifford and Cavanagh (2).

Exhibit III-9
RELATIVE PRODUCT QUALITY AND BUSINESS PERFORMANCE
Four year average return on investment*

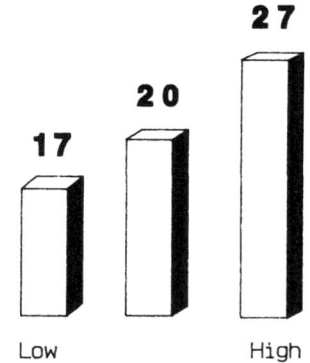

RELATIVE PRODUCT QUALITY

* Before interest expense and corporate overhead;
525 midsize businesses
Source: PIMS Program

Concerning examples of the winning companies we soon consider it to be electronics, chemistry and apparatus producers and also companies in innovation oriented areas. This is not the case at all. In all markets the winning performance is possible according to C&C: not only with sealed air "bubble wrap" packaging, but also with something familiar like chicken. The company Perdue Chickens belongs to the winners and it has a very stringent quality program which is termed "obsessive quality control" by C&C.

This is also true for Dunkin Donut, producing 350 million cups of coffee in 1984. Quality also plays a distinguishing role in the succes of coffee and donuts. The Dunkin demands on coffee suppliers serve us a good example:
- a twenty-two page specification of "what it requires in a coffee bean";
- beans are used within 10 days of their delivery;
- once the coffee is brewed it can be served for only 18 minutes;
- the coffee must be brewed between 91 and 92°C exactly; and
- use <u>real cream</u> not half and half, not milk, not the sugar based powders!

Quality management in response to the successful Japanese industries is becoming a central theme in the new spirit of the eighties and plays an increasing role in the international competition.

Is this also true for the dairy industry?
In my opinion there is more than one reason why the dairy industry has to observe quality developments made in modern industry and use them to their advantage:

1. In large parts of the dairy world companies and processing plants have increased in size and management control became intricate; then there is a greater chance that quality will fall by the way-side. This can be overcome by systematic application of quality management.

2. The distance between the milk producing cow and the consumer is becoming greater; shelflife plays a greater role because the links in the chain have become more complicated. Consistent quality management is necessary to prevent damaging results.

3. The development of new products and markets is needed more than ever. Demands are set on analysis of consumer wishes, and these have become much stricter in recent years. It also implies to the product development and quality specifications of the resulting product. This makes high demands of quality management.

4. Attention is no longer given to the product alone as was the previous practice. The spectrum of quality has expanded and now includes packaging, information on that packaging, possible uses for the product and the keeping quality information. The product has become more than the produced item, as it leaves the processing plant. It must be seen in the way the consumer receives it. These complications make systematic quality assurance necessary.

5. International competition has increased and is no longer restricted to plain bulk products competition. It aims at the creation of specific trade name markets.
    The variety of distant markets proves "quality management".

6. Some parent-companies license companies in other countries to use their product and market knowledge and trade name. In order to do so those companies must fulfil the quality requirements of the parent-company so that there is no quality difference in the final product.

7. Also in countries where the dairy industry is newly developed and is expanding, the systematic quality control is necessary. In India, for example, milk for the big cities has to be supplied from and conveyed over large distances. There quality management is also as necessary as it is in the European and Northamerican markets.

8. An increasing number of contracts contain official standards to be fulfilled, as is cited by Armstrong (3): "more contracts, particularly commercial contracts, require the supplier to meet a standard: the demands for a fully integrated assured product or service".

9. One step further is certification of the manufacturers quality assurance, so that the consumer can rely on well organised consideration of quality. Certification is already in practice in other sectors, but will be increasingly used in the food industry.

10. Developments in the terrain of product liability mainly in the U.S.A. but soon in the E.E.C., will force manufacturers to promote the introduction of certified quality assurance.

These are several of the reasons why consideration of the "quality" theme is worth while. Then we arrive automatically at the third category in connection with quality.

3. THE PROFESSIONAL

Though essential, the professionals have made the sphere of work of the quality management rather unclear, particularly the terminology is not as unequivocal as we would like it to be. This has once been characterized as: "a choral singing under the direction of hundred conductors".
    Together with the American Society for Quality Control (ASQC), the quality people united in the European Organisation for Quality Control (EOQC) have - after years of discussion - composed a "glossary" (4).

4. QUALITY

Such a terminology discussion begins with the concept "quality", what is it? The EOQC-definition is as follows:
> quality is the totality of features and characteristics of a product or service to bear on its ability to satisfy a given need.

There is a foot note explaining that quality is quite often used in reference to "grade"; but this is a category indicator for products with the same "functional use or purpose", but with diverse features.

In philosophy an early distinction is made between objective and subjective quality. "Given need" in the EOQC-definition effuses a certain objectivity, measurability. This is important in the research previous to the development of a product.

Subjectivity dominates the customers' final judgement. The product's success depends upon its fulfilment of consumer expectations; and naturally price also plays a role.

My favourite definition is:
> Quality is the measure of a products' fulfilment of consumer expectations.

A short definition which need further explanation. Although all composits can be explained in detail, the following comments are restricted to the product and the expectations.

5. THE PRODUCT

Firstly the notion of "product". In several definitions service is considered separately, but this is not really necessary if the concept of product is interpreted in the loosest sense. It is, as some economists say, a collection of attributes which are offered to the consumer; the production of something the consumer wishes to pay for.

With products such as cheese there is a definite material core, but even if the product is an airflight there is still something substantial about it. In both cases it is surrounded by an entourage of factors which influence the decision about quality.

Most foodstuffs, for example, are packaged. The protection of the product, the easy handling of the package, the possibility of reclosing, all play a role. All three are not necessarily compatible, e.g., in the United States easy handling won over product protection: gallon cartons of milk are not as easy to handle as transparent polythene bottles and the latter superseded the gallon cartons, although the milk can be damaged by light.

The packaging but also the information on the packaging is part of the product. This even too applies to shop presentation; if cooling is necessary, is the temperature really lower than 10°C? Is the margin in the shelflife adequate enough for the consumer? And so forth.

In short: product is the nucleus surrounded by a cloud of attributes which are important to the judgement of the consumer and should be recognized in a good quality management.

6. THE EXPECTATIONS

There is a great deal to write about the consumers and their expectations; the following aspects are focussed on in more details:

When a manufacturer introduces a new product, the expectations of the intended consumers are influenced by a introductory advertising campaign, as well as by the existing quality reliability of the manufacturing company, and of course by several other factors.

If, after the initial introduction, continuous purchasing goes on, the consumer expects constant characteristics of that product. A contrary reaction results in negative appraisal. Even when a manufacturer does succeed, repeated use can lead to aversion.

Expectations, as in the theory of Maslow, can, in regard to the hierarchy of human needs, be subconscious.

The assumption that foodstuff is safe is considered as natural. If this assumption is not satisfied, as in the effects of the Tsjernobyl fall-out on foodproducts, then all the other features of a product are no longer countable and its appreciation falls to nothing; as the Southern German dairy industry discovered in the beginning of May 1986.

It is easy to appreciate that radio-active milk is dangerous, but not so easy to see the danger in drinking raw milk; raw is tasty, but dangerous; it is a real danger which claims victims every year, even in the western world.

The interaction of products and expectations is a fascinating cause of study; for example the difference in appreciation between pasteurized and UHT milk based on a single sampling is not so great as with constant usage, at least for a large number of consumers.

7. QUALITY INDICATORS

The way in which the expectations of consumers are formed is the subject of market research which makes analyses of the factors which play a role in the decision to buy.

In a recent study about the perception of quality in foodstuffs, written by Steenkamp, Wieringa and Meulenberg (5), the following 3 steps-structure is proposed:
1. <u>Information</u> about a product is gained from experience, advertising, from acquaintances, general information and viewing the product.

2. <u>Quality indicators</u> are formed in the way indicated by Cox: "any informational stimulus about or relating to the product quality", such as nutritional value, means of realizing shelflife, price, brand, retailer.
3. In the <u>final judgement</u> the attributes of the product itself play a role.

According to this research the most important quality indicators for pasteurized milk are:

	cited by
1. the keeping quality instructions	74% of the 1069 respondents
2. the type of packaging	40
3. the brand	30
4. the shop	25
5. the price	25
6. the nutritional value information	24

8. EVOLUTION OF THE NOTION OF QUALITY

In the previous section emphasis was placed on the subjective way consumers judge quality. The emphasis shifting from objective to subjective is also cited by Bokern of Philips International (6). He observed a shifting of:
"Fitness for specification (Crosby)
Fitness for use (Juran)
Fitness for user
to
Fitness for users expectations"
In spite of the subjective character of "quality", it is the task of the quality professional to achieve objectivity to make good quality management possible; the subjective source, however, must not be forgotten. Then they can concentrate on specifications and conformance to specifications.

9. PRODUCT DEVELOPMENT, DESIGN QUALITY

It is necessary to probe the wishes of the consumer for whom the product is intended, and to carry out market research before a product is developed. Transposing these wishes into product features is the beginning of the development. Such a product development may make use of a consumers' panel leading to a design which may be manufactured. The features and specifications of the design, the design quality, are the starting point of the production (see scheme 1).

With the introduction into the market the consumer tests the quality of the design and it can be seen if the product fulfils the expectations of the consumer.

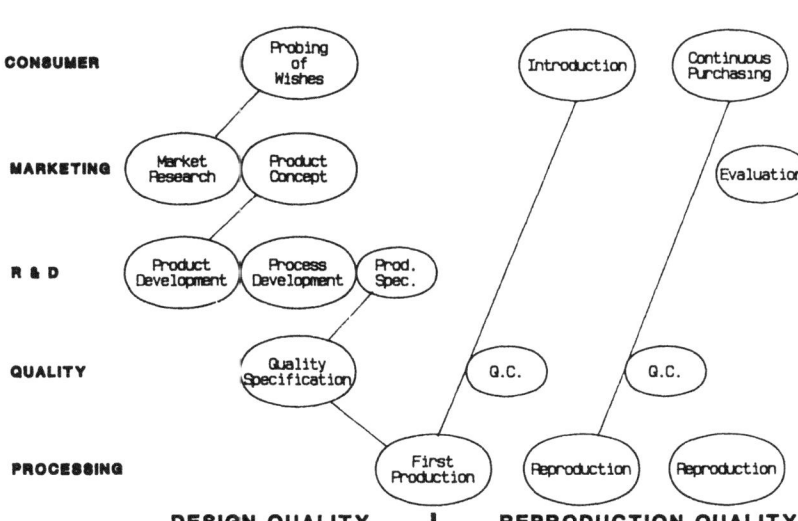

SCHEME 1
SIMPLIFIED SCHEME OF PRODUCT DESIGN AND REPRODUCTION

The next important point is the reproduction of the designed quality in full production afterwards.

Quality control of the reproduction is a rather different aspect of quality management than the control of design- and development process.

In the food industry the quality control of full production is very important because commercial success is not made with the introduction but with repeated purchasing of the new product.

"Design quality" does not meet completely the definition of quality because it cites features and characteristics which are not really tested by the consumers' judgement. Nevertheless it is a very worthy distinction of the form of quality : "conformance to specifications" in full production.

In contrast to Crosby (1) Juran (7;8) makes a clear distinction between the two areas of quality:
"The design must reflect the needs for fitness for use; the product must also conform the design".
"Quality of design versus quality of reproduction!".

In an examination of product specifications not only the new products must be considered but also the improvement of existing products needs retesting.

10. QUALITY MANAGEMENT, QUALITY ASSURANCE, QUALITY CONTROL
   (see appendix)

A different kind of confusion occurs when using the terms quality control and quality assurance, because the words assurance and control in other languages than English have a different meaning.
   This was discussed by Freund in the EOQC-Congress in 1982 (9). Freund terms the difference between quality control and quality assurance, quality system and quality management respectively "a matter of territorial challenge, generating more heat than light".

Crosby (1) states that "quality is good enough to stand by itself". The rest is management.
   It is not only Crosby who prefers the definition "quality management", but also Kano (10) has recently expressed the desire to minimize all the subdivisions of quality and to rely on the single use of quality management.
   However, the definitions quality assurance and quality control are so generally accepted, that it is necessary to discuss them yet in more detail.

Assurance has in French, German and Dutch the same meaning as the word insurance in English. Control has in those languages the same meaning as the word inspection in English. Management on the other hand has such an international recognition that the chance of misunderstandings is not very great, especially as its definition is so loose.

The EOQC-glossary gives the following definitions:
   Quality management:
   "The totality of functions involved in the determination and achievement of quality".
encompassing:
   Quality assurance:
   "All those planned or systematic actions necessary to provide adequate confidence that a product or service will satisfy given needs".
   Quality control:
   "The operational techniques and activities that are used to sustain a quality of product or service that will satisfy given needs".

In short:
   QM, the people and their piloting guidelines and equipment,
   QA, the building-up of an interrelated system: "can you live in it, or is it a house of straw?"
   QC, measuring, comparising with the set rules and readjusting when necessary: keeping everything stable and predictable. Based on a control-loop concept.

11. QUALITY MANAGEMENT

Various starting points are possible to make quality operational in daily practice.

An important starting point from Juran is: the systematic incorporation of quality in the product and the processing of that product. This has been practised in the dairy industry for a long time, but perhaps not as systematic as intended by Juran. In any case the systematic incorporation of quality in the product and processing of the product is a concept that has to be a continual centre of consideration in the present and in the future. See also Townsend (11).

The inspection of the final product is also a starting point in the quest for quality. This is applicable to all industries, but in the food industry it became a specific microbiological stress, because of perishability and danger of disseminating contagious diseases.

Christian (12) and Van Schothorst & Duke (13) have recorded and summarized this under the heading Microbiological Quality Assurance (MQA):

Good Manufacturing Practice (GMP). "Those procedures which consistently yield products of acceptable microbial quality, suitably monitored by laboratory tests."

Hazard Analysis and Critical Control Point (HACCP). "hazards, particularly microbiological ones that may exist in a process are identified and appropriate points for testing to determine whether these hazards have been controlled and also identified (Christian). Hazards: both potential to cause food poisoning as well as spoilage before reaching the end of shelf life."

"Good Manufacturing Practice" has especially become very popular for several reasons and from this some variables have arisen, such as Good Laboratory Practice, Good Hygiene Practice, Good Commercial Practice.

In the larger context of industry in general the well-known quality authorities have developed and introduced plans and systems for the realisation of an active quality management.
Some renowned ideas are:

The managerial break-through of Juran (7).
He was a pioneer of quality management in Japan, U.S.A. and numerous other countries.

Total Quality Control introduced by Feigenbaum (14).
He uses "quality cost"-analysis as a starting point followed by a systematic grid analysis through the whole organisation.

Crosby (1) is renowned for his Zero-Defects approach:
"Make it right the first time".

Ishikawa (15) is the founder of <u>Quality Circles</u>:
Round table discussions on the shop floor level. It is a success in Japan, but is also beginning to be successful in a number of western industries, but with variable results, Mohr (16).

Ishikawa believes that, with some of the other renowned programs there is too much attention given to people and he wants to avoid this by focussing attention of the quality circles on systems.

12. INTEGRATION

Finally the notion of "integration".
Integration is the threedimensional attuning and linking of the different composites of quality management.

In length: from the cow to the consumer; it is sometimes termed as integrated chain control.
In width: not only the product but also the packaging, the distribution, the information on the packaging, the treatment of complaints and so forth.
In height: through the whole organization, beginning with the topmanagement, as well as the other levels; in other words: top down and bottom up.

13. BUREAUCRACY

Once again "the winning performance". A serious threat to the success of a potential good approach is increasing bureaucracy. Quality management requires systems, but systems can easily deteriorate in bureaucracy. Continuous testing on the tendencies of bureaucracy is essential, because it is deadly to the success of companies and to quality. "Bureaucracy is the mortal enemy", Clifford and Cavanagh (2).

14. QUOTATIONS

The most important quality authorities have given their opinions about the underlining aspects of quality management, e.g. a small selection of comments:
Crosby (1):
 "Quality is too important to leave it to the professionals";
 "People still probe continually to measure the depth of top management commitment";
 "The results of communication are real and long lasting; the results of motivation are shallow and short-lived".

Feigenbaum (14):
"Genuine product quality of leadership is based upon recognition that quality is a customer determination to-day, not a marketing man's determination, or an engineer's, or a general manager's. It recognizes that quality satisfaction is based on the consumer's actual experience with the product, objective or sensed, factual or instinctive, and always representing a moving target in a competitive market".

Juran (7; 8):
"Quality cannot be inspected into a product, but must be built in";
"80% are management controlable defects, not operator controlable".

Crosby (1):
"Good things only happen when planned, bad things happen on their own".

15. REFERENCES

1. Crosby, Philip B.; 'Quality is free'.
    309 pp., Mc Graw-Hill, New York 1979.
2. Clifford, D.K. & Cavanagh, R.E.; 'The winning performance'.
    292 pp., Toronto 1985.
3. Armstrong, C.W.; 'Assessment of management/quality systems'.
    EOQC Congres C14, Amsterdam 1982.
4. European Organisation for Quality Control (EOQC); Glossary of terms used in the management of quality. Fifth edition 1981.
5. Steenkamp, J.E.B.M., Wieringa, B. and Meulenberg, M.T.G., 'Kwaliteitsperceptie van voedingsmiddelen I'. (in dutch) Swoka the Hague, 169 pp.
6. Bokern, J.H.R., Interview (in dutch);
    Specifiek 70, 5/1986.
7. Juran, J.M., Gryna, F.M. & Bingham, R.S.; Quality Control Handbook, third ed. 1800 pp. Mc Graw-Hill, New York, 1974.
8. Juran, J.M.; 'A prescription from the West ... four years later'.
    29th EOQC congress, vol. 1, 55-66, 1985.
9. Freund, R.A.; 'Progress in international quality definitions'.
    EOQC Congress B 3, Amsterdam 1982.
   Freund, R.A.; 'Definitions and basic quality concepts'.
    Sigma 4, 1986, 5-9.
10. Kano, N.; Newsletter to Quality Organisations, 1983.
11. Townsend, P.L.; 'Quality is everybody's business'. ASQC Congress, Transaction 31-35, 1986.
12. Christian, J.H.B.: 'MQA and GMP'. In Microbiological Quality Assurance in Industry. Conference Manual University of Sussex, 1983.

13. Schothorst, M. van & Duke, A.M.; 'How does MQA relate to standards?'
    In Microbiological Quality Assurance in Industry. Conference Manual University of Sussex, 1983.
14. Feigenbaum, A.V.; 'Management, modern industrialisation and quality'. EOQC Congress A1, Amsterdam 1982.
15. Ishikawa, K., 'Total quality control and quality circles'. Interview in Sigma 29 (1983) 4 (in dutch).
16. Mohr, W.: 'Quality circles', an American overview. 29th EOQC Congress, vol. 3, 73-83, Estoril 1985.

APPENDIX:

QUALITY, QUALITY ASSURANCE & QUALITY CONTROL IN 17 EUROPEAN LANGUAGES

	Quality	Quality Assurance	Quality Control
1. Norway	Kvalitet	Kvalitetssikring	Kvalitetsstyring
2. Sweden	Kvalitet	Kvallitetssäkring	Kvalitetsstyrning
3. Danmark	Kvalitet	Kvalitetssikring	Kvalitetsstyring
4. Netherlands	Kwaliteit	Kwaliteitsborging	Kwaliteitsbeheersing
5. Germany	Qualität	Qualitätssicherung	Qualitätslenkung
6. United Kingdom	Quality	Quality assurance	Quality control
7. France	Qualité	Assurance de la qualité	Maitrise technique de la qualité
8. Italy	Qualità	Assicurazione della qualità	Controllo della qualità
9. Portugal	Qualidade	Garantia da qualidade	Controle da qualidade
10. Spain	Calidad	Aseguramiento de la calidad	Control de la calidad
11. Rumania	Calitate	Asigurarea calităţii	Controlul calităţii
12. Turkey	Kalite	Kalite Güvencesi	Kalite kontrol
13. Poland	Jakosc	Zapewnienie jakości	Kierowanie jakoścía
14. Czechoslovakia	Jakost	Zajišťováni jakosti	Řizeni jakosti
15. Finland	Laatu	Laadunvarmistus	Laadunohjaus
16. Greec	Ποιότητα	Διασφάλιση ποιότητάς	Ἔλεγχος ποιότητας
17. Sovjet Union	Качество	Обеспечение качества	Управление качеством

RAW MILK QUALITY, PREREQUISITE FOR EXCELLENT MILK AND DAIRY PRODUCTS
FARM INSPECTION AND QUALITY CONTROL

Harold Wainess, Consultant
Northfield, Illinois, USA

A little over a hundred and fifty years ago, the discovery was made that pathogenic organisms were the cause of many diseases such as typhoid, tuberculosis, cholera, streptococcal sore throat and others. From that point, it took considerable time before some investigators linked various outbreaks, particularly those affecting children, with the milk supply.

Although individuals in many countries had recognized that raw milk was the causative agent for a number of these diseases, it wasn't until 1893 that Morel in France inaugarated a maternity and child welfare program, where, in order to provide clean milk for the children, the community maintained a herd of its own. This was followed in Germany, England, France and the USA by the establishment of "clean" milk stations for children. The most famous were those established in New York City, Hamburg and Paris. These examples were followed in New York by the philanthropist Nathan Strauss, who established a system of milk stations which he supported for 26 years. The milk was modified according to formula, pasteurized and dispensed in nursing bottles, and mothers instructed on feeding their young children. In 1902, these stations distributed 250,000 bottles monthly. This was one of the impulses for government action, and soon official milk stations offering pasteurized milk were established in the UK, Germany and France.

It was not until 1902 that the New York City Health Department actually assigned inspectors to visit dairy farms supplying the city. Their purpose was to investigate the conditions under which milk was produced and to "endeavor to educate farmers to the proper idea of sanitary milk production". All of this led, at the turn of the century, to the U.S. Public Health Service (USPHS) activities in the area of milk hygiene to study and conduct research, and to identify and evaluate hygienic measures for prevention of milk-borne diseases. These studies led to the conclusion that effective public health control of milk-borne disease requires the application of hygienic measures throughout the production, handling, pasteurization and distribution of milk. In 1938, milk-borne outbreaks constituted 25% of all disease outbreaks due to infected foods and contaminated water, and slowly the incidence of milk-borne disease in the United States and throughout Europe and Japan was sharply reduced. By 1984, this had dropped to less than 20%, but since that time, with the incidence of outbreaks of pathogenic strains of salmonella, listeria, yersinia and campylobacter, this percentage may have increased drastically.

The USPHS began devising a program (1924)[1] for prevention of milk-borne disease. This was on a very limited scale, but delineated the hygienic requirements for dairy farms and pasteurization plants.

The Grade A Public Health Service Milk Ordinance is divided into a series of parts, sections and appendices. One is devoted mainly to definitions for various products and specifications for chemical, microbiological and temperature requirements, and includes permits, labeling, laboratory examination, etc. Another gives the Public Health reason for each requirement and administrative procedures that are designed to unify the interpretation of the ordinance. The final part contains 11 appendices with explanatory material on various aspects of milk hygiene technology, such as individual water supplies, sewage disposal systems, pasteurization equipment, specification and tests, dairy farm inspector certification procedures and milk production methods. Of interest here is the section entitled "Standards for Milk and Milk Products: Sanitation Requirements for Grade A Milk for Pasteurization".

Section 7 specifically requires that each dairy farm shall be inspected by the regulatory agency prior to the issuance of a permit. Following that, each bulk milk pickup tanker and its appurtenances used by a milk hauler who hauls milk from the dairy plant to the milk plant or transfer station shall be inspected at least once every twelve months, and each dairy farm and transfer station shall be inspected at least once every six months.

There are 21 basic hygienic requirements for Grade A raw milk[2] covering the following fields. Adhering to these requirements, together with similar requirements for dairy plants, will not only produce a safe pasteurized product, but one with a shelf life of 14 days or more.

1. Animal health
2. Construction of milking barn and/or parlor
3. Cleanliness of milking barn or parlor
4. Cow yard
5. Milk house or milk room construction and facilities
6. Milk house or milk room cleanliness
7. Toilet
8. Water supply
9. Utensils and equipment construction
10. Utensils and equipment cleaning
11. Utensils and equipment sanitizing
12. Utensils and equipment storage
13. Utensils and equipment handling
14. Milking -- flanks, udders and teats
15. Milking -- surcingles, milk stools
16. Protection from contamination
17. Personnel -- hand washing facilities

18. Personnel cleanliness
19. Cooling
20. Vehicles
21. Insect and rodent control

Each of these requirements is divided into three categories. The first specifies the legal requirements; the second, the Public Health reason for these requirements; and the third, the administrative procedures which are the methods that the dairy farmer must use in order to comply with hygienic requirements. Compliance with all these requirements is one of the reasons why the maximum bacterial limit from individual producers is not permitted to exceed 100,000 per ml and why many farmers consistently are able to produce raw milk for pasteurization with total counts of less than 10,000 per ml. Obviously, the result has been a better milk supply for pasteurization and an increase in the shelf life of pasteurized milk.

A detailed discussion of all the requirements for hygienic dairy farms would require a considerable amount of time and a complete reading of the Grade A Pasteurized Milk Ordinance. However, some brief comments can be made describing each of the requirements. It is interesting to point out that all the requirements have been given a hygienic value from 1 to 10 (out of 100), based on their significance as far as the health of the milk is concerned. Most dairy farms are able to maintain an average of over 90% compliance, indicating that the degree of difficulty is not as hard as it appears on the surface.

The first section on abnormal milk specifically states that cows showing evidence of the secretion of abnormal milk (based upon various tests, which are defined), shall be milked last or with separate equipment and the milk discarded. This also applies to cows treated with or exposed to chemical and medicinal agents or insecticides not approved for use by the Environmental Protection Agency. Pertaining to antibiotics, the Standard states "no zone equal to or greater than 16 mm with a Bacillus stearothermophilus disk assay method or similar system". For individual producer milk, the somatic cell count is not to exceed one million per ml.

There are then specific requirements for the construction of the milking barn and milk parlor. Floors are to be constructed of concrete or equally impervious material. Walls and ceilings are smooth, painted or finished in an approved manner, and the ceiling is to be dust-tight. Sufficient light, either natural or artificial, is to be distributed throughout for day and night milking. Overcrowding is not permitted, and all feed is to be stored in dust-tight covered boxes or bins. It is specifically required that all milking must be done in an approved barn, stable or parlor. Field milking is not permitted because of the potential contamination not only from the environment, but from birds, other animals and grasses. Swine and fowl are to be kept out of the milking barn or parlor. In addition, the milking barn or parlor, including pipelines and equipment, must be kept clean. This requires daily cleaning to minimize chances of contamination, and it is recommended, where barns are provided with water under pressure, that the floors be scrubbed after each milking. In cases where running water is not available, the floors may be brushed dry and limed.

The cow-yard (which is the enclosed or unenclosed area adjacent to the milking barn in which the cows congregate) is graded and drained, and waste from the barn or the cows not allowed to pool in the cow-yard. Furthermore, manure, soil bedding and waste feed are not permitted to accumulate in the area. Brushing of cows is completed prior to milking, and all flanks, bellies, tails and udders are clipped as often as necessary to facilitate the cleaning of these areas and keep them free from dirt. In addition, the udders and teats of all milking cows are cleaned and treated with a sanitizing solution and are relatively dry just prior to milking. In any case, milking must be done in the milking barn, table or parlor. The milk house is a separate building of sufficient size to provide for cooling, handling and storage of milk and the washing, sanitizing and storing of milk containers and utensils. There are specific requirements covering the design and construction of the floors, drains, walls and ceilings.

A minimum of 220 lux is provided at all working areas whether from natural or artificial light. During dusty weather, windows are closed and the milk house adequately ventilated. Furthermore, the milk house can be used for no other purpose than the necessary dairy farm operations and cannot open directly into any barn, stable or domestic room. Water under pressure is required in the milk house, plus a hot water heather and two-compartment sink of sufficient size to be effective for the cleaning of all equipment. Where milk is transferred from a bulk holding/cooling tank to a transport tank, it must be done through a hose port located in the milk house wall. This port is to be fitted with a tight cover and closed, except when the port is in use. The milk house itself must be kept clean at all times. All water used for milk and milking operations must be approved as safe by the state water control authority and meet minimum microbiological standards. Cross connections of any type, submerged inlets, and improper water well construction are not permitted. Samples of the water supply are to be taken upon the initial approval of the physical structure and at least once every three years thereafter.

Hand washing facilities are located convenient to the milk house, milking barn or parlor. These facilities include soap, running water, individual hygienic towels and a sink. The importance of clean hands, particularly among those doing the milking or those performing any milk house function, is emphasized.

There are very strict construction requirements for multi-use containers, equipment and utensils used on the dairy farm. Basically, these requirements are similar to those for equipment used in the dairy plants. Where 3A Standards[3] have been developed for specific equipment, these standards also apply.

To depict the detailed requirements for all aspects of dairy farm hygiene, the section covering "Utensils and Equipment -- Construction" quoted below is exactly as it appears in the Code[2]. This is an example that applies to all other sections of the Code.

Item 9r. UTENSILS AND EQUIPMENT — CONSTRUCTION

All multiuse containers, equipment, and utensils used in the handling, storage or transportation of milk shall be made of smooth, nonabsorbent, corrosion-resistant, nontoxic materials, and shall be so constructed as to be easily cleaned. All containers, utensils and equipment shall be in good repair. All milk pails used for hand milking and stripping shall be seamless and of the hooded type. Multiple-use woven material shall not be used for straining milk. All single-service articles shall have been manufactured, packaged, transported, and handled in a sanitary manner and shall comply with the applicable requirements of Item 11p of this section. Articles intended for single-service use shall not be reused.

Farm holding/cooling tanks, welded sanitary piping, and transportation tanks shall comply with the applicable requirements of Items 10p and 11p of this section.

PUBLIC-HEALTH REASON

Milk containers and other utensils without flush joints and seams, without smooth, easily cleaned, and accessible surfaces, and not made of durable, non-corrodible material, are apt to harbor accumulations in which undesirable bacterial growth is supported. Single-service articles which have not been manufactured and handled in a sanitary manner may contaminate the milk.

Milk pails of small-mouth design, known as hooded milk pails, decrease the possibility of hairs, dust, chaff, and other undesirable foreign substances getting into the milk at the time of milking.

ADMINISTRATIVE PROCEDURES

This item is deemed to be satisfied when:
1. All multiuse containers, equipment, and utensils which are exposed to milk or milk products, or from which liquids may drip, drain or be drawn into milk or milk products are made of smooth, impervious, nonabsorbent, safe materials of the following types:
   A. Stainless steel of the AISI (American Iron and Steel Institute) 300 series; or
   B. Equally corrosion-resistant, nontoxic metal; or
   C. Heat-resistant glass; or
   D. Plastic or rubber and rubberlike materials which are relatively inert, resistant to scratching, scoring, decomposition, crazing, chipping and distortion, under normal use; are nontoxic, fat resistant, relatively nonabsorbent, relatively insoluble, do not release component chemicals or impart flavor or odor to the product, and which maintain their original properties under repeated use conditions.
2. Single-service articles have been manufactured, packaged, transported and handled in a sanitary manner and comply with the applicable requirements of Item 11p.
3. Articles intended for single-service use are not reused.

4. All containers, equipment and utensils are free of breaks and corrosion.
5. All joints in such containers, equipment, and utensils are smooth and free from pits, cracks, or inclusions.
6. Clean-in-place milk pipelines and return-solution lines are self-draining. If gaskets are used, they shall be self-positioning and of material meeting specifications described in 1-d above, and shall be of such design, finish, and application as to form a smooth, flush interior surface. If gaskets are not used, all fittings shall have self-positioning faces designed to form a smooth, flush interior surface. All interior surfaces of welded joints in pipelines shall be smooth and free of pits, cracks and inclusions.
7. Detailed plans for clean-in-place pipeline systems are submitted to the regulatory agency for written approval prior to installation. No alteration or addition shall be made to any milk pipeline system without prior written approval of the regulatory agency.
8. Strainers, if used, are of perforated metal design, or so constructed as to utilize single-service strainer media.
9. Seamless hooded pails having an opening not exceeding one-third the area of that of an open pail of the same size are used for hand milking and hand stripping.
10. All milking machines, including heads, milk claws, milk tubing, and other milk-contact surfaces can be easily cleaned and inspected. Pipelines, milking equipment, and appurtenances which require a screwdriver or special tool shall be considered easily accessible for inspection providing the necessary tools are available at the milk house.
11. Milk cans have umbrella-type lids.
12. Farm holding/cooling tanks, welded sanitary piping, and transportation tanks comply with the applicable requirements of Item 10p and 11p of this section.

Obviously, the cleaning of all milk house equipment is important and it is required that such equipment be thoroughly cleaned and sanitized before each usage.

All milk containers, utensils and equipment, including milking machine hoses, are stored in the milk house, either in a sanitizing solution or on racks, until used. Pipeline milkers which are designed for mechanical cleaning may be stored in the milking barn or parlor, provided that the equipment is properly designed for that purpose. If any manual cleaning of product surfaces has to be done, it must be done in the milk house. The design of in-place equipment must be such as to create complete drainage of equipment. Furthermore, after sanitization, all equipment has to be handled in such a manner as to prevent contamination of the milk or any other product surfaces.

Whenever air under pressure is used for agitation or movement of milk, or is directed at a milk contact surface, this air must be free of oil, dust, rust, excessive moisture, extraneous materials and odor, and must meet a set of standards described in the Code.[2]

One highly important requirement is that raw milk for pasteurization be cooled to 7°C. or less within two hours after milking. All vehicles used to transport milk from the dairy farm to the milk plant are to be constructed and operated to protect their contents from the sun, freezing and contamination, and be kept clean both inside and out. There are specific requirements for the cleaning and sanitizing of these vehicles.

In order to prevent contamination from insects and rodents, all surroundings are kept neat and clean and free of any conditions which might harbor or be conducive to the breeding of such insects and rodents. Excess storage of manure is not permitted during fly seasons, and milk rooms must be kept free of insects and rodents by effective screening or other similar protective devices.

These are not static requirements! Since their introduction in 1924, they have been revised 16 times and partial revisions and interpretations made on a continuous basis. Add all these assurances of quality together with good hygienic practices in the pasteurization plant, followed by proper package and packaging, and then supplement with 7°C. maximum storage, until consumed. Thus, it is common for Grade A pasteurized milk in the USA to have acceptable shelf life of 14 days.

References:
1. 1924 -- USPHS -- Standard Milk Ordinance
2. 1982 -- USPHS (FDA) Grade A Pasteurized Milk Ordinance
3. 1986 -- 3A Sanitary Standards Committee

MILK INSPECTION AND QUALITY PAYMENT

F. Harding
Milk Marketing Board
Thames Ditton
Surrey
England

ABSTRACT. Milk is progressively mixed into large volumes as it moves down the chain from the cow to the consumer. The closer to the cow quality faults are identified, the lower is the financial loss attributable to poor quality milk. There is a need for quick, simple, inexpensive tests which a farmer can use in order to control the quality of the milk he consigns. Similarly, simple rapid quality control tests are needed for bulk milk at the point of collection and tanker milk at the point of reception. Quality Payment Schemes have two functions. The first is to ensure that farmers are differentially paid according to the market value of their milk. The second is to seek longer term improvements in quality trends. The payment schemes in England and Wales are cited as an example of the success which can be achieved in terms of improving milk quality through payment incentives.

1. QUALITY CONTROL

Milk costed at the point of production, contributes significantly to the value of the final milk product cost. Quality control of milk and improvement of quality through payment systems are therefore, of financial necessity, widely applied in the dairy industry.

Quality control tests currently being used are however not ideal. Our industry needs simpler, quicker methods in order to identify milk of poor marketability and the nearer to the cow that these apply the better with respect to minimising financial loss. This is even more true in times of quotas since if a farmer's supply is financially penalised (rather than rejected) for antibiotics, for example, and that milk is part of his quota, his loss could have been reduced had he been aware of the short-comings of that milk and discarded it in favour of milk of good quality.

In simple terms, farmers cannot afford to suffer financial reductions for part of their "in quota" milk nor can Industry afford to have large volumes of milk downgraded because of contamination due to milk from one cow or one farm supply. The prime objective, therefore, in

an ideal world, ought to be a milk inspection system which enables the farmer to identify
(a) faults in a cow's milk before it is commingled with milk from other cows
(b) milk of poor quality from a farm before it is blended with milk from other farms and
(c) milk of poor quality from a tanker before it is blended with milk from other tankers.

Failure to identify quality defects at each of these stages brings increases in financial losses.

Table I Increase in Financial Loss

	Average Volume	Financial Value	Factor Increase
Cow ↓	15.5 litres	£2.48	(1)
Vat ↓	925 litres	£148	(60)
Tanker ↓	9,000 litres	£1,440	(580)
Silo	90,000 litres	£14,400	(5800)

It is important, therefore, to have a disincentive so that at each stage of mixing of milk there is an accountability - preferably a financial accountability. To enforce such an accountability it may be necessary to sample each supply each day and hold samples for test should a quality problem occur.

1.1 Quality Control at the Cow

The commercial pressure of high throughput per man leaves little time for quality surveillance. However, poor quality assessment of milk "at the cow" risks contamination of the bulk supply resulting in possible rejection or a financial penalty for the whole supply due to any of the parameters given in Table II.

Table II Quality Problems Risking Rejection

Faults	Action
1. Antibiotics	- keep records of treatment, identify treated cows clearly, avoid carry over by milking treated cows last or using separate equipment.
2. Taints	- avoid using phenolic or chlorophenol compounds near milk.
3. Extraneous Water	- see IDF Document 154: 1983
4. Blood	- check teats for cuts, sores, etc.
5. Dirt	- wash and dry udders.
6. Colostrum	- withhold milk for prescribed period after calving

Care exercised at milking should limit risks of contamination. The health of the cow should be checked and where mastitis is suspected, milk should be withheld. Milking equipment should be subjected to regular checks for cleanliness and CIP systems should be regularly checked. A thorough understanding by the cowman of possible sources of contamination and the financial consequences of failure to maintain quality is also desirable if farmers are to minimise the risk of contamination.

In the UK, for example, we have had problems with chlorophenol taints caused by the use of chlorophenolic compounds themselves or phenolic compounds which have become chlorinated by hypochlorite. These chemicals are the active components of commercial disinfectants. Phenols themselves do not cause taint problems in milk unless present at high levels (1,000 to 2,000ppb). Chlorophenols, however, may be detected by taste at levels as low as 1 part per thousand million (1ppb). Education of farm staff explaining sources of chlorophenols and the problems they create is important if contamination and dramatic financial losses are to be prevented. The threat of financial penalty plays its part. Producers contaminating tankers or even silos have been made to pay for all of the milk in that tanker or silo when they have been shown to be responsible. Here the sampling and testing of each consignment enables the source of a problem to be traced hence ensuring that the party responsible can be held accountable.

With antibiotic treatments the producer currently relies on the drug manufacturers or veterinarians giving withholding times which, if followed, should prevent antibiotic contamination. A cow side "dip in" test (rather like the bedside tests used in clinical chemistry) if developed could be used to show whether milk from antibiotic treated cows is positive or negative before it is put into the bulk vat. Modern developments in analysis make the development of such a technique possible in the near future.

1.2 Quality Control at the Point of Collection

With respect to the collection of milk - the tanker driver should inspect milk by sight and smell to ensure that any obviously contaminated milk is not collected. Such checks will, however, only identify fairly obvious contaminants such as dirt. Tanker drivers, like farmers, would also benefit from a simple dip-in test for antibiotics. Sensitive rapid "dip in" tests might also be envisaged for bacterial count and contaminants in the future. Automatic recording of temperature on tankers, coupled with automatic refusal to accept milk over a pre-set temperature, also provide a control which might ensure that set standards are met.

1.3 Quality Control at the Point of Delivery

The receiver of milk should also test incoming supplies using simple quick test methods. Rapid antibiotic and bacteriological test methods should be used where possible. Smell and taste of tanker samples (after laboratory pasteurisation) should be part of the acceptance routine to ensure that silos are not contaminated by tainted milk and temperature measurement should also be routinely made.

2. QUALITY PAYMENT

The control procedures for milk mentioned earlier are usually complemented by quality improvement encouraged through quality payment schemes. Quality payment schemes have a dual purpose:
  (a) to ensure that farmers are paid differentially according to market value of their milk
  Table III shows the effect of butterfat content of milk on butter yield. Similar calculations can be used for the impact of fat and protein on cheese yield or of solids-not fat on skimmed milk powder yield.

Table III The Effect of Compositional Quality on Product Yield

Producer A with milk of 3.00% fat requires 30,355 litres of milk to make 1 tonne butter whereas
Producer B with 3.90% fat requires only 23,350 litres of milk or 23% less milk than A to make the same weight of butter

  and
  (b) to encourage longer term improvements in milk quality through payment incentives.

Quality payment systems vary widely from country to country. However, the normal pattern involves payment on compositional quality, antibiotics and hygienic quality of milk.

We in England and Wales operate payment schemes through six Central Testing Laboratories (1), (2), (3).

(i) Compositional Quality - We pay farmers on the fat, protein and
lactose content of their milk. We try to reflect the market
value of each constituent in the payment to farmers (Table IV).

Table IV England and Wales Compositional Quality Scheme 1986

	pence/litre per one %	Average Quality	Payment	%
Fat	1.924	3.96%	7.619	49.6
Protein	1.956	3.27%	6.396	41.7
Lactose	0.288	4.63%	1.333	8.7
Total:			15.348	

Constituent values are changed as market values change, thus
as the market value of protein has increased and that for fat
has declined, this has been reflected in the price the
producer receives for each constituent. In some countries
penalties are made for extraneous water. We do not operate
such penalties as extraneous water would simply proportion-
ately lower a producer's quality and his financial returns.

(ii) Hygienic Quality - There is no simple quantitative relationship between bacterial count (total, thermoduric or psychrotrophic) and product quality. The tendency therefore is to
work towards low total bacterial counts. In England and Wales
we penalise milk in excess of 100,000 bacteria/ml and pay a
bonus for milk of 20,000 bacteria/ml or less.

Table V England and Wales Hygienic Quality Scheme
Total Bacterial Count Month
Average x '000

		1986	1987
		ppl	ppl
Band A	< 21	+ 0.205	+0.205 (+ 1.3%)
Band B	21 to 100	nil	nil
Band C	101 to 250	- 0.6	-2.4, -3.6 or -4.8
Band D	>250,000	- 2.4	(-15.6%, -23.5% or -31.3%)

The results of introducing this payment scheme which includes
a bonus payment have been most impressive in that the TBC of
the national milk supply was reduced from 92,000 to 24,000 in
six months with a similar percentage reduction in thermoduric
organisms.

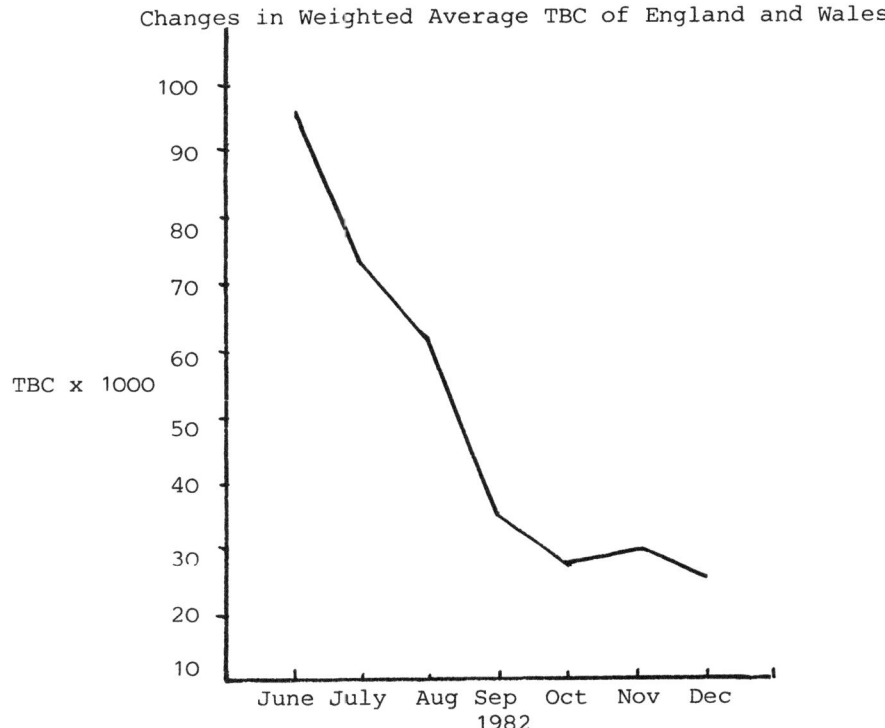

Changes in Weighted Average TBC of England and Wales Milk

Such results clearly show that payment schemes do have an impact on quality.

(iii) Antibiotics - Many countries apply retrospective penalties in order to reduce levels of contamination.

Table VI England and Wales Scheme: Trends in Failure Rates

	1962	1984	1985	1986*
Fails at 0.02 iu/ml	9%	0.40%	0.37%	-
Fails at 0.01 iu/ml	-	0.62%	0.57%	0.33%

* Based on first six months results
All milk in the consignment is paid at 5p, 3p or 1ppl (depending on frequency of failures)
Such schemes reduce the incidence of antibiotics in milk. The development of a rapid 'dip in' test would ensure that contaminated milk was kept out of supplies going to human consumption.

(iv) Somatic Cell Counts - Some countries apply quality payment schemes based on somatic cell count. We in the UK do not. We were surprised, in applying our hygiene scheme, to note a 20% reduction in somatic cell counts. This was achieved because farmers wished to obtain a bonus on bacterial counts. To achieve this they kept mastitic milk out of the milk they sold, in order to protect the bacterial count, and they practised better hygiene during milking.

Quality payment schemes provide ongoing information to farmers who often work in isolation and who value regular information on the quality of their milk. Speedy communication of results is important when there are signs of declining quality and our laboratories have a sophisticated communication network to provide for this.

3. QUALITY ASSURANCE

Lastly - not only is there a need for raw milk quality control and for incentives through quality payments but there is also a need for surveillance of milk to establish levels of contamination and trends with respect to trace contaminants (4) such as:
- Aflatoxins
- Toxic Trace Metals
- Pesticides/PCB's

As an industry this information is needed to satisfy consumers of the "purity" of our products.

4. CONCLUSION

The UK market for dairy products is £4.5 billion per annum. A significant part of the cost is represented by the raw material - milk. Simple, rapid quality control test methods and good quality payment schemes with a rapid feedback of results to producers are necessary in achieving high standards for milk inspection and quality payment in order that we can maximise the quality and value of our raw material.

Quality is sometimes seen as an immovable donkey. Donkeys are known, however, to be enticed to move in the right direction when offered a carrot - payment incentives. Those reluctant to move can be encouraged by the use of a stick - financial penalties or the risk of rejection. We have developed a balanced use of both and I believe that our industry has a raw material quality it can be proud of.

5. REFERENCES

(1) Harding F — "Milk Quality Central Testing" J. Soc. Dy. Technol. Vol. 35. No. 1 Jan 1982

(2) Longstaff G W — "Central Testing in England and Wales and its impact on Milk Quality" J. Soc. Dy. Technol. Vol. 38 No. 1 Jan 1985

(3) Harding F — "Short Shelf Life Products" 1984 Soc. Dy Technol. 72 Ermine Street, Huntingdon, Cambs.

(4) Harding F — XXI International Dairy Congress, Moscow July 1982

THE EFFECT OF RAW MILK HYGIENIC QUALITY ON THE QUALITY
OF DAIRY PRODUCTS

D. I. Jervis
St. Ivel Technical Centre
Abbey House
Church Street
Bradford on Avon, Wiltshire

ABSTRACT. The paper reviews the effect of raw milk thermoduric bacteria
and psychrotrophes on the quality and yield of dairy products. The
implications of extending the shelf life of certain products when raw
milk Total Bacterial Count is $>10^6$ are considered. The effect of
cheese making is also noted, with some indication that Total Bacterial
Count of $>10^6$ cfu/ml may reduce yields and quality.

INTRODUCTION

The influence of raw milk hygienic quality in the manufacture of dairy
products is of considerable commercial significance, with both shelf
life and efficiency of the dairy as important factors. For the convenience of the dairy operator and for reasons of economy there is a trend
to move away from the traditional 7 day working week to 6 or 5 days
and also to move bulk loads of raw milk between dairies. These factors
have the effect of extending the time raw milk can be held before
processing and therefore bring into focus the importance of hygienic
quality of raw milk as this relates to both product quality and process
efficiency. A further factor to be taken into consideration is the
trend for longer shelf life demands for fresh dairy products which
again represents the need for efficiency in distribution.
    It is the intention of this paper to briefly review the consider-
able amount of information available in order to suggest where the
industry should concentrate to maintain standards in the light of
these trends.
    The hygienic quality of raw milk is generally considered in terms
of Total Bacterial Count (TBC) and by an estimate of pasteurisation
resistant bacteria (Thermoduric). Hygienic quality also includes
parameters for antibiotics derived from animal therapy as these are
important to both human health and in the manufacture of fermented
dairy products. It is not intended to discuss antibiotics further.
Bacteriological parameters can be considered under three headings:

1. Specifications established for raw milk
2. Survival of pasteurisation affecting product specification limits

3. Metabolic effects in raw milk that affect product quality and process efficiency

1. SPECIFICATIONS ESTABLISHED FOR RAW MILK

Specifications are set to reflect the desired quality of raw milk and often form the basis of quality payment schemes rather than be of a statutory nature. The USA and Canada set a TBC limit of 100,000 cfu/ml for Grade A raw milk. In other countries bonus payments are made when TBCs are less than 100,000 cfu/ml and in some instances additional bonus payment is triggered at lower values (e.g. 20,000 cfu/ml, 30,000 cfu/ml). Penalty deductions on milk price are imposed when counts exceed 100,000 cfu/ml in most instances where payment schemes are influenced by hygienic quality.

The Council of European Communities [1] specifies TBC limits for raw milk intended for use in manufacture of Pasteurised, UHT and Sterilised milk. The standards apply to milk on receipt at the processing plant and are <300,000 cfu/ml (step 1, to be applied in 1989), reducing to <100,000 cfu/ml (step 2, to be applied in 1993). This directive also requires that milk held prior to pasteurisation at the processing plant (5°C or less) for more than 36 hours should have a TBC of <600,000 cfu/ml (step 1) reducing to 200,000 cfu/ml (step 2) immediately before process.

2. SURVIVAL OF PASTEURISATION AFFECTING PRODUCT SPECIFICATION LIMITS

Statutory limits for TBCs on the day of processing for pasteurised milk range from 5,000 to 50,000 cfu/ml. The Council of European Communities [1] specifies TBC limits of 50,000 cfu/ml (step 1) reducing to 30,000 cfu/ml (step 2). The ability to achieve these standards depends principally on the level of pasteurisation resistant bacteria in raw milk and this reflects hygiene standards in raw milk handling at all stages, from farm to processing plant. There is relatively little published information on thermoduric bacteria in raw milk but the following information (personal communication) illustrates the levels expected where TBC is achieving standards of <100,000 cfu/ml for 95% of samples.

TABLE I   Thermoduric count (cfu/ml) in raw milk supplies

Milk Source	Percentage distribution		
	<2,000	2 - 10,000	>10,000
Farm Tank	70.4	18.5	11.1
Collection Tanker	58.0	31.9	10.4
Silo	47.3	39.6	13.1

Examination of the data in Table I suggests that in some instances it would be difficult to achieve the stricter TBC limits for pasteurised milk, particularly as some of the data included in the >10,000 category represents values in excess of 50,000 cfu/ml for thermoduric bacteria.

In addition to pasteurised milk other dairy products might be expected
to be influenced in this way by extreme levels of thermoduric bacteria
in raw milk supplies. This would include statutory specifications
such as the Intervention standard for milk powders (<40,000 cfu/ml) as
well as many commercial specifications for fresh dairy products (e.g.
pasteurised cream <1,000 cfu/ml).

An additional point to consider is that a proportion of the thermoduric bacteria derived from raw milk can grow at refrigeration temperatures. Cousin [2] in a comprehensive review of published data on the
presence and activity of psychrotrophic microorganisms in milk and
dairy products concluded that the thermoduric psychrotrophes can grow
in milk and affect shelf life by production of a variety of off
flavours (bitter, rancid, fruity, unclean). Such faults can develop
in as little as six days at 7.2°C in pasteurised milk. **Bacillus cereus**
strains have been implicated in bitty cream and sweet curdling defects
in pasteurised milk. Other species of Bacillus, including **Bacillus
circulans** and **B.coagulans** have also been implicated, along with a
variety of nonsporing Gram's positive bacteria.

Psychrotrophic sporeforming bacteria usually have relatively
extended lag periods (8-14 days) and long mean generation times (22-26
hours) at 7°C. However, a strain of **Bacillus coagulans** with a generation time of 24-30 hours at 2°C is reported to show growth in pasteurised milk in 13-17 days at that temperature. From this and other data
cited by Cousin it can be concluded that psychrotrophic sporeforming
bacteria grow relatively slowly in pasteurised milk and creams if
refrigeration is maintained at 7°C or less. However, recent trends in
the dairy industry for longer shelf life in pasteurised milks and
creams are leading to more extensive application of aseptic filling
procedures to avoid post-pasteurisation contamination of product by
rapidly growing Gram's negative psychrotrophes (mean generation time 7
hours at 5-7°C). With spoilage of this type eliminated it can be
speculated that the shelf life of these products will be ultimately
restricted by growth of thermoduric psychrotrophes derived as part of
the thermoduric population of raw milk received for processing.
Perhaps the industry will need to pay more attention to this area in
the future and undertake more research to identify levels of thermoduric psychrotrophes in raw milk supplies and see how existing specification parameters control this aspect of raw milk hygiene.

3. METABOLIC EFFECTS IN RAW MILK THAT AFFECT PRODUCT QUALITY
    AND PROCESS EFFICIENCY

The introduction of refrigerated bulk tanks on the farm and the associated refrigerated transport and storage of raw milk has effectively
eliminated acid development and associated off flavours due to lactic
streptococci and other mesophilic bacteria. However, the extended
storage of raw milk at refrigeration temperatures (7°C or less) has
created new problems due to the selection of psychrotrophic bacteria.
The psychrotrophic bacteria growing in raw milk at 7°C or less are heat
sensitive and therefore effectively eliminated by pasteurisation (72°C
15 seconds). However many of the raw milk psychrotrophes, which are

usually dominated by Pseudomonas spp., produce heat stable proteinases and lipases, which can degrade protein and fat respectively, both in the raw milk and in dairy products.

Fairbairn and Law [3] have reviewed the proteinases of psychrotrophic bacteria in relation to milk and milk products. This review article draws several significant conclusions from a broad selection of published data. The optimum temperature for proteinase production by a raw milk isolate of Pseudomonas fluorescens was 20°C, although up to 55% of the maximum production occurred at 5°C. Further, significantly higher levels of proteinase per unit growth were obtained at 5°C compared with 20°C. This and other data led the authors to conclude that milk psychrotrophic bacteria synthesise increased quantities of proteinase at low temperatures at the expense of cell yield. Stepaniak and Fox [4] showed that whilst Pseudomonas proteinases gave maximum activity at 45-47.5°C, 19-27% of the maximum proteolytic activity was retained at 5°C. The heat stability of microbial proteinases is well established. Griffith et al [5] showed that the proteinases of 13 species of psychrotrophic pseudomonads isolated from raw milk retained 55-65% activity after heating to 77°C for 17 seconds and 20-40% after exposure to 140°C for 5 seconds. Fairbairn and Law [3] discuss other supporting data that confirms the extreme heat resistance of proteinases produced by psychrotrophic Pseudomonas spp. in milk although they also note the apparently anomalous increase in sensitivity to low heat treatment (55°C) in some cases. It is concluded, however, that this phenomenon is too variable to make practicable the application of a low pre-heat treatment to reduce the effect of preformed enzymes on dairy products.

In addition to proteinases, raw milk psychrotrophic bacteria also elaborate lipases. The conditions of production of these enzymes; their temperature range for operation and their heat stability are broadly similar to those for bacterial proteinases [2].

It is clear, therefore, that the psychrotrophic flora of bulk refrigerated raw milk is capable of elaborating proteolytic and lipolytic enzymes. The extent to which milk and milk products may be affected by these enzymes is now briefly discussed.

3.1 Heat Treated Milk

Muir and Philips [6] demonstrated that at 6°C lipolysis did not become apparent in raw milk until the psychrotrophic count exceeded 5×10^6 cfu/ml. Table 2 summarises their data by comparing psychrotrophe count with free fatty acid development in the milk as an indicator of lipolytic breakdown of milk fat. It is considered that the organoleptic threshold for lipolytic taint equates with 3.0 milliequivalents free fatty acid/100g fat.

TABLE II   The relationship between psychrotrophe count
               and lipolysis in raw milk

Psychrotrophe count (cfu/ml)	Frequency of occurrence (%) for free fatty acid (meg./100g fat)			
	n.	0-1.5	1.5-3.0	>3.0
less than 10^6	159	9	91	0
$10^6 - 5 \times 10^6$	47	4	92	4
more than 10^7	56	0	70	30

Adapted from Muir and Philips [6]

Should free fatty acid be developed in raw milk to a level exceeding the organoleptic threshold then heat treated milk (pasteurised and UHT) produced from it would be similarly tainted. Further, any lipase elaborated in raw milk might be expected to survive heat treatment and, in the absence of a dominant bacterial spoilage flora, give rise to a lypolytic (rancid) taint in stored product. Muir & Philips [6] also presented information showing the relationship between psychrotrophe count and the organoleptic assessment of stored cream. This is summarised in Table III.

TABLE III   The relationship between psychrotrophe count
               and taste panel score in stored, pasteurised
               double cream

Psychrotrophe count (cfu/ml)	Frequency of occurrence (%) Taste Panel score			
	n.	10	10-8	< 8
$10^5 - 10^6$	15	60	34	7
$10^6 - 10^7$	20	65	15	20
more than 10^7	66	6	21	73

Adapted from Muir and Philips [6]

The data in Table III is based on psychrotrophic growth in the stored cream as a result of post-pasteurisation contamination but the authors conclude that similar growth in raw milk or raw cream would produce the same organoleptic deterioration in stored pasteurised product through the mechanisms discussed.

Baker [7] associates the development of off flavours in pasteurised milk held at 3-4°C with proteolysis brought about by heat stable proteinases elaborated in raw milk. The author concludes that the prepasteurisation history of the milk determines the subsequent shelf life of pasteurised milk in the absence of post-pasteurisation contamination.

There is considerable evidence that proteinases elaborated in raw milk can cause gelation in UHT and sterilised milks. Data collected by Law et al [8] and summarised by Muir and Philips [6] illustrates this effect. Raw milk was inoculated with a strain of <u>Pseudomonas fluorescens</u> and after storage at 7.5°C for various times the milk was UHT sterilised, aseptically packed and stored under controlled conditions. Data is summarised in Table IV and shows that as counts exceed 8×10^6 cfu/ml in the raw milk the shelf life (time to gelation) is progressively decreased.

TABLE IV The relationship between time to onset of gelation in UHT milk and initial bacterial count

Count (cfu/ml)	Time to gelation (days)
less than 8.0×10^6	More than 140
8.0×10^6	Approximately 63
5.0×10^7	Approximately 12

Adapted from Law [8]

Adams et al [9] showed that a milk psychrotrophe produced up to 20 units of protease in skim milk over 2 days at 4°C, whilst cell numbers increased from 10^3 to 10^4 cfu/ml. These authors also showed that UHT sterilised skim milk spoiled in as little as 3 days with 8.9 units/ml of the protease present.

3.2 Cheese

Cousin [2] in reviewing the effect of cold stored milk on cheese manufacture noted that several authors indicated no detrimental effect with raw milk psychrotrophe counts of up to 10^6 cfu/ml. However many other authors were cited indicating that cheese made from milk of inferior quality did suffer from flavour and textural defects due to the activity of heat resistant proteinases and lipases derived from raw milk psychrotrophes. Thus Cheddar cheese made from milk stored at 3.5°C for 48 hours showed significant organoleptic defects including rancid taint associated with a high content of free fatty acids. Ohren and Tuckey [10] demonstrated that pasteurisation resistant lipases derived from raw milk psychrotrophes gave an excessive level of free fatty acids in mature Cheddar cheese and claim that the best cheeses were manufactured from raw milks where the psychrotrophe counts were between 10^3 and 2×10^4 cfu/ml. However, a careful examination of the data in this paper suggests that the threshold for excessive free fatty acid development might be between 5×10^5 and 10^6 cfu/ml. Law et al [11] investigated the relationship between the level of psychrotrophes in raw milk and rancidity in Cheddar cheese at 15-16 weeks age. The data from this work is summarised in Table V.

TABLE V The relationship between bacterial count in raw milk inoculated with Pseudomonas spp. and rancidity in Cheddar Cheese at 15-16 weeks

Initial count (cfu/ml)	Rancidity Score (0-4, 0 = no rancidity)
6×10^5	0.1
8.8×10^5	0.8
1.2×10^6	0.2
2.0×10^6	0.1
8.0×10^6	2.0
9.1×10^6	2.1
1.0×10^7	1.6
1.1×10^7	2.5

Adapted from Law et al [11]

The data in Table V suggests that the threshold for detection of flavour defect was associated with a raw milk psychrotrophe count of between 2×10^6 and 8×10^6 cfu/ml. Bitter flavours in cheese have been attributed to proteinases derived from raw milk psychrotrophes and in addition to rancidity, soapy, yeasty, butyric and fruity flavours have been reported as a result of lipolytic activity (Cousin [2]).

There are numerous reports concerning the effect of growth of psychrotrophes during milk storage on cheese yields. Fairbairn and Law [3] in reviewing this literature conclude that, although results vary, a reduced yield was usually associated with increased milk storage time. They cite one report where use of raw milk with 10^6 cfu/ml psychrotrophes resulted in a 5% loss of yield of soft cheese as estimated from the excess of nitrogen in whey. Also cited is work by Mohamed and Bassette [12] showing the effect of high psychrotrophe counts in raw milk on cottage cheese manufacture. Cottage cheese was manufactured experimentally from raw milk with a range of initial counts and incorporating three separation temperatures. Data is summarised in Table VI.

TABLE VI Vat failures with inoculated raw milk in making cottage cheese by the direct acid set method at three separate temperatures

Psychrotrophic count - cfu/ml	Separation temperatures (°C)		
	10	32	49
2.2×10^6	Failure	Failure	Failure
1.2×10^6	Failure	Failure	Failure
5.9×10^5	OK	OK	Failure
3.0×10^5	OK	OK	OK

Adapted from Mohamed and Bassett [12]

This work showed that as the psychrotrophic count in raw milk before separation reached 5.9×10^5 vat failures occurred at normal separation temperatures. Vat failure was typified by extensive disintegration and shattering of curd that made recovery impossible. Loss of yield when there was no vat failure (that is psychrotrophe count less than 5.9×10^5) was 0.4% based on 20% solids curd. Similar results have been obtained with cottage cheese manufacture using lactic cultures rather than direct set methods (unpublished). The ability to successfully make cottage cheese when high psychrotrophe counts are present in raw milk appears to be in part starter culture dependent. In one trial, one of five starters failed to make curd when the psychrotrophe count reached 10^6 cfu/ml whilst 4 of 5 starters failed to make cheese when counts reached 7.5×10^7 cfu/ml. However, at 10^5 cfu/ml a weak curd was noted for all starter cultures in the trial and at 10^6 cfu/ml sediment was noted in the cheese vats in each case, indicating some loss of yield. It is concluded that some loss of cottage cheese yield occurs when the psychrotrophic count in raw milk reaches 10^5–10^6 cfu/ml.

It is generally accepted that soluble casein concentration increases substantially during storage of raw milk at 2-6°C with up to 42% of total casein dissolved after storatge at 4°C for 48 hours. (Reimerdes [13]). β casein dissociates readily at low temperatures whilst α_s/κ casein is also dissolved to some degree with prolonged storage. Heating stored raw milk to 60°C results in the reconstitution of the casein micelle and the return of the protein content of serum to that of fresh milk. Under these circumstances cheese making should not be influenced by prolonged cold storage of raw milk. However proteinases released by psychrotrophic bacteria would be expected to degrade dissociated casein so preventing or impairing the reconstitution of the casein micelle and therefore affecting cheesemaking efficiency. Some preliminary work has demonstrated a reduction of β casein in stored raw milk as psychrotrophic count increases with a 25% reduction being recorded at 10^7 to 10^8 cfu/ml. Law [6] concluded that 10^4-10^6 cfu/ml psychrotrophes in raw milk are unlikely to have an adverse effect on the manufacture or quality of Cheddar cheese. Richter [14] pointed out that this conclusion is based on work with milk less than 72 hours old and may not be universally valid. The solubility of β casein and degredation by bacterial proteinases over longer periods of refrigerated storage might be relevant to this variation in observations. Also the significance of thermisation applied to reduce TBC and so allow for prolonged storage of raw milk with existing specifications should not be overlooked as preformed microbial enzymes will not be affected by the heat treatment given.

SUMMARY

There are established specifications for TBC of raw refrigerated milk received by the dairy industry.

The level of thermoduric bacteria may in some instances prevent product specification with respect to TBC being met. A more significant point is that part of the thermoduric flora of raw milk is also psychrotrophic and as aseptic packaging of non-sterile dairy products is extended to achieve longer shelf life the thermoduric psychrotrophes will be of increased importance to the industry.

Of greatest significance to the industry is the effect of proteinases and lipases released into raw milk by psychrotrophic bacteria to affect manufacturing efficiency and degrade stored product. Cheese quality and yield can be affected with significant economic implications and defects develop in stored sterile products. This latter effect might be expected to take on a greater significance in non-sterile products as aseptic packaging is extended to give longer shelf life. The majority of published data reviewed suggests that dairy products are not significantly affected by microbial enzymes until psychrotrophic levels reach 5×10^6 cfu/ml in raw milk. However there are indications that levels of 10^5-10^6 cfu/ml can be significant in some cases. The significance of extended storage of raw milk by reducing temperature and/or thermisation should be carefully considered in this context.

REFERENCES

[1] Council Directive on Health and Animal Health Problems affecting Intra Community Trade in Heat Treated Milk : 85/397/EEC

[2] M. A. Cousin (1982) Journal of Food Protection : <u>45</u> No. 2 : 172

[3] D. J. Fairbairn and B. A. Law (1986) Journal of Dairy Research :
<u>53</u> : 139

[4] L. Stepaniak and P. F. Fox (1984) Journal of Dairy Research (1985)
<u>52</u> : 77

[5] M. W. Griffiths, J. D. Phillips & D. D. Muir (1981) Journal of Applied Bacteriology <u>50</u> : 289

[6] D. D. Muir and J. D. Philips (1984) Milchwissenschaft
<u>39</u> : (1) : 7

[7] S. K. Baker (1983) The Australian Journal of Dairy Technology
September 1983 : 124

[8] B. A. Law, A. T. Andrews and M. E. Sharpe (1977) Journal of Dairy Science <u>44</u> : 145

[9] D. M. Adams, J. T. Barach, M. L. Speck (1975) Journal of Dairy Science <u>58</u> : No. 6 : 828

[10] J. A. Ohren and S. L. Tuckey (1969) Journal of Dairy Science
52 : No. 5 : 598

[11] B. A. Law, A. T. Andrews, A. J. Cliffe, M. E. Sharpe, H. R. Chapman (1979) Journal of Dairy Research <u>46</u> : 479

[12] F. O. Mohamed and R. Bassett (1979) Journal of Dairy Science
<u>62</u> : 222

[13] E. H. Reimerdes (1982) Developments in Dairy Chemistry I :
Edited P. F. Fox, published Applied Science Publishers

[14] R. Richter (1981) Journal of Food Protection <u>44</u> : 471

TOTAL QUALITY CONTROL

Introduction by Mr. Peter van de Wetering
Quality Manager, Scania Nederland B.V.
Thursday, 2nd October 1986

Quality is of vital importance to us. It is the beginning and end of all we do. The conditions behind continuous quality improvement are constant adjustment of our objectives, sound management and complete involvement of every employee.

The subject of my address today is "Total Quality Control". I want to deal with this subject first by giving an outline of several developments in the world of quality over the years. Then by answering the question of how to define the word quality and how to apply this concept in our company. Finally, I will try to visualize the future of a few current trends.

Technical development moves fast, especially if we look at what has happened over a longer period. To do that, I'll take a few strides back to the beginning of the Fifties because I think the future is decided by what happened in the past.

The early Fifties saw continuation of the tradition whereby the customer took care of maintenance himself. Short maintenance intervals, unexpected failures, lack of a service network and availability of technical know-how in the transport companies all promoted in-house maintenance. In short, the customer was forced to undertake repairs himself, yet the requirements set for industrial vehicles - covering reliability, service life, fuel consumption, performance, safety, comfort and environmental protection - increased slowly but surely during the course of the years. Meeting this growing package of requirements was achieved through quality improvement.

Today we produce the modern 9-litre engine. Compared with the engine of the Fifties, it is 15 to 20% more economical, much more powerful, weighs half as much, but lasts 2 to 3 times longer.

Improving quality brings about a drastic cut in repairs and
maintenance. Our workshop statistics showed the average need for
repairs expressed in manhours per year per vehicle dropped by a good
60% from 1967 to 1985, and this trend still prevails. The time taken
in the average vehicle park for preventive maintenance - all the jobs
from major to minor - dropped to one quarter in that same period, due
mainly to extending the so-called primary maintenance interval from
2,500 km to 5,000 km, to 10,000 km, then to the 20,000 km at present.

The drastic cut in the need for maintenance and repair is a direct
result of rising design and manufacturing quality. The quality of
maintenance and repairs must also be geared to this high level of
quality in the design and manufacture of the vehicle.

THE NEED FOR QUALITY IS RISING

So much for now on development over the past decades. I move on now to
my subject "Total Quality Control". In my opinion this approach is
sensible only if it is applied across the board, in other words over
the whole range of operations - from the first design sketch - through
to delivery of the finished product and service to the customer. We
are in fact talking about three different aspects: quality in design,
in manufacture and in service.

In the case of manufacture, I'm thinking not only of in-house
manufacture of parts but also of the constant contact with subcontrac-
tors which must result in an equal level of quality. But I'll come
back to that later.

Quality in research and development

I'd like to start with quality in research and development. I'll take
as an example development of the new 9-litre diesel engine. During
the 7 to 8 year development of this engine, we were able to make
remarkable cuts in the development period by using computer controlled
design systems. Now we can develop parallel operations much faster and
move away from the traditional sequence of research, development,
design, manufacture and investment in machines and equipment. It meant
the input for this new design had to be described accurately and as
fully as possible. During the design phase we were also busy
researching the right manufacturing method. This means the Design,
Development, Production and Productivity Departments were already
working in close co-operation with the so-called CAD/CAM system. This
method not only cuts costs but also offers quality thoroughly
integrated in the system. This led to the manufacture of an entirely
new engine of very modern concept. This engine is built in a new
factory making the best possible use of the latest methods and
technology.
It is very important that parts manufactured by others fully satisfy
the Scania quality requirements. So before a supplier is allowed to
deliver, clear-cut agreements are reached on how quality control will
be carried out at the supplier's works. The fact that products meet

the requirements set by Scania must also be demonstrated. Whenever new developments are involved, Scania works in close co-operation with the relevant supplier.

Quality in manufacturing

The next logical step is quality in manufacturing. At its factories in Zwolle and Meppel, Scania Nederland manufactures heavy trucks for delivery in the Netherlands, Belgium, Luxemburg, France, West-Germany and Italy. Axles, chassis, engines and cabins are manufactured in-house to the finished product stage. To be able to meet the given specifications and design requirements, checking the finished product regularly against these requirements is extremely important. Several aids developed to this end are now used in various operations.

POP

One of the major action programmes we can mention in this context is the POP project which has been running since 1978. POP stands for Process Optimalization Project. The aim is <u>further improvement of manufacturing quality and cutting unnecessary costs</u>.

QUIS

A major part of the POP project is the <u>Quality Information System</u>. This means that specific groups of parts are represented in a fixed manner. By combining this with types of possible deviations, a mixed system was created in which various possible error causes were combined with various possible truck parts. This was the basis for an information system set up to facilitate adjustment of manufacturing results using fast and reliable data processing, where so-called check cards are used for axles, chassis, engines and cabs. These data carriers are carried along with each vehicle produced right from the start of the process and are retained for 12 years after delivery of the vehicle. The data are entered into computers. The information required can be supplied very fast by means of various analysis programs, including pareto analysis. The information collected this way can be analyzed in different ways. The range offers a choice between: the entire company - that is, Zwolle with Meppel - or separately, one specific section or one specific station within that section where about 3 to 4 mechanics are working. This means a selection is made from the 20% of causes which are generally correct for 80% of the consequences. These "big fish" are the input for the problem approach. Since 1980 we have been using for this a 6-stage analysis model which actually shows the path from symptom - to cause - to solution. These 6 steps are: the 1st phase - detection, 2nd phase - define the problem, 3rd phase - trace the causes, 4th phase - find folutions, 5th phase - introduce solutions and finally the 6th phase - after-care.
A project once solved can be monitored automatically by the computer and reported as soon as the phenomenon emerges again. With this information it is possible to apply quality improvements in a dedicated manner.

Quality Circles

At Scania we deliberately opted for the group approach. In 1980 we named these groups "Quality Circles". A combination of the 6-stage model I just illustrated and a control group of the people directly involved make it possible to find a dedicated solution to any problem. But before I go into this further, I would first like to emphasize the importance of sound training and instruction.

Training and Instruction

Training and instruction are essential to turning such a project into a success. We started with a large-scale training programme for quality. All members of staff have taken part in one training course or another. These can range from the "Management of Quality Control" course given by the renowned Dr. Juran to a training course consisting of nine one-hour meetings for members of the quality circles. Since 1981, a total of some 1,100 people have spent 25,000 hours on quality training courses.

Individual Responsibility

In this contect I would like to emphasize once again how important it is for every employee to be entirely committed. The responsibility of the individual man or woman within the work organization plays a central part here. At Scania, every employee works on the quality of his or her work. The principles and practice behind quality could be freely translated as "improve the world - start with yourself". Quality is an attitude of mind.

But now a little more about working with quality circles. The definition we use is: a quality circle consists of a group of interested people who work on the same product or in the same discipline. Using the system model, they work on quality improvement in brief regular meetings once a week under the direction of their own section head or foreman. After the approach to a problem has been formulated, it is presented to immediate management. The results are clearly measurable product improvements in terms of quality. An increase in productivity and, I would like to emphasize this here, a growing awareness of quality. Meantime some 240 projects have been completed using this quality circle approach and the directly measurable savings amount to more than 2 million guilders a year. The experience we have gained with this approach over some 6 years I regard as very positive. That is why we have started a project in the office with the same nature. The target is to introduce our system for all other employees without exception. Meantime 10 quality circles are in operation.

Quality in Customer Service

I now come to quality in customer service. For this kind of quality control, our sales organization in the Netherlands adopted the quality manual for workshops. This quality manual is a guide, a check list and a source of information for monitoring all quality aspects of the operations. We are thinking here not only of training mechanics and sales personnel in the widest sense, but also of tools, machinery, measuring equipment, organization, administration and planning. In short, all aspects which have an effect on the quality of the service. Last but not least, I mention personal advice to the customer, especially if the purchase of new equipment is involved. It is important to choose a Scania truck, in close consultation with the customer, which exactly suits the transport need. Factors playing a part here are reliability, minimum unscheduled downtime, safety, comfort and service life.

Finally, I would say it is essential to obtain fast up-to-date information from the market, enabling us to adjust our product or parts of it to the requirements of the market and of society - requirements which are carried along in the continuous process of research and development.

Prospects

So what do we expect of the future? Let me look first at developments in engineering. There is, of course, a limit to increasing the power of industrial vehicle engines. That limit is dictated by developments in the legal overall loading capacity and what is technically feasible. We will probably not see any real changes in Europe in the near future. As I see it, the maximum engine power the customer wants for a truck will end up at about 500 HP or 368 kW.

We will push specific fuel consumption down even further. We will try to improve thermal efficiency by using, among others, ceramic materials, in which case the cooling system as we know it now will disappear.

Progress in electronics opens up new perspectives for complete engine management. This development will make accurate fuel economy possible.

The new computer controlled gearshift - CAG (Computer Aided Gear-shifting) - makes shifting gears much easier. The conventional gearshift lever has disappeared from the cab. This means the driver is not limited in his choice for shifting gears. On the contrary, he can now choose one of three different ways. First, he can accept a computer proposed gearshift, which appears on the instrument panel display. All he has to do is depress the clutch to change gear. If he thinks the current gear is correct, he can disregard the computer proposal and continue in the same gear. Thirdly, he can use a small lever to shift to the higher or lower gear he thinks is needed in the given circumstances.

The use of electronics and microcomputers in truck driving will gain more and more ground. CAG is a fine example, but systems electronics will play an ever increasing part in braking and steering as well. One of the major tasks of new equipment and complicated electronic systems is to build reliability and low maintenance into every new design. It might sound paradoxical, but with more complicated engineering we achieve greater vehicle reliability.

Ladies and gentlemen,
among other things, I have discussed here the cycle from customer to factory. His experience and wishes lead to a continuous stream of product modifications or improvements. New manufacturing methods develop as a result of modifications to the design. These have to be implemented both in-house and externally, at the subcontractor's works.
The need for quality control is increasing on all sides. The quality policy we have developed over the years at Scania is geared primarily to fast and effective translation of demand from the market into products which are delivered on time at reasonable prices and also completely satisfy the requirements both market and society.

And to achieve that, application of an effective and efficient quality policy is essential.

DISCUSSION

K. Salminen (Finland) : What is the role of Scania top management in IOC?
P. van de Wetering (the Netherlands) : Is has to be informed on qualitative and quantitative aspects. Indications have to be given as how to improve (for example break through concept). After a certain time, give a progress report and suggest further improvements.

K. Samshuijzen (the Netherlands) : You started quite a few years ago with your quality system, more or less according to a Japanese example. Now you have experience in this matter. If you had to start again, would you do it the same way?
P. van de Wetering (the Netherlands) : Yes, we think that it is very important that all people are involved and know their responsibility. We tried to start with trouble shooters and learned very quick that we missed vital information. For that reason we have recently started the same programme in our offices.

R. Grijpma (the Netherlands) : How did Scania achieve to get all people involved in this organization development (mentality change towards quality)? Especially middle management and the different functional organizations such as: design/development, manufacturing and service. Means of getting involved: quality circles in regard to shop floor level, quality costs in regard to top management and break-through theory (Juran) to top management.
P. van de Wetering (the Netherlands) : We started on a small scale in a pilot department, and after succes, we sold the concept through other departments. We have a suggestion box system that receives much more ideas now than before. Difficulties may arise when a middle manager is working in the traditional role as Boss with power to decide. We must accept a participating way of management and sometimes that takes time. It is the role of higher management to introduce that.

M.C. van der Haven (the Netherlands) : Can you explain more about the requirements for the somatic cell count. In the slide a number of 1500 000 was mentioned. This seems rather high to me.
H. Wainess (USA) : The somatic cell count requirement is 1000 000, and is based on recommendations from the National Mastitis Council and other groups to the US Food & Drug Administration. Over a period of time this has been reduced and will be further reduced.

S.Y. Ali (Pakistan) : One of the slides showed that the bacterial payment scheme was started in October and that the farmers were warned earlier in July. There was a dramatic reduction of bacterial load from July to October. How do the atmospheric temperature and humidity affect the bacterial counts in your area and what has been your experience this year in June and July; have the bacterial counts remained as low as October to December last year?

F. Harding (U.K.) : We find a slight seasonal variation in bacterial counts with counts very high in summer, probably due to bacteria multiplying on milking machine surfaces. However, nationally this only results in a slight increase for example from 17 000 to 19 000/ml.

A. Gilmour (U.K.) : With reference to Dr. Jervis's paper I would like to point out that there may not always be a good correlation between psychrotrophic bacterial numbers and enzyme levels in raw milk. This is because:
- the organisms may not be in the late lag phase of growth which is visually required for enzyme production
- and some strains present may not be capable of enzyme production.

Thus it is possible that <u>high</u> count milk may have a low level of enzyme while <u>low</u> count milk (perhaps resulting from mixing fresh with older milk) might have a <u>high</u> enzyme level.

D.I. Jervis (U.K.) : The comments made by Dr. Gilmour are valid. It is because of the lack of correlations between TBC and enzyme effect that a change to measuring enzyme to judge suitability of stored raw milk for processing is proposed in this paper.

G. Odet (France) : J'ai bien noté les différentes classes de qualité du lait cru, qui sont basées sur les numérations bactériennes totales. Prend ou également en compte, aux USA et au Royaume-Uni, l'intervalle de temps, la durée, entu la récolte du lait à la ferme et le traitement à l'usine? Y-a-t'il une réglementation sur ce sujet?

D.I. Jervis (U.K.) : There is no specification for this age of raw milk for processing. In this paper it was suggested that TBC might not always reflect suitability of milk for processing and that for this reason a change to measuring enzyme content might be considered.

K.J. Kirkpatrick (New Zealand) : Do tight microbiological standards lead to an increase in chemical sanitiser residues in milk? Has this been monitored for and if so what are the results?

F. Harding (U.K.) : Not in our experience. We find that farmers can regularly meet a standard of less than 10 000 <u>without</u> residues of sanitiser found in their milk.

L.P.M. Langeveld (the Netherlands) :
1. What is the temperature Mr. Jervis thinks of when he speaks about a keeping quality of pasteurized milk of about three weeks?
2. Mr. Jervis has stressed that the thermoduric bacteria are important if a good keeping quality on pasteurized milk is wished. I think special tests will be necessary to estimate the concentration of spores of psychrotrophic sporeforming bacteria which are able to grow in pasteurized milk at 4-7° C.

D.I. Jervis (U.K.) :
1. 5 - 7° C;
2. I agree that specific notice of psychrotrophic thermoduric bacteria will become necessary when shelf life of pasteurized milk and its products is extended by exclusion of post pasteurization contaminants.

W.H.J. Bakker (the Netherlands) : Can you explain why the USA have implemented farm inspections. Our experience with a system of farm inspections of about 25 years ago were such that we have rejected that system in favour of a system for payment of milk on its hygienic quality.

H. Wainess (USA) : The system in the USA is based on strict enforcement of the farm inspection requirements and the bacterial count. Individual dairy co-operatives and dairy plants pay premiums on the basis of total hygienic quality. If a farmer neglects to meet the Grade A requirements after appropriate warnings (both inspection and bacterial counts) he is then removed from the market until he complies. Our government acts solely as a public health agency and is not involved in paying farmers.

H.R. Lehmann (F.R.Germany) : I am surprised to see that the interest in raw milk quality is almost exclusively focussed on bacteriological aspects and cell count. Especially for machinery manufacturers, milk quality in relation to sensivity to physical damage and heat stability, is very important. My question to the dairy technologists is to compose a thorough specification of raw milk quality with respect to physical aspects.

M.G. van den Berg (the Netherlands) : It is indeed an important aspect of milk quality. It is impossible to discuss it at this moment, but I will bring this subject forward within the IDF commission B.

Seminar IX: Integrated Quality Control and Assurance

Session 2

Chairman: Dr. D. B. Gammack (U.K.)
Secretary: Ir. B. M. A. Delsing (The Netherlands)

Quality marks

QUALITY MANAGEMENT AND MANAGEMENT INFORMATION

Kari Salminen
Valio Finnish Co-operative Dairies' Association
Research and Development Department
P.O.Box 176
00181 Helsinki
Finland

It is well known that the world wide success of many Japanese companies is based on their having adopted the revolutionary pattern of thought of the 50's. We now call it Integrated Quality Control (IQC) or Total Quality Control (TQC) or Company Wide Quality Control (CWQC). These management philosophies emphasize, as implied by their very names, the central role of quality in a company's overall strategy. Along with the ever-increasing competition this kind of thinking is now also conquering the companies of the western world.

Quality. A short word with a meaning that seems to change considerably depending on with whom it is discussed in a company. The quality concept of marketing may be different from that of production, and a third shade may be introduced by R & D. If these quality concepts differ from that desired by the customer, the results are dramatic.

What in fact is quality? In 1985 the International Organization for Standardization has defined quality as follows: "The totality of features and characteristics of a product, process or service that bear on its ability to satisfy stated or implied needs".

Quality can be defined even more briefly: "Quality is the customer's satisfaction" or "Quality is the company's ability to satisfy the needs of the customers". I personally think the latter is more to the point. It emphasizes IQC's inner essence and starting point: quality is managed from the customer's needs, expectations, motives and attitudes, not on the producer's or manufacturer's terms. There is a risk this is not always understood and accepted in traditional production oriented companies owned by dairy farmers.

In this connection I would, however, like to point out straightaway that within the food industries it is just in the dairy industry that quality has occupied an emphasized position. Let us only think of the hygienic passage of milk from milking to the collection and further on to the processing plants and from there as many hygienic, tasty and nutritious products to the shop shelves to reach the consumers. This system has been so excellent that even at present milk and milk products play a central role in human nutrition in many countries. In Finland for example their total consumption, calculated as milk, is still about 400 l/capita/year. The consumption is, however, decreasing. In particular the consumption of traditional volume products, milk, butter milk and butter is decreasing alarmingly. In many countries margarine has got the better of butter. Both of them are again being threatened by new low fat spreads.

Cheese and ice cream e.g. are being threatened by imitation products. Hard as it may be, we working in the dairy industry have to admit that the quality of many dairy products does not quite keep up with the needs of the modern consumer.

So once again, what is quality?

Quality in itself implies two dimensions. We speak about essential (or must be) and attractive quality elements. The difference is by no means razor-sharp, and quality elements seem to be changing continuously. In my youth the customer had a milk can of his own into which a desirable amount of fat-standardized, pasteurized, non-homogenized milk was measured off. Then the dairy industry added new attractive quality elements to milk: it was homogenized and packed into glass bottles, plastic sachets and then cartons. Nowadays these items are already of essential quality, but at the same time new attractive quality elements were futher included in milk. So there are milks of different fat contents on the market. In Finland for example milks with a fat content of 3.9 % and 1.9 % as well as skimmed milk are marketed. In some countries the attractivity of milk has been increased by flavourings etc. Sometimes complete about-turns of quality approach take place. Thus certain consumer groups find the essential quality familiar in Finland, that is homogenization, unattractive. So we are at present test-marketing for these consumer groups a non-homogenized milk which means that in a way we are returning back to the 60's! As far as I know, in Sweden the non-homogenized, full fat "old times' milk" has gained a market share of as high as 7 % in some districts at least.

So, management has to follow continuously the changing needs of the consumer and modify them into quality targets. The essential quality, like hygiene for instance, <u>always</u> has to be first-rate, although the consumers do not even notice it often. The essential quality does not in fact increase the consumer's satisfaction in regard to the product. On the other hand, lack of this quality element in a product is an immediate cause for dissatisfaction to the consumers and may lead to complaints, measures taken by the authorities, falling-off of the markets as well as a tarnished product image and even company image. However, it is not sufficient to take care only of the essential quality. In the ever-increasing competition it is important to gather into a product as many attractive quality elements as possible and so increase its consumer value. Generally speaking it can be said that attractive quality elements are factors difficult to define, based often on feelings and beliefs. To a great extent such are conceptions of the nutritive value of a product and its beneficial effect on human health. Thus the partly tarnished belief of the consumer on the beneficial effects of milk, destroyed by the fat hypothesis, should be revived for example by emphasizing the value of the high calcium content of the low fat milk with respect to man's great demand for calcium. Regarding fermented milk products the conception of the beneficial effects of lactic acid bacteria on human health might be just that attractive factor required in struggling towards a better future.

I have been speaking such a long time about quality to show that <u>the first task of the dairy top management</u> is to define the quality targets of the products, or still better, <u>the real needs of the customers</u>. That is the very beginning for creating the company's quality policy.

The basis of IQC is <u>market in</u>, that is to say to develop products that are sought by the consumers. In a company the function which has the closest contact with the consumers is marketing. Marketing is best equipped to discern and discover the needs of the consumers. The market research must catch the trends and discover the needs of the consumers ahead of the competitors. I would like to stress the central role of strategic market research in a successful company. In cooperation with production and R & D division the consumer's needs must be converted into new ideas and new quality products. After these have been accepted by the top management, it has to lead the whole organization to produce the desired quality. That is called quality management: The aspect of the overall management function that determines and implements the quality policy (ISO). This is a deepgoing event concerning all company levels. So, in order to produce Finnish emmental cheese of a high and uniform quality we have to manage quality from the very moment when silagemaking starts on the farm. The pH of the silage is lowered instantly below 4. It is only with this kind of silage that we can produce cheese milk which has a butyric acid bacterial spore content below 2000/l. This again is a condition for making Finnish emmental which is produced without an addition of nitrate or other preservatives, this being one of our main quality targets. When all working and manufacturing stages from the field to the barn, from the barn to the cheese plant and from there to delivery and to shops are performed carefully, strictly following the instructions, inspection can be minimized. <u>Quality must be built into each process. It cannot be created through inspection.</u>

In creating the above operation chain the management has at the same time built an uninterrupted information system from the markets to each level of the company and even from there to the production of milk and silage. An important part of the system is to clearly share the quality responsibility. A cheesemaker cannot be responsible for a poor quality cheese, if the initial reason for the failure is to be found on the milk producing farm. The phrase "Next process is our customer" well describes clear-cut delegation of responsibility. Thus quality management is management by quality objectives.

Strictly speaking quality is an abstraction which in itself cannot be managed or directed. It is always man who in the end creates quality into a product and whom it is possible to manage to make quality. This takes place with quality targets. For reaching them the staff has to have at its disposal the right materials, machines and equipment, manufacturing and working procedures as well as instructions and training to use them.

One cannot define quality without knowing <u>the cost.</u> No matter how high the quality, if the product is overpriced, it cannot gain the customer's satisfaction. Nobody wants bad quality even at a low price, either.

So each product has an optimum quality, that is the best possible quality/price ratio from the point of view of the sales and production costs. This means that parallel with market research and quality information system there has to be the company's cost accounting, analysis and information system. In general the companies have a well developed cost accounting system.

An excellent example of the quality/price thinking in the dairy industry is the quality pricing of producer milk.

At present the overproduction of milk in industrialized countries has led to restrictive actions and so caused great cost pressure in dairy business. In reducing costs, which is essential, it has to be remembered that this must not take place at the expense of quality. The first task of a company is to impose the quality targets on the basis of the consumer's real needs, and secondly to carry them out in all the working stages at the lowest possible cost.

I have here tried to describe the thought patterns and priorities of Quality Management. I am afraid the picture is static. The deepest nature of IQC is, however, dynamic, an object for continuous development and learning. As soon as the top management has become convinced of that the operative management will take care of everything in an excellent way and as has been agreed, it has to ask: "Have the real needs of the customers changed? Are our quality targets proportional to the changing needs? Would it be possible to do everything in a different, better way?

HAZARD ANALYSIS AND CRITICAL CONTROL POINTS

J.H.B.Christian
CSIRO Division of Food Research
PO Box 52
North Ryde. NSW 2113
Australia.

All quality control is costly and an important objective must be to make it as cost effective as possible. This may be achieved through a policy of preventative quality assurance by a procedure known as Hazard Analysis Critical Control Points (HACCP). The HACCP system provides a more specific and critical approach to the control of microbiological hazards in a process than that achievable by traditional inspection and quality control procedures. It involves identifying the hazards and their severity, determining the critical points where the hazards can be controlled, specifying criteria to be met at the critical control points, establishing monitoring procedures at these points, and taking any appropriate corrective action.
    The systematic HACCP approach is highly applicable to the dairy manufacturing industry and examples are given. It should also be used in the formulation of Codes of Practice or regulations by regulatory authorities.

INTRODUCTION

The relative ineffectiveness, and particularly the cost ineffectiveness, of traditional methods of control based largely on inspection and microbiological testing, has long been recognized in the food industry. However, quality control procedures have been incorporated into quality assurance programs and these in turn have been modified to become what may be termed preventative quality assurance. There is nothing particularly new in this. It is a rational and logical approach to preventing defects, generally microbiological in our context, by considering first the hazards that need to be controlled and then the identification of the best means to control these hazards. A name given to this form of preventative quality assurance is Hazard Analysis Critical Control Points (HACCP).
    The HACCP concept was first presented at the 1971 National Conference on Food Protection (United States Department of Health, Education and Welfare, 1972), and was subsequently developed in a number of

publications (Bauman, 1974; Peterson and Gunnerson, 1974; Ito, 1974; Kaufmann and Schaffner, 1974), the implication being that it was in wide use in the food processing industry in the USA. However, informal surveys in 1980 revealed that the concept was not in common use elsewhere.

The World Health Organization believed that HACCP was a concept that could be applied in many countries, developing as well as developed, to improve the microbiological safety and quality of foods. To this end, a panel of members of the International Commission on Microbiological Specifications for Foods (ICMSF) was convened by WHO to prepare a report on HACCP and its general application (WHO/ICMSF, 1982). Much of the present paper is taken, with permission, from that Report.

A recommendation in the Report was that a food be selected which, on the basis of firm epidemiological evidence, is a hazard in national or international trade and a pilot program for the application of the HACCP system to the selected product be initiated. In the event, WHO decided that rather than focus on a food it would focus on a hazard, and a further ICMSF *ad hoc* Committee was formed to report on the use of HACCP to prevent and control foodborne salmonellosis (WHO/ICMSF, 1986).

In this paper, only microbiological hazards are considered.

HAZARD ANALYSIS CRITICAL CONTROL POINT (HACCP) SYSTEM

The HACCP system consists of: (1) an assessment of hazards associated with growing, harvesting, processing/manufacturing, distribution, marketing, preparation and/or use of a given raw material or food product; (2) determination of critical control points required to control any identified hazard(s); and (3) establishment of procedures to monitor critical control points. Basically, the HACCP system provides a more specific and critical approach to the control of microbiological hazards than that achievable by traditional inspection and quality control procedures.

Hazard analysis

'Hazards' include contamination of food with unacceptable levels of foodborne disease-causing microorganisms and/or contamination with spoilage organisms to the extent that hazards occur within the expected shelf-life or use of the product.

A hazard analysis consists of an evaluation of all procedures concerned with the production, distribution, and use of raw materials and food products: (1) to identify potentially hazardous raw materials and foods that may contain poisonous substances, pathogens, or large numbers of food spoilage microorganisms, and/or that can support microbial growth; (2) to find sources and specific points of contamination by observing each step in the food chain; and (3) to determine the potential for microorganisms to survive or multiply during production, processing, distribution, storage, and preparation for consumption.

Such analyses should be carried out on all existing products and on any new products that a processor intends to manufacture. Changes in raw materials used, product formulation, processing, packaging, distribution, or intended use of the product should indicate the need for re-analysis of hazards, because such changes could adversely affect safety or shelf-life. Microbiological hazards will vary from one product to another, depending upon the raw materials, the processing procedures, the manner in which the finished product is marketed, and its ultimate use. They may vary from one food processing plant to another producing the same product and therefore must be determined by observations and investigations of the particular processing plant.

Hazards caused by microbiological contamination of raw materials must be evaluated. In some cases, microbiological safety and shelf-life depend almost entirely on the selection of microbiologically suitable raw materials. For example, in the manufacture of dry blended products which are reconstituted without further heating, processing cannot be relied upon to eliminate contamination present in raw materials.

Many of the processes used in food manufacture, such as heat treatment, acidulation, fermentation and salting, will destroy or inhibit the growth of harmful microorganisms. Hazards associated with these procedures must be evaluated, and the consequences of failure of processing steps designed to destroy or inhibit the microorganisms must be understood. For example, the failure of a starter culture to initiate acid production promptly may permit the growth of staphylococci and enterotoxin production during the manufacture of cheese.

The physicochemical characteristics of a finished product that influence growth, death, or survival of microorganisms should be identified. These include such factors as water activity (a_w), pH, the presence of preservatives, the packaging system, and the gaseous environment within it. If interactions between various physical and chemical agents are relied on for safety, e.g. a_w and pH, pH and preservatives, packaging and gas atmosphere, fermentation and pH reduction, then these factors must be defined in terms of their influence on the microbial flora during processing, distribution, storage, and use by the consumer.

Hazard analysis should include an evaluation of the potential of the food processing plant environment as a source of contamination to the finished product. For example:
- To what extent is there opportunity for cross-contamination between contaminated raw materials and finished goods?
- Is air movement away from finished goods and toward raw materials?
- Are there steps, such as the manual handling of products that are eaten without further cooking, where employees could contaminate the finished product with pathogenic microorganisms?

Critical control points

A critical control point is a location, practice, procedure or a process which, if controlled, could minimize or prevent contamination with foodborne pathogens or spoilage microorganisms, or their survival or unacceptable growth (WHO/ICMSF, 1986).

Incoming raw materials may constitute critical control points, depending upon their origin and use. If one or more steps in a process can be depended upon to eliminate harmful microorganisms in a particular raw material, that raw material does not constitute a critical control point. If, on the other hand, neither processing techniques nor consumer use can be depended upon to eliminate harmful microorganisms from raw materials, then these constitute critical control points. Particularly important in this respect are sensitive raw materials, e.g. dried eggs and milk which may contain salmonellae.

Processing time-temperature combinations are frequently the most critical control points. For example, if a heat process is depended upon to destroy microorganisms, then the required combinations of processing time and temperature must be carefully established and followed. Similarly, the temperatures at which products are held prior to and during cooling and freezing and the length of time they are held are frequently critical control points.

Amongst other factors that can adversely affect safety and quality is improper sanitation in the plant, and packaging materials. If poor sanitation in a particular process step is likely to affect adversely safety of the finished product, this would constitute a critical control point. Products may be subjected to environmental contamination from such sources as air, water, insects, rodents and personnel. Incoming packaging materials do not usually constitute a critical control point, except in the case of containers used for canned foods, where lack of integrity of the finished package may affect the safety or quality of the end-product.

Monitoring

After analysing the hazards presented by a particular product and identifying critical control points, it is necessary to establish monitoring systems to ensure that these points are under control. Such monitoring may involve only visual inspection - for example, the preoperational inspection of a temperature recorder to determine that the chart has been properly installed. Similarly, hand-washing procedures should be observed. Although such observations do not involve measurements, they should be recorded on suitable check lists. More commonly, chemical, physical or microbiological tests are used for monitoring.

For each critical control point the appropriate monitoring test must be determined, the procedures documented and the frequency of testing specified. Applicable statistically-sound sampling plans must be employed. Ideally, monitoring is by tests giving rapid results - measurement of temperature, time, pH, a_w, and other physical or chemical criteria. For sterilized foods all of the checks will generally be of this type, with no microbiological testing. However, monitoring systems for non-sterilized foods may involve microbiological examinations, for example, the production of non-fat dried milk. The prime hazard presented here is the danger of post-processing contamination with *Salmonella*. Such products should therefore meet recommended microbiological criteria of the International Dairy Federation or of ICMSF (1986), and those required by regulatory agencies. Assuming monitoring of the pasteurization process indicates proper time and temperature relationships, the most likely source of contamination of the finished product would be the environment. It has been repeatedly shown that when potential for such environmental contamination exists, a continuing environmental sampling program is more likely to detect a problem than is finished product analysis. Accordingly, in such operations well-selected points in the environment that constitute critical control points should be constantly monitored. If salmonellae are detected in samples from such points, negative results of tests on finished products should be interpreted with extreme care. An analogous situation would be presented by a dry-blended product composed of multiple ingredients, each of which had been pre-tested and found negative for *Salmonella*. Here, raw materials, the environment and the finished product are critical control points which must be subject to constant monitoring.

APPLICATION OF THE HACCP SYSTEM

Approaches to hazard analysis

When considering hazard analysis, both food-poisoning and spoilage microorganisms are of concern. Knowledge that a food is hazardous may derive from one of two sources: (1) epidemiological information indicating that a product is potentially a health hazard or is microbiologically unstable may derive from effective surveillance programs that collect data on the incidence of foodborne disease and assess significance; (2) technical information may indicate that the product poses a health hazard or is subject to spoilage. Here, reaching a decision with respect to hazard is far more difficult than in the first situation.

In a hazard analysis, the following types of questions should be addressed:
1. What are the conditions of intended distribution and use?
   Is the product to be distributed under ambient or cold storage temperatures?
   What is the expected shelf-life both during distribution and storage and in the hands of the person who will ultimately use the product?
   How will the product be prepared for consumption?
   Is it likely to be cooked and then held for a period of time before consumption?
   What mishandling of the product is likely to occur in the hands of the consumer or during marketing?
2. What is the product formulation?
   What is the pH?
   What is the water activity?
   Are preservatives used?
   What packaging is used, and is this integral to product stability, e.g. the vacuum packaging of perishable foods?
3. What is the intended process?
   Consideration should be given to those steps that lead to the destruction, inhibition, or growth of foodborne disease or spoilage

microorganisms.

Based on answers to the above, and other available information, the expert food microbiologist is able to give a preliminary assessment of the potential hazard(s) involved in the manufacture, distribution and use of the product. However, it is desirable and in many cases necessary to check the assessment by inoculation of the product with appropriate foodborne pathogens and potential spoilage organisms. The inoculated food must be packaged under intended marketing conditions and then be subjected to tests under expected storage, distribution and consumer use conditions. Such tests should include evaluation of the effects of mishandling on product safety and stability.

Approaches to critical control point determination

Sometimes critical control points are obvious from the hazard analysis. Epidemiological data collected during investigations of outbreaks that occurred in similar places can also be used as a guide. At other times, more extensive research on the food or the process, including microbiological investigations, may be necessary to establish appropriate control points. Of particular value is a determination of temperatures and times at which the product is held during processing.

Approaches to the establishment of monitoring system

The type of monitoring system depends upon the nature of the critical control point under consideration.
1. If a raw material is a critical control point, a specification should be set for that raw material detailing the microbiological tests, sampling plans and limits to be employed.
2. Monitoring of process critical control points may involve microbiological tests but may be best achieved by physical and chemical tests, because the results of these are more rapidly available. There are, however, situations where in-process microbiological monitoring is necessary as, for example, in the production of highly sensitive foods for infants, children or malnourished persons. It may also be necessary to monitor the effectiveness of sanitation measures by the use of microbiological tests.
3. Visual observation, although it may appear to be a mundane activity, is often the key means of monitoring critical control points. Personnel responsible for such monitoring require considerable training and expertise.
4. End-product monitoring by microbiological testing is generally very limited. More often, determination of product attributes, such as pH, water activity, preservative level and salt content will give far more information about safety and stability. There are situations where microbiological examination of the finished product is mandatory, e.g. the examination of certain high-risk foods for *Salmonella*. For this purpose the sampling plans and analytical procedure recommended by ICMSF should be followed (ICMSF, 1974, 1978).

Check lists should be employed for monitoring critical control points. These should show details of the location of the points, the monitoring procedures, the frequency of monitoring and satisfactory compliance criteria.

EXAMPLES

It must be repeated that no two processes or plants are identical. However, there are usually general considerations that apply broadly to the processing of a particular product. Some examples, drawn from WHO/ICMSF (1986) follow.

Pasteurized milk: Adequate and rapid chilling of the raw milk is essential to prevent multiplication of pathogens, but the first critical control point (CCP) is pasteurization. HTST plants have flow diversion valves to ensure that milk is recirculated if the required pasteurizing conditions are not achieved. The temperature and time of this process should be measured and recorded. The second CCP is prevention of contamination in subsequent stages by ensuring that equipment and materials are clean and sanitary. This CCP is controlled by visual monitoring of pipelines and observing that processing and cleaning instructions are followed. Analysis for the presence of pathogens in the final product is not necessary.

Non-fat dried milk: Again the first CCP is pasteurization, monitored as above. The second CCP is the environment in which the milk is heated, dried, cooled, sifted and packaged. It is monitored visually to observe, for instance, source and direction of airflow, condition of air filters and presence of standing water. Such observations should be supplemented by *Salmonella* tests. Because of the time taken to obtain results, *Salmonella* assays do not provide ongoing control of this CCP, but they do indicate, retrospectively, whether the visual monitoring was effective. In spite of these control measures, which should minimize *Salmonella* contamination, end-product should be tested for *Salmonella* with an appropriate sampling plan.

Cheddar cheese: Although the production of Cheddar cheese involves many steps, only three are generally considered CCP's to ensure the safety of the product. The first is the heating of the raw milk either under pasteurization or sub-pasteurization (e.g. 64-70°C for 15-20 seconds) conditions, to be monitored for temperature and time to ensure the destruction of vegetative pathogenic bacteria.

The second CCP is the controlled fermentation, which should produce acid at a rate which prevents or strongly inhibits the growth of any contaminating food-poisoning bacteria. Monitoring includes checking the temperature of the milk to ensure that it permits optimal activity of the starter culture and measuring the rate of increase in titratable acidity.

The final CCP is the aging of the cheese. Sufficient aging at appropriate temperature will generally lead to death of contaminating pathogens in a cheese that has had a successful fermentation. However many factors influence the rate at which die-off occurs.

CONCLUSION

The implementation of the HACCP approach to all dairy manufacturing processes, and indeed to pre-manufacturing stages, can provide a more efficient and cost effective control than traditional procedures. Regulatory authorities should also apply HACCP in the drafting of Codes of Practice and the framing of regulations. Although it does not do away with microbiological testing for all processes, it does reduce it and in so doing depends largely on physical and chemical methods where quick responses make it possible to take corrective action, where necessary, with a minimum loss of time and product.

REFERENCES

Bauman, H.E. (1974). 'The HACCP concept and microbiological hazard categories'. *Food Technol.* **28**, 30-34,74.
ICMSF (International Commission on Microbiological Specifications for Foods) (1978). Microorganisms in Foods. 1. Their significance and methods of enumeration. 2nd edition. University of Toronto Press, Toronto.
ICMSF (International Commission on Microbiological Specifications for Foods) (1986). Microorganisms in Foods 2. Sampling for Microbiological Analysis: Principles and Specific Applications. 2nd Edition. University of Toronto Press, Toronto.
Ito, K. (1974). 'Microbiological critical control points for canned foods'. *Food Technol.* **28**, 46,48.
Kaufmann, F.L. and Schaffner, R.M. (1974). 'Hazard analysis, critical control points and good manufacturing practices regulations (sanitation) in food plant inspections'. *Proc.IV Int.Congress Food Science and Technol.* pp.402-407.
Peterson, A.C. and Gunnerson, R.E. (1974). 'Microbiological critical control points in frozen foods'. *Food Technol.* **28**, 37-44.
WHO/ICMSF (World Health Organization/International Commission on Microbiological Specifications for Foods) (1982). Report of the WHO/ICMSF meeting on hazard analysis: critical control point system in food hygiene. VPH 82.37. World Health Organization, Geneva.
WHO/ICMSF (World Health Organization/International Commission on Microbiological Specifications for Foods) (1986). Prevention and control of foodborne salmonellosis through application of the Hazard Analysis Critical Control Point System. (in press).

QUALITY CONTROL AND QUALITY ASSURANCE OF PROCESSING A PRODUCT AT DIFFERENT LOCATIONS.

G. ODET
DELEGUE SCIENTIFIQUE SODIMA
C.L.C.P. - B.P. 221
94203 IVRY-SUR-SEINE CEDEX
France

ABSTRACT. La gestion de la qualité est située comme une partie intégrante de la gestion générale de l'entreprise. Le schéma général de l'assurance de la qualité comprend : la définition des produits, les cahiers des charges, la formation des hommes, la réalisation des contrôles, la présentation et la communication des résultats, les actions correctives.
    L'auto-contrôle est imposé par les distances entre usines fabriquant les mêmes produits, et ses résultats doivent être exprimés d'une façon claire et immédiatement compréhensible.
    L'exemple cité de Sodima-Yoplait montre d'abord la sensibilité particulière des responsables de sociétés réparties sur les cinq continents à la qualité de leur production. La définition du yoghourt Yoplait, produit frais, et l'interdiction de tout traitement physique après la fermentation autre que le refroidissement à température positive, sont intégralement respectées et ne peuvent faire l'objet d'aucune transaction. L'audit qualité, examen approfondi de tous les éléments intervenant dans la qualité finale des produits, est mis en œuvre exceptionnellement en coopération avec la société concernée.

Le sujet de cette intervention est titré en langue anglaise par "Quality control and quality assurance of processing a product at different locations". La traduction française donne "Contrôle et maîtrise de la qualité d'un produit fabriqué dans des lieux géographiques différents".
    A noter que les emplacements de fabrication peuvent appartenir à un même pays, ou se trouver au contraire très éloignés, appartenant même à des continents différents. Il y a au départ un problème général de gestion de la qualité.

GESTION DE LA QUALITE

L'Afnor, Association Française de Normalisation, précise que "la gestion de la qualité est située comme une partie intégrante de la gestion générale de l'entreprise". C'est le principe essentiel à retenir, dont la conséquence immédiate est que la direction générale et les services qui participent à la gestion de l'entreprise sont concernés par la qualité des produits.

ASSURANCE DE LA QUALITE

L'Afnor définit également l'assurance de la qualité par "la mise en œuvre d'un ensemble approprié de dispositions préétablies et systématiques destinées à donner confiance en l'obtention de la qualité requise". Ce qui veut dire que le fait d'avoir le souci de maintenir un niveau de qualité n'est pas suffisant, il faut encore disposer d'un schéma préétabli pour assurer les contrôles.

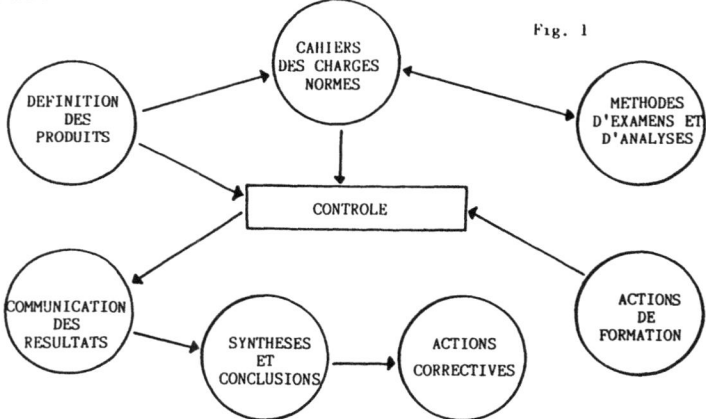

Fig. 1

Un tel schéma (Fig. 1) fait apparaître les titres suivants. En préalable
1. La définition des produits ;
2. Les cahiers des charges comprenant les normes des matières premières et produits finis, produits et emballages et méthodes d'analyses ;
3. La formation des personnes directement concernées à l'approvisionnement, à la fabrication, aux services de contrôle ;
   Au stade suivant :
4. Le contrôle - réalisation pratique - mode et fréquence d'échantillonnage ;
5. La présentation et la communication des résultats - les destinataires et le circuit d'information sont précisés ;
6. Les actions correctives auxquelles sont imposés des délais de réalisation.
Quelques précisions sont nécessaires concernant les points 1 et 2.

LA DEFINITION DES PRODUITS

Elle est en accord avec la réglementation nationale.
   Elle définit les caractéristiques majeures du produit, sur lesquelles aucune déviation n'est admise. Ces caractéristiques font

l'objet d'un accord général sur un même concept. Cependant, le concept étant adopté, les habitudes locales et les goûts particuliers des consommateurs sont pris en considération.

LES CAHIERS DES CHARGES

Guides pratiques des fabricants, des distributeurs, des contrôleurs, les cahiers des charges contiennent les spécifications :
- Formulations
- Technologies
- Equipements industriels
- Fournitures d'ingrédients et de matériaux d'emballages
- Définition de l'emballage (graphisme, étiquetage, datage)
- Contrôle et méthodes d'analyses

Les cahiers des charges définissent les moyens à mettre en œuvre pour répondre aux trois impératifs de la qualité :
- l'aptitude des produits à l'usage auquel ils sont destinés
- la conformité à des spécifications
- l'engagement de la marque.

Les cahiers des charges sont les pièces maîtresses de l'assurance de la qualité. Il est vérifié dans chaque cas que leur application, aux niveaux des équipements, des technologies et des ingrédients, permet d'obtenir des produits conformes. La preuve étant faite, il reste à maintenir les paramètres de qualité d'un jour à l'autre, d'un mois sur l'autre, pendant l'exploitation normale de l'usine. C'est cette constance des caractéristiques des produits qui témoigne de la maîtrise de la qualité. L'assurance, c'est-à-dire la confiance nécessaire, a pour base l'auto-contrôle.

L'AUTO-CONTROLE

L'auto-contrôle n'est pas considéré comme l'auto-critique, en ce sens qu'il s'intègre dans la gestion de chaque entreprise, comme un contrôle systématique de la qualité de la production réalisé par le fabricant lui-même, conformément à son engagement.

L'auto-contrôle est imposé par les distances entre usines fabricant les mêmes produits.

Le contenu de l'auto-contrôle comprend tous les aspects d'un produit dans sa présentation commerciale :
- Présentation extérieure

Conformité de l'emballage - propreté extérieure - qualité de la fermeture - facilité de l'ouverture dans le cas de film thermosoudé - étiquetage - datage visible.
- Le produit

Apparence, texture, goût et odeur - composition physico-chimique - normes bactériologiques, normes de conservation (à température précisée).

L'EXPLOITATION DES RESULTATS DE L'AUTO-CONTROLE

Les résultats peuvent être exprimés d'une façon simple par la proportion de résultats hors normes pour une caractéristique donnée, par exemple, une mesure physique ou bactériologique (extrait sec, poids, acidité, viscosité, germes de contamination...).

Une présentation simple, soit sous forme graphique, soit sous forme de tableau (Fig. 2) est immédiatement comprise par les personnes concernées à tous niveaux. Il y a lieu d'insister sur ce dernier point. En effet, de nombreux résultats de contrôles restent fréquemment "lettre morte", passent inaperçus, parce qu'aucun responsable ne les a analysés. Combien d'analyses faites très sérieusement et très régulièrement par le laboratoire restent inexploitées, parce que personne ne prend le temps d'en faire une synthèse significative pour le directeur ou les cadres responsables ? Il y a donc lieu d'étudier soigneusement de quelle façon les résultats de l'auto-contrôle seront rassemblés, présentés et diffusés.

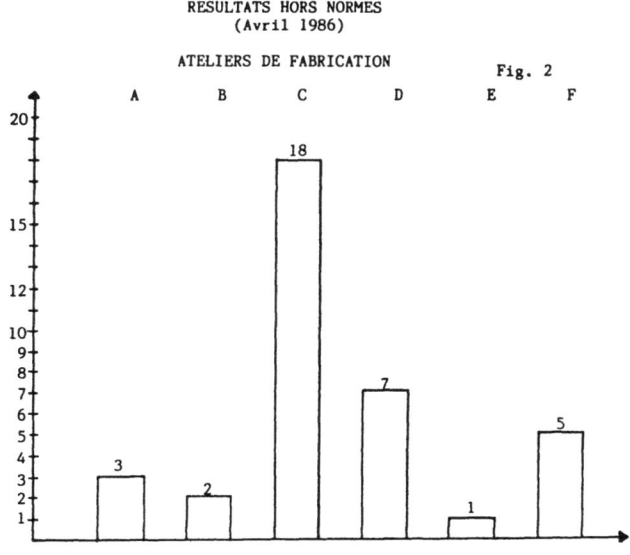

Le contrôle systématique et permanent a l'inconvénient de devenir une routine, en perdant une partie de sa valeur d'expression. On remarque aussi que la qualité des produits fabriqués, jour après jour, dans une même usine, varie peu, sauf accident. Il y a une constance de certaines qualités comme de certains défauts dans chaque usine, qui tient à la qualité du lait et des matières premières, aux équipements industriels, et à la personnalité des hommes.

C'est la raison pour laquelle il est de temps en temps nécessaire de "faire le point" par un examen approfondi du fonctionnement d'une usine.

L'AUDIT QUALITE

Défini par l'Afnor, l'audit qualité est "l'examen méthodique d'une situation relative à un produit, un procédé, une organisation en matière de qualité, réalisé en coopération avec les intéressés en vue de vérifier la conformité de cette situation aux dispositions préétablies et l'adéquation de ces dernières à l'objectif recherché".

L'examen peut être confié à une personne appartenant à la société concernée, ou étrangère à celle-ci. Ce n'est jamais un contrôle permanent, mais une appréciation approfondie correspondant à une durée limitée.

Le domaine d'action de l'audit qualité est plus vaste que celui du contrôle permanent, en ce sens, qu'il s'intéresse à toutes les données techniques et d'organisation qui ont une influence sur la qualité des produits. Exemples : disposition d'un bâtiment, réalisation d'un circuit de tuyauterie, concentration de solutions utilisées pour les nettoyages, etc...

Il faut bien noter que l'audit qualité est réalisé en plein accord avec les intéressés, qui y participent eux-mêmes et apportent leurs propres observations.

UN EXEMPLE DE GESTION DE LA QUALITE : SODIMA ET LE YOGHOURT YOPLAIT

Le yoghourt sous la marque Yoplait est présent dans de nombreux pays situés dans les cinq continents dans une gamme de produits laitiers frais (desserts, formages frais). La maîtrise de la qualité dans des emplacements aussi éloignés est fondée sur les hommes et les techniques.

Les hommes : la motivation est absolument essentielle. Chaque responsable doit se sentir concerné. L'un des premiers symposia annuels, qui s'est tenu à Lausanne en 1977, avait pour thème : "Le succès par la qualité". Et, chaque année la qualité reste une préoccupation majeure.

La sensibilisation des hommes est le préalable indispensable. Des actions dans ce sens sont réalisées au bénéfice des techniciens et des forces de vente. Ces actions font appel dans une large mesure à la présentation audio-visuelle.

Chacun sait que la qualité est un engagement de la marque.

La qualité n'est pas une affaire de spécialiste. Chacun est concerné dans la société, depuis l'ouvrier jusqu'au directeur général, en passant par les services d'approvisionnement, de contrôle, de distribution et de vente.

Cette motivation constante doit faire intervenir l'aspect qualité dans la plupart des décisions prises à l'intérieur des sociétés. Cela concerne en particulier les décisions commerciales. La distribution des produits dans une nouvelle aire géographique s'accompagne parfois de contraintes nouvelles, en relation avec les caractéristiques locales, les conditions de température de conservation des produits, les exigences du transport, les rotations de stocks, etc...

Un exemple concerne des yoghourts placés dans des cartons enveloppants avec couvercles et transportés à longues distances. Les cartons de yoghourts étaient placés dans des camions très bien réfrigérés, avec d'autres marchandises, mais étant de dimensions relativement réduites, ils étaient placés dans toutes les positions et utilisés par le chauffeur pour boucher les trous dans son chargement. Lorsqu'après le transport et le déchargement, les pots se retrouvaient dans leur position normale, du produit restait présent entre le bord du pot et le couvercle, et au premier réchauffement, on constatait des suintements qui souillaient les pots extérieurement. Le transport des pots de yoghourt dans des casiers ouverts, donc maintenus à l'endroit, ne pouvait pas mettre en évidence ce grave défaut.

Un deuxième exemple fait intervenir le groupage de plusieurs pots de yoghourt sous film rétractable. On avait observé l'apparition de moisissures sur la face externe des pots après 2 ou 3 semaines de conservation.

Après examen méthodique, l'explication de ce grave défaut était établie : quelques gouttelettes de lait provenant de la machine de conditionnement restaient fréquemment à l'extérieur des pots. Ces gouttelettes séchaient au cours de l'incubation en chambre chaude, et leur trace n'était pratiquement plus visible.

Après refroidissement en tunnel, le groupage sous-film rétractable, qui s'accompagne du passage dans un four de rétraction, provoquait une condensation d'humidité à l'extérieur des pots, et cette humidité était suffisante pour permettre le développement de moisissures à l'emplacement des taches de lait.

On comprend par ces deux exemples, la nécessité d'intéresser chaque maillon de la ligne de fabrication à la qualité du produit final, c'est-à-dire, la qualité que reçoit le consommateur.

Les techniques : le schéma exposé précédemment et illustré dans la Figure 1 est repris intégralement.

La définition du produit : yoghourt Yoplait. Elle s'inspire de la réglementation française. Le yoghourt est un lait fermenté par deux espèces bactériennes spécifiques : S. thermophilus - L. bulgaricus. Aucune autre espèce ne peut être présente (hors la présence de quelques contaminants).

Les bactéries spécifiques sont vivantes et abondantes. Leur dénombrement est supérieur à 10^8 par g.

De plus, le yoghourt ne subit aucun traitement physique tel que le chauffage ou la congélation après la fermentation. Cette exigence est fondamentale et ne peut admettre aucune discussion.

Le yoghourt est un produit frais, distribué et commercialisé sous réfrigération à température positive (4-6° C). Il a une durée de vie limitée à 4 semaines, et la date limite de consommation est inscrite visiblement sur l'emballage.

Enfin, aucun additif de texture n'est utilisé au cours de la fabrication.

Le cahier des charges : il définit le procédé, les formulations, la composition et les caractéristiques analytiques du produit, les emballages, les graphismes. Les fournisseurs de fruits et d'arômes font l'objet d'un agrément.

L'auto-contrôle : il comprend les examens physico-chimiques, bactériologiques, organoleptiques et les tests de conservation.

Les résultats sont rassemblés sur un tableau de bord, et comparés systématiquement aux normes. Ces présentations sont simples et immédiatement compréhensibles, sous forme de tableaux ou de graphiques.

Le contrôle note, non seulement, la qualité du produit, mais aussi, les éléments de l'emballage (propreté du pot, netteté du datage, étanchéité, facilité d'ouverture).

Les actions correctives : chaque fabricant s'est engagé au respect du cahier des charges. Lorsque les résultats des contrôles font apparaitre des défauts, des actions correctives sont nécessaires. Ces actions correctives utilisent largement l'expérience rassemblée par l'ensemble des fabricants. Des délais sont précisés pour la mise en œuvre des actions correctives.

L'audit qualité : c'est une intervention exceptionnelle, qui exige le déplacement de une ou plusieurs personnes dans une entreprise et une analyse minutieuse de tous les éléments qui interviennent sur la qualité des produits chez les consommateurs.

CONCLUSIONS

Le contrôle qualitatif est fondé sur : une définition suffisamment précise des produits, la détermination des normes auxquelles les produits doivent être conformes, la définition des méthodes d'examens et d'analyses.

La maîtrise de la qualité exige : la réalisation des contrôles, la présentation et la diffusion des résultats des contrôles, la mise en œuvre d'actions correctives.

Le contrôle par le fabricant lui-même (auto-contrôle) donne les résultats les plus rapides et permet la mise en œuvre d'actions correctives immédiates.

Réf. AFNOR - Gérer et Assurer la Qualité
Tour Europe - Cedex 7
92080 Paris La Défense

QUALITY ASSURANCE and IMAGE in RELATION to the CONSUMER

A M Duke
Central Quality Assurance Laboratory
Nestec Limited
Avenue Nestlé 55
1800 Vevey
Switzerland

ABSTRACT. Consumer satisfaction is attained if the food is above a "plateau" of acceptability, which relates to a period after manufacture, during which the food remains sensorially pleasing. To remain above the "plateau", quality assurance must extend beyond the factory through distribution up to the consumer. Although for stable products there are less risks at the interface with distribution and the consumer, for fresh and chilled products, this becomes a very critical point.

INTRODUCTION

Although drying is one of the oldest methods of conserving foods, cold conservation of perishable products represents one of the best and most natural ways of maintaining their original nutritional and sensorial characteristics, in the eyes of the consumer. Cold conservation, which includes both refrigeration and freezing, is becoming more and more important as consumers demand more healthy, natural and less-processed foods. In particular, refrigeration is a method of conservation which lends itself to this overall image of healthy and natural. It is also becoming more critical to use refrigeration as the current direction of healthier eating demands removal of preservatives from dairy products when, in addition, longer shelf-lives are expected.

This consumer association of refrigeration with natural has not just been self-evolved. A review of some advertising slogans highlights how often there is an attempt to portray an image of natural, good quality to the consumer, by convincing him that the chilled products (the popular name now for refrigerated foods) are 'as fresh as the day the cow was milked', 'natures own product', 'maintained for you in peak condition', and all the other descriptions that can currently be found. These types of claims can create that initial interest in, and image of the product. They will influence greatly the decision as to whether the product meets the first expectations of the consumer.

In fact, evaluating some dairy products straight off the production line or after handling, at the point of leaving the food processing plant, appearance and taste can meet exactly these claims. However, consumers must accept that all foods, whether fresh or processed, begin to deteriorate from the moment they are first handled and eventually packaged, and that this deterioration naturally continues through distribution, in the shops, up to when the crucial test is made at the time of consumption. The speed of such natural deterioration will depend on the characteristics of the product, 'shelf-stable' cheeses and UHT milks being at one extreme and some fresh dairy products, such as creams and some desserts, at the other. Recently the detection of Listeria in, for instance, soft cheeses, has raised some questions about the safety aspects of so-called "natural deterioration" following traditional commercial practices. This will not, however, be discussed in the context of this paper as this quality aspect is not directly linked to loss of image due to faults in the "chill-chain".

The opinion of the quality at the time of consumption will confirm, or produce deception about that initial image created maybe in the first place by the advertising. The consumer will thus be satisfied and may plan to buy the product again or will take note to avoid that product; perhaps even going so far as to complain to the processing company about the divergence from expectations. The task of the processor must therefore be to try to ensure that, having surveilled product quality during handling or manufacture, any further quality deterioration during distribution and up to consumption, must be as far as possible within natural limits. This means that he must be aware of any recently discovered areas where he can do more to assure quality maintenance after processing, for example, monitoring of the dairy products for psychrotrophs. Then, by experience, limits after which natural deterioration make the dairy product unacceptable and maybe unsaleable, can be defined by sell-by and/or use-by dates.

For 'shelf-stable' products, and even in most cases for frozen dairy foods, this maintenance of quality up to the expected expiry date has not been such a problem for the processor. The major product group of concern includes some of those products falling into the chilled category being stored at temperatures, for example, between 0-8 °C. (Note: 8 °C was chosen for safety reasons, but for various products and in various countries the upper temperature limit varies). It is clear that this grey zone between frozen and ambient conditions presents the most problems concerning maintenance of recommended temperature.

This chilled category is large and includes as the most successful, yoghurts and other acidified dairy products, which are not so effected by non-respect of temperature of storage. However, the safety aspects of that smaller sector of dairy products, which could deteriorate and even become a risk at temperatures] 8 °C, cannot be ignored.

The aim of this paper will be to investigate current problems being experienced by processors in assuring that these chilled foods, when consumed, are in the best condition possible, taking into consideration natural deterioration; in other words, that they remain above the plateau of acceptability established, surveilled and verified by the processor, starting from the point that the products leave the factory. Using the passage of the product through distribution, retail and up to the consumer, some ideas will be given on what can be done to retain the product's good image in the opinion of the consumer, maintaining its reputation as a top-quality food.

DISTRIBUTION

The use of chilled distribution to transport perishable foods from the point of harvesting, or first handling, to the consumers was developed by the Romans. They carried to Rome, from all parts of their Empire, fish covered with ice in metal containers. However, although there are other reports of chilling with ice being used to preserve the wholesomeness of foods, the real development of chilled distribution began in 1873, by James Harrison, a Scottish printer who had emigrated to Australia. Seeing the abundance of meat in Australia he tried, first without success, to bring chilled meat to England. Eventually, in 1880 the first acceptable meat arrived on board the Strathleven.

Following on from that date, technology advanced quite rapidly and refrigerators became a feature of many homes. First attempts to distribute chilled foods by van over short distances was achieved by suspending dry ice in a net, although the efficiency of this system was not too good.

Since these early days, refrigerated distribution of chilled foods has expanded enormously due, of course, to the rapidly growing production in the chilled-food sector. Delivery vehicles have become better equipped to maintain required chilled temperature. Particularly, in the last 12 years, according to results published at that time of an English study group (1), improvements have been made. Their results indicated that in 1974 only about one-third of all the vehicles were insulated and a quarter chilled. Specifically for dairy products at least 79.9 % were refrigerated, but when delivered with other products, this percentage was much inferior.

However, this is not to say that refrigerated delivery has now reached its optimal point. Problems still exist, even though reports and adverts arrive in food magazines concerning totally reliable chill-chains. These do exist of course, but in a recent survey (results not published), of seven retail delivery vans studied, it was found that only four were considered to have refrigeration units in proper working order.

One van, even when stationary in the depot with its refrigeration unit operating, allowed product temperature to rise. Greater problems were experienced when products were stored over a period of a few hours, and certainly rapid product deterioration was indicated if these vans were used for storage, particularly after loading Saturday for Monday delivery. The temperature of the van, at time of loading, was critical and obviously problems were experienced if the optimal chill temperature had not been initially achieved.

Although not reported by this survey, it has been stated that temperatures of distribution are greatly influenced by the numbers of deliveries made and the care taken by the driver to ensure that the products are exposed as little as possible to the ambient temperature. Therefore concluding, it would appear that with this situation many chilled products, during distribution, are going to suffer more than the natural deterioration predicted at temperatures of [8 °C.

Unfortunately the consumer, when forming his image of the product, does not know that maybe it has suffered abuse temperatures in distribution which have caused texture changes, presence of disagreeable odours and flavours, rancidity and acidity, etc. The first reaction would probably be to blame the processor for having released a sub-standard product, particularly if the product was within the sell-by and/or use-by date.

The question that therefore must be posed, and one that has been asked by many processors of chilled foods, is what can be done to carry out surveillance of quality during distribution. This is necessary so that, for example, in cases of complaints substantiated evidence can be provided if the distribution was at fault. However, more important is that action could be taken to improve the distribution based on such results of surveillance.

The obvious answer is the use of temperature recorders in the vans. Such devices, particularly based on microprocessors, are becoming available, although once again the recent survey previously mentioned (results unpublished) showed that of the seven retail delivery vans studied, only one appeared to have its temperature indicator operating correctly.

In a recent symposium (2) two papers were presented concerning temperature recording devices based on processors, ie, Autolog TTM (Remonsys Ltd) and Cool Cat (Systematic Micro Ltd). In the quoted survey I-point time temperature strips from Sweden were employed, but no definite conclusions were drawn about their efficiency. There had been ideas to label every individual unit with a temperature sensitive indicator, but this is as yet impractical.

Whatever the system, such devices are desirable for all vans, being simple to instal, relatively cheap and maybe tamper-proof, although it would be good to think that the latter is unnecessary, relying upon the honesty and sense of responsibility of the delivery man. In fact, the Cool Cat System requires that the delivery man has this sense of responsibility as it has a built-in alarm which permits immediate action in the event of deviation. There should be communication between the processor and the delivery group (whether internal or 3rd party) concerning all records of temperature. This is a part of the extended quality assurance system of the processor.

Finally, the retailer can also help to check distribution. This is in his interests as well as the processor. The retailer can:

- verify the temperature of the products delivered, from several parts of the batch
- refuse products which do not meet the agreed temperature, communicating such information to the processor
- help to limit time during which the products wait at the delivery bay

Summing up, there are many ways in which product quality, and thus good image, in the eyes of the consumer can be assured during distribution, but certainly it requires collaboration between all groups involved.

RETAIL

There have been as many reports concerning the problems of optimal temperature maintenance during retailing of chilled foods, as during distribution (2). Recently, in particular, there is a new flood of reports, coincident with the increased marketing of chilled culinary products, but such information is obviously still also relevant for the susceptible chilled dairy products.

In addition to the uneven temperatures in the cabinets, other problems can be experienced such as poor stock rotations. Fortunately, in retail, the consumer is already closer to the real conditions under which the product is being handled. He or she, maybe, would realise that it is not altogether the processor's fault that a product is unacceptable, when the temperature of the refrigerated unit is 'too' warm or even, as once noted, switched off. The consumer can also be a source of problem to maintaining quality in retail due to the habit of taking from the back of stocks, or taking products, then putting them back in more exposed, uncooled places.

Inspite of this chance that the consumer would not blame the processor, the latter must still be concerned about his product image and, ideally, should try to help to educate the retailers in how to handle the chilled products. One successful experiment with the ideal situation was carried out in Mexico where the shop assistants were trained to assure maintenance of stock at optimal conditions in the storage room, and in the displays, to quickly transfer products between store and display, including rapid price marking, to assure stock rotation and return of product after date limit.

Unfortunately, this is not practical for all situations, but within chain stores education can be promoted. Processors must try to encourage such education via various trade or consumer groups. This education is, in effect, carried out by one larger chain such as stated in an article from their Food Tech. Division (3). Based on the motto of 'keep it clean, keep it cold, keep it moving' practical programmes are developed, relying on the Market Manager to be personally committed and to motivate personnel with regular monitoring. However, one area where the least control is possible is the small village store, which often has less turnover, is over-stocked, with older display units and is maybe less susceptible to be controlled by official inspectors. No answer to this latter problem is evident, but maybe consumer complaints from such outlets can be used to advantage to highlight poorly operating shops.

Another area where processors and even better, retailers, can help is to work with manufacturers to make optimal display cabinets. Some large well-known chain stores have obviously achieved great success in this field.

Summarising, education and assurance of good equipment seems to be the answer to maintaining good product quality and image in the retail.

CONSUMER

Finally, the product after distribution and retail, will arrive at the consumer, who will make the product image assessment of appearance and taste, plus all the other factors including value for money, packaging appearance and ease to open. If the consumers have some doubts or are displeased, they will either discard the product or complain but, whatever, there will be a loss of image. As previously stated, it will quite often be the processor who will be blamed.

Maybe as a side-line, but actually linked with an important concluding point, are the details concerning consumer perception of good quality and foods to be rejected, which were recently reported by Woodburn (4). In a study of why consumers discarded foods, it was concluded that 62 % of householders did not make the correct decision (this 62 % could be equated with poor image judgement). 18 % of food discarded was with respect to expiry date, though there was generally a lack of understanding of the labelling. Many decisions to discard were not related to excess temperature or time of storage and required temperature of storage of many products was unknown.

Maybe some of the above points were exaggerated, but it would appear that consumer education is necessary in order to increase awareness of what should be perceived as representing good quality and what relates to maintaining good quality. This could particularly help processors of fragile chilled dairy foods, but the links between processor and consumer in terms of their education are still very tenuous.

LEGAL IMPLICATIONS

When reviewing all the possible quality problems due to failure of distribution, retail and consumer handling, in relation to the responsibility of the processor, it is essential to mention the legal implications. Not only is loss of product image important, but also a legal affair, if the quality of the product gave rise to sickness of the consumer.

The law varies from country to country, but in some places, such as France, the distributor and retailer can also possibly be held responsible (5). However, in Europe, the latest EEC documents cause rise for concern, when there is doubt about being able to assure good quality during distribution and retail (6). In particular, article 7 concerning producer liability is important, quoting 'that the producer shall not be liable if, having regard to the circumstances, it is probable that the defect which caused the damage did not exist at the time when the product was put into circulation by him or that this defect came into being afterwards.'

The question which could be posed regarding chilled products concerns the following situation. Although 'the bacteria X' was present after processing and would have been no problem if temperatures were maintained below Y°C, 'these bacteria' became a defect due to temperature abuse during distribution. Was the defect or not due to processor, who knew 'bacteria X' was in the product?

Another legal point that might be raised, which caused one processor to almost give up efforts with improving retail outlets, was related from France. A small local store had been identified by various means as causing abuse of chilled products, but when action was to be taken, the local people prevented closure of the shop as 'their favourite Mr. Z' was being persecuted by 'the grand officials and large companies'.

CONCLUSION

Inspite of all that has been said, the chilled dairy food business is increasing and this would suggest that products must have a very good image with the consumers. Some work has to be done in the areas highlighted to prevent those cases where, especially, temperature abuse can occur, particularly if it leads to risks for the consumer. The processor himself can help, as initially stated, by being aware of recently discovered areas where he can exercise even better quality monitoring of, for example, significant micro-organisms, before distribution begins.

As the image of chilled cairy food is good, it is assumed that consumers have concurrently developed a reliance on their safe quality, rejecting them only if sensorially, or evidently, due to yeast or mould growth something went wrong. Unfortunately, sensorial changes do not always proceed or accompany growth of pathogens and, therefore, it is essential to try and assure good commercial practices, which comprise not only the good manufacturing practice, but also the subsequent distribution and sale. A spoiled product may do some damage to the brand image, but a food poisoning case could destroy it.

REFERENCES

(1) MAAF (UK) Steering Group Food Freshness Final Report (1976)

(2) Proceedings of Chilled Food Symposium (1985) organised by Campden Food Preservation Research Association

(3) Microbial Control of Meat - A Retailer's Approach
Winslow RL J. of Food Protection (1982) 1169 - 1171

(4) Microbial and Quality Assessment of Household Food Discards
Vandoreits Jol Woodburn M J
J of Food Protection (1985) 924 - 931

(5) La chaine du Froid
M.P. Estoup
Option Qualité (1986) No 28 12-19

(6) Official J. of European Communities 29 July '85 L210/29

BRITISH FOOD MARKS

D.W. Atkinson
National Dairy Council
5/7 John Princes Street
London W1M OAP
ENGLAND

INTRODUCTION

In recent years in the UK a proliferation of quality marks have appeared on anything from wool through to food stuffs. On food alone 20 marks exist under the Food from Britain scheme which started in March 1985.
  Why is this occuring?
  Firstly, consumers appear to be more discriminating in what they purchase. They need an indication of quality through mainly visual means, because very rarely can food stuffs be tasted prior to purchase and it is often impractical. The trend is certainly towards a demand for, and willingness to pay for, higher quality. The success of one of Britain's leading retailers Marks and Spencer's food department most strikingly demonstrates this fact.
  Secondly, the marks appear primarily on what can be called commodity products and allow differentiation from direct competition on the basis of a minimum standard.
  Until comparatively recently, the cheeses of England and Wales were not subject to a uniform grading scheme with the result that consumers experienced some confusion over how to select from the varieties of Cheddar from both Britain and overseas markets. All this changed in July 1983 when the Cheese Mark scheme was introduced, initially only on Cheddar, to rectify this situation. The scheme has subsequently been extended to nearly all the English territorial cheeses. Its objectives were threefold:
1. To improve the quality and consistency of English cheese.
2. To help support the value of English and Welsh cheese.
3. To provide a symbol to convey to the consumer the high quality of English cheese.

THE SCHEME

For perishable foods, providing premium quality products to the

consumer depends on more than simply controlling quality at the time
of manufacture. High standards must be established at each stage of
the distribution chain. For these to be effectively introduced each
area must also understand the benefits of the scheme to themselves.
In the main this benefit is the ultimate advantage to be gained from
supplying a superior product which is consequently differentiated from
its competition, and which can command a higher price. In most cases
secondary advantages are also evident, such as better stock
management.

MANUFACTURE AND PACKING

Currently 18 manufacturers, representing 31 creameries, and 25
packing points are registered to use the English Cheese Mark. This
covers the majority of cheese produced in England and Wales.

Government regulations place stringent controls on food
production in respect of health and safety. The Cheese Mark scheme is
therefore principally concerned with the quality of the manufactured
product. Under the scheme manufacturers own in-house graders assess
the eligibility of any batch of cheese to carry the Mark at a given
age band as laid down in the Cheese Mark specifications. For example,
Cheddar is normally graded at 8 weeks and cannot be graded at under 6
weeks of age. The cheese is graded based on a sensory judgement of
aroma/flavour (maximum 45 points), body/texture (maximum 35 points),
finish (maximum 10 points) and colour (maximum 10 points). The cheese
may only receive the Mark if it gains over 85 points overall with at
least 38 for aroma and flavour. Inspectors with years of knowledge and
experience in selecting cheese are employed by the National Dairy
Council, to visit manufacturers and packers to take random samples and
ensure the cheese is being graded correctly. Consistency of grading is
ensured through 6-monthly grading sessions involving company graders,
the Cheese Mark Inspectors and members of the retail trade, where
individual borderline samples of cheese are discussed, to highlight
potential inconsistencies in grading.

Anyone found to be abusing the scheme can have their licence
removed. To date this final sanction has only been applied once.

The power of this threat is very strong as many retailers specify
that the cheese they buy must be of Cheese Mark standard and they are
willing to pay more for it. Exclusion from the scheme would
effectively mean a manufacturer would be unable to supply the multiple
retailers who dominate the market.

Careful administration and identification in cheese production as
demanded by the Cheese Mark scheme in itself reaps benefits of better
stock control, less wastage and fewer returns from retailers.

DISTRIBUTION

The speed and quality of the distribution system will affect the quality
of the product on the retailers shelf. The use of refridgerated vans,

speedy transfer from the delivery bay into storage, good storage facilities in wholesalers and each outlet are all needed. In most cases all this is within the retailers control. Good distribution costs money but maintains quality and reduces wastage.

RETAILERS

As so much can happen to the product between manufacture and purchase any quality mark, to be truly effective, must be able to monitor the product at this final stage. Retail surveys can identify obvious problems; mould; bad packaging, poor stocking of chilled cabinets, the visibility of the Mark. Much more is needed to monitor the real quality of what is a perishable product, whose properties change as it continues to age and mature during distribution. Both storage temperature and size of block will affect the changes that occur and these can be quite marked in terms of both texture and flavour.

It would be possible to take samples and have these retested by the Inspectors, but a major aim of this analysis is to influence the retailer on the basis of consumers reaction. We therefore generate a monitoring programme, through which samples of cheese are purchased at random from a range of retail outlets and passed to a housewife panel for trial. Obtaining their views plays a double role, a) as a verification that the Inspectors are providing the consumer with what she wants, and b) supplying a unbiased voice that the powerful retailer is willing to respond to.

The panel is selected to take account of the regional preferences in terms of flavour and visual characteristics of cheese. Cheeses are judged on a seven point scale from excellent to very poor. If one product is persistently substandard the results are presented to the retailer and guidance given as to the cause of the problem and how it can be overcome. It is then the responsibility of the retailers to take the appropriate action - usually in conjunction with his suppliers. Apart from the obvious problem areas mentioned previously, quality can be adversely affected by poor stock handling, rotation and storage, long residual shelf lives, overfilling or incorrect temperature of chilled cabinets.

In summary; although retailers request Cheese Mark cheese when they are buying they must be interested in maintaining the quality for it to be good. The only people who can stimulate this interest are the public. If they are interested in selecting cheese on the basis of quality there will be pressure on all concerned to improve quality.

WHAT DOES THE CONSUMER THINK ABOUT QUALITY MARKS?

Awareness of the different marks varies mainly dependant on the amount of promotional support provided for individual marks. Cheese Mark advertising has been extensive, starting in July 1983 and continuing through to the end of 1985, appearing for a total of 45 weeks and

accumulating over 3,400 television ratings. It now appears as an
important supporting element to all generic advertising for English
cheese.

The Cheese Mark is the best known of the British Food Marks with
over 90% of housewives having seen the symbol, and nearly 40%
understanding that it represents quality tested.

But do they want or use marks? Qualitative research has clearly
shown that the Cheese Mark acts as a real reassurance to the housewife who has been worried in the past by the inconsistency of cheeses.
Trends are certainly moving to demand for better quality, although
there will always be a proportion who will buy on price alone.

Even given this obvious need to help the consumer identify
quality, the Cheese Mark is recognised from packaging by only one-
third of respondents. Retailers are concerned to maintain their own
corporate identity, which in the UK in some cases may be very strong -
to the extent that they regard a widely-available quality mark as
something which could position them no better than any 'other' Cheese
Mark retailers. Hence they tend to limit the prominence of the Mark
itself on packaging. This is exacerbated by the growth of nutritional
labelling which limits the space available on the pack. In some cases
this can be overcome by secondary or integral labelling featuring the
Mark, or by point of sale promotion.

WHAT HAS THE CHEESE MARK ACHIEVED?

By monitoring cheese at both ends of the distribution chain, the
quality has been influenced throughout.

As a result the achievements of the Cheese Mark are significant:
Firstly, Cheese Mark cheese is better quality. Taking the
consumer panel data since January 1985, cheese carrying the Cheese
Mark has performed significantly better than non-Mark cheese. This
is given the perviously mentioned problem where some of the higher
quality retailers, although specifying Mark standard in their buying
requirements, do not display it.

Secondly, a premium price has been maintained for English Cheddar.
Thirdly, research tells us that the housewife is reassured by the
presence of the Cheese Mark and understands that it means quality.

THE FUTURE

As indicated so far the Cheese mark scheme has achieved widespread
recognition among consumers and has been welcomed by many retailers.
Since the inception of the scheme the method by which the cheeses
have been inspected and graded has meant efficient and reliable quality
control.

The National Dairy Council will be strengthening its commitment
and allocating resources to further enhance the scheme. This will
guarantee an even stronger assurance of quality and consistency for

those cheeses carrying the Cheese Mark.

Firstly, the regularity of inspection of the packers is to be doubled to once a fortnight bringing the frequency of packers' inspections up to the same level as they required for the manufacturers.

Secondly, manufacturers and packers are also being encouraged to demonstrate their commitment to upholding the standards of the Cheese Mark by allowing Cheese Mark inspectors to make unannounced visits thus waiving the current licence requirement of 24 hours notice.

More work needs to be done to get retailer commitment to the Cheese Mark, but this will rely on continued education of the consumer about cheese, and continued promotion of the quality of English cheese as embodied in the mark. In the end all those involved in the marketing of cheese - be they manufacturers, packers, wholesalers or retailers - are bound to respond to a consumer demand for quality.

AUSTRALIAN & NEW ZEALAND QUALITY MARKS

Dr K.J. Kirkpatrick
New Zealand Dairy Board
P.O. Box 417
Wellington
New Zealand

There are relatively few quality marks in use in Australia and New Zealand and fewer yet specifically in the food and dairy industries. Before discussing those that are in use, it is useful to consider the range of product marks that have quality connotations and which are in general trade use. It is particularly pertinent to review and analyse the differing perceptions of these various types of quality mark, and what they may mean to the consumer. Some examples are to be found illustrated in the figure.

WOOLMARK

The well known Woolmark is perhaps one of the most conspicuous examples of an internationally recognised quality mark, carrying with it identity of raw material origin; close integration with promotional activities; and appropriate and policed product quality standards. In effect the Woolmark is an internationally recognised "commodity brand name" although not in fact not in quite the specific proprietary ownership sense that is commonly associated with a brand.

REAL SEAL

A further distinct form of quality mark, which does not appear in Australia and New Zealand but is in active use in the United States dairy industry, is the Real Seal used to emphasise authentic dairy products rather than imitation products. In a sense it is analogous to the Woolmark but national in its application. The development of the Real Seal arose from a particular set of circumstances applying in the United States market where the encroachment of imitation products into the dairy sector is a cause for concern.

STANDARDS MARKS

Quite distinct from proprietary brand names or international promotional symbols, there are in many countries national standards association marks which relate particularly to products that comply with nationally established and registered product quality standards. With these standard marks, there is no particular connotation as to quality assurance programmes or quality control procedures that may be in operation in manufacture except insofar as the use of the standard mark and standard number may be policed by a National Standards Association.

The quality attributes as perceived by the consumer, relate principally to compliance with a finished product quality standard. Such standards generally speaking apply to non-food products of a relatively mature and unchanging nature. The machinery for establishing and agreeing standards within an industry is time-consuming and more appropriate to stable and well established products long past the stage of rapid development and evolution. As an example the mark of the New Zealand Standards Association is illustrated in the figure.

DESIGN MARK

Yet a further and different type of quality mark is represented in New Zealand by the Design Mark standard promoted by the Design Council of New Zealand. Products bearing the Design Mark have been judged subjectively and to an extent objectively, to represent excellent industrial and consumer product design and fitness for purpose, with intelligent use of materials and attractive features, generally representing good design. In many cases the products are unique and it is the excellence of the design which is being highlighted rather than compliance with any particular objective standard. Indeed products bearing the Design Mark are often quite individual for which there may be no nationally or internationally established standard.

GOVERNMENT APPROVALS

A further type of quality mark, exemplified by that applying to the New Zealand dairy industry, is the stamp of the Ministry of Agriculture & Fisheries, indicating that all aspects of the manufacture, testing and certification of products concerned has been in compliance with regulations governing the dairy industry in New Zealand. This stamp of approval is in the nature of a quality mark, where the attributes conveyed to the consumer are of an integrated comprehensive quality assurance system from raw material through to finished product with independent Government authorised and controlled inspection procedures.

Similar sorts of schemes and authentication stamps exist in many other dairy and food industries around the world.

LABORATORY REGISTRATION

Another form of quality mark, although in this case relating more to laboratory and quality assurance systems, is in use in both Australia and New Zealand as well as a few other countries. This is the laboratory registration system in New Zealand known by the acronym Telarc, representing the Testing Laboratory Registration Council, while in Australia the equivalent is NATA, standing for the National Association of Testing Associations Australia. These organisations operate systems for comprehensive and detailed inspection of laboratories covering a very wide range of areas of product testing including food and biological products. The inspections concentrate on competence and ability to undertake a defined range of tests, and as well focus on the laboratory management and reporting systems that assure accurate test reports. Test reports may be annotated with the Telarc or the NATA symbols, only under stringently devised and controlled circumstances.

In New Zealand the Telarc system has been extended to include appraisal and certification of quality assurance systems in industry, where standards for the definition and management of quality systems have been developed. In some other countries similar activities are taken within the bounds of a Standards Association.

TRIM PORK

A form of promotional and food product oriented quality mark was recently devised and used in New Zealand to expand the marketing of pork products. The so-called "Trim Pork" campaign included the establishment of product quality standards, the development of promotional material and a quality mark to confirm that products complied with the standards established. Participation by producers and retailers required compliance with these standards. This was a successful campaign which depended to an extent on a recognisable symbol or quality mark that linked the standards and promotional activity and point of sale material. The symbol is illustrated in the figure.

PROPRIETARY BRANDS

Many companies through the operation of a consistent commerical policy including quality assurance have developed brands (which may sometimes also be the company name) for their products, which have the attributes of a quality mark so far as the consumer perception is concerned.

GENERAL REVIEW

From this brief review, it can be seen that the term "quality mark" can embrace to a greater or lesser extent, a rather wide range of concepts. These include compliance with rigorously defined technical standards of both national and international currency; can include connotations of quality control or quality assurance systems, which ensures a high level of compliance with standards; has varying degrees of proprietary brand content ranging perhaps from 0 to 100%; may contain a substantial element of commercial and promotional activity; and may include compliance with laboratory testing standards and quality assurance systems. The relative balance of each of these factors and the varying connotations to the consumer, differ considerably between the different types of quality mark.

In Australia and New Zealand, there are relatively few quality marks associated with food products, other than those referred to earlier. Since the primary objective of a quality mark is to provide a distinquishable quality reference readily recognised by the consumer or potential buyer of produce, it is at first surprising that quality marks seem to be less widely used than might otherwise be expected.

Perhaps however this is more readily understandable given the strong tendency over the last several decades for the development of nationally and internationally recognised product brand names in all classes of product. In effect brands have developed to a major extent on the basis of the quality of product consistently delivered to the consumer. Given the major commitment and expenditure required to achieve such a consumer franchise, it is not surprising that in many respects well established brand names have become the most common form of quality mark.

The trend long term seems to be from non-branded commodity based quality standards and marks towards branded proprietary products, where there is comprehensive application by the brand owner of quality management systems, to all aspects of the production and marketing of the product in order to deliver to the consumer a consistent reliable product meeting customer expectations.

Given these considerations, the introduction of a new quality mark which did not carry with it the ability to assure comprehensive quality management systems were in operation throughout the supply chain to support the consumer perception of the quality mark, there would be a strong likelihood of disappointing consumer expectations and failing to achieve the consumer franchise that is the successful result of internationally and nationally promoted brand name products.

Examples of Quality Marks

SCANDINAVIAN AND DUTCH QUALITY MARKS

J.M. van der Bas
Netherlands Controlling Authority for milk and milk products
P.O. Box 250
3830 AG Leusden
Holland

ABSTRACT. Quality marks for dairy produce in Western Europe are about as old as the century. They guarantee that the produce has been subjected to a strict control on quality by an outside authority. The guarantee given has extended in the course of the years as has the concept of quality. The value of the marks may have diminished due to the growing importance of trademarks. Quality marks continue to perform a function as indicators of origin. In the future quality marks might get a new meaning and give a guarantee that quality labelled products have been produced under the strict observance of internationally agreed quality assurance principles.

HOW IT ALL BEGAN

At the first International Dairy Congress in 1903 in Brussels a heated debate took place about the question how to stop the adulteration of butter with foreign fats. It was a debate about "fraudulent practice" and "guaranteed products". The 700 participants, representing 15 countries held different views.
A majority was in favour of the method of Soxhlet, who wanted to make the addition of small amounts of a harmless tracer obligatory to all margarines and foreign fats used to adulterate butter. The tracer should not change the organoleptic characteristics of margarine and foreign fats, but should be easily detectable by simple means even in butter mixed with only 10% of foreign fats. Starch, sesame oil and phenolphthalein were considered to be suitable tracers.
A minority at the Congress had little confidence in the method of Soxhlet. The delegation of the Netherlands, e.g. emphasized that the adulteration of butter with foreign fats not spiked with tracers, might become very profitable under the Soxhlet regime. The Dutch advocated strongly both the foundation of controlling institutions to guarantee the genuiness of the butter in the respective countries and the introduction of national control marks for butter.
It was not before the third International Dairy Congress in 1907 in Scheveningen, that an agreement was reached. The countries represented

at this Congress agreed to prohibit the import of butter from
countries where the addition of tracers to margarine and foreign fats
was not obligatory or where an efficient controlling authority with
regard to butter was not effective. Butter produced under the supervision of a controlling authority should bear an official control
mark.
As is often the case in international disputes, there was no winner,
nor a looser. The principle of an official control mark for butter was
however accepted by the International Dairy Federation.

QUALITY MARKS TODAY

Having said this it will not surprise that the Danish, Dutch, Swedish
and perhaps also the other Scandinavian quality marks have been used
since the beginning of this century.
The Danish mark is called the LURMARK and consists of two pairs of
intertwined lurs. A lur is an ancient nordic musical instrument.
The Swedish mark is called the RUNEBRAND. The six shields on the ship
of the vikings contain the initials in the runic alphabet of the name
of the Swedish Government Control Board.
The Finish mark is quite clear and doesn't need any explanation.
In the past in Norway butter and cheese were often carried in a special wooden container, called a TINE, and Norway adopted the TINE as
its dairy quality symbol.
The Dutch quality marks show the arms of the state. In The Netherlands
there are slightly different marks for butter, milkpowder and other
products. For butter and milkpowder two different quality grades
exist, Extra Quality with a blue coloured and Standard Quality with a
green coloured mark.
Cheese has a specially prepared mark made from casein. Before this
casein mark was developed in 1911 the cheese was tattood with an
instrument of torture, which certainly did not improve the quality of
the cheese rind.

Dutch and Scandinavian quality marks have more in common than the time
of origin. They were introduced with the aim of protecting the good
reputation of the products concerned and hence increasing the sale and
in particular the export of milk products. In most countries the
quality mark is obligatory in case of export.
The marks serve an economic purpose. Introduced in the first place as
a guarantee for genuiness the marks gradually developed into quality
marks. The requirements to be fulfilled by quality labelled products
go far beyond the level of protection of public health. Appearance,
flavour and texture are important quality criteria taken into account
in granting the use of a quality mark to an individual producer.
Testing and sensory evaluation of samples is carried out by Government
Inspection Services or by private Controlling Authorities under
Government supervision.
In all countries there is a strong link with the Ministry of Agriculture, which is understandable in relation to the economic aim of

the marks.
In some countries the quality marks have a special meaning with regard to payment according to quality or with regard to intervention programmes.
It should be noticed that quality marks became of importance especially for bulk dairy products, such as butter and cheese in a period of time when brand names were hardly used for these products. Nowadays it can be observed that international enterprises with well established brand names are not always eager to display an official national quality mark on their products. They like to have exactly the same labels, regardless of the country of origin of their products. Are we going to survive the time of national quality marks ? Perhaps, although at present some quality marks are still connected with main price differences.

LOOK INTO THE FUTURE

Quality marks were introduced in the beginning of this century as a means to facilitate international trade. As such they have lost a great deal of their meaning.
Today it has been emphasized that quality control is more than the inspection of end-products. It's managing the manufacturing process in such a way that everything goes right from the start. But who is going to guarantee that everything has gone right from the start ? Will this be a new task for controlling authorities ? Should quality marks get a new meaning and give a new guarantee that the products labelled with these marks have been prepared under the strict observance of internationally agreed quality assurance principles ? Should controlling authorities from different countries come together to find new ways to facilitate international trade in dairy products ? I am raising the questions, you may have the answers.

Scandinavian and Dutch Quality marks

DANMARK

FINLAND

SWEDEN

NORWAY

DUTCH MILKPOWDER MARK

DUTCH BUTTER MARK

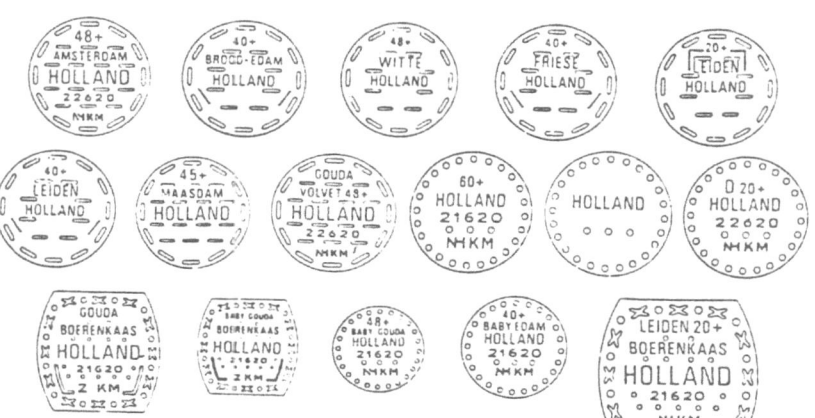

DUTCH CHEESEMARKS

DISCUSSION

J.M. van der Bas (the Netherlands) : Mr. Christian has said that end-product testing can be reduced in case quality assurance principles are applied correctly. How far can this reduction go? Do you think that end-product testing can be withdrawn completely?
J.H.B. Christian (Australia) : End-product control can not be avoided. Statistically, when quality-assurance principles are applied correctly, the results of end product checks have much less significance, but companies will keep doing end-product checks as a faith, because they see it as an extra safety.

A.E. Penning (the Netherlands) : Are maintaining of cold chain and doorstep delivery not conflicting?
M. Duke (Switzerland) : The Nestlé company is not involved in door-step delivery. Temperature control is more difficult when the products are left for the consumer to pick up.

SYNOPSIS OF THE CONGRESS

E. Mann, Ph.D.
International Dairy Federation
Square Vergote 41
B-1040 Brussels
BELGIUM

ABSTRACT. The composition of the programme of the XXII International Dairy Congress is compared with those of previous congresses. The main aspects of what has been presented during the former congress are summarized in seven challenges.

We are coming close to the end of the XXII International Dairy Congress, this having been the third one held in the Netherlands. Unfortunately, owing to circumstances entirely beyond my control, I was unable to attend the III Congress held in this country in 1907. However, 33 years ago, in 1953, I did attend the XII Congress in The Hague, a truly splendid Congress held under the shadow of the great North Sea floods which had threatened to engulf large parts of the Netherlands but which the ingenious and hard-working Dutch people managed to turn into profit by the construction of large, protective dykes and the reclamation of large areas of land from the sea.
    This year's International Dairy Congress was held under the shadow of a different kind of flood for many countries represented here, not least our host country, and will require all the ingenuity of the Dutch and their fellow sufferers to solve in order to save the dairy farmers and the dairy industries in many countries from the most serious consequences. I shall be returning to that subject shortly but, in the first place, it fives me great pleasure to be able to report that Dutch ingenuity and innovation has shown itself to good effect in the congress programme which did, in many ways, break away from the more traditional concept of using Dairy Congresses as platforms for progress reports on all developments in dairy science and technology during the preceding 4 years, backed up during the preceding 4 Congresses by the publications of Brief Communications attempting to provide up-to-date progress reports on current research in the world's dairy and food laboratories.
    In its place, our Dutch friends devised a programme of high scientific and technical excellence, whose centre piece was based on 9 seminars the subjects of which having been selected carefully in order to reflect current major world interests and trends in the

production, processing and marketing of milk and milk products but which, it must be said, did not by any means attempt to cover progress in the whole area of dairy science and technology.

However, the careful choise of the most eminent en suitable speakers for each subject ensured Seminars and discussions of outstanding quality, a fact which will become apparent from the published proceedings.

Around this centre piece of 9 seminars, we had 5 plenary lectures by world authorities on important subjects of current interest, as well as 9 round table discussions, some in simultaneous sessions in order to accommodate the obvious interest in such round table discussions enabling specialists to exchange views and experiences on topical subjects on a more informal basis than is possible in the larger seminars.

The brief communications of the four most recent congresses were replaced by poster presentations, a congress technique of growing importance, although on this occasion, the number of such presentations was surprisingly low.

Amongst the many other activities were a number of technical excursions enabling many delegates to witness for themselves the enormous progress made in Dutch dairy farming, dairy research and technical dairy development, the great Machevo exhibition, held to coincide with the Dairy Congress, and a workshop on the possibility of utilizing dairy ingredients in the non-food industries, organized by the European Federation of Biotechnology.

Whether the new congress pattern set by our Dutch colleagues and executed with great dedication and efficiency, will be the pattern for future Dairy Congresses, or whether we shall revert to the old-style Congresses or someting entirely different, is something which is now under discussion between officials of the International Dairy Federation and the countries which are going to host future Dairy Congresses. The problems relating to the organization and form of future Dairy Congresses do, of course, sink into insignificance when compared with the problems which face world dairying and against the background of which this Congress was held.

Rather than present you with a list of recommendations of conclusions which tend to become carved in stone to be looked at rather than to be acted upon, I want to leave you with a series of challenges, some more important than others, some applying more to one region than another but all of them, I believe, crucial for the survival of the dairy industry world-wide.

(1) Urgent steps must be taken to reduce the growing imbalance between supply of and demand for milk and milk products in many parts of the world and between different regions of the world. There are probably as many suggestions to a solution to this paramount problem, as there are politicians and economists. However, the challenges here clearly include: (a) the taking of appropriate measures to limit milk production in the Western world whilst taking the necessary steps at the same time to mitigate the effects of the resulting structural adaptation of dairy farming to these changes; (b) providing support, including food aid where appropriate, for the development of the

dairy industry in the Third World; (c) initiating an international dairy agreement within the pattern of the G.A.T.T. negotiations leading, among other things, to a more stable price structure (on these markets).

(2) Increasing efforts should be made to decrease the costs of milk production by all the modern genetic, scientific and technical means available to us, thereby maintaining or improving the competitiveness of milk and milk products in relation to other foods.

(3) The theme of milk as a tool for rural and socio-economic development, pioneered with such outstanding success in India during the last 20 - 25 years, should be extended to other regions of the world where it is applicable, leading, hopefully, to a more integrated approach between food aid, food trade and dairy development in the regions concerned, than has been achieved so far. In this connection, it is worth elaborating that modern developments in dairy farming and milk processing seem to be creating a widening technology gap between the western and much of the Third World. For those regions in the Third World where it is appropriate, for instance where milk is a well-accepted food in highly urbanized societies, great efforts should be made to bridge this gap through international cooperations and, here again, the Indian experience could be most valuable.

(4) Both traditional and new milk and milk products should be promoted in a more positive manner than at present through more effective nutritional labelling, better consumer education, as well as that of the medical profession on diet and health matters, and improving the information flow between the dairy industry and the consuming public on the unique and positive properties of milk and milk products in the context of a healthy way of life. In order to enhance this image still further, we must ensure that we are providing a completely safe product through hygienic production, handling and processing methods.

(5) While there can be no doubt that the dairy industry has been the father of the food industry world-wide, as regards research and development, industrial processing and quality assurance, to mention only a few aspects, it is time for the dairy industry to recognize that the son has now grown up and stands alongside us as a full equal. The challenge here is to break down the very real barriers, physiological, technical, organizational and economical, which still exists between us and the food industry and to recognize that we are an integrated part, albeit a very important part of the food industry.

(6) If we can take that fital step, the next and perhaps most important challenge becomes easier to face. Namely that milk contains a treasure chest of some 2000 ingredients, many with unique functional and nutritional properties, which modern scientific and technological processes are capable of isolating and refining for a multiplicity of users in the food and related industries. This is, of course, not a simple challenge and depends on our ability to produce the right product at the right price in the right market place, often in competition with lower priced products claiming to have similar properties. However, we have already seen much progress in this area and take courage, for example, by what is happening to whey, until

not so many years ago regarded as a disposable by-product of cheese making, yet today an increasingly important source of valuable ingredients for the food and dairy industries.

(7) Perhaps the final challenge in this connection is to make sure that dairy farmers and processors get their fair share of the value added to milk as the result of the generation of such milk ingredients for use in other industries.

I have kept this synopsos of the Congress deliberately short in the hope that you are more likely to listen to it and weigh up the pros and cons of the challenges which I have presented to you, than if I had given an hour long dissertation. In any event, you can rest assured that those challenges which are within the capacity of the IDF to act upon, will be dealt within the various Commissions of the IDF.

It remains for me to thank you for your attention and to wish you a safe journey home.

MIX
Papier aus verantwortungsvollen Quellen
Paper from responsible sources
FSC® C105338

If you have any concerns about our products,
you can contact us on
ProductSafety@springernature.com

In case Publisher is established outside the EU,
the EU authorized representative is:
Springer Nature Customer Service Center GmbH
Europaplatz 3, 69115 Heidelberg, Germany

Printed by Libri Plureos GmbH
in Hamburg, Germany